Lecture Notes in Computer Science 1196

Edited by G. Goos, J. Hartmanis and J. van Leeuwen

Springer
Berlin
Heidelberg
New York
Barcelona
Budapest
Hong Kong
London
Milan
Paris
Santa Clara
Singapore
Tokyo

Lubin Vulkov Jerzy Waśniewski
Plamen Yalamov (Eds.)

Numerical Analysis and Its Applications

First International Workshop, WNAA'96
Rousse, Bulgaria, June 24-26, 1996
Proceedings

Volume Editors

Lubin Vulkov
Plamen Yalamov
University of Rousse, Center for Applied Mathematics and Informatics
7017 Rousse, Bulgaria
E-mail: vulkov@ami.ru.acad.bg
yalamov@iscbg.acad.bg

Jerzy Waśniewski
Danish Computing Centre for Research and Education
DTU, UNI-C, Building 304, DK-2800 Lyngby, Denmark
E-mail: jerzy.wasniewski@uni-c.dk

Cataloging-in-Publication data applied for

Die Deutsche Bibliothek - CIP-Einheitsaufnahme

Numerical analysis and its applications : first international workshop ; proceedings / WNAA '96, Rousse, Bulgaria, June 24 - 26, 1996. Lubin Vulkov ... (ed.). - Berlin ; Heidelberg ; New York ; Barcelona ; Budapest ; Hong Kong ; London ; Milan ; Paris ; Santa Clara ; Singapore ; Tokyo : Springer, 1997
(Lecture notes in computer science ; 1196)
ISBN 3-540-62598-4

NE: Vulkov, Lubin [Hrsg.]; WNAA <1, 1996, Ruse>; GT

CR Subject Classification (1991): G.1, F.2.1, G.4, I.6

ISSN 0302-9743
ISBN 3-540-62598-4 Springer-Verlag Berlin Heidelberg New York

Printed in Germany

Typesetting: Camera-ready by author
SPIN 10548987 06/3142 – 5 4 3 2 1 0 Printed on acid-free paper

Preface

This volume of the Lecture Notes in Computer Science series is the Proceedings of the First Workshop on Numerical Analysis and its Applications, which was held at the University of Rousse, Bulgaria, June 24-27, 1996. The workshop attracted more than 90 participants from 22 countries. The volume includes 14 invited lectures and 57 contributed papers presented at the workshop. The workshop was organized by the Department of Numerical Analysis and Statistics at the University of Rousse. Support and help were also received from the Bulgarian Mathematical Society, the Technical University of Gabrovo, the Parallel Computing Laboratory of the Bulgarian Academy of Sciences, the University of Texas at Permian Basin, the ACM (Association for Computing Machinery) Special Interest Group on Numerical Mathematics, SIAM (Society for Industrial and Applied Mathematics), Bulgaria - Insurance and Re-insurance Holding Group, DEC of Bulgaria, MAPLESOFT, MATHWORKS for Central and Eastern Europe, and the Numerical Algorithms Group Ltd.

The purpose of the workshop was to bring together scientists working in the area of numerical analysis, and also people from its applications in physics, chemistry, and other natural and engineering sciences. The interaction between these groups seems to be quite useful for development of new algorithms and ideas in numerical methods, and for applying some of already existing methods in practice.

We are indebted to our colleagues who helped us in the preparation and organisation of the workshop, during the meeting, and in editing this proceedings.

November 1996

Lubin Vulkov
Jerzy Waśniewski
Plamen Yalamov

Table of Contents*

---

* Italic style indicates the speaker.
Bold style indicates the title of the paper and the invited speaker.

The Newtonian Continuation Method for Numerical Study of 3D Polaron Problem

P. G. Akishin, I. V. Puzynin, Yu. S. Smirnov

Laboratory of Computing Techniques and Automation
E-mail addresses: akishin@lcta28.jinr.dubna.su
puzynin@lcta15.jinr.dubna.su
smirnov@lcta6.jinr.dubna.su

Abstract. The numerical approach for study of the 3D-nonlinear problem for the spherically-nonsymmetric polaron is considered. An expansions in spherical harmonics are used for the approximation of the solution. The iterative Newton's scheme with an additional parametrization of the initial equation for the solving of the nonlinear problem is proposed. The results of numerical modelling are discussed. The comparison of the obtained eigenvalues with the known ones confirms the efficiency of the elaborated algorithms.

1 Introduction

We consider the problem of numerical solving of the spherically-nonsymmetrical model of polaron [1]. Such a model of polaron describes the behavior of a non-relativistic particle (electron) in the field, created by the interaction with medium.

Let us formulate a mathematical statement of the problem for a finite constant of coupling. The wave function ψ and potentials u, v satisfy the following system of partial differential equations in the space R^3 :

$$\Delta\psi(\mathbf{x}) + (u(\mathbf{x}) - v(\mathbf{x}))\psi(\mathbf{x}) - \lambda\psi(\mathbf{x}) = 0, \tag{1}$$

$$\Delta u(\mathbf{x}) + 4\pi\psi^2(\mathbf{x}) = 0, \tag{2}$$

$$\Delta v(\mathbf{x}) - c^2 v(\mathbf{x}) + 4\pi\psi^2(\mathbf{x}) = 0, \tag{3}$$

where Δ is the Laplace operator and c is the constant of coupling.

The wave function $\psi(\mathbf{x})$ satisfy the normalization condition:

$$\int_{R^3} \psi^2(\mathbf{x})dV_{\mathbf{x}} = 1. \tag{4}$$

In spherically symmetrical case (when $\psi(\mathbf{x})$ depends on $r = |\mathbf{x}|$ only), the system (1)-(2) may be reduced to a boundary problem for the system of ordinary differential equations on semi-axis [2],[3]. This problem has been studied quite well. In paper [4] the authors consider the axi-symmetrical solutions of the polaron model. The papers [5],[7] are devoted to numerical investigations of the problem for the nonsymmetrical case.

In the present paper to approximate equations (1)-(2) we use the approach proposed in [5],[6]. To solve the nonlinear discrete problem, a parametrization of initial equations is introduced by means of the additional continuous parameter t [8]. At the initial moment $t = 0$ the problem is reduced to the sufficiently simple spectral Helmholtz problem for a ball. All eigenvalues and eigenfunctions of this problem may be easily found. After that, the solution of the original nonlinear problem (1)-(2) may be obtained if we use a movement with respect to continuous parameter t.

2 Discretization

In this section we consider the problem of discretization of the system of equations (1)-(2). For this aim we use the Bubnov - Galerkin method [9]. Let functions $\psi(\mathbf{x})$, $u(\mathbf{x})$ and $v(\mathbf{x})$ be expanded in spherical harmonic series [5]:

$$\psi(\mathbf{x}) = \sum_{k=0}^{\infty}\sum_{l=-k}^{k} \psi_{kl}(r)Y_{kl}(\theta,\varphi), \tag{5}$$

$$u(\mathbf{x}) = \sum_{k=0}^{\infty}\sum_{l=-k}^{k} u_{kl}(r)Y_{kl}(\theta,\varphi), \qquad v(\mathbf{x}) = \sum_{k=0}^{\infty}\sum_{l=-k}^{k} v_{kl}(r)Y_{kl}(\theta,\varphi), \tag{6}$$

where Y_{kl} are spherical harmonics [10].

Taking into account a finite number K of terms in the expansions (5),(6) for approximations of functions $\psi(\mathbf{x}), u(\mathbf{x}), v(\mathbf{x})$, multiplying (1)-(2) by the spherical harmonics and integrating over the angle variables, we get the following approximate system of equations for the functions $\tilde{\psi}_{kl} = \psi_{kl}r$, $\tilde{u}_{kl} = u_{kl}r$, $\tilde{v}_{kl} = v_{kl}r$:

$$\tilde{\psi}''_{kl} - \frac{k(k+1)}{r^2}\tilde{\psi}_{kl} + \frac{1}{r}\sum_{k_1,l_1}\sum_{k_2,l_2} W^{ll_1l_2}_{kk_1k_2}\tilde{\psi}_{k_1l_1}\tilde{u}_{k_2l_2} = \lambda\tilde{\psi}_{kl} \tag{7}$$

$$\tilde{u}''_{kl} - \frac{k(k+1)}{r^2}\tilde{u}_{kl} + 4\pi\frac{1}{r}\sum_{k_1,l_1}\sum_{k_2,l_2} W^{ll_1l_2}_{kk_1k_2}\tilde{\psi}_{k_1l_1}\tilde{\psi}_{k_2l_2} = 0, \tag{8}$$

$$\tilde{v}''_{kl} - (c^2 + \frac{k(k+1)}{r^2})\tilde{v}_{kl} + 4\pi\frac{1}{r}\sum_{k_1,l_1}\sum_{k_2,l_2} W^{ll_1l_2}_{kk_1k_2}\tilde{\psi}_{k_1l_1}\tilde{\psi}_{k_2l_2} = 0, \tag{9}$$

$$k = 0,1,2,...,K; l = 0,\pm1,\pm2,...,\pm k.$$

Here coefficients $W^{l_1l_2l_3}_{k_1k_2k_3}$ equal to the integral over the unit sphere for the product of spherical harmonics:

$$W^{l_1l_2l_3}_{k_1k_2k_3} = \int_0^{2\pi}\left[\int_0^{\pi} Y_{k_1l_1}(\theta,\varphi)Y_{k_2l_2}(\theta,\varphi)Y_{k_3l_3}(\theta,\varphi)sin\theta d\theta\right]d\varphi. \tag{10}$$

The normalization condition (4) may be written in the following form:

$$\sum_{k,l}\int_0^\infty \tilde{\psi}_{k,l}^2 dr = 1. \tag{11}$$

Let $[0,R]$ be an interval of the variation of r. We assume that R is large enough, so functions $\tilde{\psi}_{kl}(r), \tilde{u}_{kl}(r), \tilde{v}_{kl}(r)$ have a behavior, which may be approximated analytically for $r \geq R$. We choose a uniform grid of nodes for variable $r : \{r_i; i = \overline{1,N}\}\left(r_i = (i-1)h, h = \frac{R}{N-1}\right)$ to discretize the system (7)-(8). The second order accuracy finite difference scheme with respect to step h is used for the approximation of equations (7)-(8). Let us introduce the notations $\psi_{kl}^i = \tilde{\psi}_{kl}(r_i), u_{kl}^i = \tilde{u}_{kl}(r_i), v_{kl}^i = \tilde{v}_{kl}(r_i)$. Then the discrete system of equations has the following form:

$$\frac{\psi_{kl}^{i+1} - 2\psi_{kl}^i + \psi_{kl}^{i-1}}{h^2} - \frac{k(k+1)}{r_i^2}\psi_{kl}^i + \frac{1}{r_i}\sum_{k_1,l_1}\sum_{k_2,l_2} W_{kk_1k_2}^{ll_1l_2}\psi_{k_1l_1}^i u_{k_2l_2}^i = \lambda\psi_{kl}^i \tag{12}$$

$$\frac{u_{kl}^{i+1} - 2u_{kl}^i + u_{kl}^{i-1}}{h^2} - \frac{k(k+1)}{r_i^2}u_{kl}^i + 4\pi\frac{1}{r_i}\sum_{k_1,l_1}\sum_{k_2,l_2} W_{kk_1k_2}^{ll_1l_2}\psi_{k_1l_1}^i \psi_{k_2l_2}^i = 0 \tag{13}$$

$$\frac{v_{kl}^{i+1} - 2v_{kl}^i + v_{kl}^{i-1}}{h^2} - (c^2 + \frac{k(k+1)}{r_i^2})v_{kl}^i + 4\pi\frac{1}{r_i}\sum_{k_1,l_1}\sum_{k_2,l_2} W_{kk_1k_2}^{ll_1l_2}\psi_{k_1l_1}^i \psi_{k_2l_2}^i = 0 \tag{14}$$

$k = 0,1,2,...K; l = 0,\pm1,\pm2,...,\pm k; i = \overline{2,N-1}$. For $r_1 = 0$ we have the following left-hand boundary conditions: $\psi_{kl}^1 = 0, u_{kl}^1 = 0, v_{kl}^1 = 0$. Taking into account exponential vanishing of the function ψ_{kl} on infinity, we get the right-hand boundary condition for ψ_{kl}^i:

$$\frac{\psi_{kl}^N - \psi_{kl}^{N-2}}{2h} = (-\sqrt{\lambda} + \frac{k}{r_{N-1}})\psi_{kl}^{N-1}. \tag{15}$$

Analogously, for u_{kl}^i and v_{kl}^i we have

$$\frac{u_{kl}^N - u_{kl}^{N-2}}{2h} = \frac{k}{r_{N-1}}u_{kl}^{N-1}, \qquad \frac{v_{kl}^N - v_{kl}^{N-2}}{2h} = (-c + \frac{k}{r_{N-1}})v_{kl}^{N-1}. \tag{16}$$

By using some numerical approximation for (11) the normalization conditions may be written in the form:

$$\sum_{k,l}\sum_{i=1}^{N}\alpha_i(\psi_{kl}^i)^2 = 1, \tag{17}$$

where α_i are the coefficients of the quadrature formula for calculation of integrals.

To find the approximate solutions of the original problem (1)-(2), it is necessary to solve the nonlinear system of algebraic equations (12)-(17) for the unknown variables $\{\psi_{kl}^i, u_{kl}^i, v_{kl}^i, i = \overline{1,N}\}$ and the spectral parameter λ.

3 The method for solving the nonlinear discrete systems

Let $\boldsymbol{\Psi}_i$ be the vector with the components ψ^i_{kl} , $\mathbf{U}_i$ and $\mathbf{V}_i$ be the vectors with the components u^i_{kl} and v^i_{kl} correspondingly. Analogously we define the vectors corresponding to the nonlinear terms (double sums) in the equations (12) and (13), as $\mathbf{F}_i = \mathbf{F}_i(\boldsymbol{\Psi}_i, \mathbf{U}_i)$, $\mathbf{G}_i = \mathbf{G}_i(\boldsymbol{\Psi}_i, \boldsymbol{\Psi}_i)$. Then the system (12)-(13) can be written in the form:

$$\frac{\boldsymbol{\Psi}_{i+1} - 2\boldsymbol{\Psi}_i + \boldsymbol{\Psi}_{i-1}}{h^2} + [D_i]\boldsymbol{\Psi}_i + \mathbf{F}_i(\boldsymbol{\Psi}_i, \mathbf{U}_i) = \lambda \boldsymbol{\Psi}_i, \tag{18}$$

$$\frac{\mathbf{U}_{i+1} - 2\mathbf{U}_i + \mathbf{U}_{i-1}}{h^2} + [D_i]\mathbf{U}_i + \mathbf{G}_i(\boldsymbol{\Psi}_i, \boldsymbol{\Psi}_i) = 0, \tag{19}$$

$$\frac{\mathbf{V}_{i+1} - 2\mathbf{V}_i + \mathbf{V}_{i-1}}{h^2} - c^2\mathbf{V}_i + [D_i]\mathbf{V}_i + \mathbf{G}_i(\boldsymbol{\Psi}_i, \boldsymbol{\Psi}_i) = 0, \tag{20}$$

where $i = \overline{2, N-1}$; the matrices $[D_i]$ are diagonal matrices with coefficients on the main diagonal depending on the number k of the correspondent harmonic. The normalization condition (17) may be written in new variables as

$$\sum_{i=1}^{N} \alpha_i |\boldsymbol{\Psi}_i|^2 - 1 = 0, \tag{21}$$

The left-hand boundary conditions reduce to the following equations:

$$\boldsymbol{\Psi}_1 = 0, \qquad \mathbf{U}_1 = 0, \qquad \mathbf{V}_1 = 0. \tag{22}$$

Analogously, we rewrite the right-hand boundary conditions:

$$\frac{\mathbf{U}_N - \mathbf{U}_{N-2}}{2h} = [B_1]\mathbf{U}_{N-1}, \qquad \frac{\mathbf{V}_N - \mathbf{V}_{N-2}}{2h} = (-c[E] + [B_2])\mathbf{V}_{N-1}, \tag{23}$$

$$\frac{\boldsymbol{\Psi}_N - \boldsymbol{\Psi}_{N-2}}{2h} = (-\sqrt{\lambda}[E] + [B_3])\boldsymbol{\Psi}_{N-1}, \tag{24}$$

where $[E]$ is an identity matrix, $[B_1]$, $[B_2]$, $[B_3]$ are diagonal matrices.

Let us consider an auxiliary problem. We can write the right boundary condition for the wave functions in the following form:

$$\boldsymbol{\Psi}_N = 0. \tag{25}$$

Then the system of equations (18)-(23),(25) also approximates the problem (7)-(11), but with the lesser accuracy. To solve the nonlinear problem (18)-(23),(25), we use the continuation method from [8]. In equations (18)-(20) we introduce a parameter $t, (t \in [0, 1])$ by the following way:

$$\mathbf{F}_i(t) = t\mathbf{F}_i(\boldsymbol{\Psi}_i, \mathbf{U}_i), \qquad \mathbf{G}_i(t) = t\mathbf{G}_i(\boldsymbol{\Psi}_i, \boldsymbol{\Psi}_i).$$

After this substitution the system (18)-(20) may be transformed to the following system of equations:

$$\frac{\boldsymbol{\Psi}_{i+1} - 2\boldsymbol{\Psi}_i + \boldsymbol{\Psi}_{i-1}}{h^2} + D_i\boldsymbol{\Psi}_i + t\mathbf{F}_i(\boldsymbol{\Psi}_i, \mathbf{U}_i) = \lambda\boldsymbol{\Psi}_i \tag{26}$$

$$\frac{\mathbf{U}_{i+1} - 2\mathbf{U}_i + \mathbf{U}_{i-1}}{h^2} + D_i\mathbf{U}_i + t\mathbf{G}_i(\boldsymbol{\Psi}_i, \boldsymbol{\Psi}_i) = 0, \tag{27}$$

$$\frac{\mathbf{V}_{i+1} - 2\mathbf{V}_i + \mathbf{V}_{i-1}}{h^2} - c^2\mathbf{V}_i + D_i\mathbf{V}_i + t\mathbf{G}_i(\boldsymbol{\Psi}_i, \boldsymbol{\Psi}_i) = 0, \tag{28}$$

with boundary conditions (22)-(23),(25). In this case the solutions $\boldsymbol{\Psi}_i, \mathbf{U}_i, \mathbf{V}_i, \lambda$ of the problem (21)-(23),(25)-(28) will be the functions of the parameter t ($t \in [0,1]$). Obviously, for $t = 1$ we have the original system of equations (18)-(23),(25), and for $t = 0$ we obtain the sufficiently simple non-connected linear spectral problem for each harmonic separately.

To find the values of functions $\boldsymbol{\Psi}(t), \mathbf{U}(t), \mathbf{V}(t), \lambda(t)$ at the moment $t = 0$ for the fixed orbital momentum k_0, we solve the one-dimensional spectral problem for one equation only:

$$\frac{y_{i+1} - 2y_i + y_{i-1}}{h^2} - \frac{k_0(k_0+1)}{r_i^2}y_i = \lambda y_i, i = \overline{2, N-1} \tag{29}$$

with boundary and normalization conditions

$$y_1 = 0, \qquad y_N = 0; \qquad \sum_{i=1}^{N} \alpha_i y_i^2 = 1. \tag{30}$$

Let $\{y_i^*\}, \lambda_i^*$ be a solution of the problem (29)-(30). For the fixed number $l_0 : -k_0 \le l_0 \le k_0$ we assume

$$\psi_{kl}^i(t)|_{t=0} = \begin{cases} y_i^*/\sqrt{2}, & l = l_0, k = k_0 \\ y_i^*/\sqrt{2}, & l = -l_0, k = k_0 \\ 0, & l \ne l_0 \\ 0, & k \ne k_0. \end{cases} \tag{31}$$

This case corresponds to the real initial approximation for determing the solutions which are even functions with respect to angle φ. The case corresponding to the odd initial approximation can be written by the following way:

$$\psi_{kl}^i(t)|_{t=0} = \begin{cases} y_i^*/\sqrt{2}, & l = l_0, k = k_0 \\ -y_i^*/\sqrt{2}, & l = -l_0, k = k_0 \\ 0, & l \ne l_0 \\ 0, & k \ne k_0. \end{cases} \tag{32}$$

Accordingly for the spectral parameter we have $\lambda(t) = \lambda^*$ for $t = 0$.

So, for every $t \in [0,1]$ we have obtained the nonlinear boundary problem. Let $\{t_j; j = \overline{0, M}\}(t_0 = 0, t_M = 1)$ be some network of the interval [0,1]. To find the solutions $\psi_{kl}(t_j)$, $u_{kl}(t_j)$, $v_{kl}(t_j)$, $\lambda(t_j)$ of the problem (29)-(30) in the point t_j

the Newton method is used. Supposing the difference $|t_j - t_{j+1}|$ is small enough, we have the good initial approximation from previous step t_{j-1} for Newtonian iterative procedure.

Let $\boldsymbol{\Psi}_i^m$ be the vector of components ψ_{kl}^i on $m-$th step of Newtonian iterative process. Analogously let $\mathbf{U}_i^m$ and $\mathbf{V}_i^m$ be the vectors with components u_{kl}^i and v_{kl}^i. Then we have the following boundary problem for the vectors of corrections $\delta\boldsymbol{\Psi}_i^m = \boldsymbol{\Psi}_i^{m+1} - \boldsymbol{\Psi}_i^m, \delta\mathbf{U}_i^m = \mathbf{U}_i^{m+1} - \mathbf{U}_i^m, \delta\mathbf{V}_i^m = \mathbf{V}_i^{m+1} - \mathbf{V}_i^m$:

$$\frac{\delta\boldsymbol{\Psi}_{i+1}^m - 2\delta\boldsymbol{\Psi}_i^m + \delta\boldsymbol{\Psi}_{i-1}^m}{h^2} + [D_i]\delta\boldsymbol{\Psi}_i^m + [A_{11}](\mathbf{u}_i^m)\delta\boldsymbol{\Psi}_i^m + [A_{12}](\boldsymbol{\Psi}_i^m)\delta\mathbf{U}_i^m - \lambda_m\delta\boldsymbol{\Psi}_i^m +$$

$$+\delta\lambda_m\Psi_i^m = -\left(\frac{\boldsymbol{\Psi}_{i-1}^m - 2\boldsymbol{\Psi}_i^m + \boldsymbol{\Psi}_{i+1}^m}{h^2} + [D_i]\boldsymbol{\Psi}_i^m + \mathbf{F}_i(\boldsymbol{\Psi}_i^m, \mathbf{U}_i^m) - \lambda_m\boldsymbol{\Psi}_i^m\right),$$

$$\frac{\delta\mathbf{U}_{i+1}^m - 2\delta\mathbf{U}_i^m + \delta\mathbf{U}_{i-1}^m}{h^2} + [D_i]\delta\mathbf{U}_i^m + [A_{21}](\mathbf{U}_i^m)\delta\mathbf{U}_i^m + [A_{22}](\mathbf{U}_i^m)\delta\mathbf{U}_i^m =$$

$$= -\left(\frac{\mathbf{U}_{i-1}^m - 2\mathbf{U}_i^m + \mathbf{U}_{i+1}^m}{h^2} + [D_i]\mathbf{U}_i^m + \mathbf{G}_i(\boldsymbol{\Psi}_i^m, \boldsymbol{\Psi}_i^m)\right),$$

$$\frac{\delta\mathbf{V}_{i+1}^m - 2\delta\mathbf{V}_i^m + \delta\mathbf{V}_{i-1}^m}{h^2} + ([D_i] - c^2[E])\delta\mathbf{V}_i^m + [A_{31}](\mathbf{V}_i^m)\delta\mathbf{V}_i^m + [A_{32}](\mathbf{V}_i^m)\delta\mathbf{V}_i^m =$$

$$= -\left(\frac{\mathbf{V}_{i-1}^m - 2\mathbf{V}_i^m + \mathbf{V}_{i+1}^m}{h^2} + ([D_i] - c^2[E])\mathbf{V}_i^m + \mathbf{G}_i(\boldsymbol{\Psi}_i^m, \boldsymbol{\Psi}_i^m)\right),$$

where $[A_{11}], [A_{12}], [A_{21}], [A_{22}], [A_{31}], [A_{32}]$ are the matrices corresponding to differential operators for functions $\mathbf{F}_i$ and $\mathbf{G}_i$ from (18)-(20). After that, the boundary conditions can be written in the following form:

$$\delta\boldsymbol{\Psi}_0^m = 0, \delta\boldsymbol{\Psi}_N^m = 0; \qquad \delta\mathbf{U}_0^m = 0, \delta\mathbf{U}_N^m = 0; \qquad \delta\mathbf{V}_0^m = 0, \delta\mathbf{V}_N^m = 0.$$

To find the correction $\delta\lambda_m$ for spectral parameter, we have the equation:

$$(\delta\boldsymbol{\Psi}^m, \boldsymbol{\Psi}^m) = 0$$

To solve this boundary problem, we apply the matrix sweep method [11]. The found solutions $\{\boldsymbol{\Psi}_i^*(t_M), \mathbf{U}_i^*(t_M)\,\mathbf{V}_i^*(t_M)\}, \lambda^*(t_M)$ are used as the initial approximations for Newtonian iterative process to solve the problem (18)-(23) with the non-zero boundary conditions (24). Initial approximations $\{y_i^*\}, \lambda^*$ for $t = 0$ have been found by using the procedure from [12].

4 The results of numerical simulation

On the base of the proposed techniques the FORTRAN code to calculate the three-dimensional egenvalues and eigenfunctions for the polaron model has been created. These programs allow one to find as spherically-symmetrical and axi-symmetrical solutions as essentially three-dimensional solutions.

Now we define some classification of solutions. Let the class of solutions $\Omega_{k_0 l_0}$ correspond to the solutions obtained from initial guesses with $k = k_0, l = l_0$ in (31),(32). It should be noted that from iteration process we have functions ψ_{kl}, u_{kl} equal to 0 for $k < k_0$. So the class Ω_{00} includes the spherically-symmetrical solutions of the polaron problem. The case with $l_0 = 0$ corresponds to the axial symmetry of the problem. In Table 1 the summary of the obtained results for polaron model in the limit of strong coupling ($c = \infty$) and the comparison with results from [5] are shown. From this Table we see that the fast convergence of solutions in respect to the number of harmonics takes place near the ground state, but there is a dynamics of the numerical results for the low part of spectrum. In order to obtain the reliable results in this case it is necessary to increase the dimension of the solving discrete problem. In Table 2 the numerical results in dependence on the constant of coupling c for different classes Ω_{kl} are given. From this Table one can see the fast convergence of the solutions for large value c to the solutions in limit of strong coupling.

This investigation has been supported by the Russian Foundation for Fundamental Research, grant N $94 - 01 - 01119$.

References

1. Pekar, S.I.: Issledovania po elektronnoi teorii kristallov. Moscow, Gostekhizdat, 1951 (in Russian)
2. Amirkhanov, I.V. et.al.: Report of Science Investigation Biological Center of USSR Science Academy. Pushchino 1988
3. Amirkhanov, I.V. et.al.: Report of Science Investigation Biological Center of USSR Science Academy. Pushchino 1990
4. Amirkhanov, I.V., Zemlyanaya, E.V., Puzynina, T.P.: JINR Report. P-11-91-139 Dubna 1991
5. Gabdullin, R.R.: Report of Science Investigation Biological Center of USSR Science Academy. Pushchino 1991
6. Amirkhanov, I.V., Puzynin, I.V., Puzynina, T.P., Zemlyanaya, E.V.: Preprint of JINR. E-11-92-205 Dubna 1992
7. Gabdullin, R.R. - DAN Rossii, 1993, v.333, 1, .23 (in Russian)
8. Ortega, J.M., Rheinboldt, W.C.: Iterative solution of nonlinear equations in several variables. Academic Press. New York and London 1970
9. Fletcher, C.A. : J. Computational Galerkin Methods, Springer-Verlag, New York, Berlin, Heidelberg, Tokyo, 1984
10. Korn, G., Korn, T.: Mathematical handbook for scientists and engineer. McGraw-Hill book comp., New York, Toronto, London, 1961
11. Samarsky, A.A.: Teorija raznosthykh skhem. Moscow, Nauka, 1977 (in Russian)
12. Akishin P.G.,Puzynin I.V. JINR Report. 5-10992, Dubna, 1977.

Table 1. The calculated eigenvalues λ in dependence on the number of harmonics K.

Eigenvalue	Class	$K=1$	$K=3$	$K=5$	results from [5]
λ_0	Ω_{00}	0.08139	-	-	0.0814
λ_1	Ω_{00}	0.01540	-	-	0.0154
λ_2	Ω_{00}	0.00626	-	-	0.0062
λ_3	Ω_{00}	0.00337	-	-	-
λ_0	Ω_{10}	0.02705	0.03392	0.03443	0.0343
λ_1	Ω_{10}	0.00880	0.00833	0.01327	0.0126
λ_2	Ω_{10}	0.00432	0.00407	0.00415	-
λ_3	Ω_{10}	0.00255	0.00263	0.00276	-
λ_0	Ω_{21}	-	0.01490	0.01642	0.0159
λ_1	Ω_{21}	-	0.00618	0.00573	-
λ_2	Ω_{21}	-	0.00342	-	-

Table 2. Eigenvalues λ in dependence on the value of the constant of coupling c.

Eigenvalue	Class	$C=0.1$	$C=1.0$	$C=10.0$	$C=\infty$
λ_0	Ω_{00}	0.0347194	0.0769554	0.0813419	0.0813949
λ_1	Ω_{00}	0.0127952	0.0153239	0.0153978	0.0153986
λ_2	Ω_{00}	0.0058969	0.0062565	0.0062632	0.0062638
λ_0	Ω_{10}	0.0215822	0.0333973	0.0339144	0.0339199
λ_1	Ω_{10}	0.0078265	0.0083211	0.0083275	0.0083276
λ_2	Ω_{10}	-	0.0040679	0.0040688	0.0040688
λ_3	Ω_{10}	-	0.0026337	0.0026339	0.0026339
λ_0	Ω_{21}	0.0126877	0.0148575	0.0148967	0.0148970
λ_1	Ω_{21}	0.0059044	0.0061771	0.00618088	0.00618088
λ_2	Ω_{21}	-	-	0.0034152	0.0034152

CANM in Numerical Investigation of QCD Problems

I.V.Amirkhanov, I.V.Puzynin, T.P.Puzynina, T.A.Strizh, E.V.Zemlyanaya

Laboratory of Computing Techniques and Automation
Joint Institute for Nuclear Research,
141980 Dubna, Russia

Abstract. The iterative schemes based on the Continuous analogue of the Newton's method (CANM) have been applied to the solving a number of the QCD problems.
A mathematical statement of boundary problems is formulated for the QCD-inspired potential quarkonium model on the basis of Schwinger - Dyson and Bethe - Salpeter equations. The explicit form of these equations depends on a chosen effective potential. In the different cases we have the systems of nonlinear integral, integrodifferential or differential equations. The cases of the oscillator, Yukawa, Gauss, Coulomb and linear potentials are considered.

1 Introduction

Physical programs for experiments on the meson-factories set for the particle physics theory a problem of constructing a unified description for the spectra and formfactors of light and heavy mesons. In recent works [1] the quark-potential model, based on the QCD effective Hamiltonian in Coulomb gauge has been developed. In this approach the description of mesons spectra reduces to the solution of boundary value problems for three-dimensional Schwinger - Dyson (S-D) and Bethe - Salpeter (B-S) equation. The S-D equation describes the quark characteristics 'in' meson. The solution of the B-S equations (eigenvalues and eigenfunctions) is associated to the masses and the wave functions of the free mesons. The aim of numerical analyses of these equations is to investigate characteristics of the model as well as to describe experimental data for the meson masses and decay constants.

The effective method for the numerical solving of these equations is the continuous analogue of the Newton's method (CANM) [2].

Particularly, Newtonian modified iterative schemes are successfully used for numerical investigation of SD and BS equations for the most popular potentials, such as oscillator [3], Yukawa [4], Gauss [5], Coulomb and linear [6], as well as for the generalization of this model for finite temperature [7].

A mathematical statement of these problems and a generalized CANM iterative scheme are presented below.

2 Schwinger - Dyson Equation

The SD equation for an arbitrary potential $V(|\bar{p}-\bar{q}|)$ can be written as the coupled equations [3]:

$$\begin{cases} E(p)cos(2v(p)) = m_0 + \frac{1}{2}\int d\bar{q}V(|\bar{p}-\bar{q}|)cos(2v(q))/(2\pi)^3 \\ E(p)sin(2v(p)) = p + \frac{1}{2}\int d\bar{q}V(|\bar{p}-\bar{q}|)\xi sin(2v(q))/(2\pi)^3, \end{cases} \tag{1}$$

where the integration over the three - dimensional vector space of $\bar{q}$ is supposed, $\xi = (\bar{p}/p, \bar{q}/q)$ is the scalar product of unique vectors, m_0 is the given constant (the current quark mass). $E(p)$ and $v(p)$ are the quark energy and wave function, respectively, that should be founded.

Integrating over angle $\Omega\bar{q}$ in these equations, we arrive at

$$\begin{cases} E(p)cos(2v(p)) = m_0 + I_1 \\ E(p)sin(2v(p)) = p + I_2, \end{cases} \tag{2}$$

where

$$I_1 = \int_0^\infty dqV_1(p,q)cos(2v(q)), \qquad I_2 = \int_0^\infty dqV_2(p,q)sin(2v(q)), \tag{3}$$

$$V_1 = \frac{1}{2}\frac{1}{(2\pi)^3}q^2\int d\Omega V(|\bar{p}-\bar{q}|), \qquad V_2 = \frac{1}{2}\frac{1}{(2\pi)^3}q^2\int d\Omega\xi V(|\bar{p}-\bar{q}|). \tag{4}$$

The asymptotical conditions for the energy function $E(p)$ and the mass function $m(p) = E(p)cos(2v(p))$ are:

$$\lim_{p\to\infty} E(p) = p, \qquad \lim_{p\to 0} E(p) = const, \tag{5}$$

$$\lim_{p\to\infty} m(p) = m_0, \qquad \lim_{p\to 0} m(p) = const. \tag{6}$$

The asymptotical behavior of functions $v(p)$ and $\phi(p)$ has the form:

$$\lim_{p\to 0} v(p) = 0, \quad \lim_{p\to\infty} v(p) = \pi/4, \quad \lim_{p\to 0}\phi(p) = \pi/2, \quad \lim_{p\to\infty}\phi(p) = 0. \tag{7}$$

To provide the performance of Eqs.(5)-(6) some subtractions schemes are used

$$\begin{cases} I_1(p) = \int_0^\infty dqV_1(p,q)cos(2v(q))f_1(p,q), \\ I_2(p) = \int_0^\infty dqV_2(p,q)sin(2v(q))f_2(p,q). \end{cases} \tag{8}$$

Here $f_1(p,q)$ and $f_2(p,q)$ are some functions.

Let us consider the effective potential as the sum of the Gaussian and oscillator ones [3,5]:

$$V = V_G + V_O, \qquad V_G = v_g \exp(-\mu^2 r^2) + C, \qquad V_O = -v_o r^2, \tag{9}$$

where C is a constant, $v_g > 0$, $\mu > 0$ and $v_o > 0$ are parameters of the potential. Using the Fourier transformations, we obtain the potential in the momentum space

$$V(|\bar{p} - \bar{q}|) = \frac{v_g}{(\sqrt{\pi})^3} R^3 \exp(-R^2|\bar{p} - \bar{q}|^2) + C(2\pi)^3 \delta(|\bar{p} - \bar{q}|) -$$
$$- v_o(2\pi)^3 \Delta_{\bar{p}} \delta(|\bar{p} - \bar{q}|), \quad R = 1/(2\mu). \tag{10}$$

Substituting the effective potential (10) into Eqs.(2)-(4) and turning to the dimensionless values we obtain:

$$\begin{cases} (\tilde{E}(\tilde{p}) - \tilde{C}/2) cos(2v(\tilde{p})) = \tilde{m}_0 + \tilde{I}_1 \\ (\tilde{E}(\tilde{p}) - \tilde{C}/2) sin(2v(\tilde{p})) = \tilde{p} + \tilde{I}_2. \end{cases} \tag{11}$$

Here

$$\begin{cases} \tilde{I}_1 = [(sin(\phi(\tilde{p})))'' + \frac{2}{p}(sin(\phi(\tilde{p})))'] + \tilde{J}_1, \\ \tilde{I}_2 = [(cos(\phi(\tilde{p})))'' + \frac{2}{p}(cos(\phi(\tilde{p})))' - \frac{2}{p^2} cos(\phi(\tilde{p}))] + \tilde{J}_2, \end{cases} \tag{12}$$

$$\tilde{J}_1 = \hat{\alpha} \int_0^\infty d\tilde{q} \tilde{V}_1(\tilde{p}, \tilde{q}) cos(2v(\tilde{q})), \qquad \tilde{J}_2 = \hat{\alpha} \int_0^\infty d\tilde{q} \tilde{V}_2(\tilde{p}, \tilde{q}) sin(2v(\tilde{q})), \tag{13}$$

$$\tilde{V}_1 = \tilde{R} \frac{\tilde{q}}{\tilde{p}} [\exp(-\tilde{R}^2(\tilde{p}^2 + \tilde{q}^2)) sh(2\tilde{R}^2 \tilde{p}\tilde{q})], \tag{14}$$

$$\tilde{V}_2 = \frac{1}{2} \frac{1}{\tilde{R}\tilde{p}^2} \{\exp(-\tilde{R}^2(\tilde{p}^2 + \tilde{q}^2)) [2\tilde{R}^2 \tilde{p}\tilde{q} ch(2\tilde{R}^2 \tilde{p}\tilde{q}) - sh(2\tilde{R}^2 \tilde{p}\tilde{q})]\}, \tag{15}$$

For Yukawa potential in coordinate space we have [4]:

$$V = \frac{\alpha}{r} \{\exp(-\mu_1 r) - \exp(-\mu_2 r)\} + C, \tag{16}$$

where α, μ_1, μ_2, C are the potential parameters, $\mu_1 \neq \mu_2$, and in the momentum space this potential takes form:

$$V(\bar{p} - \bar{q}) = 4\pi\alpha \left\{ \frac{1}{{\mu^2}_1 + |\bar{p} - \bar{q}|^2} - \frac{1}{{\mu^2}_2 + |\bar{p} - \bar{q}|^2} \right\} + + C(2\pi)^3 \delta(|\bar{p} - \bar{q}|) \tag{17}$$

Eq.(4) takes form

$$V_1 = \frac{\alpha q}{2p} \left[\ln \left| \frac{(p+q)^2 + \mu_1^2}{(p-q)^2 + \mu_1^2} \right| - \ln \left| \frac{(p+q)^2 + \mu_2^2}{(p-q)^2 + \mu_2^2} \right| \right] \tag{18}$$

$$V_2 = \frac{\alpha q}{2p} \left[\frac{p^2 + q^2 + \mu_1^2}{4pq} \ln \left| \frac{(p+q)^2 + \mu_1^2}{(p-q)^2 + \mu_1^2} \right| - \frac{p^2 + q^2 + \mu_2^2}{4pq} \ln \left| \frac{(p+q)^2 + \mu_2^2}{(p-q)^2 + \mu_2^2} \right| \right]. \tag{19}$$

For the linear and coulomb potential we have coordinate representation [6]

$$V(r) = -\alpha\frac{1}{r} + \sigma r, \tag{20}$$

where α and σ are parameters. The momentum representation is:

$$V(|\bar{p}-\bar{q}|) = -\alpha\frac{4\pi}{|\bar{p}-\bar{q}|^2} - \sigma\frac{8\pi}{|\bar{p}-\bar{q}|^4}. \tag{21}$$

Substituting (21) into (3) we have integrals I_1 and I_2 which are nonconvergent if $|\bar{p}-\bar{q}| \to 0$. Using identity

$$\frac{1}{|\bar{p}-\bar{q}|^4} = \frac{1}{6}\triangle_p\frac{1}{|\bar{p}-\bar{q}|^2} \tag{22}$$

we can rewrite system (2) as

$$\begin{cases} E(p)cos(2v(p)) = m_0 + I_{1K} + I_{1L} \\ E(p)sin(2v(p)) = p + I_{2K} + I_{2L}, \end{cases} \tag{23}$$

where

$$\bar{\alpha} = -\frac{\alpha}{2\pi}, \qquad \bar{\sigma} = -\frac{\sigma}{\pi},$$

$$I_{1C} = \frac{\bar{\alpha}\hat{I}_{1C}}{p}, I_{2C} = \frac{\bar{\alpha}\hat{I}_{2C}}{p}, I_{1L} = \frac{\bar{\sigma}\hat{I}''_{1C}}{2p}, I_{2L} = \frac{\bar{\sigma}}{2p}[\hat{I}''_{2C} - \frac{2\hat{I}_{2C}}{p^2}], \tag{24}$$

$$\begin{cases} \hat{I}_{1C}(p) = \int\limits_0^\infty dqq ln|\frac{p+q}{p-q}|cos(2v(q))f_1(p,q), \\ \hat{I}_{2C}(p) = \int\limits_0^\infty dqq[-1 + \frac{p^2+q^2}{2pq}ln|\frac{p+q}{p-q}|]sin(2v(q))f_2(p,q). \end{cases} \tag{25}$$

3 Bethe - Salpeter Equation

The BS equation for the pseudoscalar meson as the quark-antiquark bound state has been obtained in ref.[5] for the spherically symmetric case, and can be written as following coupled equations:

$$MU_{\binom{2}{1}}(p) = (E_t(p) - \frac{C}{2})U_{\binom{1}{2}}(p) - \tag{26}$$

$$-2\int\limits_0^\infty dq[C_p^{\binom{-}{+}}C_q^{\binom{-}{+}}\hat{V}_1(p,q) + S_p^{\binom{-}{+}}S_q^{\binom{-}{+}}\hat{V}_2(p,q)]U_{\binom{1}{2}}(q)$$

The normalization condition is:

$$\frac{4N_C}{M}\frac{1}{(2\pi)^3}\int dqU_1(q)U_2(q) = 1, \tag{27}$$

where $N_C = 3$ is the quantum number and

$$\hat{V}_1(p,q) = \frac{p}{q}V_1(p,q), \qquad \hat{V}_2(p,q) = \frac{p}{q}V_2(p,q), \tag{28}$$

$V_1(p,q)$ и $V_2(p,q)$ are determined by the Eqs.(4),

$$C_p^{(\pm)} = cos(v_1(p) \pm v_2(p)), \qquad S_p^{(\pm)} = sin(v_1(p) \pm v_2(p)),$$

$v_1(p), v_2(p)$ and E_1, E_2 are solutions of the SD equation for quark and antiquark, $E_t(p) = E_1(p)+E_2(p)$ is the total energy of the meson, M is the eigenvalue (mass of the coupled state), $U_{\binom{1}{2}}$ is the wave function.

Using the solutions of the system, we can calculate the lepton decay constant for the pseudoscalar mesons:

$$F_\pi = \frac{4N_C}{M}\frac{1}{(2\pi)^3}\sqrt{4\pi}\int\limits_0^\infty dqU_2(q)cos(v_1(q)+v_2(q))q. \tag{29}$$

The solution of the system (26) must satisfy the asymptotic conditions:

$$\lim_{p\to 0} U_{\binom{1}{2}}(p) = 0, \qquad \lim_{p\to\infty} U_{\binom{1}{2}}(p) = 0. \tag{30}$$

For the potential (10) with $C = 0$ we obtain [3,5]:

$$MU_{\binom{2}{1}}(p) + U''_{\binom{1}{2}}(p) + W_{\binom{1}{2}}(p)U_{\binom{1}{2}}(p) = \tag{31}$$

$$= -2\int\limits_0^\infty dq[C_p^{(\mp)}C_q^{(\mp)}\hat{V}_1(p,q) + S_p^{(\mp)}S_q^{(\mp)}\hat{V}_2(p,q)]U_{\binom{1}{2}}(q),$$

$$\frac{4N_C}{M}\frac{1}{(2\pi)^3}\int\limits_0^\infty dqU_1(q)U_2(q) = 1, \tag{32}$$

where

$$W_1 = -\{E_t + \frac{1}{4}(\phi_1' + \phi_2')^2 + \frac{2}{p^2}cos^2(\frac{\phi_1+\phi_2}{2})\}, \tag{33}$$

$$W_2 = -\{E_t + \frac{1}{4}(\phi_1' - \phi_2')^2 + \frac{2}{p^2}sin^2(\frac{\phi_1-\phi_2}{2})\}, \tag{34}$$

$$\phi_i = -2v_i + \pi/2, \quad i = 1,2, \tag{35}$$

$$U_{\binom{2}{1}00} = U_{\binom{2}{1}}, \tag{36}$$

$$\hat{V}_1(p,q) = \frac{p}{q}V_1(p,q), \quad \hat{V}_2(p,q) = \frac{p}{q}V_2(p,q), \tag{37}$$

$V_1(p,q)$ and $V_2(p,q)$ are determined by relations (14) and (15).

For Yukawa potential (17) we have the system (26) with

$$\hat{V}_1(p,q) = \frac{p}{q}V_1(p,q), \qquad \hat{V}_2(p,q) = \frac{p}{q}V_2(p,q). \tag{38}$$

$V_1(p,q)$ и $V_2(p,q)$ are determined by the eqs.(23-24) correspondingly.

In ref.[6] it is shown that for the linear and coulomb potential BS equations can be reduced to the next system

$$\begin{cases} V_{13}U_1'' + V_{12}U_1' + V_{11}U_1 - MU_2 - 2J_{12} = 0 \\ V_{23}U_2'' + V_{22}U_1' + V_{21}U_2 - MU_1 - 2J_{22} = 0 \end{cases} \tag{39}$$

where

$$V_{11} = E_t(p) - 2\{C_p^{(-)}[\bar\alpha CV_1^{(-)} + \frac{\bar\sigma}{2}(CV_1^{(-)})''] + S_p^{(-)}[\bar\alpha SV_2^{(-)} + + \frac{\bar\sigma}{2}((SV_2^{(-)})'' - \frac{2}{p^2}SV_2^{(-)})]\}, \tag{40}$$

$$V_{12} = -2\bar\sigma[C_p^{(-)}(CV_1^{(-)})' + S_p^{(-)}(SV_2^{(-)})'], \tag{41}$$

$$V_{13} = -\bar\sigma[(C_p^{(-)}CV_1^{(-)} + S_p^{(-)}SV_2^{(-)}], \tag{42}$$

$$V_{21} = E_t(p) - 2\{C_p^{(+)}[\bar\alpha CV_1^{(+)} + \frac{\bar\sigma}{2}(CV_1^{(+)})''] + S_p^{(+)}[\bar\alpha SV_2^{(+)} + + \frac{\bar\sigma}{2}((SV_2^{(+)})'' - \frac{2}{p^2}SV_2^{(+)})]\}, \tag{43}$$

$$V_{22} = -2\bar\sigma[C_p^{(+)}(CV_1^{(+)})' + S_p^{(+)}(SV_2^{(+)})'], \tag{44}$$

$$V_{23} = -\bar\sigma[C_p^{(+)}CV_1^{(+)} + S_p^{(+)}SV_2^{(+)}], \tag{45}$$

$$J_{12} = C_p^{(-)}[\bar\alpha W_{12C}^{(-)}(p) + \frac{\bar\sigma}{2}(W_{12C}^{(-)}(p))''] + S_p^{(-)}[\bar\alpha W_{22C}^{(-)}(p) + + \frac{\bar\sigma}{2}\{(W_{22C}^{(-)}(p))'' - \frac{2}{p^2}W_{22C}^{(-)}(p)\}], \tag{46}$$

$$J_{22} = C_p^{(+)}[\bar\alpha W_{12C}^{(+)}(p) + \frac{\bar\sigma}{2}(W_{12C}^{(+)}(p))''] + S_p^{(+)}[\bar\alpha W_{22C}^{(+)}(p) + + \frac{\bar\sigma}{2}\{(W_{22C}^{(+)}(p))'' - \frac{2}{p^2}W_{22C}^{(+)}(p)\}], \tag{47}$$

$$CV_1^{(\pm)} = C_p^{(\pm)}SV_1, \qquad SV_2^{(\pm)} = S_p^{(\pm)}SV_2.$$

$$SV_1(p) = \int dq ln|\frac{p+q}{p-q}|, \qquad SV_2(p) = \int dq[-1 + \frac{p^2+q^2}{2pq}ln|\frac{p+q}{p-q}|],$$

$$\tilde W_{11C}^{(\pm)} = C_p^{(\pm)}\tilde{SV}_1(p)U_{\binom{2}{1}}(p),$$

$$\tilde W_{12C}^{(\pm)} = \int_0^\infty dq ln|\frac{p+q}{p-q}|[C_q^{(\pm)}U_{\binom{2}{1}}(q) - \frac{q}{p}C_p^{(\pm)}U_{\binom{2}{1}}(p)],$$

$$\tilde W_{21C}^{(\pm)} = S_p^{(\pm)}\tilde{SV}_2(p)U_{\binom{2}{1}}(p),$$

$$\tilde W_{22C}^{(\pm)} = \int_0^\infty dq[-1 + \frac{p^2+q^2}{2pq}ln|\frac{p+q}{p-q}|][S_q^{(\pm)}U_{\binom{2}{1}}(q) - \frac{q}{p}S_p^{(\pm)}U_{\binom{2}{1}}(p)].$$

4 Description of the Iterative Scheme on the Base of CANM

It can be seen that the explicit form of the S-D and B-S equations depends on the chosen effective potential. The simulation of nonlinear effects in the QCD-inspired field models results in the singular, nonlinear boundary value problems for the differential, integral and integrodifferential equations depending on the "external" physical parameters of a model. Each of these equations can be considered as a nonlinear functional equation.

According to the generalized continuous analogue of the Newton method for a nonlinear functional equation in B - space

$$\phi(z) = 0 \tag{48}$$

which represents an investigated problem we consider the evolution equation with respect to continuous parameter t

$$\frac{d}{dt}\phi(t, z(t)) = -\phi(t, z(t)), \qquad 0 \le t < \infty \tag{49}$$

with the initial condition $z(0) = z_0$.

A parametrization of $\phi = \phi(t, z(t))$ is performed so that for $t = 0$ we have a simple equation

$$\phi(0, z(0)) \equiv \phi_0(z_0) = 0$$

which can be solved easily and $\lim_{t\to\infty} \phi(t, z(t)) = \phi(z)$.

From Eq.(49) designating $A(t) = \phi_z^{'}(t, z(t))$ we obtain

$$\frac{dz}{dt} = -A^{-1}(t)[\phi(t, z(t)) + \phi_t^{'}(t, z(t))]. \tag{50}$$

Since the integral of Eq.(49) is $\phi(t, z(t)) = e^{-t}\phi(0, z_0)$, then for $t \to \infty$, $||\phi(t, z(t))|| \to 0$, and $z(t)$ converges to unknown solution z^*.

The discretization of the continuous parameter for Eq.(50) $t : (t_0, t_1, ...t_k)$; $t_0 = 0$, $t_{k+1} - t_k = \tau_k$ is performed in the framework of the Eiler scheme approximation

$$z(t_{k+1}) = z(t_k) + \tau_k v_k \tag{51}$$

where

$$v_k = -B(t_k)[\phi(t_k, z(t_k)) + \phi_t^{'}(t_k, z(t_k))], \qquad B(t_k) = A^{-1}(z(t_k)).$$

Calculating an iteration correction v_k and step τ_k for each t_k, we receive a new approximation $z(t_{k+1})$ to solution z^*.

The iteration process should be continued until the next relation is performed:

$$\| \phi(t_k, z(t_k)) \| \le \epsilon \tag{52}$$

where $\epsilon > 0$ is a given small number.

The convergence of this iteration process is justified in ref.[2].

There are different modifications of CANM allowing to develop iterative schemes which take into account specialties of problems considered. In ref.[8] one of such modifications applied for solving considered problems is developed.

Note that the combination of the CANM and Continuation method [9] allows to investigate effectively the solution dependence upon the problem parameters including the physical problem "external" parameters and the computational schemes "internal" parameters.

Authors are grateful to Russian Foundation for Basic Research RFBR (Grant 94 - 01 - 01119) for support.

References

1. Kalinovsky Yu.L., Kallies W., Kaschluhn L., Münchow L., Pervushin V.N. and Sarikov N.A.: Mesons in the Low - Energy Limit of QCD. Fortschr. Phys. **38** (1990) 333–343
2. Puzynin I.V. and Zhanlav T.: On convergency of iteration on the basis of Continuous analogue of Newton method. Comput. Math.and Math.Phys. **32** (1992) 846–856
3. Amirkhanov I.V., Zemlyanaya E.V., Pervushin V.N., Puzynin I.V., Puzynina T.P., Sarikov N.A. and Strizh T.A.: Numerical investigation of Schwinger - Dyson and Bethe - Salpeter equations with Gauss and oscillator potentials at the framework of the quarkonium model. JINR Communication **E11-94-509** (1994) 1–20
4. Amirkhanov I.V., Davlatov H.F., Zemlyanaya E.V., Pervushin V.N., Puzynin I.V., Puzynina T.P., Sarikov N.A. and Strizh T.A.: Numerical investigation of modification of QCD-inspired quarkonium model with the Yukawa potential. JINR Communication **P11-94-523** (1994) 1–12
5. Amirkhanov I.V., Zemlyanaya E.V., Pervushin V.N., Puzynin I.V., Puzynina T.P., Sarikov N.A. and Strizh T.A.: Numerical investigation of Schwinger - Dyson and Bethe - Salpeter equations with Gauss potential at the framework of the quarkonium model. Math. Model. **6** (1994) 55–70
6. Amirkhanov I.V., Zemlyanaya E.V., Puzynin I.V., Puzynina T.P. and Strizh T.A.: On some problems of numerical investigation of the quarkonium model with the coulomb and linear potentials, Math. Model. **7** (1995) 34–48
7. Amirkhanov I.V., Zemlyanaya E.V., Pervushin V.N., Puzynin I.V., Puzynina T.P.and Strizh T.A.: Numerical analyzes of a quarkonium model at finite temperature. Math. Model. **8** (1996) (to appear)
8. Puzynin I.V., I.V.Amirkhanov I.V., Puzynina T.P. and Zemlyanay E.V.: The newtonian iterative scheme with simultaneous calculating the inverse operator for the derivative of nonlinear function. JINR Rapid Comm. **5[62]-93** (1993) 63–73
9. Ortega J.M. and Rheinboldt W.C.: Iterative solution of nonlinear equations in several variables. Academic Press (1970) N.Y.

Parallel Iterative Solvers for Banded Linear Systems

Pierluigi Amodio[1] and Francesca Mazzia[1]

Dipartimento di Matematica, Università di Bari.
Via E. Orabona 4, I-70125 Bari, Italy. Fax: +39 80 5442722.
E-mail: labor@alphamath.dm.uniba.it

Abstract. A parallel implementation of the SOR iterative method is presented for the solution of block banded linear systems. The algorithm is based on the block reordering of the coefficient matrix used by the domain decomposition methods. It is proved that the obtained iteration matrix maintains the same spectral properties of the corresponding sequential method and also the same optimal parameter of relaxation. The parallel SOR algorithm is then applied to the solution of linear systems arising from the discretization of elliptic partial differential equations in order to obtain an interesting comparison with the coloring schemes.

Key words and phrases. Parallel algorithms, iterative solvers, SOR iteration.

1 Introduction

Numerical linear algebra is the kernel of most of the existing algorithms for the solution of problems arising from physics, chemistry, and engineering. The use of parallel computers has reduced the execution time and has allowed the solution of quite difficult problems such as, for example, those derived from the meteorology.

In this paper we are interested in iterative methods for the parallel solution of large banded linear systems. Banded systems (in particular tridiagonal systems) arise from the discretization of several PDEs and ODEs [2, 4, 7]. Their parallel solution by means of the classical iterations (Gauss-Seidel, SOR) has received particular attention especially when the coefficient matrices arise from the discretization of elliptic partial differential equations on rectangular domains. In particular, most of the approaches are based on the SOR iteration applied to different orderings, called coloring, of the grid nodes (see [1, 2]).

The easiest and most famous ordering is called red/black, since it corresponds to perform a red/black (or odd/even) permutation to the problem given by the natural rowwise ordering. An important property of this ordering is that its iteration matrix has the same eigenvalues of that corresponding to the natural rowwise. Therefore, the algorithm associated to the red/black has the same asymptotic convergence rate of the sequential SOR.

In this paper we analyze a class of parallel iteration schemes which preserves the convergence properties of the natural rowwise scheme. The algorithms are based on the partitioning of the coefficient matrix among the processors used by the domain decomposition methods [3, 7]. They are obtained by applying classical iterative methods to a permuted problem, where the permutation matrix is chosen in order to emphasize parallel computations.

In Section 2 we present the parallel block and point SOR algorithms applied to block banded matrices, without considering the internal structure of each block. In Section 3 we analyze the parallel solution of elliptic boundary value problems by exploiting the internal band structure of each block. In this way some new iteration schemes, similar to the classical multicoloring schemes, are derived.

2 The parallel SOR method

Let us consider the solution of the linear system

$$Ax = f, \tag{1}$$

where the coefficient matrix A is block banded

$$A = \begin{pmatrix} A_1 & C_{1,1} & \dots & C_{r,1} & & \\ B_{1,2} & A_2 & C_{1,2} & & \ddots & \\ \vdots & \ddots & \ddots & \ddots & & C_{r,n-r} \\ B_{s,s+1} & & \ddots & \ddots & \ddots & \vdots \\ & \ddots & & \ddots & A_{n-1} & C_{1,n-1} \\ & & B_{s,n} & \dots & B_{1,n} & A_n \end{pmatrix}, \tag{2}$$

and the blocks A_i, B_{ij} and C_{ij} are $m \times m$ matrices.

Let $p \le \min(n/(2r), n/s)$ be an integer that exactly divides n, consider the following partitioning of A:

$$A = \begin{pmatrix} M_1 & -U'_1 & & \\ -L'_2 & M_2 & \ddots & \\ & \ddots & \ddots & -U'_{p-1} \\ & & -L'_p & M_p \end{pmatrix},$$

where $M_i = D_i - U_i - L_i$ are $(n/p) \times (n/p)$ block banded matrices with D_i as their main block diagonal and $-L_i$ and $-U_i$ as their lower and upper block triangular part; L'_i and U'_i are block matrices respectively with s off-diagonals in the upper right corner and r off-diagonals in the lower left corner.

Moreover, let

$$D = \begin{pmatrix} D_1 & & \\ & \ddots & \\ & & D_p \end{pmatrix}, \quad L_p = \begin{pmatrix} L_1 & & \\ & \ddots & \\ & & L_p \end{pmatrix}, \quad U_p = \begin{pmatrix} U_1 & & \\ & \ddots & \\ & & U_p \end{pmatrix},$$

$$L'_p = \begin{pmatrix} O & & & \\ L'_2 & O & & \\ & \ddots & \ddots & \\ & & L'_p & O \end{pmatrix}, \quad U'_p = \begin{pmatrix} O & U'_1 & & \\ & O & \ddots & \\ & & \ddots & U'_{p-1} \\ & & & O \end{pmatrix}.$$

Then:

$$A = D - L_p - L'_p - U_p - U'_p.$$

The block SOR iteration is expressed by

$$(D - \omega(L_p + L'_p))x^{(k)} = \omega f + ((1-\omega)D + \omega(U_p + U'_p))x^{(k-1)} \qquad k = 1, 2, \ldots$$

and the iteration matrix is:

$$G = (D - \omega(L_p + L'_p))^{-1}((1-\omega)D + \omega(U_p + U'_p)). \tag{3}$$

We define as the parallel block SOR iteration the algorithm given by the following recursion (see also Fig. 1):

$$(D - \omega(L_p + U'_p))x^{(k)} = \omega f + ((1-\omega)D + \omega(U_p + L'_p))x^{(k-1)} \qquad k = 1, 2, \ldots \tag{4}$$

This algorithm derives from the parallel Gauss-Seidel iteration presented in [4] and may be efficiently implemented on at most p processors. Its theoretical properties are based on the following:

Lemma 1. *Let $A = M - N$ be a splitting for the matrix A, with M nonsingular. Then λ is an eigenvalue of $M^{-1}N$ iff $det(\lambda M - N) = 0$.*

Proof. If λ is an eigenvalue of $M^{-1}N$, then

$$0 = \det(M^{-1}N - \lambda I) = \det(M^{-1}(N - \lambda M) = \det(N - \lambda M). \qquad \square$$

Let

$$G_p = (D - \omega(L_p + U'_p))^{-1}((1 - \omega D) + \omega(U_p + L'_p)) \tag{5}$$

be the iteration matrix associated to (4) (obviously $G_1 \equiv G$), then the following theorems hold:

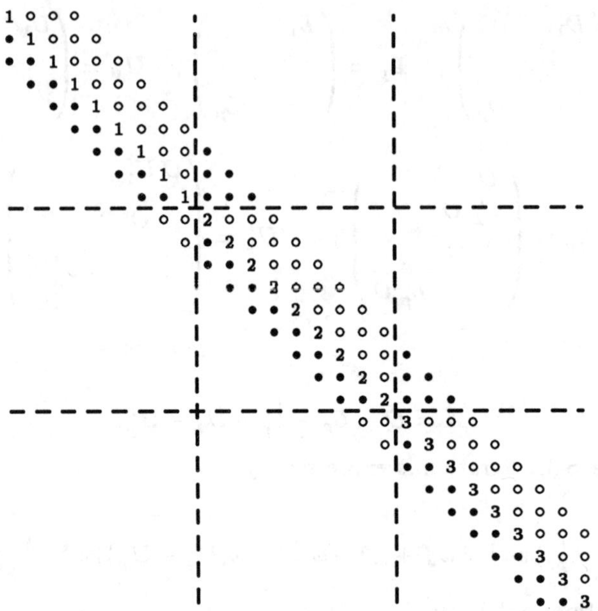

Fig. 1. *Parallel splitting on 3 processors for the matrix (2). The elements of L_p and U'_p are represented by black points, the elements of L'_p and U_p by white points. The numbers on the main diagonal denote the processor that handles the corresponding block row.*

Theorem 2. *The parallel block SOR iteration with $p \leq \min(n/(2r), n/s)$ converges iff the block SOR iteration converges.*

Proof. Let $\lambda \neq 0$ be an eigenvalue of the block SOR iteration, we define the nonsingular matrix

$$S(\lambda) = \begin{pmatrix} I & & & \\ & \lambda I & & \\ & & \ddots & \\ & & & \lambda^{p-1} I \end{pmatrix}, \tag{6}$$

where the identity matrix I has the same size of the blocks M_i. From the previous lemma and the equivalence

$$\begin{aligned} &\lambda(D - \omega(L_p + L'_p)) - (1-\omega)D - \omega(U_p + U'_p) = \\ &\quad S(\lambda)\left(\lambda(D - \omega(L_p + U'_p)) - (1-\omega)D - \omega(U_p + L'_p)\right) S(\lambda)^{-1} \end{aligned}$$

we have that λ is an eigenvalue of the parallel block SOR iteration. □

Theorem 3. *If $p \leq \min(n/(2r), n/s)$, then the iteration matrices of the block SOR iteration and of the parallel block SOR iteration have the same optimum parameter of relaxation ω.*

Proof. Since ω depends on the spectrum of the iteration matrices of the two algorithms, the thesis easily derives from the previous Theorem 2. □

It is possible to obtain a parallel point SOR by considering a different splitting of the matrix A. If the matrix D in (3) and (4) contains only the main diagonal elements of each block A_i and the matrices L_p and U_p are opportunely changed, then (3) becomes the point SOR iteration and (4) is a parallel implementation of the point SOR. Theorem 2 may be generalized to emphasize the spectral properties of this iteration scheme:

Theorem 4. *The parallel point SOR iteration with* $p \leq \min(n/(2r), n/s)$ *converges iff the point SOR iteration converges.*

Proof. It is sufficient to choose the matrices D, L_p and U_p in order to obtain the splitting of the point SOR algorithm and follow the proof of Theorem 2. □

3 Parallel solution of elliptic problems

In this section we generalize the parallel point SOR iteration to the solution of the linear systems arising from the discretization of elliptic partial differential equations on rectangular domains. In this case, by ordering the grid nodes in the natural rowwise fashion (left to right, bottom to top) we obtain a block tridiagonal matrix with sparse blocks. The structure of each block depends on the kind of discretization used and on the complexity of the problem. If no mixed derivatives are presented, a 5-point discretization leads to a block tridiagonal matrix with the main diagonal blocks tridiagonal and the other blocks diagonal. The 9-point approximation leads to a block tridiagonal matrix with tridiagonal blocks; a matrix of the same structure may be also obtained when mixed derivative are presented.

The parallel solution of elliptic problems and of the linear systems with this structure has been treated by many authors [1, 2, 6, 7, 8] and some interesting results have been derived by grouping the algorithms into equivalence classes that are known to have the same convergence behavior. Some model problems such us the 5-point and 9-point approximation to the Laplacian have been extensively studied and also the corresponding optimal relaxation parameter has been found [2, 5]. Moreover the most important results are derived when the matrix resulting after the discretization is symmetric and positive definite. In the following we will refer only on the structure of each block and the properties of the parallel algorithms will also hold for nonsymmetric matrices.

The parallel block and point SOR in (4) applied to elliptic problems correspond to decompose the domain by rows among the processors (see Fig. 2a). If we want to use a larger number of processors, we can divide the original domain in subrectangles (see Fig. 2b); the corresponding iterative scheme is obtained by exploiting the structure of each block.

Let us consider a $n \times n$ block tridiagonal matrix with tridiagonal blocks of size m. The iteration scheme can be derived by applying a procedure similar

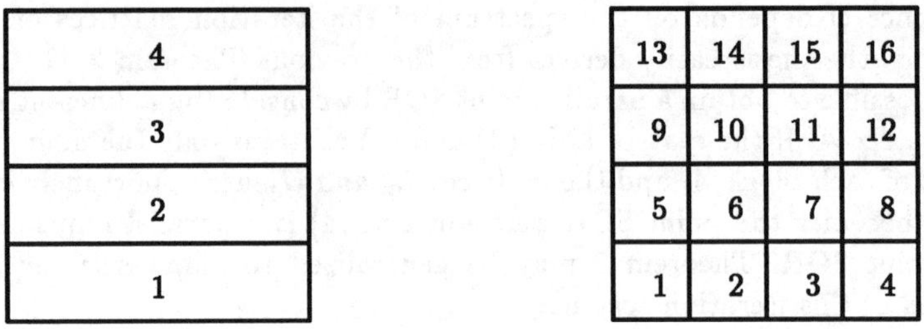

4
3
2
1

(a)

13	14	15	16
9	10	11	12
5	6	7	8
1	2	3	4

(b)

Fig. 2. Parallel decomposition of the domain among the processors.

to that proposed in the previous section. The substantial difference is that, for $\lambda \neq 0$, instead of the matrix $S(\lambda)$ in (6) we use

$$S(\lambda) = \begin{pmatrix} R(\lambda) & & & \\ & \lambda R(\lambda) & & \\ & & \ddots & \\ & & & \lambda^{p-1} R(\lambda) \end{pmatrix}, \tag{7}$$

where $R(\lambda)$ is a suitable nonsingular matrix and $p \leq n/2$ is the number of rows in the domain decomposition. By choosing opportunely the matrix $R(\lambda)$, we may derive parallel SOR schemes with the same spectral properties of the point SOR scheme. For example, we may use the matrix $R(\lambda)$ defined as

$$R(\lambda) = \begin{pmatrix} I_{m(n/p-2)} & & \\ & T_1(\lambda) & \\ & & T_2(\lambda) \end{pmatrix},$$

where, let $w = (1, 1, \lambda, \lambda, \lambda^2, \lambda^2, \ldots)$,

$$W_1 = \begin{pmatrix} w_1 & & \\ & \ddots & \\ & & w_q \end{pmatrix} \quad \text{and} \quad W_2 = \begin{pmatrix} w_2 & & \\ & \ddots & \\ & & w_{q+1} \end{pmatrix},$$

$T_1(\lambda)$ and $T_2(\lambda)$ are defined as

$$T_1(\lambda) = W_1 \otimes I_{m/q}, \qquad T_2(\lambda) = W_2 \otimes I_{m/q}$$

and $q \leq m/2$ is the number of columns in the domain decomposition (see Fig. 2b).

When applied to the 9-point discretization, we have

$$\begin{aligned} & S(\lambda)^{-1} \left(\lambda(D - \omega(L_p + L'_p)) - (1-\omega)D - \omega(U_p + U'_p)\right) S(\lambda) = \\ & \lambda(D - \omega L_{pq}) - (1-\omega)D - \omega U_{pq}, \end{aligned} \tag{8}$$

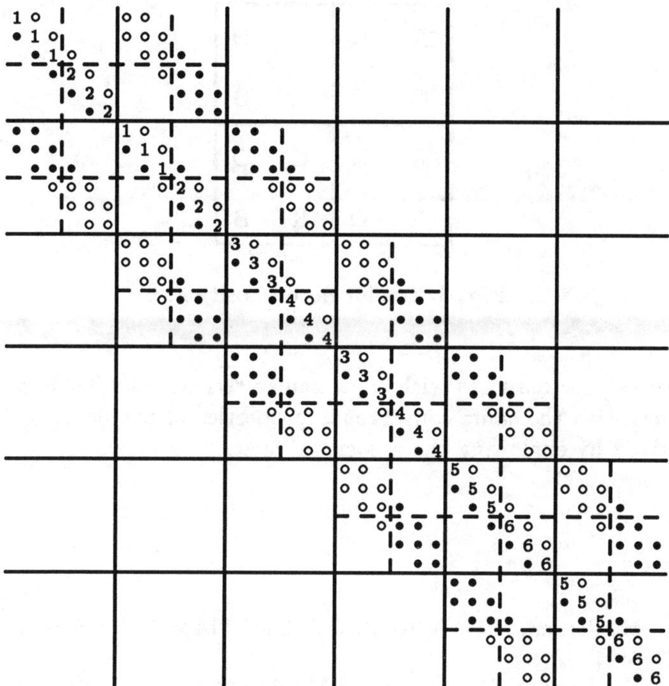

Fig. 3. *Parallel splitting on 6 processors (2 inside each block) for a matrix with 9 nonnull diagonals. The elements of L_{pq} are represented by black points, the elements of U_{pq} by white points. The numbers on the main diagonal denote the processor that handles the corresponding row.*

where the matrices $\boldsymbol{L}_{pq}$ and $\boldsymbol{U}_{pq}$ have a structure similar to that in Fig. 3. From Theorem 2 we derive that the parallel scheme:

$$(\boldsymbol{D} - \omega \boldsymbol{L}_{pq})\boldsymbol{x}^{(k)} = \omega \boldsymbol{f} + ((1-\omega)\boldsymbol{D} + \omega \boldsymbol{U}_{pq})\boldsymbol{x}^{(k-1)} \qquad k = 1, 2, \ldots \tag{9}$$

will converge with the same asymptotic behaviour of the point SOR scheme. This scheme may be implemented in parallel on at most pq processors. If $p = n/2$ and $q = m/2$ it is equal to the classical R/B/G/O scheme as presented in Fig. 4. In [1] this 4-color SOR iteration has been proved to have the same convergence behavior of the natural rowwise scheme only in the case of $\boldsymbol{A}$ symmetric and positive definite.

If the trasformation matrix $S(\lambda)$ is applied to the 5-point discretization, because of the particular structure of the off-diagonals, the choice $p = n/2$ and $q = m$ gives rise to a parallel iteration scheme which is equivalent to the classical red/black.

The presented results suggest that the convergence properties of other multi-coloring schemes could be investigated by using the same approach with different

R	B	G	O
G	O	R	B
R	B	G	O
G	O	R	B

Fig. 4. 4-color decomposition.

$R(\lambda)$. Moreover, for matrices with a particular structure, suitable parallel iterative schemes with the same convergence properties of the point or block SOR may be derived by exploiting the associated sparsity pattern.

References

1. Adams, L.M., Jordan, H.F.: Is SOR color-blind? SIAM J. Sci. Statist. Comput. **7** (1986) 490–506
2. Adams, L.M., Leveque, R.J., Young, D.M.: Analysis of the SOR iteration for the 9-point Laplacian. SIAM J. Numer. Anal. **25** (1988) 1156–1180
3. Amodio, P., Brugnano, L., Politi, T.: Parallel factorizations for tridiagonal matrices. SIAM J. Numer. Anal. **30** (1993) 813–823
4. Amodio, P., Mazzia, F.: A parallel Gauss-Seidel method for block tridiagonal linear systems. SIAM J. Sci. Comput. (to appear)
5. Chan, T.F., Elman, H.C.: Fourier Analysis of iterative methods for elliptic problems. SIAM Rev. **31** (1989) 20–49
6. Evans, D.J.: Parallel S.O.R. iterative methods. Parallel Comput. **1** (1984) 3–18
7. Ortega, J.M.: Introduction to parallel and vector solution of linear systems. Plenum Press, New York, 1988
8. Patel N.R., Jordan, H.F.: A parallelized point rowwise successive over-relaxation method on a multiprocessor. Parallel Comput. **1** (1984) 207–222

Finite Element Approximations of Some Central Curves *

Andrey B. Andreev and Todor D. Todorov

Department of Mathematics
Technical University of Gabrovo
Bulgaria

Abstract. The aim of this paper is to obtain the conditions of applicability of FLAC3 finite elements in the plane. Some approximate properties and applications are discussed.

1 Introduction.

Using curved elements is an efficient procedure to increase the accuracy of approximation of a boundary value problem in numerical analysis. The triangular isoparametric elements are the elements most often used by engineers when solving boundary value problems in two variables. It is natural to apply the corresponding one - dimensional elements for approximation of boundary. We adopt notations of Argyris and Mlejnek [1], and we consider FLAC3 finite element approximations of the next central curves:

$$\Gamma : \begin{cases} x = cos^{2n+1}t \\ y = sin^{2n+1}t, \quad t \in [0, 2\pi) \end{cases}$$

where n is an integer and $n \geq 1$.

There are number of reasons for considering this problem:

i) FLAC3 curvilinear finite elements are closely associated with six - node isoparametric elements (TRIC6 elements). The good approximation of boundary, using TRIC6 elements, leads to the phenomenon of superconvergence [2] and supraconvergence [3]. From the computational point of view the curved elements make it possible to obtain the same order of accuracy as in the case when original boundary is polygon [4, 5].

ii) This curves may be boundary of domain in $\Re^2$ with turning points [6];

iii) Bounded domain Ω in $\Re^2$ with piecewise smooth boundary $\partial\Omega$ consisting of parts of central curves;

iv) Arch problem (see [7]).

The curve Γ can be derived into a finite number of arcs, each of which has the parametric representation with smooth functions.

* This work is partially supported by the Bulgarian Ministry of Science and Technologies, Contract MM-524/95.

2 Finite Element Transformations.

Let $(\hat{K}, \hat{P}, \hat{\Sigma})$ be the finite element of reference defined as follows:
$-\hat{K} = \{\hat{x}; 0 \leq \hat{x} \leq 1\}$ is the space [0,1].
$-\hat{P} = P_2$, where P_2 is the space of all polynomials of degree not exeeding second on $\hat{K}$.
$-\hat{\Sigma} = \{\hat{x};\ \hat{x} = \frac{i}{2}, i = 0, 1, 2\}$ is the set of all Lagrangian interpolation nodes.

We consider a set Γ_h approximating Γ and let τ_h be finite element triangulation with parameter h defined as follows :

$$\tau_h = \{K = F_K(\hat{K}) \mid \textit{the nodes } A_i \in \Gamma \ \ i = 1, 2, 3; \ \ meas(K) < h\}$$

where $F_K : \hat{K} \to K \subset \Re^2$ is a finite element transformation:

$$F_K : \begin{pmatrix} x \\ y \end{pmatrix} = \hat{x}(2\hat{x} - 1)\begin{pmatrix} x_1 \\ y_1 \end{pmatrix} + (1 - \hat{x})(1 - 2\hat{x})\begin{pmatrix} x_2 \\ y_2 \end{pmatrix} + 4\hat{x}(1 - \hat{x})\begin{pmatrix} x_3 \\ y_3 \end{pmatrix} \quad (1)$$

and $A_i(x_i, y_i)$, i = 1, 2, 3.

We denote by h_K the length of K and assume that all h_K are bounded by h.

The Jacobian of transformation (1) is :

$$J(\hat{x}) = (x\prime(\hat{x}), y\prime(\hat{x}))^T.$$

We introduce the functionals E and H from C[0,1] to $\Re$ with:

$$Ez = -z(1) - 3z(0) + 4z(0.5), \quad Hz = z(0) + z(1) - 2z(0.5).$$

Then, we can write :

$$F_K : \begin{cases} x = \varphi_K(\hat{x}) = 2Hx.\hat{x}^2 + Ex.\hat{x} + x_2 \\ y = \psi_K(\hat{x}) = 2Hy.\hat{x}^2 + Ey.\hat{x} + y_2 \end{cases} \quad (2)$$

3 Properties of FLAC3 elements with nodes on Γ.

An important property is invertibility of finite element transformations. The problem about invertibility of eight - node two dimensional quadratic isoparametric transformations is considered in [8]. Necessary and sufficient conditions for invertibility of six - node quadratic isoparametric finite element transformations, associated with finite elements with one curved side, are proved in [9], and with two curved sides in [10]. We will prove a more general result conserning isoparametric transformation.

Theorem 1 *Let T be a transformation from [0,1] to $\Re^2$ defined by:*

$$T : \begin{pmatrix} x \\ y \end{pmatrix} = B\begin{pmatrix} \hat{x}^2 \\ \hat{x} \end{pmatrix} + b,$$

where

$$B = \begin{pmatrix} b_{11} & b_{12} \\ b_{21} & b_{22} \end{pmatrix}, \quad b = \begin{pmatrix} b_{13} \\ b_{23} \end{pmatrix}.$$

Then T is noninvertible if and only if there exist a point $\hat{x}_s \in (0, 1)$ such that $T'(\hat{x}_s) = (0, 0)^T$.

Proof. Sufficiency. Let there exist $\hat{x}_s \in (0,1)$ for which

$$T'(\hat{x}_s) = B\begin{pmatrix} 2\hat{x}_s \\ 1 \end{pmatrix} = \begin{pmatrix} 0 \\ 0 \end{pmatrix}. \tag{3}$$

We will show that T is noninvertible.

a) We suppose that $0 < \hat{x}_s \leq 0.5$ Considering the difference:

$$T(2\hat{x}_s) - T(0) = B\begin{pmatrix} 4\hat{x}_s^2 \\ 2\hat{x}_s \end{pmatrix} = 2\hat{x}_s B\begin{pmatrix} 2\hat{x}_s \\ 1 \end{pmatrix} = \begin{pmatrix} 0 \\ 0 \end{pmatrix}.$$

We obtain $T(2\hat{x}_s) = T(0)$ and $0 < 2\hat{x}_s \leq 1$. Therefore, the transformation T is noninvertible.

b) Let $0.5 < \hat{x}_s < 1$. We consider the difference :

$$T(2\hat{x}_s - 1) - T(1) = B\begin{pmatrix} (2\hat{x}_s - 1)^2 \\ 2\hat{x}_s - 1 \end{pmatrix} - B\begin{pmatrix} 1 \\ 1 \end{pmatrix}$$

$$= B\begin{pmatrix} 4\hat{x}_s(\hat{x}_s - 1 \\ 2(\hat{x}_s - 1) \end{pmatrix} = 2(\hat{x}_s - 1)B\begin{pmatrix} 2\hat{x}_s \\ 1 \end{pmatrix} = \begin{pmatrix} 0 \\ 0 \end{pmatrix}.$$

We have $T(2\hat{x}_s - 1) = T(1)$ and $0 < 2\hat{x}_s - 1 < 1$. Therefore, the transformation T is noninvertible.

Necessity. Let the transformation T be noninvertible. We will prove the existence of $\hat{x}_s \in (0,1)$ such that (3) be fulfilled. We take $\hat{x}_I \neq \hat{x}_{II}, \quad \hat{x}_I, \hat{x}_{II} \in (0,1)$ such that $T(\hat{x}_{II}) - T(\hat{x}_I) = 0$, having in mind that T is noninvertible. Hence,

$$T(\hat{x}_{II}) - T(\hat{x}_I) = B\begin{pmatrix} (\hat{x}_{II})^2 \\ \hat{x}_{II} \end{pmatrix} - B\begin{pmatrix} (\hat{x}_I)^2 \\ \hat{x}_I \end{pmatrix} = (\hat{x}_{II} - \hat{x}_I)B\begin{pmatrix} \hat{x}_I + \hat{x}_{II} \\ 1 \end{pmatrix} = \begin{pmatrix} 0 \\ 0 \end{pmatrix}.$$

But $\hat{x}_I \neq \hat{x}_{II}$, therefore $B\binom{\hat{x}_I + \hat{x}_{II}}{1} = \binom{0}{0}$. We put $\hat{x}_s = \frac{\hat{x}_I + \hat{x}_{II}}{2}$. We obtain $B\binom{2\hat{x}_s}{1} = \binom{0}{0}$. where $\hat{x}_s \in (0,1)$ and (3) is proved.

Corollary 1 *The transformation T is invertible if and only if for every $\hat{x} \in (0,1), T\prime(\hat{x}) \neq (0,0)^T$.*

We have the folloing corollary when the nodes of K belong to Γ

Corollary 2 *The transformation F_K, $K \in \tau_h$ is invertible if and only if the nodes of K are three different points.*

Proof. The matrix B of transformation F_K which is defined with (2) is :

$$B = \begin{pmatrix} 2Hx & Ex \\ 2Hy & Ey \end{pmatrix}, \quad b = \begin{pmatrix} x_2 \\ y_2 \end{pmatrix}.$$

We easily calculate:

$$det(B) = (-4). \begin{vmatrix} x_1 & y_1 & 1 \\ x_2 & y_2 & 1 \\ x_3 & y_3 & 1 \end{vmatrix}$$

Taking into account that A_i $i = 1,2,3$ are different points on Γ, then $det(B) \neq 0$, and we prove the invertibility of F_K.

We cosider for the sake of symmetry $x \geq 0, \;\; y \geq 0$. Then

$$\Gamma_1 : \begin{cases} x = cos^{2n+1}t \\ y = sin^{2n+1}t, \quad t \in [0, \frac{\pi}{2}]. \end{cases}$$

We can present the equation of the curve in the explicit form :

$$\Gamma_1 : y_n(x) = (1 - x^{\frac{2}{2n+1}})^{\frac{2n+1}{2}}, \; x \in [0,1], n \in N. \tag{4}$$

Obviously $y_n(x)$ is a monotone decreasing function. We denote

$$B_K : \begin{cases} x_1 + \frac{1}{4}h_1 \leq x \leq x_1 + \frac{3}{4}h_1 \\ y_2 + \frac{1}{4}h_2 \leq y \leq y_2 + \frac{3}{4}h_2 \end{cases},$$

where $h_1 = x_2 - x_1, \;\; h_2 = y_1 - y_2, \;\; h_1, h_2 > 0$.

Theorem 2 *The transformation $F_K, \;\; K \in \tau_h$ gives monotone decreasing function $y = f_K(x)$ if and only if the node $A_3 \in B_K$.*

Proof. We derive from (2):

$$y\prime(\hat{x}) = 4Hy.\hat{x} + Ey, \quad y\prime(0) = -y_1 - 3y_2 + 4y_3, \quad y\prime(1) = 3y_1 + y_2 - 4y_3$$

a) let $y_2 \leq y_3 < y_2 + 0.25h_2$. We obtain

$$y\prime(0) < -y_1 - 3y_2 + 4y_2 + h_2 = 0, \quad y\prime(1) \geq 3y_1 + y_2 - 4y_2 = 3y_1 - 3y_2 > 0.$$

Then we have $y\prime(0).y\prime(1) < 0$.

b) if $y_2 + \frac{3}{4}h_2 < y_3 \leq y_1$, it follows

$$y\prime(0) > -y_1 - 3y_2 + 4y_2 + 3h_2 = 2y_1 - 2y_2 > 0$$

$$y\prime(1) < 3y_1 + y_2 - 4y_2 - 3h_2 = 3(y_1 - y_2) - 3h_2 = 0$$

and we have $y\prime(0).y\prime(1) < 0$.

c) if $y_2 + 0.25h_2 \leq y_3 \leq y_2 + 0.75h_2$, it follows

$$y\prime(\hat{x}) = y_1(4\hat{x} - 1) + y_2(4\hat{x} - 3) + 4y_3(1 - 2\hat{x})$$

c1) if $0 \leq \hat{x} \leq 0.5$ then

$$y\prime(\hat{x}) \geq y_1(4\hat{x} - 1) + y_2(4\hat{x} - 3) + 4(y_2 + \frac{1}{4}h_2)(1 - 2\hat{x}) = 2\hat{x}h_2 \geq 0$$

c2) if $0.5 < \hat{x} \leq 1$ then

$$y\prime(\hat{x}) \geq y_1(4\hat{x} - 1) + y_2(4\hat{x} - 3) + 4(y_2 + \frac{3}{4}h_2)(1 - 2\hat{x}) = 2h_2(1 - \hat{x}) \geq 0$$

The case c) gives us $y\prime(\hat{x}) \geq 0, \;\; \hat{x} \in [0,1]$. Hence $y(\hat{x}), \;\; \hat{x} \in [0,1]$ is monotone incresing function if and only if $y_3 \in [y_2 + 0.25h_2, y_2 + 0.75h_2]$. Analogically, the function $x(\hat{x})$ is monotone decreasing if and only if $x_3 \in [x_1 + 0.25h_1, x_1 + 0.75h_1]$, which complete the proof.

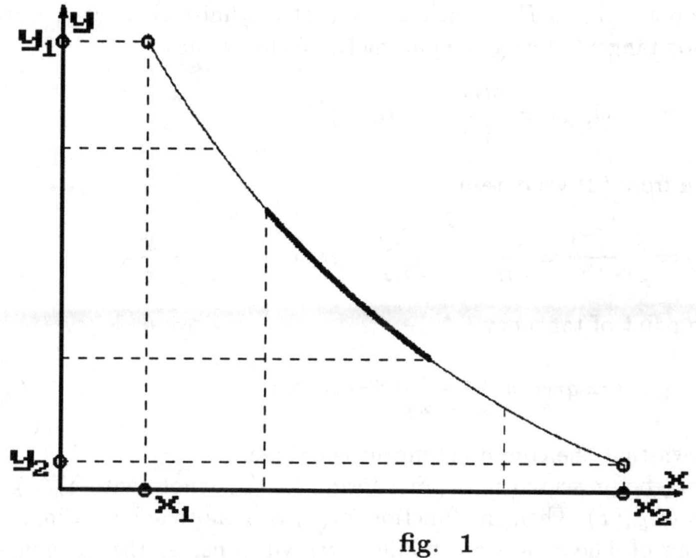

fig. 1

Remark 1 There are h_1 and h_2 sufficiently small such that $B_K \bigcap \Gamma \neq \emptyset$ for any $n \in N$. In other words, if we have a given point $A_1(x_1, y_1)$ on Γ, we choose h_1 and h_2 (or $A_2(x_2, y_2)$ on Γ) such that

$$(x_1 + \frac{1}{4}h_1)^{\frac{2}{2n+1}} + (y_1 - \frac{3}{4}h_2)^{\frac{2}{2n+1}} < 1$$

with

$$x_1^{\frac{2}{2n+1}} + y_1^{\frac{2}{2n+1}} = 1.$$

We present a case where $B_K \bigcap \Gamma = \emptyset$ in the fig. 2. We have a point of selfintersection in this case.

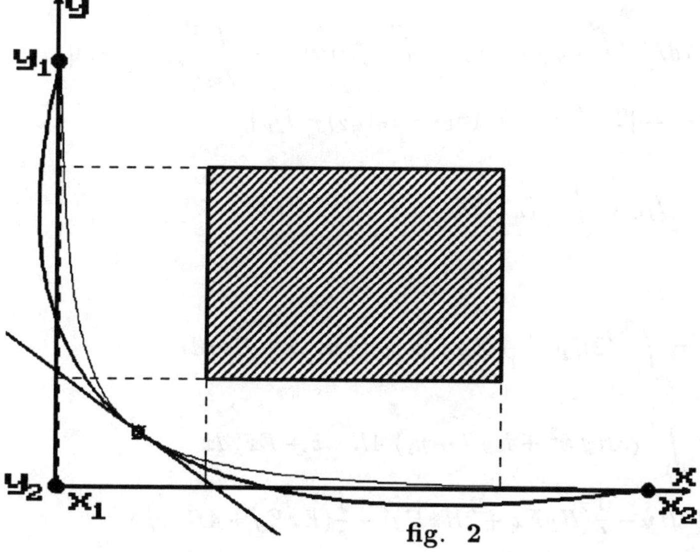

fig. 2

Let us choose the point A_3 on Γ_1 in such a way that the finite element $K \in \tau_h$ and Γ_1 are common tangent. Using the parametric form, we have

$$y\prime_n(x) = \frac{\dot{y}(t)}{\dot{x}(t)} = -(tan\ t)^{2n-1}.$$

On the other hand from (2) we obtain :

$$f\prime_K(x) = \frac{\varphi\prime_K(\hat{x})}{\psi\prime_K(\hat{x})} = \frac{4Hy.\hat{x}^2 + Ey}{4Hx.\hat{x}^2 + Ex}, \quad f\prime_K(x_3) = \frac{y_1 - y_2}{x_1 - x_2}.$$

We determine the point of tangency:

$$t = arctan(\frac{y_2 - y_1}{x_1 - x_2})^{\frac{1}{2n-1}}, n \geq 1. \tag{5}$$

Thus we have shown that the common tangent is unique.

Let the curve Γ_1 be presented in explicit form (4). We denote with $Y_n(x)$ a primitive function of $y_n(x)$. Then the function $Y_n(x)$ is monotone increasing.

If the conditions of Theorem 2 are fulfilled, we will consider the difference between the curve Γ_1 and its finite element approximation in L_1 norm.

Case 1 The curve Γ_1 and the finite element K have unique point of intersection A_3 (fig. 1). Then

$$\|y_n(x) - f_K(x)\|_{L_1,[x_1,x_2]} = \int_{x_1}^{x_2} |y_n(x) - f_K(x)|\, dx$$

$$= \int_{x_1}^{x_3} |y_n(x) - f_K(x)|\, dx + \int_{x_3}^{x_2} |y_n(x) - f_K(x)|\, dx$$

$$= |\int_{x_1}^{x_3} (y_n(x) - f_K(x))\, dx| + |\int_{x_3}^{x_2} (y_n(x) - f_K(x))\, dx|$$

$$= |(\int_{x_1}^{x_3} y_n(x)\, dx - \int_{x_1}^{x_2} y_n(x)\, dx) - (\int_{x_1}^{x_3} f_K(x)\, dx - \int_{x_3}^{x_2} f_K(x)\, dx)|$$

$$= |2Y_n(x_3) - Y_n(x_1) - Y_n(x_2) - I_{f_K}|$$

where

$$I_{f_K} = \int_{x_1}^{x_3} f_K(x)\, dx - \int_{x_3}^{x_2} f_K(x))\, dx$$

Using (2) we have

$$I_{f_K} = \int_0^{\frac{1}{2}} (2Hy.\hat{x}^2 + Ey.\hat{x} + y_2)(4Hx.\hat{x} + Ex)\, d\hat{x}$$

$$+ \int_1^{\frac{1}{2}} (2Hy.\hat{x}^2 + Ey.\hat{x} + y_2)(4Hx.\hat{x} + Ex)\, d\hat{x}$$

$$= -\frac{7}{4}HxHy - \frac{1}{2}(HyEx + 2HxEy) - \frac{1}{4}(ExEy + 4Hx.y_2)$$

We obtain

$$I_{f_k} = \frac{1}{2}(-x_1y_1 - x_1y_3 - x_2y_2 - x_2y_3 + x_3y_1 + x_3y_2 + 2x_3y_3)$$

Finally:

$$\|y_n(x) - f_K(x)\|_{L_1,[x_1,x_2]} = |2Y_n(x_3) - Y_n(x_1) - Y_n(x_2)$$

$$+\tilde{x}y_3 - \tilde{y}x_3 - x_3y_3 + \frac{x_1y_1 + x_2y_2}{2}|, \tag{6}$$

where

$$\tilde{x} = \frac{x_1 + x_2}{2}, \tilde{y} = \frac{y_1 + y_2}{2}$$

Case 2 Let the curve Γ_1 and the element $K \in \tau_h$ have contact in such a way that K is in one side of Γ_1 (fig.2). Clearly, the both curves have a common tangent on the node $A_3(x_3, y_3)$. We easily calculate:

$$\|y_n(x) - f_K(x)\|_{L_1,[x_1,x_2]} = \int_{x_1}^{x_2} |y_n(x) - f_K(x)|\,dx = |\int_{x_1}^{x_2} (y_n(x) - f_K(x))\,dx|$$

$$= |Y_n(x_2) - Y_n(x_1) - \int_{x_1}^{x_2} f_K(x)\,dx|.$$

Using (2) we obtain:

$$\int_{x_1}^{x_2} f_K(x)\,dx = \int_0^1 (2Hy.\hat{x}^2 + Ey.\hat{x} + y_2)(4Hx.\hat{x} + Ex)\,d\hat{x}$$

$$= 2HxHy + \frac{2}{3}(2HxEy + HyEx) + \frac{1}{2}(4Hx.y_2 + ExEy) + Ex.y_2$$

$$= \frac{1}{3}(\frac{3}{2}x_1y_1 - \frac{1}{2}x_1y_2 + 2x_1y_3 + \frac{1}{2}x_2y_1 - \frac{3}{2}x_2y_2 - 2x_2y_3 - 2x_3y_1 + 2x_3y_2).$$

Finally, we have :

$$\|y_n(x) - f_K(x)\|_{L_1,[x_1,x_2]} = |Y_n(x_2) - Y_n(x_1)$$

$$+\frac{4}{3}x_3(y_1 - \tilde{y}) + \frac{4}{3}y_3(x_2 - \tilde{x}) + \frac{x_1y_2 - x_2y_1}{6} - \frac{x_1y_1 - x_2y_2}{2}|. \tag{7}$$

Thus, we have proved the following

Theorem 3 *Let the conditions of theorem 2 be fulfilled. There exists one and only one point $A_3(x_3, y_3)$ on Γ_1 such that the finite element $K \in \tau_h$ and Γ_1 have common tangent. The parameter t corresponding to this point is given by (5). The relations (6) and (7) give approximation properties of FLAC3 finite elements in the plane when the considered central curves are used.*

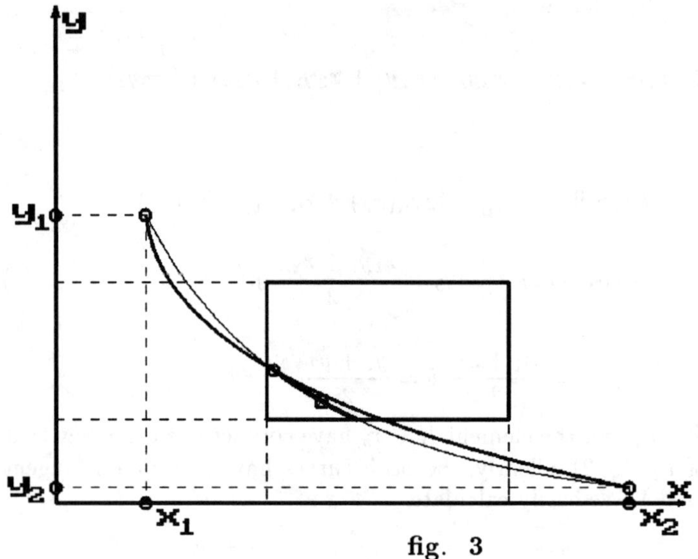

fig. 3

Remark 2 Similarly to the above cases one can research the case where there are two intern points of intersection. But the numerical expiriments show that the case 2 is prefereble (fig. 3).

4 An application

Let us consider the Laplace equation

$$\Delta u = f \tag{8}$$

in a plane domain Ω with a bondary

$$\partial\Omega \;:\; (x-1)^{\frac{2}{2n+1}} + y^{\frac{2}{2n+1}} = 1, \quad x \in [0,2];$$

Thus the origin O(0,0) is a turning point of the boundary.

Theorem 4 *If the equation (8) has the Dirichlet (or Neumann) condition on $\partial\Omega$ and $f \in L_2(\Omega)$, there exists a unique solution $u \in H^2(\Omega) \bigcap {H_0}^1(\Omega)$, (or $u \in H^2(\Omega)$), where $H^2(\Omega)$ and ${H_0}^1(\Omega)$ are standard Sobolev spaces [7].*

Proof. Tthis is consequence of theorem 3.3.1 from [6]. We denote:

$$\varphi_1(x) = -(1-(x-1)^{\frac{2}{2n+1}})^{\frac{2n+1}{2}}$$

$$\varphi_2(x) = (1-(x-1)^{\frac{2}{2n+1}})^{\frac{2n+1}{2}}$$

. We will prove

$$\lim_{x\to 0} \frac{(|\varphi_1''(x)| + |\varphi_2''(x)|)(\varphi_2(x)-\varphi_1(x))}{(\varphi\prime_2(x)-\varphi\prime_1(x))^2} < 2$$

We have:

$$\varphi_1 = -\varphi_2; \quad \varphi_1{}' = -\varphi_2{}' \quad and \quad \varphi_1'' = -\varphi_2''$$

We put

$$x - 1 = X; \quad \frac{2n+1}{2} = N \quad X \in [-1, 1]$$

Then

$$\varphi_2(X) = (1 - X^{\frac{1}{N}})^N,$$

$$\varphi\prime_2(X) = -X^{\frac{1-N}{N}}(1 - X^{\frac{1}{N}})^{N-1},$$

$$\varphi_2''(X) = \frac{N-1}{N} X^{\frac{1-2N}{N}}(1 - X^{\frac{1}{N}})^{N-2}.$$

We obtain:

$$\lim_{X\to 1} \frac{\varphi_2''(X)\varphi_2(X)}{(\varphi\prime_2(X))^2} = \frac{N-1}{N} = \frac{2n-1}{2n+1} < 1$$

We have the same arguments for the other three turning points of $\partial\Omega$.

References

1. Argyris, J., Mlejnek, H.-P.: Finite element methods. Friedr. Vieweg & Sohn, Braunschweig / Wiesbaden, 1986 (In German)
2. Andreev, A.B., Lazarov, R.D.: Superconvrgence of the gradient for quadratic triangular finite elements. Numer. Meth. for PDE, USA, **4** (1988) 15-32
3. Lebaud, M.P.: Error estimate in an isoparametric finite element eigenvalue problem. Math. of Computation **63** (1994) 19-40
4. Zlamal, M: Curved elements in the finite element method I. SIAM J. Numer. Analysis **10** (1973) 229-240
5. Zlamal, M.: Curved elements in the finite element method II. SIAM J. Numer. Analysis **11** (1974) 347-368
6. Grisvard, P.: Eliptic problems in nonsmooth domains. Pitman Advanced Publishing Program, 1985
7. Ciarlet, P.G.: The finite element method for eliptic problems. North - Holland, Amsterdam, 1978
8. Frey, A.E., Hall, C.A., Porsching, T.A.: Some results on the global inversion of bilinear and quadratic isoparametric finite element transformations. Mathematics of computation **32** (1978) 725-749
9. Mitchell, A.R., Wait, R.: The finite element method in partial differential equations. Weley, New York, 1977
10. Todorov, T.D.: Invertibility of isoparametric finite element transformations associated with finite elements with two curved sides. Fifth international conference on differential equations and applications -CDEV́, Rousse, Bulgaria, August 24-29, 1995

Utilization of CANM for Magnetic Field Calculation

E.A.Ayrian, A.V.Fedorov

Laboratory of Computing Techniques and Automation
Joint Institute for Nuclear Research,
141980 Dubna, Russia

Abstract. Iterative schemes based on continuous analogue of the Newton's method have been successfully used in computational modeling of various nonlinear physical problems. The paper concerns the utilization of this method to solve magnetostatic field equations. The extended convergence domain of the method makes possible to construct fast, reliable, robust iteration scheme for magnetic field calculations. Data illustrated convergence of the scheme are presented.

1 Introduction

Iterative schemes based on continuous analogue of the Newton's method (CANM) have been successfully used in computational modeling of various nonlinear physical problems [1]. These schemes are characterized by such properties as adaptation to concrete physical problem, tacking account of a priori and computational information to improve schemes characteristic, evolution from simple equation to complicated one.

Computational modeling by using CANM was implemented for numerical analysis of the blow up regimes of combustion of two–component nonlinear heat–conducting medium [2], for solution of coupled FEM and BEM equations [3], for magnetic field calculation [see e.g.4]. This paper concerns the utilization of CANM to solve a magnetostatic field equations. These equations in 2D may be reduced to boundary value problem for elliptic equation. Computational domain includes a ferromagnetic material with nonlinear properties. The shape of ferromagnetic usually has a complex structure, so a finite elements are used for discretization. For high level field calculations some problems arise with the convergence of utilized iteration processes. It means that iteration parameters should be specified for each problem. The extended convergence domain of CANM [1,5] makes possible to construct fast, reliable, robust iteration scheme for magnetic field calculation independent of field level, domain geometry, material characteristic.

2 Problem Formulation

Weak Galerkin formulation of magnetostatic field problem in 2D is reduced to equation

$$\int_{\Omega} \nu(|\nabla A|^2)\nabla A \cdot \nabla u d\Omega = \int_{\Omega} j u d\Omega, \ \forall u \in H_0^1(\Omega), \tag{1}$$

where unknown z component of magnetic vector potential A is a function of space variables x and y, j is known current density, $\nu = \nu(|\nabla A|^2)$ is magnetic reluctivity of ferromagnetic obtained from experimental data, $\nabla u = \left[\frac{\partial u}{\partial x}, \frac{\partial u}{\partial y}\right]$, $H_0^1(\Omega) = \left\{u | u \in L^2(\Omega), \frac{\partial u}{\partial x}, \frac{\partial u}{\partial y} \in L^2(\Omega), u|_{\Gamma} = 0\right\}$.

Now let us do some important but rather practical assumptions on function ν and its derivative

$$0 < d_1 \leq \nu \leq d_2, \tag{2}$$

$$0 < d_3 \leq \nu + 2\frac{d\nu}{d|\nabla A|^2}|\nabla A|^2 \leq d_4. \tag{3}$$

The assumption (3) has an important role in convergence of Newton-like iteration methods applied to such problem. We must be sure that this assumption satisfies for ν function approximation from experimental measurment.

Under assumptions (2), (3) the problem (1) has unique solution $A \in H_0^1(\Omega)$, bilinear form in lefthand side is symmetrical and positive definite.

3 CANM Iteration Scheme, Discretization

Now apply CANM iteration scheme [1] to equation (1). This yields

$$\int_{\Omega} (\nu(|\nabla A^n|^2) + 2\nu'(|\nabla A^n|^2)|\nabla A^n|^2)\nabla p^n \cdot \nabla u \, d\Omega =$$

$$= \int_{\Omega} (ju - \nabla A^n \cdot \nabla u) \, d\Omega, \ \forall u \in H_0^1(\Omega) \tag{4}$$

$$A^{n+1} = A^n + \tau_n p^n. \tag{5}$$

The lefthand side of equation (4) is the Frechet derivative or Jacobian of (1), the righthand side is the residual. Potential A^{n+1} for next iteration is defined as the sum of the potential A^n and the correction term p multiplied by iteration parameter τ, p is a solution of problem (4).

Equations (4) and (5) set CANM iteration scheme. To discretize equations the domain Ω is divided into finite elements - linear triangles in our calculations on which the vector potential is supposed to have a linear variation

$$A^e(x, y) = \sum_{r=1}^{3} A_r N_r^e(x, y).$$

Applying the weighted residual process to equation (4) with the shape functions $N_r(x,y)$ as weighting functions yields nonlinear system of algebraic equations

$$J(A_h^n)p_h^n = f - K(A_h^n)A_h^n \tag{6}$$
$$A_h^{n+1} = A_h^n + \tau_n p_h^n \tag{7}$$

with matrixes $J(A_h^n)$ and $K(A_h^n)$ are obtained by ensembling element matrixes

$$K(A_h^n) = \nu^e M^e$$
$$J(A_h^n) = 2\nu'^e M^e [A_e] [A_e]^T M^e + J(A_h^n)$$
$$M_{rq}^e = \int\limits_{\Omega^e} (\frac{\partial N_r}{\partial x}\frac{\partial N_q}{\partial x} + \frac{\partial N_r}{\partial y}\frac{\partial N_q}{\partial y}) d\Omega$$
$$f_r = \int\limits_{\Omega^e} j N_r(x,y) d\Omega, \quad r,q = 1,2,3,$$

$[A_e]$ – is a column vector of potential value in element nodes.

Assumptions (2) and (3) on ν and ν' guarantee the positive definiteness of matrixes J and K.

4 System Solution and Results

Linear system of equations (6) was solved by the conjugate gradients method with incomplete Cholessky preconditioner.

As it was mentioned above for high level field calculations it is important to choose accurately an iteration parameter to obtain a convergent process. We determined the parameter τ^n by minimization along Newtonian direction to fulfill the inequality

$$\left\|residual^{n+1}\right\| < \left\|residual^{n}\right\|.$$

Iteration process stopped when residual norm and norm of correction p^n were less than 10^{-4}.

In the table we present number of iterations (it) neccessary to reach the accuracy above, values of iteration parameter τ on each iteration in dependence on current (I) in coils.

it/I(kA)	1	3	5	7	9	11
1	1.	1.	0.655	0.501	0.397	0.330
2	1.	1.	0.857	0.479	0.317	0.238
3	1.	1.	0.590	0.114	0.053	0.053
4		1.	1.	0.206	0.140	0.133
5		1.	1.	0.195	0.237	0.184
6			1.	0.668	0.248	0.286
7			1.	0.773	0.395	0.290
8			1.	0.947	0.400	0.320
9			1.	1.	0.830	0.461
10				1.	0.874	0.618
11				1.	1.	0.818
12					1.	1.
13					1.	1.
14					1.	1.

The result is that CANM based iteration scheme converges for any field level. For low field there is rapid convergence of Newton's method with $\tau_n = 1$. For high field convergence is achieved by appropriate choosing of τ_n and last iterations again are rapid Newton's iterations.

Authors are grateful to Russian Foundation for Basic Research RFBR (Grant 95 - 01 - 01467a) for support.

References

1. Zhidkov E.P., Makarenko G.I., Puzynin I.V.: A continuous analog of Newton's method in nonlinear problem of physics. J.Part.Nucl. **4** 1 (1973) 127–166
2. Dimova S.N., Kaschiev M.S., Koleva M.G., Vasileva D.P.: Numerical analysis of the blow up regimes of combustion of two–component nonlinear heat–conducting medium. Comput. Math.and Math.Phys. **35** (1995) 380–399
3. Ayrian E.A., Fedorov A.V., Yuldashev O.I., Zhidkov E.P.: One approach for coupling FEM and BEM. Proceedings of ISNA'92, Part III. Contributed Papers (1992) 12–19
4. Ayrian E.A., Zhidkov E.P., Fedorov A.V., Khoromskii B.N., Shelaev I.A., Yudin I.P., Yuldashev O.I.: Numerical algorithms for particle accelerator magnetic system calculations. J.Part.Nucl. **21** 1 (1990) 251–307
5. Puzynin I.V. and Zhanlav T.: On convergency of iteration on the basis of Continuous analogue of Newton method. Comput. Math.and Math.Phys. **32** (1992) 846–856

Numerical Algorithm for Simulation of Coupled Heat-Mass Transfer and Chemical Reaction in Two-Phase Steady Flow

I.B. Bazhlekov, M.G. Koleva, D.P. Vasileva

Department of Computational Mathematics,
Institute of Mathematics and Informatics - BAS,
Acad. G. Bontchev str., bl.8, 1113 Sofia, Bulgaria

Abstract. An iterative algorithm for numerical simulation of coupled heat-mass transfer and chemical reaction around flat interface in two-phase steady laminar flow is presented. The mathematical model is based on the boundary layer approximation of 2D Navier-Stokes equations and corresponding convection-diffusion equations for heat and concentration in both phases. The obtained system of nonlinear partial differential equations is discretized by means of explicit and implicit difference schemes on nonuniform grids. The central or upwind differencing of the convective terms is used. The resulting systems of algebraic equations are linearized and solved by TDMA method. The obtained values of the velocity, concentration and temperature are compared with existing theoretical results at different values of the parameters.

1 Introduction

The interest in the investigations of coupled heat-mass transfer and chemical reactions in fluid flows has grown considerably recently because ot their importance for many industrial processes. For a long time the most of the theoretical investigations in this area have been performed emphasizing on one, as the main part of the process, without taking into account the others or using their simple models. So, a large number of investigations have been devoted to fluid flows [1], as in some cases [2, 3] the influence of heat or concentration on the flow has been considered. Others [4, 5] have studied heat-mass transfer at given fluid flows. Thermocapillary effects in falling liquid films resulting from heat-mass transfer, accounting for the effects of chemical reaction have been studied by [6]. However, coupled hydrodynamics and heat-mass transfer, accounting for the effects of chemical reactions in two-phase flows have been less investigated.

The main goal of the present paper is to perform a numerical algorithm for the simulation of such processes taking into account the mutual influence between hydrodynamics, heat and mass transfer and chemical reactions more completely. For this purpose an 2D tow-phase steady flow with flat interface between both phases is considered here. From a mathematical point of view the influence of heat and mass transfer on hydrodynamics is taken into account by means of additional terms in the boundary conditions at the interface, which introduces additional non-linearity [7]. The first of these non-linear effects, Marangoni effect, is a consequence of the rise of secondary flows due to the irregular distribution of the substance or temperature along the interface. The second effect is a result of the concentration gradient directed normally to the interface.

The mathematical formulation of the above problem is given in the next Section and the numerical algorithm is described in Section 3. The algorithm is based on Finite Difference Method (FDM) [8] and simple iterations for solving of the non-linear problems. Numerical tests and comparisons with existing theoretical data, as well as some results are presented in Section 4.

2 Mathematical model

The mathematical model is based on the boundary layer approximation of 2D Navier-Stokes equations and corresponding convection-diffusion equations for heat and concentration. The equations are considered in both phases $\Omega_1 = \{0 \le x \le 1;\ y_1 \ge 0\}$ and $\Omega_2 = \{0 \le x \le 1;\ y_2 \le 0\}$ respectively. The interface is $\{y_1 = y_2 = 0\}$. For convenience the transformation: $y_2 = -y_2$ and $v_2 = -v_2$ is applied and the equations are written below at $y_1,\ y_2 > 0$:

– Equations for the velocity (u_1, v_1) in the first phase $\Omega_1^f = \{0 \le x \le 1;\quad 0 \le y_1 \le Y_1\}$:

$$u_1(x,y_1)\frac{\partial u_1}{\partial x}(x,y_1) + v_1(x,y_1)\frac{\partial u_1}{\partial y_1}(x,y_1) = \frac{\partial^2 u_1}{\partial y_1^2}(x,y_1); \tag{1}$$

$$\frac{\partial u_1}{\partial x}(x,y_1) + \frac{\partial v_1}{\partial y_1}(x,y_1) = 0; \tag{2}$$

with boundary conditions:

$$u_1(0,y_1) = 1; \tag{3}$$

$$u_1(x,0) = \theta_1 u_2(x,0); \tag{4}$$

$$v_1(x,0) = -\theta_4 \frac{\partial C_1}{\partial y_1}(x,0); \tag{5}$$

$$u_1(x,Y_1) = 1; \tag{6}$$

– Equations for the velocity (u_2, v_2) in the second phase $\Omega_2^f = \{0 \le x \le 1;\quad 0 \le y_2 \le Y_2\}$:

$$u_2(x,y_2)\frac{\partial u_2}{\partial x}(x,y_2) + v_2(x,y_2)\frac{\partial u_2}{\partial y_2}(x,y_2) = \frac{\partial^2 u_2}{\partial y_2^2}(x,y_2); \tag{7}$$

$$\frac{\partial u_2}{\partial x}(x,y_2) + \frac{\partial v_2}{\partial y_2}(x,y_2) = 0; \tag{8}$$

with boundary conditions:

$$u_2(0,y_2) = 1; \tag{9}$$

$$\frac{\partial u_2}{\partial y_2}(x,0) = -\theta_2 \frac{\partial u_1}{\partial y_1}(x,0) - \theta_3 \frac{\partial T_2}{\partial x}(x,0) - \theta_8 \frac{\partial C_2}{\partial x}(x,0); \tag{10}$$

$$v_2(x,0) = 0; \tag{11}$$

$$u_2(x,Y_2) = 1; \tag{12}$$

– Equations for the concentration C_1 in the first phase $\Omega_1^C = \{0 \le x \le 1; \quad 0 \le y_1 \le \bar{Y}_1\}$:

$$u_1(x,y_1)\frac{\partial C_1}{\partial x}(x,y_1) + v_1(x,y_1)\frac{\partial C_1}{\partial y_1}(x,y_1) = \frac{1}{Sc_1}\frac{\partial^2 C_1}{\partial y_1^2}(x,y_1); \tag{13}$$

with boundary conditions:

$$C_1(0,y_1) = 1; \tag{14}$$

$$C_1(x,0) = -C_2(x,0); \tag{15}$$

$$C_1(x,\bar{Y}_1) = 1; \tag{16}$$

– Equations for the concentration C_2 in the second phase $\Omega_2^C = \{0 \le x \le 1; \quad 0 \le y_2 \le \bar{Y}_2\}$:

$$u_2(x,y_2)\frac{\partial C_2}{\partial x}(x,y_2) + v_2(x,y_2)\frac{\partial C_2}{\partial y_2}(x,y_2) = \frac{1}{Sc_2}\frac{\partial^2 C_2}{\partial y_2^2}(x,y_2) - Da\, C_2(x,y_2); \tag{17}$$

with boundary conditions:

$$C_2(0,y_2) = 0; \tag{18}$$

$$\frac{\partial C_2}{\partial y_2}(x,0) = \theta_5\frac{\partial C_1}{\partial y_1}(x,0); \tag{19}$$

$$C_2(x,\bar{Y}_2) = 0; \tag{20}$$

– Equations for the temperature T_1 in the first phase $\Omega_1^T = \{0 \le x \le 1; \quad 0 \le y_1 \le \bar{\bar{Y}}_1\}$:

$$u_1(x,y_1)\frac{\partial T_1}{\partial x}(x,y_1) + v_1(x,y_1)\frac{\partial T_1}{\partial y_1}(x,y_1) = \frac{1}{Pr_1}\frac{\partial^2 T_1}{\partial y_1^2}(x,y_1); \tag{21}$$

with boundary conditions:

$$T_1(0,y_1) = 0; \tag{22}$$

$$T_1(x,0) = T_2(x,0); \tag{23}$$

$$T_1(x,\bar{\bar{Y}}_1) = 0; \tag{24}$$

– Equations for the temperature T_2 in the second phase $\Omega_2^T = \{0 \le x \le 1; \quad 0 \le y_2 \le \bar{\bar{Y}}_2\}$:

$$u_2(x,y_2)\frac{\partial T_2}{\partial x}(x,y_2) + v_2(x,y_2)\frac{\partial T_2}{\partial y_2}(x,y_2) = \frac{1}{Pr_2}\frac{\partial^2 T_2}{\partial y_2^2}(x,y_2) - \theta_6 C_2(x,y_2); \tag{25}$$

with boundary conditions:

$$T_2(0,y_2) = 0; \tag{26}$$

$$\frac{\partial T_2}{\partial y_2}(x,0) = -\theta_7\frac{\partial T_1}{\partial y_1}(x,0); \tag{27}$$

$$T_2(x,\bar{\bar{Y}}_2) = 0. \tag{28}$$

3 Numerical algorithm

The boundary layer equations for velocity (1), (7); concentration (13), (17) and temperature (21), (25) are written in a general form below:

$$u_i(x,y_i)\frac{\partial w_i}{\partial x}(x,y_i) = p_i\frac{\partial^2 w_i}{\partial y_i^2}(x,y_i) - v_i(x,y_i)\frac{\partial w_i}{\partial y_i}(x,y_i) - \\ q_i w_i(x,y_i) - s_i C_i(x,y_i), \quad i=1,2. \tag{29}$$

The pairs of the boundary conditions (4), (10); (15), (19) and (23), (27) can be generalized as

$$\alpha_{1i}w_1(x,0)+\beta_{1i}\frac{\partial w_1}{\partial y_1}(x,0) = \alpha_{2i}w_2(x,0)+\beta_{2i}\frac{\partial w_2}{\partial y_2}(x,0)+\gamma_i(x), \quad i=1,2; \tag{30}$$

and the pairs of the boundary conditions (6), (12); (16), (20) and (24), (28) as

$$w_i(x,\tilde{Y}_i) = \delta_i, \quad i=1,2. \tag{31}$$

In the above equations w_i stands for u_i, C_i or T_i and $\tilde{Y}_i$ stands for Y_i, $\bar{Y}_i$ or $\bar{\bar{Y}}_i$. The constants p_i, q_i, s_i, α_{ij}, β_{ij} and δ_i are different for different w_i. The function $\gamma_i(x)$ appears in (30) only in the case when $w_i = u_i$, $i=2$ and depends on C_2 and T_2 (see (10)).

The equations (2) and (8) are

$$\frac{\partial v_i}{\partial y_i}(x,y_i) = -\frac{\partial u_i}{\partial x}(x,y_i), \quad i=1,2; \tag{32}$$

and the boundary conditions (5) and (11) are of the kind

$$v_i(x,0) = \mu_i(x), \quad i=1,2, \tag{33}$$

where μ_i might depend on C_i.

Let $\hat{Y}_i = \max\{Y_i, \bar{Y}_i, \bar{\bar{Y}}_i\}$ and let us introduce the nonuniform grid $\omega_\tau \times \omega_h$,

$$\omega_\tau = \{x_k,\ x_k = x_{k-1}+\tau_k;\ k=1,2,\ldots,m;\ x_0=0;\ \sum_{k=1}^{m}\tau_k = 1\};$$

$$\omega_h = \{y_{ij},\ y_{ij} = y_{ij-1}+h_{ij};\ i=1,2,\ j=1,2,\ldots,n_i;\ y_{i0}=0;\ \sum_{j=1}^{n_i}h_{ij} = \hat{Y}_i\}.$$

Using FDM [8] and denoting by w_{ij}^k the approximate solution for $w_i(x_k,y_j)$, the following explicit difference scheme for determining w_{ij}^k on the k-th layer $(1 \le k \le m)$

$$u_{ij}^{k-1}\frac{w_{ij}^k - w_{ij}^{k-1}}{\tau_k} = \frac{2p_i}{h_{ij}+h_{ij+1}}\left(\frac{w_{ij+1}^{k-1}-w_{ij}^{k-1}}{h_{ij+1}} - \frac{w_{ij}^{k-1}-w_{ij-1}^{k-1}}{h_{ij}}\right) - \\ v_{ij}^{k-1}\frac{w_{ij+1}^{k-1}-w_{ij-1}^{k-1}}{h_{ij}+h_{ij+1}} - q_i w_{ij}^{k-1} - s_i C_{ij}^{k-1}, \quad i=1,2,\ j=1,2,\ldots,n_i-1; \tag{34}$$

$$\alpha_{1i} w_{10}^k + \beta_{1i} \frac{w_{11}^k - w_{10}^k}{h_{11}} = \alpha_{2i} w_{20}^k + \beta_{2i} \frac{w_{21}^k - w_{20}^k}{h_{21}} + \gamma_i^k, \quad i = 1,2; \tag{35}$$

$$w_{in_i}^k = \delta_i, \quad i = 1,2 \tag{36}$$

could be considered. For this scheme the stability condition:

$$\tau_k \le \frac{u_{ij}^{k-1} h_{ij} h_{ij+1}}{2p_i + q_i h_{ij} h_{ij+1}}, \quad i = 1,2, \ \ j = 1,2,\ldots,n_i - 1 \tag{37}$$

must be satisfied.

To find directly the values of the unknowns w_{ij}^k, first the equations for C_i, $i = 1,2$, then for T_i, $i = 1,2$ and after that for u_i, $i = 1,2$ have to be solved.

For determining the values of v_i, $i = 1,2$ the forward Euler's method is used:

$$v_{i0}^k = \gamma_i^k, \quad v_{ij+1}^k = v_{ij}^k - h_{ij+1} \frac{u_{ij+1}^k - u_{ij+1}^{k-1}}{\tau_k}, \quad j = 0,1,\ldots,n_i - 1. \tag{38}$$

However, for the problem (1)–(28) the explicit scheme (34)–(38), because of the stability condition (37), requires very small steps τ_k and that is why this scheme is practically unusable. This is commented in Section 4.

The next implicit difference schemes could be also used:

– with central differencing of convective terms

$$\begin{aligned} u_{ij}^k \frac{w_{ij}^k - w_{ij}^{k-1}}{\tau_k} &= \frac{2p_i}{h_{ij} + h_{ij+1}} \left(\frac{w_{ij+1}^k - w_{ij}^k}{h_{ij+1}} - \frac{w_{ij}^k - w_{ij-1}^k}{h_{ij}} \right) - \\ v_{ij}^k \frac{w_{ij+1}^k - w_{ij-1}^k}{h_{ij} + h_{ij+1}} &- q_i w_{ij}^k - s_i C_{ij}^k, \quad i = 1,2, \ j = 1,2,\ldots,n_i - 1; \end{aligned} \tag{39}$$

– with upwind differencing of convective terms

$$\begin{aligned} u_{ij}^k \frac{w_{ij}^k - w_{ij}^{k-1}}{\tau_k} &= \frac{2p_i}{h_{ij} + h_{ij+1}} \left(\frac{w_{ij+1}^k - w_{ij}^k}{h_{ij+1}} - \frac{w_{ij}^k - w_{ij-1}^k}{h_{ij}} \right) - \\ (v_{ij}^{k\,-}) \frac{w_{ij+1}^k - w_{ij}^k}{h_{ij+1}} &- (v_{ij}^{k\,+}) \frac{w_{ij}^k - w_{ij-1}^k}{h_{ij}} - q_i w_{ij}^k - s_i C_{ij}^k, \end{aligned} \tag{40}$$

where $v_{ij}^{k\,-} = \min\{v_{ij}^k, 0\}$, $v_{ij}^{k\,+} = \max\{v_{ij}^k, 0\}$, $i = 1,2$, $j = 1,2,\ldots,n_i - 1$.

The boundary conditions at the point (0,0) are not consistent and the schemes based on (39) or (40) are not conservative. Because of that, to obtain a sufficiently accurate numerical solution small steps τ_k have to be used near the origin.

To get a conservative difference scheme the equation (29) is written in the divergent form

$$\frac{\partial(u_i w_i)}{\partial x}(x, y_i) = p_i \frac{\partial^2 w_i}{\partial y_i^2}(x, y_i) - \frac{\partial(v_i w_i)}{\partial y_i}(x, y_i) - q_i w_i(x, y_i) - s_i C_i(x, y_i) \tag{41}$$

and then the following approximation of (41):

$$\frac{u_{ij}^k w_{ij}^k - u_{ij}^{k-1} w_{ij}^{k-1}}{\tau_k} = \frac{2p_i}{h_{ij}+h_{ij+1}} \left(\frac{w_{ij+1}^k - w_{ij}^k}{h_{ij+1}} - \frac{w_{ij}^k - w_{ij-1}^k}{h_{ij}} \right) -$$
$$\frac{v_{ij+1}^k w_{ij+1}^k - v_{ij-1}^k w_{ij-1}^k}{h_{ij}+h_{ij+1}} - q_i w_{ij}^k - s_i C_{ij}^k, \quad i=1,2,\ j=1,2,\ldots,n_i-1 \tag{42}$$

is obtained using the balance method [8].

For the above three implicit difference schemes the boundary conditions are approximated using (35), (36). The values of v_{ij}^k are also sought by means of (38).

Because the coefficients u_{ij}^k and v_{ij}^k are unknown, the simplest iterative process is used – their values are taken from the previous iteration. The values, found using the explicit difference scheme (34)-(38) (without the stability condition (37)) are taken as an initial approximation for this iterative process.

The system of linear algebraic equations (39), (35), (36); (40), (35), (36) or (41), (35), (36) is solved on each iteration using the counter three diagonal matrix algorithm (counter sweep method) [9].

4 Numerical tests, comparison and results

The difference schemes described in the previous section are tested on different meshes and compared with existing theoretical results at some values of the parameters. The results considered below are obtained at $\theta_1 = 0.1$, $\theta_2 = 0.145$, $\theta_5 = 18.3$, $\theta_7 = 0.034$, $\theta_8 = 0$, $Sc_1 = 0.735$, $Sc_2 = 564$, $Pr_1 = 0.666$, $Pr_2 = 6.54$, $Y_1 = 6$, $Y_2 = 6$, $\bar{Y}_1 = 7$, $\bar{Y}_2 = 0.26$, $\bar{\bar{Y}}_1 = 7.4$ and $\bar{\bar{Y}}_2 = 2.4$. ¿From a chemical point of view the following quantitative characteristics of the solution are interesting – the velocity and its second derivatives at $x = 1$, as well as the integrals:

$$I_1 = \int_0^1 \left(\frac{\partial C_1}{\partial y_1}\right)_{y_1=0} dx, \qquad I_2 = \int_0^1 \left(C_1 \frac{\partial C_1}{\partial y_1}\right)_{y_1=0} dx,$$

$$I_3 = \int_0^1 \left(T_1 \frac{\partial C_1}{\partial y_1}\right)_{y_1=0} dx, \qquad I_4 = \int_0^1 \left(\frac{\partial T_1}{\partial y_1}\right)_{y_1=0} dx.$$

The derivatives are calculated using formulas for numerical differentiation on nonuniform grids. The integrals are obtained by means of the trapezoidal rule.

To investigate the accuracy, convergence and reliability of the proposed numerical algorithm meshes $\omega^{lr} = \omega_\tau^l \times \omega_h^r$ with different steps τ_k^l and h_{ij}^r are used. The grid ω_τ^l is such that $\tau_k^l = \tau_k^0/2^l$, $\tau_k^0 = 0.001$. The grid ω_h^r is such that $h_{i1}^r = h_{\min}^r$; $h_{in_i^r}^r = h_{\max}^r$; $h_{ij}^r = t_i^r h_{ij-1}^r$, $i=1,2$, $j=2,3,\ldots,n_i^r$; $h_{\min}^0 = 0.001$; $h_{\max}^0 = 0.05$; $h_{\min}^r = h_{\min}^0/2^r$; $h_{\max}^r = h_{\max}^0/2^r$; t_i^r and n_i^r are determined uniquely from $h_{\min}^r$ and $h_{\max}^r$. For the mesh $\tilde{\omega}^0 = \tilde{\omega}_\tau \times \omega_h^0$ the submesh $\tilde{\omega}_\tau$ is such that $\tau_k = 10^{-5}$, $k = 1,2,\ldots,1000$ and $\tau_k = 10^{-3}$, $k > 1000$.

Example 1. In order to compare the different implicit difference schemes numerical results for the integral I_1 at $Da = 0$, $\theta_3 = 0$, $\theta_4 = 0$ and $\theta_6 = 0$ are shown in Table 1. In this case the value of I_1 has been obtained by means of asymptotic theory in [7] and it is 0.45547.

Table 1. The values of I_1 at $Da = 0$, $\theta_3 = 0$, $\theta_4 = 0$, $\theta_6 = 0$, obtained using implicit difference schemes, based on different approximations of (29).

ω	Approximation of (29) with		
	(40)	(39)	(42)
ω^{00}	0.43445	0.43545	0.45462
ω^{10}	0.44001	0.44101	0.45455
ω^{20}	0.44402	0.44502	0.45451
ω^{30}	0.44687	0.44788	0.45448
$\tilde{\omega}^{0}$	0.45184	0.45291	0.45556

ω	Approximation of (29) with		
	(40)	(39)	(42)
ω^{00}	0.43445	0.43545	0.45462
ω^{01}	0.43512	0.43562	0.45508
ω^{02}	0.43547	0.43573	0.45529
ω^{03}	0.43565	0.43579	0.45540
ω^{04}	0.43574	0.43581	0.45544

It is seen from the above table that:

– there is not an essential difference between results obtained using upwind (40) and central differencing (39) of the convective terms;

– in order to obtain sufficiently good results the non-conservative schemes (39) and (40) require a smaller step τ_k especially near the origin than the conservative difference scheme (42);

– the value of I_1 calculated using (42) on the mesh ω^{04} differs from that have been obtained in [7] with less than 0.02%.

The numerical results are computed on IBM PC 486DX4-100 Mhz. The CPU time for solving the problem on ω^{00} is 7′30″. For the finer meshes it increases proportionally to the number of the nodes in the mesh. Let us note, for comparison, that if the explicit difference scheme (35) is used on $\omega_\tau \times \omega_h^0$, because of the stability condition (37) (the steps τ_k have to be less than 4×10^{-8}) the corresponding CPU time would be more than 300 hours.

Example 2. The numerical algorithm is also tested at $Da = 0$, $\theta_3 = 0$, $\theta_4 = 0.2$ and $\theta_6 = 0$. At these values of the parameters the value of I_1 obtained by means of asymptotic theory in [7] is 0.48844. The computed value of I_1 using (42) on ω^{03} is 0.48041.

Example 3. Some numerical results at $Da = 100$, $\theta_3 = 2.08 \times 10^{-4}$, $\theta_4 = 0.2$ and $\theta_6 = 86$ are shown in Table 2, where

$$M_1 = \max_{y_1 \in [0, Y_1]} \frac{\partial^2 u_1}{\partial y_1^2}.$$

It is seen that further refinement of the mesh ω_{02} does not change essentially the shown values.

Table 2. Results, obtained using implicit difference scheme (42) at $Da = 100$, $\theta_3 = 2.08 \times 10^{-4}$, $\theta_4 = 0.2$, $\theta_6 = 86$

ω	I_1	I_2	I_3	I_4	$u_1(1,0)$	$v_1(1,Y_1)$	M_1
ω^{00}	0.67930	0.06470	0.02435	-0.02710	0.10924	0.58046	-0.12294
ω^{01}	0.68129	0.06168	0.02453	-0.02720	0.10924	0.58266	-0.12260
ω^{02}	0.68224	0.06019	0.02462	-0.02726	0.10924	0.58372	-0.12243
ω^{03}	0.68270	0.05944	0.02466	-0.02728	0.10924	0.58429	-0.12235
ω^{12}	0.68221	0.05992	0.02458	-0.02725	0.10924	0.58388	-0.12251

Acknowledgements. The authors are highly appreciative to Prof. Chr. Boyadjiev for the posed problem and the helpful discussions.

This work is partially supported under Grants MM-443/94 and MM-501/95 by the Ministry of Education and Science - Bulgaria.

References

1. Iliev, O., Makarov, M.: A block-matrix iterative numerical method for coupled solving 2D Navier-Stokes equations. J. Comput. Phys. **121** (1995) 324–330
2. Shankar, N., Subramanian, S.: The slow axisymmetric termocapillary migration of an eccentricaly placed bubble inside a drop in zero gravity. J. Coll. Interface Sci. **94** (1983) 258–275
3. Mok, L., Kim, K.: Motion of a gas bubble inside a spherical liquid container with a vertical temperature gradient. J. Fluid. Mech. **176** (1987) 521–531
4. Maron Moalem, D., Miloh, T.: Theoretical analysis of heat and mass transfer through eccentric spherical fluid shells at large Peclet number. Appl. Sci. Res. **32** (1976) 395–414
5. Rasmussen, R., Levizzani, V., Pruppacher, H.: A numerical study of the heat transfer through a fluid layer with recirculating between concentric and eccentric spheres. Pure Appl. Geophys. **120** (1982) 702–720
6. Boyadjiev, Chr.: Thermocapillary effects in falling liquid films, I. General theory. Hung. J. Industrial Chem. **9** (1988) 195–201
7. Boyadjiev, Chr.: The theory of non-linear mass transfer in system with intensive interphase mass transfer. Bulg. Chem. Comm. **26** (1993) 33–58
8. Samarskii, A.: Theory of the difference schemes. Moskow, Nauka (1987) (in Russian)
9. Samarskii, A., Nikolaev E.: Methods for solving of difference schemes. Moskow, Nauka (1978) (in Russian)

Basic Techniques for Numerical Linear Algebra on Bulk Synchronous Parallel Computers

Rob H. Bisseling

Department of Mathematics, Utrecht University
P. O. Box 80010, 3508 TA Utrecht, the Netherlands
http://www.math.ruu.nl/people/bisseling

Abstract. The bulk synchronous parallel (BSP) model promises scalable and portable software for a wide range of applications. A BSP computer consists of several processors, each with private memory, and a communication network that delivers access to remote memory in uniform time.
Numerical linear algebra computations can benefit from the BSP model, both in terms of simplicity and efficiency. Dense LU decomposition and other computations can be made more efficient by using the new technique of two-phase randomised broadcasting, which is motivated by a cost analysis in the BSP model. For LU decomposition with partial pivoting, this technique reduces the communication time by a factor of $(\sqrt{p}+1)/3$, where p is the number of processors.
Theoretical analysis, together with benchmark values for machine parameters, can be used to predict execution time. Such predictions are verified by numerical experiments on a 64-processor Cray T3D. The experimental results confirm the advantage of two-phase randomised broadcasting.

1 Introduction

The field of parallel numerical linear algebra has rapidly evolved over the last decade. A major development has been the acceptance of the two-dimensional cyclic data distribution as a standard for dense matrix computations on parallel computers with distributed memory. The two-dimensional cyclic distribution of an $m \times n$ matrix A over a parallel computer with $p = M \cdot N$ processors is given by

$$a_{ij} \longmapsto P(i \bmod M, j \bmod N), \text{ for } 0 \le i < m \text{ and } 0 \le j < n \ , \tag{1}$$

where $P(s,t)$, with $0 \le s < M$ and $0 \le t < N$, is a two-dimensional processor number. To the best of our knowledge, this distribution was first used by O'Leary and Stewart [12] in 1985. It has been the basis of parallel linear algebra libraries such as PARPACK [1, 3], a prototype library for dense and sparse matrix computations which was developed for use on transputer meshes, and ScaLaPack [4], a public domain library for dense matrix computations which is available for many different parallel computers. (ScaLaPack uses a generalisation of (1), where each matrix element a_{ij} is replaced by a submatrix A_{ij}.) Distribution (1) has acquired a variety of names, such as 'torus-wrap mapping', which is used in [8], and 'scattered square decomposition' [5]. Following PARPACK, we use the term $M \times N$ *grid distribution.* In many cases, it is best to choose $M \approx N \approx \sqrt{p}$. For an optimality proof of the grid distribution with respect to load balancing and a discussion of its communication properties, see [3]. Application of this distribution in a wide range of numerical linear algebra computations (LU decomposition, QR factorisation, and Householder tridiagonalisation) is discussed in [8].

The chosen data distribution need not have any relation with the physical architecture of a parallel computer. Even though terms as 'processor row $P(s,*)$', or 'processor column $P(*,t)$' are used in this paper and elsewhere, these terms just describe a collection of processors with particular processor identities and not a physical submachine. Although many parallel linear algebra algorithms were originally developed for rectangular meshes or hypercubes, today it is recognised that these algorithms can often be used on other architectures as well, simply by taking the processor numbering as a logical numbering.

Parallel algorithms are always developed within a given programming model, whether it is explicitly specified or not. A simple parallel programming model that leads to portable and scalable software can clearly be of great benefit in applications, including numerical linear algebra. The Bulk Synchronous Parallel (BSP) model by Valiant [15] is such a simple model; it will be explained briefly in Sect. 2. The BSP model allows us to analyse the time complexity of parallel algorithms using only a few parameters. On the basis of such an analysis, algorithms can be better understood and possibly improved. One improvement is the technique of two-phase randomised broadcasting, which is presented in Sect. 3 and tested in Sect. 4. The technique is based on the idea of using intermediate randomly chosen processors, which originates in routing algorithms [14].

Within the BSP framework, there have been a few studies of broadcasting as part of parallel numerical linear algebra. Gerbessiotis and Valiant [6] present and analyse an algorithm for Gauss-Jordan elimination with partial pivoting. Their algorithm broadcasts matrix rows and columns in $\log_2 \sqrt{p}$ phases. In preliminary work with Timmers [13], we implemented the technique of two-phase randomised broadcasting as part of a parallel Cholesky factorisation. Although our theoretical analysis revealed major benefits, we did not observe them in practice, for the simple reason that we used a parallel computer with only four processors. To reap the benefits of this technique, more processors must be used. This is done in Sect. 4 of the present work, where 64 processors are used to study two-phase randomised broadcasting as part of a parallel LU decomposition. Recent theoretical work on communication primitives for the BSP model such as broadcast and parallel prefix can be found in [9].

2 BSP model

The BSP model was proposed by Valiant in 1990 [15]. It defines an architecture, a type of algorithm, and a function for charging costs to algorithms. We use the variant of the cost function proposed in [2]. For a recent survey of BSP computing, see [10].

A *BSP computer* consists of p processors, each with private memory, and a communication network that allows processors to access private memories of other processors. Each processor can read from or write to any memory cell in the entire machine. If the cell is local, the read or write operation is relatively fast. If the cell belongs to another processor, a message must be sent through the communication network, and this takes more time. The access time for different non-local memories is the same.

A *BSP algorithm* consists of a sequence of supersteps, each ended by a global barrier synchronisation. In a *computation superstep*, each processor performs a sequence of operations on locally held data. In numerical linear algebra, these operations are mainly floating point operations (flops). In a *communication superstep*, each processor sends and receives a number of messages. The messages do not synchronise the sender with the receiver and they do not block progress. Synchronisation takes place only at the end of a superstep.

The *BSP cost function* is defined as follows. An *h-relation* is a communication superstep where each processor sends at most h data words to other processors and receives at most h data words. We denote the maximum number of words sent by any processor by h_{s}, and the maximum number received by h_{r}. Therefore,

$$h = \max\{h_{\mathrm{s}}, h_{\mathrm{r}}\} \ . \tag{2}$$

This equation reflects the assumption that a processor can send and receive data simultaneously. Charging costs on the basis of h is motivated by the assumption that the bottleneck of communication lies at the entry or exit of the communication network, so that simply counting the maximum number of sends and receives per processor gives a good indication of communication time.

The cost of an h-relation is

$$T_{\mathrm{comm}}(h) = hg + l \ , \tag{3}$$

where g and l are machine-dependent parameters. The cost unit is the time of one flop. Cost function (3) is chosen because of the expected linear increase of communication time with h. The processor that sends or receives the maximum number of data words determines h and hence the communication cost. Since $g = \lim_{h\to\infty} T_{\mathrm{comm}}(h)/h$, the value of g can be viewed as the time (in flops) needed to send one word into the communication network, or to receive one word from it, in a situation of continuous message traffic. The linear cost function includes a nonzero constant l because each h-relation incurs a fixed cost. This fixed cost includes: the cost of global synchronisation; the cost of ensuring that all communicated data have arrived at their destination; and startup costs of sending messages.

The cost of a computation superstep with an amount of work w is

$$T_{\mathrm{comp}}(w) = w + l \ . \tag{4}$$

The amount of work w is defined as the maximum number of flops performed in the superstep by any processor. The value of l is taken to be the same as that of a communication superstep, despite the fact that the fixed cost is less: global synchronisation is still necessary, but the other costs disappear. The advantage of having one parameter l is simplicity; the total synchronisation cost of an algorithm can be determined by simply counting the supersteps. As a consequence of (3) and (4), the total cost of a BSP algorithm becomes an expression of the form $a + bg + cl$.

A BSP computer can be characterised by four parameter: p is the number of processors; s is the single-processor speed measured in flop/s; g is the communication cost per data word; and l is the synchronisation cost of a superstep. (We slightly abuse the language, because l also includes other costs.) We call a computer with these four parameters a BSPC(p, s, g, l). The execution time of an algorithm with cost $a+bg+cl$ on a BSPC(p, s, g, l) is $(a+bg+cl)/s$ seconds.

Estimates for g and l of a particular machine can be obtained by benchmarking a range of *full* h-relations, i.e., h-relations where each processor sends and receives exactly h data. In the field of numerical linear algebra, it is appropriate to use 64-bit reals as data words. The measured cost of a full h-relation will be an upper bound on the cost of an arbitrary h-relation. For a good benchmark, a representative full h-relation must be chosen which is unrelated to the specific architectural characteristics of the machine. A cyclic pattern will often suffice: processor $P(i)$, $0 \leq i < p$, sends its first data word to $P((i+1) \bmod p)$, the next to $P((i+2) \bmod p)$, and so on. The source processor $P(i)$ is skipped as the destination of messages, since local assignments do not require communication. (As an alternative, an average for a set of random full h-relations can be measured.)

Figure 1 shows the results of benchmarking cyclic full h-relations on 64 processors of a Cray T3D. The nodes of the Cray T3D used in this experiment consist of a 150 MHz Dec Alpha processor and a memory of 64 Mbyte. The communication network is a three-dimensional torus. The Cray T3D is transformed into a BSP computer by using the Cray T3D implementation of the Oxford BSP library [11], version 1.1. The measured data points of the figure lie close to the straight line of a least-squares fit, so that the behaviour is indeed linear as modelled by (3). The fitted BSP parameters are $g = 11.5$ and $l = 61.4$. The sequential computing speed $s = 6.0$ Mflop/s is obtained by measuring the time of a DAXPY operation $\mathbf{y} := \alpha\mathbf{x} + \mathbf{y}$, where $\mathbf{x}$ and $\mathbf{y}$ are vectors of length 1024 and α is a scalar.

Communication in the BSP model does not require any form of synchronisation between the sender and the receiver. Global barrier synchronisations guarantee memory integrity at the start of every superstep. The absence of sender/receiver synchronisation makes it possible to view the communications as one-sided: the processor that initiates the communication of a message is active and the other processor is passive. If the initiator is the sender, we call the send operation a *put*; if it is the receiver, we call the receive operation a *get*. Conceptually, puts and gets are single-processor operations. More and more, puts and gets are supported in hardware, which makes them very efficient. An

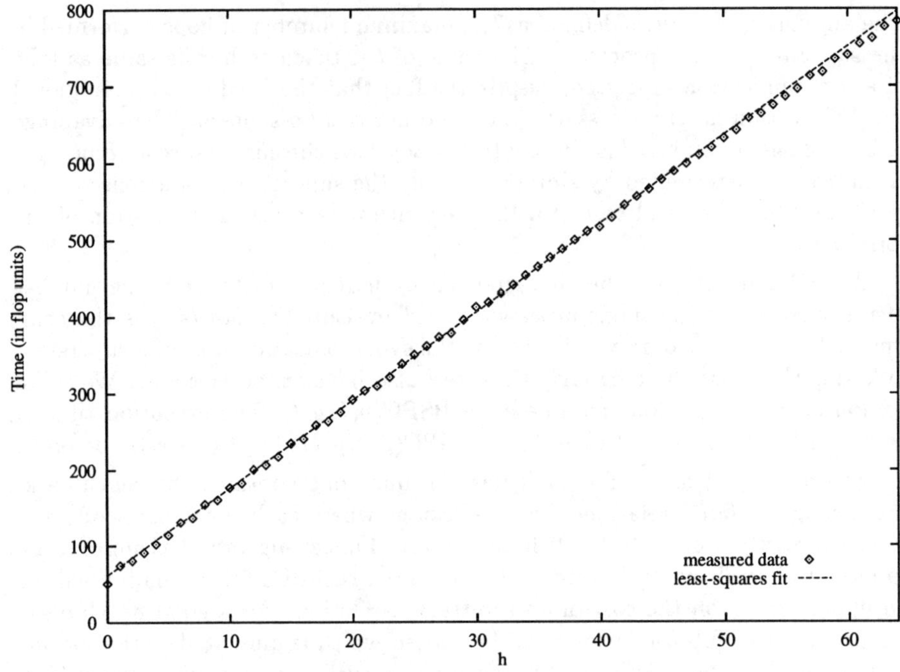

Fig. 1. Time of an h-relation on a 64-processor Cray T3D

early example of such support is the SHMEM facility of the Cray T3D. The use of one-sided communications makes programs simpler: the program text only includes a put or get statement for the initiator. This is in contrast to traditional message-passing programs where both parts of a matching send/receive pair must be included. In dense matrix computations, puts are usually sufficient. In sparse matrix computations, gets are sometimes needed because the receiver of the data knows which data it needs, for instance because of its local sparsity pattern, but the sender does not have this knowledge. Note that we use the terms 'send' and 'receive', even if we only perform puts. Furthermore, we still analyse algorithms counting both sent and received data words. The advantage of one-sided communications lies mainly in simpler program texts of algorithms and their implementations.

3 Two-Phase Randomised Broadcasting

An important communication operation in parallel algorithms for numerical linear algebra is the broadcast of a matrix row or column. Let us examine a column broadcast in detail. The $M \times N$ grid distribution assigns matrix element a_{ij} to processor $P(i \bmod M, j \bmod N)$. Suppose we have to broadcast column k, i.e., each matrix element a_{ik} must be communicated to all processors that contain elements from matrix row i. For the grid distribution, these processors are contained in processor row $P(i \bmod M, *)$. Often, a column broadcast and a similar row broadcast prepare the ground for a subsequent rank-1 update of the matrix. The broadcasts communicate the elements a_{ik} and a_{kj} that are required in assignments of the form $a_{ij} := a_{ij} - a_{ik}a_{kj}$.

Usually, only part of a column k must be broadcast, such as e.g. the elements a_{ik} with $i > k$ in stage k of an LU decomposition. In our explanation, however, we assume without loss of generality that all elements a_{ik}, $0 \leq i < m$, must be broadcast, where $m \geq 1$. A simple algorithm that performs this column broadcast is given by Fig. 2.

```
if k mod N = t then
        for all i : 0 ≤ i < m ∧ i mod M = s do
                put a_ik in P(s, *);
sync;
```

Fig. 2. Program text for processor $P(s,t)$ of one-phase broadcast of column k

This broadcast consists of one phase, which is a communication superstep. The cost of this superstep can be analysed as follows. The only processors that send data are the M processors from $P(*, k \bmod N)$. These processors send $N-1$ copies of at most $R = \lceil m/M \rceil$ matrix elements, so that the maximum number of elements sent per processor is $h_s = R(N-1)$. In general, all processors except the senders receive elements from column k. The maximum number of elements received per processor is $h_r = R$. It follows that $h = h_s = R(N-1)$. (Note that this count also holds in the special case $N = 1$, where $h = 0$.) The resulting cost of the one-phase broadcast is

$$T(\text{one-phase broadcast}) = \left\lceil \frac{m}{M} \right\rceil (N-1)g + l \ . \tag{5}$$

In the common case $M = N = \sqrt{p}$ with $m \gg \sqrt{p} \gg 1$, the cost becomes $T \approx mg + l$.

This cost analysis reveals a major disadvantage of the straightforward algorithm. The maximum number of sends is $N-1$ times larger than the maximum number of receives. In the ideal situation, h_s and h_r would both be equal to the average number of data communicated per processor, which is the communication volume divided by p. In the one-phase broadcast, however, the communication work is badly balanced. The senders have to copy the column elements and send all these copies out, whereas the receivers only receive one copy of each column element. For $M = N = \sqrt{p}$ with $m \gg \sqrt{p} \gg 1$, we are far from the ideal situation: $h_s \approx m$, $h_r \approx m/\sqrt{p}$, and the average number of communicated data is about $m/\sqrt{p}$.

The communication imbalance can be eliminated by first sending one copy of each data element to a randomly chosen intermediate processor and making this processor responsible for copying and for sending the copies to the destination. This method is similar to two-phase randomised routing [14], which sends packets from source to destination through a randomly chosen intermediate location, to avoid congestion in the routing network. The new broadcasting method splits the original communication superstep into two communication supersteps: an unbalanced h-relation which randomises the location of the data elements to be broadcast; and a balanced h-relation which performs the broadcast itself. In analogy with the routing case, we call the resulting pair of h-relations a *two-phase randomised broadcast.*

In general, the intermediate processor in a two-phase randomised broadcast is chosen randomly. However, in certain situations with regular communication patterns, for instance in dense matrix computations, the intermediate processor can also be chosen deterministically. (For sparse matrix computations, which are often irregular, a random choice may be more appropriate.) Processor $P(s, k \bmod N)$, $0 \le s < M$, is the source of the matrix elements a_{ik} with $i \bmod M = s$. The row index i of a local element a_{ik} of this processor can be written as $i = \mathtt{i}M + s$, where $\mathtt{i}$ is a local index. Note that $\mathtt{i} = i \operatorname{div} M$, and that the local indices of the local elements are consecutive. The intermediate processor for an element a_{ik} of $P(s, k \bmod N)$ can be chosen within processor row $P(s, *)$. This is a natural choice, because the broadcast involves a nearly equal number of elements for each processor row and because source and destination are always in the same processor row. The consecutive local indices $\mathtt{i}$ can be used to assign intermediate addresses in a cyclic fashion (similar to the cyclic method of the grid distribution itself). This can be done by assigning elements with local index $\mathtt{i}$ to processor $P(s, \mathtt{i} \bmod N)$. The resulting two-phase broadcast is given by Fig. 3.

```
if k mod N = t then
        for all i : 0 ≤ i < m ∧ i mod M = s do
                put a_ik in P(s, (i div M) mod N);
sync;
for all i : 0 ≤ i < m ∧ i mod M = s ∧ (i div M) mod N = t do
        put a_ik in P(s, *);
sync;
```

Fig. 3. Program text for processor $P(s,t)$ of two-phase broadcast of column k

The cost analysis of the two-phase broadcast is as follows. The broadcast consists of two communication supersteps. Let $R = \lceil m/M \rceil$. The first superstep has $h_s = R - \lfloor R/N \rfloor$, since at least $\lfloor R/N \rfloor$ puts are local, and $h_r = \lceil R/N \rceil$. For simplicity, we write $h \le R$. The second superstep has $h_s = \lceil R/N \rceil (N-1)$

and $h_r = R - \lfloor R/N \rfloor$. An upper bound for h_s can be obtained from $h_s = \lceil R/N \rceil (N-1) \leq (R/N+1)N = R+N$. We write $h \leq R+N$. The resulting cost of the two-phase broadcast is

$$T(\text{two-phase broadcast}) \leq (2\left\lceil \frac{m}{M} \right\rceil + N)g + 2l \ . \tag{6}$$

For $M = N = \sqrt{p}$ with $m \gg p$, the cost becomes $T \approx 2(m/\sqrt{p})g + 2l$. Compared to the one-phase broadcast, the communication cost decreases by the considerable factor of $\sqrt{p}/2$, at the modest expense of doubling the synchronisation time.

The cost of the two-phase broadcast is close to minimal because a lower bound for every broadcast is

$$T(\text{broadcast}) \geq \left\lceil \frac{m}{M} \right\rceil g + l \ . \tag{7}$$

The lower bound follows from the fact that the processor with most words to broadcast has to send $\lceil m/M \rceil$ data words at least once and this takes at least one superstep.

Figure 4 illustrates the implementation of a two-phase broadcast as part of an LU decomposition program. The program fragment performs the broadcast of elements a_{ik} with $k < i < n$, for an $n \times n$ matrix A. The program is written in ANSI C extended with primitives from the proposed BSP Worldwide standard library [7]. The program uses two-dimensional processor coordinates `(s,t)` and the corresponding one-dimensional identity `s+t*M`. The first `bsp_put` primitive writes the double a_{ik} = `a[i][kc]` into `ak[i]` of processor `s+(i mod N)*M`. The value of `i0` is the smallest integer `i` $\geq$ `kr1` with `i` mod $N = t$.

```
#define SZD (sizeof(double))
/* nr =  number of local rows of matrix A
   kr1 = first local row with global index > k */

if (k%N==t){
     /* kc = local column with global index = k */
     for(i=kr; i<nr; i++){
          bsp_put(s+(i%N)*M,&a[i][kc],ak,i*SZD,SZD);
     }
}
bsp_sync();
for(i=i0; i<nr; i +=N){
     for(t1=0; t1<N; t1++){
          bsp_put(s+t1*M,&ak[i],ak,i*SZD,SZD);
     }
}
bsp_sync();
```

Fig. 4. Implementation of two-phase broadcast from LU decomposition in ANSI C with BSP extensions

4 Experimental Results

The theoretical analysis of Sect. 3 shows that two-phase randomised broadcasting is advantageous. In this section, we put this claim to the test by performing numerical experiments on a parallel computer. We choose LU decomposition with partial pivoting as our test problem and we perform our tests on 64 nodes of a Cray T3D. This parallel computer with parameters $p = 64$, $s = 6.0\,\mathrm{Mflop/s}$, $g = 11.5$, and $l = 61.4$ has good BSP characteristics, because of the low values of g and l and because of its scalable and predictable behaviour, see Sect. 2. For the experiments, we implemented LU decomposition with one-phase and two-phase broadcasts. Except for the broadcasts, the two implementations are identical. The $n \times n$ matrix A which is decomposed into $PA = LU$ is distributed according to the 8×8 grid distribution. The test matrix is chosen such that in each stage k, $0 \leq k < n$, of the LU decomposition, row k is explicitly swapped with the pivot row r, forcing communication. This is done to generate the worst-case communication behaviour.

The communication operations of LU decomposition in stage k are: a broadcast of column k which involves elements a_{ik} with $k < i < n$; a broadcast of row k which involves elements a_{kj} with $k < j < n$; and a swap of rows k and r which involves elements a_{kj} and a_{rj} with $0 \leq j < n$. The time complexity of the broadcasts is analysed in Sect. 3; the row swap costs about $(n/\sqrt{p})g + l$ flop units. Each stage of the parallel LU decomposition algorithm with two-phase randomised broadcasting contains six supersteps. (This small number can be obtained in an implementation by combining supersteps and by performing computation and communication in the same superstep.) The total cost of the algorithms can be shown to be equal to

$$T(\text{LU with one-phase broadcast}) \approx \frac{2n^3}{3p} + \left(n^2 + \frac{n^2}{\sqrt{p}}\right) g + 5nl \ , \qquad (8)$$

and

$$T(\text{LU with two-phase broadcast}) \approx \frac{2n^3}{3p} + \frac{3n^2}{\sqrt{p}} g + 6nl \ . \qquad (9)$$

Figure 5 shows the measured time of LU decomposition with partial pivoting for matrices of order n=250–5000. Two-phase randomised broadcasting clearly improves performance for all problem sizes. The largest gain factor achieved is 2.53, for $n = 500$. Theoretically, the total communication cost of the algorithm decreases from $(n^2 + n^2/\sqrt{p})g$ to $(3n^2/\sqrt{p})g$. This is a decrease by a factor of $(\sqrt{p}+1)/3$, which equals three for $p = 64$. Synchronisation is dominant for small problems and computation is dominant for large problems; in both asymptotic cases the decrease in communication time is relatively small compared to the total solution time. In the intermediate range of problem size, however, the decrease is significant. The experiments show that in the range of n = 250–1500 the decrease of the total time is more than a factor of two, but even for $n = 5000$ it is substantial: the two-phase version takes only 143 s, whereas the one-phase version takes 202 s.

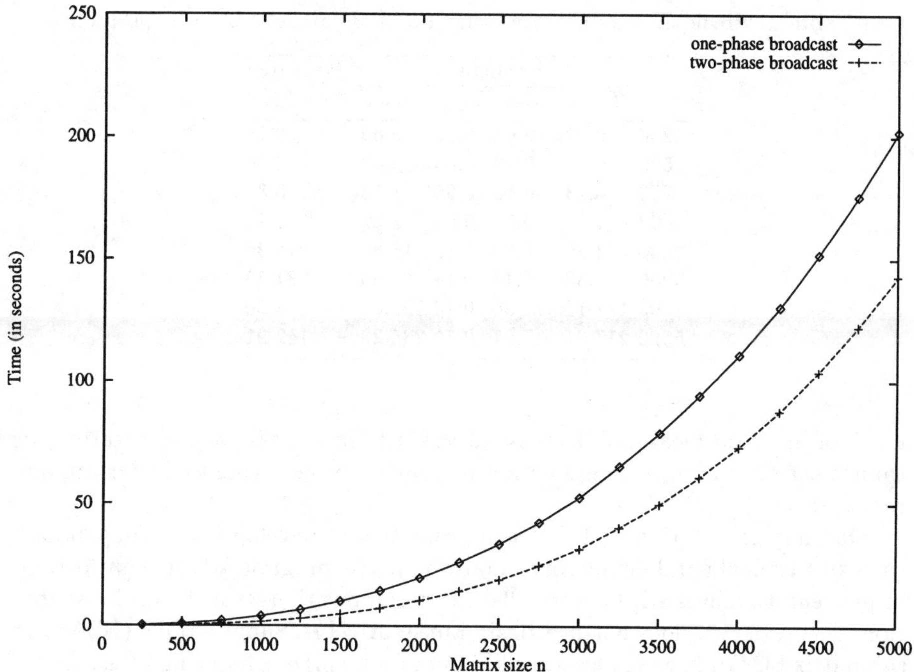

Fig. 5. Time of LU decomposition with partial pivoting on a 64-processor Cray T3D

The theoretical timing formula (9) and the measured machine characteristics p, s, g, and l can be used to predict the total execution time and its three components. Table 1 compares predicted and measured execution times. The table displays reasonable agreement for small n, but significant discrepancy for larger n. This discrepancy is mainly due to a sequential effect: it is difficult to predict sequential computing time, because the actual computing rate may differ from one application to another. Simply substituting a benchmark result in a theoretical time formula gives what may be termed an *ab initio prediction*, i.e. a prediction from basic principles, which may be useful as an indication of expected performance, but not as an accurate estimate. Of course, the prediction can always be improved by using the measured computing rate for the particular application, instead of the benchmark rate s. (For LU decomposition, we measured a sequential rate of 9.35 Mflop/s for $n = 1000$, which is considerably faster than the benchmark rate of 6.0 Mflop/s.)

A breakdown of the predicted total time in computation time, communication time, and synchronisation time gives additional insight. Synchronisation time is negligible for this machine and this problem, except in the case of very small n. This implies that the increase in synchronisation time caused by two-phase randomised broadcasting is irrelevant. The savings in communication time, however, is significant. For example, for $n = 5000$, the predicted communication time is brought down from 53.91 s to 17.97 s, and this predicted decrease accounts for

Table 1. Predicted and measured time (in s) of parallel LU decomposition

	Predicted				Measured
n	T_{comp}	T_{comm}	T_{sync}	T_{total}	T_{total}
250	0.03	0.04	0.02	0.09	0.10
500	0.22	0.18	0.03	0.43	0.36
750	0.73	0.40	0.05	1.18	0.87
1000	1.74	0.72	0.06	2.52	1.72
2000	13.89	2.88	0.12	16.89	10.33
3000	46.88	6.47	0.18	53.53	31.53
4000	111.11	11.50	0.25	122.86	73.20
5000	217.01	17.97	0.31	235.29	142.91

much of the total measured decrease of 59 s. (As in the case of computation, we cannot expect an ab initio prediction to give an exact account of all communication effects.)

Our implementation of LU decomposition was developed to study broadcasts in a typical parallel matrix computation. The program is a straightforward implementation in ANSI C of parallel LU decomposition with the grid distribution. The program does not use Basic Linear Algebra Subprograms (BLAS) or submatrix blocking, which are crucial in obtaining high computing rates. As expected, our implementation is far from optimal with respect to computing speed; it achieves 0.56 Gflop/s for $n = 8000$ and $p = 64$, whereas ScaLaPack attains 5.3 Gflop/s [4] for the same problem on the same machine. Absolute performance can of course be improved by using BLAS wherever possible and reorganising the algorithm to use matrix-matrix multiplication in the core computation. The data distribution need not be changed. Note that for higher computing rates the issue of communication will become relatively more important and hence the benefits of two-phase randomised broadcasting will grow in significance.

5 Conclusion

The main result of this work is that significant performance improvement can be obtained by using the BSP model in parallel numerical linear algebra. The new technique of two-phase randomised broadcasting has been introduced to balance the communication work in a parallel computer. This technique can be used to accelerate row and column broadcasts in LU decomposition, Cholesky factorisation, QR factorisation, and Householder tridiagonalisation. Numerical experiments on a Cray T3D show that the improvement can be observed in practice. Application is not restricted to a BSP context; the technique may be used wherever row or column broadcasts occur in parallel numerical linear algebra. The derivation of the technique is natural in the BSP model. One may speculate that it is much more difficult to obtain such techniques without the guidance of the BSP model.

Acknowledgements

This work is partially supported by the university grants programme of NCF in the Netherlands and Cray Research. Numerical experiments were performed on the Cray T3D of the Ecole Polytechnique Fédéral de Lausanne.

References

1. R. H. Bisseling and L. D. J. C. Loyens. Towards peak parallel LINPACK performance on 400 transputers. *Supercomputer*, 45:20–27, 1991.
2. R. H. Bisseling and W. F. McColl. Scientific computing on bulk synchronous parallel architectures. In B. Pehrson and I. Simon, editors, *Technology and Foundations: Information Processing '94, Vol. I*, volume 51 of *IFIP Transactions A*, pages 509–514. Elsevier Science Publishers, Amsterdam, 1994.
3. R. H. Bisseling and J. G. G. van de Vorst. Parallel LU decomposition on a transputer network. In G. A. van Zee and J. G. G. van de Vorst, editors, *Parallel Computing 1988*, volume 384 of *Lecture Notes in Computer Science*, pages 61–77. Springer-Verlag, Berlin, 1989.
4. J. J. Dongarra and D. W. Walker. Software libraries for linear algebra computations on high performance computers. *SIAM Review*, 37(2):151–180, 1995.
5. G. C. Fox, M. A. Johnson, G. A. Lyzenga, S. W. Otto, J. K. Salmon, and D. W. Walker. *Solving Problems on Concurrent Processors: Vol. I, General Techniques and Regular Problems.* Prentice Hall, Englewood Cliffs, NJ, 1988.
6. A. V. Gerbessiotis and L. G. Valiant. Direct bulk-synchronous parallel algorithms. *Journal of Parallel and Distributed Computing*, 22(2):251–267, 1994.
7. M. W. Goudreau, J. M. D. Hill, K. Lang, B. McColl, S. B. Rao, D. C. Stefanescu, T. Suel, and T. Tsantilas. A proposal for the BSP Worldwide standard library. Technical report, Oxford Parallel, Oxford, UK, Apr. 1996.
8. B. A. Hendrickson and D. E. Womble. The torus-wrap mapping for dense matrix calculations on massively parallel computers. *SIAM Journal on Scientific Computing*, 15(5):1201–1226, 1994.
9. B. H. H. Juurlink and H. A. G. Wijshoff. Communication primitives for BSP computers. *Information Processing Letters*, to appear, 1996.
10. W. F. McColl. Scalable computing. In J. van Leeuwen, editor, *Computer Science Today: Recent Trends and Developments*, volume 1000 of *Lecture Notes in Computer Science*, pages 46–61. Springer-Verlag, Berlin, 1995.
11. R. Miller. A library for bulk synchronous parallel programming. In *General Purpose Parallel Computing*, pages 100–108. British Computer Society Parallel Processing Specialist Group, 1993.
12. D. P. O'Leary and G. W. Stewart. Data-flow algorithms for parallel matrix computations. *Communications of the ACM*, 28(8):840–853, 1985.
13. P. Timmers. Implementing dense Cholesky factorization on a BSP computer. Master's thesis, Department of Mathematics, Utrecht University, Utrecht, the Netherlands, June 1994.
14. L. G. Valiant. A scheme for fast parallel communication. *SIAM Journal on Computing*, 11:350–361, 1982.
15. L. G. Valiant. A bridging model for parallel computation. *Communications of the ACM*, 33(8):103–111, 1990.

Convergence of a Crank-Nicolson Difference Scheme for Heat Equations with Interface in the Heat Flow and Concentrated Heat Capacity

Ilia Braianov

Center of Applied Mathematics and Informatics
University of Rousse, 7017 Rousse, Bulgaria.
e-mail: Ilia.Braianov@ami.ru.acad.bg

Abstract. In this paper we consider initial value problems for heat equation with discontinuous heat flow and concentrated heat capacity in interior points or at the boundary. Convergence of the Crank-Nicolson scheme is analyzed via the concept of elliptic projection. Namely, second order convergence is proved for the corresponding elliptic problems in special norms. Then, splitting the error of the heat problem into two errors we prove second order estimates in space and time in modified L_2 norm.

1 Introduction

We begin with the problem:

$$\frac{\partial u(x,t)}{\partial t} = \frac{\partial}{\partial x}\left(a(x)\frac{\partial u(x,t)}{\partial x}\right) + b(x,t)u, \; in\, [0,\xi)\cup(\xi,1]\times[0,T], \tag{1}$$

$$[u]_{x=\xi} = u(\xi+0,t) - u(\xi-0,t) = 0, \; in\, [0,T], \tag{2}$$

$$C\frac{\partial u(\xi,t)}{\partial t} = \left[a(x)\frac{\partial u(x,t)}{\partial x}\right]_{x=\xi}, \; 0<\xi<1, \; in\, [0,T], \tag{3}$$

$$u(0,t) = 0, \; in\, [0,T], \tag{4}$$

$$u(1,t) = 0, \; in\, [0,T], \tag{5}$$

$$u(x,0) = \varphi(x), \; in\, [0,1]. \tag{6}$$

Here $C = const > 0$ is the heat capacity concentrated at point $x = \xi$, $a(x)$ is a positive function on $[0,1]$, smooth on $[0,\xi]$ and on $[\xi,1]$, with a possible discontinuity at ξ, and $b(x,t)$ is nonpositive function smooth on $[0,\xi]\times[0,T]$ and on $[\xi,1]\times[0,T]$, with a possible discontinuity on $\xi\times[0,T]$. We assume that the coefficients and the initial value φ are such that the problems we shall consider, possess a unique solutions continuous on $[0,1]\times[0,T]$ and sufficiently smooth for our purposes on $[0,\xi]\times[0,T]$ and on $[\xi,1]\times[0,T]$. If the capacity C_0 is concentrated on the boundary $x=0$, then the following boundary condition instead of (4) is stated:

$$C_0\frac{\partial u(0,t)}{\partial t} = a(0)\frac{\partial u(0,t)}{\partial x}, \; C_0 = const > 0. \tag{7}$$

The flow may have discontinuity in a point ξ :

$$\rho a(\xi+0)\frac{\partial u(\xi+0,t)}{\partial x}-a(\xi-0)\frac{\partial u(\xi-0,t)}{\partial x}=0,\ \rho=const>0. \tag{8}$$

The problem (1), (4), (5)-(7) with $b = 0$ and $a(x)$-continuous function is solved numerically in ([3], p. 424). We shall prove second order convergence of the following problem (P_1) given by: (1)-(3), (7), (5), (6). In a similar way second order convergence for the problem (P_2): (1)- (6) and (P_3): (1), (2), (8), (7), (5), (6) can be established.

The convergence question for differential equations with discontinuous coefficients has been studied from many authors. To obtain optimal order error estimates in space we use a technique which consists in comparing the finite difference approximations of (P_1) to the solution of associated discrete stationary problem (S_1). This idea originates from Samarskii ([2], p. 199) and is developed at the error analysis of Galerkin finite element methods [4]. The method is applied in [1] to the equation (1) at zero Dirichlet boundary conditions with discontinuous heat flow.

Multiplying (1) by $u(x,t)$, integrating by parts over $[0,1]$ taking into account (2), (3), (5) and (7), we get for problem (P_1)

$$\frac{\partial}{\partial t}\left(\int_0^1 u^2(x,t)dx+Cu^2(\xi,t)+C_0u^2(0,t)\right)=$$

$$=-2\int_0^1 a(x)\left(\frac{\partial u(x,t)}{\partial x}\right)^2dx+2\int_0^1 b(x,t)u^2(x,t)dx\le 0.$$

It follows from here the energy stability of solution to (P_1):

$$\|u(.,s)\|_{L_{2,\xi,0}}\le\|u(.,t)\|_{L_{2,\xi,0}},\ 0\le t\le s\le T, \tag{9}$$

and in particular,

$$\|u(.,t)\|_{L_{2,\xi,0}}\le\|\varphi\|_{L_{2,\xi,0}}, \tag{10}$$

where

$$\|v(.,t)\|^2_{L_{2,\xi,0}}=\int_0^1 v^2(x,t)dx+C^2v^2(\xi,t)+C_0v^2(0,t). \tag{11}$$

In a similar way one can prove the following inequalities for (P_2) and (P_3), respectively:

$$\|u(.,s)\|_{L_{2,\xi}}\le\|u(.,t)\|_{L_{2,\xi}},\ 0\le t\le s\le T,$$

where

$$\|v(.,t)\|^2_{L_{2,\xi}}=\int_0^1 v^2(x,t)dx+C^2v^2(\xi,t), \tag{12}$$

and

$$\|u(.,s)\|_{L_{2,0,\xi,\rho}}\le\|u(.,t)\|_{L_{2,0,\xi,\rho}},\ 0\le t\le s\le T,$$

where

$$\|v(.,t)\|^2_{L_{2,0,\xi,\rho}}=\int_0^\xi v^2(x,t)dx+\rho\int_\xi^1 v^2(x,t)dx+C_0v^2(0,t). \tag{13}$$

In section 2 we state three point boundary value problems (namely (S_1), (S_2), (S_3)) associated with (P_1), (P_2), (P_3), respectively. More in detail is studied (S_1). We discretize (S_1) by the analog of the difference scheme used in the discretization of problem (P_1), namely (31), (32), (33). Then, we prove second order convergence of the resulting approximation to solution of (S_1) in discrete maximum norm and the discrete analog of (11). We use these estimates in the proof of the error bound (38) in Theorem 1. In section 3 the elliptic approximation technique is applied to obtain optimal error estimates for the discretization of (P_1). Analogous estimates can be proved for (P_2) and (P_3), in discrete maximal norm and norms which are discrete analogy of (12) and (13) respectively.

2 Stationary problems

In this section we consider the following elliptic problem:

$$(S_1)\quad \begin{aligned} &(a(x)w'(x))' = f(x),\ in\,[0,\xi)\cup(\xi,1], && (14)\\ &[a(x)w'(x)]_{x=\xi} = C_1, && (15)\\ &[w(x)]_{x=\xi} = 0, && (16)\\ &a(0)w'(0) = C_2, && (17)\\ &w(1) = 0. && (18) \end{aligned}$$

Here C_1 and C_2 are some positive constants. The stationary problem corresponding to (P_2) is as follows: (14)-(16), (18), (19) and to (P_3) : (14), (16)-(18), (20) and

$$w(0) = 0, \tag{19}$$

$$\rho a(\xi+0)w'(\xi+0,t) - a(\xi-0)w'(\xi-0,t) = 0. \tag{20}$$

The stationary problem (S_1) will be only treated. We partition the interval $[0,1]$ into N+1 parts, where $x_0 = 0, x_{N+1} = 1, x_m = \xi$, $(1 < m < N)$, $x_n = x_{n-1}+h_-$ if $1 \le n \le m$ and $x_n = x_{n-1}+h_+$ if $m+1 \le n \le N+1$. The right and left difference derivatives in x_n are denoted by $u_{x,n}$ and $u_{\bar{x},n}$ respectively. Consequently, centered second difference quotients that approximate the function $(a(x)u'(x))'$ at $x_n \ne x_m$ will be denoted by $(a_{n+\frac12}u_{x,n})_{\bar{x},n}$, where we put $a_{n+\frac12} = a(\frac{x_n+x_{n+1}}{2})$ for $n = 0, ..., N$. At the interface we let:

$$\delta\left(a_{m+\frac12}u_{x,m}\right) = \left(a_{m+\frac12}\frac{u_{m+1}-u_m}{h_+} - a_{m-\frac12}\frac{u_m-u_{m-1}}{h_-}\right)/\hat{h},$$

where $\hat{h} = (h_+ + h_-)/2$.

We approximate (S_1) by the following discrete problem:

$$(D_1)\quad \begin{aligned} &\left(a_{n+\frac12}W_{x,n}\right)_{\bar{x},n} = f(x_n),\ if\ 1 \le n \le N,\ n \ne m,\\ &\delta\left(a_{m+\frac12}W_{x,m}\right) = \hat{f}(x_m) + \frac{C_1}{\hat{h}},\\ &\frac{a_{\frac12}W_{x,0}}{h_-} = \frac{f_0}{2} + \frac{C_2}{h_-},\ W_{N+1} = 0, \end{aligned}$$

where $\hat{f}(x_m) = (h_+ f(x_m + 0) + h_- f(x_m - 0))/2\hat{h}$. The $N+1 \times N+1$ tridiagonal matrix of the linear system (D_1) is obviously invertible i.e. W is uniquely defined. Let us introduce in $\mathcal{R}_{1,0}^{N+2}(w = (w_0, \dots, w_{N+1}), w_{N+1} = 0, w_n \in \mathcal{R})$ the bilinear form $a_h(.,.)$:

$$a_h(v,u) = -a_{\frac{1}{2}} v_{x,0} u_0 - h_- \sum_{n=1}^{m-1} (a_{n+\frac{1}{2}} v_{x,n})_{\bar{x},n} u_n -$$

$$-\hat{h}\delta(a_{m+\frac{1}{2}} v_{x,m}) u_m - h_+ \sum_{n=m+1}^{N} (a_{n+\frac{1}{2}} v_{x,n})_{\bar{x},n} u_n. \tag{21}$$

It is easily seen that:

$$\forall v \in R_{1,0}^{N+2},\ a_h(v,v) = h_- \sum_{n=0}^{m-1} a_{n+\frac{1}{2}} |v_{x,n}|^2 + h_+ \sum_{n=m}^{N} a_{n+\frac{1}{2}} |v_{x,n}|^2, \tag{22}$$

$$\forall v \in R_{1,0}^{N+2},\ a_h(v,v) \le \frac{4|a|_\infty [|v||_h^2}{[min(h_-, h_+)]^2}, \tag{23}$$

and that $a_h(.,.)$ is an inner product in $\mathcal{R}_{1,0}^{N+2}$. Here by $[|v||_h^2$ we denote the following norm:

$$[|v||_h^2 = \frac{h_-}{2} v_0^2 + \sum_{n=1}^{m-1} h_- v_n^2 + \frac{\hat{h} v_m^2}{2} + \sum_{n=m+1}^{N} h_+ v_n^2, \tag{24}$$

and $|a|_\infty = \max_{x \in (0,1)} a(x)$. Further, letting $\underline{a} = \min_{0 \le x \le 1} a(x)$, we see that the following discrete Sobolev type inequality holds:

$$\forall v \in R_{1,0}^{N+2},\ \max_n |v_n|^2 \le \frac{2}{\underline{a}} a_h(v,v). \tag{25}$$

A trivial consequence of (25) is that:

$$\exists M_1 > 0,\ \forall v \in R_{1,0}^{N+2},\ a_h(v,v) \ge M_1 [|v||_h^2. \tag{26}$$

Let us denote by (D_1') the problem obtained from (D_1) by replacing the right-hand sides with $\zeta_{x,n}, \zeta_{x,m}, \zeta_{x,0}$ and 0, respectively. An a priori estimate for the solution of problem (D_1') is given in the following

Lemma 1 *Let v be a solution of problem (D_1'). Then there exists a constant $M_2 > 0$ such that*

$$a_h(v,v)^{\frac{1}{2}} \le M_2 [|\zeta||_h. \tag{27}$$

Proof. Multiplying the first equation in (D_1') by $h_- v_n$ if $1 \le n \le m-1$ and by $h_+ v_n$ if $m+1 \le n \le N$, the second by $\hat{h} v_m$ and the third one by $h_- v_0$ we obtain

$$-a_h(v,v) = \sum_{n=0}^{m-1} (\zeta_{n+1} - \zeta_n) v_n + (\zeta_{m+1} - \zeta_m) v_m + \sum_{n=m+1}^{N} (\zeta_{n+1} - \zeta_n) v_n =$$

$$= -\zeta_0 v_0 - \sum_{n=0}^{m-1} \zeta_{n+1}(v_{n+1} - v_n) - \sum_{n=m}^{N} \zeta_{n+1}(v_{n+1} - v_n).$$

Using the Cauchy-Schwarz inequality (22) and (25) we have,

$$a_h(v,v) \le M_2 a_h(v,v)^{\frac{1}{2}}[|\zeta||_h,$$

and (27) follows. □

Let $e_n = w_n - W_n, n = 0, ..., N+1$ and $h = max(h_-, h_+)$. Two approximation results for W are given in the following

Lemma 2 *Let the solution of (S_1) be continuous in $[0,1]$ and smooth in $[0,\xi]$ and in $[\xi, 1]$. Then there exists a constant M_3, independent of h and w such that:*

$$\max_{0\le n\le N+1} |e_n| \le M_3 h^2 \sum_{j=1}^{4} |w^{(j)}|_\infty, \tag{28}$$

and

$$[a_h(e,e)]^{\frac{1}{2}} \le M_3 h^2 \sum_{j=1}^{4} |w^{(j)}|_\infty, \tag{29}$$

where $|w^{(j)}|_\infty = \max\left(\sup_{0\le x<\xi} |w^{(j)}(x)|_\infty, \sup_{\xi<x\le 1} |w^{(j)}(x)|_\infty\right)$.

Proof. Let us first note that (28) follows immediately from (29) in view of (25). Therefore it remains to show (29). Using (D_1) we have for $1 \le n \le N, n \ne m$,

$$\left(a_{n+\frac{1}{2}} e_{x,n}\right)_{\bar{x},n} = \left(a_{n+\frac{1}{2}} w_{x,n}\right)_{\bar{x},n} - f(x_n) = \begin{cases} -h_-^2 g_n, \, if \ 1 \le n \le m-1, \\ -h_+^2 g_n, \, if \ m+1 \le n \le N, \end{cases}$$

and for $n = m$, $n = 0$

$$\hat{h}\delta(a_{m+\frac{1}{2}} e_{x_m}) = -h_+^2 g_m^{'} - h_-^2 g_m^{''}, \ a_{\frac{1}{2}} e_{x,0} = -h_-^2 g_0.$$

Then

$$a_h(e,e) = h_-^2 g_0 e_0 + h_-^3 \sum_{n=1}^{m-1} g_n e_n + \left(h_+^2 g_m^{'} + h_-^2 g_m^{''}\right) e_m + h_+^3 \sum_{n=m+1}^{N} g_n e_n.$$

We introduce an auxiliary vector $\eta \in \mathcal{R}_{1,0}^{N+2}$ by the relations:

$$\eta_{n+1} - \eta_n = h_-^3 g_n, \, n = 1, \ldots, m-1, \ \eta_{n+1} - \eta_n = h_+^3 g_n, \, n = m+1, \ldots, N,$$
$$\eta_{m+1} - \eta_m = h_+^2 g_m^{'} + h_-^2 g_m^{''}, \ \eta_1 - \eta_0 = h_-^2 g_0.$$

Obviously

$$\eta_0 = -h_-^2 g_0 - h_-^3 \sum_{n=1}^{m-1} g_j - \left(h_+^2 g_m^{'} + h_-^2 g_m^{''}\right) - h_+^3 \sum_{n=m}^{N} g_n,$$

$$\eta_s = -h_-^3 \sum_{n=s}^{m-1} g_n - \left(h_+^2 g_m^{'} + h_-^2 g_m^{''}\right) - h_+^3 \sum_{n=m+1}^{N} g_n, \, s = 1, \ldots m,$$

$$\eta_s = -h_+^3 \sum_{n=s}^{N} g_n, \, s = m+1, \ldots, N.$$

Then

$$[|\eta||_h \leq M_2^{'} h^2 \sum_{j=1}^{4} |w^{(j)}|_\infty, \tag{30}$$

and (29) now follows from (27) and (30). □

3 The heat problems

In this section we consider the discretization of problem (P_1) by the Crank-Nicolson method modified at the interface node $x_m = \xi$. We partition the interval $[0, T]$ into J parts and the interval $[0, 1]$ into $N+1$ parts. We shall approximate u_n^j by U_n^j, where $U^j = (U_0^j, \ldots, U_{N+1}^j)^t \in \mathcal{R}_{1,0}^{N+2}(U_{N+1}^j = 0; U_n^j \in \mathcal{R})$. Starting with $U_n^0 = \varphi_n$, we denote $U_n^{j+\frac{1}{2}} = (U_n^{j+1} + U_n^j)/2$, $t^{j+\frac{1}{2}} = t^j + \tau/2$, $b_n^{j+\frac{1}{2}} = b(x_n, t^{j+\frac{1}{2}})$ and put $U_n^{\tau,j} = (U_n^{j+1} - U_n^j)/\tau$. Then for $0 \leq j \leq J-1$ our scheme becomes (see [3]),

$$U_n^{\tau,j} - \left(a_{n+\frac{1}{2}} U_{x,n}^{j+\frac{1}{2}}\right)_{\bar{x},n} - b_n^{j+\frac{1}{2}} U_n^{j+\frac{1}{2}} = 0,\ 1 \leq n \leq N,\ n \neq m, \tag{31}$$

$$(H_1) \qquad \left(1 + C/\hat{h}\right) U_m^{\tau,j} - \delta\left(a_{m+\frac{1}{2}} U_{x,m}^{j+\frac{1}{2}}\right) - \hat{b}\left(x_m, t^{j+\frac{1}{2}}\right) U_m^{j+\frac{1}{2}} = 0, \tag{32}$$

$$(0.5 + C_0/h_-) U_0^{\tau,j} - a_{\frac{1}{2}} U_{x,0}^{j+\frac{1}{2}}/h_- - 0.5 b_0^{j+\frac{1}{2}} U_0^{j+\frac{1}{2}} = 0, \tag{33}$$

$$U_{N+1}^j = 0,\ U_n^0 = \varphi_n, \tag{34}$$

where by analogy to $\hat{f}$, $\hat{b}(x_m, t) = (h_- b(x_m - 0, t) + h_+ b(x_m + 0, t))/(2\hat{h})$.

Let (H_1') is obtained from (H_1) by replacing the right-hand sides with $\eta_n(j\tau)$, $(1 + C/\hat{h})\eta_m(j\tau)$, $(0.5 + C_0/h_-)\eta_0(j\tau)$, 0 and φ_n, respectively. The energy stability of the solution to problem (P_1) has been shown in section 1. The discrete analog of (9), (10) for scheme (H_1) is proved in the next Lemma. Denote by

$$[|v||_{\star,h}^2 = [|v||_h^2 + C_0 v_0^2 + C v_m^2. \tag{35}$$

Lemma 3 *Let Z be a solutuon of (H_1') then*

$$[|Z^{j+1}||_{\star,h} \leq [|Z^j||_{\star,h} + \tau[|\eta(j\tau)||_{\star,h}, \tag{36}$$

and in particular

$$[|Z^{j+1}||_{\star,h} \leq [|Z^0||_{\star,h} + \sum_{j=0}^{j} \tau[|\eta(j\tau)||_{\star,h}. \tag{37}$$

Proof. Multiplying (31) by $h_- Z_n^{j+\frac{1}{2}}$ if $n \leq m-1$ and by $h_+ Z_n^{j+\frac{1}{2}}$ if $n \geq m+1$, (32) by $\hat{h} Z_m^{j+\frac{1}{2}}$ and (33) by $Z_0^{j+\frac{1}{2}}$, summing from $n = 0$ to $n = N$, using (21), (22), (24), (35), and the sign conditions of a, b, apllying the Cauchy-Schwarz inequality we obtain,

$$[|Z^{j+1}||_{\star,h}^2 - [|Z^j||_{\star,h}^2 \leq \tau[|\eta(j\tau)||_{\star,h}[|Z^{j+1} + Z^j||_{\star,h}.$$

Hence,

$$[|Z^{j+1}\|_{\star,h} \leq [|Z^j\|_{\star,h} + \tau[\|\eta(j\tau)\|_{\star,h}. \qquad \square$$

From (36) and (37) follows that U^{j+1} exists uniquely as the solution of $J \times J$ tridiagonal system represented by (H_1).

Let $u^j = u(x, t^j)$, $U^j = U(x, t^j)$. The main result in this paper is given in the following

Theorem 1 *Let u^j satisfy (P_1) and U^j satisfy (H_1). Then there exists a constant $C(u)$ independent of h and τ and dependent of the derivatives of u up to order 4, such that*

$$\max_{0\leq j\leq J} [\|u^j - U^j\|_{\star,h} \leq C(u)(\tau^2 + h^2). \tag{38}$$

Proof. For $0 \leq t \leq T$, let $W(t) = (W_0, \ldots, W_{N+1})^t \in \mathcal{R}^{N+2}_{1,0}$ be the elliptic approximation to $u(.,t)$ defined by

$$(E_1) \quad \begin{aligned} &\left(a_{n+\frac{1}{2}} W_{x,n}\right)_{\bar{x},n} = (Lu)(x_n, t),\ 1 \leq n \leq N,\ n \neq m, \\ &\delta\left(a_{m+\frac{1}{2}} W_{x,m}\right) = \left(\hat{L}u\right)(x_m, t) + \frac{C}{\hat{h}} u_t(x_m, t), \\ &\frac{a_{\frac{1}{2}} W_{x,0}}{h_-} = \frac{1}{2}(Lu)(0,t) + \frac{C_0 u_t(0,t)}{h_-},\ W_{N+1} = 0, \end{aligned}$$

where $Lu := (a(x)u_x)_x$ with $W^j = W(t^j)$. Denote $\zeta^j = u^j - W^j$, $\theta^j = W^j - U^j$, $j = 0, \ldots, J$. So that $u^j - U^j = \zeta^j + \theta^j$. According to (28), (24), and (35) we have

$$\max_{0\leq j\leq J} [\|\zeta^j\|_{\star,h} \leq Ch^2. \tag{39}$$

Hence, it remains to estimate $[\|\theta^j\|_{\star,h}$. Using (H_1), (P_1) and (E_1) and letting $\theta_n^{j+\frac{1}{2}} = (\theta_n^{j+1} + \theta_n^j)/2$, we obtain for $1 \leq n \leq N, n \neq m$

$$\theta_n^{\tau,j} = \left(a_{n+\frac{1}{2}} \theta_{x,n}^{j+\frac{1}{2}}\right)_{\bar{x},n} + b_n^{j+\frac{1}{2}} \theta_n^{j+\frac{1}{2}} + w_n^j, \tag{40}$$

where the term w_n^j is of order $\mathcal{O}(\tau^2 + h^2)$ (see [1]).

Using again (H_1) and (E_1) we have on the interface

$$\left(1 + \frac{C}{\hat{h}}\right) \theta_m^{\tau,j} = \delta\left(a_{m+\frac{1}{2}} \theta_{x,m}^{j+\frac{1}{2}}\right) + \hat{b}_m^{j+\frac{1}{2}} \theta_m^{j+\frac{1}{2}} + \left(1 + \frac{C}{\hat{h}}\right) w_m^j, \tag{41}$$

where $w_m^j = w_{m_1}^j + \ldots + w_{m_7}^j$

$$w_{m_1}^j = \left(W_m^{\tau,j} - u_m^{\tau,j}\right),\ w_{m_2}^j = \left(u_m^{\tau,j} - u_t\left(x_m, t^{j+\frac{1}{2}}\right)\right),$$

$$\left(1 + \frac{C}{\hat{h}}\right) w_{m_3}^j = \frac{C}{\hat{h}} \left(u_t\left(x_m, t^{j+\frac{1}{2}}\right) - \frac{1}{2}\left(u_t\left(x_m, t^j\right) + u_t\left(x_m, t^{j+1}\right)\right)\right),$$

$$\left(1+\frac{C}{\hat{h}}\right) w_{m_4}^j = \hat{L}u\left(x_m, t^{j+\frac{1}{2}}\right) - \frac{1}{2}\left(\hat{L}u\left(x_m, t^j\right) + \hat{L}u\left(x_m, t^{j+1}\right)\right),$$

$$\left(1+\frac{C}{\hat{h}}\right) w_{m_5}^j = -\frac{1}{2}\hat{b}_m^{j+\frac{1}{2}}\left(W_m^j + W_m^{j+1} - u\left(x_m, t^j\right) - u\left(x_m, t^{j+1}\right)\right),$$

$$\left(1+\frac{C}{\hat{h}}\right) w_{m_6}^j = -\frac{1}{2}\hat{b}_m^{j+\frac{1}{2}}\left(\frac{1}{2}\left(u\left(x_m, t^j\right) + u\left(x_m, t^{j+1}\right)\right) - u\left(x_m, t^{j+\frac{1}{2}}\right)\right),$$

$$\left(1+\frac{C}{\hat{h}}\right) w_{m_7}^j = u_t\left(x_m, t^{j+\frac{1}{2}}\right) - \hat{L}u\left(x_m, t^{j+\frac{1}{2}}\right) - \hat{b}_m^{j+\frac{1}{2}} u\left(x_m, t^{j+\frac{1}{2}}\right).$$

Obviously the terms $w_{m_2}^j$, $\left(1+\frac{C}{\hat{h}}\right) w_{m_4}^j$, $\left(1+\frac{C}{\hat{h}}\right) w_{m_6}^j$ and $\left(1+\frac{C}{\hat{h}}\right) w_{m_7}^j$ are $\mathcal{O}\left(\tau^2+h^2\right)$ and the term $\left(1+\frac{C}{\hat{h}}\right) w_{m_3}^j$ is $\mathcal{O}\left(\frac{C}{\hat{h}}(\tau^2+h^2)\right)$. It follows from (28) and the commutativity of the elliptic approximation operator and time differentiation that $w_{m_1}^j$ is $\mathcal{O}\left(\tau^2+h^2\right)$ and from (28) follows that the term $\left(1+\frac{C}{\hat{h}}\right) w_{m_5}^j$ is $\mathcal{O}(\tau^2+h^2)$. Hence, w_m^j is $\mathcal{O}(\tau^2+h^2)$.

At the left bound $x=0$ using again (E_1) and $(H_1\ )$ we have

$$\left(\frac{C_0}{h_-}+\frac{1}{2}\right)\theta_0^{\tau,j} = \frac{a_{\frac{1}{2}}\theta_{x,0}^{j+\frac{1}{2}}}{h_-} + \frac{1}{2}b_0^{j+\frac{1}{2}}\theta_0^{j+\frac{1}{2}} + \left(\frac{C_0}{h_-}+\frac{1}{2}\right) w_0^j. \tag{42}$$

As in the interface we can prove that w_0^j is $\mathcal{O}(\tau^2+h^2)$.

From (36), (40), (41) and (42) we obtain,

$$[\|\theta^{j+1}\|_{\star,h} - [\|\theta^j\|_{\star,h} \le \tau[\|w^j\|_{\star,h}.$$

Therefore

$$\max_j[\|\theta^j\|_{\star,h} \le C(u)(\tau^2+h^2), \tag{43}$$

and (38) now follows from (39) and (43). □

Acknowledgment

This work is supported by Ministry of Science and Education of Bulgaria under Grant # MM–524/95.

References

1. Akrivis, G.D., Dougalis, V.A.: Finite difference discretizations of Some initial and boundary value problems with interface. Math. Comp. **56** (1991) 505–522
2. Samarskii, A.A.: Introduction in difference scheme theory. Nauka, Moscow 1971 (Russian)
3. Samarskii, A.A.: Theory of difference schemes. Nauka, Moscow 1977 (Russian)
4. Thomée, V.: Galerkin finite element methods for parabolic problems. Lecture notes in Math. **1054**, Springer-Verlag, Berlin, Heidelberg, New York, Tokyo 1984

Treatment of Large Air Pollution Models

J. Brandt[1], J. Christensen[1], I. Dimov[2], K. Georgiev[2], I. Uria[3] and Z. Zlatev[1]

[1] National Environmental Research Institute,
Frederiksborgvej 399, P. O. Box 358, DK-4000 Roskilde, Denmark
e-mail: lujbr@sun4.dmu.dk, lujc@sun1.dmu.dk, luzz@sun2.dmu.dk
[2] Central Laboratory for Parallel Information Processing, Bulgarian Academy of Sciences, Acad. G. Bonchev str., Bl. 25-A, 1113 Sofia, Bulgaria;
e-mail: ivdimov@iscbg.acad.bg, georgiev@iscbg.acad.bg
[3] LABEIN, Bilbao, Spain, Cuesta de Olabeaga, 16
e-mail jtauragn@bi.ehu.es

Abstract. Large ozone concentrations have harmful effects on forests and crops when these exceed some critical levels. It is believed that the damages in USA due to high ozone concentrations exceed several billions dollars. Therefore it is worthwhile to investigate different actions that could be applied in the attempts to reduce the harmful effects. One needs reliable mathematical models in such studies. Reliable models are normally very big and it is difficult to treat them numerically, because they lead, after some kind of discretization and after the implementation of some appropriate splitting procedure, to several very huge systems of ordinary differential equations (up to order of 10^6). Moreover, these systems have to be treated numerically during many time-steps (typically several thousand time-steps per run are necessary). The use of modern parallel and/or vector machines is an important condition in the efforts to handle successfully big air pollution models. If the numerical algorithms are both sufficiently fast and sufficiently accurate, then different simulations can be carried out.

1 Mathematical description of an LRTAP model

The long range transport of air pollutants (**LRTAP**) over Europe is studied, [23], by a system of partial differential equations (**PDE's**):

$$\frac{\partial c_s}{\partial t} = -\frac{\partial(uc_s)}{\partial x} - \frac{\partial(vc_s)}{\partial y} - \frac{\partial(wc_s)}{\partial z} \tag{1}$$

$$+\frac{\partial}{\partial x}\left(K_x\frac{\partial c_s}{\partial x}\right) + \frac{\partial}{\partial y}\left(K_y\frac{\partial c_s}{\partial y}\right) + \frac{\partial}{\partial z}\left(K_z\frac{\partial c_s}{\partial z}\right)$$

$$+E_s + Q_s(c_1, c_2, \ldots, c_q) - (\kappa_{1s} + \kappa_{2s})c_s, \quad s = 1, 2, \ldots, q.$$

The number of equations q is equal to the number of species that are studied by the model and has been varied from 10 to 168 in our studies. The different quantities involved in the mathematical model are defined as follows: (i) the concentrations are denoted by c_s, (ii) u, v and w are wind velocities, (iii) K_x, K_y and K_z are diffusion coefficients, (iv) the emission sources are described by E_s, (v) κ_{1s} and κ_{2s} are deposition coefficients, (vi) the chemical reactions are denoted by $Q_s(c_1, c_2, \ldots, c_q)$.

1.1 Splitting the model to sub-models

It is difficult to treat the system of **PDE's** (1) directly. This is the reason for using different kinds of splitting. A simple splitting procedure, based on ideas proposed in [13] and [14], can be defined, for $s = 1, 2, \ldots, q$, by five sub-models, representing respectively the horizontal advection, the horizontal diffusion, the chemistry, the deposition and the vertical exchange:

$$\frac{\partial c_s^{(1)}}{\partial t} = -\frac{\partial (u c_s^{(1)})}{\partial x} - \frac{\partial (v c_s^{(1)})}{\partial y} \tag{2}$$

$$\frac{\partial c_s^{(2)}}{\partial t} = \frac{\partial}{\partial x}\left(K_x \frac{\partial c_s^{(2)}}{\partial x}\right) + \frac{\partial}{\partial y}\left(K_y \frac{\partial c_s^{(2)}}{\partial y}\right) \tag{3}$$

$$\frac{d c_s^{(3)}}{dt} = E_s + Q_s(c_1^{(3)}, c_2^{(3)}, \ldots, c_q^{(3)}) \tag{4}$$

$$\frac{d c_s^{(4)}}{dt} = -(\kappa_{1s} + \kappa_{2s}) c_s^{(4)} \tag{5}$$

$$\frac{\partial c_s^{(5)}}{\partial t} = -\frac{\partial (w c_s^{(5)})}{\partial z} + \frac{\partial}{\partial z}\left(K_z \frac{\partial c_s^{(5)}}{\partial z}\right) \tag{6}$$

1.2 Space discretization

If the model is split into sub-models as in the previous paragraph, then the discretization methods will lead to five systems of ordinary differential equations (ODE's):

$$\frac{dg^{(i)}}{dt} = f^{(i)}(t, g^{(i)}), \qquad g^{(i)} \in R^{N_x \times N_y \times N_z \times N_s}, \qquad f^{(i)} \in R^{N_x \times N_y \times N_z \times N_s}, \tag{7}$$

where $i = 1, 2, 3, 4, 5$, N_x, N_y and N_z are the numbers of grid-points along the coordinate axes and $N_s = q$ is the number of chemical species. The functions $f^{(i)}, i = 1, 2, 3, 4, 5$, depend on the discretization methods used in the numerical treatment of the different sub-models.

2 Numerical methods

The mathematical terms in the five sub-models have different properties. Therefore it is natural to apply different numerical methods in the treatment of these sub-models in an attempt to optimize globally the computational process.

2.1 Numerical treatment of the horizontal advection

The horizontal advection causes a lot of difficulties during the numerical treatment. Different methods have been proposed: pseudospectral algorithms, finite differences, finite elements, special algorithms producing non-negative solutions, semi-Lagrangian algorithms. It may be appropriate to try wavelets.

The pseudospectral algorithm used in the Danish Eulerian Model is based on the use of trigonometric interpolation to calculate approximations to the spatial derivatives. The algorithm is accurate and a good implementation is not very expensive. However, it requires periodic boundary conditions and extra efforts are needed in order to deal with this problem. Other pseudospectral algorithms can be found in [7].

The **finite differences** are still very popular. It is easy to implement such methods. However, the horizontal advection deals very often with sharp gradients. This causes problems for the finite difference methods. One must apply high-order differences to resolve the gradients. The simpler low-order differences (used in the up-wind methods) produce smooth solutions, but also introduce a lot of artificial diffusion; see Fig.1.

Finite elements algorithms are becoming more and more popular. Different finite elements can be used in the different parts of the space domain, and this is a great advantage of the finite elements methods. Smaller and/or more accurate finite elements could be used in difficult sub-domain. However, it is rather difficult to implement such a method in conjunction with the other physical processes. The fact is that the finite element algorithms in air pollution models are still used with the same finite elements in the whole region (an exception being the use of finite elements to handle the vertical exchange; see below). Linear one-dimensional finite elements are popular. Such methods have been used in some of our experiments.

Algorithms producing non-negative solutions are also very popular. The previous three groups of methods could produce negative concentrations, which leads to a disaster when the chemical reactions are handled. There are different algorithms which produce non-negative concentrations. Some experiments with the methods developed by Bott [1] and Holm [10] have been carried out. These methods produce smooth concentration fields with no negative concentrations, but they are in general more expensive than many of the methods from the previous three groups.

The semi-Lagrangian methods are accurate and have good stability properties. They have been used in several air pollution models. We have experimented with a semi-Lagrangian scheme, which has been more expensive than the other methods we used. However, our implementation of this scheme can probably be improved considerably.

We have not carried out experiments with methods based on the use of **wavelets**. It would be interesting to see how such methods will perform.

Many of the algorithms listed above may produce **negative** concentrations If this is the case, then some kind of smoothing has to be applied before the treatment of the chemical reactions. It is difficult to justify theoretically the application of a smoothing procedure. The selective filter proposed by Forester [6] seems to work rather well in practice. This filter performs smoothing only in regions around detected oscillations. Therefore, the high concentrations are normally not affected by the use of this filter.

2.2 Numerical treatment of the horizontal diffusion

The same algorithms as those used in the treatment of the horizontal advection can be carried out to handle the horizontal diffusion (excepting the semi-Lagrangian algorithms). A semi-analytical approach has also been applied in the experiments. The concentrations are expanded in a Fourier series which are truncated after some term and contributions due to the horizontal diffusion are calculated by using the truncated series. The success of this algorithm is due to the fact that the horizontal advection dominates over the horizontal diffusion. Therefore the desired approximations can be calculated by keeping relatively small number of terms in the truncated series.

2.3 Numerical treatment of the chemical transformations

The quasi-steady-state-approximation, QSSA, is commonly used in air pollution modelling since 1978, [9]. Recently updated versions of the QSSA have been proposed, [21], [2]. This method may have difficulties in the periods around sun-rises and sun-sets when the photochemical reactions are activated or disactivated, but it is very cheap computationally.

Classical methods (mainly the Backward Euler Formula, the Trapezoidal Rule and some Runge-Kutta methods) have also been used; see [3], [16], [17], [19]. These methods produce more accurate results than the QSSA, but they are more expensive. Recently the classical methods have been used with different kinds of partitioning ([8], [20]). The use of partitioning improves the performance of the methods without affecting too much the accuracy.

Extrapolation techniques are very popular in other fields of science and engineering. It may be useful to try such techniques also in the treatment of the chemical parts of large air pollution problems; these methods have been used for solving some other chemical problems in [5].

The chemical sub-models are sometimes described by a system of differential-algebraic equations. Techniques similar to those used in [18] can be applied in this situation.

2.4 Numerical treatment of the deposition terms

The deposition sub-model consists of a system of **linear** ODE's. Moreover, the equations of this system are decoupled. Every equation can be solved exactly, and this is the most straight-forward approach. It should be stressed, however, that other methods must be used when the deposition sub-model is described by more advanced physical mechanisms.

2.5 Numerical treatment of the vertical exchange

The grid in the vertical direction is not equi-distant. The use of finite elements is a natural solution when this sub-model is handled. Classical methods can be used to solve the ODE system obtained after the application of a finite element algorithm. The algorithms actually used in the Danish Eulerian Model, [23], are the linear one-dimensional finite elements and the well-known θ-method, [12].

2.6 Numerical algorithms used at lower levels

The major algorithms used to treat the sub-models are described in the previous paragraphs. Many other algorithms are used at lower levels. FFT's are used in the pseudospectral algorithm, [23]. Predictor-corrector schemes with several different correctors are used in the time-discretization of the ODE systems arising in the horizontal advection sub-model (such schemes have improved stability properties; [22]). BLAS's and solvers for systems of linear algebraic equations are used in different parts of the model, [23].

3 Need for high speed computers

The number of equations in the five ODE systems (7) grows very quickly when the number of grid-points and/or the number of chemical species is increased; see Table 1. The systems of ODE's are treated successively at each time-step. The typical number of time-steps is 3456 (when meteorological data covering a period of one month + five days to start up the model are to be handled). The number of time-steps for the chemical sub-model is sometimes even larger, because smaller step-sizes have to be used in this sub-model; [23].

Number of species	$(32 \times 32 \times 10)$	$(96 \times 96 \times 10)$	$(192 \times 192 \times 10)$
1	10240	92160	368640
2	20480	184320	737280
10	102400	921600	3686400
35	358400	3225600	12902400
56	573440	5160960	21381120
168	1720320	15482880	61931520

Table 1
The numbers of equations per ODE system treated at every time-step.

Such large problems can be solved **only** if new and modern high-speed computers are used. Moreover, it is necessary to select the right numerical algorithms (which are most suitable for the high speed computers available) and to perform the programming work very carefully in order to exploit fully the great potential power of the vector and/or parallel computers; see [23].

4 Numerical experiments

The following simple test-example (proposed originally in [11]; see also [23]) can be used to test the accuracy of (a) the advection algorithms, (b) the chemical algorithms and, what is most important, (c) the coupling of advection with chemistry.

$$\frac{\partial c_s}{\partial t} = -(1-y)\frac{\partial c_s}{\partial x} - (x-1)\frac{\partial c_s}{\partial y} + Q_s(c_1, c_2, \ldots, c_q), \tag{8}$$

$$s = 1, 2, \ldots, q, \qquad 0 \le x \le 2 \qquad \textit{and} \qquad 0 \le y \le 2. \tag{9}$$

4.1 Testing the accuracy of some advection schemes

Remove the chemical terms and set $q = 1$. Then (8)-(9) is reduced to the classical Molenkampf-Crowley rotation test ([15], [4]). The wind velocities are defined so that (i) the wind trajectories are concentric circles with centre in $x = 1$, $y = 1$ and (ii) the motion is with a constant angular velocity. At some time t_{end} a full rotation of the concentration field around $x = 1$, $y = 1$ will be accomplished and, thus, at t_{end} the concentrations must be distributed in the same way as at the beginning (at t_{start}). Results, obtained after one rotation with (i) the up-wind method, (ii) the finite elements algorithm, (iii) the Bott's scheme ([1]) and (iv) the scheme proposed by Holm ([10]), are given in Fig. 1. These results should be compared with the upper left plot in Fig. 2 (in the ideal case, where there are no errors, the results obtained by the numerical methods should be identical with the initial distribution). The Holm's scheme is the best one among these methods. However, it is also the most time-consuming. Holm has improved the performance of his scheme and now it is much more efficient (private communication). The Bott's scheme gives also good overall accuracy, however, this scheme seems to have some problems on the base of the cone. The finite elements algorithm produces some oscillations, but it is considerably cheaper than the other two algorithms. The simple and cheap up-wind method has problems with preserving the high concentrations at the top of the cone. The pseudospectral algorithm gives much more accurate results for this test (see the upper right plot in Fig. 2), However, it requires periodic boundary conditions. The conclusion is that all methods have both advantages and disadvantages and no method is perfect. The problem of finding an optimal advection method is still open, The solution of this problem depends both on the computer that is to be used and on the particular implementation of the numerical method chosen in the air pollution model under consideration.

4.2 Testing the accuracy of the QSSA

Consider the case where only the chemical terms are retained in (8). In this case the chemistry is carried out independently for each grid-point. Moreover, at many grid-points the calculations are identical. Therefore, only a few points are important, and the chemical transformations in these points can be performed (with a small time-step) in a very accurate manner. After that the whole solution field at the end-point of the time interval can be reproduced. This allows us to check the accuracy of the results obtained by the chemical algorithm chosen (at least at the end-point of the time-interval). This procedure has been used to determine a time-stepsize by which good results can be produced by using QSSA. The time-stepsize so found was 30 seconds (the same time-stepsize has been recommended in [9]).

4.3 Coupling advection with chemistry

Consider now the coupling of the chemical sub-model with the other sub-models in a large air pollution model. To simplify the discussion only the coupling of chemistry and advection will be considered. Similar discussion can be carried out when the other processes (the diffusion, deposition and vertical exchange processes) are added to the advection-chemistry combination.

Assume that both the advection algorithm and the chemistry algorithm have been carefully tested. Assume also that the tests indicate that both algorithms perform satisfactorily well. Then the big question is: **Will also the combination of these two algorithms perform well?** It will be shown that, unfortunately, the answer to this question is in some cases negative.

Consider Fig. 2 and Fig. 3. The upper plot on the left hand side in each of these two figures represents the initial distribution of the concentrations. The upper plot of the right hand side represents the distribution of the concentrations after a full rotation in the case when a pure advection test is carried out. **In the ideal case, when there are no errors, the upper two plots must be identical.** The lower plot on the left hand side represents the case where only a pure chemical test is run over the time-interval needed to perform a full rotation. The lower plot on the right hand side represents the distribution of the concentrations after one rotation for the most general case that can be treated by this test; the case where both the transport and the chemical reactions are activated. **In the ideal case, when there are no errors, the lower two plots must be identical.** The results shown on Fig. 2 are very good. However, the results shown on Fig. 3 indicate that problems could appear. In this particular case the difficulties are due to the sharp gradients caused by the chemical reactions. Note too, that the change of the shape of the plot is dramatical.

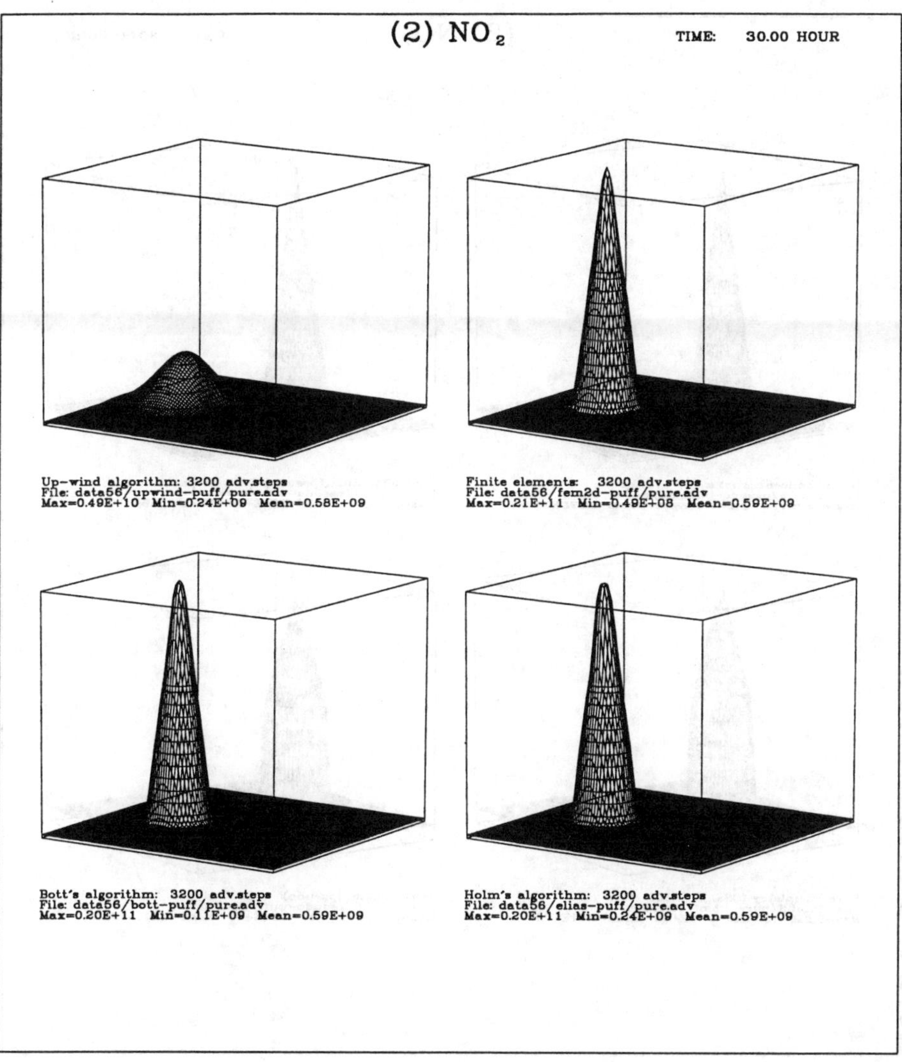

Figure 1

Pure advection test: (a) the up-wind method (upper, left), (b) finite elements (upper, right), (c) Bott's algorithm (lower, left) and (d) Holm's algorithm (lower, right).

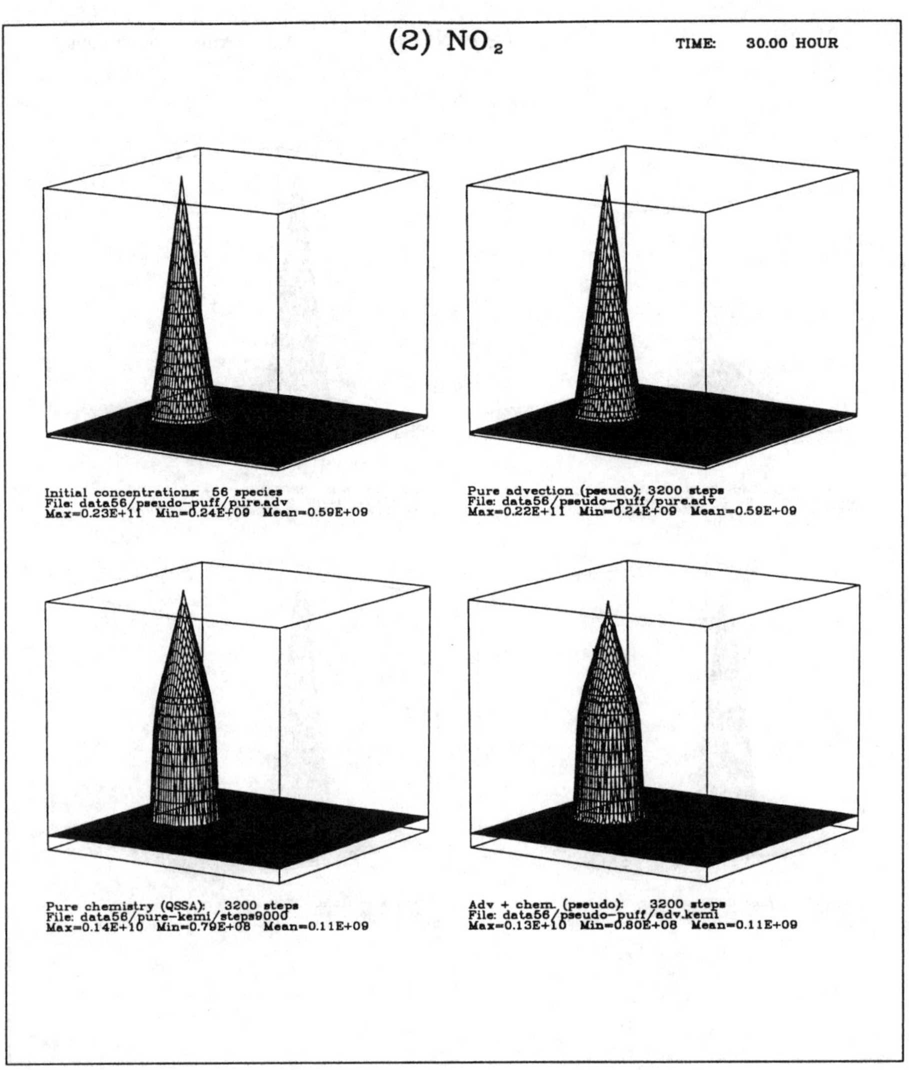

Figure 2

Nitrogen di-oxide concentrations at: (a) the beginning (upper, left), (b) the end of the pure advection test (upper, right), (c) the end of the pure chemistry test (lower, left) and (d) the end of the advection-chemistry test (lower, right).

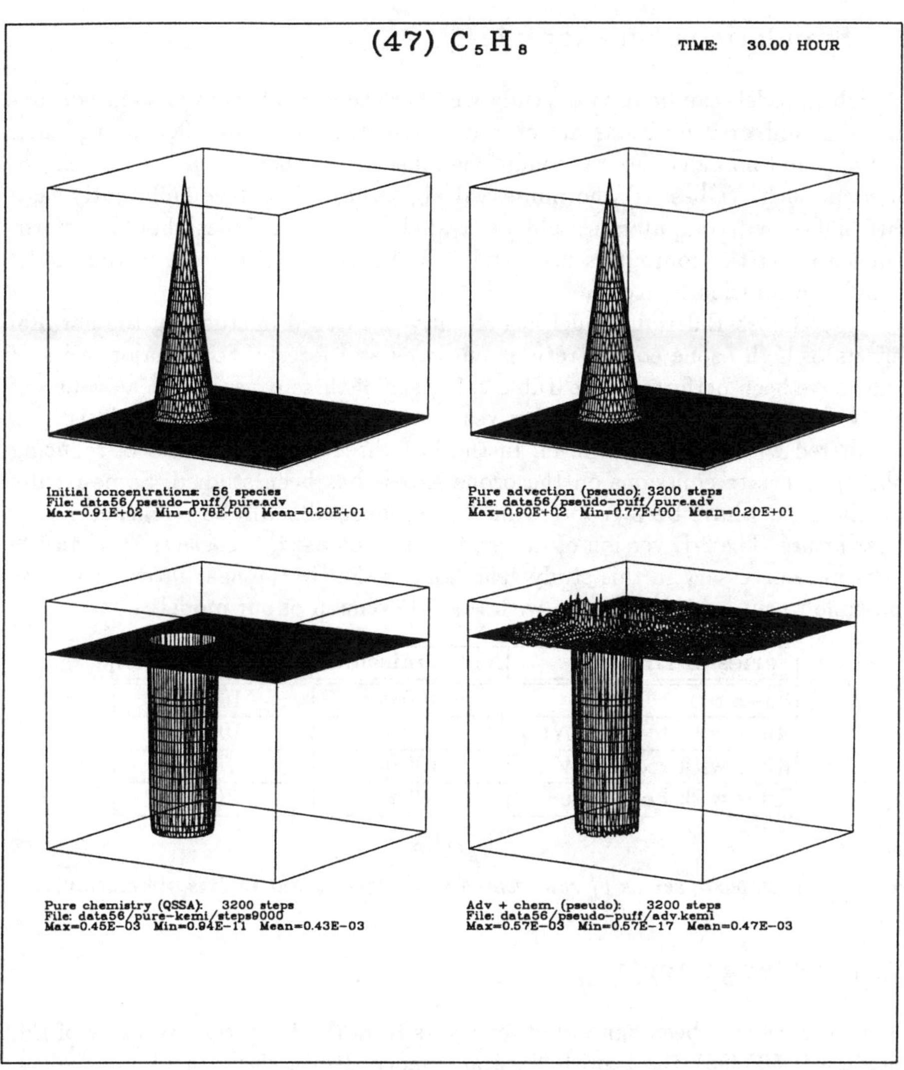

Figure 3

Isoprene concentrations at: (a) the beginning (upper, left), (b) the end of the pure advection test (upper, right), (c) the end of the pure chemistry test (lower, left) and (d) the end of the advection-chemistry test (lower, right).

5 Simulations: an example

Reliable models can be used to study whether the critical levels of some concentrations and/or depositions are exceeded and, thus, harmful effects on plants, animals and humans. Several conditions must be satisfied when the models are used in such studies: (i) the numerical algorithms should be sufficiently fast, (ii) high-speed computers should be available, (iii) the codes should perform efficiently on the computers used and (iv) the results produced by the model should be sufficiently accurate.

The Danish Eulerian Model has recently been used in studying the harmful effects of high ozone concentrations on crops and forests. Four major series of runs have been performed; see Table 2. In each of these series the mean values of the excess ozone for five successive years, 1989-1993, have been calculated and compared with the critical level. In the last three series the effect of reducing the appropriate emissions on the ozone excess has been studied. Some results on these runs have been reported in [24]; more results will be published in the near future. The 2-D version of the model has been used, because more than 720 runs were necessary in this study. The hope is that in the near future it will be possible to run such long series with the 3-D version of our model.

Series of runs	NO_x **emissions**	VOC **emissions**
Basic runs	100%	100%
Runs with reduced NO_x	70%	100%
Runs with reduced VOC	100%	70%
Runs with both reduced	70%	70%

Table 2

The four basic series of runs which were carried out in this simulation.

Acknowledgements

This research has been supported by grants from the Bulgarian Ministry of Education (I-505/95), the Danish Natural Sciences Research Councill and NATO.

References

1. Bott, A., "A positive definite advection scheme obtained by nonlinear renormalization of the advective fluxes", Mon. Weather Rev., 117(1989), 1006-1015.
2. Brandt, J., Wasniewski, J. and Zlatev, Z., "Handling the chemical part in large air pollution models", Appl. Math. and Comp. Sci., 6 (1996), 101-121.
3. Chock, D. P., Winkler, S. L. and Sun, P, "Comparison of stiff chemistry solvers for air quality models", Environ. Sci. Technol., 28 (1994), 1882-1892.
4. Crowley, W. P., "Numerical advection experiments", Mon. Weath. Rev., 96 (1968), 1-11.
5. Deuflhard, P., Nowak, U. and Wulkow, M., "Recent development in chemical computing", Computers Chem. Engng., 14 (1990), 1249-1258.

6. Forester, C. K., "Higher order monotonic convective difference schemes", J. Comput. Phys., 23(1977), 1-22.
7. Fornberg, B., "A practical guide to pseidospectral methods". Cambridge University Press, Cambridge, 1996.
8. Hertel, O., Berkowicz, R., Christensen, J. and Hov, Ø., "Test of two numerical schemes for use in atmospheric transport-chemistry models", Atmos. Environ., 27A (1993), 2591-2611.
9. Hesstvedt, E., Hov, Ø. and Isaksen, I. A., "Quasi-steady-state approximations in air pollution modelling: comparison of two numerical schemes for oxidant prediction", Internat. J. Chem. Kinetics, 10 (1978), 971-994.
10. Holm, E. V., "High-order numerical methods for advection in atmospheric models", PhD Thesis, Department of Meteorology, Stockholm University, 1994.
11. Hov, Ø., Zlatev, Z., Berkowicz, R., Eliassen, A. and Prahm, L. P., "Comparison of numerical techniques for use in air pollution models with non-linear chemical reactions" , Atmos. Environ., 23 (1988), 967-983.
12. Lambert, J. D., "Numerical methods for ordinary differential equations", Wiley, Chichester-New York-Brisbane-Toronto-Singapore, 1991.
13. Marchuk, G. I., "Mathematical modeling for the problem of the environment", Studies in Mathematics and Applications, No. 16, North-Holland, Amsterdam, 1985.
14. McRae G. J., Goodin, W. R. and Seinfeld, J. H., "Numerical solution of the atmospheric diffusion equations for chemically reacting flows", J. Comp. Phys., 45 (1984), 1-42.
15. Molenkampf, C. R., "Accuracy of finite-difference methods applied to the advection equation", J. Appl. Meteor., 7 (1968), 160-167.
16. Odman, M. T., Kumar, N. and Russell, A. G., "A comparison of fast chemical kinetic solvers for air quality modeling", Atmos. Environ., 26A (1992), 1783-1789.
17. Peters, L. K., Berkowitz, C. M., Carmichael, G. R., Easter, R. C., Fairweather, G., Ghan, S. J., Hales, J. M., Leung, L. R., Pennell, W. R., Potra, F. A., Saylor, R. D. and Tsang, T. T., "The current state and future direction of Eulerian models in simulating the tropospherical chemistry and transport of trace species: A review", Atmos. Environ., 29 (1995), 189-221
18. Petzold, L. R., "Order results for implicit Runge-Kutta methods applied to differential-algebraic systems", SIAM J. Numer. Anal., 23 (1986), 837-852.
19. Shieh, D. S., Chang, Y. and Carmichael, G. R., "The evaluation of numerical techniques for solution of stiff ordinary differential equations arising from chemical kinetic problems", Environ. Software, 3 (1988), 28-38.
20. Skelboe, S. and Zlatev, Z., "Exploiting the natural partitioning in the numerical solution of ODE systems arising in atmospheric chemistry", Springer, Berlin, to appear.
21. Verwer, J. G. and Simpson, D., "Explicit methods for stiff ODE's from atmospheric chemistry", Appl. Numer. Math., to appear.
22. Zlatev, Z., "Application of predictor-corrector schemes with several correctors in solving air pollution problems", BIT, 24 (1984), 700-715.
23. Zlatev, Z., "Computer treatment of large air pollution models", Kluwer Academic Publishers, Dordrecht-Boston-London, 1995.
24. Zlatev, Z., Fenger, J. and Mortensen, L., "Relationships between emission sources and excess ozone concentrations", Comput. Math. Applics., to appear.

Boundary Value Methods for the Numerical Approximation of Ordinary Differential Equations

Luigi Brugnano

Dipartimento di Energetica, Università di Firenze
Via C. Lombroso 6/17, 50134 Firenze, Italy

Abstract. Many numerical methods for the approximation of ordinary differential equations (ODEs) are obtained by using Linear Multistep Formulae (LMF). Such methods, however, in their usual implementation suffer of heavy theoretical limitations, summarized by the two well known Dahlquist barriers. For this reason, Runge-Kutta schemes have become more popular than LMF, in the last twenty years. This situation has recently changed, with the introduction of Boundary Value Methods (BVMs), which are methods still based on LMF. Their main feature consists in approximating a given continuous initial value problem (IVP) by means of a discrete boundary value problem (BVP). Such use allows to avoid order barriers for stable methods. Moreover, BVMs provide several families of methods, which make them very flexible and computationally efficient. In particular, we shall see that they allow a natural implementation of efficient mesh selection strategies.

1 Introduction

Usually, the solution of an initial value ODE problem,

$$y' = f(t,y), \quad t \in [t_0, T], \qquad y(t_0) = \eta, \tag{1}$$

is obtained by using a k-step LMF,

$$\sum_{i=0}^{k} \alpha_i y_{n+i} = h \sum_{i=0}^{k} \beta_i f_{n+i}. \tag{2}$$

In the previous equation, y_n denotes, as usual, the discrete approximation of the solution $y(t)$ at $t = t_n \equiv t_0 + nh$, $h = (T - t_0)/N$, and $f_n \equiv f(t_n, y_n)$. Since (2) is a kth order difference equation, then k conditions need to be imposed to obtain the discrete solution. Usually, such conditions are obtained by fixing the first k values, $y_0, \ldots, y_{k-1}$, of the discrete solution. That is, the continuous IVP (1) is approximated by means of a discrete IVP. This approach is very straightforward. However, it suffers of heavy theoretical limitations, summarized by the two Dahlquist barriers.

It is possible to get rid of such limitations by suitably modifying the use of LMF. This is, in fact, the idea on which Boundary Value Methods rely. Early references on such methods can be found in [4, 12]. However, only in the last three years such methods have been systematically studied, starting from [14]. In particular, a linear stability theory has been recently devised [6], which has made possible the derivation of several families of methods, each containing stable methods of arbitrarily high order. In this paper, a brief review on BVMs is presented, along with a mesh selection strategy which is very efficient for such methods.

In Sect. 2 the main facts about BVMs will be recalled, and in Sect. 3 the principal families of methods are sketched. In Sect. 4 the block version of the methods is presented, along with the mesh selection strategy. Finally, in Sect. 5 some numerical examples on difficult stiff problems are reported, showing the effectiveness of BVMs.

2 Boundary Value Methods

Suppose, when approximating (1) by means of (2), to fix the first $k_1 \le k$ values of the discrete solution, $y_0, \ldots, y_{k_1-1}$, and the last $k_2 \equiv k - k_1$ ones, $y_{N-k_2+1}, \ldots, y_N$. In this way, the discrete problem becomes

$$\begin{array}{ll} \sum_{i=-k_1}^{k_2} \alpha_{i+k_1} y_{n+i} = h \sum_{i=-k_1}^{k_2} \beta_{i+k_1} f_{n+i}, & n = k_1, \ldots, N-k_2, \\ y_0, \ldots, y_{k_1-1}, \quad y_{N-k_2+1}, \ldots, y_N, & \text{fixed.} \end{array} \tag{3}$$

That is, the continuous IVP (1) is approximated by means of a discrete BVP. This approach defines a BVM with (k_1, k_2)-*boundary conditions.* Observe that, for $k_1 = k$ and, therefore, $k_2 = 0$, problem (3) becomes an IVP, so that BVMs contain as a proper subclass the usual initial value methods for ODEs based on LMF.

In order to completely exploit all the advantages of this new approach, that is, to derive effective BVMs, we need to generalize the known notions of stability. This is done by introducing the following polynomials [6].

Definition 1. Let $p(z)$ be a polynomial of degree k, and let $|z_1| \le \ldots \le |z_k|$ be its roots. We say that $p(z)$ is a

- $S_{k_1k_2}$-*polynomial* if $|z_{k_1}| < 1 < |z_{k_1+1}|$;
- $N_{k_1k_2}$-*polynomial* if $|z_{k_1}| \le 1 < |z_{k_1+1}|$, with simple zeros of unit modulus.

Observe that for $k_1 = k$ and $k_2 = 0$, one obtains the usual Schur polynomials and von Neumann polynomials, respectively.

Now, let us denote by $\rho(z)$ and $\sigma(z)$ the two polynomials associated with the LMF (2), and, as usual, let $\pi(z, q) = \rho(z) - q\sigma(z)$ denote the stability polynomial. The following definitions are then stated [6].

Definition 2. A BVM with (k_1, k_2)-boundary conditions is

– $0_{k_1k_2}$*-stable* if the corresponding polynomial $\rho(z)$ is a $N_{k_1k_2}$-polynomial;
– (k_1, k_2)*-absolutely stable*, for a given $q \in \mathbb{C}$, if the polynomial $\pi(z,q)$ is a $S_{k_1k_2}$-polynomial. The region

$$D_{k_1k_2} = \{q \in \mathbb{C} : \pi(z,q) \text{ is a } S_{k_1k_2}\text{-polynomial}\}$$

is said *region of* (k_1, k_2)*-absolute stability*;
– $A_{k_1k_2}$*-stable* if $\mathbb{C}^- \subseteq D_{k_1k_2}$.

Observe that the previous definitions reduce to the usual stability notions, when $k_1 = k$ and $k_2 = 0$.

The problem of finding the $k-1$ additional values

$$y_1, \ldots, y_{k_1-1}, \quad y_{N-k_2+1}, \ldots, y_N,$$

in (3) is easily solved by treating them as unknowns. This is done by introducing a set of $k-1$ *additional equations*, independent of those provided by the *main formula* (3). Such equations are conveniently derived by a set of $k_1 - 1$ *initial additional methods*,

$$\sum_{i=0}^{r} \alpha_i^{(j)} y_i = h \sum_{i=0}^{r} \beta_i^{(j)} f_i, \qquad j = 1, \ldots, k_1 - 1, \tag{4}$$

and k_2 *final* ones,

$$\sum_{i=0}^{r} \alpha_{r-i}^{(j)} y_{N-i} = h \sum_{i=0}^{r} \beta_{r-i}^{(j)} f_{N-i}, \qquad j = N - k_2 + 1, \ldots, N. \tag{5}$$

Such additional methods have to be chosen with the same order of the main formula (3), in order to have the same order for the whole composite method (3)–(5). As we shall see in the next section, for almost all BVMs the number r of steps of the additional methods is the same of the main formula, that is $r = k$. Moreover, each BVM will be coupled with the most appropriate set of additional methods.

3 Families Of BVMs

In this section, we present the most important families of BVMs. Of such families, one is here introduced for the first time. A common feature for all these families is that all of them contain $0_{k_1k_2}$-stable, $A_{k_1k_2}$-stable methods of arbitrary high order. This, in turn, confirms that there are no more order barriers for stable BVMs.

3.1 Generalized BDF

The *Generalized Backward Differentiation Formulae (GBDF)* have the following form [6], for all $k \geq 1$,

$$\sum_{i=-\nu}^{k-\nu} \alpha_{i+\nu} y_{n+i} = h f_n, \qquad n = \nu, \ldots, N-k+\nu, \tag{6}$$

where

$$\nu = \begin{cases} \frac{k+2}{2}, & \text{for even } k, \\ \frac{k+1}{2}, & \text{for odd } k. \end{cases}$$

The coefficients $\{\alpha_i\}$ are uniquely determined by imposing a $O(h^{k+1})$ truncation error. Such formulae are to be used with $(\nu, k-\nu)$-boundary condition or, equivalently, they are conveniently coupled with the following set of initial additional methods,

$$\sum_{i=0}^{k} \alpha_i^{(j)} y_i = h f_j, \qquad j = 1, \ldots, \nu - 1,$$

and final additional ones,

$$\sum_{i=0}^{k} \alpha_{k-i}^{(j)} y_{N-i} = h f_j, \qquad j = N-k+\nu+1, \ldots, N.$$

The coefficients of the additional formulae are uniquely determined in order to have the same truncation error of the main formula (6).

Observe that the formulae obtained for $\nu = k$ are the usual BDF, which are 0-unstable, for $k \geq 7$, while GBDF are $0_{\nu,k-\nu}$-stable, $A_{\nu,k-\nu}$-stable, and with order of convergence k, for all $k \geq 1$.

3.2 Generalized Adams Methods

The *Generalized Adams Methods (GAMs)* are BVMs in the following form [10],

$$y_n - y_{n-1} = h \sum_{i=-\nu}^{k-\nu} \beta_{i+\nu} f_{n+i}, \qquad n = \nu, \ldots, N-k+\nu. \tag{7}$$

where ν is defined according to

$$\nu = \begin{cases} \frac{k+1}{2}, & \text{for odd } k, \\ \frac{k}{2}, & \text{for even } k, \end{cases} \tag{8}$$

and the coefficients $\{\beta_i\}$ are uniquely determined in order to have a method of order $k+1$. For each $k \geq 1$, they must be used with $(\nu, k-\nu)$-boundary conditions, and are $0_{\nu,k-\nu}$-stable, $A_{\nu,k-\nu}$-stable methods. They are conveniently used with the following set of additional initial methods,

$$y_j - y_{j-1} = h\sum_{i=0}^{k} \beta_i^{(j)} f_i, \qquad j = 1, \ldots, \nu - 1,$$

and final ones,

$$y_j - y_{j-1} = h\sum_{i=0}^{k} \beta_{k-i}^{(j)} f_{N-i}, \qquad j = N - k + \nu + 1, \ldots, N.$$

The coefficients of the additional methods are uniquely determined by imposing each formula to have the same order, $k+1$, of the main method (7).

The formulae obtained in correspondence of odd values of k are also called *Extended Trapezoidal Rules (ETRs)* [3], since the formula obtained for $k=1$ is the trapezoidal rule. Such formulae belong to the class of *symmetric schemes*, that we shall consider later.

3.3 Extended Trapezoidal Rules Of Second Kind (ETR$_2$s)

Let us consider the methods having the following general form,

$$\sum_{i=-\nu}^{k-\nu} \alpha_{i+\nu} y_{n+i} = h(\beta f_n + (1-\beta) f_{n-1}), \qquad n = \nu, \ldots, N - k + \nu, \tag{9}$$

where ν is chosen according to (8), and the coefficients $\{\alpha_i\}$ and β are uniquely determined by imposing a $O(h^{k+2})$ truncation error. The formulae obtained for k even will be called *unsymmetric ETR$_2$s*, while those obtained for k odd have been called *ETR$_2$s* [5, 8].

In particular, in the latter case one obtains $\beta = 1 - \beta = \frac{1}{2}$, and the corresponding formulae belong to the class of symmetric schemes.

All (unsymmetric) ETR$_2$s are $0_{\nu,k-\nu}$-stable, $A_{\nu,k-\nu}$-stable formulae, and must be used with $(\nu, k-\nu)$-boundary conditions. The following set of additional initial methods,

$$\sum_{i=0}^{k} \alpha_i^{(j)} y_i = h(\beta^{(j)} f_j + (1-\beta^{(j)}) f_{j-1}), \qquad j = 1, \ldots, \nu - 1,$$

and final ones,

$$\sum_{i=0}^{k} \alpha_{k-i}^{(j)} y_{N-i} = h(\beta^{(j)} f_j + (1-\beta^{(j)}) f_{j-1}), \qquad j = N - k + \nu + 1, \ldots, N,$$

is conveniently associated with the main formula (9). The coefficients $\{\alpha_i^{(j)}\}$ and $\beta^{(j)}$ of each additional method are uniquely determined by imposing the same order, $k+1$, of the main formula.

3.4 Symmetric Schemes

We have collected as *symmetric schemes* [5, 8], BVMs having the following general properties,

- they have an odd number of steps, $k = 2\nu - 1$, and must be used with $(\nu, \nu-1)$-boundary conditions (i.e., they require $\nu - 1$ initial and $\nu - 1$ final additional methods);
- the corresponding polynomials $\rho(z)$ have skew-symmetric coefficients. That is, $z^k \rho(z^{-1}) = -\rho(z)$;
- the corresponding polynomials $\sigma(z)$ have symmetric coefficients. That is, $z^k \sigma(z^{-1}) = \sigma(z)$.
- $D_{\nu,\nu-1} \equiv \mathbb{C}^-$.

Such schemes are conveniently used for either approximating continuous BVPs [7], or Hamiltonian problems [5, 8].

Both ETRs and ETR$_2$s are symmetric schemes. Another important family of BVMs fits this class, namely *Top Order Methods (TOMs)* [1]. Such method have the following general form,

$$\sum_{i=0}^{\nu-1} \alpha_i (y_{n-\nu+i} - y_{n+\nu-1-i}) = h \sum_{i=0}^{\nu-1} \beta_i (f_{n-\nu+i} + f_{n+\nu-1-i}), \qquad (10)$$
$$n = \nu, \ldots, N - \nu + 1,$$

where the coefficients $\{\alpha_i\}$ and $\{\beta_i\}$ are determined in order to have the maximum possible order for a k-step formula, that is $p = 2k$. They can be conveniently used with the following initial

$$y_j - y_{j-1} = h \sum_{i=0}^{2k-1} \beta_i^{(j)} f_i, \qquad j = 1, \ldots, \nu - 1,$$

and final additional methods,

$$y_j - y_{j-1} = h \sum_{i=0}^{2k-1} \beta_{2k-1-i}^{(j)} f_{N-i}, \qquad j = N - \nu + 2, \ldots, N.$$

The unknown coefficients $\{\beta_i^{(j)}\}$ of the additional methods are uniquely determined by imposing that they have the same order $p = 2k$ of the main formula (10).

4 Block BVMs And Mesh Selection

The arguments presented in the last section allow to consider a BVM, together with its corresponding additional methods, as a composite method. Having fixed a suitable N, such method allows to pass from the approximation at $t = t_0$ to the one at $t = t_N$. It is then possible to discretize the interval $[t_0, T]$ by using two different meshes: a coarser one containing the $p + 1$ points

$$\tau_i = \tau_{i-1} + \hat{h}_i, \quad i = 1, \ldots, p, \qquad \tau_0 \equiv t_0, \quad \tau_p \equiv T,$$

and a finer one, which discretizes each subinterval $[\tau_{i-1}, \tau_i]$, $i = 1, \ldots, p$, with a constant finer stepsize $h_i = \hat{h}_i / N$.

In more detail, by using the initial condition $y_0 = \eta$ provided by the continuous problem (1), we can apply the (composite) BVM over the first subinterval $[\tau_0, \tau_1]$, with constant finer stepsize $h_1 = \hat{h}_1 / N$. The discrete approximation of the solution at the points

$$t_j = \tau_0 + j h_1, \qquad j = 1, \ldots, N,$$

is then obtained. One then uses the approximated value at $t_N \equiv \tau_1$ for computing the discrete approximation over the second subinterval $[\tau_1, \tau_2]$, by using the same BVM with finer stepsize $h_2 = \hat{h}_2 / N$.

It is evident that the process can be iterated $p-2$ more times, thus providing a discrete approximation over the entire interval $[t_0, T]$.

The resulting procedure defines the block version of BVMs [8], which has two important practical implications:

1. the stepsize variation becomes very simple, since inside each block the used stepsize is constant, while one may vary the stepsize in the coarser mesh;
2. it allows a very efficient parallel implementation of such methods [2].

In [9] a novel mesh selection strategy, which is very effective for the approximation of continuous BVPs, has been introduced. We here present a different mesh selection strategy, which is very efficient for the approximation of continuous IVP. Essentially, it is based on deferred correction [15, 19, 20], which assumes, for BVMs, a very natural implementation. For simplicity, we assume that a single block implementation of BVMs is used for approximating the continuous problem (1), which will be assumed to be scalar. Then, let us denote by

$$F_p(\mathbf{y}) = \mathbf{0} \tag{11}$$

the discrete problem obtained by applying a BVM of order p, where $\mathbf{y} = (y_0, \ldots, y_N)^T$ is the discrete solution. It is not difficult to verify that

$$F_p(\mathbf{y}) \equiv A_p \mathbf{y} - h B_p f(\mathbf{y}) - \begin{pmatrix} \eta \\ \mathbf{0} \end{pmatrix},$$

where $f(\mathbf{y}) = (f(t_0, y_0), \ldots, f(t_N, y_N))^T$, h is the used stepsize, and A_p, B_b are $(N+1)\times(N+1)$ matrices, whose rows contain the coefficients of the considered method,

$$A_p = \begin{pmatrix} 1 & 0 & \ldots & & \ldots & \ldots & 0 \\ \alpha_0^{(1)} & \ldots & \alpha_k^{(1)} & & & & \\ \vdots & & \vdots & & & & \\ \alpha_0^{(k_1-1)} & \ldots & \alpha_k^{(k_1-1)} & & & & \\ \alpha_0 & \ldots & \alpha_k & & & & \\ & \ddots & & & \ddots & & \\ & & \ddots & & & \ddots & \\ & & & & \alpha_0 & \ldots & \alpha_k \\ & & & & \alpha_0^{(N-k_2+1)} & \ldots & \alpha_k^{(N-k_2+1)} \\ & & & & \vdots & & \vdots \\ & & & & \alpha_0^{(N)} & \ldots & \alpha_k^{(N)} \end{pmatrix}_{(N+1)\times(N+1)},$$

$$B_p = \begin{pmatrix} 0 & 0 & \ldots & & \ldots & \ldots & 0 \\ \beta_0^{(1)} & \ldots & \beta_k^{(1)} & & & & \\ \vdots & & \vdots & & & & \\ \beta_0^{(k_1-1)} & \ldots & \beta_k^{(k_1-1)} & & & & \\ \beta_0 & \ldots & \beta_k & & & & \\ & \ddots & & & \ddots & & \\ & & \ddots & & & \ddots & \\ & & & & \beta_0 & \ldots & \beta_k \\ & & & & \beta_0^{(N-k_2+1)} & \ldots & \beta_k^{(N-k_2+1)} \\ & & & & \vdots & & \vdots \\ & & & & \beta_0^{(N)} & \ldots & \beta_k^{(N)} \end{pmatrix}_{(N+1)\times(N+1)}.$$

In the above expression, we have assumed the main formula, and the corresponding additional methods, to have the same number, k, of steps. Moreover, we shall also assume that the matrix B_p has unit row sums. In the following, this assumption will hold for all the considered methods.

Let now $\hat{\mathbf{y}} = (y(t_0), \ldots, y(t_N))^T$ be the restriction of the continuous solution to the mesh. Then, since all the formulae defining the BVM have the same order p, it will be

$$F_p(\hat{\mathbf{y}}) = \boldsymbol{\tau}_p \equiv \begin{pmatrix} 0 \\ \tau_1 \\ \vdots \\ \tau_N \end{pmatrix}, \qquad \tau_i = O(h^{p+1}), \quad i = 1, \ldots, N. \tag{12}$$

From the previous equation, one readily obtains the following first order approximation,

$$\hat{\mathbf{y}} - \mathbf{y} \approx M_p^{-1} \boldsymbol{\tau}_p, \tag{13}$$

where M_p is the Jacobian matrix of F_p evaluated at $\hat{\mathbf{y}}$,

$$M_p = A_p - hB_p J_f(\hat{\mathbf{y}}),$$

and, by denoting with f_y the Jacobian of $f(t, y)$,

$$J_f(\hat{\mathbf{y}}) = \begin{pmatrix} f_y(t_0, y(t_0)) & & \\ & \ddots & \\ & & f_y(t_N, y(t_N)) \end{pmatrix}.$$

Let now put the discrete solution $\mathbf{y}$ inside the discrete problem obtained by using a different method, of order $q > p$, over the same mesh. It will be

$$F_q(\mathbf{y}) = -\mathbf{u}, \tag{14}$$

where, in general, the vector $\mathbf{u}$ will not be the zero vector. The following result then holds true.

Theorem 3. *Provided that the continuous solution is suitably smooth, one has (see (12) and (14))* $\mathbf{u} = \boldsymbol{\tau}_p + O(h^{p+2})$.

Proof. Since the method defining F_q has order q, it will be

$$F_q(\hat{\mathbf{y}}) = \boldsymbol{\tau}_q \equiv O(h^{q+1}).$$

From (13) one then obtains

$$F_q(\mathbf{y}) \approx F_q(\hat{\mathbf{y}} - M_p^{-1}\boldsymbol{\tau}_p) \approx F_q(\hat{\mathbf{y}}) - M_q M_p^{-1}\boldsymbol{\tau}_p = \boldsymbol{\tau}_q - M_q M_p^{-1}\boldsymbol{\tau}_p,$$

where $M_q = A_q - hB_q J_f(\hat{\mathbf{y}})$ denotes the Jacobian matrix of F_q evaluated at $\hat{\mathbf{y}}$. Since in the above approximations the neglected terms are $O(h^{2p})$, the thesis then follows by proving that

$$M_q \mathbf{v} = \boldsymbol{\tau}_p + O(h^{p+2}), \qquad \mathbf{v} = M_p^{-1}\boldsymbol{\tau}_p.$$

The entries of $\boldsymbol{\tau}_p$ are $O(h^{p+1})$ and, consequently, those of the vector $\mathbf{v}$ will be $O(h^p)$. That is,

$$\mathbf{v} = h^p \mathbf{c} \equiv h^p \begin{pmatrix} c(t_0) \\ \vdots \\ c(t_N) \end{pmatrix},$$

where $c(t)$ is a suitably smooth function, under the made hypothesis on $y(t)$. Since the first entry of $\boldsymbol{\tau}_p$ is zero (see (12)), then $c(t_0) = 0$. Moreover, the function $c(t)$ is not uniquely defined, so that we may assume $c'(t_0) = 0$ as well. From the previous equations, one then obtains

$$\tau_p = M_p\mathbf{v} = h^p(A_p\mathbf{c} - hB_pJ_f(\hat{\mathbf{y}})\mathbf{c}) = h^p(A_p\mathbf{c} - hB_p\mathbf{c}') + h^{p+1}B_p(\mathbf{c}' - J_f(\hat{\mathbf{y}})\mathbf{c}),$$

where the vector $\mathbf{c}' = (c'(t_0), \ldots, c'(t_N))^T$. Because of the following trivial equalities,

$$A_p\mathbf{c} - hB_p\mathbf{c}' = O(h^{p+1}), \qquad B_p(\mathbf{c}' - J_f(\hat{\mathbf{y}})\mathbf{c}) = (\mathbf{c}' - J_f(\hat{\mathbf{y}})\mathbf{c}) + O(h),$$

one concludes that $\tau_p = h^{p+1}(\mathbf{c}' - J_f(\hat{\mathbf{y}})\mathbf{c}) + O(h^{p+2})$. Consequently, by means of similar arguments, one finally obtains

$$\begin{aligned} M_q\mathbf{v} &= h^p(A_q\mathbf{c} - hB_qJ_f(\hat{\mathbf{y}})\mathbf{c}) = h^p(A_q\mathbf{c} - hB_q\mathbf{c}') + h^{p+1}B_q(\mathbf{c}' - J_f(\hat{\mathbf{y}})\mathbf{c}) \\ &= O(h^{p+q+1}) + h^{p+1}(\mathbf{c}' - J_f(\hat{\mathbf{y}})\mathbf{c}) + O(h^{p+2}) = \tau_p + O(h^{p+2}). \end{aligned}$$

□

From the previous result and equations (11), (13), (14), the following estimate then easily follows,

$$\hat{\mathbf{y}} - \mathbf{y} \approx M_p^{-1}(F_p(\mathbf{y}) - F_q(\mathbf{y})). \tag{15}$$

We observe that, when computing (15), the matrix M_p has already been factored for solving (11).

When the methods defining F_p and F_q are both GBDF, then $B_p = B_q$ and, therefore, $\tau_p \approx F_p(\mathbf{y}) - F_q(\mathbf{y}) \equiv (A_p - A_q)\mathbf{y}$. As consequence, the estimate of the local error does not explicitly depend on f. As observed in [13, page 134], this feature makes the estimate suitable for approximating stiff problems.

Finally, we observe that, having fixed a tolerance for the error, the estimate (15), together with the usual extrapolation procedure, can be easily used for determining the appropriate stepsize.

5 Numerical Examples

We here report a few numerical examples on severe stiff test problems taken from the literature. For comparisons, we also report the performance of LSODE [16] (one of the most popular ODE solvers) and of the Matlab stiff ODE solver ODE23S [18] on the same problems.

We first consider the Robertson's problem,

$$\begin{aligned} y_1' &= -.04y_1 + 10^4y_2y_3, & y_1(0) &= 1, \\ y_2' &= .04y_1 - 10^4y_2y_3 - 3\cdot 10^7y_2^2, & y_2(0) &= 0, \\ y_3' &= 3\cdot 10^7y_2^2, & y_3(0) &= 0. \end{aligned}$$

We use the fifth-order GBDF with tolerance 10^{-7} for the error. The estimate of the error is obtained from (15) by considering the GBDF of order seven. The interval $[0, 2 \cdot 10^{20}]$ is covered with 540 steps, and stepsizes which monotonically increase from 10^{-4} up to $5.2 \cdot 10^{19}$. LSODE, with parameters $mf = 21$ and $atol = rtol = 10^{-7}$, requires 637 steps to cover the same interval. On the other hand, ODE23S, with parameters $atol = rtol = 10^{-7}$ and analytic Jacobian, fails to cover the whole integration interval.

Let us now consider the following problem due to Curtis [17, page 409],

$$y' = A_\nu(t)y + f(t), \quad t \in [0, 10\pi], \qquad y(0) = \begin{pmatrix} 1 \\ 0 \end{pmatrix},$$

where $\nu = 10^3$, and

$$f(t) = \begin{pmatrix} -\sin(t) \\ \cos(t) \end{pmatrix} - A_\nu(t) \begin{pmatrix} \cos(t) \\ \sin(t) \end{pmatrix},$$

$$A_\nu(t) = M_\nu(t) \begin{pmatrix} -1001 & 0 \\ 0 & -1 \end{pmatrix} M_\nu^T(t), \qquad M_\nu(t) = \begin{pmatrix} \cos(\nu t) & \sin(\nu t) \\ -\sin(\nu t) & \cos(\nu t) \end{pmatrix}.$$

The solution of the problem is given by $y(t) = (\cos(t)\ \sin(t))^T$, independently of the value of the parameter ν, but the problem is very stiff, despite the fact that the solution is smooth. However, the latter feature suggests that stable high order methods should perform well. For this reason, we consider the GBDF of order twenty on this problem. The stepsize is changed by using the estimate of the global errors obtained by means of the GBDF of order twenty two. Having fixed a tolerance 10^{-5} for the error, the integration interval is covered with 56 mesh points, stepsizes ranging from 0.5 to 0.64, and a maximum absolute error $2.7 \cdot 10^{-6}$. To get an idea of the performance, consider that LSODE, with parameters $mf = 21$ and $atol = rtol = 10^{-5}$, needs 2356 steps to cover the integration interval, with a maximum error $1.7 \cdot 10^{-4}$. By using the default parameters $rtol = 10^{-3}$ and $atol = 10^{-6}$, ODE23S requires more than 12000 steps to cover the integration interval, with a maximum error $2.4 \cdot 10^{-3}$.

Finally, we consider the Van der Pol's equations,

$$\begin{aligned} y_1' &= y_2, & y_1(0) &= 2, \\ y_2' &= -y_1 + \mu y_2(1 - y_1^2), & y_2(0) &= 0, \end{aligned}$$

with the parameter $\mu = 10^3$. Such equations have an attractive limit cycle, which is readily reached from the chosen starting point. We integrate up to $T = 2 \cdot 10^3$, so that a whole limit cycle is covered. By using the fifth-order GBDF, coupled with the GBDF of order seven for the error estimate, and a tolerance 10^{-7} for the error, the integration interval is covered with 1930 steps. The selected stepsizes range from $7.5 \cdot 10^{-6}$, where the solution has the most rapid variations, to $3.5 \cdot 10^1$, where it is very smooth. By using the parameters $mf = 21$ and $atol = rtol = 10^{-7}$, LSODE requires 1328 steps to cover the integration interval, while 5081 steps are required by ODE23S with parameters $atol = rtol = 10^{-7}$.

Acknowledgements. The author is very indebted with professor D. Trigiante for the helpful discussions.

References

1. Amodio, P.: *A*-stable *k*-step linear multistep formulae of order $2k$ for the solution of stiff ODEs. (submitted)
2. Amodio, P., Brugnano, L.: Parallel implementation of block Boundary Value Methods for ODEs. (submitted)
3. Amodio, P., Mazzia, F.: A boundary value approach to the numerical solution of ODEs by multistep methods. J. of Difference Eq. and Appl. **1** (1995) 353–367
4. Axelsson, A. O. H., Verwer, J. G.: Boundary value techniques for initial value problems in ordinary differential equations. Math. Comp. **45** (1985) 153–171
5. Brugnano, L.: Essentially symplectic Boundary Value Methods for linear Hamiltonian systems. J. Comput. Math. (to appear)
6. Brugnano, L., Trigiante, D.: Convergence and stability of Boundary Value Methods for ordinary differential equations. J. Comput. Appl. Math. (to appear)
7. Brugnano, L., Trigiante, D.: High order multistep methods for boundary value problems. Appl. Num. Math. **18** (1995) 79–94
8. Brugnano, L., Trigiante, D.: Block Boundary Value Methods for linear Hamiltonian systems. J. of Appl. Math. (to appear)
9. Brugnano, L., Trigiante, D.: A new mesh selection strategy for ODEs. (submitted)
10. Brugnano, L., Trigiante, D.: Boundary Value Methods: the third way between Linear Multistep and Runge-Kutta methods. Computer and Math. with Appl. (to appear)
11. Brugnano, L., Trigiante, D.: *Solving Differential Problems by Multistep Initial and Boundary Value Methods.* (in preparation)
12. Cash, J. R.: *Stable Recursions.* Academic Press, London, 1979
13. Hairer, E., Wanner, G.: *Solving ordinary differential equations II.* Springer-Verlag, Berlin, 1991
14. Lopez, L., Trigiante, D.: Boundary Value Methods and BV-stability in the solution of initial value problems. Appl. Numer. Math. **11** (1993) 225–239
15. Pereyra, V. L.: On improving an approximate solution of a functional equation by deferred corrections. Numer. Math. **8** (1966) 376–391
16. Radhakrishnan, K., Hindmarsh, A. C.: Description and use of LSODE, the Livermore Solver for Ordinary Differential Equations. NASA Reference Publication 1327, LLNL Report UCRL-ID-113855, 1993
17. Shampine, L. F.: *Numerical Solution of Ordinary Differential Equations.* Chapman & Hall, New York, 1994
18. Shampine, L. F., Reichelt, M. W.: The Matlab ODE Suite. Report, 1995
19. Stetter, H. J.: Economical global error estimation. In "Stiff Differential Systems", R. A. Willoughby ed., Plenum Press, New York, 1974
20. Zadunaisky, P.: On the estimation of errors propagated in the numerical solution of ordinary differential equations. Numer. Math. **27** (1976) 21–39

Fast Algorithms for Problems on Thermal Tomography

Raymond H. Chan* , Chun-pong Cheung** and Hai-wei Sun***

Abstract. In this paper, we study an ill-posed, nonlinear inverse problem in heat conduction and hydrology applications. In [2], the problem is linearized to give a linear integral equation, which is then solved by the Tikhonov method with the identity as the regularization operator. We prove in this paper that the resulting equation is well-condition and has clustered spectrum. Hence if the conjugate gradient method is used to solve the equation, we expect superlinear convergence. However, we note that the identity operator does not give good solution to the original equation in general. Therefore in this paper, we use the Laplacian operator as the regularization operator instead. With the Laplacian operator, the regularized equation is ill-conditioned and hence a preconditioner is required to speed up the convergence rate if the equation is solved by the conjugate gradient method. We here propose to use the Laplacian operator itself as preconditioner. This preconditioner can be inverted easily by fast sine-transforms and we prove that the resulting preconditioned system is well-conditioned and has clustered spectrum too. Hence the conjugate gradient method converges superlinearly for the preconditioned system. Numerical results are given to illustrate the fast convergence.

1 Introduction

In this paper, we consider the inverse problem of finding a pair (u, a) for the Cauchy problem

$$\begin{aligned} u_t - \operatorname{div}(a\nabla u) &= \delta(x - x^*)\delta(t) && \text{on } \mathbb{R}^m \times (0, T), \\ u &= 0 && \text{on } \mathbb{R}^m \times \{0\}, \end{aligned} \tag{1}$$

where δ is the Dirac delta function. The unknown coefficient function $a = a(x)$ is assumed to be bounded and measurable and has the expansion $a(x) = 1 + f(x)$, where $f(x)$ is zero outside a bounded domain $\Omega \subset \mathbb{R}^m$. The solution $u(x, t; x^*)$ is assumed to be given for $x, x^* \in \Omega^*$ and $t \in (0, T)$ where Ω^* is a bounded domain in $\mathbb{R}^m$ whose closure does not intersect Ω. This kind of problems arises in the

* Department of Mathematics, The Chinese University of Hong Kong, Shatin, Hong Kong. Research supported by CUHK Direct Grant 220600680.

** Department of Mathematics, The Chinese University of Hong Kong, Shatin, Hong Kong. Research supported by CUHK Grants for Overseas Academic Activities.

*** Department of Mathematics and Physics, Guangdong University of Technology, Guangzhou, People's Republic of China.

heat conduction and in hydrology applications. The diffusion coefficient $a(x)$ is to be recovered from the boundary measurements of the solution $u(x, t; x^*)$ in (1). In real applications, x^* is the source and x is the sensor.

Since (1) is ill-posed and nonlinear, it is difficult to solve. Recently, Elayyan and Isakov have derived in [2] a linearized equation for (1). After linearization, the unknown function f is given as the solution of a linear convolution integral equation. Since the integral equation is also ill-posed, in [2], the equation is solved by using the Tikhonov regularization method with the identity operator as the regularization operator.

In this paper, we first show that if the identity operator is used as the regularization operator, then the resulting operator is well-conditioned with clustered spectrum. Hence if the equation is solved by the conjugate gradient method, then we have superlinear convergence. However, we will illustrate by examples that the identity operator does not give good solution for the original integral equation. In this paper, we therefore use the Laplacian operator as the regularization operator instead. Our numerical examples show that it is a better regularization operator than the identity operator in that the solutions thus obtained are more accurate.

The main drawback in using the Laplacian operator is that the regularized equation becomes ill-conditioned. If the conjugate gradient method is used to solve the equation, the iteration number required for convergence will increase with the size of the discretization matrix. To speed up the convergence rate, we propose to use the Laplacian itself as the preconditioner for the equation. For all 1-dimensional problems or 2-dimensional problems on rectangular domains, the inverse of the Laplacian can easily be found by using fast sine-transforms. We prove that the preconditioned system is well-conditioned and has clustered spectrum. Therefore the convergence rate of the conjugate gradient method for the preconditioned system is expected to be superlinear. Numerical results are given to illustrate this fast convergence.

The outline of the paper is as follows. In §2, we briefly recall the linearization scheme in [2]. In §3, we discuss the identity regularization methods used in [2] and show that the resulting operator has clustered spectrum. Numerical results are given in §4 to illustrate that the Laplacian operator will give much better results than the identity operator. Then in §5, we show that with the Laplacian operator as preconditioner, the resulting preconditioned operator is still well-conditioned and has clustered spectrum. Finally, concluding remarks are given in §6.

2 Linearization of the Inverse Problem

In this section, we review the linearization method proposed by Elayyan and Isakov [2] for the inverse problem (1). We first consider $\tilde{u}$, the zeroth order approximation to u, which satisfies

$$\begin{aligned} \tilde{u}_t - \Delta \tilde{u} &= \delta(x - x^*)\delta(t) && \text{on } \mathbb{R}^m \times (0, T), \\ \tilde{u} &= 0 && \text{on } \mathbb{R}^m \times \{0\}. \end{aligned} \tag{2}$$

We note that (2) can be obtained from (1) by replacing $a(x)$ there by 1, the zeroth order approximation of $a(x)$. We remark that there is a closed form solution in integral form for $\tilde{u}$ in (2), see [2].

Now we let $\tilde{v} = u - \tilde{u}$ be the residual of the approximation. Substituting this expansion into (1), we get

$$\begin{aligned} \tilde{v}_t - \Delta\tilde{v} &= \operatorname{div}(f\nabla\tilde{u}) + \operatorname{div}(f\nabla\tilde{v}) \text{ on } \mathbb{R}^m \times (0,T), \\ \tilde{v} &= 0 \qquad\qquad\qquad\qquad\qquad\quad \text{ on } \mathbb{R}^m \times \{0\}. \end{aligned} \tag{3}$$

We note that the nonlinear part of (3) is the second term in the right hand side, which can be dropped out as it is small when compared with the first term; see Elayyan and Isakov [2]. Thus we have, to the first order approximation,

$$\begin{aligned} \tilde{v}_t - \Delta\tilde{v} &= \operatorname{div}(f\nabla\tilde{u}) \text{ on } \mathbb{R}^m \times (0,T), \\ \tilde{v} &= 0 \qquad\qquad\quad \text{ on } \mathbb{R}^m \times \{0\}. \end{aligned} \tag{4}$$

Representing the solution $\tilde{u}$ in (2) in integral form, we can express $\tilde{v}$ in (4) as

$$\tilde{v}(x^*,t) = -\frac{1}{4(4\pi)^m}\int_0^t\int_{\mathbb{R}^m} f(x)\frac{|x^*-x|^2}{(\tau(t-\tau))^{m/2+1}}\exp\left(-\frac{|x^*-x|^2 t}{4\tau(t-\tau)}\right)dx d\tau. \tag{5}$$

Since $u(x,t;x^*)$ is given in $\Omega^* \times (0,T)$ and $\tilde{u}$ in (2) can be solved exactly, the left hand side $\tilde{v} = u - \tilde{u}$ in (5) is known in $\Omega^* \times (0,T)$. Since f is zero outside Ω, we can rewrite (5) as

$$-\frac{1}{4(4\pi)^m}\int_0^t\int_{\Omega} f(x)\frac{|x^*-x|^2}{(\tau(t-\tau))^{m/2+1}}\exp\left(-\frac{|x^*-x|^2 t}{4\tau(t-\tau)}\right)dx d\tau = \tilde{v}(x^*,t), \tag{6}$$

where the right hand side $\tilde{v}(x^*,t)$ is known in $\Omega^* \times (0,T)$.

For simplicity, we set $t = T$. Then the integral equation (6) can be written as

$$\mathcal{K}f = b, \tag{7}$$

with $f \in L^2(\Omega)$, $b \in L^2(\Omega^*)$, and

$$(\mathcal{K}f)(y) = \int_{\Omega} k(y-x)f(x)dx.$$

Here the convolution kernel function $k(x)$ is given by

$$k(x) = -\frac{1}{4(4\pi)^m}\int_0^T \frac{|x|^2}{(\tau(T-\tau))^{m/2+1}}\exp\left(-\frac{|x|^2 T}{4\tau(T-\tau)}\right)d\tau. \tag{8}$$

Thus $\mathcal{K}$ is an operator from $L^2(\Omega)$ into $L^2(\Omega^*)$. The uniqueness of the solution f in (7) has been shown in [2].

Equation (7) is a Fredholm integral equation of the first kind. Since $\mathcal{K}$ maps any Sobolev space on Ω into the space of functions that are analytic in a neighborhood of Ω^*, (7) is ill-posed, see [2]. Regularization method is therefore needed to solve (7). In the next two sections, we will study the Tikhonov regularization method for this problem.

3 Regularizations by the Identity Operator

In the Tikhonov approach for solving the ill-posed equation (7), the equation is replaced by the regularized equation

$$(\alpha\mathcal{Q} + \mathcal{K}^*\mathcal{K})f = \mathcal{K}^*b, \tag{9}$$

where α is the regularization parameter and $\mathcal{Q}$ is the regularization operator. In [2], $\mathcal{Q}$ is chosen to be the identity operator $\mathcal{I}$.

Since $\mathcal{K}$ is a convolution kernel, the discretized matrix $\mathbf{K}$ of $\mathcal{K}$ by the rectangular quadrature rule is a Toeplitz matrix for 1-dimensional problem and a block-Toeplitz-Toeplitz-block matrix for 2-dimensional problem, see for instance Chan and Ng [1]. Thus the matrix-vector multiplication of $\mathbf{K}$ with any vector is of order $O(n \log n)$ operations, where n is the size of the matrix $\mathbf{K}$, see for instance Chan and Ng [1]. Hence if the conjugate gradient method is applied to solving the discrete equation of (9), the cost per iteration is of order $O(n \log n)$ operations, see for instance Stoer and Bulirsch [4, p. 606]. As for the convergence rate, we have

Theorem 1. *The spectrum of the regularized operator $\alpha\mathcal{I} + \mathcal{K}^*\mathcal{K}$ is clustered around α. In particular, if the conjugate gradient method is applied to solving the regularized system, then the convergence rate will be superlinear.*

Proof. We first note that any operator $\mathcal{K}$ with kernel function $k \in L^2(\Omega \times \Omega^*)$ is a compact operator from $L^2(\Omega)$ to $L^2(\Omega^*)$, see for instance Hackbusch [3, Theorem 3.2.7]. Since the kernel function $k(x)$ in (8) is smooth, $\mathcal{K}$ is a compact operator from $L^2(\Omega)$ to $L^2(\Omega^*)$. Hence $\mathcal{K}^*\mathcal{K}$ is also compact from $L^2(\Omega)$ to $L^2(\Omega)$. Therefore its spectrum is clustered around 0. In particular, the spectrum of the regularized operator $\alpha\mathcal{I} + \mathcal{K}^*\mathcal{K}$ is clustered around α. Thus if the conjugate gradient method is applied to solving the regularized system, then the convergence rate will be superlinear. □

Thus the conjugate gradient method for solving (7) will converge in a fixed number of iterations independent of n, the size of the discretization matrix. Since the cost per iteration of the conjugate gradient method is of order $O(n \log n)$ operations, the total cost of solving (7) is also of order $O(n \log n)$ operations.

4 Regularization by the Laplacian Operator

Although the identity-regularized equation converges superlinearly, we find that the regularized solution so obtained is not good in general. In the following, we propose to use the Laplacian operator as the regularization operator, i.e., the regularization equation is given by

$$(\alpha\Delta + \mathcal{K}^*\mathcal{K})f = \mathcal{K}^*b. \tag{10}$$

We now illustrate the differences between the two regularization operators by examples. In the computation, the discretization matrix $\mathbf{K}$ is obtained by applying the rectangle rule to $\mathcal{K}$. The regularized equations are solved by the conjugate gradient method with stopping criteria $\|\mathbf{r}_q\|_\infty/\|\mathbf{r}_0\|_\infty < 10^{-7}$, where $\mathbf{r}_q$ is the residual after the qth iteration. The initial guess is chosen to be the zero vector. All computations were done by Matlab on an IBM 43P-133 workstation. The best regularization parameter α were obtained experimentally by trying different values of α. We found that it is rather robust – change in the first decimal place usually gives the same result. We note that since the entries of $\mathbf{K}$ are very small (of magnitude smaller than 10^{-4} and 10^{-7} respectively for the 1-dimensional and 2-dimensional problems we tried), α thus computed are small too.

We start with two one-dimensional problems with $\Omega = [0, 2]$, $\Omega^* = [3, 5]$, $T = 4$ and $n = 256$. The true solutions $f(x)$ are chosen to be $\sin(\pi x/2)$ and $x^3 + 3x^2 + 2x$. The right hand side vector $\mathbf{b} = \mathbf{K}\mathbf{f}$ of the discrete equation is obtained accordingly. Then 10% random noise vector $\mathbf{n}$ is added to the right hand side $\mathbf{b}$. More precisely, $\|\mathbf{n}\|_\infty/\|\mathbf{b}\|_\infty = 0.1$ and we solve

$$(\alpha\mathbf{Q} + \mathbf{K}^*\mathbf{K})\mathbf{f} = \mathbf{K}^*(\mathbf{b} + \mathbf{n}), \tag{11}$$

where $\mathbf{Q}$ is either the identity matrix or the discrete Laplacian.

In Figure 1, the solid line, the dash line and the dot line show the original image, the result from the identity regularization and the result from Laplacian regularization respectively. In the first example, the value of α for both regularization operators are 10^{-7}. In the second example, $\alpha = 10^{-9}$ for the identity regularization and $\alpha = 10^{-10}$ for Laplacian regularization.

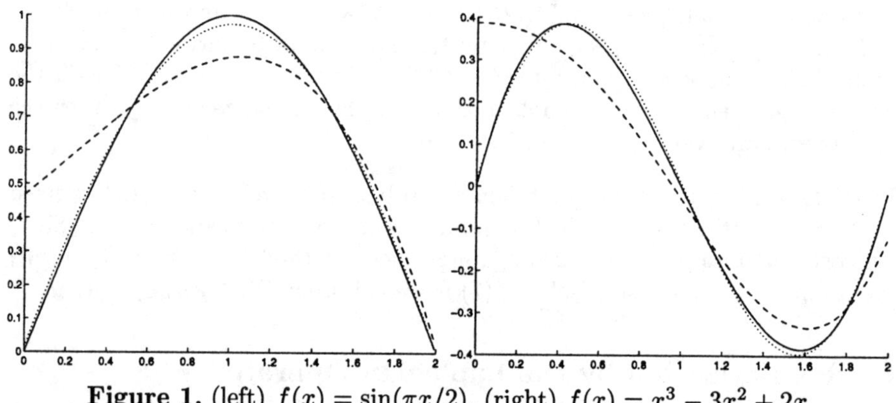

Figure 1. (left) $f(x) = \sin(\pi x/2)$, (right) $f(x) = x^3 - 3x^2 + 2x$

Next we consider the two-dimensional case. We choose $\Omega = [0, 2]^2$, $\Omega^* = [3, 5]^2$, $T = 4$ and $n = 64^2$. The test functions are $f(x, y) = \sin(\pi x)\sin(\pi y)$ and $f(x, y) = (x^3 - 3x^2 + 2x)(e^{2y^3 - 4y^2} - 1)$. Ten percent relative noise vector $\mathbf{n}$ is added to the right hand side vector $\mathbf{b}$ as in the 1-dimensional case, see (11). Figure 2 shows the original images for both test functions. Figures 3 and 4 are the images recovered by the identity regularization method and the Laplacian regularization method respectively.

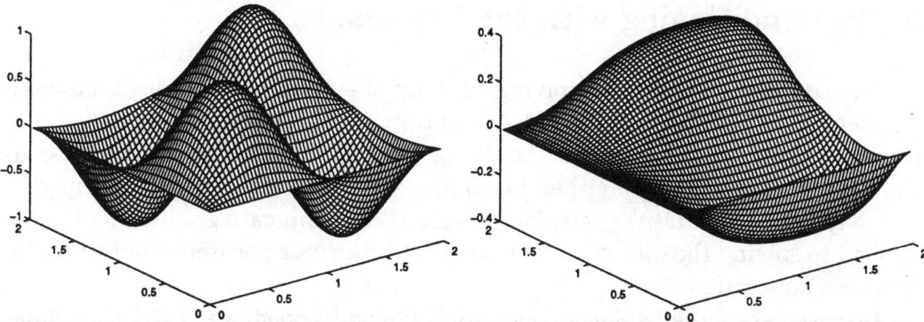

Figure 2. Original Image: (left) $\sin(\pi x)\sin(\pi y)$, (right) $(x^3 - 3x^2 + 2x)(e^{2y^3-4y^2} - 1)$

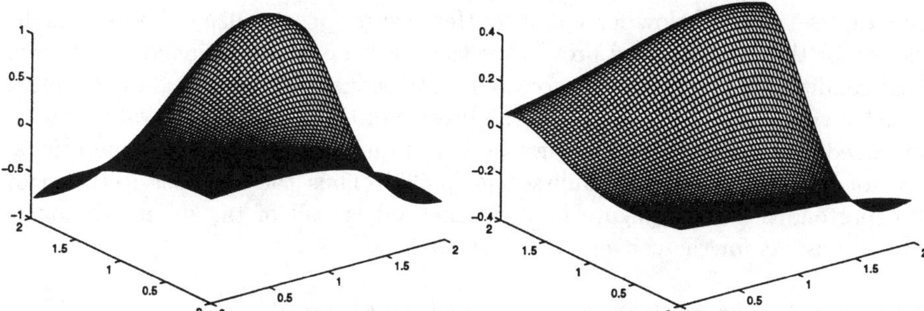

Figure 3. Identity Regularization with (left) $\alpha = 5 \cdot 10^{-14}$, (right) $\alpha = 10^{-13}$

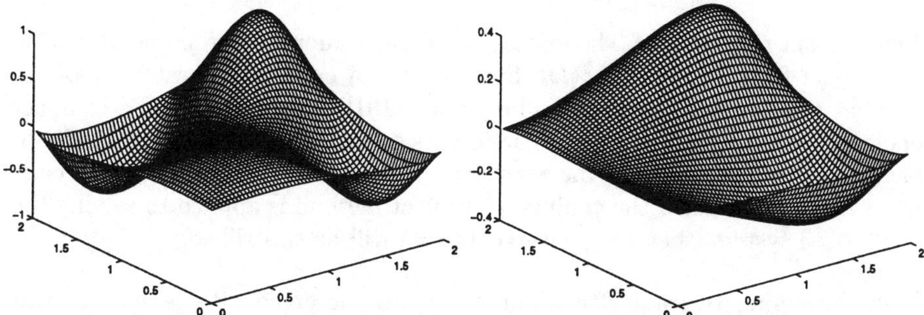

Figure 4. Laplacian Regularization with (left) $\alpha = 3 \cdot 10^{-15}$, (right) $\alpha = 10^{-13}$

From the examples, we see that the solutions obtained by the Laplacian regularization are much better than those from the identity regularization. Although we have shown only two examples here, we emphasize that we have similar results for other examples we tested.

5 Preconditioning with the Laplacian

In this section, we consider solving (10) by the conjugate gradient method. As mentioned previously, the cost of multiplying **K** to any vector is of order $O(n \log n)$ operations where n is the size of the matrix. Since the discrete Laplacian is a band-matrix with either 3 or 5 non-zero bands, the cost of multiplying it to any vector is of $O(n)$ operations. Hence if the conjugate gradient method is applied to solving the discrete equation of (10), the cost per iteration is of order $O(n \log n)$ operations.

However, since the Laplacian operator is ill-conditioned, we expect the Laplacian regularized equation (10) to be ill-conditioned. In particular, the number of iterations required for convergence will increase with increasing matrix size, see Tables 1 and 2 below for instance. Here we propose to use $\alpha\Delta$ as preconditioner for the equation and prove that the resulting preconditioned equation is well-conditioned. We note that for all 1-dimensional problems or 2-dimensional problems on rectangular domains, the inverse of the Laplacian operator can be obtained quite efficiently by using fast sine-transforms in $O(n \log n)$ operations, see for instance Stoer and Bulirsch [4, p.592]. Thus the cost per iteration of the preconditioned conjugate gradient method is still of the order $O(n \log n)$ operations. As for the convergence rate, we have

Theorem 2. *The spectrum of the preconditioned operator $\mathcal{I} + \alpha^{-1}\Delta^{-1}\mathcal{K}^*\mathcal{K}$ is clustered around* 1. *In particular, if the conjugate gradient method is applied to solving the preconditioned system, then the convergence rate will be superlinear.*

Proof. From the proof of Theorem 1, we already know that the operator $\mathcal{K}^*\mathcal{K}$ is compact from $L^2(\Omega)$ to $L^2(\Omega)$. Since Δ^{-1} is a bounded operator on $L^2(\Omega)$ (see for instance, Gilbarg and Trudinger [5, p.101]), $\Delta^{-1}\mathcal{K}^*\mathcal{K}$ is still a compact operator from $L^2(\Omega)$ to $L^2(\Omega)$. Therefore its spectrum is clustered around 0. In particular, the spectrum of the regularized operator $\mathcal{I} + \alpha^{-1}\Delta^{-1}\mathcal{K}^*\mathcal{K}$ is clustered around 1. Thus if the conjugate gradient method is applied to solving the regularized system, then the convergence rate will be superlinear. □

Thus the conjugate gradient method for solving the preconditioned system will converge in a fixed number of iterations independent of n. Since the cost per iteration of the preconditioned conjugate gradient method is of order $O(n \log n)$ operations, the total cost of solving (10) is also of order $O(n \log n)$ operations.

In the following, we illustrate the fast convergence of our preconditioned systems by using examples considered in §4. We apply the conjugate gradient method to (10) with or without the preconditioners. The numbers of iterations required for convergence are listed in Tables 1 and 2, where "No" and "$\alpha\Delta$" indicate whether preconditioning is used. The regularization parameter α is as given in §4 and is not changed with n. (We find from our numerical results that the best α is basically unchanged w.r.t n.) The stopping tolerance is again set to $\|\mathbf{r}_q\|_\infty / \|\mathbf{r}_0\|_\infty < 10^{-7}$. The initial guess is chosen to be the zero vector. All computations were done by Matlab on an IBM 43P-133 workstation. We note that the number of iterations is roughly constant if $\alpha\Delta$ is used as the preconditioner whereas the number is increasing for the non-preconditioned system.

$f(x)$	$\sin(\pi x/2)$		$x^3 - 3x^2 + 2x$	
n	No	$\alpha\Delta$	No	$\alpha\Delta$
32	33	3	28	5
64	64	3	63	5
128	128	3	116	5
256	256	3	230	5
512	512	3	441	5
1024	1024	3	849	5

Table 1. Numbers of the Iterations for 1-D Examples.

$f(x,y)$	$\sin(\pi x)\sin(\pi y)$		$(x^3 - 3x^2 + 2x)(e^{2y^3-4y^2} - 1)$	
n	No	$\alpha\Delta$	No	$\alpha\Delta$
8^2	18	12	16	10
16^2	31	10	29	10
32^2	58	13	52	10
64^2	92	10	95	13
128^2	185	10	195	10

Table 2. Numbers of the Iterations for 2-D Examples.

6 Conclusing Remarks

In this preliminary report, we consider solving the integral equation (6) which is obtained by linearizing (1). We find that the Laplacian operator is a better regularization operator than the identity operator. To speed up the convergence rate of the regularized equations, we propose to use the Laplacian operator itself as preconditioner and prove that the resulting preconditioned systems have clustered spectra. We will compare our proposed operator with other regularization operators (such as the TV-norm operator) and methods in future work.

References

1. R. Chan and M. Ng: Conjugate Gradient Method for Toeplitz Systems. SIAM Review (to appear)
2. A. Elayyan and V. Isakov: On Thermal Tomography. SIAM J. of Appl. Math. (to appear)
3. W. Hackbusch: Integral Equations. ISNM Vol 120, Birkhäuser Verlag, Basel, 1995.
4. J. Stoer and R. Bulirsch: Introduction to Numerical Analysis. Springer-Verlag, Berlin, 1991.
5. D. Gilbarg and N. Trudinger: Elliptic Partial Differential Equations of Second Order. 2nd Ed., Springer-Verlag, Berlin, 1983.

Convergence in Iterative Polynomial Root-Finding

Tien Chi Chen[1]

United College, The Chinese University of Hong Kong, Hong Kong
Present address: 6566 Tam O'Shanter Drive, San Jose CA 95120, USA.
e-mail: Tcchenuccu@aol.com

Abstract. For target α of the Nth-degree polynomial $P(z)$, $|\delta^*/\delta| \equiv |(z^* - \alpha)/(z - \alpha)| = O\left[|\rho\delta|^{q-1}\right] < 1$ if $q > 1$ and $|\rho\delta| << 1$, regardless of $|\delta|$ itself. Even if α is not a zero but the centroid of a cluster, the recomputed multiplicity estimate $m(z)$ could lead to a component zero. In global iterations, popular methods proved inadequate, yet for symmetric clusters the CLAM formula $z^* = z - (NP/P')\left(1 - Q^{m/N}\right)/(1-Q)$, where $Q = \left[N\left(1 - PP''/P'^2\right) - 1\right]/(N/m - 1)$, converges in principle to an m-fold zero in one iteration, using any finite guess outside the cluster centroid. Equipped with countermeasures against rebounds caused by local clusters, the formula has never been found to fail for general polynomials, and with an initial guess based on zeros of a symmetric cluster, usually converge in a few iterations.

1 An iteration theory based on finite deviations

Given an Nth-degree polynomial $P(z) \equiv \sum b_i z^i$, with $b_N = 1$, having zeros $\{\alpha_j\}$ and associated multiplicities $\{m_j\}$, we study iteration functions for multiplicity m:

$$z^* = F(z; m, s_1, \mu) = z - G_m(\mu)/s_1 \tag{1}$$

with

$$s_k \equiv \left[(-1)^{k-1}/(k-1)!\right] d^k \ln P/dz^k = \sum m_j/(z - \alpha_j) \tag{2}$$

$$\mu \equiv s_1^2/s_2 = 1/\left(1 - PP''/P'^2\right) \tag{3}$$

$$\begin{aligned} G_m(\mu) &= m \text{ in Newton's method} \\ &= 2m\mu/(m+\mu) \text{ in Halley's method} \\ &= N/\left\{1 + \left[(N/m-1)(N/\mu-1)\right]^{1/2}\right\} (m \neq 0, \mu \text{ or } N) \\ &\quad \text{in Laguerre's method} \\ &= N\left(1 - Q^{m/N}\right)/(1-Q), \text{ with } Q \equiv (N/\mu - 1)/(N/m - 1) \\ &\quad (m \neq 0, \mu \text{ or } N) \text{ in CLuster-Adapted Method (CLAM) [1] [2]} \end{aligned} \tag{4}$$

$G_m(\mu)$ is rational in μ for Halley's method, also for Newton's method in degenerate form. In the irrational methods we compute the fractional power thus:

$$(x+iy)^{u/v} = \left(re^{i\phi}\right)^{u/v} = \left(r^{u/v}\right) e^{i\phi u/v} = a+ib; \quad \phi \in (-\pi, \pi] \tag{5}$$

Analytical behavior being our main concern, all effects of roundoff are ignored in the discussions unless stated explicitly.

2 Superlinear deviation reduction

For the implied target α, we aim to reduce the magnitude of the *deviation ratio* $\delta^*/\delta \equiv (z^* - \alpha)/(z-\alpha)$ by comparing s_k with v/δ^k where v could be either m or N in the iteration formula. Using the pure-number measure

$$\beta_{kv}\delta^k \equiv s_k/\left(v/\delta^k\right) - 1 \tag{6}$$

we have

$$\mu = v\left[1 + 2\beta_{1v}\delta + \left(\beta_{1v}^2 - \beta_{2v}\right)\delta^2 + O\left(\beta_{1v}\beta_{2v}\delta^3\right)\right] \tag{7}$$

2.1 Aiming at zeros

If (v, α) equals, say, (m_1, α_1), we have, by identifying v with $m = m_1$

$$\begin{aligned}
(\delta^*/\delta)_{v=m} &= \beta_{1m}\delta + O\left(\beta_{1m}^2\delta^2\right) && \text{(Newton)}\\
&= \frac{1}{2}\left(\beta_{1m}^2 + \beta_{2m}\right)\delta^2 + O\left(\beta_{1m}^3\delta^3\right)\\
&\quad + O\left(\beta_{1m}\beta_{2m}\delta^3\right) && \text{(Halley)}\\
&= \frac{1}{2}\left[-\beta_{1m}^2/(N/m-1) + \beta_{2m}\right]\delta^2 + O\left(\beta_{1m}^3\delta^3\right)\\
&\quad + O\left(\beta_{1m}\beta_{2m}\delta^3\right) && \text{(Laguerre)}\\
&= \frac{1}{2}\left[(N/m-5)\,\beta_{1m}^2/3\,(N/m-1) - \beta_{2m}\right]\delta^2 + O\left(\beta_{1m}^3\delta^3\right)\\
&\quad + O\left(\beta_{1m}\beta_{2m}\delta^3\right) && \text{(CLAM)}\,(8)
\end{aligned}$$

The above formulas still hold if $v = m = N$ for the rational methods. Yet in the two irrational formulas N is already present, and this choice merely degrades the iteration function to Newton's formula. Nethertheless, for $v = N \neq m$

$$\begin{aligned}
(\delta^*/\delta)_{v=N\neq m} &= (N/m-1)\left[-2\beta_{1N}\delta/(N/m-1)\right]^{1/2} + O\left(\beta_{1N}\delta\right) && \text{(Laguerre)}\\
&= \left[-2\beta_{1N}\delta/(N/m-1)\right]^{m/N} + O\left(\beta_{1N}\delta\right) && \text{(CLAM)}(9)
\end{aligned}$$

For all above cases, $\delta^*/\delta = O\left[(\rho\delta)^{q-1}\right]$, or more familiarly, $\delta^* = O\left(\rho^{q-1}\delta^q\right)$, and the order of the iteration is said to equal q. Further, since $q > 1$, when $|\rho\delta| << 1$, true ***superlinear deviation reduction*** is achieved for the iteration, yielding $|\delta^*/\delta| < 1$; the persistence of this condition leads to ultimate convergence to α. As $\rho\delta$ generally has coefficients of order unity, reduction may already occur with $|\rho\delta| < 1$.

For convergence to $\alpha = \alpha_1$, classical theory often requires $m = m_1$ and $|\delta| << 1$, yet coordinate rescaling can shrink $|\delta|$ as much as one pleases without changing $|\delta^*/\delta|$. We have shown that $|\delta^*/\delta|$ actually depends on the *pure number* $\rho\delta$, invariant under coordinate rescaling. The iterate is in the "classical domain", converging towards the intended zero α_1, undistracted by the other zeros, if

$$m = m_1 \text{ and } |\delta| << \text{Min}\left\{|z - \alpha_j|\right\} \text{ for all } \alpha_j \neq \alpha_1 \tag{10}$$

2.2 Aiming at a centroid

Interestingly, α need not be confined to $\{\alpha_j\}$ nor any fixed target, and the iterates could shift its focus to new targets uncovered along the way, even when $|\delta|$ greatly exceeds the diameter of the zero collection as long as $|\rho\delta|$ is small.

When viewed from afar, the collection of all zeros of P looks like an N-fold zero at the centroid $c \equiv \sum m_j\alpha_j / \sum m_j = -b_{N-1}/Nb_N$. Let $\delta = z - c$ and $\varepsilon_j \equiv (c - \alpha_j)/\delta$, we have $s_k = \sum m_j/\delta^k (1 + \varepsilon_j)^k = \left[N + O\left(\sum m_j\varepsilon_j^2\right)\right]/\delta^k$. With $v = m = N$, $\beta_{1N}\delta = \sum m_j\varepsilon_j^2/N + O\left(\sum m_j\varepsilon_j^3/N\right)$, having the same order as $\beta_{2N}\delta^2 = 1.5\sum m_j\varepsilon_j^2/N + O\left(\sum m_j\varepsilon_j^3/N\right)$. We have

$$\begin{aligned}
(\delta^*/\delta)_{v=N=m} &= \sum m_j\varepsilon_j^2/N + O\left(\sum m_j\varepsilon_j^3/N\right) && \text{(Newton)}\\
&= 1.5\sum m_j\varepsilon_j^2/N + O\left(\sum m_j\varepsilon_j^3/N\right) && \text{(Halley)}\\
&= O\left(\rho\delta\right)^2 = O\left(1/\delta^2\right) \text{ with}\\
&\quad \rho^2 = O\left[\sum m_j\left(c - \alpha_j\right)^2/N\right]/\delta^4 && \text{(for both cases) (11)}
\end{aligned}$$

For the two irrational formulas again we require $N \neq m$

$$\begin{aligned}
(\delta^*/\delta)_{v=N\neq m} &= (N/m - 1)\left[-2\sum m_j\varepsilon_j^2/N\,(N/m - 1)\right]^{1/2}\\
&\quad + O\left(\sum m_j\varepsilon_j^2/N\right) = O\left(\rho\delta\right) && \text{(Laguerre)}\\
&= \left[-2\sum m_j\varepsilon_j^2/N\,(N/m - 1)\right]^{m/N}\\
&\quad + O\left(\sum m_j\varepsilon_j^2/N\right) = O\left(\rho\delta\right)^{2m/N} && \text{(CLAM) (12)}
\end{aligned}$$

If $\{\varepsilon_j << 1\}$, then $\sum m_j\varepsilon_j^2/N << 1$, and the farther z is from c, the nearer is z^*. We shall continue to call such deviation reduction "superlinear" for want of a better, all-inclusive term. δ^*/δ is $O(1/\delta)$ in the two rational methods, usually bringing z^* extremely close to the centroid, but only $O(1/\delta^2)$ in Laguerre's method and $O(1/\delta^{m/N})$ in CLAM. But the cenroid, easily computable, is not where most zeros are, and proximity to it may bring uncertainty or even catastrophes, as shall be seen presently.

3 Estimating the multiplicity

It is gratifying that, whenever $s_1 \approx m/(z-\alpha)$, uncannily $\mu \approx m$. Thus m need not be known, but can be estimated from μ as $m(z)$, and can even adapt to the iterate environment. In iteration functions already using μ the extra cost is minimal.

3.1 A compromise estimate

One is tempted to substitute μ for m directly, yet this again degrades third-order methods into second order. Surprisingly, degradation does not occur if m is *rounded* from a real form of μ. Just rounding the real part is not enough, as the imaginary part, often sizable, do contribute to m. Also, the choice $m = 1$ may mean slow convergence, but is usually safe in the face of uncertainty; $m = N$ is not allowed for the irrational formulas and (as shall be seen) may mean trouble with the rational ones. The following is a compromise for the ith iteration:

Let

$$w = \text{Floor } [0.5 + N/(1 + |N/\mu - 1|)] \tag{13}$$

then

$$\begin{aligned} m(z) &= w \text{ if } (1 \le w < N) \text{ and } (i \bmod 10 \ne 0) \\ &= 1 \text{ otherwise} \end{aligned} \tag{14}$$

3.2 Navigating by $m(z)$

The steering by $m(z)$ performs very well in practice, and it is recommended that $m(z)$ be recomputed at every iteration to take full advantage of refocusing. $m(z)$ should give the best multiplicity when superlinear deviation reduction is at hand, and elsewhere it seems to do as well as the other simple alternatives.

The iterates often describe a zig-zag path, reflecting the shifts in focus. While it is unrealistic to expect superlinear deviation reduction at every iteration, reductions by orders of magnitude are not uncommon, even for large $|\delta|$ despite significance loss, though small iterate displacements appear from time to time, often before large advances. The target aimed at is often neither a zero nor an easily recognizable centroid until near the end, and we have not found a good way to direct the iterates towards a zero in some preferred region after planting the first guess.

For example, $P(z) = (z-3)(z-1)(z-1.1-0.1i)^2(z-1.1+0.1i)^2$, the last 3 zeros form a local cluster with total multiplicity 5 and centroid at $(1.08+0i)$. CLAM with $(1+3i)$ as first guess gives the $m(z)$ sequence $(5,1,2,2)$, and the iterates $(1.107+0.009i)$, $(1.108+0.08i)$, $(1.10007+0.10006i)$, and $[(1.1-4\cdot10^{-10})+(0.1+4\cdot10^{-10})\,i]$, showing a shift of focus from the cluster centroid to a double zero within. The next $m(z)$ is 1, probably due to roundoff with small $|P|$. In Iteration 2 with z inside the cluster, $\mu = (-0.46-0.23i)$, and $N/(1+|N/\mu-1| = 0.44$; $m(z)$ was set to 1.

4 Global iterations

Since superlinear reduction does not really demand $|\delta| << 1$, one can contemplate *global iterations*, with initial guess z almost anywhere in the complex plane [1]. Aside from its own intrinsig interest, a study of global iterations could also help to delimit the applicability of known methods. We exclude from consideration infinity and zeros of s_1 as points unsuited for automatic computation, and expect precision loss when $|\delta|$ is several orders of magnitude greater than the diameter of the zeros.

We shall use as test polynomial the *symmetric cluster function* with centroid c, (complex) radius r and a fanout of n zeros, all of multiplicity N/n :

$$S_n(z;c,r,N) \equiv [(z-c)^n - r^n]^{N/m} \tag{15}$$

Here

$$s_1 = N(z-c)^{n-1}/[(z-c)^n - r^n] \tag{16}$$

$$\mu = N/[1+(n-1)r^n/(z-c)^n] \tag{17}$$

4.1 Rational methods

For z far from c, $m = N$ is required for superlinear reduction, giving z^* very close to c. But the next iterate z^{**} is discouraging for all integer $m \leq N$. Newton's method gives $|(z^{**}-c)/(z-c)| \geq 1$, and z^{**} is no neare to the zeros than z two iterations ago [1]. On the other hand, in Halley's method the iterate is held fast by c, with $z^{**} \approx z^*$, giving a false signal for convergence based on an iterate displacement criterion.

It turn out that, when z^* lies at the intersection of reflective symmtry axes, the two rational methods can only yield iterates also at an intersection, namely either z or ∞, and this effect is already felt even when z is near the intersection [3]. In addition, z on a single reflective symmetry axis cannot converge to zeros outside the axis. These shortcomings disqualify both methods for global iterations; they should be used only if z is already fairly close to a zero, say in the "classical domain".

4.2 Laguerre's method

Laguerre's method with $m = N/n$ converges globally for $S_n(z)$ with $n \leq 4$. Yet for $n \geq 5$, oscillating divergence occurs unless z lies within an annulus enclosing the circle passing through all zeros, its width shrinking as N increases. We conclude that it is not suited for global iterations either.

For $S_{10}(z; 0, 1, 10) = z^{10} - 1$, using $m = 1$ and $z = (0.5 + 0i)$, the nearest zero is $(1 + 0i)$ with $|\delta| = 0.5$. The Laguerre iterates vacillate about the circle $r = 1$ with increasing amplitude, and the (δ^*/δ) sequence is $(-2.54, -0.55, -19.7, -0.07, -9.6 \cdot 10^{13} \ldots)$. Incidentally, upon reducing the coordinates by a factor of 10^6, we have $|\delta_{new}| = 0.5 \cdot 10^{-6} << 1$, yet $\delta^*_{new}/\delta_{new} = \delta^*/\delta$ still.

4.3 CLAM

By far the most promising candidate found is CLAM, which converges globally with $m = N/n$ to a zero of $S_n(z)$ in just *one* iteration, for any finite guess z not coinciding with the centroid. Indeed, CLAM can be derived through approximating the given $P(z)$ by $S_n(z)$, matching s_1, μ, and $m = N/n$.

For $S_{10}(z; 0, 1, 10)$, to reach $|P| < 2 \cdot 10^{-14}$ with $m = 1$, CLAM takes only one iteration for $z = (0.5 + 0i)$; five for $z = (1 + i/3) \cdot 10^6$ due to rounoff, but one iteration again using an asymptotic version of the CLAM formula even for $z = (1 + i/3) \cdot 10^{15}$.

5 The rebound and countermeasures

5.1 The phenomenon

When z is close to a local yet far from other zeros, z^* may be close to the cluster centroid. If this cluster has high enough symmetry to make little or no contribution to the total s_1, its component zeros would become transparent to the iteration method. Attraction by external targets could then cause z^{**} to be flung away from the cluster, possibly towards another cluster centroid. This *rebound* phenomenon happens for all superlinear methods; it could recur indefinitely, possibly involving different clusters, and often interspersed by normal iterate behavior.

5.2 Countermeasures

Rebounds are characterized by sudden increases in $|P|$ and/or excessive iterate movement relative to the centroid. It is considered detected if

$$m(z) > 1 \text{ and } |P(z^{**})/P(z^*)| \geq 20 \qquad \text{(Type 1)} \quad (18)$$

$$\text{and/or } m(z) > 1 \text{ and } |z^{**} - z^*|/|z^* - c| \geq m(z)\pi/N$$

$$\text{with } |P(z^{**})/P(z^*)| \geq \frac{3}{4} \qquad \text{(Type 2)} \quad (19)$$

the remedy assumes the local cluster to be $S_m(z; z^*, r, m)$ with $r = [-P(z^*)]^{1/m}$. z^{**} is assigned a new value on the assumed periphery of the cluster, away from z^{**} :

$$(z^{**})_{new} = z^* + |r| e^{i\phi} \tag{20}$$

with

$$\phi = \tan^{-1}\{[\mathrm{Im}(z^*) - \mathrm{Im}(z^{**})] / [\mathrm{Re}(z^*) - \mathrm{Re}(z^{**})]\} \tag{21}$$

and (N, c) is replaced by (m, z^*) until the next rebound, if any.

One could also use $m = 1$ to break a sequence of undetected rebounds at the risk of first-order convergence. Our multiplicity estimate embodies this safety tactic.

CLAM was used with initial guess $(-2 - 2i)$ for the 8th-degree polynomial with zeros $(\pm 1 \pm 0.01 \pm 0.01i)$, forming two symmetric clusters centered at $(\pm 1 + 0i)$. A type-2 rebound was detected during the 4th iteration. If left alone, the iterates would jump between the two local centroids indefinitely; correction leads to convergence in 3 more iterations $\left(|P| = 4 \cdot 10^{-15}\right)$.

6 Using an initial guess

6.1 An initial guess

An excellent initial guesses is $z = c + [-P(c)/b_N]^{1/N}$, where b_i is the coefficient of z^i in $P(z)$ and c the centroid. If $P(z) = S_N(z; c, r, N)$, it is just the zero $(c + r)$, reachable also in one CLAM iteration for arbitrary guess using $m = 1$. This guess could save several iterations for CLAM; it can avoid the initial oscillating divergence in Laguerre's method, though local symmetry could still cause later ones.

6.2 Finding one or more zeros

A CLAM program package CLAM96, improved from the earlier SCARFS [2], has been written in the APL language. The package accept coefficients of P, and starts with the computed initial guess, recomputing $m(z)$ for every iteration. For numerical accuracy, the iteration functions are actually evaluated in terms of $(1/s_1)$, $(1/\mu)$ and $(1 - Q)$. Several iterations usually suffice to stabilize z; and further iterations might reduce $m(z)$ to unity, even reduce accuracy.

For more zeros, the polynomial is deflated after each zero is found. As $m(z)$ tends to become 1 when the iterate gets too close to a zero, the newly found zero is used as initial guess for the deflated polynomial, hoping to catch the missed members of multiple zeros, if any. The penultimata zero takes only one iteration, and the last zero requires no polynomial evaluation.

6.3 Examples using CLAM96

All 16 zeros $(\pm 1 \pm 0.01 \pm 0.01i)(\pm 0.01 \pm i \pm 0.01i)$ in $P(z)$ were computed to $|P| < 2 \cdot 10^{-14}$ in 70 iterations (210 horners), averaging 4.38 iterations (13.0 horners) per zero. Rebounds were detected and corrected twice, for $N = 16$ and $N = 8$.

All zeros of $(z - 1 - i)^2 (z - 4 + 3i)(z - 4 - 3i)(z - 3.999 - 3i)$ were computed in 10 CLAM iterations (30 horners; to $|P| < 6 \cdot 10^{-13}$), comparing with 151 horners by the Newton-based method proposed by Ralston and Rabinowitz, and 83 horners by the Traub-Jenkins method [4]. Just 3 CLAM iterations (9 horners) yielded all zeros of $(z^2 + 9)(z - 3)^4$ to $|P| < 10^{-12}$;Muller's method took 108 horners [5].

7 Summary and discussions

Superlinear reduction occurs if $|\rho\delta| << 1$, even when $|\delta|$ is large. A computed $m(z)$ can seek the target automatically, shifting its focus from a cluster centroid to a component zero when needed. In global iterations, an iterate from afar causes precision loss, but the greatest challenge occurs when the iterate is within clusters with untoward symmetry; CLAM, equipped with rebound countermeasures, appears to be failure-free here. The proposed initial guess usually leads to convergence in just a few iterations.

Nevertheless, much remains to be done, especially in steering the iterate towards specific zeros, in simplifying the algorithms, and in making other methods, most notably Laguerre's method, fast and reliable.

8 Acknowledgment

The writer acknowledges stimulation received from Dr. Ralph Willoughby, also Professors Tony Chan, Franklin Luk, Omar Wing and Wang Yuan. The work had its inception while the writer was at IBM Corporation.

References

1. Chen T.C.: Iterative zero-finding revisited, in W.I. Hogarth and B.J. Noye (Eds.) Computational Techniques and Applications:CTAC-89 (Proc. Computaional Techniques and Applications Conf., Brisbane, Australia, July 1989), New York. Hemisphere Pub. Corp. (1990) 583–590
2. Chen, T.C.: SCARFS, an efficient polynomial zero-finder system. APL 93 Conf. Proc. (APL Quote Quard **24** (1993) 47–54
3. Chen, T.C.: Symmetry and nonconvergence in iterative polynomial zero-finding. (To be published)
4. Ralston, A., Rabinowitz, P.: A First Course in Numerical Analysis. 2nd Ed., New York: McGraw-Hill, 1978
5. Atkinson, K.E.: An Introduction to Numerical Analysis. New York:Wiley, 1978

Splitting Time Methods and One Dimensional Special Meshes for Reaction-Diffusion Parabolic Problems

C. Clavero[1], J.C. Jorge[2], F. Lisbona[3], G.I. Shishkin[4]

[1] Departamento de Matemática Aplicada. Universidad de Zaragoza. Zaragoza. Spain. e-mail Clavero@posta.unizar.es

[2] Departamento de Matemática e Informática.Universidad Pública de Navarra. Pamplona. Spain. e-mail Jcjorge@si.upna.es

[3] Departamento de Matemática Aplicada. Universidad de Zaragoza. Zaragoza. Spain. e-mail Lisbona@posta.unizar.es

[4] Institute of Mathematics and Mechanics, Ekaterinburg, Russia. e-mail Sgi@eqmph.imm.intec.ru

Abstract. A numerical method is developed for a time dependent reaction diffusion two dimensional problem. This method is deduced by combining an alternating direction technique and the central finite difference scheme on some special piecewise uniform meshes. We prove that this method is uniformly convergent with respect to the diffusion parameter ε, achieving order 1 in both spatial and time variables. The theoretical results are confirmed by the numerical experiences performed.

1 Introduction

Let us consider the singularly perturbed parabolic problem

$$\begin{aligned} &\frac{\partial u}{\partial t} - \varepsilon \Delta u + ku = f \text{ in } \Omega \times [0,T] \equiv [0,1] \times [0,1] \times [0,T], \qquad (1)\\ &u(x,y,0) = u_0(x,y) \text{ in } \Omega,\\ &u(0,y,t) = u(1,y,t) = 0 \text{ in } [0,1] \times [0,T],\\ &u(x,0,t) = u(x,1,t) = 0 \text{ in } [0,1] \times [0,T], \end{aligned}$$

which models some linear reaction-diffusion process in fluid mechanics. We shall assume that $\varepsilon > 0$, $k \equiv k(x,y,t) > 0$ and $f \equiv f(x,y,t)$ are sufficiently smooth functions and it is verified the compatibility condition

$$f(0,0,t) = f(0,1,t) = f(1,0,t) = f(1,1,t) = 0. \qquad (2)$$

When the diffusion parameter ε is small with respect to the reaction coefficient k, the solutions of these problems, even for smooth data, present rapid variations

* This research has been partially supported by the CICYT project. num. AMB94-0396 and by a project of Gobierno de Navarra

in certain narrow regions closed to the boundary of the domain Ω. In this paper we are interested in uniform convergent methods, i.e., methods in which the rate of convergence and the error constant must be independent of the singular perturbation parameter ε. Many contributions in this direction have been carried out in two main ways: exponential fitting and non-uniform special meshes, see [2], [5], [7]. Nevertheless, for time dependent multidimensional problems, only a few contributions have been done, see [3], [4]. To solve these problems we shall use an alternating direction technique (ADI). In an earlier paper [1], we study some exponential fitting finite difference schemes for the resulting singularly perturbed one dimensional problems on each line, obtaining uniform convergence in some cases. In order to improve these results, in this paper we study a scheme which use the same ADI technique and the standard central finite difference scheme on some special meshes introduced by Shishkin, [8], [9], which are easily defined and have a number of points independent of ε. So, we obtain a first order uniformly convergent method, which besides have a low computational cost in comparison with the classical two dimensional discretizations.

Throughout the paper we shall denote C a generic positive constant that is not dependent on both ε and on the meshes. Let us consider the operators

$$L_{x,\varepsilon} \equiv -\varepsilon \frac{\partial^2}{\partial x^2} + k_1, \quad L_{y,\varepsilon} \equiv -\varepsilon \frac{\partial^2}{\partial y^2} + k_2, \tag{3}$$

with $k_i(x,y,t) > \gamma_i > 0, i = 1,2$ and $k_1 + k_2 = k$. We shall also consider decompositions for the source term $f = f_1 + f_2$ where f_1, f_2 will be chosen satisfying

$$f_1(x,0,t) = f_1(x,1,t) = 0, \quad f_2(0,y,t) = f_2(1,y,t) = 0. \tag{4}$$

2 The Time Semidiscretization

A first stage towards defining the method consists of a time discretization process by means of the following fractional steps scheme (see [6]):

$$a) \quad u^0 = u_0(x,y), \tag{5}$$
$$b) \quad (I + \Delta t L_{x,\varepsilon})u^{n+\frac{1}{2}} = u^n + \Delta t f_1(t_{n+1}), \tag{6}$$
$$u^{n+\frac{1}{2}}(0,y) = u^{n+\frac{1}{2}}(1,y) = 0, \tag{7}$$
$$c) \quad (I + \Delta t L_{y,\varepsilon})u^{n+1} = u^{n+\frac{1}{2}} + \Delta t f_2(t_{n+1}), (f_1 + f_2 = f), \tag{8}$$
$$u^{n+1}(x,0) = u^{n+1}(x,1) = 0, \tag{9}$$

and we could obtain semidiscrete approximations $u^n(x,y)$ to the solution $u(x,y,t)$ of (1) at the time levels $t_n = n\Delta t$, if we solved exactly the steps (6–9). To study the scheme (5–9), we consider the local error, defined by $e_{n+1} = u(t_{n+1}) - \hat{u}^{n+1}$, where $\hat{u}^{n+1}$ is the approximation to $u(t_{n+1})$ given after a time step by the semidiscrete method (5–9), taking $u^n = u(t_n)$. Then, we have

Lemma 1. *Let us assume that* $u, \frac{\partial u}{\partial t}, \frac{\partial^2 u}{\partial t^2} \in C^0[\bar{\Omega} \times [0,T]]$ *and*

$$|\frac{\partial^{i_0}}{\partial t^{i_0}} u(x,y,t)| \leq C, \quad (x,y,t,) \in \overline{\Omega} \times [0,T], \, i_0 \leq 2. \tag{10}$$

Then, the local error satisfies $\|e_{n+1}\|_\infty \leq C(\Delta t)^2$.

The combination of this result and a stability result for (5–9) based on a maximum principle for the operators $(I + \Delta t L_{i,\varepsilon})$, $i = x, y$, gives the following convergence result.

Theorem 2. *Under assumptions of Lemma 1,*

$$sup_{n \leq \frac{T}{\Delta t}} \|u(t_n) - u^n\|_\infty \leq C\Delta t. \tag{11}$$

Hence, the time semidiscretization process is convergent.

The proofs of the previous results can be seen in [1].

Note. The smoothness assumptions in Lemma 1 for the exact solution $u(t)$ of (1), are the natural ones for analyzing time-discretization of parabolic problems via the consistency and stability in the maximum norm. In the same direction, the ε–independence of bounds for $|\frac{\partial^i u}{\partial t^i}|$, $i = 0,1,2$ in $\Omega \times [0,T]$ is a natural restriction to obtain a uniform convergence result in time. Sufficient conditions for u to fulfill the smoothness hypotheses of Lemma 1, are the following:

1) *The ε–independent smoothness of data*, i.e., $u_0, (L_{x,\varepsilon} + L_{y,\varepsilon})u_0, (L_{x,\varepsilon} + L_{y,\varepsilon})^2 u_0$, $f, \frac{\partial f}{\partial t}, \frac{\partial^2 f}{\partial t^2}, (L_{x,\varepsilon} + L_{y,\varepsilon})f(x,y,0)$, are continuous and bounded independently of ε,

2) *Compatibility among data*, i.e., $u_0 = 0$ in $\partial\Omega, (L_{x,\varepsilon} + L_{y,\varepsilon})u_0 = f(x,y,0)$ in $\partial\Omega$, $(L_{x,\varepsilon} + L_{y,\varepsilon})^2 u_0 = (L_{x,\varepsilon} + L_{y,\varepsilon})f(x,y,0) - \frac{\partial f}{\partial t}(x,y,0)$ in $\partial\Omega$, and (2).

These conditions can be easily obtained by differentiation of the differential equation in (1), and proving sufficient conditions for the continuity and ε-independently estimates of the solutions of resulting problems in $\bar{\Omega} \times [0,T]$.

3 The Spatial Discretization

In this section we study the totally discrete scheme, deduced after the space discretization of (5–9). In this way, we define a non necessarily uniform rectangular mesh $\Omega_{\varepsilon,h}$ as the tensor product $I_{x,\varepsilon,h} \times I_{y,\varepsilon,h}$ of one-dimensional special meshes, which will be generated as follows:

$$I_{x,\varepsilon,h} \equiv \{0 = x_0, x_1, \ldots, x_{\frac{N}{4}} = \sigma, \ldots, x_{\frac{3N}{4}} = 1 - \sigma, \ldots, x_N = 1\}, \tag{12}$$

with $x_i = i\frac{4\sigma}{N}, i = 0, \ldots, \frac{N}{4}$, $x_i = \sigma + (i - \frac{N}{4})\frac{2(1-2\sigma)}{N}, i = \frac{N}{4} + 1, \ldots, \frac{3N}{4}$ and $x_i = 1 - \sigma + (i - \frac{3N}{4})\frac{4\sigma}{N}, i = \frac{3N}{4} + 1, \ldots, N$, where $\frac{N}{4} \in \mathbb{N}$, $h = \frac{1}{N}$ and

$$\sigma = \min(\frac{1}{4}, m\sqrt{\varepsilon}\log N), \tag{13}$$

where m is a positive constant, which satisfies $m\gamma_1 \geq 1$, independent of ε and N. In similar way can be defined $I_{y,\varepsilon,h}$.

On the rectangular grid $\Omega_{\varepsilon,h}$, if we denote by $[.]_h$ the restriction of a function defined in $[0,1]\times[0,1]$ to $\Omega_{\varepsilon,h}$, the totally discrete approximations u_h^n to $[u(t_n)]_h$ are defined by

$$a)\quad u_h^0 = [u_0]_h, \tag{14}$$

$$b)\quad (I + \Delta t L_{x,\varepsilon,h})u_h^{n+\frac{1}{2}} = u_h^n + \Delta t[f_1(x,y,t_{n+1})]_h, \tag{15}$$

$$u_h^{n+\frac{1}{2}}(0,y) = u_h^{n+\frac{1}{2}}(1,y) = 0, \quad y \in I_{y,\varepsilon,h}, \tag{16}$$

$$c)\quad (I + \Delta t L_{y,\varepsilon,h})u_h^{n+1} = u_h^{n+\frac{1}{2}} + \Delta t[f_2(x,y,t_{n+1})]_h, \tag{17}$$

$$u_h^{n+1}(x,0) = u_h^{n+1}(x,1) = 0, \quad x \in I_{x,\varepsilon,h}, \tag{18}$$

where $L_{x,\varepsilon,h}$ (and analogously $L_{y,\varepsilon,h}$) is the discretization of the differential operator $L_{x,\varepsilon}$ ($L_{y,\varepsilon}$) using the one-dimensional central difference scheme on $I_{x,\varepsilon,h}$ ($I_{y,\varepsilon,h}$). For (14–18) we set and prove an intermediate uniform convergence result.

Theorem 3. *Let us assume that $\hat{u}^{n+1/2}$ and $\hat{u}^{n+1}$ have the asymptotic behaviour given in section 4 by (33) and let $\hat{u}_h^{n+1}$ be the numerical solution of (14–18) taking $u_h^n \equiv [u(t_n)]_h$. Then*

$$\|[\hat{u}^{n+1}]_h - \hat{u}_h^{n+1}\|_\infty \leq C\,\Delta t\, h. \tag{19}$$

Proof. To compare $\hat{u}^{n+1}$ and $\hat{u}_h^{n+1}$ the following process is done. In the first half step of the algorithm (5–9), we have the family of one-dimensional stationary singularly perturbed problems

$$(I + \Delta t L_{x,\varepsilon})\hat{u}^{n+\frac{1}{2}}(x,y) = u(x,y,t_n) + \Delta t f_1(x,y,t_{n+1}), \tag{20}$$

$$\hat{u}_h^{n+\frac{1}{2}}(0,y) = \hat{u}_h^{n+\frac{1}{2}}(1,y) = 0, \quad y \in I_{y,\varepsilon,h}, \tag{21}$$

and its discrete version

$$(I + \Delta t L_{x,\varepsilon,h})\hat{u}_h^{n+\frac{1}{2}} = [u(x,y,t_n)]_h + \Delta t[f_1(x,y,t_{n+1})]_h, \tag{22}$$

$$\hat{u}_h^{n+\frac{1}{2}}(0,y) = \hat{u}_h^{n+\frac{1}{2}}(1,y) = 0, \quad y \in I_{y,\varepsilon,h}. \tag{23}$$

Under the hypotheses of the theorem, we can apply the results of section 4 to obtain the estimates

$$\|[\hat{u}^{n+\frac{1}{2}}]_h - \hat{u}_h^{n+\frac{1}{2}}\|_\infty \leq C\,\Delta t\, h.$$

For the second fractional step, we have

$$(I + \Delta t L_{y,\varepsilon})\hat{u}^{n+1}(x,y) = \hat{u}^{n+1/2}(x,y) + \Delta t f_2(x,y,t_{n+1}), \tag{24}$$

$$\hat{u}^{n+1}(x,0) = \hat{u}^{n+1}(x,1) = 0, \quad x \in I_{x,\varepsilon,h}, \tag{25}$$

in the algorithm (5–9) and its discrete version

$$(I + \Delta t L_{y,\varepsilon,h})\hat{u}_h^{n+1} = \hat{u}_h^{n+1/2} + \Delta t[f_2(x,y,t_{n+1})]_h, \tag{26}$$

$$\hat{u}_h^{n+1}(x,0) = \hat{u}_h^{n+1}(x,1) = 0, \quad x \in I_{x,\varepsilon,h}, \tag{27}$$

in the algorithm (14–18). To find the relation between $\hat{u}^{n+1}$ and $\hat{u}_h^{n+1}$, we introduce the auxiliar problem

$$(I + \Delta t L_{y,\varepsilon,h})\tilde{u}_h^{n+1} = [\hat{u}^{n+\frac{1}{2}}]_h + \Delta t[f_2(t_{n+1})]_h, \tag{28}$$
$$\tilde{u}_h^{n+1}(x,0) = \tilde{u}_h^{n+1}(x,1) = 0, \quad x \in I_{x,\varepsilon,h}. \tag{29}$$

Using the same arguments that in the first fractional step, we obtain

$$\|[\hat{u}^{n+1}]_h - \tilde{u}_h^{n+1}\|_\infty \le C\,\Delta t\; h.$$

If we take into account also that

$$\tilde{u}_h^{n+1} - \hat{u}_h^{n+1} = (I + \Delta t L_{y,\varepsilon,h})^{-1}([\hat{u}^{n+\frac{1}{2}}]_h - \hat{u}_h^{n+\frac{1}{2}}),$$

and recall the stability result, (19) follows. □

To prove the uniform convergence of the totally discrete scheme we have split the global error $\|[u(t_n)]_h - u_h^n\|_\infty$ into the contributions from the time and space discretizations. Combining the properties of these two processes, we have the following result, proved in [1].

Theorem 4. *Let u be the solution of (1) and $\{u_h^n\}_n$ the solution of (14–18). Under the hypotheses of Lemma 1 and Theorem 3, there exists a constant C such that*

$$\|[u(t_n)]_h - u_h^n\|_\infty \le C(\Delta t + h). \tag{30}$$

4 Uniform Analysis of the Spatial Discretization

In this section we study the discretization of problems (20–21) and (24–25) given by (22–23) and (26–27) respectively. We show only the analysis for (20–23) and similarly can be done for (24–27). For simplicity, we denote $T_{x,\varepsilon,\Delta t}w \equiv -\varepsilon\Delta t w''(x) + (1 + \Delta t k_1(x,y))w(x)$, then, problem (20–21) can be considered as a family of 1D singularly perturbed problems

$$T_{x,\varepsilon,\Delta t}w = u(x,y,t_n) + \Delta t f_1(x,y,t_{n+1}), \quad 0 < x < 1, \tag{31}$$
$$w(0) = 0, \quad w(1) = 0, \tag{32}$$

depending on the parameter $y \in [0,1]$, $k_1(x,y) > \gamma_1 > 0$ and k_1, f_1 are sufficiently smooth functions. It is known (see [8], [9]) that $w \equiv w(x,\varepsilon)$ satisfies

$$\left|\frac{\partial^i w(x,y,t_n,\Delta t)}{\partial x^i}\right| \le C\big[1+\varepsilon^{-i/2}\big(\exp(-\gamma\varepsilon^{-1/2}x)+\exp(-\gamma\varepsilon^{-1/2}(1-x))\big)\big]. \tag{33}$$

On the mesh $I_{x,\varepsilon,h}$ defined by (12–13), the central finite difference scheme can be written as

$$T_{x,\varepsilon,h,\Delta t}W_j \equiv r_j^- W_{j-1} + r_j^c W_j + r_j^+ W_{j+1} = f_j, \quad j = 1,\dots,N-1, \tag{34}$$
$$W_0 = 0, \quad W_N = 0, \tag{35}$$
$$r_j^- = \frac{-\varepsilon\Delta t}{h_j \tilde{h}_j}, \quad r_j^+ = \frac{-\varepsilon\Delta t}{h_{j+1}\tilde{h}_j} \quad r_j^c = 1 + \Delta t k_{1,j} - r_j^- - r_j^+, \tag{36}$$
$$k_{1,j} = k_1(x_j,y), \quad f_j = u(x_j,y,t_n) + \Delta t f_1(x_j,y,t_{n+1}), \tag{37}$$

where $h_j = 4\sigma/N$, $j = 1,\ldots,N/4$ and $j = (3N/4)+1,\ldots,N$ $h_j = 2(1-2\sigma)/N$, $j = (N/4)+1,\ldots,3N/4$, $\tilde{h}_j = (h_j + h_{j+1})/2$, $j = 1,\ldots,N$. We only present the case $\sigma = m\sqrt{\varepsilon}\log N$ (otherwise the problem can be studied in classical way).

Using Taylor expansions, with the remainder in the classical or integral form, and taking account the special meshes considered for the numerical discretization, we can obtain the following estimates for the local error $\tau_j = T_{x,\varepsilon,h,\Delta t}(w(x_j)) - (T_{x,\varepsilon,\Delta t}w)(x_j)$ for $j = 1,\ldots,N-1$:

1) If $0 < x_j < \sigma$ or $1-\sigma < x_j < 1$, then

$$|\tau_j| \leq C\Delta t \frac{\log^2 N}{N^2}. \tag{38}$$

2) If $\sigma \leq x_j \leq 1-\sigma$, then

$$|\tau_j| \leq \frac{C\Delta t}{N}. \tag{39}$$

From (38) and (39) it follows that the scheme (34–37) is uniformly consistent. As the operator $T_{x,\varepsilon,h,\Delta t}$ is also uniformly stable, we have that

$$\max_j |w(x_j) - W_j| \leq C\Delta t\, h. \tag{40}$$

5 Numerical results

Finally, we show the numerical results obtained in the integration of a problem of type (1). We have considered an example with known exact solution in order to compute exactly the pointwise errors $e_\varepsilon^{N,\Delta t}(n,i,j) = |u(t_n,x_i,y_j) - u^N(t_n,x_i,y_j)|$, where the superscript indicates the number of mesh points used in the x-direction and in the y-direction (for simplicity we take the same number of points in both directions), and $t_n = n\Delta t$. For each ε the maximum nodal error is given by

$$E_{\varepsilon,N,\Delta t} = \max_{n,i,j} e_\varepsilon^{N,\Delta t}(n,i,j), \tag{41}$$

and for each N and Δt, the ε-uniform maximum nodal error is defined by $E_{N,\Delta t} = \max_\varepsilon E_{\varepsilon,N,\Delta t}$. Computed values of $E_{\varepsilon,N,\Delta t}$ and $E_{N,\Delta t}$ are given in Table 1 for various values of ε, N and Δt. To obtain a numerical ε-uniform rate of convergence, we proceed as in [10] and the results are given in Table 2. In the example below, we choose the following decomposition of the function f: $f_1(x,y,t) = f(x,y,t) - f_2(x,y,t)$, $f_2(x,y,t) = f(x,0,t) + y(f(x,1,t) - f(x,0,t))$, which satisfies (4).

The example that we solve is given by

$$\begin{aligned}
&u_t - \varepsilon\Delta u + (2 + x^2 + y^2 + xy)u = f(x,y,\varepsilon,t), \quad (x,y,t) \in \Omega \times [0,5],\\
&u(0,y,t) = u(1,y,t) = 0, \quad y \in [0,1], t \in [0,5]\\
&u(x,0,t) = u(x,1,t) = 0, \quad x \in [0,1], t \in [0,5]\\
&u(x,y,0) = g(x,y,\varepsilon), \quad x,y \in [0,1],
\end{aligned}$$

where f and g are such that the exact solution is given by $u_\varepsilon(x,y,t) = (2 - e^{-t})(h(x)-1)(h(y)-1)$, with

$$h(\zeta) = \frac{e^{-\frac{\zeta}{\sqrt{\varepsilon}}} + e^{-\frac{1-\zeta}{\sqrt{\varepsilon}}}}{1+e^{-\frac{1}{\sqrt{\varepsilon}}}}. \tag{42}$$

Table 1. Maximum Nodal Errors $E_{\varepsilon,N,\Delta t}$ and $E_{N,\Delta t}$

ε	N=8 Δt= 0.1	N=16 Δt= 0.05	N=32 Δt= 0.025	N=64 Δt=0.0125	N=128 Δt=0.00625
1	5.555E-4	3.437E-4	1.981E-4	1.076E-4	5.625E-5
10^{-1}	2.437E-2	1.242E-2	6.349E-3	3.199E-3	1.607E-3
10^{-2}	7.722E-2	4.082E-2	1.999E-2	9.873E-3	4.903E-3
10^{-3}	1.071E-1	5.168E-2	2.797E-2	1.404E-2	7.088E-3
10^{-4}	1.779E-1	8.237E-2	3.201E-2	1.642E-2	8.320E-3
10^{-5}	2.043E-1	1.047E-1	4.961E-2	2.118E-2	8.901E-3
10^{-6}	2.130E-1	1.125E-1	5.658E-2	2.722E-2	1.230E-2
10^{-7}	2.158E-1	1.151E-1	5.894E-2	2.943E-2	1.433E-2
$E_{N,\Delta t}$	0.2158	0.1151	0.0589	0.0294	0.0143

Table 2. Numerical order of convergence

ε	N=8	N=16	N=32	N=64
1	0.692	0.794	0.880	0.935
10^{-1}	0.972	0.968	0.988	0.993
10^{-2}	0.919	1.029	1.017	1.009
10^{-3}	1.051	0.885	0.994	0.986
10^{-4}	1.110	1.365	0.963	0.980
10^{-5}	0.964	1.077	1.227	1.250
10^{-6}	0.920	0.991	1.055	1.146
10^{-7}	0.906	0.965	1.001	1.038
Min.	0.692	0.794	0.880	0.935

References

1. Clavero, C., Jorge, J.C., Lisbona, F.: Uniformly convergent schemes for singular perturbation problems combining alternating directions and exponential fitting techniques. In Applications of Avanced Computational Methods for Boundary and Interior Layers (J.J.H. Miller, Ed). Boole Press. Dublin (1993) 33–52
2. Farrell, P.A.: Sufficient conditions for uniform convergence of a class of difference schemes for a singularly perturbed problem. IMA J. Numer. Anal. **7** (1987) 459–472
3. Hegarty, A.F., Miller, J.J.H., O'Riordan, E., Shishkin, G.I.: Special numerical methods for convection-dominated laminar flows at arbitrary Reynolds number. East-West J. Numer. Math. **2** (1994) 65–74
4. Hegarty, A.F., Miller, J.J.H., O'Riordan, E., Shishkin, G.I.: Special meshes for finite difference approximations to an advection-diffusion equation with parabolic layers. J. Comp. Phys. **117** (1995) 47–54
5. Herceg, D.: Uniform fourth order difference scheme for a singular perturbation problem. Numer. Math. **56** (1990) 675–693
6. Jorge, J. C., Lisbona, F.: Contractivity results for alternating direction schemes in Hilbert spaces. App. Numer. Math. **15** (1994) 65–75
7. Kellogg, R.B., Tsan, A.: Analysis of some difference approximations for a singular perturbation problem without turning points. Math. Comp. **32** (1978) 1025–1039
8. Shishkin, G.I.: Grid pproximation of singularly perturbed boundary value problems with convective terms. Sov. J. Numer. Anal. Math. Modelling **5** (1990) 173–187
9. Shishkin, G.I.: Grid approximations of singularly perturbed elliptic and parabolic equations (in Russian). Russian Academy Science, Ural Branch, Ekaterinburg (1992)
10. Stynes M., Roos, H.G.: The Midpoint Upwind Scheme. Preprint of Dep. of Math., University College, Cork (1996)

High-Order Stiff ODE Solvers via Automatic Differentiation and Rational Prediction

G. F. Corliss[1], A. Griewank[2], P. Henneberger[2], G. Kirlinger[3], F. A. Potra[4], H. J. Stetter[3]

[1] Dept of Math, Stat and Comp Sci, Marquette University, Milwaukee, WI, 53233-2214, USA
[2] TU Dresden, Institut fuer Wissenschaftliches Rechnen, 01062 Dresden, Germany
[3] TU Wien, Wiedner Hauptstr. 8-10, A-1040 Wien, Austria
[4] Dept of Mathematics, The University of Iowa, Iowa City, Iowa, 52242, USA

Abstract. A class of higher order methods is investigated which can be viewed as implicit Taylor series methods based on Hermite quadratures. Improved automatic differentiation techniques for the claculation of the Taylor-coefficients and their Jacobians are used. A new rational predictor is used which can allow for larger step sizes on stiff problems.

1 Introduction

The design and implementation of numerical methods for the stepwise approximate solution of initial value problems for systems of ordinary differential equations

$$\begin{aligned} y'(t) &= f(y(t),t)\ , \quad y(t) \in \mathbb{R}^n \\ y(0) &= y_0 \end{aligned} \tag{1}$$

is rather well understood, but there is still a good deal of active research in this area.

For non-stiff systems, through the availability of efficient implementations of automatic differentiation (cf. e.g. [2]), the use of local Taylor expansions along trajectories has become feasible as a basis for the numerical solution of (1). One may also use integration steps which are of the order of the radius of convergence of the Taylor expansion, see [2]. Due to their explicit nature, these algorithms are not well suited for stiff systems so various combinations are used together. These considerations have led the authors to the methods discussed in this paper. They may be viewed as a generalization of various approaches: Taylor series methods, the implicit Euler method, and collocation methods. These methods may attain any given order of consistency, and they are L-stable with an appropriate choice of parameters. A natural continuous approximation of the solution trajectory which is uniformly of the same high-order is associated with the pointwise approximate solution.

The methods, following ideas of Obrechkoff [9, 10], Milne [7] and Wanner [12, 13] have been known since 1940 and are discussed in [6, 8, 3]. However, high-order methods have rarely been explored. The difficulty is the system of nonlinear equations for the approximation y_ν at t_ν involves the values of higher Taylor-coefficients of the solution at (t_ν, y_ν) with respect to t. Further, to solve the resulting nonlinear system by a Newton iteration one needs the partial derivatives of these Taylor-coefficients with respect to the next solution point. Automatic differentiation provides a critical enabling technology for solving these problems.

2 Description of the Methods

By far the most popular and successful approach to the design of stepwise methods for the numerical solution of systems of ordinary differential equations (1) is the following: Consider the *identity*

$$y(t_\nu) = y(t_{\nu-1}) + \int_{t_{\nu-1}}^{t_\nu} y'(t)\,dt, \tag{2}$$

and replace y' in $[t_{\nu-1}, t_\nu]$ by some approximation which depends parametrically on values of y' at nodes $t_{n,\zeta}$. Identification of these values with the values of $f(t_{n,\zeta}, y_{n,\zeta})$ leads to a relation solely between values of y. This approach is often called collocation.

With initial value problems for (2), the best-known examples for this approach are the Adams-Bashforth and Adams-Moulton methods which replace y' in (2) by its interpolation polynomial. This approach provides a continuous approximation $\tilde{y}(t)$ for the solution.

Using automatic differentiation, Chang and Corliss [2] (and others) have replaced the derivative of the trajectory through $(y_{\nu-1}, t_{\nu-1})$ by its Taylor polynomial $T_{\nu-1,p}$ at $t_{\nu-1}$ of a given high degree $p-1$:

$$T_{\nu-1,p}(t) = y'_{\nu-1} + \sum_{j=1}^{p-1} \frac{1}{j!}\, y_{\nu-1}^{(j+1)}(t-t_{\nu-1})^j, \tag{3}$$

where the values of the vectors $y_{\nu-1}^{(j)}$ are calculated using automatic differentiation of (1).

Integrating the Adams-Bashforth method in (2) then leads to Taylor series methods (see [2]) with the continuous extension

$$\tilde{y}(t) = y_{\nu-1} + \sum_{j=1}^{p} y_{\nu-1,j}(t-t_{\nu-1})^j. \tag{4}$$

The coefficient vectors

$$F_{\nu,j}(y_\nu) \equiv\; y_{\nu,j} := \frac{1}{j!} y_\nu^{(j)} \in \mathbb{R}^n, \tag{5}$$

arise naturally in the automatic differentiation procedures so that these represent the simplest algorithmic form, cf. [2]. The vector-functions $F_{\nu,j} : \mathbb{R}^n \mapsto \mathbb{R}^n$ are uniquely defined by the right hand side f and the time t_ν.

Although the classical Taylor series approach permits the attainment of arbitrarily high orders p and of a step size h_ν which is close to the radius of convergence of the Taylor series (cf.[2]), its use for stiff systems is restricted by its lack of stability. The modulus of any fixed Taylor polynomial for $\exp(-t)$ goes to infinity with an increase of the step $h_\nu > 0$, contrary to the behaviour of the approximated exponential function.

To overcome this difficulty we consider a Hermite-type interpolation polynomial for y'. The availability of the values of y' and a number of its derivatives at $t_{\nu-1}$ and t_ν is required. The increment function of the forward step formula obtained after substitution of the polynomial into (2) and integration will depend on the $y_\nu^{(j)}$, and thus, via (1), on y_ν. This implicitness is at least for derivative-free methods a prerequisite for the suitability of the method for stiff systems.

The remainder of the paper we will be concerned only with one-step methods. We may therefore always consider the analytic solution running through the last solution $y_{\nu-1}$ so the initial value problem can be rewritten as

$$\begin{aligned} y'(t) &= f(y(t),t)\ , \quad y(t) \in \mathbb{R}^n \\ y(t_{\nu-1}) &= y_{\nu-1} \end{aligned} \tag{6}$$

Moreover, let $y_{\nu-1}^{(j)}$ and $y_\nu^{(j)}$, $j = 0, 1, 2, \ldots$, be the values ($\in \mathbb{R}^n$) of the j-th derivatives of the solution $y(t)$ of (6) at $t_{\nu-1}$ and t_ν, respectively :

$$y_\mu^{(j)} := \frac{d^j}{dt^j} y(t_\mu), \ \text{for } j \in \mathbb{N}, \ \text{and } \mu = \nu - 1, \ \nu \ . \tag{7}$$

Consider the unique polynomial $P_{pq}(s) \in \mathbb{P}^{p+q-1}$ which interpolates the function $g(s) \equiv y'(t_\nu + s \cdot h_\nu)$ and its first $p-1$ and $q-1$ derivatives at 0 and 1, respectively ($p, q \geq 1$): $P_{pq}^{(j-1)}(t) = g^{(j-1)}(t) = y_{\nu-1}^{(j)} h_\nu^{j-1}$. The Lagrangian representation of P_{pq} has the form

$$P_{pq}(s) = \sum_{j=1}^{p} l_j^{pq}(s) y_{\nu-1}^{(j)} h_\nu^{j-1} + \sum_{j=1}^{q} (-1)^{j-1} l_j^{qp}(1-s) y_\nu^{(j)} h_\nu^{j-1} \tag{8}$$

with Lagrangian basis polynomials $l_j^{pq} \in \mathbb{P}^{p+q-1}$, for $j = 1(1)p$. Now we substitute P_{pq} for g in

$$y(t_\nu) = y(t_{\nu-1}) + h_\nu \int_0^1 g(s)\, ds + \mathcal{O}(h_\nu^{p+q+1}) \tag{9}$$

the resulting *approximate* relation between $y(t_\nu)$ and $y(t_{\nu-1})$ defines our class of methods

$$\sum_{j=0}^{q} (-1)^j c_j^{qp} F_{\nu,j}(y(t_\nu)) h_\nu^j = \sum_{j=0}^{p} c_j^{pq} F_{\nu-1,j}(t_{\nu-1}) h_\nu^j + \mathcal{O}(h_\nu^{p+q+1}), \tag{10}$$

with

$$c_j^{pq} := j! \int_0^1 l_j^{pq}(s)\, ds. \tag{11}$$

Here, the $F_{\mu,j}$ are defined as in (5).

Neglecting the approximation error we define the next numerical solution point $y_\nu = y_{\nu,0}$ as solution to the implicit equation

$$\begin{aligned} H_\nu^{pq}(y_\nu) &= y_{\nu-1/2} \quad \text{with} \\ H_\nu^{pq}(y_\nu) &\equiv \sum_{j=0}^{q} c_j^{qp} F_{\nu,j}(y_\nu)(-h)^j \quad \text{and} \\ y_{\nu-1/2} &\equiv \sum_{j=0}^{p} c_j^{pq} F_{\nu-1,j}(y_{\nu-1})h^j \quad . \end{aligned} \tag{12}$$

This system of equations has the components of y_ν as unknowns since the initial vector $y_{\nu-1}$ is known. The right hand side vector $y_{\nu-1/2}$ in (12) may be interpreted as a match point of two modified Taylor expansions, one forward from $y_{\nu-1}$ and one backward from y_ν. Both expansions cover half the step length since we will see that roughly $c_j^{pq} \approx (0.5)^j$ when $p \approx q$, which holds for the most interesting schemes. The match point $y_{\nu-1/2}$ may be viewed as a mid-point between $y_{\nu-1}$ and y_ν. On stiff problems both expansions may look divergent, so that $y_{\nu-1/2}$ is likely to lie off the map and consideration needs to be taken with regards to scaling and numerical cancelation.

We now state two important facts about the methods (See also [10]).

Theorem 1. *For arbitrary nonnegative integers p and q, and for* $y \in C^{p+q+1}$ $[t_{\nu-1}, t_\nu]$, *the method defined by* (12) *has the order of consistency* $p+q$; *the order of the local error is* $p+q+1$. *Furthermore, the coefficients* (11) *of* (12) *are identical with the coefficients of the rational* (q,p)*-Padé approximation* $P_{pq}(z)$ *of* $\exp(z)$

$$\exp(z) = P_{pq}(z) + O(z^{p+q+1}) = \frac{\sum_{j=0}^{p} c_j^{pq}(z^j/j!)}{\sum_{j=0}^{q}(-1)^j c_j^{qp}(z^j/j!)} + O(z^{p+q+1}) \tag{13}$$

which are known to be (cf. e.g. [11])

$$c_j^{pq} = \binom{p}{j} \Big/ \binom{p+q}{j} = \frac{p!(p+q-j)!}{(p+q)!(p-j)!}. \tag{14}$$

Theorem 2. *The HOP methods defined by equation* (12) *are A-stable for* $q-2 \le p \le q$ *and L-stable for* $q-2 \le p \le q-1$.

Our primary interest in HOP-methods is for stiff systems, so we will use them in the range specified by Theorem 2.2 because A- and L-stability are desirable properties in this context. On the other hand, A- or L-stability is not sufficient in the case of stiff systems (1); cf. e.g. [4]. Since our methods use higher derivatives of the solution, further reaching stability concepts cannot immediately be applied to it.

With a nontrivial HOP method (i.e. with $q \geq 1$), the computation of the approximation y_ν at $t_\nu = t_{\nu-1} + h_\nu$ requires the solution of the nonlinear algebraic system

$$H_\nu^{qp}(y_\nu) \equiv \sum_{j=0}^{q} c_j^{qp} F_{\nu,j}(y_\nu)(-h)^j \;=\; y_{\nu-1/2}$$

with $y_{\nu-1/2}$ as defined in (12).

The Taylor coefficients $y_{\mu,j} = F_{\nu,j}(y_\nu)$ for $j \geq 1$ and $\mu = \nu - 1, \nu$ are computed through automatic differentiation of the right hand side $f(y,t)$ as described in Section 6. Because of the stiffness of (1), we must use a Newton-like method to solve (2) for y_ν. To this end we need to evaluate or approximate the Jacobian

$$R_\nu^{pq} \;\equiv\; [H_\nu^{pq}(y_\nu)]' \;=\; I + \sum_{j=1}^{q} c_j^{qp} F'_{\nu,j}(y_\nu)(-h)^j$$

In Sections 5 and 6 we will show show how the total derivatives

$$B_{\nu,j} \;\equiv\; \frac{dy_{\nu,j}}{dy_\nu} \;=\; F'_{\nu,j}(y_\nu) \;\in\; \mathbb{R}^{n\times n}$$

can be computed using automatic differentiation.

3 The Time-Varying Linear Case

Here, we consider the comparatively simple situation where $y'(t) = A(t)\,y(t) + b(t)$ and $y(t_{\nu-1}) = y_{\nu-1}$ for at least $q+1$ times differentiable coefficient functions $A : \mathbb{R} \mapsto \mathbb{R}^{n\times n}$ and $b : \mathbb{R} \mapsto \mathbb{R}^n$ In this setting, certain relations are easily derived and intuitively clear and the prediction $\tilde{y}_\nu = 0$ requires one Newton step to an explicit formula for the next point y_ν.

With the matrix $A_{\nu,j} \in \mathbb{R}^{n\times n}$ the j-th Taylor-coefficient of $A(t)$ at $t = t_\nu$ we have for given $q > 0$

$$A(t) \;=\; \sum_{j=0}^{q-1} A_{\nu,j}(t-t_\nu)^j + O(t-t_\nu)^q$$

and similarly with vectors $b_{\nu,j}$

$$b(t) \;=\; \sum_{j=0}^{q-1} b_{\nu,j}(t-t_\nu)^j + O(t-t_\nu)^q \;.$$

Substituting these expansions into the affine ODE and identifying coefficients for the powers of $(t - t_\nu)$ we obtain the recurrence

$$y_{\nu,j+1} \;=\; \frac{1}{j+1}\left[b_{\nu,j} + \sum_{i=0}^{j} A_{\nu,j-i}\, y_{\nu,i}\right] \qquad (15)$$

Differentiating with respect to $y_\nu = y_{\nu,0}$ we obtain

$$B_{\nu,j} \equiv \frac{d\,y_{\nu,j+1}}{d\,y_\nu} = \frac{1}{j+1}\left[A_{\nu,j} + \sum_{i=1}^{j} A_{\nu,j-i} B_{\nu,i-1}\right] \quad . \tag{16}$$

Now we can compute the Jacobian of the algebraic system (2) as the linear combination

$$R_\nu^{qp} \equiv [H_\nu^{qp}]' = \left[I + \sum_{j=1}^{q} (-1)^j h_\nu^j c_j^{qp} B_{\nu,j-1}\right] \quad , \tag{17}$$

which depends only on t_ν but not y_ν because of the assumed linearity. Provided this square matrix is nonsingular, we obtain the explicit solution.

Autonomous versus Time-dependent

The HOP methods are mathematically invariant with respect to the usual trick of appending the time as an extra component to the state vector. The resulting approximations $y_\nu \approx y(t_\nu)$ will be the same, provided the algebraic systems (12) is solved exactly at each step. However, from the point of numerical efficiency and accuracy there may be a significant difference. In particular, it is clear that linearity is always lost (except when the original system was already autonomous). Strictly speaking, this means that Newton's method no longer converges in one step. However, one can easily see that the Jacobian R_ν^{qp} of the extended system still has a special structure in that it is block triangular with the blocks in the diagonal being constant. On algebraic systems with such Jacobian Newton's method reaches the exact solution after a number of steps equal to the number of blocks in the diagonal. Here this number is two and if a predictor is at least smart enough to get the time component of the solution vector right a single correction step by Newton's method will already suffice. Even then there remains an inefficiency since the partial derivatives of the extended right-hand side with respect to the time-component are evaluated or approximated but then play absolutely no role in computing the full solution-vector. This observation applies not only to the HOP method considered here but also other implicit single- or multi-step methods. Therefore, we will continue to develop our method in the general, non-autonomous framework.

Solutions to Approximating Problems

Consider the situation where $A(t)$ and $b(t)$ are approximated by equally smooth paths $\tilde{A}_{\nu-1}(t)$ and $\tilde{b}_{\nu-1}(t)$ such that for $j = 0, 1, \ldots q-1$ we have $\partial^j/\partial t^j(A(t) - \tilde{A}_{\nu-1}(t))$ and $\partial^j/\partial t^j(b(t) - \tilde{b}_{\nu-1}(t))$ are $\mathcal{O}(t - t_{\nu-1})^{(q-j)}$. This condition is met with taylor interpolating polynomials. The perturbed ODE is biven by $\tilde{y}'(t) = \tilde{A}_{\nu-1}(t)\,y(t) + \tilde{b}_{\nu-1}(t)$ where $\tilde{y}(t_{\nu-1}) = y_{\nu-1}$. The transition from the original system to the approximate system effects the solution at time $t_\nu = t_{\nu-1} + h_\nu$ by a quantity of order $\mathcal{O}(h_\nu^{q+1})$.

4 Rational Predictors and the Derived ODE

The Obrechkoff methods require a good predictor so we derive a rational function of all available information. After convergence to the point $y_{\nu-1}$ we know not only the Taylor coefficients but also those of the matrix path and the vector path. The Jacobian is abbreviated as $J(t) \equiv (A(t), a(t)) = f'(y(t), t)$.

We will make extensive use of the **derived ODE** $z'(t) = A(t)\, z(t) + a(t)$ with $z_{\nu-1} = y'_{\nu-1}$ for the state derivative $z(t) \equiv y'(t)$. By definition the Taylor-coefficients satisfy the identity. $z_{\nu,j} = (j+1) y_{\nu,j+1}$. This is then substituted into (15).

Although the coefficient functions of the derived ODE are usually transcendental and unknown at general t, we can form approximations using their known Taylor coefficients. They are of order $\mathcal{O}(t - t_{\nu-1})^{(q-j)}$. Integrating the resulting ODE by the (q,p) HOP method we obtain an approximation of order h^{q+1} to $z_\nu = y_{\nu,1}$. The computation of this point involves according to (17) the forming and factoring of the Jacobian

$$\tilde{R}_\nu^{q,p} = \left[I + \sum_{j=1}^{q} (-1)^j h_\nu^j c_j^{qp} \tilde{B}_{\nu,j-1} \right] \quad . \tag{18}$$

Here, the $\tilde{B}_{\nu,j-1}$ are obtained according to (16) from the $\tilde{A}_{\nu,j}$ representing the Taylor coefficients of $\tilde{A}_{\nu-1}(t)$ at $t = t_\nu$. This so-called Taylor shift from $t = t_{\nu-1}$ to $t = t_\nu$ can be achieved by the extended Horner scheme, whose complexity is quadratic in q.

Applying the recurrence (15) for the derived ODE to the approximate z-quantities we then obtain $\tilde{z}_{\nu,j}$ of order h^{q+1-j} to $z_{\nu,j} = (j+1) y_{\nu,j+1}$ so that after rescaling for $j = 1, \ldots q+1$

$$\tilde{y}_{\nu,j} \equiv \frac{1}{j} \tilde{z}_{\nu,j-1} = y_{\nu,j} + \mathcal{O}(h^{q+2-j}) \quad . \tag{19}$$

Thus we have obtained fairly high-order approximations to the derivatives of the analytical solution $y(t)$ of (6) at $t = t_\nu$. Substituting these into the desired relation (10) we finally obtain the actual predictor value

$$\tilde{y}_\nu \equiv \sum_{j=0}^{p} h_\nu^j c_j^{pq} y_{\nu-1,j} - \sum_{j=1}^{q} (-1)^j h_\nu^j c_j^{qp} \tilde{y}_{\nu,j} \; .$$

Combining the analysis in the section we can formulate the following result:

Theorem 4.1: For sufficiently smooth f and $p > 0$, the predicted value satisfies

$$\tilde{y}_\nu = y(t_\nu) + O(h_\nu^{q+2}) \; .$$

The order of this predictor is quite satisfactory since squaring the residual by one Newton correction will theoretically reduce it to order h_ν^{2q+4}. This error is at least h_ν^3 times smaller than the local discretization error $\mathcal{O}(h_\nu^{q+p+1})$ since $p \leq q$ for the A-stable schemes. The same effect can still be achieved if the correction step is done with an order h_ν^p approximation of the Jacobian R_ν^{qp} of the algebraic system. This quasi-Newton approach may save the effort to first evaluate the A_j, then form the B_j, and finally factor the resulting approximation $\tilde{R}_\nu^{qp}$ at the predicted point $\tilde{y}_\nu$. However, there are alternatives with similarly low complexities, which we will discuss in Section 5.

The mere order of the predictor discussed above is not the main consideration, since it could also be achieved with polynomial expansion or interpolation. The key advantage is that the predictor is rational in the function and derivative data due to the inversion of the matrix R_ν^{qp} for the derived ODE. To highlight this property we note the following interesting fact.

Corollary 4.2: If the original ODE is affine with $A(t) = A_0$ constant and $b(t)$ a polynomial of degree at most $q + 1$, then the predicted value $\tilde{y}_\nu$ is equal to the discrete solution y_ν.

This result cannot be generalized to cases where $A(t)$ is non-constant. Our numerical experiments with the rational predictor were quite good. A central question remains whether a polynomial approximation to the Jacobian path $A(t)$ makes sense in the nonlinear situation where it is likely to be as *stiff* as the underlying solution $y(t)$. There is some reason for optimism since the Taylor coefficients of $A(t)$ do not enter directly into the predictor but as rational function with the denominator having a higher rank than the numerator. We are currently investigating the idea of defining $\tilde{A}_{\nu-1}(t)$ as Padé approximation to $A(t)$ in the non-commutative algebra of square matrices. The vector function $a(t)$ can be approximated as though it was the last row in the Jacobian of the corresponding autonomous system.

5 Error Estimation and Correction

Error Prediction and Estimation

The information available at the old point $y_{\nu-1}$ yields an approximation $\tilde{y}_{\nu,q+1} = y_{\nu,q+1} + O(h_\nu)$, which we have not used. Thus we can compute the predicted residual $r_\nu^{p+1,q+1}$ of the $(p+1, q+1)$ scheme at the predicted point $\tilde{y}_\nu$. Now we need an approximation $\tilde{R}_\nu$ to the Jacobian $R_\nu^{q+1,p+1}$ of this higher order scheme to compute the Newton-step $\tilde{R}_\nu^{-1}\tilde{r}_\nu^{p+1,q+1}$ as an estimate for the discrepancy between $\tilde{y}_\nu$ and the actual solution. For simplicity one may want to use simply the matrix $\tilde{R}_\nu$ as defined by the residual, which has already been formed and factorized. On very stiff problems this may not be a good idea since $R_\nu^{q+1,p+1}$ contains the matrix $B_{\nu,q}$ which has the leading term $A_{\nu,0}^q$ whereas $B_{\nu,q-1}$ is only of order $A_{\nu,0}^{q-1}$. In the linear constant coefficient case this is exactly the case since all $A_{\nu,j}$ with $j > 0$ vanish. In general it may make sense to compute $\tilde{B}_{\nu,q}$ with $\tilde{A}_{\nu,q}$ set to zero since $A_{\nu-1,q}$ is not known either. The resulting $R_\nu^{q+1,p+1}$ will then be exact for constant coefficient problems so that the a priori error estimate discussed so far will coincide with the a posteriori error estimate to be discussed later.

After convergence to y_ν has been achieved with satisfactory accuracy the higher order residual can be computed with actual information in the sense that the $\tilde{y}_{\nu,j}$ are replaced by the $F_{\nu,j}(y_\nu)$. In many cases we will then also have a new matrix R_ν^{pq} formed and factored, which we may use to precondition the residual. Again, it may be necessary to include the highest term $B_{\nu,q}$ or an approximation and therefore obtain a satisfactory pre-conditioner for the residual.

If the predicted point is rather poor, y_ν^{q+2} can be inaccurate and therefore cause the $(R_\nu^{q+1,p+1})^{-1} r^{q+1,p+1}$ to be a bad local error estimate. To avoid this phenomenon, the higher Newton-step $(R_\nu^{q,p+1})^{-1} r^{q,p+1}$ is preferred because no extra coefficients need to be recalculated, but $(R_\nu^{q,p+1})^{-1}$ and $r^{q,p+1}$ do need to be reformed.

Computing a Newton-Step

Now suppose, we have settled on a suitable step-size h_ν and computed a predictor $\tilde{y}_\nu$. In order to perform a Newton correction we need the Jacobian $[H_\nu^{qp}]'$ which was found to be equal to R_ν^{qp} as defined in (17) in the linear time-varying case. In contrast to (15) that identity is still true in the general case for the following reason.

Given any smooth curve t

$$y(t) \equiv \sum_{j=0}^{d-1} y_j (t-t_0)^j + O((t-t_0)^d \in C^d(\mathbb{R}, \mathbb{R}^n), \tag{20}$$

one obtains a smooth image

$$z(t) \equiv f(y(t),t) \;=\; \sum_{j=0}^{d-1} z_j (t-t_0)^j + O(t-t_0)^d, \tag{21}$$

provided the underlying right hand side

$$f \;:\; \mathbb{R}^m \mapsto \mathbb{R}^n$$

is sufficiently smooth. Since we have not yet imposed any differential relationship between $y(t)$ and $z(t)$ we may consider the coefficient vectors y_j as independent variables and obtain the resulting coefficient vectors z_j as functions

$$z_j \;=\; z_j(y_0, y_1, \ldots, y_j) \quad \text{for} \quad j = 0, \ldots, d-1 \quad .$$

The partial derivatives of the z_j with respect to the y_i for $0 \leq i \leq j$ are $n \times n$ matrices with the following remarkable invariance property.

Lemma 1 *If f is d times continuously differentiable on some neighborhood of a point $(y_0, t_0) \in \mathbb{R}^n$, then for all $0 \leq i \leq j < d$, $\partial z_j / \partial y_i = A_{j-i}$, with A_i the i-th Taylor coefficient of the Jacobian of $f'(y(t),t)$ at $t = t_0$, i.e.*

$$f'(y(t),t) = \sum_{i=0}^{d-1} A_i (t-t_0)^i + O(t-t_0)^d. \tag{22}$$

This result was already established in [1] for the autonomous case where f does not explicitly depend on t. In the previous two sections we have always viewed the matrices A_j as Taylor coefficients of the Jacobian $A(t) = f'(y(t), t)$ and used them in that capacity for the prediction process. Now we have verified that they represent also the Jacobians of the Taylor coefficient vectors z_j and y_i with respect to each other. In that role we can use them for the correction process. Imposing now the relation

$$z(t) = y'(t) \quad \Rightarrow \quad y_{j+1} = \frac{1}{j+1} z_j$$

it follows by the chain rule that the total derivatives $B_j = dy_{j+1}/dy_0$ still satisfy the recurrence (16) and the Jacobian of the nonlinear system is still given by

$$R_\nu^{qp} = \left[I + \sum_{j=1}^{q} (-1)^j h_\nu^j c_j^{qp} B_{\nu,j-1} \right] \quad , \tag{23}$$

A straight forward application of Newton's method would require the evaluation of the A_j, the computation of the B_j, the formation of the Jacobian R_ν and its factorization at the predictor and each subsequent iterate. The corresponding matrices at the finally accepted iterate would then be used for the predictor over the next step. Except possibly in the time-varying linear case, we must expect that the effort for one such composite matrix evaluation is much bigger than that for recursively evaluating the Taylor-coefficients $\tilde{y}_{\nu,j} = F_{\nu,j}(\tilde{y}_\nu)$ at an iterate $\tilde{y}_\nu$. Hence one would certainly want to perform no more than one composite Jacobian evaluation per time-step.

Outline of Algorithm

If the predictor works well, one can assume that the correction step(s) are much smaller than the prediction step. Therefore we recommend that the Jacobians be evaluated and factorized at the predicted point rather than the main iterates. This means that the correction is done with up-to-date information, whereas the prediction is done with information from the previously predicted point. Below is an algorithm for the transition from the given point $y_{\nu-1}$ to the next y_ν.

1. Select a trial step size h_ν, calculate $y_{\nu-1/2}$ as in (12) and for $j \leq q$ compute $A_{\nu,j}$ and $b_{\nu,j}$ as Taylor-shifts of the available $A_{\nu-1,j}$ and $b_{\nu-1,j}$.
2. Compute the $B_{\nu,j}$ from the $A_{\nu,j}$ by (16), form their linear combination R_ν according to (17), and factor it.
3. Compute the predicted point $\tilde{y}_\nu$ as in (4) with its Taylor-coefficients $\tilde{y}_{\nu,j}$ for $j \leq q+1$. If R_ν is singular or $\tilde{y}_\nu$ unsuitable restart with smaller h_ν.
4. Evaluate $\tilde{y}_{\nu,j} = F_{\nu,j}(\tilde{y}_\nu)$ and compute the residual $r_\nu = H_\nu(\tilde{y}_{\nu-1}) - y_{\nu-1/2}$. If $\tilde{y}_\nu$ is the predictor point evaluate A_j's, compute B_j's, form and factor R_ν.
5. Calculate $\Delta y_\nu \equiv R_\nu^{-1} r_\nu$ and depending on the size of its norm
   a: terminate iteration and skip to 6. if $\|\Delta y_\nu\|$ is small

b: correct $y_\nu = \tilde{y}_\nu - \Delta y_\nu$ and goto 4. if $||\Delta y_\nu||$ is medium
c: restart step with smaller h_ν if $||\Delta_\nu||$ is large

6. Perform a posteriori discretization error estimate and start new step $h_\nu = h_{\nu-1}(K * E/T)^{1/p+q+1}$ where K=constant, E = local error, T = tolerance, and go to 3. if acceptable and otherwise restart old step with reduced h.

6 Numerical Results

The two test problems, Robertson's Problem and a linear time dependent ODE, were compared to Radau5 ($s = 3$) and Radaup for higher orders ($s = 5, s = 7$). Robertsons problem,a prominent of stiff literature, was integrated at the transient phase $t = [0.0, 0.1]$.

$$\left.\begin{array}{l} y_1' = -0.04y_1 + 10^4 y_2 y_3 \\ y_2' = 0.04y_1 - 10^4 y_2 y_3 - 3 * 10^7 y_2 y_2 \\ y_3' = 3 * 10^7 y_2 y_2 \end{array}\right\} \quad \text{with} \quad \left\{\begin{array}{l} y_1(0) = 1 \\ y_2(0) = 0 \\ y_3(0) = 0 \end{array}\right.$$

and $h_0 = 2.0 * 10^{-3}$. The tolerance depends on the order of the method; Tol $= 10^{-(2s)} = 10^{-(q+p+1)}$. The table 1 represents the number of steps (NS), function evaluations (NF), Jacobian evaluations (NJ), and rejected steps (rej) for the HOP and IRK methods at orders of (3,2), (5,4), (7,6).

method	order	NS	NF	NJ	rej
HOP	(3,2)	6	8	7	0
IRK	(3,2)	7	76	7	0
HOP	(5,4)	13	22	16	1
IRK	(5,4)	14	209	10	0
HOP	(7,6)	18	37	23	2
IRK	(7,6)	23	450	13	0

The table shows that the HOP method requires fewer steps and fewer functions evaluations but more Jacobian evaluations. The HOP method takes larger steps over the transient phase $t = [0, 0.02]$, but thereafter the IRK integrates over larger steps. This is due to the extrapolation technique used in the HOP-methods in order to the predict the Jacobian at the next grid point. This explicit technique causes the (q,p) HOP method to integrate over a restricted step size. A rational approximation of the Jacobian at the next point should hinder this step size restriction. The second test problem is a linear time varying ODE.

$$\left.\begin{array}{l} y_1' = .2(y_2 - y_1) \\ y_2' = 10y_1 - (60 - 0.125y_3)y_2 + .125y_3 \\ y_3' = 1 \end{array}\right\} \quad \text{with} \quad \left\{\begin{array}{l} y_1(0) = 0 \\ y_2(0) = 0 \\ y_3(0) = 0 \end{array}\right.$$

and $h_0 = 1.0 * 10^{-4}$. Time has been represented as y_3, and therefore $y_3' = 1$. Table 2 displays the results of the number of steps, function evaluations, Jacobian evaluations and rejected steps.

method	order	NS	NF	NJ	rej
HOP	(3,2)	6	8	7	0
IRK	(3,2)	7	76	7	0
HOP	(5,4)	13	22	16	1
IRK	(5,4)	14	209	10	0
HOP	(7,6)	18	37	23	2
IRK	(7,6)	23	450	13	0

Due to the linear time dependency, the extrapolation technique was not nearly as restricting as in the Robertson problem. The HOP-methods take fewer steps than the IRK (Radaup) method.

References

1. Christianson, B.: Reverse accumulation and accurate rounding error estimates for Taylor series coefficients. Technical Report No.239, Hatfield Polytechnic, July 1991
2. Chang, Y.F., Corliss, G.: Solving ordinary differential equations using Taylor series. ACM Trans. Math. Software **8** (1982) 114–144
3. Hairer, E., Norsett, S.P., Wanner, G.: Solving Ordinary Differential Equations I. Springer-Verlag, Berlin, 1987
4. Hairer, E., Wanner, G.: Solving Ordinary Differential Equations II. Springer-Verlag, Berlin, 1991
5. Kirlinger, G., Corliss, G.F.: On Implicit Taylor Series Methods for Stiff ODEs. in Computer Arithmetic and Enclosure Methods, ed. L. Atanassova and J. Herzberger, North Holland, Amsterdam, London, New York, Tokyo, (1992) 371-380
6. Lambert, J.D.: Computational Methods in Ordinary Differential Equations. John Wiley & Sons, London, New York, Sydney, Toronto, 1973
7. Milne, W.E.: A note on the numerical integration of differential equations. J. Res. Nat. Bur. Standards **43** (1949) 537-542
8. Milne, W.E.: Numerical Solution of Differential Equations. John Wiley & Sons, London, New York, Sydney, Toronto, 1953
9. Obrechkoff, N.: Neue Quadraturformeln. Abh. Preuss. Akad. Wiss. Math. Nat. Kl.**4** (1940)
10. Obrechkoff, N.: Sur le quadrature mecaniques (Bulgarian, French summary). Spisanie Bulgar. Akad. Nauk. **65** (1942) 191-289
11. Padé, H.: Sur la représentation approachée d'une fonction par des fractions rationelles. Thesis, Ann. de l'Éc. Nor. (3) **9** (1892)
12. Wanner, G.: On the integration of stiff differential equations. Technical Report, October 1976, Université de Genéve Section de Mathematique, 1211 GENEVE 24th, Suisse
13. Wanner, G.: STIFFI, A Program for Ordinary Differential Equations. Technical Report, October 1976, Université de Genéve Section de Mathematique, 1211 GENEVE 24th, Suisse

The Use of Discrete Sine Transform in Computations with Toeplitz Matrices

Fabio Di Benedetto

Dipartimento di Matematica, Università di Genova.

Abstract. The $\mathcal{T}$ algebra, related to the discrete sine transform, is an efficient tool for approximating Toeplitz matrices arising in image processing. We present two applications concerning the computation of singular values and the preconditioning of least squares problems.

1 Introduction

In several fields of image restoration (e.g. in presence of distortion by the atmospheric turbulence or the refraction generated by a lens) it is necessary to solve a linear system of the form

$$Tx = b, \tag{1}$$

which is the discretization of a Fredholm integral equation of the first kind. The usual ill-posedness of the continuous problem makes the system (1) heavily ill-conditioned and very sensitive to the (noise) errors affecting b.

In order to ensure the reliability of the computed solution, a standard approach is based on the definition of a *regularization operator* R_μ, depending on the parameter μ, such that $\lim_{\mu\to 0} R_\mu b = x$ and R_μ is uniformly bounded with respect to the dimension.

Instead of the true solution x, one computes the approximation $x_\mu = R_\mu b$; μ should be chosen in order to make x_μ sufficiently close to x, and so that computing x_μ becomes a well-conditioned problem.

Several techniques of regularization have been proposed in literature: among all we recall truncated singular value decomposition (SVD: μ represents the threshold parameter of truncation), iterative methods (μ is related to the number of iterations) and Tikhonov regularization, where $R_\mu b$ solves the minimization problem

$$\min_x \left(\|Tx-b\|_2^2 + \mu^2\|Lx\|_2^2\right) \quad ; \tag{2}$$

L usually discretizes a $2s$-th order differentiation operator.

The explicit or approximate knowledge of the singular values of T is of crucial importance in the choice of the parameter μ for many regularization techniques; moreover, the large size of the problem (2) requires the use of fast iterative methods in order to solve it.

In this paper we consider the case where the matrix T has the *Toeplitz structure*, $T = (t_{j-i})_{i,j=1}^n$: this arises when the continuous integral operator has a displacement kernel. In the context of 2-dimensional image processing, the blocks

$\{t_k\}$ are $p \times p$ Toeplitz matrices themselves: in this case we will call T a *2-level Toeplitz matrix*.

We review in Sect. 2 the properties of Toeplitz matrices and of the algebra $\mathcal{T}$ related to the discrete sine trasform, which has two main applications: computation and/or estimation of the singular values of a Toeplitz matrix (presented in Sect. 3), and preconditioning of the least squares problem (2) (recalled in Sect. 4).

2 Toeplitz and $\mathcal{T}$ Matrices

It is often useful to associate a given $n \times n$ Toeplitz matrix T to a suitable function $f(z)$ defined on the unit circle S^1.

Definition 1. Let $T = (t_{j-i})_{i,j=1}^n$ with $\sum_{k=-\infty}^{+\infty} |t_k| < +\infty$; we say that T is *generated* by $f(z) : S^1 \to \mathbb{C}$ if the Laurent expansion of $f(z)$ defines the entries of T:

$$f(z) = \sum_{k=-\infty}^{+\infty} t_k z^k \qquad \forall z \in S^1.$$

In this case, we will use the notation $T = T_n(f)$ or simply $T(f)$.

Two important examples are given by *banded Toeplitz matrices* for which $f(z)$ is a finite sum, whose length is related to the bandwidth of T, and by *rational Toeplitz matrices*, that are generated by a rational function with respect to z.

In the 2-level case, each block of T equals $t_k = (t_{k,r-s})_{r,s=1}^p$; under the assumption $\sum_{k,h \in \mathbb{Z}} |t_{k,h}| < +\infty$, we construct the Laurent expansion

$$\phi(z, w) = \sum_{k,h \in \mathbb{Z}} t_{k,h} z^k w^h$$

with respect to the variables $z, w \in S^1$ and we use the notation $T = T[\phi]$.

A powerful tool for the numerical treatment of symmetric Toeplitz matrices is represented by the so-called $\mathcal{T}$ algebra, firstly introduced by Bini and Capovani [1] and later independently rediscovered in other works [9, 2, 4].

Theorem 2. *Let $\tau = (\tau_{i,j})_{i,j=1}^n$ be a $n \times n$ symmetric matrix. The following conditions are equivalent:*

1. *τ satisfies the* cross sum rule, *that is*

$$\tau_{i,j+1} + \tau_{i,j-1} = \tau_{i+1,j} + \tau_{i-1,j} \quad \textit{for } 1 \le i, j \le n,$$

where $\tau_{i,0} = \tau_{0,j} = \tau_{n+1,j} = \tau_{i,n+1} = 0$.

2. *τ is diagonalized by the discrete sine transform:*

$$E_n = \left(\sqrt{\frac{2}{n+1}} \sin \frac{\pi i j}{n+1} \right)_{i,j=1}^n \Longrightarrow E_n \tau E_n = \Lambda_T,$$

where Λ_T is diagonal and E_n is symmetric and orthogonal.

3. There exists a symmetric Toeplitz matrix $T = (t_{j-i})_{i,j=1}^{n}$ *such that*

$$\tau = T - H, \quad H = \begin{pmatrix} t_2 & \dots t_{n-1} & 0 & 0 \\ \vdots & \cdot^{\cdot^{\cdot}} & & 0 \\ t_{n-1} & & 0 & t_{n-1} \\ 0 & & \cdot^{\cdot^{\cdot}} & \vdots \\ 0 & 0 & t_{n-1} \dots & t_2 \end{pmatrix} . \qquad (3)$$

Definition 3. $\mathcal{T}$ is the set of all matrices τ satisfying one of the conditions 1, 2 or 3 of Theorem 2.

It follows immediately from condition 2 that $\mathcal{T}$ is a matrix algebra; the diagonalization of $\tau \in \mathcal{T}$ can be performed in $O(\log n)$ parallel steps with $O(n)$ processors through the fast sine transform.

Moreover, it is evident that if $T(f)$ is a given symmetric Toeplitz matrix generated by $f(z)$, then the relation (3) defines a corresponding $\tau(f) \in \mathcal{T}$. If f is positive or $T(f)$ is banded, for large n $T(f)$ is positive definite iff $\tau(f)$ is.

The 2-level extension $\mathcal{T}^{(2)}$ of the class $\mathcal{T}$ leads to an algebra of quadrantally symmetric matrices: that is, the matrices are symmetric at the block level and each inner block is symmetric itself.

More precisely, if $E_{n,p}$ is the Kronecker product $E_n \otimes E_p$, involving the discrete sine transform in two dimensions, the new definition is

$$\mathcal{T}^{(2)} = \{P \in \mathbb{R}^{np \times np} : E_{n,p}\, P\, E_{n,p} \text{ is diagonal}\} .$$

Now, the diagonalization in $\mathcal{T}^{(2)}$ can be solved with $O(np\log(np))$ operations or $O(\log n + \log p)$ parallel steps and $O(np)$ processors.

A matrix $\tau = (\tau_{i,j})_{i,j=1}^{n} \in \mathcal{T}^{(2)}$ satisfies the cross sum rule at the block level; in addition, every inner block $\tau_{i,j}$ belongs to $\mathcal{T}$.

Finally, if $T = T[\phi]$ is a quadrantally symmetric 2-level Toeplitz matrix, a natural approximation $\tau[\phi] \in \mathcal{T}^{(2)}$ can be constructed as follows:

- for every block t_k of T, compute the approximation $\tau_k \in \mathcal{T}$ by using (3);
- define the block Toeplitz matrix T_τ whose blocks are the τ_k's;
- $\tau[\phi]$ is defined as a block correction of T_τ like (3), with the t_k's replaced by the τ_k's.

Most of our results are based on the following theorem [5] investigating the structure of the product of two Toeplitz matrices.

Theorem 4. *Let* $T(f) \in \mathbb{R}^{m \times n}$ *and* $T(g) \in \mathbb{R}^{l \times m}$ *be generated by the functions* $f(z)$ *and* $g(z)$*. If* $\epsilon > 0$*, then there exists* M *only depending on* ϵ, f *and* g *such that, for sufficiently large* l, m, n*, matrices* E', E'' *can be found in order to verify the following conditions:*

1. $T(g)T(f) - T(g\,f) = E' + E''$ *where* $T(g\,f) \in \mathbb{R}^{l \times n}$*;*
2. $\|E''\|_2 \le \epsilon$*;*

3. $\mathrm{rk}(E') \leq M$, *where* $\mathrm{rk}(X)$ *denotes the rank of the matrix* X.

In the special case where one of $T(f)$ and $T(g)$ is banded with maximal bandwidth independent on the dimensions, M is related to this bandwidth and we can choose $E'' = 0$. Moreover, block and 2-level versions of the previous theorem are also valid; see [5] for all details.

3 Eigenvalue and Singular Value Computation

In order to compute the eigenvalues of a symmetric Toeplitz matrix T, an efficient method consists of retrieving the spectrum of T from the eigensystem of suitable $\mathcal{T}$ matrices, through the application of an updating procedure.

The idea is very simple in the banded case, but it can be also applied to rational Toeplitz matrices: assume $T = T(f)$ with

$$f(z) = \frac{A(z)}{B(z)} + \frac{C(z^{-1})}{D(z^{-1})}; \tag{4}$$

A, B, C and D are polynomials in z, and in the symmetric case we have $C(z) = A(z)$ and $D(z) = B(z)$.

A classical result due to Dickinson [8] states the relation

$$T = T(B)^{-1}T(A) + T(C)^{*}T(D)^{-*}, \tag{5}$$

where the Toeplitz matrices arising in the right-hand side are banded and upper triangular ($*$ denotes the conjugate transpose).

An eigenvalue λ of the $n \times n$ symmetric matrix T solves the determinant of

$$T - \lambda I = T(B)^{-1}\left[T(A)T(\tilde{B}) + T(B)T(\tilde{A}) - \lambda T(B)T(\tilde{B})\right]T(\tilde{B})^{-1}, \tag{6}$$

where $\tilde{A}$ and $\tilde{B}$ are abbreviations of $A(z^{-1})$ and $B(z^{-1})$, respectively.

The bracketed expression of (6) differs from a Toeplitz matrix by a number of elements independent of n, in view of Theorem 4: the result is of the form

$$T(\alpha) - \lambda T(\mu) + E_\alpha - \lambda E_\mu,$$

where $T(\alpha)$ and $T(\mu)$ are banded Toeplitz matrices generated by the functions

$$\alpha(z) = A(z)B(z^{-1}) + A(z^{-1})B(z), \quad \mu(z) = B(z)B(z^{-1}),$$

and E_α, E_μ are shown in [7] to have the only nonzero entries in the $K^* \times K^*$ south-east corner, where $K^* := \max(\deg A, \deg B)$.

Moreover, for both $T(\alpha)$ and $T(\mu)$ the approximations in $\mathcal{T}$ can be constructed: the corrections H_α and H_μ can add new nonzero elements to E_α and E_μ only in the north-west corners of size $K^* - 1$.

Applying a congruence relation to (6) yields that λ solves the perturbed generalized problem

$$(\tau(\alpha) + E_1)y = \lambda(\tau(\mu) + E_2)y, \tag{7}$$

where the rank of $E_1 = E_\alpha + H_\alpha$ and $E_2 = E_\mu + H_\mu$ is not greater than $2K^* - 1$.

Once the vector y in (7) is known, the original eigenvector u of T relative to λ can be recovered through the relation $u = T(\tilde{B})y$.

It is now possible to compute a common expansion of E_1 and E_2 as sums of rank-one terms, of the form

$$E_1 = \sum_{k=1}^{K} \delta_\alpha^k q_k q_k^*, \quad E_2 = \sum_{k=1}^{K} \delta_\mu^k q_k q_k^*; \tag{8}$$

for example, if E_1 and E_2 were simultaneously diagonalizable, then the length K would be equal to $2K^* - 1$, the vectors q_k and the scalar pairs $(\delta_\alpha^k, \delta_\mu^k)$ being recovered from the generalized eigenpairs associated to the pencil (E_1, E_2). In the general case, the expansion (8) is redundant with respect to the rank of E_1 and E_2, but the length K can always be made less than $3K^*$.

The computation of the eigenvalues of T is based on the transition from the simple problem

$$\tau(\alpha)y = \lambda\tau(\mu)y,$$

whose explicit solution is known thanks to Theorem 2, point 2, to the perturbed problem (7) through a sequence of K successive rank-one updates, each of them involving the k-th terms in (8).

A suitable re-ordering in the expansion (8) allows to obtain a sequence of pencils (A_k, M_k) such that

1. $A_0 = \tau(\alpha)$, $M_0 = \tau(\mu)$;
2. every M_k is symmetric positive definite;
3. $A_k = A_{k-1} + \delta_\alpha^k q_k q_k^*$, $M_k = M_{k-1} + \delta_\mu^k q_k q_k^*$;
4. $A_K = \tau(\alpha) + E_1$, $M_K = \tau(\mu) + E_2$.

The diagonalization of each intermediate pencil is performed by using the algebraic properties of the rank-one update of the generalized eigenvalue problem: the topic is studied in detail in [7], where the classical results concerning the standard eigenvalue problem are generalized. Here we summarize the most important conclusions.

Definition 5. Let $\Lambda_\alpha = \mathrm{diag}\{\alpha_1, \ldots, \alpha_n\}$, $\Lambda_\mu = \mathrm{diag}\{\mu_1, \ldots, \mu_n\}$, $\lambda_i = \alpha_i/\mu_i$ and $z = (\zeta_1 \; \ldots \; \zeta_n)^*$. The problem

$$(\Lambda_\alpha + \alpha zz^*)x = \lambda(\Lambda_\mu + \mu zz^*)x, \quad x \neq 0 \tag{9}$$

is *non-defective* if the following conditions hold:

1. $\zeta_i \neq 0$ for each $i = 1, \ldots, n$;
2. $\lambda_i \neq \lambda_j$ if $i \neq j$;
3. $\alpha \neq \mu\lambda_i$ for each i.

Every generalized updated problem can be deflated, in order to get a non-defective problem of reduced dimension (see [7] for details).

Theorem 6. *If Definition 5 applies, then every eigenvalue of (9) solves the* generalized secular equation

$$w(\lambda) = 1 + (\alpha - \lambda\mu)\sum_{j=1}^{n} \frac{\zeta_j^2}{\alpha_j - \lambda\mu_j} = 0;$$

a set of separators consists of the values $\{\lambda_i\}$ and, if $\mu > 0$, of the ratio α/μ.

Moreover, an eigenvector x associated to the eigenvalue λ is given by $x = (\Lambda_\alpha - \lambda\Lambda_\mu)^{-1}z$.

The previous theorem allows us to solve a generalized non-defective problem by searching the roots of $w(\lambda)$; knowledge of the separators is useful for any rootfinding procedure.

The approach described in this section can be easily applied to the block or 2-level case, even though the number of updating steps depends on the dimension.

A further application involves the singular value decomposition of a nonsymmetric $m \times n$ matrix $T(f)$.

Theorem 7. *Let σ be a singular value of $T(f)$, with $f(z)$ defined as in (4), and let $u \in \mathbb{R}^n$ be its associated right singular vector. Define*

$$\alpha(z) = A(z)D(z^{-1}) + B(z)C(z^{-1}), \quad \mu(z) = B(z)D(z^{-1});$$

if $\tilde{f}$ is the abbreviation of $f(z^{-1})$ for every function f, there exist $n \times n$ matrices E_1, E_2 such that $\mathrm{rk}(E_1), \mathrm{rk}(E_2) \leq M$ only depending on f and the relation

$$(\tau(\alpha\tilde{\alpha}) + E_1)y = \sigma^2(\tau(\mu\tilde{\mu}) + E_2)y \tag{10}$$

holds with $u = T_n(\tilde{D})T_n(B)y$.

Proof. A first application of Theorem 4 to the expression (5) gives

$$T(f) = T_m(B)^{-1}T(\alpha)T_n(D)^* + \hat{E}, \tag{11}$$

the rank of $\hat{E}$ being independent of the size of $T(f)$. Since $T_m(B)$ is upper triangular, $T_m(B)^{-1} = T_m(1/B)$, while $T(\alpha)$ is a banded $m \times n$ Toeplitz matrix; applying again Theorem 4 twice, we get

$$T_m(B)^{-1}T(\alpha) = T(\alpha/B) + E_l, \; T(\alpha)T_n(B)^{-1} = T(\alpha/B) + E_r,$$

whence $T_m(B)^{-1}T(\alpha) = T(\alpha)T_n(B)^{-1} + \tilde{E}$ with $\tilde{E}$ having low rank. Substituting in (11) and considering $R = T(f)^*T(f)$ yields

$$R = T_n(D)^{-1}T_n(B)^{-*}[T(\alpha)^*T(\alpha) + E_0]T_n(B)^{-1}T_n(D)^{-*};$$

since σ and u satisfy the relation $Ru = \sigma^2 u$, by applying the definition of y given in the statement and the above expression of R we have

$$(T(\alpha)^*T(\alpha) + E_0)y = \sigma^2 T_n(B)^*T_n(D)T_n(D)^*T_n(B)y \; . \tag{12}$$

We further invoke Theorem 4 to the Toeplitz products appearing in (12) and we approximate in the $\mathcal{T}$ class the resulting band Toeplitz matrices, obtaining the desired relation (10) with two low-rank corrections E_1 and E_2. □

Hence, the computation of the singular values too can be reduced to a suitable updating procedure as we have seen before.

We just mention here that the eigenvalues of $\tau(f)$ are good approximations for the spectrum of the symmetric Toeplitz matrix $T(f)$.

For example, the eigenvalues of $T(f)$ and $\tau(f)$ are equidistributed (the result is generalizable to the block case: see [11]), this can be exploited in order to derive fine estimates of the extremal eigenvalues of $T(f)$ [10].

Such informations could be very useful in order to design effective regularization strategies for the given problem, without performing a large amount of computations.

4 Iterative Solution of a Regularized Problem

When Tikhonov regularization is used, the least-squares problem (2) can be rewritten as

$$\min \|\hat{T}x - \hat{b}\|_2, \quad \hat{T} = \begin{pmatrix} T \\ \mu L \end{pmatrix}, \; \hat{b} = \begin{pmatrix} b \\ 0 \end{pmatrix} \; . \tag{13}$$

The normal equations associated to (13) have the form

$$(T^*T + \mu^2 L^* L)x = T^* b; \tag{14}$$

the matrix $R := T^*T + \mu^2 L^* L$ can be related, thanks to Theorem 4, to a Toeplitz matrix.

For instance, in the bidimensional case we have $T = T[\phi]$ and $L = T[\delta_s]$, where $\delta_s(z,w) = d_s(z) + d_s(w)$ with $d_s(z) = (2 - z - z^{-1})^s$; Theorem 4 implies

$$R = T[|\phi|^2 + \mu^2 |\delta_s|^2] + E' + E'' \; .$$

It turns out that $P := \tau[|\phi|^2 + \mu^2|\delta_s|^2] \in \mathcal{T}^{(2)}$ is positive definite and can be chosen as an effective preconditioner for the solution of (14) by a Conjugate Gradient method; in fact, the following clustering theorem holds.

Theorem 8. *For every $\epsilon > 0$, the spectrum $\sigma(\Delta)$ of the matrix $\Delta = R - P \in \mathbb{R}^{np \times np}$ satisfies the property*

$$\#\{\lambda \in \sigma(\Delta) : \; |\lambda| \geq \epsilon\} = O(n + p) \; .$$

In the one-dimensional case and in many practical situations the above number of outliers reduces to $O(1)$, resulting in a superlinear rate of convergence. The definition of P we have considered is comparable (and in some cases outperforms) to the alternative choice of P among circulant matrices [3]. Some numerical examples may be found in [5, 6].

The preconditioning step is performed through the diagonalization of P, yielding a computational cost per iteration of $O(np \log(n + p))$ operations; in a parallel model of computation, the time complexity is logarithmic with respect to the size of the problem.

References

1. Bini, D., Capovani, M.: Spectral and computational properties of band symmetric Toeplitz matrices. Linear Algebra Appl. **52** (1983) 99–126.
2. Boman, E., Koltracht, I.: Fast transform based preconditioners for Toeplitz equations. SIAM J. Matrix Anal. Appl. **16** (1995) 628–645.
3. Chan, R. H., Nagy, J. G., Plemmons, R. J.: FFT-based preconditioners for Toeplitz-block least squares problems. SIAM J. Numer. Anal. **30** (1993) 1740–1768.
4. Chan, R. H., Ng, M. K., Wong, C. K.: Sine transform based preconditioners for symmetric Toeplitz systems. Linear Algebra Appl. **232** (1996) 237–260.
5. Di Benedetto, F.: Solution of nonsymmetric Toeplitz systems by preconditioning of the normal equations. Preprint # 268, Dipartimento di Matematica, Università di Genova, 1994.
6. Di Benedetto, F.: Iterative solution of Toeplitz systems by preconditioning with the discrete sine transform. Proceedings of SPIE International Symposium, San Diego, 1995.
7. Di Benedetto, F.: Generalized updating problems and computation of the eigenvalues of rational Toeplitz matrices. (to appear)
8. Dickinson, B. W.: Solution of linear equations with rational Toeplitz matrices. Math. Comp. **34** (1980) 227–233.
9. Mertens, L., Van de Vel, H.: A special class of structured matrices constructed with the Kronecker product and its use for difference equations. Linear Algebra Appl. **106** (1988) 117–147.
10. Serra, S.: On the extreme spectral properties of Toeplitz matrices generated by L^1 functions with several minima/maxima. BIT **36** (1996) 135–143.
11. Miranda, M., Tilli, P.: On the asymptotic spectrum of Hermitian block Toeplitz matrices generated by matrix-valued functions. (to appear)

A Finite State Stochastic Minimax Optimal Control Problem with Infinite Horizon

Silvia C. Di Marco and Roberto L.V. González

CONICET - Universidad Nacional de Rosario
Rosario, Argentina.
e-mail:dimarco@unrctu.edu.ar

Abstract. We consider here a stochastic discrete minimax control problem with infinite horizon. We prove the existence of solution, we characterize it and we present iterative methods to compute it numerically.

1 The stochastic optimization problem

We consider a stochastic process Y which at each time $n \in \mathbb{N}$ is in one of the N states x_i belonging to $\Omega = \{x_i : i = 1, \ldots N\}$. Moreover, we assume that, at each time n an action a can be chosen in a finite set A, with $|A| = N_A$, (controlled finite Markov chain (see [11])). The transition probabilities of the process Y are given by

$$P\{Y(n+1) = j \,/\, Y(n) = i, \alpha(n) = a\} = p_{i,j}(a), \tag{1}$$

where $p_{i,j}(a)$ denotes the transition probability from the state x_i to the state x_j when the control a is used in x_i and the process is controlled by the policy α.

We denote by $\mathcal{A}$ the set of non-anticipative or progressively measurable control policies (see [5, 6, 9]). We say that a non-anticipative policy α is stationary, in case this policy is defined by a map $\Omega \mapsto A$. Since there exists a bijection between the stationary policies (of a N states problem) and the set A^N, we consider $\alpha \in A^N$ when α is stationary.

We are interested in minimizing, with respect to the non-anticipative policies, the expectation of the maximum of a function f ($f : \Omega \times A \mapsto \mathbb{R}^+$), on the chain along the time. So, the functional to be minimized is given by J

$$J(i, \alpha(\cdot)) = \mathbb{E}_\alpha \left\{ \max_{n \in \mathbb{N}_0} f(Y_\alpha(n), \alpha(n)) / Y_\alpha(0) = i \right\}, \tag{2}$$

where $\mathbb{E}_\alpha$ denotes the expectation of the process when the policy α has been chosen. The maximum of the right side of (2) is well defined because it is computed on a finite number of states and actions.

The optimal cost function is given by

$$u(i) = \min \{J(i, \alpha(\cdot)) : \alpha(\cdot) \in \mathcal{A}\} . \tag{3}$$

Remark. Even when the chosen policy is stationary, the stochastic process

$$\zeta_n = \left\{ \max_{\nu \leq n} f(Y_\alpha(\nu), \alpha(\nu)) \right\}$$

is not a Markov one. At time n, it depends not only on the state x_n but also on the previous ones. To apply dynamic programming techniques to the problem (3), we transform it in a Markov process through the introduction of an additional variable.

2 Transformation in a Markov problem

2.1 Generalized auxiliary states

We consider x the initial state of the process Y_α, with $\alpha(\cdot)$ an arbitrary non-anticipative policy. The auxiliary process Z is given by a sequence of random variables z_n with values in F, which evolves according to

$$z_{n+1} = z_n + (f(Y_\alpha(n), \alpha(n)) - z_n)^+ = \max\{f(Y_\alpha(n), \alpha(n)),\ z_n\},$$

with initial value $z = p$, $p \in F$. F is the set of admissible states of the process Z,

$$F = \{0\} \bigcup \{f(i,a) : i \in \Omega, a \in A\}, \quad |F| = N \times N_A + 1.$$

Remark. It is easy to see that

$$z_n = \max\left\{p, \max_{k=1,\dots,n} f(Y_\alpha(k), \alpha(k))\right\}.$$

In consequence it is immediate to check that the variable z_n "stores" the maximum of f up to time n, on the trajectory Y when the policy α is employed.

For each $a \in A$, we define the operator $T_a : F \times \Omega \to F$ such that

$$T_a(p,i) = p + (f(i,a) - p)^+. \tag{4}$$

We consider now the stochastic process (Z, Y) whose state at time $n \in \mathbb{N}$ is given by

$$(z_n, Y(n)) \in F \times \Omega. \tag{5}$$

The transition probabilities of the stochastic process (Z, Y) are

$$P\{(z_{n+1}, Y_\alpha(n+1)) = (q,j) \,/\, (z_n, Y_\alpha(n)) = (p,i), \alpha(n) = a\} = p_{i,j}(a)\, \mathcal{X}_{q=T_a(p,i)}.$$

In consequence, (Z, Y) is a Markov chain.

Definition 1. Let α be a non-anticipative policy. We say that α is a *generalized feedback* if it is given by a function $\alpha : F \times \Omega \mapsto A$.

Definition 2. We give the following definition of recurrent state for the chain (Z, Y); (see [1, 10, 11]). We say that (p, i) is a *recurrent state* under the action of the feedback control policy α, if there exists $C_p(i) \subset \Omega$, such that $\forall\, j \in C_p(i)$,

- $T_a(p, j) = p$;
- there exists $m, k \in \mathbb{N}$ such that $p^m_{i,j} > 0$ and $p^k_{j,i} > 0$;
- $p_{j,l} = 0$, $\forall\, l \notin C_p(i)$.

2.2 Auxiliary problem

We consider in the chain (Z, Y) the following stochastic optimal control problem with infinite horizon,

$$v_\alpha(p,x) = \mathbb{E}_\alpha \left\{ \sum_{n=0} (f(Y_\alpha(n), \alpha(n)) - z_n)^+ / Y_\alpha(0) = x, z \; = p \right\}, \tag{6}$$

whose associated optimal cost function is

$$v(p,x) = \min \{ v_\alpha(p,x) : \alpha(\cdot) \in \mathcal{A} \}. \tag{7}$$

Proposition 3. *The following properties hold*

1. $v(p,x) \geq 0, \; \forall \, (p,x) \in F \times \Omega$.
2. *If* $\tilde{F} = \max \{ f(i,a) : i \in \Omega, a \in A \}$, *then* $v(\tilde{F}, x) = 0, \; \forall \, x \in \Omega$.

Proposition 4. *Let g be a generalized feedback control policy and (p,i) be a recurrent state of the chain determined by g. Then $v_g(p,i) = 0$.*

Now we give a relation concerning the minimax problem (3) and the stochastic control problem with cumulative cost (7).

Proposition 5. $u(x) = v(0,x) = \max \{ v(p,x) : p \in F \}$.

2.3 Dynamical programming. The operator M.

The stochastic control problem (7) has been widely studied in the literature of this subject. See e.g. [11, 12]. We refer to chapter 6 of [11] for the proofs of the following Proposition and Corollary.

Proposition 6. *The function v satisfies the following dynamical programming equation:*

$$v(p,i) = \min_{a \in A} \left\{ (f(i,a) - p)^+ + \mathbb{E}_a \left(v(T_a(p,i), Y_\alpha(1)) \right) \right\}. \tag{8}$$

Corollary 1 *Let π be a generalized feedback policy such that $\pi(p,x)$ produces the minimum of the right side of (8). Then, π is an optimal policy.*

Corollary 2 *Let π be a generalized feedback policy. For all (p,i) recurrent state of the chain determined by π, it results $v(p,i) = 0$.*

Operators and their properties

Definition 7. We define the operator $M : \mathbb{R}^{(NN_A+1)N} \mapsto \mathbb{R}^{(NN_A+1)N}$ such that

$$(Mw)(p,i) = \min_{a\in A}\left\{(f(i,a)-p)^{+} + \mathbb{E}_a\left(w(T_a(p,i),Y_\alpha(1))\right)\right\}. \tag{9}$$

Remark. From the dynamical programming principle proved in the Proposition 6, we have that

$$M\,v = v. \tag{10}$$

Definition 8. For each generalized feedback g, we define the operator $M_g : \mathbb{R}^{(NN_A+1)N} \mapsto \mathbb{R}^{(NN_A+1)N}$ such that

$$(M_g\,w)\,(p,i) = (f(i,g(p,i))-p)^{+} + \mathbb{E}_{g(p,i)}\left(w(T_{g(p,i)}(p,i),Y_g(1))\right). \tag{11}$$

Remark. If g is a given generalized feedback, it is easy to see that v_g is a fixed point of M_g.

Proposition 9. *Let g be a generalized feedback. Then, the following properties hold:*

1. *M_g and M are monotone.*
2. *$\left(M_g^{(n)}0\right)(p,i) \to v_g(p,i)$ when $n\to\infty$.*

Characterization of the optimal cost v

Definition 10. In relation to equation (10), we define the associated sets of supersolutions and subsolutions:

$$S = \{s : F\times\Omega \mapsto \mathbb{R}^{+} \,/\, M\,s \le s\},$$
$$W = \{w : F\times\Omega \mapsto \mathbb{R} \,/\, M\,w \ge w\}.$$

It is easy to see that $v \in S\bigcap W$.

In the following Theorem the value function v is characterized in the set of supersolutions as the minimum element and as the limit of a particular sequence of subsolutions.

Theorem 11. *The following properties are valid*

- $v = \min\{s : s\in S\}$,
- $\left(M^{(n)}0\right)(p,i) \to v(p,i)$ *when* $n\to\infty$.

3 Iterative computation of the function u

3.1 Value iteration

From the previous Theorem we have that the optimal cost can be computed as the limit of the sequence of subsolutions generated by 0. This suggests the following iterative scheme.

Algorithm

Step 0: $w = 0$, $n = 0$.

Step 1: Compute $w^{n+1} = M w^n$.

Step 2: If $w^{n+1} = w^n$, stop.

Step 3: Set $n = n + 1$ and go to Step 1.

From Theorem 11, the algorithm converges to the solution v. The convergence can be dismally slow as it is shown in the following example.

Example 1 We consider a chain with N nodes, with $|A| = 1$ and the instantaneous cost f given by

$$f_i = \left| \begin{array}{ll} 0 & i = 1, \ldots, N-1, \\ \\ 1 & i = N. \end{array} \right. \tag{12}$$

The transition probabilities are given by

$$p_{i,j} = \left| \begin{array}{ll} 1 - 2\rho & j = i, \\ \\ \rho & j = i - 1, \\ \\ \rho & j = i + 1, \\ \\ 0 & \text{otherwise}, \end{array} \right. \tag{13}$$

where the equality between nodes is equality module N.

The error $v - w^n$ converges to 0 verifying in addition the relation

$$\lim_{n \to} \frac{\|v - w^n\|}{\lambda^n} = 1, \tag{14}$$

where $\lambda = 1 - 4\,\rho\,\sin^2(\pi/2N)$. Then, $\lambda \to 1$ when $\rho \to 0$. From (14) we see that the convergence is very slow when $\rho \sim 0$.

3.2 Policy iteration

To avoid the slowness of the convergence of the value iteration algorithm, we try to apply the Newton method. Since the fixed point problem $w = M\,w$ is not a contraction, we do not use the Newton method directly (which is known as the policy iteration method in this type of problems). We apply it on a sequence of perturbed problems that converges to the original problem and whose associated fixed point operators are contractions.

We consider the following discounted optimal control problem with λ the discount coefficient, $0 < \lambda < 1$.

$$V_\lambda(p,x) = \min\left\{V_{\pi,\lambda}(p,x) : \pi(\cdot) \in \mathcal{A}\right\}, \tag{15}$$

where

$$V_{\pi,\lambda}(p,x) = \mathbb{E}_\pi\left(\sum_{n=0} \lambda^n \left(f(Y_\pi(n),\pi(n)) - z_n\right)^+ / Y_\pi(0) = x, z_o = p\right). \tag{16}$$

It is easy to see that V_λ verifies the following dynamical programming equation.

$$V_\lambda(p,i) = \min_{a\in A}\left\{(f(i,a)-p)^+ + \lambda\sum_{j=1}^{N} p_{i,j}(a)\cdot V_\lambda(T_a(p,i),j)\right\}. \tag{17}$$

Definition 12. Given $\lambda \in (0,1)$ we define $Q_\lambda : \mathbb{R}^{(NN_A+1)N} \mapsto \mathbb{R}^{(NN_A+1)N}$ such that

$$(Q_\lambda\, w)(p,i) = \min_{a\in A}\left\{(f(i,a)-p)^+ + \lambda\sum_{j\in\Omega} p_{i,j}(a)\cdot w(T_a(p,i),j)\right\}. \tag{18}$$

Remark. The operator Q_λ is a contraction and V_λ is its unique fixed point. Besides, $V_\lambda \to v$, when $\lambda \to 1$.

Algorithm

Step 0: Let $\{\lambda_n\}_{n\in\mathbb{N}}$ be an increasing positive sequence, which converges to 1,

$\varepsilon > 0$, $n = 1$.

Step 1: Compute $V_{\lambda_n} = Q_{\lambda_n} V_{\lambda_n}$.

Determine the stationary policy $\hat{\pi}$ whose cost is V_{λ_n}.

Step 2: Solve the linear system

$$\left|\begin{array}{ll} w = 0 & \forall\ (p,x) \text{ recurrent,} \\ & \\ w = M_{\hat{\pi}} w & \text{otherwise.} \end{array}\right.$$

Step 3: If $w - V_{\lambda_n} \leq \varepsilon$, stop.

Set $n = n + 1$ and go to Step 1.

Remark. The Step 1 can be executed using the algorithms developed in [7]. They converge to the solution of the discounted problem in a finite number of steps.

Convergence of the algorithm

Remark. Let us see that the stopping rule is verified after a finite number of iterations. En effect, let us assume that $\forall\, n \in \mathbb{N}$ we have that

$$w - V_{\lambda_n} > \varepsilon\,. \tag{19}$$

It is easy to see that there exists $K > 0$ (only dependent on the data of the problem but independent on the policy used in (20)) such that

$$w - V_{\lambda_n} \leq K(1 - \lambda_n)\,. \tag{20}$$

Then, from (19) and (20), it results that $\forall n \in \mathbb{N}$,

$$\varepsilon < K(1 - \lambda_n)\,, \tag{21}$$

which implies that $\varepsilon = 0$ because $\lambda_n \to 1$. Then, (19) can not hold $\forall\, n \in \mathbb{N}$ and in consequence, the stopping rule is verified for a finite value of n.

Theorem 13. *When the algorithm finishes, a sub-optimal policy is available. The cost of this policy is w and it is valid that*

$$v \leq w \leq v + \varepsilon\,. \tag{22}$$

Example 2 Again, let us consider the problem described in Example 1 to compare both algorithms. In the value iteration method we have used the stopping test $||Mw - w|| \leq 10^{-6}$.

Algorithm	Time
Value iteration	29,65 s
Policy iteration	0,27 s

Table 1. Comparative table, $N = 5, N_A = 1, \rho = 0.1$

4 Conclusions

We have studied here a discrete minimax stochastic problem defined on a finite Markov chain and we have devised two algorithms to compute numerically the value function of the problem. This function is a natural approximation of the optimal cost corresponding to the continuous time minimax problem (with infinite horizon) studied in [2, 4]. In that case, the Markov chain is obtained via Kushner's procedure to discretize the dynamics of the system appearing in the continuous problem. We want to remark that it seems necessary to use additional hypotheses in order to obtain the convergence of the discrete solutions towards the continuous solution. Among these ones,

- conditions which assure that the finite horizon problems converges to the infinite horizon problem,
- existence of an attractor state or of an attractive cycle.

The analysis of these issues is the object of [3].

References

1. Borkar V.S.: Topics in controlled Markov chains. Longman–Pitman, London, (1991)
2. Di Marco S.C.: Sobre la optimización minimax y tiempos de detención óptimos. Thesis University of Rosario, Argentine (1996)
3. Di Marco S.C.: On the numerical solution of minimax optimal control problem with infinite horizon. (Work in progress)
4. Di Marco S.C., González R.L.V.: Minimax optimal control problem – Infinite horizon case. (Work in progress)
5. Fleming W.H., Rishel R.W.: Deterministic and stochastic optimal control. Springer–Verlag, New York, (1975)
6. Fleming W.H., Soner H.M.: Controlled Markov processes and viscosity solutions. Springer–Verlag, New York, (1993)
7. González R.L.V., Sagastizábal C.A: Un algorithme pour la résolution rapide d'équations discrètes de Hamilton–Jacobi–Bellman. Comptes Rendus Acad. Sc. Paris, Série I Tome 311 (1990) 45–50
8. Kushner H.J.: Probability methods for approximations in stochastic control and for elliptic equations. Academic Press, New York, (1977)
9. Kushner H.J., Dupuis P.G.: Numerical methods for stochastic control problems in continuous time. Springer–Verlag, New York, (1992)
10. Romanovsky V.I.: Discrete Markov chains. Walters–Noordhoff, Groningen, The Netherlands, (1970)
11. Ross S.M.: Applied probability models with optimization applications. Holden–Day, San Francisco, (1970)
12. Ross S.M.: Introduction to stochastic dynamic programming. Academic Press, New York, (1983)

Verified Solving of Linear Systems with Uncertainties in Maple

Nelli S. Dimitrova[1] and Christian P. Ullrich[2]

[1] Institute of Biophysics, Bulgarian Academy of Sciences, 1113 Sofia, Bulgaria ***
[2] Institute for Computer Science, University of Basel, 4056 Basel, Switzerland

Abstract. An algorithm with result verification for linear systems involving uncertainties in the input data is realized for Maple. We give an overview on the collection of procedures designed as a package and report on numerical experiments which confirm the feasibility of implementing such algorithms in computer algebra systems. The package will be made available under the name `Velisy` for the Maple share library.

1 Introduction

The computer algebra system Maple provides exact rational arithmetic which avoids round-off errors. Integers can be very large – the length limit is system dependent but generally much larger than users will encounter [1]. To represent the result exactly as many digits as nessecary are carried along the computations. In many applications this may lead to exponential growth in the storage requirements and pessimistic runtimes. In numerical analysis floating-point arithmetic is the fast way of performing computations. Like other computer algebra systems, Maple also offers efficient facilities for floating-point arithmetic with arbitrary precision. The global variable `Digits` which has 10 as its default value controls the accuracy of the floating-point calculations [2]. As mentioned in [4] each individual floating-point operation or single function evaluation is of accuracy 0.6 ulp (unit in the last place) but there are no facilities in Maple (as distributed) for studying the effect of rounding errors or of data uncertainties. Let us consider the (10×10)-Hilbert matrix $A = (a_{ij})$, defined by $a_{ij} = 1/(i+j-1)$. Given the exact solution $x = (1, 1, \ldots, 1)$, construct the exact right-hand side $b = Ax$ and evaluate it to floats. The Maple procedure `linsolve` from the `linalg` package in the Maple library displays the following result s:

$$s = [1.000125309\ .993269074\ 1.066366880\ .992524270\ -1.374494199$$
$$13.17498454\ -26.93555894\ 34.66042059\ -19.72680448\ 6.149413769].$$

Remember that the exact solution is $x = (1, 1, \ldots, 1)$! An experienced user knows that Hilbert matrices are very sensitive to rounding errors. The Maple

*** This work has been partially supported by the National Science Fund under grant No. MM 521/95 and by the Swiss National Science Foundation

`linsolve` routine neither provides any estimation on the correctness of the result nor displays a (warning) message about it – s is given as the "true" result.

In recent years new techniques – algorithms with result verification – have been developed in numerical analysis which allow to verify the correctness of the computed results on the computer (see e.g. [7], [8], [9]). Moreover, the computer establishes the existence and uniqueness of a solution of the numerical problem. The output has the quality of a mathematical proof for statements of the form "the computed result contains the exact solution" or "the problem does not possess a solution"; sometimes a failure message is returned, that is the algorithm can not claim inclusion for the exact solution (set) of the problem; wrong information is never passed to the user. Algorithms with result verification enable us to solve many mathematical problems involving uncertain (interval-valued and more generally set-valued) input data and have strong impact on mathematical modeling in natural and engineering sciences.

This paper demonstrates how to implement an algorithm with result verification for linear systems in the computer algebra system Maple. The algorithm used is known from [5]–[7]. For the sake of convenience we briefly present it in section 2. Section 3 contains the description of the Maple procedures realizing the algorithm. Numerical results are reported in section 4.

2 Theoretical Background

Denote by $IR^{n\times n}$ the set of interval matrices $[A] = (A_{ij})$, i.e. matrices with real compact intervals $A_{ij} = [A_{ij}^-, A_{ij}^+]$, $A_{ij}^- \leq A_{ij}^+$, $i,j = 1,2,\ldots,n$ in the components. The real matrix $\omega([A]) = (\omega(A_{ij})) = (A_{ij}^+ - A_{ij}^-)$ is called width of $[A]$; obviously $\omega([A]) \geq 0$ holds with componentwise order relation. The real matrix $\mu([A]) = (\mu(A_{ij})) = ((A_{ij}^- + A_{ij}^+)/2)$ is called midpoint of $[A]$. For an interval vector $[b]$ the real vectors $\omega([b])$ and $\mu([b])$ can be defined similarly. Consider the interval linear system

$$[A]x = [b]. \tag{1}$$

Definition 1. [10] The set $\Sigma([A],[b]) = \{x \in R^n : Ax = b, A \in [A], b \in [b]\}$ is called a solution set of the interval linear system (1).

The solution set $\Sigma([A],[b])$ can have very complicated structure [9]. If every real matrix $A \in [A]$ is nonsingular, then the interval matrix $[A]$ is called nonsingular and $\Sigma([A],[b])$ is bounded. Any interval vector $[x] \in IR^n$ satisfying $[x] \supseteq \Sigma([A],[b])$ is called solution of (1).

Definition 2. Let $[A] = (A_{ij}) \in IR^{n\times n}$ and the real matrix $\langle [A] \rangle = (\alpha_{ij})$ be defined by

$$\alpha_{ij} = \begin{cases} \max\{|a_{ij}| : a_{ij} \in A_{ij}\} & \text{if } i = j, \\ -\min\{|a_{ij}| : a_{ij} \in A_{ij}\} & \text{if } i \neq j. \end{cases}$$

If $\langle [A] \rangle$ is a M-matrix, then $[A]$ is called H-matrix. (For an introduction to the theory of these matrices we recommend [9].)

Suppose that $[A]$ is a nonsingular interval matrix. For constructing a solution $[x]$ of (1), the following approach is proposed in [10]: in a first step compute an approximation u for the exact solution $x = A^{-1}b$ with arbitrary real matrix $A \in [A]$ and real vector $b \in [b]$; in a second step, called verification phase, find a symmetric interval vector $[-z, z]$, $z \geq 0$, such that $[x] = u + [-z, z] \supseteq \Sigma([A], [b])$ is satisfied. In [10] this approach is applied to the problem $Ax = [b]$ with a real banded M-matrix A; the approximate solution u is computed using LDL^T-factorization of A; in the case of a symmetric and positive definite matrix A the LL^T-decomposition is used. In the latter case the Cholesky method is modified using a long accumulator for scalar products.

In [5]–[7] the considerations are concentrated to real (point) matrices and interval right-hand sides too. The new idea in the mentioned papers is to use existing and well-known program libraries such as LINPACK for computing the approximate solution and then to extend them by new routines for the verification phase.

We intended to exploit the last idea for Maple, i.e. to use existing Maple routines for finding the approximate solution. Unfortunately Maple procedures for linear algebra problems are not based on factorizations. So we had to implement two procedures – one for LU-decomposition and another for forward/backward substitution. We hope they will be of interest for the Maple users.

Definition 3. [10] For a nonsingular lower triangular matrix $L = (l_{ij})$ and a right-hand side $b = (b_i)$, $b \geq 0$, we define $y = \langle L \rangle^{-1} b$ by

$$y_i = (b_i + \sum_{j=1}^{i-1} |l_{ij}| y_j)/|l_{ii}|, \quad i = 1, 2, \ldots, n.$$

Analogously, $y = \langle U \rangle^{-1} b$ can be defined for a nonsingular upper triangular matrix U and $b \geq 0$.

Applying Crout's strategy [8] for triangular decomposition to a point matrix A we obtain $A = LU$ with lower triangular matrix $L = (l_{ij})$, $l_{ij} = 0$ for $i < j$, $l_{ii} = 1$, and upper triangular matrix $U = (u_{ij})$, $u_{ij} = 0$ for $i > j$, $i, j = 1, 2, \ldots, n$. Executing the decomposition in floating-point arithmetic yields $A \approx \tilde{L}\tilde{U}$. Usually $\tilde{L}$ and $\tilde{U}$ are stored in A. Partial pivoting can be used to avoid termination. By forward/backward substitution $\tilde{L}y = Pb$, $\tilde{U}u = y$, the approximate solution u is then computed (P means the permutation matrix).

Assume that componentwise bounds

$$|[A] - \tilde{L}\tilde{U}| \leq \Delta, \qquad |[b] - [A]u| \leq \varrho$$

are known with a matrix $\Delta \geq 0$ and a vector $\varrho \geq 0$. Compute according Definition 3 the vectors

$$h = \langle \tilde{U} \rangle^{-1} \langle \tilde{L} \rangle^{-1} \varrho, \qquad y = \langle \tilde{U} \rangle^{-1} \langle \tilde{L} \rangle^{-1} \Delta h$$

and let $h > y$. For a given real number $\delta > \max\{y_i/(h_i - y_i), i = 1, \ldots, n\}$ define the vector $z = (1 + \delta)h$, $z > 0$. Then $\Sigma([A], [b]) \subseteq u + [-z, z]$ is satisfied (see [10], Theorem 10).

For $A \in [A]$ and $b \in [b]$ the following inequalities hold true:

$$|[A] - \tilde{L}\tilde{U}| \leq |[A] - A| + |A - \tilde{L}\tilde{U}| \leq \omega([A]) + |A - \tilde{L}\tilde{U}|, \tag{2}$$
$$|[b] - [A]u| \leq \omega([b]) + \omega([A])|u| + |b - Au|. \tag{3}$$

Componentwise bounds for $|A - \tilde{L}\tilde{U}|$ and $|b - Au|$ can be found in [11]:

$$|A - \tilde{L}\tilde{U}| \leq \frac{\varepsilon}{1 - n\varepsilon}(|\tilde{L}|D|\tilde{U}| - |\tilde{U}|), \tag{4}$$
$$|b - Au| \leq \frac{2\varepsilon(n+1)}{1 - n\varepsilon}|\tilde{L}||\tilde{U}||u|, \tag{5}$$

where ε is the machine accuracy and $D = diag(1, 2, \ldots, n)$. The estimate (4) is valid if $n\varepsilon < 1$; (5) holds if $n\varepsilon \leq 0.5$. Both bounds are valid if no exeption (overflow, underflow, division by zero etc.) occurs during computation.

In [5] a second bound for $|b - Au|$ can be found:

$$|b - Au| \leq |b - \tilde{L}\tilde{U}u| + |A - \tilde{L}\tilde{U}||u|. \tag{6}$$

The rightmost expressions in (4)–(6) depend on known data and can be computed; they remain valid for $|PA - \tilde{L}\tilde{U}|$ and $|Pb - PAu|$. Bound (5) is less accurate then (6) but easier to evaluate [6].

To avoid storing of an additional vector $\omega([b])$, in [5] the estimate $\omega([b]) \leq m_b|b|$ with a number $m_b \geq 0$ is proposed. We could use the same idea for $\omega([A])$ but this will not avoid storing of an additional (floating-point) matrix.

Bounds (2) and (3) are valid for point matrices $A = [A]$ with $\omega(A) = 0$. Note that in this case the matrix has to be exactly representable in the system of machine numbers.

3 The Maple Routines

Let $[A] \in IR^{n \times n}$ be a nonsingular matrix. We firstly present the main steps of the algorithm for solving (1). Let $A \approx \mu([A])$ and $b \approx \mu([b])$ be approximations in floating-point arithmetic to the midpoint of $[A]$ and $[b]$ respectively. Let the floating-point number $m_b \geq 0$ be such that $\omega([b]) \leq m_b|b|$ holds true and $\omega([A])$ be the width of $[A]$. The following steps are executed:

Step 1. Compute in floating-point arithmetic the factorization $A \approx \tilde{L}\tilde{U}$; use partial pivoting if nessecary.

Step 2. Compute an approximate solution u using forward/backward substitution $\tilde{L}y = Pb$, $\tilde{U}u = y$.

Step 3. Compute by means of appropriate directed roundings the bound $\varrho \in \{\varrho_1, \varrho_2\}$, where

$$\varrho_1 = m_b P|b| + P\omega([A])|u| + \frac{2\varepsilon(n+1)}{1 - n\varepsilon}|\tilde{L}||\tilde{U}||u|,$$
$$\varrho_2 = m_b P|b| + P\omega([A])|u| + |Pb - \tilde{L}\tilde{U}u| + \frac{\varepsilon}{1 - n\varepsilon}(|\tilde{L}|D|\tilde{U}| - |\tilde{U}|)|u|.$$

Step 4. Compute according Definition 3 and using directed roundings the vectors

$$h = \langle\tilde{U}\rangle^{-1}\langle\tilde{L}\rangle^{-1}\varrho, \qquad \varrho \in \{\varrho_1, \varrho_2\},$$
$$y = \langle\tilde{U}\rangle^{-1}\langle\tilde{L}\rangle^{-1}(P\omega([A])h + \frac{\varepsilon}{1-n\varepsilon}(|\tilde{L}|D|\tilde{U}| - |\tilde{U}|)h).$$

If $h > y$ is satisfied, goto step 5. If there is an index i such that $h_i \leq y_i$ is fulfilled, print a failure message and stop – the algorithm can not compute an inclusion for the solution set.

Step 5. Compute $\delta = \max\{y_i/(h_i - y_i), i = 1, 2, \ldots, n\}$ and $z = (1+\delta)h$ by means of directed roundings. Construct the interval vector $u + [-z, z]$ as an enclosure for the solution set and stop.

Steps 1 and 2 perform the approximation phase of the algorithm, steps 3, 4 and 5 – the verification phase. In step 3 the computation of ϱ_1 or ϱ_2 is chosen by the user.

A detailed discussion about the reasons why the algorithm might fail is presented in [5]. In general this problem arises when the matrix $[A]$ does not have the H-matrix property.

The following Maple procedures realize separate steps of the algorithm and are included in the package **Velisy**.

LUdecomp(A,prm): performs in floating-point arithmetic the LU-factorization of a square matrix **A** applying Crout's algorithm with partial pivoting. If the matrix **A** is nonsingular, it is overwritten by the decomposition: the upper triangular part (including the main diagonal) of **A** contains the entries of $\tilde{U}$ and the strongly lower triangular part of **A** contains the elements of $\tilde{L}$; the diagonal entries of $\tilde{L}$ are normalized to unity. The parameter **prm** is optional. It is a n-dimensional vector, initialized by **prm**$=(1, 2, \ldots, n)$, which contains the indices for the pivot rows (permutation vector). The procedure is invoked by **LUdecomp(A,'prm')** or simply **LUdecomp(A)**. A matrix with floating-point components is returned.

LUsolve(ALU,b,p): computes the solution of the linear system **ALU**x = **Pb**, where **ALU** is the square LU-decomposed matrix returned by **LUdecomp**, **b** is the right-hand side vector, **p** is the permutation vector. The argument **p** is optional, i.e. the invokation **LUsolve(ALU,b)** is also allowed, we then solve **ALU**x = **b**. **ALU** and **b** are not changed during computation. An array is returned and displayed as a row vector.

The following procedures use **INTPAK** – the experimental interval package included in the Maple share library [4]. From **INTPAK** we use the procedures **type/interval**, **construct**, **is_in**, the directed roundings **Interval_Round_Up**, **Interval_Round_Down** as well as **Interval_midpoint**, **Interval_width** etc.

AbsDefect(ALU,u,b,mb,wA,p): computes the estimate ϱ_1. **ALU** is the decomposed matrix, **b** is a vector, **mb**> 0 is a floating-point number, **u** is the approximate solution returned from **LUsolve**, **wA** is the width of $[A]$ and **p** is the permutation vector. **wA** and **p** are optional parameters. The invokation **AbsDefect(ALU,u,b,mb,p)** or **AbsDefect(ALU,u,b,mb)** means that a point matrix $A = [A]$ is processed. All computations are performed using appropriate directed roundings. An array is returned and displayed as a row vector.

AbsExDefect(ALU,u,b,mb,wA,p): performs the computation of the estimate ϱ_2. The meaninig of the parameteres is as in **AbsDefect**. All computations are performed using appropriate directed roundings. An array is returned and displayed as a row vector.

VerifSol(ALU,u,dfct,wA,p): executes consecutively the computations presented in steps 4 and 5 of the algorithm. **ALU** is the LU-factorized matrix, **u** is the approximate solution, **dfct** is a vector returned either from **AbsExDefect** or **AbsDefect**, **wA** is the width of $[A]$ and **p** is the permutation vector. The invocation **VerifSol(ALU,u,dfct)** means that a point matrix $A = [A]$ is processed. All computations are performed using directed roundings. An array is returned and displayed as a row vector.

An additional procedure **IntBounds(B)** is implemented. It performs the computation of the midpoint **b** of the interval **B** and a floating-point number **mB**≥ 0 such that **width(B)**$\leq$ **mB** $*$ |**b**| is satisfied. Directed roundings are used during computations. A two-dimensional array of the form **[b,mB]** is returned.

4 Numerical Results

For constructing the first two examples we used the following approach. Starting from a given exactly representable matrix A and a given exact solution x we find the right-hand side vector $b = Ax$, $b = (b_1, b_2, \ldots, b_n)$. Applying the **INTPAK** procedure **construct(·,'rounded')** to each component b_i we obtain the interval right-hand side vector $[b] = ([b]_i)$. After loading the **Velisy** package we firstly call **IntBounds** with each component $[b]_i$ to derive the array $[b_i, m[b]_i]$ and the value **mb**$=$ $\max\{m[b]_i, i = 1, 2, \ldots, n\}$. Then we consecutively call **LUdecomp**, **LUsolve**, **AbsDefect** or **AbsExDefect** (which can be determined by the user in interactive input) and **VerifSol**. The so designed Maple worksheet either yields a failure message or produces the solution $u + [-z, z]$.

Example 1. As a first example consider the Hilbert matrix A from section 1. After executing the worksheet with **Digits:=10** and **AbsDefect** we obtain

$$y[1] = .4045073414\,10^{15},$$
$$h[1] = .8632068806\,10^{8},$$

that is $y_1 > h_1$ and the verification fails. The permutation vector returned by **LUdecomp** is p =[1 2 6 10 3 4 8 5 9 7], i.e. the triangular decomposition was performed with partial pivoting. This result is an indication that the matrix is (probably) ill-conditioned and forces further examinations. We increase the value of **Digits** to 25. This time the verification succeeds and an interval vector is displayed which contains the exact solution. The component widths (which present the relative errors in the solution) vary between $\approx 0.46 \times 10^{-4}$ for the largest component and $\approx 0.17 \times 10^{-8}$ for the smallest one. The output from **linalg[linsolve]** in floating-point arithmetic is

$$s = [.9999999999999999997708410\ 1.000000000000000195261784$$
$$.9999999999999995889045432\ 1.000000000000037014210546$$

.99999999999998248682240191.00000000000000478157321334
.99999999999992200690792554 1.00000000000000749903562631
.99999999999996080457431607 1.00000000000000085861251741].

Example 2 [5]. The matrix $A = (a_{ij})$ defined by

$$a_{ij} = \begin{cases} n & \text{if } i = j, \\ \frac{i-j}{i+j-1} & \text{if } i \neq j \end{cases}$$

is a dense H-matrix. Using the rational numbers in Maple the components of A are exactly representable. With the exact solution $x = (x_1, x_2, \ldots, x_n)$, $x_i = 1/i$, we execute the Maple worksheet, called **VelisyTest** with **Digits:=10** and **AbsDefect**. Table 1 compares the runtime in seconds spent by **VelisyTest** and **linsolve** for different dimensions n. The latter solves the system $Ax = b$ in rational arithmetic. All computations are performed on IBM PC 486 DX4/120 Mhz with 4 MB RAM.

dimension	10	15	20	25	30	35	40	45	50
VelisyTest	2	2	5	11	15	28	60	100	337
linsolve	0	1	3	8	30	127	?	?	?

Table 1

The sign "?" means that **linsolve** is not able to solve the corresponding problem exactly.

Various numerical experiments for different values of **Digits** with this linear system show that the verification routines guarantee a relative accuracy of order $10^{-\mathtt{Digits}+3}$.

Table 2 compares the runtimes in seconds for the approximation and the verification phase of our algorithm.

dimension	10	15	20	25	30	35	40	45	50
Approximation step	1	1	1	3	5	8	12	20	41
Verification step	1	2	5	8	10	20	38	80	296

Table 2

The results are rather pessimistic. The reason is that for each floating-point operation a directed rounding routine from **INTPACK** is used. As mentioned in [4], the interval package "was not written with efficiency in mind"; it "was intended to be exploratory and not for production use". Nevertheless, Table 1 shows that the time spent by **linsolve** to solve the problem exactly is much longer than the time used for verification even in the case of smaller systems. The results would be much more impressive if fast rounded floating-point operations could be used.

Example 3. The following small example shows how linear systems with interval matrices can be solved by our algorithm. This problem is part of a more general task which leads to solving this system. Let

$$[A] = \begin{bmatrix} [[1.359999999, 1.460000001], [-1.2, -1.2], [-1.6, -1.6]] \\ [1.2, 1.2], [1.839999999, 1.920000001], [-1.2, -1.2] \\ [.9, .9], [1.2, 1.2], [1.219999999, 1.320000001] \end{bmatrix}$$

$[b] = [[.01302499769, .01302500231]$
$[-.002400001301, -.002399998699][-.01427500221, -.01427499779]]$

With `Digits:=10` and `AbsExDefect` we obtain the following result

$$\begin{aligned}[x] = [[&-.002419102431, .0004246913160] \\ [&-.004539365870, -.004069360524] \\ [&-.006755960822, -.005574837608]]\end{aligned}$$

References

1. Char, B. W., Geddes,K. O., Gonnet, G. H., Leong, B. L., Monagan, M. B., Watt, S. M.: Maple V Language Reference Manual. Springer Verlag (1991)
2. Char, B. W., Geddes, K. O., Gonnet, G. H., Leong, B. L., Monagan, M. B., Watt, S. M.: First Leaves: A Tutorial Introduction to Maple V. Springer Verlag (1992)
3. Maple V Release 3 for DOS and Windows. Getting started. Waterloo Maple Software (1994)
4. Connel, A. E., Corless, R. M.: An Experimental Interval Arithmetic Package in Maple. Interval Computations **2** (1993) 120–134
5. Falcó Korn, C.: Die Erweiterung von Software-Bibliotheken zur effizienten Verifikation der Approximationslösung linearer Gleichungssysteme. PhD Thesis, Institut für Informatik, Universität Basel, Switzerland (1993)
6. Falcó Korn, C., Ullrich, C. P.: Extending LINPACK by Verification Routines for Linear Systems. Mathematics and Computers in Simulation **39** (1995) 21–37
7. Falcó Korn, C., Hörmann, B., Ullrich, C. P.: Verification may be Better than Estimation. To appear in SIAM Journal on Scientific Computing (1996) 6 pages
8. Hammer, R., Hocks, M., Kulisch, U., Ratz, D.: Numerical Toolbox for Verified Computing. Basic Numerical Problems; Theory, Algorithms and Pascal-XSC Programs. Springer Verlag (1993)
9. Neumaier, A.: Interval Methods for Systems of Equations. Cambridge University Press (1990)
10. Rump, S. M.: Inclusion of the Solution of Large Linear Systems with M-matrix. Interval Computations **1(3)** (1992) 22–43
11. Stoer, J.: Einführung in die numerische Mathematik I - vierte Auflage. Springer Verlag (1978)

Iterative Monte Carlo Algorithms for Linear Algebra Problems *

I.T. Dimov[1] and A.N. Karaivanova[1]

Central Laboratory for Parallel Computing, Bulgarian Academy of Sciences, Acad. G. Bonchev St.,bl. 25 A Sofia, 1113, Bulgaria

Key words: Monte Carlo method, matrix computations, convergence, eigenvalues, convergent iterative process

MSC subject classification: 65 C 05, 65 U 05

Abstract. A common Monte Carlo approach for linear algebra problems is presented. The considered problems are inverting a matrix B, solving systems of linear algebraic equations of the form $Bu = b$ and calculating eigenvalues of symmetric matrices. Several algorithms using the same Markov chains with different random variables are described.

The presented algorithms contain iterations with a resolvent matrix (used as iterative operator) of a given matrix. For inverting matrices and solving linear systems a mapping of the spectral parameter domain of convergence is established. This transformation leads to a new resolvent matrix and, respectively, to a new random variable constructed on the corresponding Markov chain, which allows the use of smaller number of iterations for reaching a given error. For calculating of eigenvalues an additional parameter in resolvent operator is involved to accelerate the algorithm convergence. The convergence of the iterative processes is proved and the convergence rate is compared with the rate of existing Monte Carlo algorithms for similar problems.

An error analysis is done.

Numerical experiments for Monte Carlo Almost Optimal (MAO) algorithms are performed on CRAY Y-MP C92A. It is shown that the accuracy and the algorithm complexity practically does not depend on the size of the matrix.

1 Introduction

Monte Carlo almost optimal (MAO) algorithms for inverting square nonsingular matrices B, solving systems of linear algebraic equations $Bu = b$ and evaluating the smallest eigenvalue of symmetric matrices B are proposed and studied.

The basic idea of Monte Carlo methods consists in the following: for the problem under consideration a random process is built with the property that the random variables created give the approximate solution of the problem. Generally speaking, such a random process can be built not in a unique way.

The probable error for the usual Monte Carlo method (which does not use any additional a priori information about the regularity of the solution) is defined in [So73] as:

$$r_N = c_{0.5} s(\theta) N^{-1/2}, \tag{1}$$

where $c_{0.5} \approx 0.6745$; $s(\theta)$ is the standard deviation.

A criterion of quality of the Monte Carlo method is the dispersion of the random variable θ, whose mathematical expectation is equal to the value of a linear functional of the solution J. Let θ be a random variable in the usual Monte Carlo method such that $J = E\theta$. Let $\hat{\theta}$

* Supported by Ministry of Science and Education of Bulgaria under Grants # I 501/95 and # MM 449/94

be another random variable for which $J = E\hat{\theta}$ and the conditions providing the existence and the finiteness of the dispersion $D\hat{\theta} = \left(E\hat{\theta}^2 - (E\hat{\theta})^2\right)^{\frac{1}{2}}$ be fulfilled.

The method for which $D\hat{\theta} < D\theta$ is called "efficient Monte Carlo method". A method of this type is proposed by Kahn [Ka50] (for evaluation of integrals) and by Mikhailov [Mi87] and Dimov [Di91], [DT93] (for evaluating integral equations). The method with the smallest $D\hat{\theta}_{MO}$ is called Monte Carlo Optimal method (MO). It is shown in [DT93a] that methods of this type are time-consuming. From an algorithmistic point of view it is more efficient to use MAO algorithm with a dispersion $D\hat{\theta}_{MAO}$ for which $D\hat{\theta}_{MAO} \geq D\hat{\theta}_{MO}$, but $|D\hat{\theta}_{MAO} - D\hat{\theta}_{MO}| \leq \varepsilon$, where ε is a small parameter and the random variable $\hat{\theta}_{MAO}$ is easily constructed and the simulation of the corresponding random process for it is less time-consuming.

The presented algorithms are algorithms of the second type (i.e. MAO algorithms). The reduction of the value of the dispersion is obtained using a special kind of transition-density matrices.

2 Method and Algorithms

2.1 Formulation

Consider a matrix A:

$$A = \{a_{ij}\}_{i,j=1}^{n}, \quad A \in I\!R^n \times I\!R^n \tag{2}$$

and a vector

$$f = (f_1, \ldots, f_n)^t \in I\!R^n \tag{3}$$

The matrix A can be consider as a linear operator $A[I\!R^n \rightarrow I\!R^n]$, so that the linear transformation

$$Af \in I\!R^n \tag{4}$$

defines a new vector in $I\!R^n$.

Since iterative Monte Carlo algorithms using the transformation (4) will be considered, the linear transformation (4) will be called "iteration". The algebraic transformation (4) plays a fundamental role in iterative Monte Carlo algorithms.

Now consider the following three problems **Pi (i=1,2,3)** for the matrix A:

Problem P1. Evaluating the inner product

$$J(u) = (h, u) = \sum_{i=1}^{n} h_i u_i \tag{5}$$

of the solution $u \in I\!R^n$ of the linear algebraic system

$$Bu = b, \tag{6}$$

where $B \in I\!R^n \times I\!R^n$ is a given matrix; $b \in I\!R^n$ and $h \in I\!R^n$ are given vectors.

Clearly it is possible to choose a matrix $M \in I\!R^n \times I\!R^n$ such that $MB = I - A$, where $I \in I\!R^n \times I\!R^n$ is the identity matrix and $Mb = f$, $f \in I\!R^n$.

Then (6) becomes

$$u = Au + f. \tag{7}$$

It will be assumed that

$$(i) \quad \begin{cases} 1.\ \text{The matrices } M \text{ and } A \text{ are both non-singular;} \\ 2.\ |\lambda(A)| < 1 \ \text{ for all eigenvalues } \lambda(A) \text{ of } A, \end{cases}$$

that is, all values $\lambda(A)$ for which

$$Au = \lambda(A)u \tag{8}$$

is satisfied.

Problem P2. Inverting of matrices, i.e. evaluating of matrix

$$C = B^{-1}, \tag{9}$$

where $B \in I\!R^n \times I\!R^n$ is a given real matrix.

It will be assumed that the following conditions are fulfilled:

$$(ii) \qquad \begin{cases} 1. \text{ The matrix } B \text{ is non-singular;} \\ 2. \ ||\lambda(B)| - 1| < 1 \text{ for all eigenvalues } \lambda(B) \text{ of } B. \end{cases}$$

Problem P3. Evaluating of eigenvalues:

$$Au = \lambda(A)u. \tag{10}$$

It is assumed that

$$(iii) \qquad \begin{cases} 1. \ A \text{ is a symmetric matrix, i.e. } a_{ij} = a_{ji} \text{ for all } i,j = 1,\ldots,n; \\ 2. \ \lambda_{min} = \lambda_n < \lambda_{n-1} \leq \lambda_{n-2} \leq \ldots \leq \lambda_2 < \lambda_1 = \lambda_{max}. \end{cases}$$

Obviously, if the condition (i) is fulfilled, the solution of **the problem P1** can be obtained using the iterations (4).

For **the problem P2** the following matrix:

$$A = I - B \tag{11}$$

can be constructed.

Since it is assumed that the condition (ii) is fulfilled, the inverse matrix $C = B^{-1}$ can be presented as

$$C = \sum_{i=0}^{\infty} A^i. \tag{12}$$

For **the problem P3** under conditions (iii) an iterative process of the type (4) can be used for calculating the largest eigenvalue:

$$\lambda_1(A) = \lim_{i\to\infty} \frac{(h, A^i f)}{(h, A^{i-1} f)}, \tag{13}$$

since for symmetric matrices $\lambda_{min}(A)$ is a real number.

We will be interested in evaluating the smallest eigenvalue $\lambda_{min}(A)$ using an iterative process of the type (4). It will be done by introducing a new matrix for realizing Monte Carlo iterations.

Let $A = \{a_{ij}\}_{i,j=1}^n$ be a given matrix and $f = \{f_i\}_{i=1}^n$ and $h = \{h_i\}_{i=1}^n$ are vectors.

For the problems $P_i (i = 1, 2, 3)$ we create a stochastic process using the matrix A and vectors f and h.

Consider an initial density vector $p = \{p_i\}_{i=1}^n \in I\!R^n$, such that $p_i \geq 0, i = 1, \ldots, n$ and $\sum_{i=1}^n p_i = 1$.

Consider also a transition density matrix $P = \{p_{ij}\}_{i,j=1}^n \in I\!R^n \times I\!R^n$, such that $p_{ij} \geq 0$, $i,j = 1,\ldots,n$ and $\sum_{j=1}^n p_{ij} = 1$, for any $i = 1,\ldots,n$.

Define sets of permissible densities P_h and P_A. The initial density vector $p = \{p_i\}_{i=1}^n$ is called **permissible** to the vector $h = \{h_i\}_{i=1}^n \in I\!R^n$, i.e. $p \in P_h$, if

$$p_i > 0, \text{ when } h_i \neq 0 \text{ and } p_i = 0, \text{ when } h_i = 0 \text{ for } i = 1, \ldots, n. \quad (14)$$

The transition density matrix $P = \{p_{ij}\}_{i,j=1}^n$ is called **permissible** to the matrix $A = \{a_{ij}\}_{i,j=1}^n$, i.e. $P \in P_A$, if

$$p_{ij} > 0, \text{when } a_{ij} \neq 0 \text{ and } p_{ij} = 0, \text{when } a_{ij} = 0 \text{ for } i, j = 1, \ldots, n. \quad (15)$$

Consider the following Markov chain:

$$k_0 \to k_1 \to \ldots \to k_i. \quad (16)$$

where $k_j = 1, 2, \ldots, n$ for $j = 1, \ldots, i$ are natural random numbers.

The rules for constructing the chain (16) are:

$$Pr(k_0 = \alpha) = p_\alpha, \quad Pr(k_j = \beta | k_{j-1} = \alpha) = p_{\alpha\beta}. \quad (17)$$

Assume that

$$p = \{p_\alpha\}_{\alpha=1}^n \in P_h, \quad P = \{p_{\alpha\beta}\}_{\alpha,\beta=1}^n \in P_A. \quad (18)$$

Now define the random variables W_j using the following recursion formula:

$$W_0 = \frac{h_{k_0}}{p_{k_0}}, \quad W_j = W_{j-1} \frac{a_{k_{j-1}k_j}}{p_{k_{j-1}k_j}}, \; j = 1, \ldots, i. \quad (19)$$

The random variables W_j, $j = 1, \ldots, i$ can also be considered as weights on the Markov chain (17).

From all possible permissible densities we choose the following

$$p = \{p_\alpha\}_{\alpha=1}^n \in P_h, \; p_\alpha = \frac{|h_\alpha|}{\sum_{\alpha=1}^n |h_\alpha|}; \quad (20)$$

$$P = \{p_{\alpha\beta}\}_{\alpha,\beta=1}^n \in P_A, p_{\alpha\beta} = \frac{|a_{\alpha\beta}|}{\sum_{\beta=1}^n |a_{\alpha\beta}|}, \alpha = 1, \ldots, n. \quad (21)$$

Such a choice of the initial density vector and the transition density matrix leads to MAO algorithm. The initial density vector $p = \{p_\alpha\}_{\alpha=1}^n$ is called almost optimal initial density vector and the transition density matrix $P = \{p_{\alpha\beta}\}_{\alpha,\beta=1}^n$ is called almost optimal density matrix.

Instead of the finite random trajectory T_i in our algorithms we consider an infinite trajectory with a state coordinate $\delta_m (m = 1, 2, \ldots)$. Assume $\delta_m = 0$ if the trajectory is broken and $\delta_m = 1$ in other cases. Let

$$\Delta_m = \delta_0 \times \delta_1 \times \ldots \times \delta_m. \quad (22)$$

So, $\Delta_m = 1$ up to the first break of the trajectory and $\Delta_m = 0$ after that.

It is easy to show [So73], that under the conditions (i), (ii) and (iii), the following equalities are fulfilled:

$$E\{W_i f_{k_i}\} = (h, A^i f), \quad i = 1, 2, \ldots; \quad (23)$$

$$E\{\sum_{i=0}^N W_i f_{k_i}\} = (h, u), \quad (P1), \quad (24)$$

$$E\{\sum_{i|k_i=r'} W_i\} = c_{rr'}, \quad (P2), \quad (25)$$

where $(i|k_i = r')$ means a summation only for weights W_i for which $k_i = r'$ and $C = \{c_{rr'}\}_{r,r'=1}^n$.

$$\frac{E\{W_i f_{k_i}\}}{E\{W_{i-1} f_{k_{i-1}}\}} \approx \lambda_1(A), \text{ for sufficiently large "i"} \quad (P3). \tag{26}$$

2.2 Algorithms for inverting of matrices and solving linear systems

Consider problem **P1** where A (see (7)) is a symmetric matrix with real eigenvalues λ_k, $\lambda_k \in (-\infty, -a]$, and $a > 0$ and resolvent $R_\lambda = \sum_{k=0}^{\infty} c_k \lambda^k$, where $c_k = A^{k+1}$. In [DK96] the interval $(-\infty, -a]$ is mapped onto the interval $[-1, 1]$ using the following transformation

$$\lambda = \psi(\alpha) = \frac{4a\alpha}{(1-\alpha)^2}. \tag{27}$$

and a new random variable with a smaller dispersion (compared to the dispersion of the r.w. in (24) and (25)) is defined

$$\theta_m^*[h] = \sum_{j=0}^{m} g_j^{(m)} W_j f_{k_j} \quad , \tag{28}$$

where $W_0 = \frac{h_{k_0}}{p_0}$, $g_0^{(m)} = 1$, $g_k^{(m)} = \sum_{j=k}^{m} (4a)^k C_{k+j-1}^{2k-1} \alpha_*^j$, $W_j = W_{j-1} \frac{a_{k_{j-1},k_j}}{p_{k_{j-1},k_j}}$, $j = 1, 2, \ldots, m$.

($k_0, k_1, \ldots, k_m$ is a Markov chain with initial density function p_{k_0} and transition density function p_{k_{j-1},k_j}).

The following theorem is proved ([DK96]):

Theorem 1. *Given the above matrix A, whose resolvent R_λ does not converge, the map (27) and the random variable (28), the following assertion*

$$E\left\{\lim_{m\to\infty} \sum_{\nu=0}^{m} g_j^{(m)} W_j f_{k_j}\right\} = (h, u)$$

holds. This theorem allows the use of the random variable $\theta_m^*[h]$ for evaluating the inner product (5) of **P1**.

It is proved that this modification of the known algorithms leads to accelerating the convergence and decreasing the probable error.

For calculating one component of the solution , for example the "r"th component of u, we must choose

$$h = e(r) = (0, ..., 0, 1, 0, ..., 0), \tag{29}$$

where "1" is in the "r"th position. It follows that

$$(h, u) = \sum_{\alpha}^{n} e_\alpha(r) u_\alpha = u_r$$

and the corresponding Monte Carlo method is given by

$$u_r \approx \frac{1}{N} \sum_{s=1}^{N} \theta_m^*[e(r)]_s, \tag{30}$$

where N is the number of chains.

Consider **Problem P2**. To find the inverse $C = \{c_{rr'}\}_{r,r'=1}^n$ of some matrix B we must first compute the elements of the matrix

$$A = I - B, \tag{31}$$

where I is the identity matrix. Clearly the inverse matrix is given by (12) which converges if the condition (ii) holds. If the last condition is not fulfilled or the corresponding Neumann series converges slowly we can use the same mapping technique for accelerating the convergence of the method.

Estimate the element $c_{rr'}$ of the inverse matrix C

Let the vector f be the following unit vector

$$f_{r'} = e(r'). \tag{32}$$

Theorem 2. *Given the above matrix A, whose resolvent R_λ does not converge, the map (27) and the random variable (28), the following assertion*

$$E\left\{ \lim_{m\to\infty} \sum_{j=0}^{m} g_j^{(m)} \frac{a_{rk_1}a_{k_1k_2}\dots a_{k_{j-1}k_j}}{p_{rk_1}p_{k_1k_2}\dots p_{k_{j-1}p_j}} f_{r'} \right\} = c_{rr'}$$

holds.

This theorem permits the use of the following Monte Carlo method for calculating elements of the inverse matrix C:

$$c_{rr'} \approx \frac{1}{N}\sum_{s=1}^{N}\left[\sum_{(j|k_j=r')}^{m} g_j^{(m)} \frac{a_{rk_1}a_{k_1k_2}\dots a_{k_{j-1}k_j}}{p_{rk_1}p_{k_1k_2}\dots p_{k_{j-1}p_j}}\right], \tag{33}$$

where $(j|k_j = r')$ means that only the variables $W_j^{(m)} = g_j^{(m)} \frac{a_{rk_1}a_{k_1k_2}\dots a_{k_{j-1}k_j}}{p_{rk_1}p_{k_1k_2}\dots p_{k_{j-1}p_j}}$ for which $k_j = r'$ are included in the sum (33).

Observe that since $W_j^{(m)}$ is only calculated in the corresponding sum for $r' = 1, 2, \dots, n$ then the same set of N chains can be used to compute a single row of the inverse matrix.

2.3 The Resolvent Monte Carlo algorithm (RMC) for the smallest eigenvalue

Now consider the **Problem P3**. Here an algorithm based on Monte Carlo iterations by the matrix A resolvent operator $R_q = [I - qA]^{-1}$ is presented.

The following presentation

$$[I - qA]^{-m} = \sum_{i=0}^{\infty} q^i C_{m+i-1}^i A^i, \ |q|\lambda < 1; \tag{34}$$

is valid because of behaviours of binomial expansion and the spectral theory of linear operators (the matrix A is a linear operator [KA64]. The eigenvalues of the matrices R_q and A are connected with the equality $\mu = \frac{1}{(1-q\lambda)}$, and the eigenfunctions coincide. According to (13), the following expression

$$\mu^{(m)} = \frac{([I - qA]^{-m} f, h)}{([I - qA]^{-(m-1)} f, h)} \to_{m\to\infty} \mu = \frac{1}{1 - q\lambda}, \quad f \in R^n, h \in R^n. \tag{35}$$

is valid. For a negative value of q, the largest eigenvalue μ_{max} of R_q corresponds to the smallest eigenvalue λ_{min} of the matrix A. Now, for constructing the method it is sufficient to prove the following theorem.

Theorem 3. *Let λ'_{max} be the largest eigenvalue of the matrix $A' = \{|a_{ij}|\}_{i,j=1}^n$ If q is choosen such that $|\lambda'_{max} q| < 1$, then*

$$([I - qA]^{-m}f, h) = E\{\sum_{i=0}^{\infty} q^i C^i_{m+i-1} W_i h(x_i)\}. \tag{36}$$

Proof. Since the expansion (34) converges in uniform operator topology [KA64] it converges for ani vector $f \in I\!R^n$:

$$([I - qA]^{-m}f, h) = \sum_{i=0}^{\infty} q^i C^i_{m+i-1} (A^i f, h). \tag{37}$$

For obtaining (37) from (36) one needs to apply (23) and to average every term of the presentation (37). Such everiging will be correct if A, f, h and q in (36) are replaced by their absolute values. If it is done the sum (36) will be finite since the condition λ'_{max} is fulfilled. Thus, for a finite sum (36) there is a finite majorant summed over all terms and the expansion can be average over all terms. The theorem is proved.

After some calculations one can obtain

$$\lambda \approx \frac{1}{q}(1 - \frac{1}{\mu^{(m)}}) = \frac{(A[I - qA]^{-m}f, h)}{([I - qA]^{-m}f, h)} =$$

$$\frac{E\sum_{i=1}^{\infty} q^{i-1} C^{i-1}_{i+m-2} W_i h(x_i)}{E\sum_{i=0}^{\infty} q^i C^i_{i+m-1} W_i h(x_i)}. \tag{38}$$

The coefficients C^n_{n+m} are calculated using the presentation

$$C^i_{i+m} = C^i_{i+m-1} + C^{i-1}_{i+m-1}.$$

From the representation

$$\mu^{(m)} = \frac{1}{1 - |q|\lambda^{(m)}} \approx \frac{(h, [I - qA]^{-m}f)}{(h, [I - qA]^{-(m-1)}f)}, \tag{39}$$

we obtain the following Resolvent Monte Carlo (RMC) algorithm for evaluating the smallest eigenvalue:

$$\lambda \approx \frac{1}{q}\left(1 - \frac{1}{\mu^{(m)}}\right) \approx \frac{E\sum_{i=0}^{l} q^i C^i_{i+m-1} W_{i+1} h(x_i)}{E\sum_{n=0}^{l} q^i C^i_{i+m-1} W_i h(x_i)}, \tag{40}$$

where $W_0 = \frac{h_{k_0}}{p_{k_0}}$ and W_i are defined by (19).

The parameter $q < 0$ has to be chosen so that to minimize the following expression

$$J(q, A) = \frac{1 + |q|\lambda_1}{1 + |q|\lambda_2}, \tag{41}$$

or if $\lambda_1 = \alpha\lambda_2, \quad (0 < \alpha < 1)$,

$$J(q, A) = 1 - \frac{|q|\lambda_2(1 - \alpha)}{1 + |q|\lambda_2}. \tag{42}$$

We choose

$$q = -\frac{1}{2||A||} \tag{43}$$

but sometimes a slightly different value of q might give better results when a number of realizations of the algorithm is considered.

Since the initial vector f can be any vector $f \in R^n$ (in particular, a unit vector), the following formula for calculating λ_{min} is used

$$\lambda \approx \frac{E\{W_1 + qC_m^1 W_2 + q^2 C_{m+1}^2 W_3 + \ldots + q^l C_{l+m-1}^l W_{l+1}\}}{E\{1 + qC_m^1 W_1 + q^2 C_{m+1}^2 W_2 + \ldots + q^l C_{l+m-1}^l W_l\}}, \quad (44)$$

that is

$$\lambda \approx \frac{\frac{1}{N}\sum_{s=1}^{N}\{\sum_{i=0}^{l} q^i C_{i+m-1}^i W_{i+1}\}_s}{\frac{1}{N}\sum_{s=1}^{N}\{\sum_{i=0}^{l} q^i C_{i+m-1}^i W_i\}_s}$$

2.4 The Inverse Monte Carlo Iterative algorithm (IMCI) for the smallest eigenvalue

Here an **Inverse Monte Carlo Iterative algorithm (IMCI)** is also considered.

This algorithm can be applied when A is a non-singular matrix. The algorithm has a high efficiency when the smallest by modules eigenvalue of A is much smaller then other eigenvalues. This algorithm can be realized as follow:

1. Calculate the inversion of the matrix A.

2. Starting from the initial vector $f_0 \in R^n$ calculate the sequence of Monte Carlo iterations:

$$f_1 = A^{-1}f_0,\ f_2 = A^{-1}f_1 \ldots,\ f_i = A^{-1}f_{i-1}, \ldots$$

The vectors $f_i \in R^n$ converge to the eigenvector which corresponds to the smallest by modules eigenvalue of A. In fact, we calculate the functionals

$$\frac{(Af_i, h_i)}{(f_i, h_i)} = \frac{(f_{i-1}, h_i)}{(f_i, h_i)}.$$

It is not necessary to calculate A^{-1} because the vectors f_k can be evaluated solving the following system of equations:

$$Af_1 = f_0$$

$$Af_2 = f_1$$

$$\ldots$$

$$Af_i = f_{i-1}.$$

For the Monte Carlo methods it is more efficient first to evaluate the inverse matrix using the algorithm proposed in and after that to apply the Monte Carlo iterations.

3 Some numerical tests

Numerical tests are performed for a number of test matrices - general symmetric dense matrices, sparse symmetric matrices (including band sparse symmetric matrices) with different behaviors. Results are obtained for solving systems of linear algebraic equations, matrix inversion and finding of eigenvalues. Since the volume of the paper is limited we will discuss only some numerical results obtained for the most difficult from algorithmistic point of view problem - the problem of evaluating the smallest eigenvalue.

The results for matrices of size $n = 512$ and $n = 1024$ are shown in tables 1, 2.

The experimental results show that both IMCI and RMC algorithms give good results even in case of small values of the parameters m and N. An information about efficiency of vectorization of the algorithms is received. This information shows that the studied algorithms are well-vectorized.

Table 1. Inverse Monte Carlo Iterative algorithm (IMCI) for MS512.2 ($\lambda_{min} = 0.2736$). (A general symmetric matrix of size 512.)

a) The number of Markov chains is fixed $N = 80$.

m	Calculated λ_{min}	Error, %
2	0.2736	0.0
3	0.2733	0.11
4	0.2739	0.11
5	0.2740	0.15
10	0.2732	0.15
50	0.2738	0.07
100	0.2757	0.76

b) The number of iterations (number of moves in every Markov chain) m is fixed - $m = 50$.

N	Calculated λ_{min}	Error, %	$CP - time$, s	$HWM-memory$
20	0.2729	0.26	5.356	1137378
40	0.2742	0.22	5.396	1137378
60	0.2748	0.44	5.468	1137378
80	0.2739	0.11	5.524	1137378
100	0.2736	0.00	5.573	1137378
500	0.2737	0.04	6.666	1137378
1000	0.2739	0.11	8.032	1137378

Remark: The values for CP-time and HWM-memory are for CRAY-YMP-C92A.

4 Conclusion

Monte Carlo algorithms for inverting matrices, solving systems of linear algebraic equations and calculating eigenvalues are presented. Markov chain, which define an almost optimal transition density function is constructed and used in all presented algorithms. The results obtained for matrix inversion and solving systems of linear algebraic equations show that the presented modifications of the known Monte Carlo algorithms lead to acceleration the convergence and decreasing the stochastic error. Two different parallel and vectorizable Monte Carlo algorithms - RMC and IMCI for calculating eigenvalues of symmetric matrices are presented and studied. The both studied algorithms are "almost optimal" from statistical point of view, i.e. the dispersion of the probable error of the random variable which is equal to the value of λ_{min} is "almost minimal" in the meaning of definition given in [Di91]. An information about efficiency of vectorization of the algorithms is received. This information shows that the studied algorithms are well-vectorized.

Table 1. c) The number of iterations (number of moves in every Markov chain) m is small and fixed - $m = 4$.

N	$Calculated$ λ_{min}	$Error,$ %	$CP - time,$ s	$HWM-$ $memory$
20	0.2737	0.04	5.296	1137378
40	0.2749	0.58	⋆	1137378
60	0.2754	0.66	⋆	1137378
80	0.2739	0.11	⋆	1137378
100	0.2736	0.00	⋆	1137378
500	0.2737	0.04	⋆	1137378
1000	0.2738	0.07	5.514	1137378

Remarks:
1. The values of CP-time and HWM-memory are for CRAY-YMP-C92A.
2. "⋆" - no estimated CP-time; the values of CP-time are between 5.296 s and 5.514 s.
3. In comparison with case b), CP-time decreases very slowly for more then 10-times decreasing of the number of moves m.
4. The corresponding NAG-routine for solving the same problem needs CP-time = 5.452 s and HWM-mem = 1 220 676.

Table 2. Inverse Monte Carlo Iterations algorithm (IMCI) for a symmetric band matrix of size 1024 ($\lambda_{min} = -0.0001376$).

N	m	$Calculated$ λ_{min}	$Error,$ %	$CP - time,$ s	$HWM-$ $memory$
5	10	$-2.5178.10^{-4}$	34.2	⋆	1137378
10	10	$-2.2573.10^{-4}$	20.3	⋆	1137378
20	10	$-1.9562.10^{-4}$	4.3	23.30	1137378
30	10	$-1.9307.10^{-4}$	2.9	23.34	1137378
40	10	$-1.9165.10^{-4}$	2.2	⋆	1137378
50	10	$-1.8942.10^{-4}$	1.0	⋆	1137378
80	10	$-1.901.10^{-4}$	1.3	23.36	1137378
10^4	5	$-2.276.10^{-4}$	21.3	⋆	1137378
10^3	5	$-2.206.10^{-4}$	17.6	⋆	1137378
100	5	$-2.194.10^{-4}$	16.9	⋆	1137378
100	4	$-2.102.10^{-4}$	12.0	⋆	1137378
100	3	$-2.736.10^{-4}$	45.8	⋆	1137378

Remarks:
1. "⋆" - no estimations for CP-time.
2. In this case the results for small number of moves in every Markov chain m are not so good lake in the case of general symmetric matrix of size 512.
3. The corresponding NAG-routine for solving the same problem with the same matrix needs CP-time = 22.467 s and HWM-mem = 1 220 679 on CRAY-YMP-C92A, but IMCI algorithm is parallelized and can be used for massively parallel machines.

References

[Cu54] Curtiss, J.H.: Monte Carlo methods for the iteration of linear operators. J. Math Phys. **32**, No 4 (1954) 209–232.

[Cu56] Curtiss, J.H.: A Theoretical Comparison of the Efficiencies of two classical methods and a Monte Carlo method for Computing one component of the solution of a set of Linear Algebraic Equations. Proc. Symposium on Monte Carlo Methods , John Wiley and Sons (1956) 191–233.

[Di91] Dimov, I.: Minimization of the Probable Error for Some Monte Carlo methods. Proc. Int. Conf. on Mathematical Modeling and Scientific Computation, Varna (1991).

[DT93] Dimov, I, Tonev, O.: Random walk on distant mesh points Monte Carlo methods. Journal of Statistical Physics **70(5/6)** (1993) 1333 - 1342.

[DT93a] Dimov, I., Tonev, O.: Monte Carlo algorithms: performance analysis for some computer architectures. Journal of Computational and Applied Mathematics **48** (1993) 253–277.

[DK96] Dimov, I., Karaivanova, A.: A Fast Monte Carlo Method for Matrix Computations, in *Iterative Methods in Linear Algebra II, IMACS Series in Computational and Applied Mathematics* (S. Margenov and P.S. Vassilevski eds.), (1996) 204–213.

[Ka50] Kahn, H.: Random sampling (Monte Carlo) techniques in neutron attenuation problems. Nucleonics **6** No 5 (1950), 27–33 ; **6**, No 6 (1950) 60–65.

[KA64] Kantorovich, L.V., Akilov, S.P.: Functional analysis in normed spases. *Pergamon Press*, New York, 1964.

[MAD94] Megson, G., Aleksandrov, V., Dimov, I.: Systolic Matrix Inversion Using a Monte Carlo Method. Journal of Parallel Algorithms and Applications **3**, No 1 (1994) 311–330.

[Mi87] Mikhailov, G.A.: Optimization of the "weight" Monte Carlo methods. *Nauka*, Moscow, 1987.

[So73] Sobol, I.M.: Monte Carlo numerical methods. *Nauka*, Moscow, 1973.

Operator Problems in Strengthened Sobolev Spaces and Numerical Methods for Them

Eugene G. D'yakonov

Department of Computer Mathematics and Cybernetics
Moscow State University
Moscow, 119899, Russia.
e-mail: dknv@cmc.msk.su

Abstract. The strengthened Sobolev spaces are naturally connected, e.g., with such important (two or three-dimensional) problems of mathematical physics as those in theory of plates and shells with stiffeners or in the capillary hydrodynamics involving the surface tension. These nonstandard Hilbert spaces allow also to set variational and operator problems on composed manifolds of different dimensionality. Spectral (eigenvalue) problems can be considered as well.
Special attention is paid to numerical methods based on the use of projective-grid methods and effective iterative methods such as multigrid and cutting methods; under natural conditions on the smoothness of the solution, it can be shown that the strengthened variant of the Kolmogorov-Bakhvalov hypothesis about asymptotically optimal algorithms for elliptic problems holds also for the above mentioned problems on composed manifolds.

1 Strengthened Sobolev spaces

In what follows, we assume, for simplicity, that Ω is a bounded domain in the Euclidean space $\mathbf{R}^d$ with Lipschitz piecewise smooth boundary $\Gamma \equiv \partial\Omega$ and $\bar{\Omega} \equiv \Omega \cup \Gamma$. We write

$$(u,v)_{0,\Omega} \equiv (u,v)_{L_2(\Omega)}, \quad |u|_{0,\Omega} \equiv (u,u)^{1/2}_{0,\Omega}, \quad |u|_{1,\Omega} \equiv (|\nabla u|^2, 1)^{1/2}_{0,\Omega}$$

and make use of the classical Sobolev space $W_2^1(\Omega) \equiv V$ (see [1–4]) with the norm

$$\|u\|_V \equiv [|u|^2_{1,\Omega} + |u|^2_{0,\Omega}]^{1/2}. \tag{1.1}$$

We start by considering the case $d = 2$ and the model strengthened Sobolev space

$$G_{1,m} \equiv G \equiv W_2^1(\Omega) \cup W_2^m(E), m = [m] \geq 1,$$

where $E \subset \bar{\Omega}$ consists of straight line segments (edges, stiffeners) $E_1, \ldots, E_{r^*}$. More precisely, if s is the respective arclength parameter and D_s refers to the differentiation along E_r, we define our pre-Hilbert space G as a subset of functions in V such that their traces on each E_r belong to $W_2^m(E_r)$, so we may define the inner product by

$$(v, v')_G \equiv (v, v')_{W_2^1(\Omega)} + \sum_{r=1}^{r*} (v, v')_{W_2^m(E_r)}. \tag{1.2}$$

Here

$$\|v\|^2_{W_2^m(E_r)} \equiv \sum_{k=0}^{m} |D_s^k v|^2_{0,E_r}, \quad \|v\|^2_{W_2^m(E)} \equiv \sum_{r=1}^{r*} \|v\|^2_{W_2^m(E_r)}.$$

For simplicity, we assume that the end points of each E_r belong to Γ. Thus all E_r (considered as cuttings lines) define a partition of $\bar{\Omega}$ into a set of blocks (panels) $P_1, \ldots, P_{r'}$.

Theorem 1.1. *The space G is complete.*

Proof. From (1.1) and (1.2), it follows that each fundamental sequence u^n in G is also fundamental in V. Hence, there exists $u \in V$ such that $u = \lim_{n\to\infty} u^n$ in the sense of V. Therefore, on each E_r traces of u^n converge to the trace of u in the sense of $L_2(S_r)$ (see [2,3]). Since the traces of u^n were uniformly bounded in $W_2^m(E_r)$, the trace of u on each E_r belongs to $W_2^m(E_r)$ and $u \in G$.

On the other hand, the sequence of traces u^n was fundamental in the sense of each $W_2^m(E_r)$; therefore, these traces converge in $W_2^m(E_r)$ to some limit functions. Since the convergence in the sense of $L_2(E_r)$ was already established, $u = \lim_{n\to\infty} u^n$ in the sense of G. □

Theorem 1.2. *Let $u \in G$ and $f \equiv u\,|_E \equiv Tr\ u$ denote the trace of u on E. Then almost everywhere on E, f coincides with a continious on E and $m-1$ times continiously differentiable on each E_r function; moreover, the operator Tr, considered as a mapping of G into $C(E)$, is compact.*

Proof. For the respective restriction, we have $f \in W_2^m(E_r)$. Therefore, we can consider it as a $m-1$ times continiously differentiable on E_r function.

Let the point M_0 correspond to the intersection of any two stiffeners. Consider a triangle $T \equiv \Delta M_0M_1M_2 \subset \bar{\Omega}$, with two sides belonging to the chosen stiffeners. For the respective restriction, we have $u \in W_2^1(T)$, and hence it has the trace f on ∂T in the sense of the space $W_2^{1/2}(\partial T)$ (see [1–4]). Therefore $f \in W_2^{1/2}(M_0M_i), i = 1, 2$, and must satisfy the condition (see [4])

$$\int_0^{s_0} \frac{|w(s)|^2}{s} ds < \infty, \tag{1.3}$$

where $w(s) \equiv f(s) - f(-s)$, s denotes the respective arclength parameter for ∂T, $s = 0$ corresponds to the point M_0, and s_0 is small enough (e.g., $s_0 = min\{|M_0M_1|; |M_0M_2|\}$). ¿From (1.3) and the continuity of f on M_0M_1 and M_0M_2, we obtain the desired continuity of f at the point M_0.

Due to the compactness of embedding of $W_2^1(E_r)$ into $C(E_r)$, $r \in [1, r^*]$, we conclude that Tr is compact. □

It is clear that Tr is linear and bounded ($Tr \in \mathcal{L}\{G; C(E)\}$).

Theorems 1.1 and 1.2 can be generalized for spaces G with inner products of type

$$(v, v')_G \equiv (v, v')_{W_2^1(\Omega)} + \sum_{r=1}^{r^*} (v, v')_{W_2^{m_r}(E_r)},$$

where $m_r > 1/2, r \in [1, r^*]$ and $W_2^{m_r}(E_r)$ denotes the corresponding Sobolev-Slobodetckii space (see [1–4]). Moreover, we can even deal with normed linear spaces G such that

$$||v||_G \equiv ||v||_{W_p^l(\Omega)} + \sum_{r=1}^{r^*} ||v||_{W_{p_r}^{m_r}(E_r)},$$

$1 < p < \infty, l \geq 1, m_r \geq 0, 1 < p_r < \infty, r \in [1, r^*]$.

We stress that a pre-Hilbert space is called Hilbert space if it is complete and separable.

Theorem 1.3. *Let the space G be defined as in Theorem 1.1 and suppose that $m \leq 2$ (see* (1.2)*). Then G is a Hilbert space.*

This theorem holds probably for all m in (1.2), but even for $m \leq 2$ the proof is rather complicated because the standard Sobolev averaging is of small use. By this reason for $m = 1$ and a given $u \in G$, we apply approximations $\hat{u}_h \in \hat{G}_h \subset G$ typical for theory of spline approximations connected with quasi-uniform triangulations $T_h(\bar{\Omega})$ (see [1,5]; for simplicity, we may restrict ourselves to a polygonal $\bar{\Omega}$ when $\hat{G}_h$ consists of continious on $\bar{\Omega}$ and linear on each triangular cell $T \in T_h(\bar{\Omega})$ functions; our triangulations must be consistent with the geometry of E and yield the respective triangulations of each panel P_j). As indicated in [1], it is possible to construct a family of projection operators $\hat{I}_h$ (mapping G into $\hat{G}_h$) such that

$$||\hat{I}_h||_{W_2^1(\Omega) \mapsto W_2^1(\Omega)} \leq K, \quad ||\hat{I}_h||_{W_2^m(E_r) \mapsto W_2^m(E_r)} \leq K$$

and the convergence

$$\lim_{h \to 0} \hat{I}_h u = u, \quad \forall u \in G \tag{1.4}$$

follows from the respective convergences in $W_2^1(\Omega)$ and $W_2^m(E_r), r \in [1, r^*]$ for smooth functions u in these relatively simple spaces (due to the classical Banach Theorem about the pointwise convergence of a family of linear bounded operators (see [6])). The operator $\hat{I}_h$ is determined by linear functionals φ_i where the index i corresponds to a vertex M_i of an elementary triangle $T \in T_h(\bar{\Omega})$. The most important point in constructing φ_i is the use of one-dimensional Steklov's averaging along the stiffener E_r (when M_i belongs to only one E_r) and the use of a sum of such averagings when M_i is a common point of several stiffeners. For the remaining points, we can use various choices including the standard two-dimensional Steklov averaging and functionals from [5]. If $m = 2$, then the construction of $\hat{G}_h$ involves the use of Hermite cubic polynomials on each E_r.

It is noteworthy that the above theorems hold for smooth arcs E_r (it is even possible to take $E_r \subset \Omega$).

As an example of more involved two-dimensional manifold, we consider

$$U^{(2)} \equiv F \cup E, \quad F \equiv \cup_{i=1}^{6} F_i, \quad E \equiv \cup_{j=1}^{12} E_j, \tag{1.5}$$

where $F_i, i \in [1,6]$, are faces of the unit cube Q and $E_i, i \in [1,12]$ are its edges. For the space $W_2^1(F) \equiv V$ with the square of the norm

$$\|v\|_V^2 \equiv \sum_{i=1}^{6} \|v\|_{W_2^1(F_i)}^2,$$

we consider a subset of functions with special traces on each edge E_j so we can define a pre-Hilbert space $G \equiv G(U^{(2)})$ with the square of the norm

$$\|v\|_{G(U^{(2)})}^2 \equiv \|v\|_V^2 + \sum_{j=1}^{12} \|v\|_{W_2^1(E_j)}^2. \tag{1.6}$$

Theorem 1.4. *Let the space $G \equiv G(U^{(2)})$ be defined by* (1.5) *and* (1.6). *Then G is a Hilbert space.*

Theorem 1.5. *Let instead of F in* (1.5) *we take $F \equiv F_1 \equiv \bar{\Omega}$ and strengthen the space $W_2^1(F_1)$ as in Theorem* 1.1 *with 4 stiffeners, that coincide with 4 edges of F_1. Let $m = 1$ (see* (1.2)*) and the traces of elements of this space (continious on the boundary of F_1) be extended to continious on E (see* (1.5)*) functions with their respective restrictions in $W_2^1(E_j)$ for each $E_j, j \in [1,12]$. Suppose that the defined in this way functions on U (see* (1.5)*) are considered as elements of a pre-Hilbert space G with the norm* (1.6). *Then G is a Hilbert space.*

As an example of three-dimensional composed manifold, we consider

$$U^{(3)} \equiv Q \cup F \cup E. \tag{1.7}$$

Theorem 1.6. *Suppose, in the space $W_2^1(Q)$, we consider only such functions that their traces on $F \equiv \partial Q$ belong to the Hilbert space $G(U^{(2)})$ defined by* (1.5) *and* (1.6). *Suppose we consider them as elements of a pre-Hilbert space $G \equiv G(U^{(3)})$ (see* (1.3)*) with the square of the norm*

$$\|v\|_{G(U^{(3)})}^2 \equiv \|v\|_{W_2^1(Q)}^2 + \|v\|_{G(U^{(2)})}^2. \tag{1.8}$$

Then G is a Hilbert space.

It is not difficult to study the case when in (1.5)–(1.8) we deal with more general blocks F and E, e.g., when some parts of them not belong to $\bar{Q}$. We stress that geometries of composed manifolds can be fairly general and include, e.g., a ball, its boundary, and a net of arcs on the boundary.

2 Examples of variational problems

In what follows, $l \in G^*$ (l is a bounded linear functional defined on a Hilbert space G), $b_L(u;v)$ is a bilinear form defined on $G \times G$; we assume that $b_L(u;v)$ is symmetric, bounded, and that the quadratic form $b_L(v;v) \equiv \bar{I}_2(v)$ is positive definite, that is, there exists $\nu_0 > 0$ such that

$$\bar{I}_2(v) \geq \nu_0 \|v\|_G^2, \quad \forall v \in G. \tag{2.1}$$

Then the variational problem of finding

$$u = \arg\min_{v \in G}[\bar{I}_2(v) - 2l(v)] \tag{2.2}$$

is correctly posed (see, e.g., [1,7,9]).

We present here a number of examples of such variational problems that are associated with the strengthened Sobolev spaces considered in the above theorems; first use of analogous problems in pre-Hilbert spaces dates back to the paper of S. Timoshenko in 1915 (see [1,9]); note that a reduction of such problems for plates with stiffeners to those for Stokes type problems was studied in [1] along with asymptotically effective numerical methods; note also that the main mathematical problem was to show that the well-known inf-sup condition (normal solvability of the divergence operator) holds for new special Hilbert spaces.

We start by indicating G from Theorem 1 and

$$\bar{I}_2(v) \equiv I_2(v) + \sum_{r=1}^{r^*} \int_{E_r} c_r^{(1)} (D_s v)^2 ds + \sum_{j=1}^{j^*} c_j^{(0)} (v(A_j^*))^2, \tag{2.3}$$

$$I_2(v) \equiv \|v\|^2_{W_2^1(\Omega)}; \tag{2.4}$$

here and in what follows, all $c_i^{(r)}$ are positive constants; $A_j^* \in E, j \in [1, j^*]$.

More interesting is the case when in constructing G (see Theorem 1.1) we start from a subspace of V, e.g., like the subspace $V_0 \equiv W_2^1(\Omega; \Gamma_0)$ of functions that vanish on Γ_0, where $\Gamma_0 \subset \Gamma$ and consists of several simple arcs (the case $\Gamma_0 = \Gamma$ is allowed). If we strengthen V_0 in the same way as V, then we obtain a Hilbert space $G[\Gamma_0]$ that is a subspace of the old space G; we remark only that if the end points of a stiffener E_r belong to Γ_0, then the trace on E_r of any $u \in G_0$ must belong to $\overset{0}{W_2^1}(E_r)$ (the case with only one end point of E_r on Γ_0 is fairly similar). Moreover, we can add points $A_i \in E,\ i \in [1, i^*]$ to the set Γ_0; it is even possible to define a subspace G_0 of G by the conditions

$$v(A_i) = 0, \quad A_i \in E, i \in [1, i^*]$$

with $\Gamma_0 \equiv \{A_i\}$.

Theorem 2.1. Suppose we consider variational problem (2.2), (2.3) *with G replaced by G_0 and*

$$I_2(v) \equiv \sum_{i=1}^{m'} c_i^{(2)}(1, |\nabla v|^2)_{0,P_i}$$

(instead of (2.4)*) in* (2.3). *Then this problem is correct.*

Theorem 2.1 implies that for variational problems in spaces like G, it is possible to set the Dirichlet boundary condition even at a single point $A_i \in E$ what makes no sense in the standard Sobolev space $W_2^1(\Omega)$. It is also easy to show that conditions like $u(A_i) \geq 0,\ A_i \in E, i \in [1, i^*]$, define a nonempty, convex and closed subset in G; thus the classical theory of variational inequalities (see [1,7,8]) can be applied.

Now we consider subspaces $G_0(U^{(2)}) \subset G(U^{(2)})$ (see Theorems 1.4 and 1.5) and $G_0(U^{(3)}) \subset G(U^{(3)})$ (see Theorems 1.6) defined in the same way as the subspace G_0.

Theorem 2.2. *Suppose we consider variational problem* (2.2), (2.3) *with* G *replaced by* $G_0(U^{(2)})$ *and*

$$I_2(v) \equiv \sum_{i=1}^{6} c_i^{(2)}(1, |\nabla v|^2)_{0,F_i}$$

or even $I_2(v) \equiv c_1^{(2)}(1, |\nabla v|^2)_{0,F_1}$ *(instead of* (2.4)*) in* (2.3). *Then this problem is correct.*

Theorem 2.4. *Suppose we consider variational problem* (2.2) *with* G *replaced by* $G_0(U^{(3)})$ *and*

$$\bar{I}_2(v) \equiv c^{(3)}(1, |\nabla v|^2)_{0,Q} + \sum_{i}^{6}(1, |\nabla v|^2)_{0,F_i} + \sum_{j=1}^{12} \int_{E_j} c_j^{(1)}(D_s v)^2 ds. \tag{2.5}$$

Then this problem is correct.

Note that in (2.5) the vector ∇v is a three-dimensional vector only in the term with $(1, |\nabla v|^2)_{0,Q}$.

Any of the above variational problems is equivalent to a correct linear operator equation

$$Lu = f$$

in the respective Hilbert space G with symmetric and positive definite linear operator L such that $(Lu, v)_G = b_L(u; v),\ \forall u \in G, \forall v \in G$ (see (2.1)). It is possible to give examples of more general correct operator equations in G including the case of nonlinear equations. In the case of several unknown function, one of the most interesting examples relates to three-dimensional elasticity problems with plates as stiffeners.

Among possible spectral problems, we mention those that are reduced to the problems $Mu = \lambda Lu$ with the above L and symmetric and compact operators M (for such problems in our Hilbert spaces G, the classical Hilbert-Schmidt theorem holds (see [1])). For example, if $\bar{I}_2(v)$ is defined by (2.5) then, in the role of M, we can take operators M such that

$$(Mv,v)_G = \alpha^{(3)}|v|^2_{0,Q} + \sum_{j=1}^{j^*} \alpha_j^{(0)}(v(A_j^*))^2,$$

where $A_j^* \in E, j \in [1, j^*]$ (see (1.5)).

3 Numerical methods and their optimization

It seems natural to assume that the solution of problem (2.2) from Theorem 2.1 satisfy conditions

$$\|u\|_{W_2^{1+\gamma}(P_i)} \leq K_{1,i}^* \quad i \in [1, r'], \tag{3.1}$$

$$\|u\|_{W_2^{1+\gamma}(E_r)} \leq K_r, \quad r \in [1, r^*], \tag{3.2}$$

where $\gamma > 0$.

Theorem 3.1. *Let M be a set of functions in G from Theorem* 2.1 *such that* (3.1) *and* (3.2) *hold. Then M is a compact set in G and, for the N-width in the sense of Kolmogorov for M, we have*

$$N(\varepsilon; M) \equiv N(\varepsilon) \asymp \varepsilon^{-2/\gamma}, \tag{3.3}$$

where $\varepsilon > 0$ is a prescribed tolerance.

Note that the N-width of M is usually defined by

$$\pi_N \equiv \inf_{G_N} \sup_{u \in M} \|u - Pu\|,$$

where G_N denotes an arbitrary subspace of G with $\dim G_N \leq N$ and P denotes the orthoprojector of G onto G_N; in our case, $\pi_N \asymp N^{-\gamma/2} \asymp \varepsilon$, which yields (3.3).

Theorem 3.2. *Let the solution u of problem from Theorem* 2.1 *be such that* (3.1) *and* (3.2) *hold. Then, given a prescribed tolerance $\varepsilon > 0$, it is possible to indicate a computational algorithm $a \equiv a_\varepsilon$ such that it yields an ε-approximation to u with the computational work*

$$W_a(\varepsilon) \asymp N(\varepsilon). \tag{3.4}$$

Theorem 3.2 implies that the strengthened variant of the Kolmogorov-Bakhvalov hypothesis about asymptotically optimal computational algorithms (see [1]) holds not only for standard elliptic problems but also for variational problems in some new Hilbert spaces like G (this remarkable hypothesis has attracted much attention of mathematicians over the last half of this century (see [1])).

Similar results are valid for all above examples of variational problems but in the case of the problem from Theorem 2.4, instead of (3.1)–(3.4) we have

$$\|u\|_{W_2^{1+\gamma}(Q} \leq K, \quad \|u\|_{W_2^{1+\gamma}(F_i)} \leq K, \quad \|u\|_{W_2^{1+\gamma}(E_j)} \leq K,$$

(for all i and j) and

$$N(\varepsilon) \asymp \varepsilon^{-3/\gamma}.$$

In describing projective-grid (finite element) methods, we confine ourselves to the case $\gamma \leq 1$ and the original problem from Theorem 2.1 with a domain Ω such that Γ is a closed broken line and all $c_j^{(0)} = 0$. We apply triangulations of $\bar{\Omega}$ and make use of spline spaces $\hat{G}_h$, where $\hat{G}_h$ consists of piecewise linear with respect to the triangles $T \in T_h(\bar{\Omega})$ functions (we assume that $T_h(\bar{\Omega})$ are consistent with geometry of E and yield triangulations of each panel; we also assume that Γ_0 is a union of some sides of triangles $T_h \in T_h(\bar{\Omega})$). Then it is possible to obtain the asymptotically optimal estimate of accuracy

$$||\hat{u} - u||_G \leq Kh^\gamma.$$

Our projective-grid method yields grid systems

$$L_h \bar{u} = \bar{f} \tag{3.5}$$

(we write them here as operator equations in the Euclidean space H with $L_h = L_h^* > 0$).

Theorem 3.3. *Suppose $T_h(\bar{\Omega}) \equiv T^{(p)}(\bar{\Omega})$ is obtained as a result of a refinement procedure that is applied recursively p times for an initial coarse triangulation $T^{(0)}(\bar{\Omega})$ with $p \asymp |\ln h|$. Then for L_h in (3.5), there exists an asymptotically optimal model operator $B_h \asymp L_h$ such that the constants of spectral equivalence and the estimates of the required computational work in solving systems with B_h are independent of all numbers $c_j^{(2)}$ and $c_j^{(1)}$.*

The desired B_h is constructed in accordance with theory of model cooperative operators (see [1,10,11]) based on proper multigrid splittings of the spline space $\hat{G}_h$.

It is also possible to obtain model operators related to domain decomposition methods, e.g., considered in [1,12,13] for standard elliptic problems.

References

1. D'yakonov, E.G: Optimization in Solving Elliptic Problems. CRC Press, Boca Raton, 1996
2. Besov, O.V., Il'in, V.P., Nikol'skii, S.M.: Integral Representation of Functions and Embedding Theorems. **1** Winston and Sons, Washington, 1978; **2** A Halsted Press Book, John Wiley, New York, 1979
3. Grisvard, P: Elliptic Problems in Nonsmooth Domains. Pitman, Boston, 1985
4. G. N. Yakovlev, G.N.: Traces of functions in the space W_p^l on piecewise smooth surfaces. Math. USSR Sb. **3** (1967) 481–498
5. Scott, L.R., Zhang, S.: Finite element interpolation of nonsmooth functions satisfying boundary conditions. SIAM J. Numer. Anal. **54** (1990) 483–493
6. Kantorovich, L.V., Akilov, G.P.: Functional Analysis in Normed Spaces. Pergamon, London, 1964
7. Ciarlet, P.: The Finite Element Method for Elliptic Problems. North-Holland Publishing Company, Amsterdam, 1975

8. Lions, J.L.: Quelque Methodes de Resolution des Problem aux Limites non Lineare. Dunod, Paris, 1969
9. Courant, R.: Variational methods for the solution of problems of equilibrium and vibrations. Bull. of Amer. Math. Soc. **49** (1943) 1–23
10. D'yakonov, E.G.: About an iterative method for solving discretized elliptic systems. Compte rendue de l' Academie Bulgare des Sciences **28** (1975) 295–297 (in Russian)
11. Axelson, O., Vassilevski, P.S.: A survey of multilevel preconditioned iterative methods. BIT **29** (1989) 769–793
12. Bramble, J.H., Pasciak, J.E., Schatz, A.H.: The construction of preconditioners for elliptic problems by substructuring. I. Math. Comput. **47** (1986) 103–134
13. Dryja, M., Smith, B.F., Widlund, O.B.: Schwarz analysis of iterative substructuring algorithms for elliptic problems in three dimensions. SIAM J. Numer. Anal. **31** (1994) 1662–1694

Convex Combinations of Matrices – Nonsingularity and Schur Stability

Ludwig Elsner[1] and Tomasz Szulc[2]

[1] Fakultät für Mathematik, Universität Bielefeld, Postfach 100 131, D-33501 Bielefeld, Germany
[2] Faculty of Mathematics and Computer Science, Adam Mickiewicz University, Matejki 48/49, 60-769 Poznań, Poland

Abstract. Using the notion of a block P-matrix, introduced previously by the authors, a characterization of the nonsingularity (Schur stability, resp.) of all convex combinations of three nonsingular (Schur stable, resp.) real matrices is derived.

1 Introduction

Consider the following problem: *For given n-by-n matrices* $\mathbf{A}_1, \mathbf{A}_2, \ldots, \mathbf{A}_k$ *having the same property "w" do characterize the inheritance of "w" by all convex combinations of* $\mathbf{A}_1, \mathbf{A}_2, \ldots, \mathbf{A}_k$.

The above problem arises in stability theory and control theory. In particular, it also arises in studies of numerical solving of systems of linear algebraic equations, and in this case, the investigation subjects are the nonsingularity inheritance and Schur stability inheritance (recall that a square matrix is called Schur stable if its spectral radius is less than one). These variants of our problem for k = 2 and complex $\mathbf{A}_1, \mathbf{A}_2$ have been solved in [3] (the nonsingularity inheritance) and in [6] (Schur stability inheritance). Our gool in this paper is to solve the mentioned variants for k = 3 and real $\mathbf{A}_1, \mathbf{A}_2, \mathbf{A}_3$. We note that all convex combinations of three real matrices $\mathbf{A}_1 = (a_{ij}^{(1)}), \mathbf{A}_2 = (a_{ij}^{(2)})$ and $\mathbf{A}_3 = (a_{ij}^{(3)})$ are contained in the interval matrix

$$\mathbf{A} = ([min\{a_{ij}^{(1)}, a_{ij}^{(2)}, a_{ij}^{(3)}\}, max\{a_{ij}^{(1)}, a_{ij}^{(2)}, a_{ij}^{(3)}\}]), 1 \leq i, j \leq n.$$

So, our goal could be achieved using interval methods (proposed, for example, in [4] and [5]). However, roughly speaking, interval tools are very "expensive" which motivates our interest in looking for other (and "cheaper") tools.

2 Preliminaries

For a square matrix $\mathbf{X}$, the spectrum and the spectral radius of $\mathbf{X}$, respectively, will be denoted by $\sigma(\mathbf{X})$ and $\rho(\mathbf{X})$, respectively. By $N(k)$ we will denote a partition of $N = \{1, \ldots, n\}$ into $k, 1 \leq k \leq n$, pair-wise disjoint subsets N_j of the cardinality n_j and by $\mathcal{T}^{(k)}$ the set of all diagonal n-by-n matrices $\mathbf{T}$ such that, for $i = 1, \ldots, k, \mathbf{T}[N_i] = t_i \mathbf{I}_{n_i}$, where $\mathbf{T}[N_i]$ is the principal submatrix of

$\mathbf{T}$ with row and column indicies in $N_i, t_i \in [0,1]$ and $\mathbf{I}_{n_i}$ is the n_i-by-n_i identity matrix (we set $\mathbf{I}_n = \mathbf{I}$).

In [1] the following definition has been introduced.

Definition 1. Let $N(k)$ be a partition of N. A matrix $\mathbf{A} \in \mathbf{R}_{n\times n}$ is called a block P-matrix with respect to the partition $N(k)$, if for any $\mathbf{T} \in \mathcal{T}^{(k)}$

$$det(\mathbf{TA} + (\mathbf{I} - \mathbf{T})) \neq 0.$$

For $\mathbf{A}, \mathbf{B}, \mathbf{C} \in \mathbf{R}_{n\times n}$ we define the block matrices $\tilde{\mathbf{A}}, \tilde{\mathbf{B}}, \tilde{\mathbf{C}}, \tilde{\mathbf{I}} \in \mathbf{R}_{3n^2\times 3n^2}$ by

$$\tilde{\mathbf{A}} = \begin{bmatrix} \hat{\mathbf{A}} & 0 & 0 \\ \hat{\mathbf{I}} & \hat{\mathbf{I}} & 0 \\ 0 & 0 & \hat{\mathbf{I}} \end{bmatrix}, \tilde{\mathbf{B}} = \begin{bmatrix} \hat{\mathbf{B}} & \hat{\mathbf{D}} & 0 \\ 0 & \hat{\mathbf{I}} & 0 \\ \hat{\mathbf{I}} & 0 & \hat{\mathbf{I}} \end{bmatrix}, \tilde{\mathbf{C}} = \begin{bmatrix} \hat{\mathbf{C}} & \hat{\mathbf{E}} & \hat{\mathbf{F}} \\ 0 & \hat{\mathbf{I}} & 0 \\ 0 & 0 & \hat{\mathbf{I}} \end{bmatrix}, \tilde{\mathbf{I}} = \begin{bmatrix} \hat{\mathbf{I}} & 0 & 0 \\ 0 & \hat{\mathbf{I}} & 0 \\ 0 & 0 & \hat{\mathbf{I}} \end{bmatrix},$$

where

$\hat{\mathbf{A}} = \mathbf{I}\otimes\mathbf{I} - \mathbf{A}\otimes\mathbf{A}$, $\hat{\mathbf{B}} = \mathbf{I}\otimes\mathbf{I} - \mathbf{B}\otimes\mathbf{B}$, $\hat{\mathbf{C}} = \mathbf{I}\otimes\mathbf{I} - \mathbf{C}\otimes\mathbf{C}$,
$\hat{\mathbf{D}} = \mathbf{A}\otimes\mathbf{B} + \mathbf{B}\otimes\mathbf{A} - \mathbf{A}\otimes\mathbf{A} - \mathbf{B}\otimes\mathbf{B}$, $\hat{\mathbf{E}} = \mathbf{A}\otimes\mathbf{C} + \mathbf{C}\otimes\mathbf{A} - \mathbf{A}\otimes\mathbf{A} - \mathbf{C}\otimes\mathbf{C}$,
$\hat{\mathbf{F}} = \mathbf{B}\otimes\mathbf{C} + \mathbf{C}\otimes\mathbf{B} - \mathbf{B}\otimes\mathbf{B} - \mathbf{C}\otimes\mathbf{C}$, $\hat{\mathbf{I}} = \mathbf{I}\otimes\mathbf{I}$

and $\mathbf{X}\otimes\mathbf{Y}$ denotes the Kronecker product of matrices $\mathbf{X}$ and $\mathbf{Y}$.

3 Results

We start with the fundamental result for our considerations.

Theorem 2. *Let* $\mathbf{A}, \mathbf{B} \in \mathbb{R}_{n\times n}$. *The following are equivalent.*
(i) $\sigma(\alpha\mathbf{A} + (1-\alpha)\mathbf{B}) \cap (-\infty, 0] = \emptyset$ *for all* $\alpha \in [0,1]$.
(ii) The matrix

$$\begin{bmatrix} \mathbf{I} & \mathbf{A}-\mathbf{B} \\ -\mathbf{I} & \mathbf{B} \end{bmatrix} \tag{1}$$

is a block P-matrix with respect to the partition $\{\{1,\ldots,n\}, \{n+1,\ldots,2n\}\}$ *of* $\{1,\ldots,2n\}$.

Proof. First we observe that for any numbers α and β we have

$$\begin{bmatrix} \alpha\mathbf{I} & 0 \\ 0 & \beta\mathbf{I} \end{bmatrix} \begin{bmatrix} \mathbf{I} & \mathbf{A}-\mathbf{B} \\ -\mathbf{I} & \mathbf{B} \end{bmatrix} + \begin{bmatrix} (1-\alpha)\mathbf{I} & 0 \\ 0 & (1-\beta)\mathbf{I} \end{bmatrix} \begin{bmatrix} \mathbf{I} & 0 \\ 0 & \mathbf{I} \end{bmatrix} = \begin{bmatrix} \mathbf{I} & \alpha(\mathbf{A}-\mathbf{B}) \\ -\beta\mathbf{I} & \beta\mathbf{B}+(1-\beta)\mathbf{I} \end{bmatrix}. \tag{2}$$

So, using the Schur determinantal formula, we get that the matrix (2) is nonsingular if and only if the Schur complement of $\mathbf{I}$ in (2), i.e., the matrix

$$\beta\mathbf{B} + (1-\beta)\mathbf{I} + \alpha\beta(\mathbf{A} - \mathbf{B}), \tag{3}$$

is nonsingular.

Now we proceed to show the equivalence of the mentioned statements.
(ii) $\Longrightarrow$(i): If (1) is a block P-matrix with respect to the partition { {1,..., n}, {n+1,..., 2n}} of {1, ..., 2n} then, by the above observation, the matrix (3) is nonsingular for any $(\alpha,\beta) \in [0,1]^2$. Hence, for any $\alpha \in [0,1]$ and $\beta \in (0,1]$ the matrix

$$\alpha\mathbf{A} + (1-\alpha)\mathbf{B} + \frac{(1-\beta)}{\beta}\mathbf{I}$$

is nonsingular which is equivalent to the spectral property of $\alpha\mathbf{A} + (1-\alpha)\mathbf{B}$ in question. So, (ii) $\Longrightarrow$(i) holds.
(i) $\Longrightarrow$(ii): The spectral property of $\alpha\mathbf{A} + (1-\alpha)\mathbf{B}$ implies that

$$\beta\mathbf{B} + (1-\beta)\mathbf{I} + \alpha\beta(\mathbf{A}-\mathbf{B})$$

is nonsingular for any $\alpha \in [0,1]$ and $\beta \in (0,1]$. But this is also trivially true for $\beta = 0$. Hence

$$\beta\mathbf{B} + (1-\beta)\mathbf{I} + \alpha\beta(\mathbf{A}-\mathbf{B})$$

is nonsingular for any $(\alpha,\beta) \in [0,1]^2$ and therefore (1) is a block P-matrix with respect to the partition $\{\{1,\ldots,n\},\{n+1,\ldots,2n\}\}$ of $\{1,\ldots,2n\}$. □

As a consequence of Theorem 2 we get the following result.

Theorem 3. *Let* $\mathbf{A},\mathbf{B},\mathbf{C} \in \mathbb{R}_{n\times n}$. *The following are equivalent.*
(i) All convex combinations of **A**, **B**, **C** *are nonsingular.*
(ii) **C** *is nonsingular and the matrix*

$$\begin{bmatrix} \mathbf{I} & (\mathbf{A}-\mathbf{B})\mathbf{C}^{-1} \\ -\mathbf{I} & \mathbf{B}\mathbf{C}^{-1} \end{bmatrix} \tag{4}$$

is a block P-matrix with respect to the partition { {1,..., n}, {n+1,..., 2n}} of {1, ..., 2n}.

Proof. (i) $\Longrightarrow$(ii): First we observe that, by (i), the nonsingularity of **C** is obvious. Now, because of Theorem 2, it suffices to show that, for all $\alpha \in [0,1]$,

$$\sigma(\alpha\mathbf{A}\mathbf{C}^{-1} + (1-\alpha)\mathbf{B}\mathbf{C}^{-1}) \cap (-\infty, 0] = \emptyset$$

Suppose that, for some $\tilde{\alpha} \in [0,1]$, the matrix

$$\tilde{\alpha}\,\mathbf{A}\mathbf{C}^{-1} + (1-\tilde{\alpha})\mathbf{B}\mathbf{C}^{-1}$$

has a nonpositive eigenvalue λ so that

$$\tilde{\alpha}\,\mathbf{A}\mathbf{C}^{-1} + (1-\tilde{\alpha})\mathbf{B}\mathbf{C}^{-1} - \lambda\mathbf{I} \tag{5}$$

is singular. Set $-\lambda = \frac{\alpha_3}{1-\alpha_3}$ where $\alpha_3 \in [0,1)$. Then, from the singularity of the matrix (5),

$$\tilde{\alpha}\,(1-\alpha_3)\mathbf{A} + (1-\tilde{\alpha})(1-\alpha_3)\mathbf{B} + \alpha_3\mathbf{C} \tag{6}$$

is singular. Observe that all coefficients in (6) are nonnegative and that their sum equals 1. So, (6) is a singular convex combination of **A**, **B** and **C** which contradicts (i).

(ii) $\Longrightarrow$(i): Suppose that for some nonnegative $\alpha_1, \alpha_2, \alpha_3$ such that $\alpha_1 + \alpha_2 > 0$ and $\sum_{i=1}^{3} \alpha_i = 1$ the matrix $\alpha_1 \mathbf{A} + \alpha_2 \mathbf{B} + \alpha_3 \mathbf{C}$ is singular. Hence, as $\mathbf{C}$ is nonsingular, it follows that

$$\frac{\alpha_1}{\alpha_1+\alpha_2}\mathbf{AC}^{-1} + \frac{\alpha_2}{\alpha_1+\alpha_2}\mathbf{BC}^{-1} + \frac{\alpha_3}{\alpha_1+\alpha_2}\mathbf{I} =$$

$$\frac{\alpha_1}{\alpha_1+\alpha_2}\mathbf{AC}^{-1} + \frac{\alpha_1}{\alpha_1+\alpha_2}\mathbf{AC}^{-1} + \frac{\alpha_3}{1-\alpha_3}\mathbf{I}$$

is singular. So, for some $\beta \in [0,1]$,

$$\sigma(\beta \mathbf{AC}^{-1} + (1-\beta)\mathbf{BC}^{-1}) \cap (-\infty, 0] \neq \emptyset,$$

a contradiction to Theorem 2. □

Theorem 4. *Let* $\mathbf{A}, \mathbf{B}, \mathbf{C} \in \mathbb{R}_{n\times n}$. *The following are equivalent.*
(i) All convex combinations of $\mathbf{A}$, $\mathbf{B}$, $\mathbf{C}$ *are Schur stable.*
(ii) $\mathbf{C}$ *is Schur stable and the matrix*

$$\begin{bmatrix} \tilde{\mathbf{I}} & (\tilde{\mathbf{A}} - \tilde{\mathbf{B}})\,\tilde{\mathbf{C}}^{-1} \\ -\tilde{\mathbf{I}} & \tilde{\mathbf{B}}\tilde{\mathbf{C}}^{-1} \end{bmatrix} \tag{7}$$

is a block P-matrix with respect to the partition $\{\{1,\ldots,3n^2\}, \{3n^2+1,\ldots,6n^2\}\}$ *of* $\{1,\ldots, 6n^2\}$.

(iii) $\mathbf{C}$ *is Schur stable and all convex combinations of* $\tilde{\mathbf{A}}, \tilde{\mathbf{B}}, \tilde{\mathbf{C}}$ *are nonsingular.*

Proof. (ii) $\Longrightarrow$(i): We show it in two steps. First we prove that (ii) $\Longrightarrow$(iii) and then that (iii) $\Longrightarrow$(i). Observe that the Schur stability of $\mathbf{C}$, together with Theorem 4.2.12 from [2] on eigenvalues of the Kronecker product, implies that $\hat{\mathbf{C}}$ is nonsingular. Then, from the definition of $\tilde{\mathbf{C}}$, it follows at once that $\tilde{\mathbf{C}}$ is also nonsingular. Hence, as (7) is a block P-matrix with respect to the partition $\{\{1,\ldots,3n^2\}, \{3n^2+1,\ldots, 6n^2\}\}$ of $\{1,\ldots, 6n^2\}$, all convex combinations of $\tilde{\mathbf{A}}$, $\tilde{\mathbf{B}}, \tilde{\mathbf{C}}$ are nonsingular by Theorem 3. So, (iii) is proved. Now we show (iii) $\Longrightarrow$(i). ¿From (iii) we have that for all nonnegative α_i, $i = 1,2,3$, with $\sum_{i=1}^{3} \alpha_i = 1$,

$$det(\alpha_1 \tilde{\mathbf{A}} + \alpha_2 \tilde{\mathbf{B}} + \alpha_3 \tilde{\mathbf{C}}) \neq 0$$

which, by the definitions of $\tilde{\mathbf{A}}, \tilde{\mathbf{B}}$, and $\tilde{\mathbf{C}}$, becomes

$$det \begin{bmatrix} \alpha_1 \hat{\mathbf{A}} + \alpha_2 \hat{\mathbf{B}} + \alpha_3 \hat{\mathbf{C}} & \alpha_2 \hat{\mathbf{D}} + \alpha_3 \hat{\mathbf{E}} & \alpha_3 \hat{\mathbf{F}} \\ \alpha_1 \hat{\mathbf{I}} & \hat{\mathbf{I}} & 0 \\ \alpha_2 \hat{\mathbf{I}} & 0 & \hat{\mathbf{I}} \end{bmatrix} \neq 0.$$

Hence, using the Schur determinantal formula and definitions of $\widetilde{\mathbf{A}}, \widetilde{\mathbf{B}}$, and $\widetilde{\mathbf{C}}$, we get that the nonsingularity of all convex combinations of $\widetilde{\mathbf{A}}, \widetilde{\mathbf{B}}, \widetilde{\mathbf{C}}$ is equivalent to the nonsingularity of the matrix

$$\alpha_1 \hat{\mathbf{A}} + \alpha_2 \hat{\mathbf{B}} + \alpha_3 \hat{\mathbf{C}} - \alpha_1(\alpha_2 \hat{\mathbf{D}} + \alpha_3 \hat{\mathbf{E}}) - \alpha_2\alpha_3 \hat{\mathbf{F}}$$

$$= \mathbf{I} \otimes \mathbf{I} - (\alpha_1\mathbf{A} + \alpha_2\mathbf{B} + \alpha_3\mathbf{C}) \otimes (\alpha_1\mathbf{A} + \alpha_2\mathbf{B} + \alpha_3\mathbf{C}) \tag{8}$$

(for all nonnegative $\alpha_i, i = 1, 2, 3$, with $\sum_{i=1}^{3} \alpha_i = 1$).

Suppose now that for some nonnegative $\widehat{\alpha}_i, i = 1, 2, 3$, such that $\sum_{i=1}^{3} \widehat{\alpha}_i = 1$, and $\widehat{\alpha}_1 + \widehat{\alpha}_2 > 0$ we have

$$\rho(\widehat{\alpha}_1 \mathbf{A} + \widehat{\alpha}_2 \mathbf{B} + \widehat{\alpha}_3 \mathbf{C}) \geq 1.$$

Then, as $\mathbf{C}$ is Schur stable and the matrix (8) is nonsingular for all nonnegative $\alpha_i \geq 0, i = 1, 2, 3$, with $\sum_{i=1}^{3} \alpha_i = 1$, by continuity there exists a matrix

$$\breve{\alpha}_1 \mathbf{A} + \breve{\alpha}_2 \mathbf{B} + \breve{\alpha}_3 \mathbf{C}$$

having an eigenvalue with modulus equals one, where $\breve{\alpha}_i, i = 1, 2, 3$, are such that $0 \leq \breve{\alpha}_i \leq \widehat{\alpha}_i$, $\breve{\alpha}_1 + \breve{\alpha}_2 > 0$ and $\sum_{i=1}^{3} \breve{\alpha}_i = 1$. Hence, by the mentioned above Theorem 4.2.12, the matrix

$$\mathbf{I} \otimes \mathbf{I} - (\breve{\alpha}_1 \mathbf{A} + \breve{\alpha}_2 \mathbf{B} + \breve{\alpha}_3 \mathbf{C}) \otimes (\breve{\alpha}_1 \mathbf{A} + \breve{\alpha}_2 \mathbf{B} + \breve{\alpha}_3 \mathbf{C})$$

is singular which contradicts (8). So, (iii) $\Longrightarrow$(i) holds.

(i) $\Longrightarrow$(ii): Observe that, by (i), the Schur stability of $\mathbf{C}$ is obvious. Now, as all convex combinations of $\mathbf{A}$,$\mathbf{B}$ and $\mathbf{C}$ are Schur stable, using again Theorem 4.2.12 from [2] we get that the matrix

$$\mathbf{I} \otimes \mathbf{I} - (\alpha_1\mathbf{A} + \alpha_2\mathbf{B} + \alpha_3\mathbf{C}) \otimes (\alpha_1\mathbf{A} + \alpha_2\mathbf{B} + \alpha_3\mathbf{C})$$

is nonsingular for all nonnegative $\alpha_i, i = 1, 2, 3$, with $\sum_{i=1}^{3} \alpha_i = 1$. Hence, repeating the argument used in the proof of the inverse implication, we obtain that all convex combinations of $\widetilde{\mathbf{A}}, \widetilde{\mathbf{B}}, \widetilde{\mathbf{C}}$ are nonsingular. To show the implication in question it suffices to observe that $\widetilde{\mathbf{C}}$ is nonsingular and apply Theorem 3.

□

We remark that setting $\mathbf{B} = \mathbf{A}$ *in Theorem 3 and Theorem 4, we regain (in the real case) the result of [3], while Theorem 4 leads to a characterization of Schur stability for all convex combinations of two matrices, which is different from that given in [6].*

References

1. Elsner, L., Szulc T.: Block P-matrices. Linear and Multilinear Algebra, submitted
2. Horn, R.A., Johnson C.R.: Topics in matrix analysis, Cambridge U.P.,Cambridge, 1991
3. Johnson, C.R., Tsatsomeros: Convex sets of nonsingular and P-matrices. Linear and Multilinear Algebra, **38** (1995) 233–240
4. Rohn, J.: System of linear equations. Linear Algebra Appl., **126** (1989) 39–78
5. Rohn, J.: Positive definiteness and stability of interval matrices. SIAM J. Matrix Anal. Appl., **15** (1994) 175–184
6. Soh, C.B.: Schur stability of convex combination matrices. Linear Algebra Appl., **128** (1990) 159–168

Numerical Solution of a BVP in Hydrodynamics (Thermocapillary Convection)

S.S. Filippov, Yu.V. Sanochkin, A.Y. Sogomonyan, and A. V. Tygliyan

Keldysh Institute of Applied Mathematics, Russian Academy of Sciences, 4 Miusskaya Square, Moscow 125047, Russia

Abstract. A boundary value problem describing thermocapillary convection in terms of Navier—Stokes and heat transfer equations is reduced to a b.v.p. for a system of ODE with an extra boundary condition. The latter problem is solved numerically.

1 Introduction

We study the system of ordinary differential equations

$$R^{-1}\phi_{yyy} - \phi\phi_{yy} + \phi_y^2 + b = 0, \tag{1a}$$

$$(PR)^{-1}\psi_{yy} - \phi\psi_y + 2\phi_y\psi = 0, \tag{1b}$$

$$(PR)^{-1}\theta_{yy} - \phi\theta_y - (PR)^{-1}\psi = 0 \tag{1c}$$

with the following boundary conditions:

$$\phi(0) = 0, \qquad \phi_y(0) = 0, \qquad \psi(0) = 0, \qquad \theta(0) = 0, \tag{2a}$$

$$\phi(1) = 0, \qquad \phi_{yy}(1) + \psi(1) = 0, \tag{2b}$$

and either

$$\psi(1) = \delta^2, \qquad \theta(1) = 1, \tag{2c}$$

or

$$\psi_y(1) = \delta^2, \qquad \theta_y(1) = 1. \tag{2d}$$

Here $\phi(y)$, $\psi(y)$, and $\theta(y)$ are the unknown functions to be determined in the interval $0 \le y \le 1$, and subscripts denote differentiation with respect to y. The solutions of the b.v.p. (1), (2) depend on three positive parameters R, P, and δ. Note that the system (1) is of the order 7, and there are 8 boundary conditions (2). This enables one to determine the unknown constant b in the Eq. (1a).

These problems emerged in the course of studying thermocapillary convection caused by a non-uniform heat flux on the free surface of a thin liquid layer (Sanochkin 1987; Sanochkin et al. 1987; Filippov and Sanochkin 1985). It is the temperature dependence of the surface tension that accounts for tangential forces, arising under a temperature gradient along the liquid surface and setting it into motion. The bulk of the liquid is involved into motion due to viscosity. Therefore, the Navier-Stokes equations is an appropriate tool to describe this phenomenon. We need also the equation of heat transfer in a flowing liquid. For a two-dimensional steady flow of an incompressible fluid these equations take the form

$$\begin{aligned} u_x + u_y &= 0, \\ uu_x + vu_y &= -\rho^{-1}p_x + \rho^{-1}\mu(u_{xx} + u_{yy}), \\ uv_x + vv_y &= -\rho^{-1}p_y + \rho^{-1}\mu(v_{xx} + v_{yy}), \\ uT_x + vT_y &= (c_p\rho)^{-1}\kappa(T_{xx} + T_{yy}), \end{aligned} \tag{3}$$

where $u(x,y)$ and $v(x,y)$ are the x and y components of the velocity vector, $p(x,y)$ is the pressure, $T(x,y)$ is the temperature, ρ is the density, μ is the viscosity, c_p is the heat capacity per unit mass, and κ is the heat conductivity. The subscripts denote partial differentiation with respect to x and y.

Let us place the origin at the bottom of the liquid layer and assume that its free surface remains plane despite of motion. Taking into account the symmetry of the flow with respect to the plane $x = 0$, we get a boundary value problem for the domain $0 \le x \le \infty, 0 \le y \le h$ with the following boundary conditions:

$$u(0,y) = 0, \qquad v_x(0,y) = 0, \qquad T_x(0,y) = 0 \tag{4a}$$

in the plane of symmetry $x = 0$,

$$u(x,0) = 0, \qquad v(x,0) = 0, \qquad T(x,0) = 0 \tag{4a$'$}$$

at the bottom $y = 0$,

$$v(x,h) = 0, \qquad \mu u_y(x,h) = \alpha_T T_x(x,h), \tag{4b}$$

and either

$$T(x,h) = T_s(x), \tag{4c}$$

or

$$T_y(x,h) = -\kappa^{-1}q_s(x) \tag{4d}$$

at the free surface $y = h$. Here α is the surface tension, $\alpha_T = d\alpha/dT < 0$, $T_s(x)$ is the temperature distribution, and $q_s(x)$ the heat flux on the surface. We assume the parabolic law for both of these functions:

$$T_s(x) = T_0(1 - x^2/2l^2), \tag{4c$'$}$$

$$q_s(x) = -\kappa T_0 h^{-1}(1 - x^2/2l^2). \tag{4d$'$}$$

There are eight values in Eqs (3) and (4) supposed to be constant: h, l, ρ, μ, c_p, κ, α_T, and T_0. The number of constants can be decreased substantially, if we introduce scaled variables

$$x/h, \quad y/h, \quad u/V, \quad v/V, \quad p/p_0, \quad \text{and } T/T_0,$$

where $V = |\alpha_T|T_0\mu^{-1}$ and $p_0 = \rho V^2$. Let us rewrite our equations and boundary conditions, using the same designations for the scaled variables as for unscaled:

$$\begin{aligned} u_x + u_y &= 0, \\ uu_x + vu_y &= -p_x + R^{-1}(u_{xx} + u_{yy}), \\ uv_x + vv_y &= -p_y + R^{-1}(v_{xx} + v_{yy}), \\ uT_x + vT_y &= (PR)^{-1}(T_{xx} + T_{yy}), \end{aligned} \tag{5}$$

$$u(0,y)=0, \qquad v_x(0,y)=0, \qquad T_x(0,y)=0, \tag{6a}$$

$$u(x,0)=0, \qquad v(x,0)=0, \qquad T(x,0)=0, \tag{6a'}$$

$$v(x,1)=0, \qquad u_y(x,1)=T_x(x,1), \tag{6b}$$

and either

$$T(x,1)=1-\delta^2x^2/2, \tag{6c}$$

or

$$T_y(x,1)=1-\delta^2x^2/2, \tag{6d}$$

The solutions of these b.v.p. in the domain $0 \le x \le \infty, 0 \le y \le 1$ are controlled by only three parameters. These are the Reynolds number $R = \rho h V \mu^{-1} = \rho h |\alpha_T| T_0 \mu^{-2}$, the Prandtl number $P = \mu c_p \kappa^{-1}$, and the geometric ratio $\delta = h/l$.

Now we shall look for solutions of b.v.p. (5), (6) with separated independent variables, and for this sake we put

$$\begin{aligned} u(x,y) &= -x\phi_y(y), \\ v(x,y) &= \phi(y), \\ p(x,y) &= a(y) + x^2 b(y)/2, \\ T(x,y) &= \theta(y) - x^2\psi(y)/2. \end{aligned} \tag{7}$$

The substitution of these functions into Eqs (5) and (6) yields the Eqs (1), (2), and

$$a_y = R^{-1}\phi_{yy} - \phi\phi_y, \qquad b_y = 0. \tag{8}$$

In Section 2 we reduce the b.v.p. (1), (2) to several initial value problems. Numerical solution of those problems is described in Section 3. Some computational results are discussed in Section 4. Section 5 contains concluding remarks.

2 Reduction to initial value problems

We begin with some transformations of the equations (1) and (2). First of all, we put

$$\phi(y) = R^{-1}\Phi(y) \tag{9}$$

and get from (1a), (2a), and (2b)

$$\Phi_{yyy} - \Phi\Phi_{yy} + \Phi_y^2 + b^* = 0, \tag{10}$$

$$\Phi(0)=0, \qquad \Phi_y(0)=0, \qquad \Phi(1)=0, \qquad \Phi_{yy}(1) = -R^*, \tag{11}$$

where

$$b^* = R^2 b, \qquad R^* = kR, \qquad k = \psi(1) > 0. \tag{12}$$

In this way we disjoint the dynamical part of the problems (1), (2) from the thermal one. For the function $\Phi(y)$ we have now a separate boundary value problem with a single parameter R^*. The superfluous boundary condition can be employed to relate the unknown constant b^* with R^*.

This boundary value problem is similar to that investigated by Brady and Acrivos (1981). They have studied viscid flow in a flat channel, the walls of which were moving with a velocity directly proportional to x. Our Eq. (10) is just their basic equation, though the boundary conditions (11) are different.

In the case (2c), i.e. when the temperature distribution on the surface is prescribed, the solution of b.v.p. (10), (11) is fully independent from the solution of the thermal problem, since $k = \delta^2$ is known. However, in the case (2d), i.e. with a given heat flux on the surface, the value of $k = \psi(1)$ is still unknown, and we cannot retrieve the values $R = k^{-1}R^*$ and $b = R^{-2}b^*$ without solving the equation for $\psi(y)$. Nevertheless, we use successfully the fruitful idea of Brady and Acrivos (1981) to convert a boundary value problem into an initial value one, and the techniques they had developed.

To this aim we apply the transformation

$$t = By, \qquad z = B^{-1}\Phi \tag{13}$$

to the equations (10) and (11):

$$z_{ttt} - zz_{tt} + z_t^2 + \beta = 0, \tag{14}$$

$$z(0) = 0, \qquad z_t(0) = 0, \qquad z(B) = 0, \qquad z_{tt}(B) = -B^{-3}R^*. \tag{15}$$

Here B is an arbitrary positive number, and

$$\beta = B^{-4}b^*. \tag{16}$$

Furthermore, for the equation (14) we consider a family of initial value problems, depending on parameter β $(-\infty < \beta < \infty)$, with initial values

$$z(0) = 0, \qquad z_t(0) = 0, \qquad z_{tt}(0) = z_{20}, \tag{17}$$

where z_{20} is an arbitrary but fixed real number. If for a given value of β there exists such a value of t, $t = t_1 > 0$, that $z(t_1) = 0$ and $z_{tt}(t_1) < 0$, then we put $B = t_1$ and calculate $R^* = R^*(\beta)$ from the last of the equations (15). The curve in the (R, β) plane, computed in this way, does not depend on the absolute value of z_{20}. Indeed, for $z_{20} \neq 0$ the transformation $\tau = \tilde{B}t$, $\zeta = \tilde{B}^{-1}z$, $\tilde{\beta} = \tilde{B}^{-4}\beta$ with $\tilde{B} = |z_{20}|^{1/3}$ yields the equation

$$\zeta_{\tau\tau\tau} - \zeta\zeta_{\tau\tau} + \zeta_\tau^2 + \tilde{\beta} = 0$$

and the initial value $\zeta_{\tau\tau}(0) = +1$ for $z_{20} > 0$, or -1 for $z_{20} < 0$. Hence, three values 1, 0, and -1 for z_{20} in (17) are sufficient. And vice versa, one may fix three values $+1$, -1, and 0 for β in (14) and consider three corresponding sets of initial value problems with a single parameter z_{20} $(-\infty < z_{20} < \infty)$. We have used both of these procedures in our computations.

Let us turn now to remaining equations in (1), (2). Taking advantage of the fact that the equation (1b) is linear and homogeneous with respect to $\psi(y)$, we set

$$\psi(y) = A\Psi(t) \tag{18}$$

and, making use of the transformations (9) and (13), reduce the boundary value problem (1b), (2a), and either (2c) or (2d) for ψ to that for Ψ:

$$P^{-1}\Psi_{tt} - z\Psi_t + 2z_t\Psi = 0, \tag{19}$$

$$\Psi(0) = 0, \tag{20a}$$

and either

$$\Psi(B) = A^{-1}\delta^2 \tag{20c}$$

or

$$\Psi_t(B) = A^{-1}B^{-1}\delta^2. \tag{20d}$$

It is clear that A can be found from the boundary conditions (20c) or (20d), if we solve the equation (19) with the initial values

$$\Psi(0) = 0, \qquad \Psi_t(0) = 1, \tag{21}$$

once $z(t)$ and B are found from (14)–(17). Thus, the value of $k = \psi(1)$ in (12) for the case (2d) is

$$k = A\Psi(B) = (B\Psi_t(B))^{-1}\Psi(B)\delta^2. \tag{22}$$

By substituting (9), (13), (18), and

$$\theta(y) = AB^{-2}\Theta(t) \tag{23}$$

into the equations (1c), (2a), and either (2c) or (2d) we get

$$\Theta_{tt} - Pz\Theta_t - \Psi = 0, \tag{24}$$

$$\Theta(0) = 0, \tag{25a}$$

and either

$$\Theta(B) = A^{-1}B^2 \tag{25c}$$

or

$$\Theta_t(B) = A^{-1}B. \tag{25d}$$

The equation (24) is a first order non-homogeneous equation with respect to $\Theta'(t) = \Theta_t(t)$. Hence, its common solution is

$$\Theta'(t) = C\Theta_0'(t) + \Theta_1'(t), \tag{26}$$

where $\Theta_0'(t)$ and $\Theta_1'(t)$ are solutions of two initial value problems for the associated homogeneous equation

$$\frac{d\Theta_0'}{dt} - Pz\Theta_0' = 0, \qquad \Theta_0'(0) = 1, \tag{27}$$

and for non-homogeneous equation

$$\frac{d\Theta_1'}{dt} - Pz\Theta_1' = \Psi, \qquad \Theta_1'(0) = 0, \tag{28}$$

respectively. Furthermore, taking into account (25a), we can write

$$\Theta(t) = C\Theta_0(t) + \Theta_1(t), \tag{29}$$

where $\Theta_i(t)$, $i = 0, 1$ are solutions of the initial value problems

$$\frac{d\Theta_i}{dt} = \Theta_i', \qquad \Theta_i(0) = 0 \qquad (i = 0,\ 1). \tag{30}$$

In the case (2c) of prescribed temperature distribution on the surface we have from (25c) and (29)

$$C = \frac{A^{-1}B^2 - \Theta_1(B)}{\Theta_0(B)}. \tag{31}$$

For the case (2d) with a given heat flux on the surface we get from (25d) and (26)

$$C = \frac{A^{-1}B - \Theta_1'(B)}{\Theta_0'(B)}. \tag{32}$$

Finally, we substitute (9), (13), and

$$a(y) = R^{-2}B^2\alpha(t) \tag{33}$$

into the first equation (8). This yields

$$\alpha_t = z_{tt} - zz_t. \tag{34}$$

The initial value of $\alpha(t)$ can be chosen arbitrarily, say

$$\alpha(0) = 0, \tag{35}$$

because Navier—Stokes equations (5) contain only derivatives of the pressure.

3 Computational considerations

Because of computational convenience considerations all the initial value problems described in Section 2 ought to be solved simultaneously. To this end let us introduce new variables

$$\begin{aligned} &Y_1 = z, \quad Y_2 = z_t, \quad Y_3 = z_{tt}, \quad Y_4 = \Psi, \quad Y_5 = \Psi_t, \\ &Y_6 = \Theta_0, \quad Y_7 = \Theta_0', \quad Y_8 = \Theta_1, \quad Y_9 = \Theta_1', \quad Y_{10} = \alpha, \end{aligned} \tag{36}$$

and write out the entire system of 1st-order ordinary differential equations with relevant initial conditions: $Y_3(0) = z_{20}$, $Y_5(0) = Y_7(0) = 1$, otherwise $Y_i(0) = 0$; $\beta\ (-\infty < \beta < \infty)$, z_{20} ($z_{20} = -1, 0, +1$), and P are parameters.

For the numerical integration of this system we have used DOPRI5, a code described in (Hairer et al. 1993). This routine is based on the 5th-order explicit Runge—Kutta method of Dormand and Prince (1980) with an incorporated interpolation formula.

With fixed values of β, z_{20}, and P we proceed step by step up to the value $t = B$, at which $Y_1(B) = 0$ and $Y_3(B) < 0$. Once such a value of t is found,

$R^* = -B^3Y_3(B)$ and $b^* = \beta(BY_3(B))^{-2}$ can be computed. We failed to prove that such a B exists always and is unique. Therefore, numerical integration was continued each time for $t > B$, but further zeros of $Y_1(t)$ were never found. Moreover, for certain combinations of β and z_{20} no zeros of $Y_1(t)$ were found at all for $t > 0$.

Finally, combining the equations (9), (12), (13), (16), (18), (20), (22), (23), (25), (29), (31)–(33), and (36), we can return to the original variables.

4 Numerical results and discussion

We have calculated velocity and temperature profiles in a wide range of parameter values. A detailed presentation of numerical results should be published elsewhere. Here only several typical examples will be discussed, mainly for the case of prescribed temperature distribution (4c′) along the surface (boundary conditions (2c)), and for a fixed value $\delta = 0.5$.

The function $\phi_y(y) = -u(x,y)/x$ represents the profile of the velocity component directed along the liquid layer. At low values of R^* it has a sole zero in the interval $0 < y < 1$, and $\phi_{yy}(0) > 0$. This means that the flow is in the positive direction of the x-axis near the surface, and in the negative direction near the bottom of the layer. The quantity $\phi_{yy}(0)$ changes its sign at $R^* = 6360$. For all R^* greater then this critical value, $\phi_{yy}(0)$ is negative, and $\phi_y(y)$ has two zeros between $y = 0$ and $y = 1$. The flow near the bottom now has the same direction as that near the surface, and the counterstream is located at the middle depth in the layer.

The temperature distribution $T(x,y)$ is defined by the functions $\theta(y)$ and $\psi(y)$ [see Eq. (7)]. With exception of very small values of P, the behaviour of these functions is quite different for subcritical and supercritical values of Reynolds number. At $x = 0$ the temperature varies rapidly near the surface in subcritical flows, and near the bottom otherwise.

It is of interest to examine the dependence of b^* on R^*. Negative sign and monotonous growth of b^* are characteristic at low values of R^*. The sign of b^* changes at $R^* \approx 1000$, but the absolute value of b^* remains very small for all $R^* > 1000$, and $b^* \to 0$ when $R^* \to \infty$. The curve $b^*(R^*)$ has two remarkable peculiarities. Firstly, it is not monotonous. There are two maxima and one minimum. The minimum is situated just at the critical value $R^* = 6360$. Secondly, this curve has an S-shaped portion between $R^* = 8125$ and $R^* = 8615$. Three different solutions of the b.v.p. (1), (2) exist for each value of R^* in this range.

Finally, it should be mentioned that velocity profiles are rather sensitive to the magnitude of the Prandtl number P. The effects of increasing both P and R^* on the velocity profile are alike. Indeed, this observation is in accordance with the well known fact that Marangoni number $\mathbf{Ma} = PR$ is exactly the parameter of importance in thermocapillary problems.

5 Concluding remarks

The boundary value problems (1), (2), that we have studied numerically, have their origin from a peculiar case of thermocapillary convection. Although the Marangoni number $\mathbf{Ma} = PR$ is characteristic for thermocapillary problems, we have chosen the Reynolds number R as a parameter by reason of similitude of the dynamical part in our problems to the problem investigated by Brady and Acrivos (1981). The basic equation of their work is precisely our equation (1a), but with other boundary conditions: $\phi(0) = \phi_{yy}(0) = \phi(1) = 0,\ \phi_y(1) = -1$.

Basic attention should be paid to bifurcation phenomena that are of particular interest. It is instructive to compare our results with those obtained in the problem studied by Brady and Acrivos (1981). Rebuilding of the flow structure occurs at certain critical values of R in both cases. On the other hand, the dependence of the pressure coefficient b on R is quite different. In our case there is a sole curve with an S-shaped portion, whereas Brady and Acrivos (1981) have found three disconnected, but intersecting branches (cf. Fig. 1 in their paper).

As soon as the presence of branching phenomena is established, it is naturally to question the stability of each mode of the flow. The authors hope to devote another paper to this topic.

References

1. Brady, J.F., Acrivos, A.: Steady flow in a channel or tube with an accelerating surface velocity. J. Fluid Mech. **112** (1981) 127
2. Dormand, J.R., Prince, P.J.: A family of embedded Runge—Kutta formulae. J. Comp. Appl. Math. **6** (1980) 19
3. Filippov, S.S., Sanochkin, Yu.V.: On the nature of mutual repulsion of cathode spots: a possible explanation. XVIIth Int. Conf. on Phenomena in Ionized Gases, Budapest 1985, Contributed Papers **1** 481
4. Hairer, E., Nørsett, S.P., Wanner, G.: Solving Ordinary Differential Equations, I: Nonstiff Problems. Second edition (1993), Springer-Verlag, Berlin
5. Sanochkin, Yu.V.: Thermocapillary convection and dynamics of the cathode spot at the liquid cathode. Nuovo cimento D **9D** No. 8 (1987) 941
6. Sanochkin, Yu.V., Tukhvatullin, R.S., Filippov, S.S.: Numerical simulation of thermocapillary convection in a liquid layer under local heating of its free surface. Izvestiya Akademii Nauk SSSR, Ser. Mekhanika Zhidkosti i Gaza No. 4 (1987) 108 (Russian)

Componentwise Error Bounds and Direct Linear System Solving

B. Hörmann* and C. P. Ullrich**

Universität Basel,
Institut für Informatik,
Mittlere Strasse 142,
CH-4056 Basel

Abstract. It is shown that componentwise relative error bounds can be computed efficiently in connection with direct linear system solving. No restriction is made on the data matrix. In a series of experiments the effectiveness of the componentwise approach is demonstrated and compared to the conventional normwise approach as incorporated in the numerical linear algebra library LAPACK.

1 Introduction

Consider the system of linear equations

$$Ax = b \ , \tag{1}$$

where A is a nonsingular real n-by-n matrix and x is a real n dimensional vector. We focus on those methods that compute an approximate solution $\widehat{x}$ of (1) by solving $F^{-1}\widehat{x} = b$ with backward/forward substitution (F^{-1} denotes a computed factorization of A, e.g. $F^{-1} \approx LU$ for Gaussian LU decomposition). From

$$\widehat{x} = Fb$$

it follows that

$$\frac{\|x - \widehat{x}\|_\infty}{\|x\|_\infty} \le \|I - FA\|_\infty$$

so that we assume $\|I - FA\|_\infty$ to be significantly less than 1. In the following section we derive upper bounds for the componentwise *absolute* error which serve in Section 3 as a basis for the computation of the largest *relative* error in $\widehat{x}$

$$\varepsilon = \max_{1 \le i \le n} \frac{|x_i - \widehat{x}_i|}{|x_i|} \ , \tag{2}$$

where x_i does not vanish.

For the rest of the paper all indices are thought to range from 1 to n.

* e-mail: hoermann@ifi.unibas.ch
** e-mail: ullrich@ifi.unibas.ch

2 Theorem of Brouwer and Bounds for the Absolute Error

Our starting point is the famous fixed-point theorem of Brouwer.

Theorem of Brouwer. *Let f be a continous mapping from a convex and compact subset S of the euclidian space $\mathbb{R}^n$ into itself. Then there exists $x \in S$ such that $f(x) = x$.*

An elegant proof can be found in the awarded paper of Gale [5].

The following corollary relates Brouwer's theorem to linear system solving. Absolute values and relations are taken entrywise[3].

Corollary. *Let $\Phi\colon y \mapsto My + q$ be a linear mapping with $M \in \mathbb{R}^{n\times n}$ and $y, q \in \mathbb{R}^n$. If for $u \in \mathbb{R}^n$ there is a positive vector $z \in \mathbb{R}^n$ such that*

$$|\Phi(u) - u| \le (I - |M|)z \ ,$$

then there exists a fixed-point $t \in \mathbb{R}^n$ of Φ and

$$|t - u| \le z \ .$$

For a proof see [4].

Relating the corollary to the purpose of finding error bounds, we search a linear mapping that has for its unique fixed-point the solution of the linear system. For the later discussion $\Phi\colon y \mapsto (I - FA)y + Fb$ will be of interest. It is easy to verify that Φ satisfies $x = \Phi(x)$, where x is the solution of (1).

Proposition. *Let $r = A\widehat{x} - b$ be the residual and $F \in \mathbb{R}^{n\times n}$. If there is a positive vector $z \in \mathbb{R}^n$ such that*

$$|Fr| + |I - FA|z \le z \tag{3}$$

then

$$|x - \widehat{x}| \le z \ .$$

Proof. Set $M = I - FA$, $q = Fb$, $u = \widehat{x}$ and apply the corollary. □

Since we assume $\|I - FA\|_\infty$ less than 1 (cf. Section 1) we conclude that z exists. Two feasible ways for determining z are known to us. The first one is due to Collatz [4] the second one was used by Rump [3] and Falcó Korn [7] for their derivation of so-called *verified* error bounds. Rump considered symmetric M-matrices while Falcó Korn looked at (symmetric and nonsymmetric) H-matrices.

In [4] a $z > 0$ is suggested that has all components equal to a positive number. With $z_i = \gamma$ the hypothesis of the proposition (3) requires

$$|Fr|_i + \gamma \sum_j |I - FA|_{ij} \le \gamma$$

[3] $|M| \in \mathbb{R}^{n\times n}$ denotes the matrix whose elements are $|m_{ij}|$ and $|v| \in \mathbb{R}^n$ denotes the vector whose components are $|v_i|$. Similar, $v \ge 0$ denotes $v_i \ge 0$.

which is satisfied by

$$\gamma = \max_i \frac{|Fr|_i}{1 - \sum_j |I - FA|_{ij}} \ .$$

Each component of the vector of absolute errors is therefore bounded from above by γ. For the upcoming discussion we shall be content with a less rigorous bound obtained using norms rather than single components. Of course,

$$\gamma \leq \frac{\|Fr\|_\infty}{1 - \|I - FA\|_\infty} \ .$$

Thus, with

$$\delta_\mathrm{C} = \frac{1}{1 - \|I - FA\|_\infty}$$

we conclude

$$\boxed{\|x - \hat{x}\|_\infty \leq \delta_\mathrm{C} \, \|Fr\|_\infty} \ . \tag{4}$$

For a z with different components consider $z = \mu|Fr|$, where μ can be determined similar to the case above from the hypothesis of the proposition. We have

$$\mu = \max_i \frac{|Fr|_i}{|Fr|_i - (|I - FA||Fr|)_i} \ . \tag{5}$$

Suppose $|Fr|_i > 0$ and set

$$\eta = \frac{\max_i |Fr|_i}{\min_i |Fr|_i} \ .$$

Dividing the nominator and the denominator in (5) by $|Fr|_i$ and observing that

$$\frac{(|I - FA||Fr|)_i}{|Fr|_i} \leq \eta \, \|I - FA\|_\infty$$

it follows

$$\mu \leq \frac{1}{1 - \eta \, \|I - FA\|_\infty} \ , \tag{6}$$

where it is assumed that

$$\eta \, \|I - FA\|_\infty < 1 \ . \tag{7}$$

Let δ_r denote the right-hand side of inequality (6). We conclude

$$\boxed{|x - \hat{x}| \leq \delta_\mathrm{r} \, |Fr|} \ . \tag{8}$$

In Section 4 we shall address the case where η is too large for (7) to be satisfied.

3 Bounds for the Accuracy

If some approximation has m correct significant digits then it is said to have an *accuracy* of m digits. To be precise, using (2) we mean the quantity[4]

$$\lfloor \log_{10}(\varepsilon) \rfloor \ .$$

The exact value of ε is known only if the solution x of (1) is at hand. Suppose an upper bound β for ε can be computed instead. Translating relative errors to accuracies this means that *at least* $\lfloor \log_{10}(\beta) \rfloor$ decimal digits are correct in each component of the approximation $\widehat{x}$. Actually, β can be obtained via the absolute error and the approximation $\widehat{x}$. Let c denote an upper bound of the absolute error

$$|x_i - \widehat{x}_i| \le c_i \ .$$

It follows that

$$\varepsilon = \max_i \frac{|x_i - \widehat{x}_i|}{|x_i|} \le \max_i \frac{c_i}{|\widehat{x}_i| - c_i} \ ,$$

where it is assumed that $|\widehat{x}_i| > c_i$. Overall, considering (4) and (8)

$$\varepsilon \le \max_i \frac{\delta_{\mathrm{C}} \|Fr\|_\infty}{|\widehat{x}_i| - \delta_{\mathrm{C}} \|Fr\|_\infty} \tag{9}$$

and

$$\varepsilon \le \max_i \frac{\delta_{\mathrm{r}} |Fr|_i}{|\widehat{x}_i| - \delta_{\mathrm{r}} |Fr|_i} \ . \tag{10}$$

(The denominators are assumed to be positive.)

4 Limiting the Growth of η

In Section 5 we shall see that error estimation according to (9) can be a good choice. But there are cases where restricting the vector of absolute errors to identical components gives very conservative estimations. On the other hand, error estimation according to (10) requires that the components of $|Fr|$ do not differ too much. This leads to the idea of "raising" small components of $|Fr|$. To this end we define the vector y by

$$y_i = \max \left\{ \frac{\|Fr\|_\infty}{\tau}, |Fr|_i \right\} \ ,$$

where for now τ denotes a real number greater than or equal to 1. Obviously, the components of y are found within $\|Fr\|_\infty / \tau$ and $\|Fr\|_\infty$ so that

$$\frac{\max_i y_i}{\min_i y_i} \le \tau \ . \tag{11}$$

[4] $\lfloor x \rfloor$ denotes the largest integer less than or equal to x

Consider the hypothesis of the proposition (3) for

$$z = \frac{1}{1-\omega}\, y \quad (0 < \omega < 1) \; .$$

From $|Fr| \le y$ it follows, using norms

$$|Fr|_i + \frac{1}{1-\omega}\,(|I - FA|\, y)_i \;\le\; y_i + \frac{1}{1-\omega}\,\|I - FA\|_\infty \,\|y\|_\infty$$

which is required to be less than or equal to z, hence

$$y_i - \omega y_i + \|I - FA\|_\infty \,\|y\|_\infty \le y_i$$

or equivalently

$$\|I - FA\|_\infty \frac{\|y\|_\infty}{\min_i y_i} \le \omega \; .$$

From (11) we conclude that

$$\tau \le \frac{\omega}{\|I - FA\|_\infty}$$

is sufficient for (3) to be satisfied. The quantity τ limits the range of differing elements of $|Fr|$ depending on the quality of the factorization of the data matrix. If the factorization is good (that is, if $\|I - FA\|_\infty$ is small) then the components of the vector of absolute errors are allowed to differ much in their magnitudes, whereas they are restricted to a small bandwidth in the case where F^{-1} is a bad factorization (typically if A is ill-conditioned). Figure 1 summarizes the procedure. Relative errors are computed from absolute errors as described in Section 3.

Fig. 1. Procedure that describes the computation of an upper bound of the largest relative error in an approximate solution to a system of linear equations (see (2)). The value of ν is returned.

IF $\|I - FA\|_\infty \le \omega$ THEN

  $\tau := \omega / \|I - FA\|_\infty$

  $y_i := \max\{\|Fr\|_\infty/\tau, |Fr|_i\}, \quad i = 1, \ldots, n$

  $\nu := \max_i (y_i / (|\widehat{x}_i| - \omega|\widehat{x}_i| - y_i))$

ELSE

  $\nu := 1$

END

5 Practical Experience

5.1 Implementation

In our C-implementation of the algorithm described in Fig. 1 we set ω to 0.9. For a rough estimation of $\|I - FA\|_\infty$ we took $\max_i |1 - Fv|$ where v denotes the vector of row sums of A, that is $v_i = \sum_j a_{ij}$.

To reduce the effects of numerical cancellation, we computed $|Fr|$ twice exploiting the idendity

$$F(A\widehat{x} - b) = FA\widehat{x} - \widehat{x} \ .$$

For the same reason, care has to be taken when evaluating $A\widehat{x}$. We tried to avoid summing up subsequent terms which differ much in their magnitudes.

(The only neat way to cope with catastrophic cancellation would be the consequent use of high accurate dot products according to the GAMM-IMACS proposal [10]. Despite the fact that hardware implementations of the high accurate dot product perform as well as conventional loop-programmed versions [11], their use is still "exotic".)

The overall performance of our implementation is governed by 5 operations of complexity $\mathcal{O}(n^2)$: one backward/forward substitution for Fv, two backward/forward substitutions for the computation of Fr, one matrix-vector multiplication for $A\widehat{x}$ and one operation for building the row sums of A.

In a real test environment we experienced a runtime performance which is comparable to that of LAPACK's error routines.

5.2 Numerical Results

Our testset currently comprises 28 matrices (parts of them are from the Harwell-Boeing collection [8]) and 6 different right-hand sides[5]. The largest matrix is of order 1140. Condition numbers range from 1 to 10^{14}. A detailed description of the testset is published elsewhere. Contact the authors for immediate information.

All computations are done on a SUN Sparc-Station 5. Floating-point numbers conform to the IEEE double format which corresponds to at least 16 decimal digits.

The numerical experiments are run on a whole set of samples. Each sample involves four steps: 1) Choose a matrix M and a "solution" vector s from the testset. 2) Compute the right-hand side by multiplying M with s. 3) Compute an approximate solution $\widehat{s}$ of the chosen linear system. 4) Compute the "true accuracy" and the "estimated accuracy". The approximation is calculated by LAPACK's GE routines [12]. By "true accuracy" we mean $\lfloor \log_{10}(\alpha) \rfloor$, where

$$\alpha = \max_i \frac{|s_i - \widehat{s}_i|}{|s_i|} \ .$$

In a first experiment we computed the error according to (9) where we set δ_C to 10. This is reasonable because we do not expect $\|I - FA\|_\infty$ to be larger than

[5] Actually we store the solution vectors.

0.9. The result is displayed in the leftmost graphic of Fig. 2. In all samples the computed accuracy is equal to or less than the true accuracy. No overestimations occurred. Two samples revealed an underestimation of four resp. five digits.

In the next experiment we turn our attention to the algorithm of Fig. 1 (see Section 5.1 for implementation details). The result is summarized in the second graphic of Fig. 2. Again, no overestimations occurred. Compared to the first experiment the "center of gravity" of the frequency graph has moved to the left.

It is well known that normwise relative errors like $\|x - \widetilde{x}\|_\infty / \|x\|_\infty$ have no clear connection to accuracy (see for example [1], [2] or Ex. 2.2.1, p. 54 of [6]). The following experiment demonstrates this fact again where we refrain from using samples with solution vectors that differ much in the magnitudes of their components because this almost always implies severe overestimations. LAPACK provides two different methods for calculating normwise errors: one via an estimation of the condition number, the other combined with an iterative refinement. For details refer to [12] or [1]. The rightmost graphic of Fig. 2 shows the result applying LAPACK's condition estimation method. Note the wide range of deviations between the computed accuracies and the true accuracy. Overestimation occurred in 28% of all samples from the reduced testset. We observed the same behaviour for LAPACK's DGERFS error routine.

Fig. 2. Frequency distribution of samples with same difference between true and computed accuracy. Accuracy estimation on the left is based on (9). Accuracy estimation in the middle is done according to the algorithm of Fig. 1. Accuracy estimation on the right is based on LAPACK's DGECON routine. Labels on the x-axis indicate the deviation of the computed accuracy from the true accuracy. Negative numbers reveal overestimations. Total number of samples for both graphs on the left is 168 and 112 for the graph on the right.

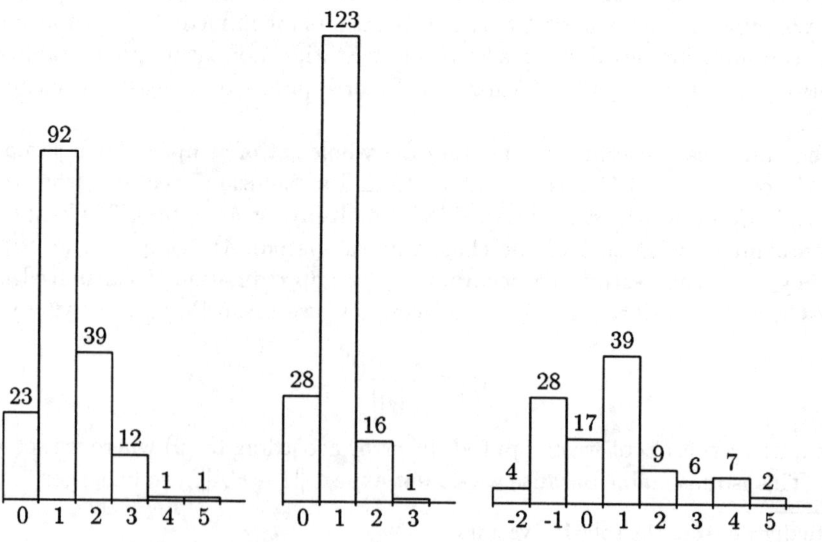

6 Conclusion

Considering the large condition numbers of some testset matrices we think that results like those in the middle of Fig. 2 are very promising especially when compared to conventional normwise error bounds.

We also made experiments with large banded systems up to dimension 10^6 using Cholesky's method and LDL^T factorization. The results are comparable to those of Fig. 2. This is no surprise since our method is based on the same assumption that holds for direct linear system solving in general: $\|I - FA\|_\infty$ less than 1.

Although we did not encounter any overestimations of the true accuracy, our method does not guarantee 100% reliability as does a verifying error routine which inherently *never* overestimates the true accuracy (see [2], [3] and [7]). But its reliability is very high and it is not restricted to H- or M-matrices.

References

1. Hörmann, B.: Notes on LAPACK and Verification. Technical Report 95–1 Universitätsrechenzentrum und Institut für Informatik Universität Basel Schweiz (1995)
2. Falcó Korn, C., Hörmann, B., Ullrich, C.: Verification May Be Better Than Estimation. SIAM J. Sci. Comput. **17** 4 (to appear 1996)
3. Rump, S.M.: Inclusion of the Solution of Large Linear Systems with Positive Definite Symmetric M-Matrix. Computer Arithmetic and Enclosure Methods Elsevier Science Publishers B. V. (1992) 339–347
4. Collatz, L.: Functional analysis and numerical mathematics. Academic Press New York (1966)
5. Gale, D.: The game of hex and the Brouwer fixed-point theorem. Amer. Math. Monthly **86** 10 (1979) 818–826
6. Golub, G., Van Loan, C.F.: Matrix Computations. Johns Hopkins University Press Baltimore MD 2nd. ed. (1989)
7. Falcó Korn, C.: Die Erweiterung von Software-Bibliotheken zur effizienten Verifikation der Approximationslösung linearer Gleichungssysteme. Ph.D. Thesis Institut für Informatik Universität Basel Schweiz (1993)
8. Duff, I.S., Grimes, R.G, Lewis, J.G.: User's Guide for the Harwell-Boeing Sparse Matrix Collection (Release I). TR/PA/92/86 CERFACS Toulouse Cedex France (1992)
9. Chatelin, F., Frayssé, V.: A statistical study of the stability of linear systems. TR/PA/91/43 CERFACS Toulouse Cedex France (1991)
10. GAMM-IMACS: Proposal for Accurate Floating-Point Vector Arithmetic. Mathematics and Computers in Simulation **35** 4 IMACS (1993)
11. Baumhof, C.: A New VLSI Vector Arithmetic Coprocessor for the PC. Proc. of the 12th Symposium on Comp. Arith. ARITH-12 '95 England (1995)
12. Anderson, E., Bai, Z., Bischof, C., Demmel, J., Dongarra, J., Du Croz, J., Greenbaum, A., Hammarling, S., McKenney, A., Ostrouchov, S., Sorensen, D.: LAPACK Users' Guide. SIAM Philadelphia 2nd. ed. (1995)

A Method for Solving the Spectral Problem for Complex Matrices*

Ivan Ganchev Ivanov[1]

Shoumen University, Shoumen 9712, Bulgaria

Abstract. An effective iterative numerical method for solving the spectral problem for an arbitrary complex matrix is described and its applicability discussed. It is a modification of Voevodin's method [5] for computing of eigenvalues and eigenvectors for complex matrix. The method presented in this paper uses similar transformations with real matrices.

1 Introduction

The problem for computing of eigenvalues and eigenvectors is one of the fundamental problems of algebra. The solving of this problem is necessary in many practical problems. Eberlein in [1], Voevodin in [5], Veselić in [4] have proposed different iterative methods for computing of eigenvalues and eigenvectors of arbitrary real and complex matrices. Let's consider the problem of computing the numbers ν and non zero vectors z satisfying the equation

$$Dz = \nu z, \tag{1}$$

where D is a complex matrix, i.e. $D = A + iB,\ A \in R^{n\times n},\ B \in R^{n\times n}, i = \sqrt{-1}$ and $R^{n\times n}$ is the set of real $n \times n$ matrices.

It is known that the problem (1) is equivalent to the spectral problem for a real block matrix

$$H = H(A, B) = \begin{pmatrix} A & -B \\ B & A \end{pmatrix}. \tag{2}$$

The matrix H is a special 2×2 block matrix. An algorithm for finding the eigenvalues and eigenvectors of H is proposed by Ivanov [2]. The existent methods for computing the eigenvalues and eigenvectors of H do not take into account the special block structure of H. The present method solves the spectral problem of H and it is a modification of Voevodin's method [5] for solving the problem (1). The method works with the elements, which are real numbers, of matrices A and B. Numerical experiments will be carried out with these algorithms for comparing the speed and the accuracy of different algorithms and for solving the problem (1) or its equivalent.

* This work is partially supported by Contract MM 521/95

2 Description of the method and convergence theorem

For the matrix H we construct the sequence

$$H_{k+1} = U_k^{-1} H_k U_k = (h_{rs}^{(k+1)}),\ H_1 = H, k = 1,2,3,\ldots, \tag{3}$$

where each $H_k = H(A_k, B_k)$ and $U_k = U_{p_k q_k}(\varphi_k)$. For each k the matrix U_k depends from three parameters p_k, q_k, φ_k.

Since the computations of each step are similar we consider the k step of algorithm. We introduce the notations

$$H_k = H(A_k, B_k) = (h_{rs}^{(k)}),\ A_k = (a_{\beta\gamma}^{(k)}),\ B_k = (b_{\beta\gamma}^{(k)}),$$
$$C_k = C(H_k) = H_k^T H_k - H_k H_k^T = H(F_k, E_k) = (c_{rs}^{(k)}),$$

where

$$F_k = F_k^T = A_k^T A_k + B_k^T B_k - A_k A_k^T - B_k B_k^T = (f_{\beta\gamma}^{(k)}),$$
$$E_k = -E_k^T = A_k^T B_k - B_k^T A_k + A_k B_k^T - B_k A_k^T = (e_{\beta\gamma}^{(k)}),$$
$$||H_k||^2 = \sum_{r,s=1}^{2n} (h_{rs}^{(k)})^2.$$

The strategy determining U_k from (3) and parameters p_k, q_k, φ_k is the following. For the matrices C_k and H_k we find the numbers

$$c^{(k)} = \max_{r\neq s} |c_{rs}^{(k)}|^{\frac{1}{2}} \quad \textit{and} \quad h^{(k)} = \max_{r\neq s} |h_{rs}^{(k)} + h_{sr}^{(k)}|.$$

If $c^{(k)} \geq h^{(k)}$ then we choose the matrix U_k to decrease Euclidean norm of H_{k+1}. If $h^{(k)} > c^{(k)}$ then we choose the matrix U_k to annihilate the off-diagonal elements $h_{p_k q_k}^{(k)} + h_{q_k p_k}^{(k)}$ and $h_{q_k p_k}^{(k)} + h_{p_k q_k}^{(k)}$ of the symmetric matrix $H_k + H_k^T$.

Then there are four possible cases

A.1. $|f_{pq}^{(k)}|^{\frac{1}{2}} = c^{(k)} \geq h^{(k)}$, $1 \leq p = p_k < q = q_k \leq n$, $\varphi = \varphi_k$. In this case $U = U_k = U_{pq}(\varphi)$ is of the type

$$U = \begin{pmatrix} S_{pq}^{-1}(\varphi) & 0 \\ 0 & S_{pq}^{-1}(\varphi) \end{pmatrix}, \tag{4}$$

where $S_{pq}(\varphi) \in R^{n\times n}$ is the matrix

$$S_{pq}(\varphi) = (s_{ij}) = \begin{cases} s_{qp} = \varphi \\ s_{ij} = \delta_{ij}, & (i,j) \notin \{(q,p)\} \end{cases}$$

From the type of $S_{pq}(\varphi)$ we see that $S_{pq}^{-1}(\varphi) = S_{pq}(-\varphi)$.

The parameter φ can be computed by the formula

$$\varphi = \frac{f_{pq}^{(k)}}{\max(|f_{pq}^{(k)}|, m_{qp}^{(k)})}, \tag{5}$$

where

$$m_{qp}^{(k)} = \sum_{i\neq q}((a_{iq}^{(k)})^2 + (b_{iq}^{(k)})^2) + \sum_{j\neq p}((a_{pj}^{(k)})^2 + (b_{pj}^{(k)})^2) + (a_{qq}^{(k)} - a_{pp}^{(k)})^2$$
$$+ (b_{qq}^{(k)} - b_{pp}^{(k)})^2 + \tau^2((a_{pq}^{(k)})^2 + (b_{pq}^{(k)})^2 + (a_{qp}^{(k)})^2 + (b_{qp}^{(k)})^2), \quad (\tau > 1).$$

A.2. $|e_{pq}^{(k)}|^{\frac{1}{2}} = c^{(k)} \geq h^{(k)}$, $1 \leq p = p_k < q = q_k \leq n$, $\varphi = \varphi_k$. Then the matrix $U = U_k = U_{pq}(\varphi)$ is of the form

$$U = \begin{pmatrix} I & -S_{pq}(\varphi) \\ S_{pq}(\varphi) & I \end{pmatrix}, \tag{6}$$

where $S_{pq}(\varphi) \in R^{n\times n}$ and

$$S_{pq}(\varphi) = (s_{ij}) = \begin{cases} s_{pq} = \varphi \\ s_{ij} = \delta_{ij}, \quad (i,j) \notin \{(p,q)\} \end{cases}$$

The matrix U^{-1} is of the type

$$U^{-1} = \begin{pmatrix} I & S_{pq}(\varphi) \\ -S_{pq}(\varphi) & I \end{pmatrix}.$$

In this case the parameter φ is defined by

$$\varphi = \frac{-e_{pq}^{(k)}}{\max(|e_{pq}^{(k)}|, m_{qp}^{(k)})}, \tag{7}$$

where

$$m_{qp}^{(k)} = \sum_{i\neq q}((a_{qi}^{(k)})^2 + (b_{qi}^{(k)})^2) + \sum_{j\neq p}((a_{jp}^{(k)})^2 + (b_{jp}^{(k)})^2) + (a_{qq}^{(k)} - a_{pp}^{(k)})^2$$
$$+ (b_{qq}^{(k)} - b_{pp}^{(k)})^2 + \tau^2((a_{qp}^{(k)})^2 + (b_{qp}^{(k)})^2 + (a_{pq}^{(k)})^2 + (b_{pq}^{(k)})^2), \quad (\tau > 1).$$

Lemma 1. *Let the matrix H_k from the sequence (3), $1 \leq p = p_k < q = q_k \leq n$, $\varphi = \varphi_k$ and $H_{k+1} = U_k^{-1} H_k U_k$ where $U_k = U_{pq}(\varphi)$ is chosen from (4) or (6) and φ is chosen from (5) or (7) respectively. Then*

$$||H_k||^2 - ||H_{k+1}||^2 \geq \frac{2}{\tau^2}\frac{(\tau^2 - \tau - 1)(c^{(k)})^4}{\max((c^{(k)})^2, m_{qp}^{(k)})} \geq \frac{1}{\tau^2}\frac{(c^{(k)})^4}{||H_k||^2}. \tag{8}$$

Proof. Let U_k be chosen from (4) and φ is computed by (5). We denote $\Delta(\varphi) = ||H_k||^2 - ||H_{k+1}||^2$ and we obtain

$$\Delta(\varphi) = 2(-G\varphi^4 - Q\varphi^3 - W\varphi^2 + 2f_{pq}^{(k)}\varphi) = 2\Delta$$

where

$$\begin{aligned}
G &= (a_{pq}^{(k)})^2 + (b_{pq}^{(k)})^2, \\
Q &= 2a_{pq}^{(k)}(a_{qq}^{(k)} - a_{pp}^{(k)}) + 2b_{pq}^{(k)}(b_{qq}^{(k)} - b_{pp}^{(k)}), \\
W &= \sum_{i\neq q}((a_{iq}^{(k)})^2 + (b_{iq}^{(k)})^2) + \sum_{j\neq p}((a_{pj}^{(k)})^2 + (b_{pj}^{(k)})^2) + \\
&\quad (a_{qq}^{(k)} - a_{pp}^{(k)})^2 + (b_{qq}^{(k)} - b_{pp}^{(k)})^2 - 2(a_{pq}^{(k)}a_{qp}^{(k)} + b_{pq}^{(k)}b_{qp}^{(k)}).
\end{aligned}$$

From W and $m_{qp}^{(k)}$ we find

$$\frac{|W|}{\max(|f_{pq}^{(k)}|, m_{qp}^{(k)})} \leq \frac{|W|}{m_{qp}^{(k)}} \leq 1.$$

We consider the inequality

$$2xy \leq \frac{t^4x^2 + y^2}{t^2} \tag{9}$$

It is true for the real numbers $t,\ x,\ y$. When $t = \sqrt{\tau},\ x = |a_{pq}^{(k)}|,\ y = |a_{qq}^{(k)} - a_{pp}^{(k)}|$ we obtain

$$2|a_{pq}^{(k)}(a_{qq}^{(k)} - a_{pp}^{(k)})| \leq \frac{1}{\tau}((\sqrt{\tau})^4(a_{pq}^{(k)})^2 + (a_{qq}^{(k)} - a_{pp}^{(k)})^2).$$

We have

$$\frac{|Q|}{\max(|f_{pq}^{(k)}|, m_{qp}^{(k)})} \leq \frac{|Q|}{m_{qp}^{(k)}} \leq$$

$$\frac{|2a_{pq}^{(k)}(a_{qq}^{(k)} - a_{pp}^{(k)}) + 2b_{pq}^{(k)}(b_{qq}^{(k)} - b_{pp}^{(k)})|}{(a_{qq}^{(k)} - a_{pp}^{(k)})^2 + (b_{qq}^{(k)} - b_{pp}^{(k)})^2 + \tau^2((a_{pq}^{(k)})^2 + (b_{pq}^{(k)})^2 + (a_{qp}^{(k)})^2 + (b_{qp}^{(k)})^2)} \leq$$

$$\frac{1}{\tau}\frac{(a_{qq}^{(k)} - a_{pp}^{(k)})^2 + (b_{qq}^{(k)} - b_{pp}^{(k)})^2 + \tau^2((a_{pq}^{(k)})^2 + (b_{pq}^{(k)})^2)}{(a_{qq}^{(k)} - a_{pp}^{(k)})^2 + (b_{qq}^{(k)} - b_{pp}^{(k)})^2 + \tau^2((a_{pq}^{(k)})^2 + (a_{qp}^{(k)})^2 + (b_{pq}^{(k)})^2 + (b_{qp}^{(k)})^2)},$$

hence

$$\frac{|Q|}{m_{qp}^{(k)}} \leq \frac{1}{\tau}.$$

By analogy we find

$$\frac{|G|}{\max(|f_{pq}^{(k)}|, m_{qp}^{(k)})} \leq \frac{|G|}{m_{qp}^{(k)}} \leq \frac{(a_{pq}^{(k)})^2 + (b_{pq}^{(k)})^2}{m_{qp}^{(k)}} \leq \frac{1}{\tau^2}$$

For Δ we obtain

$$\begin{aligned}
\Delta &\geq 2f_{pq}^{(k)}\varphi - |W|\varphi^2 - |Q|\varphi^3 - |G|\varphi^4 \\
&\geq 2\varphi^2 \max(|f_{pq}^{(k)}|, m_{qp}^{(k)}) - |W|\varphi^2 - |Q|\varphi^2 - |G|\varphi^2 \\
&= \max(|f_{pq}^{(k)}|, m_{qp}^{(k)})(2\varphi^2 - \frac{|W| + |Q| + |G|}{\max(|f_{pq}^{(k)}|, m_{qp}^{(k)})}\varphi^2) \\
&\geq \max(|f_{pq}^{(k)}|, m_{qp}^{(k)})(2\varphi^2 - (1 + \frac{1}{\tau} + \frac{1}{\tau^2})\varphi^2) \\
&= \max(|f_{pq}^{(k)}|, m_{qp}^{(k)})(2 - \frac{1 + \tau + \tau^2}{\tau^2})\varphi^2 \\
&= \max(|f_{pq}^{(k)}|, m_{qp}^{(k)})(\frac{\tau^2 - \tau - 1}{\tau^2})\varphi^2
\end{aligned}$$

Then

$$\Delta \geq \frac{\tau^2 - \tau - 1}{\tau^2} \frac{(f_{pq}^{(k)})^2}{\max(|f_{pq}^{(k)}|, m_{qp}^{(k)})}$$

Since

$$\max(|f_{pq}^{(k)}|, m_{qp}^{(k)}) < 2(\tau^2 - \tau - 1)||H_k||^2, \quad (\tau > \frac{1 + \sqrt{5}}{2}),$$

hence

$$\Delta(\varphi) = 2\Delta \geq 2\frac{\tau^2 - \tau - 1}{\tau^2} \frac{(f_{pq}^{(k)})^2}{\max(|f_{pq}^{(k)}|, m_{qp}^{(k)})} \geq \frac{1}{\tau^2} \frac{(f_{pq}^{(k)})^2}{||H_k||^2}$$

In the case when U is chosen from (6) and φ is computed from (7), the proof is analogous. The lemma is proved.

Lemma 2. *Let α, β be integer numbers and $1 \leq \alpha,\ \beta \leq 2n,\ \alpha \neq \beta,\ |\alpha - \beta| \neq n$. Let the matrix H_k from the sequence (3), $H_{k+1} = U_k^{-1} H_k U_k$ where $U = U_k$ is chosen from (4) or (6) and $\varphi = \varphi_k$ is chosen from (5) or (7) respectively. Then*

$$|h_{rs}^{(k+1)} - h_{rs}^{(k)}| \leq 2|c_{\alpha\beta}^{(k)}| \quad \textit{for all} \quad r, s.$$

This lemma is proved in an analogous way Lemma 1 from [4].

A.3. $|a_{pq}^{(k)} + a_{qp}^{(k)}| = h^{(k)} > c^{(k)},\ 1 \leq p = p_k < q = q_k \leq n,\ \varphi = \varphi_k$. In this case the matrix $U = U_{pq}(\varphi) = (u_{ij})$ differs from the $2n \times 2n$ unit matrix by the elements

$$\begin{cases} u_{pp} = u_{qq} = u_{p+np+n} = u_{q+nq+n} = \cos\varphi \\ u_{pq} = -u_{qp} = u_{p+nq+n} = -u_{q+np+n} = -\sin\varphi \end{cases}$$

From the condition $|a_{pq}^{(k+1)} + a_{qp}^{(k+1)}| = 0$ for φ we compute

$$tg2\varphi = \frac{a_{pq}^{(k)} + a_{qp}^{(k)}}{a_{pp}^{(k)} - a_{qq}^{(k)}}.$$

A.4. $|-b_{pq}^{(k)}+b_{qp}^{(k)}|=h^{(k)}>c^{(k)}$, $1\le p=p_k<q=q_k\le n$, $\varphi=\varphi_k$. In this case the matrix $U=U_{pq}(\varphi)=(u_{ij})$ differs from the $2n\times 2n$ unit matrix by the elements

$$\begin{cases} u_{pp}=u_{qq}=u_{p+np+n}=u_{q+nq+n}=\cos\varphi \\ u_{pq+n}=-u_{q+np}=u_{qp+n}=-u_{p+nq}=-\sin\varphi \end{cases}$$

From the condition $|-b_{pq}^{(k+1)}+b_{pq}^{(k+1)}|=0$ for φ we have

$$tg2\varphi=\frac{b_{qp}^{(k)}-b_{pq}^{(k)}}{a_{pp}^{(k)}-a_{qq}^{(k)}}.$$

In cases **A.3** and **A.4** the similaraty transformations $U_k^{-1}H_kU_k$ are steps from Jacobi's method for the symmetric matrix $H_k+H_k^T$ [3].

Theorem 3. *For sequence (3) we have*

I. $C(H_k)\to 0,\ k\to\infty$.

II. The symmetric matrix $\frac{1}{2}(H_k+H_k^T)$ *tends to the matrix* $diag[\lambda_1,\ldots,\lambda_n,\lambda_1,\ldots,\lambda_n]$ *where* $\lambda_s,\ s=1,\ldots,n$ *are real parts of eigenvalues of* H.

III. If $\beta\ne\gamma$ $(1\le\beta,\gamma\le n)$ *and* $\lambda_\beta\ne\lambda_\gamma$ *then*

$$a_{\beta\gamma}^{(k)}\to 0,\ k\to\infty, \qquad b_{\beta\gamma}^{(k)}\to 0,\ k\to\infty.$$

IV. If for a fixed integer number $m\in[1,n]$ *and each integer number* $t\in[1,n]$ *for which* $t\ne m$ *we have* $\lambda_t\ne\lambda_m$ *then* $b_{mm}^{(k)}\to\mu_m,\ k\to\infty$ *and* μ_m *is the imaginary part of eigenvalues with a real part* λ_m.

The theorem is proved in an analogous way with Theorems from [4], [2].

3 Numerical experiments

Our algorithm and the other algorithms compute eigenvalues and eigenvectors of different complex matrices. For experiments we have used five programs. The first program is program $G1$ for QR-algorithm [6]. The second program is $G2$ for Eberlein's algorithm [1]. The third program is program $G3$ for Voevodin's algorithm [5]. The fourth program is program $G4$ for Ivanov's algorithm [2]. The fifth program is program $G5$ for our algorithm. The programs $G1, G2, G3$ compute the eigenvalues of the complex matrix $D=A+i\,B$ in complex arithmetic. The programs $G4, G5$ compute the eigenvalues of the complex matrix $D=A+i\,B$ in real arithmetic.

Numerical experiments were made on computer Pentium using the algorithmic language Turbo Pascal and the real arithmetic having a 11 sedecimal digit mantissa. Thirty iterations are made for the computing of each eigenvalues in program $G1$. If $G1$ does not compute a eigenvalue for 30 iterations then the work of $G1$ will stop.

The programs use a cyclic choice on the pivot indices (p,q). The iterative process stops in programs $G2$, $G3$ if

$$\sum_{r\neq s} |a_{rs} + ib_{rs}| \leq 10^{-4}.$$

The iterative process stops in programs $G4$, $G5$ if

$$\sum_{r\neq s} (|a_{rs}| + |b_{rs}|) \leq 10^{-4}.$$

Let's compute eigenvalues of matrix $D = A + i\,B$ for $n = 4m+1$. The blocks A, B are of the type

$$A = (a_{rs}) = \begin{cases} a_{1s} = a_{s1} = 1 & s = 1,\ldots,n \\ a_{rs} = \cos\frac{2\pi(r-1)(s-1)}{n} & r,s = 2,\ldots,n \end{cases},$$

$$B = (b_{rs}) = \begin{cases} b_{1s} = b_{s1} = 1 & s = 1,\ldots,n \\ b_{rs} = \sin\frac{2\pi(r-1)(s-1)}{n} & r,s = 2,\ldots,n \end{cases}.$$

The exact eigenvalues of the matrix $D = A + i\,B$ are given in following table.

eigenvalues λ_s	$\sqrt{n}$	$-\sqrt{n}$	$i\,\sqrt{n}$	$-i\,\sqrt{n}$
multiplicity	$m+1$	m	m	m

The eigenvalues of the matrix D are computed when $m = 3,4,5$ with programs $G1$, $G2$, $G3$, $G4$, $G5$. Let's denote $\varepsilon = \max_i |\lambda_i - \tilde{\lambda}_i|$ where $\tilde{\lambda}_i$ are the computed eigenvalues obtained with each of the programs $G1$, $G2$, $G3$, $G4$, $G5$. The results are shown in the following table.

program	$m=3$		$m=4$		$m=5$	
	time	ε	*time*	ε	*time*	ε
$G1$	-	–	-	-	-	-
$G2$	$2''$	$4,4309\,.10^{-10}$	$4''$	$1,2136\,.10^{-9}$	$10''$	$1,6154\,.10^{-9}$
$G3$	$3''$	$2,2377\,.10^{-10}$	$9''$	$7,3752\,.10^{-10}$	$33''$	$1,4414\,.10^{-9}$
$G4$	$2''$	$2,4754\,.10^{-10}$	$5''$	$4,6841\,.10^{-10}$	$11''$	$6,4462\,.10^{-10}$
$G5$	$2''$	$2,1370\,.10^{-10}$	$6''$	$6,6897\,.10^{-10}$	$11''$	$1,7330\,.10^{-10}$

Table 1.

For the above example the program $G1$ does not compute all eigenvalues because the defined 30 iterations for each eigenvalues in the program are insufficient. The program $G5$ computes the eigenvalues faster than the program $G3$. The algorithms $G4, G5$ compute the eigenvalues faster than $G2$, $G3$ for this example. All programs compute eigenvalues with equal accuracy.

The methods in real arithmetic ($G4$, $G5$) are faster, they have simpler computational schemes and give better possibilities for parallel modifications.

References

1. Eberlein, P.: A Jacobi-like method for the automatic computation of eigenvalues and eigenvectors of an arbitrary matrix. SIAM J. **10** (1962) 74-88
2. Ivanov, I.: Algorithm for solving spectral problem of complex matrix in real arithmetic. Mathematica Balkanica **8** (1994) 51-58
3. Petkov, M., Ivanov,I.: Solution of symmetric and hermitian J-symmetric eigenvalue problem. Mathematica Balkanica **8** (1994) 337-349
4. Veselić, K.: A convergent Jacobi method for solving the eigenproblem of arbitrary real matrices. Numer. Math. **25** (1976) 179-184
5. Voevodin V.: Numerical methods of the algebra. Moscow (1966) (In Russian)
6. Wilkinson, J., Reinsch C.: Handbook for automatic computation. Linear algebra. Moscow (1976) (In Russian)

Interpolation Technique and Convergence Rate Estimates for Finite Difference Method*

Boško S. Jovanović

University of Belgrade, Faculty of Mathematics
Studentski trg 16, POB 550, 11001 Belgrade, Yugoslavia

Abstract. In this work we expose a methodology for establishing convergence rate estimates for finite difference schemes based on the interpolation theory of Banach spaces. As a model problem we consider Dirichlet boundary value problem for second order linear elliptic equation with variable coefficients from Sobolev spaces. Using interpolation theory we construct fractional-order convergence rate estimates which are consistent with the smoothness of data.

1 Introduction

Boundary–value problems with generalized solutions are of great theoretical and practical importance. One of main methods for solving theese problems is finite difference method. It is important to establish the most precise convergence rate estimates for this method.

For an elliptic boundary value problem (BVP) with solution belonging to the Sobolev space $W_p^s(\Omega)$, convergence rate estimates

$$\|u - v\|_{W_p^k(\omega)} \le C h^{s-k} \|u\|_{W_p^s(\Omega)}, \qquad s > k \tag{1}$$

are considered to be consistent with the smoothness of the solution [14]. Here u is the solution of the original BVP, v is the solution of the corresponding finite difference scheme (FDS), h is discretization parameter, $W_p^k(\omega)$ denotes the discrete Sobolev space, and C is a positive generic constant, independent of h and u.

Estimates of this type have been obtained for the broad class of elliptic problems (see [6], [12], [13], [15], [19]). Analogous results have also been obtained for parabolic and hyperbolic problems (see [5], [7], [8], [11]). As a rule, the Bramble–Hilbert lemma [2], [4] is used for proving these results.

In this paper we expose an alternative technique, based on the theory of interpolation of Banach spaces, for obtaining estimates of the form (1). The main attention is given to the problems with variable coefficients, which are multipliers in Sobolev spaces. The same technique is used in [10] for problems with constant coefficients. Convergence rate estimate for projection–difference scheme approximating second order hyperbolic equation is obtained in [22].

---

* AMS Subject Classifications (1991): 65N15, 46E35, 46B70
Supported by MST of Republic of Serbia, grant number 04M03/C

2 Interpolation of Banach Spaces

Let A_0 and A_1 be two Banach spaces, linearly and continuously imbedded in a topological linear space $\mathcal{A}$. Two such spaces are called *interpolation pair* $\{A_0, A_1\}$ (see [20], [1]). Consider also the space $A_0 \cap A_1$, with the norm

$$\|a\|_{A_0 \cap A_1} = \max\{\|a\|_{A_0}, \|a\|_{A_1}\}$$

and the space $A_0 + A_1 = \{a \in \mathcal{A} : a = a_0 + a_1,\ a_j \in A_j,\ j = 0, 1\}$, with the norm

$$\|a\|_{A_0+A_1} = \inf_{\substack{a=a_0+a_1 \\ a_j \in A_j}} \{\|a_0\|_{A_0} + \|a_1\|_{A_1}\}.$$

Obviously, $A_0 \cap A_1 \subset A_j \subset A_0 + A_1$, $j = 0, 1$.

Let us introduce category $\mathcal{C}_0$, whose objects $A, B, C, \ldots$ are Banach spaces, and morphisms – bounded linear operators $L \in \mathcal{L}(A, B)$. Let, also, $\mathcal{C}_1$ be a category whose objects are interpolation pairs $\{A_0, A_1\}, \{B_0, B_1\}, \ldots$ while morphisms are $L \in \mathcal{L}(\{A_0, A_1\}, \{B_0, B_1\})$. Here $\mathcal{L}(\{A_0, A_1\}, \{B_0, B_1\})$ denotes the set of bounded linear operators from $A_0 + A_1$ into $B_0 + B_1$, whose restrictions on A_j belong to the set $\mathcal{L}(A_j, B_j)$, $j = 0, 1$.

Functor $\mathcal{F} : \mathcal{C}_1 \to \mathcal{C}_0$ is called *interpolation functor* if

$$A_0 \cap A_1 \subset \mathcal{F}(\{A_0, A_1\}) \subset A_0 + A_1$$

for every interpolation pair $\{A_0, A_1\}$, while for every morphism $L \in \mathcal{L}(\{A_0, A_1\}, \{B_0, B_1\})$, $\mathcal{F}(L)$ is the restriction of the operator L on $\mathcal{F}(\{A_0, A_1\})$.

The corresponding Banach space $A = \mathcal{F}(\{A_0, A_1\})$ is called *interpolation space*.

Note that $A_0 \cap A_1$ and $A_0 + A_1$ are interpolation spaces.

If the inequality

$$\|L\|_{\mathcal{F}(\{A_0,A_1\}) \to \mathcal{F}(\{B_0,B_1\})} \leq C\, \|L\|^{1-\theta}_{A_0 \to B_0} \|L\|^{\theta}_{A_1 \to B_1},$$

where $0 < \theta < 1$ and $C = \text{const} \geq 1$, is satisfied for every morphism L of category $\mathcal{C}_1$, the interpolation functor $\mathcal{F}$ is called to be of the *type* θ. (Here $\|L\|_{A_j \to B_j}$ denotes standard operator norm of $L : A_j \to B_j$).

Let us consider so called complex interpolation method [3], [20]. Let us define the following sets of complex numbers: $S = \{z \in \mathbb{C} : 0 < \Re z < 1\}$ and $\bar{S} = \{z \in \mathbb{C} : 0 \leq \Re z \leq 1\}$. For the given interpolation pair $\{A_0, A_1\}$ we introduce the set $\mathcal{M}(A_0, A_1)$ of continuous functions $f : \bar{S} \to A_0 + A_1$, analytic in S, which satisfy the following conditions:

$$\sup_{z \in \bar{S}} \|f(z)\|_{A_0+A_1} < \infty,$$

$$f(j + it) \in A_j, \quad j = 0, 1, \quad t \in \mathbb{R},$$

the mapings $t \to f(j + it)$, $j = 0, 1$, are continuous on t, and

$$\|f\|_{\mathcal{M}(A_0,A_1)} = \max\left\{ \sup_{t \in \mathbb{R}} \|f(it)\|_{A_0}, \sup_{t \in \mathbb{R}} \|f(1 + it)\|_{A_1} \right\} < \infty.$$

For $0 < \theta < 1$ with $[A_0, A_1]_\theta$ we denote the set of elements $a \in A_0 + A_1$ which satisfy the conditions:

there exists a function $f \in \mathcal{M}(A_0, A_1)$ such that $f(\theta) = a$, and

$$\|a\|_{[A_0,A_1]_\theta} = \inf_{\substack{f \in \mathcal{M}(A_0,A_1) \\ f(\theta)=a}} \|f\|_{\mathcal{M}(A_0,A_1)} < \infty.$$

Defined in that way, space $[A_0, A_1]_\theta$ is an interpolation space. The following relations hold

$$\begin{aligned}
&[A_0, A_1]_\theta = [A_1, A_0]_{1-\theta}, \\
&[A, A]_\theta = A, \\
&\|a\|_{[A_0,A_1]_\theta} \le C_\theta \|a\|_{A_0}^{1-\theta} \|a\|_{A_1}^{\theta}, \qquad \forall a \in A_0 \cap A_1, \\
&A_0 \cap A_1 \quad \text{is dense in} \quad [A_0, A_1]_\theta.
\end{aligned}$$

If $A_0 \subset A_1$, then for $0 < \theta < \vartheta < 1$

$$A_0 \subset [A_0, A_1]_\theta \subset [A_0, A_1]_\vartheta \subset A_1.$$

The corresponding interpolation functor $\mathcal{F}(\{A_0, A_1\}) = [A_0, A_1]_\theta$ is of the type θ, with constant $C_\theta = 1$, i.e.

$$\|L\|_{[A_0,A_1]_\theta \to [B_0,B_1]_\theta} \le \|L\|_{A_0 \to B_0}^{1-\theta} \|L\|_{A_1 \to B_1}^{\theta}.$$

Analogous assertion holds true for bilinear operators [3], [20]:

Lemma 1. *Let $A_0 \subset A_1$, $B_0 \subset B_1$, $C_0 \subset C_1$ and let $L : A_1 \times B_1 \to C_1$ be a continuous bilinear form whose restriction on $A_0 \times B_0$ is continuous maping with values in C_0. Than L is continuous maping from $[A_0, A_1]_\theta \times [B_0, B_1]_\theta$ into $[C_0, C_1]_\theta$, and*

$$\|L\|_{[A_0,A_1]_\theta \times [B_0,B_1]_\theta \to [C_0,C_1]_\theta} \le \|L\|_{A_0 \times B_0 \to C_0}^{1-\theta} \|L\|_{A_1 \times B_1 \to C_1}^{\theta}.$$

3 Spaces $\mathbf{H}_p^s$, $\mathbf{B}_{pq}^s$ and $\mathbf{W}_p^s$

As example of interpolation function spaces let us consider the spaces of Bessel potentials H_p^s and the Besov spaces B_{pq}^s (see [1], [20]). The following assertion holds true [20]:

Lemma 2. *For $-\infty < s_0, s_1 < \infty$, $1 < p_0, p_1 < \infty$, $1 \le q_0 < \infty$, $1 \le q_1 \le \infty$ and $0 < \theta < 1$ we have*

$$\left[H_{p_0}^{s_0}(\mathbb{R}^n), H_{p_1}^{s_1}(\mathbb{R}^n)\right]_\theta = H_p^s(\mathbb{R}^n), \quad \textit{and} \tag{2}$$

$$\left[B_{p_0q_0}^{s_0}(\mathbb{R}^n), B_{p_1q_1}^{s_1}(\mathbb{R}^n)\right]_\theta = B_{pq}^s(\mathbb{R}^n), \tag{3}$$

where

$$s = (1-\theta)s_0 + \theta s_1, \qquad \frac{1}{p} = \frac{1-\theta}{p_0} + \frac{\theta}{p_1}, \qquad \frac{1}{q} = \frac{1-\theta}{q_0} + \frac{\theta}{q_1}. \tag{4}$$

The spaces H_p^s and B_{pq}^s are spaces of distributions. The following relations hold

$$\mathcal{D}(\mathbb{R}^n) \subset H_p^s(\mathbb{R}^n),\, B_{pq}^s(\mathbb{R}^n) \subset \mathcal{D}'(\mathbb{R}^n),$$

where $\mathcal{D}(\mathbb{R}^n) = C_0^\infty(\mathbb{R}^n)$ is the set of infinitely differentiable functions with compact support, and $\mathcal{D}'(\mathbb{R}^n)$ – the set of Schwartz distributions [18]. For $s = 0$

$$H_p^0(\mathbb{R}^n) = L_p(\mathbb{R}^n),$$

where L_p is the Lebesgue space of integrable functions.

For $-\infty < s < \infty$, $1 < p < \infty$, $\varepsilon > 0$ and $1 \le q_0 \le q_1 \le \infty$ the following imbeddings hold true [20]

$$B_{p,\infty}^{s+\varepsilon}(\mathbb{R}^n) \subset B_{p1}^s(\mathbb{R}^n) \subset B_{pq_0}^s(\mathbb{R}^n) \subset B_{pq_1}^s(\mathbb{R}^n) \subset B_{p,\infty}^s(\mathbb{R}^n) \subset B_{p1}^{s-\varepsilon}(\mathbb{R}^n),$$
$$H_p^{s+\varepsilon}(\mathbb{R}^n) \subset H_p^s(\mathbb{R}^n), \quad \text{and}$$
$$B_{p,\min\{p,2\}}^s(\mathbb{R}^n) \subset H_p^s(\mathbb{R}^n) \subset B_{p,\max\{p,2\}}^s(\mathbb{R}^n). \tag{5}$$

For $-\infty < t \le s < \infty$, $1 < p \le q < \infty$, $1 \le r \le \infty$ and $s - n/p \ge t - n/q$ we also have

$$B_{pr}^s(\mathbb{R}^n) \subset B_{qr}^t(\mathbb{R}^n) \qquad \text{and} \qquad H_p^s(\mathbb{R}^n) \subset H_q^t(\mathbb{R}^n).$$

For $1 < p < \infty$ the Sobolev spaces W_p^s are defined in the following manner (see [20]):

$$W_p^s(\mathbb{R}^n) = \begin{cases} H_p^s(\mathbb{R}^n), & s = 0, 1, 2, \ldots \\ B_{pp}^s(\mathbb{R}^n), & 0 < s \ne \text{integer} \end{cases} \tag{6}$$

with the norm defined as

$$\|f\|_{W_p^s} = \left(\sum_{k<s} |f|_{W_p^k}^p + |f|_{W_p^s}^p \right)^{1/p},$$

where

$$|f|_{W_p^r} = \begin{cases} \left(\sum\limits_{|\alpha|=r} \int\limits_{\mathbb{R}^n} |D^\alpha f(x)|^p dx \right)^{1/p}, & r = 0, 1, 2, \ldots \\ \left(\sum\limits_{|\alpha|=[r]} \int\limits_{\mathbb{R}^n} \int\limits_{\mathbb{R}^n} \frac{|D^\alpha f(x) - D^\alpha f(y)|^p}{|x-y|^{n+p(r-[r])}} dx dy \right)^{1/p}, & 0 < r \ne \text{integer}. \end{cases}$$

Here $\alpha = (\alpha_1, \ldots, \alpha_n)$ is a multi–index, $|\alpha| = \alpha_1 + \ldots + \alpha_n$, $x = (x_1, \ldots, x_n) \in \mathbb{R}^n$, $|x| = (x_1 + \ldots + x_n)^{1/2}$, $D^\alpha = D_1^{\alpha_1} \cdots D_n^{\alpha_n} = (\partial/\partial x_1)^{\alpha_1} \cdots (\partial/\partial x_n)^{\alpha_n}$ and $[r]$ is the integer part of r. In such a way

$$W_p^s(\mathbb{R}^n) \subset L_p(\mathbb{R}^n), \qquad s \ge 0.$$

For $s < 0$ Sobolev spaces are defined by duality:

$$W_p^s(\mathbb{R}^n) = \left(W_{p'}^{-s}(\mathbb{R}^n) \right)', \qquad 1/p + 1/p' = 1.$$

From (2), (3) and (6), for $s_0, s_1 \geq 0$, follows

$$\left[W_p^{s_0}(\mathbb{R}^n), W_p^{s_1}(\mathbb{R}^n)\right]_\theta = W_p^s(\mathbb{R}^n), \qquad s = (1-\theta)\, s_0 + \theta\, s_1, \tag{7}$$

if s_0, s_1 and s are all integer, or fractional numbers. For $p = 2$ from (5) follows

$$W_2^s(\mathbb{R}^n) = H_2^s(\mathbb{R}^n) = B_{22}^s(\mathbb{R}^n),$$

and (7) holds without restriction.

The previous results hold for the spaces H_p^s, B_{pq}^s and W_p^s in a bounded domain $\Omega \subset \mathbb{R}^n$ which satisfies the cone condition. Here we assume that $s \geq 0$ for H_p^s spaces, and $s > 0$ for B_{pq}^s spaces.

4 Multipliers in Sobolev Spaces

Let Ω be a domain in $\mathbb{R}^n$ and V and W two function spaces contained in $\mathcal{D}'(\Omega)$. A function a, defined on Ω, is called *multiplier* from V to W if, for every v in V, the product $a \cdot v$ belongs to W. The set of all multipliers from V to W is denoted by $M(V \to W)$. In particular, when $V = W$ we put $M(V) = M(V \to V)$.

In this section we shall be concerned with multipliers in Sobolev spaces. We shall restrict our attention to spaces $M(W_p^t \to W_p^s)$, where $1 < p < \infty$ and $t \geq s$. The following relations are satisfied [16]:

$$M(W_{p'}^{-s}(\mathbb{R}^n) \to W_{p'}^{-t}(\mathbb{R}^n)) = M(W_p^t(\mathbb{R}^n) \to W_p^s(\mathbb{R}^n)), \qquad 1/p + 1/p' = 1,$$
$$M(W_p^s(\mathbb{R}^n)) \subset L_\infty(\mathbb{R}^n), \qquad s \geq 0.$$

If $a \in M(W_p^t(\mathbb{R}^n) \to W_p^s(\mathbb{R}^n))$, $t \geq s \geq 0$, where t and s are both integer or fractional numbers, then:

$$\begin{aligned}
a &\in M(W_p^{t-s}(\mathbb{R}^n) \to L_p(\mathbb{R}^n)), \\
a &\in M(W_p^{t-\sigma}(\mathbb{R}^n) \to W_p^{s-\sigma}(\mathbb{R}^n)), && 0 < \sigma < s, \\
D^\alpha a &\in M(W_p^t(\mathbb{R}^n) \to W_p^{s-|\alpha|}(\mathbb{R}^n)), && |\alpha| \leq s, \\
D^\alpha a &\in M(W_p^{t-s+|\alpha|}(\mathbb{R}^n) \to L_p(\mathbb{R}^n)), && |\alpha| \leq s.
\end{aligned}$$

The following assertion holds true [16]:

Lemma 3. *If $a_\alpha \in M(W_p^{s-|\alpha|}(\mathbb{R}^n) \to W_p^{s-k}(\mathbb{R}^n))$, $s \geq k$, for every multi-index α, then the differential operator*

$$Lu = \sum_{|\alpha| \leq k} a_\alpha(x) D^\alpha u, \qquad x \in \mathbb{R}^n \tag{8}$$

defines a continuous mapping from $W_p^s(\mathbb{R}^n)$ to $W_p^{s-k}(\mathbb{R}^n)$.

Analogous result holds true for $s < 0$. If $p = 2$ then the result holds true for every s. Under certain conditions we also have the inverse result:

Lemma 4. *Let the operator* (8) *define a continuous mapping from* $W_p^s(\mathbb{R}^n)$ *to* $W_p^{s-k}(\mathbb{R}^n)$, *and* $p(s-k) > n$, $p > 1$. *Then* $a_\alpha \in M(W_p^{s-|\alpha|}(\mathbb{R}^n) \to W_p^{s-k}(\mathbb{R}^n))$, *for every multi-index* α.

The following lemma establishes sufficient conditions for a function a to be the multiplier in W_p^s (see [9]).

Lemma 5. *Let* Ω *be a bounded open domain in* $\mathbb{R}^n$ *with Lipschitz continuous boundary,* $s > 0$ *and* $1 < p < \infty$. *If* $a \in W_q^t(\Omega)$, *where*

$$q = p, \quad t = s, \quad \text{when} \quad sp > n, \quad \text{and}$$
$$q = n/s, \quad t = s + \varepsilon, \quad \varepsilon > 0, \quad \text{when} \quad sp \le n,$$

then $a \in M(W_p^s(\Omega))$.

5 Boundary Value Problem and its Approximation

As a model problem let us consider Dirichlet BVP for a second-order linear elliptic equation with variable coefficients, in the square $\Omega = (0,1)^2$:

$$-\sum_{i,j=1}^{2} D_i(a_{ij} D_j u) = f \quad \text{in} \quad \Omega, \qquad u = 0 \quad \text{on} \quad \Gamma = \partial\Omega. \tag{9}$$

We assume that the generalized solution of the problem (9) belongs to the Sobolev space $W_2^s(\Omega)$, $1 \le s \le 4$, with the right-hand side $f(x)$ belonging to $W_2^{s-2}(\Omega)$. Consequently, coefficients $a_{ij}(x)$ belong to the space of multipliers $M(W_2^{s-1}(\Omega))$. According to Lemma 5, sufficient conditions are the following:

$$\begin{array}{llll} a_{ij} \in W_p^{s-1+\delta}(\Omega), & \text{where} & & \\ p = 2, & \delta = 0, & \text{for} & s > 2, \\ p = 2/(s-1), & \delta > 0, & \text{for} & 1 < s \le 2, \\ p = \infty, & \delta = 0, & \text{for} & s = 1. \end{array}$$

We also assume that the corresponding differential operator is strongly elliptic, i.e.

$$a_{ij} = a_{ji}; \qquad \sum_{i,j=1}^{2} a_{ij} y_i y_j \ge c_0 \sum_{i=1}^{2} y_i^2, \quad x \in \Omega, \quad c_0 = \text{const} > 0.$$

Let $\overline{\omega}$ be the uniform mesh in $\overline{\Omega}$ with the step size h, $\omega = \overline{\omega} \cap \Omega$, $\gamma = \overline{\omega} \cap \Gamma$, $\gamma_{ik} = \{x \in \gamma : x_i = k, 0 < x_{3-i} < 1\}$, $k = 0, 1$, $\omega_i = \omega \cup \gamma_{i0}$ and $\omega_{12} = \omega \cup \gamma_{10} \cup \gamma_{20} \cup \{(0,0)\}$. We define finite differences in the usual manner [17]:

$$v_{x_i} = (v^{+i} - v)/h, \qquad v_{\bar{x}_i} = (v - v^{-i})/h,$$

where $v^{\pm i}(x) = v(x \pm h\, r_i)$, and r_i is the unit vector on the x_i axis.

We also define the Steklov smoothing operators (see [9])

$$T_i^+ f(x) = \int_0^1 f(x + htr_i)dt = T_i^- f(x + hr_i) = T_i f(x + 0.5hr_i).$$

These operators commute and transform derivatives to differences:

$$T_i^+ D_i u = u_{x_i}, \qquad T_i^- D_i u = u_{\bar{x}_i}.$$

We approximate problem (9) with the following FDS

$$L_h v = T_1^2 T_2^2 f \quad \text{in} \quad \omega, \qquad v = 0 \quad \text{on} \quad \gamma \tag{10}$$

where

$$L_h v = -0.5 \sum_{i,j=1}^{2} \left[(a_{ij} v_{\bar{x}_j})_{x_i} + (a_{ij} v_{x_j})_{\bar{x}_i} \right].$$

The finite difference scheme (10) is the standard symmetric FDS (see [17]) with averaged right–hand side. Note that for $s \leq 3$ the right–hand side may be discontinuous function, so FDS without averaging is not well defined.

6 Convergence of the Finite Difference Scheme

Let u be the solution of BVP (9) and v – the solution of FDS (10). The error $z = u - v$ satisfies the conditions

$$L_h z = \sum_{i,j=1}^{2} \eta_{ij,\bar{x}_i} \quad \text{in} \quad \omega, \qquad z = 0 \quad \text{on} \quad \gamma \tag{11}$$

where

$$\eta_{ij} = T_i^+ T_{3-i}^2 (a_{ij} D_j u) - 0.5\,(a_{ij} u_{x_j} + a_{ij}^{+i} u_{\bar{x}_j}^{+i}).$$

For $\varpi \subseteq \bar{\omega}$ let $(\cdot,\cdot)_{\varpi} = (\cdot,\cdot)_{L_2(\varpi)}$ and $\|\cdot\|_{\varpi} = \|\cdot\|_{L_2(\varpi)}$ denote the discrete inner product and the discrete L_2–norm on ϖ. We also define discrete Sobolev norms

$$\|v\|^2_{W_2^1(\omega)} = \|v\|^2_{\omega} + \|v_{\bar{x}_1}\|^2_{\omega_1} + \|v_{\bar{x}_2}\|^2_{\omega_2}, \qquad \text{and}$$

$$\|v\|^2_{W_2^2(\omega)} = \|v\|^2_{W_2^1(\omega)} + \|v_{x_1\bar{x}_1}\|^2_{\omega} + \|v_{x_2\bar{x}_2}\|^2_{\omega} + \|v_{x_1 x_2}\|^2_{\omega_{12}}.$$

The following assertion holds true [9]:

Lemma 6. *FDS* (11) *satisfy a priori estimates*

$$\|z\|_{W_2^1(\omega)} \leq C \sum_{i,j=1}^{2} \|\eta_{ij}\|_{\omega_i}, \qquad \textit{and} \tag{12}$$

$$\|z\|_{W_2^2(\omega)} \leq C \sum_{i,j=1}^{2} \|\eta_{ij,\bar{x}_i}\|_{\omega}. \tag{13}$$

In such a way, the problem of deriving the convergence rate estimates for FDS (10) is now reduced to estimating the right–hand side terms in (12) and (13).

Let us decompose $\eta_{ij,\bar{x}_i}$ in the following manner

$$\eta_{ij,\bar{x}_i} = T_1^2 T_2^2 D_i(a_{ij} D_j u) - 0.5\left[(a_{ij} u_{\bar{x}_j})_{x_i} + (a_{ij} u_{x_j})_{\bar{x}_i}\right] = \sum_{k=1}^{7} \zeta_{ijk},$$

where

$$\begin{aligned}
\zeta_{ij1} &= T_1^2 T_2^2 D_i(a_{ij} D_i D_j u) - (T_1^2 T_2^2 a_{ij}) \cdot (T_1^2 T_2^2 D_i D_j u),\\
\zeta_{ij2} &= (T_1^2 T_2^2 a_{ij} - a_{ij}) \cdot (T_1^2 T_2^2 D_i D_j u),\\
\zeta_{ij3} &= a_{ij}\left[T_1^2 T_2^2 D_i D_j u - 0.5\,(u_{\bar{x}_i x_j} + u_{x_i \bar{x}_j})\right],\\
\zeta_{ij4} &= T_1^2 T_2^2 (D_i a_{ij} D_j u) - (T_1^2 T_2^2 D_i a_{ij}) \cdot (T_1^2 T_2^2 D_j u),\\
\zeta_{ij5} &= \left[T_1^2 T_2^2 D_i a_{ij} - 0.5\,(a_{ij,x_i} + a_{ij,\bar{x}_i})\right] \cdot (T_1^2 T_2^2 D_j u),\\
\zeta_{ij6} &= 0.5\,(a_{ij,x_i} + a_{ij,\bar{x}_i}) \cdot \left[T_1^2 T_2^2 D_j u - 0.5\,(u_{x_j}^{-i} + u_{\bar{x}_j}^{+i})\right],\\
\zeta_{ij7} &= 0.25\,(a_{ij,x_i} - a_{ij,\bar{x}_i}) \cdot (u_{x_j}^{-i} - u_{\bar{x}_j}^{+i}).
\end{aligned}$$

The value ζ_{ij1} in the node $x \in \omega$ can be represented in the form

$$\begin{aligned}
\zeta_{ij1} = \frac{1}{2h^4} \iiiint\limits_{e\times e} &\Phi(\xi_1,\xi_2)\,\Phi(\chi_1,\chi_2)\,[a(\xi_1,\xi_2) - a(\chi_1,\chi_2)] \times \\
&\times [D_i D_j u(\xi_1,\xi_2) - D_i D_j u(\chi_1,\chi_2)]\, d\xi_1 d\xi_2 d\chi_1 d\chi_2,
\end{aligned} \tag{14}$$

where $e = (x_1 - h,\ x_1 + h) \times (x_2 - h,\ x_2 + h)$ and

$$\Phi(\xi_1,\xi_2) = \left(1 - \frac{|\xi_1 - x_1|}{h}\right)\left(1 - \frac{|\xi_2 - x_2|}{h}\right).$$

From (14) immediatly follows

$$|\zeta_{ij1}| \le \frac{C}{h}\,\|a_{ij}\|_{L_\infty(e)} \|D_i D_j u\|_{L_2(e)} \le \frac{C}{h}\,\|a_{ij}\|_{L_\infty(\Omega)} \|u\|_{W_2^2(e)}.$$

From here, summing over the mesh ω, we obtain

$$\|\zeta_{ij1}\|_\omega \le C\,\|a_{ij}\|_{L_\infty(\Omega)} \|u\|_{W_2^2(\Omega)} \le C\,\|a_{ij}\|_{W_2^{1+\varepsilon}(\Omega)} \|u\|_{W_2^2(\Omega)}, \qquad \varepsilon > 0.$$

Analogous estimates hold true also for other terms ζ_{ijk}. In such a way we obtain

$$\|\eta_{ij,\bar{x}_i}\|_\omega \le C\,\|a_{ij}\|_{W_2^{1+\varepsilon}(\Omega)} \|u\|_{W_2^2(\Omega)}, \qquad \varepsilon > 0. \tag{15}$$

From (14), using representation

$$\begin{aligned}
D_i D_j u(\xi_1,\xi_2) - D_i D_j u(\chi_1,\chi_2) &= \int_{\chi_2}^{\xi_2} D_2 D_i D_j u(\xi_1,\tau_2)\, d\tau_2 \\
&+ \int_{\chi_1}^{\xi_1} D_1 D_i D_j u(\tau_1,\chi_2)\, d\tau_1,
\end{aligned}$$

we obtain

$$|\zeta_{ij1}| \le C\, \|a_{ij}\|_{L_\infty(\Omega)} \|u\|_{W_2^3(e)}.$$

Summation over the mesh ω yields

$$\|\zeta_{ij1}\|_\omega \le Ch\, \|a_{ij}\|_{L_\infty(\Omega)} \|u\|_{W_2^3(\Omega)} \le Ch\, \|a_{ij}\|_{W_2^2(\Omega)} \|u\|_{W_2^3(\Omega)}.$$

Analogous estimates hold true for other terms ζ_{ijk}, so we obtain

$$\|\eta_{ij,\bar{x}_i}\|_\omega \le Ch\, \|a_{ij}\|_{W_2^2(\Omega)} \|u\|_{W_2^3(\Omega)}. \tag{16}$$

Finally, transforming $a(\xi_1, \xi_2) - a(\chi_1, \chi_2)$ in (14) to integral form, using Newton–Leibnitz formula, and applying Hölder's inequality, we obtain

$$|\zeta_{ij1}| \le Ch\, \|a_{ij}\|_{W_q^1(e)} \|u\|_{W^3_{2q/(q-2)}(e)}, \qquad q > 2.$$

From here, summing over the mesh ω, and using the imbeddings $W_2^3 \subset W_q^1$ and $W_2^4 \subset W^3_{2q/(q-2)}$, we obtain

$$\|\zeta_{ij1}\|_\omega \le Ch^2\, \|a_{ij}\|_{W_q^1(\Omega)} \|u\|_{W^3_{2q/(q-2)}(\Omega)} \le Ch^2\, \|a_{ij}\|_{W_2^3(\Omega)} \|u\|_{W_2^4(\Omega)}.$$

Analogous estimates hold true for other terms ζ_{ijk}, and so we obtain

$$\|\eta_{ij,\bar{x}_i}\|_\omega \le Ch^2\, \|a_{ij}\|_{W_2^3(\Omega)} \|u\|_{W_2^4(\Omega)}. \tag{17}$$

The mapping $(a_{ij}, u) \to \eta_{ij,\bar{x}_i}$ is, obviously, bilinear. From (15–17) follows that it is a bounded bilinear operator from $W_2^{1+\varepsilon}(\Omega) \times W_2^2(\Omega)$ to $L_2(\omega)$, from $W_2^2(\Omega) \times W_2^3(\Omega)$ to $L_2(\omega)$, and from $W_2^3(\Omega) \times W_2^4(\Omega)$ to $L_2(\omega)$. Applying Lemma 1, from (16) and (17) it follows that $\eta_{ij,\bar{x}_i}$ is a bounded bilinear operator from $[W_2^2(\Omega), W_2^3(\Omega)]_\theta \times [W_2^3(\Omega), W_2^4(\Omega)]_\theta$ to $L_2(\omega)$, with the norm $M \le Ch^{1+\theta}$. Accordingly to (7)

$$\left[W_2^2(\Omega), W_2^3(\Omega)\right]_\theta = W_2^{2+\theta}(\Omega) \qquad \text{and} \qquad \left[W_2^3(\Omega), W_2^4(\Omega)\right]_\theta = W_2^{3+\theta}(\Omega).$$

Setting $3 + \theta = s$, we obtain

$$\|\eta_{ij,\bar{x}_i}\|_\omega \le Ch^{s-2}\, \|a_{ij}\|_{W_2^{s-1}(\Omega)} \|u\|_{W_2^s(\Omega)}, \qquad 3 < s < 4. \tag{18}$$

From (13), (16), (17) and (18) we finally obtain the desired convergence rate estimate when the solution of BVP (9) belongs to Sobolev space with non-integer index

$$\|u - v\|_{W_2^2(\omega)} \le Ch^{s-2} \max_{i,j} \|a_{ij}\|_{W_2^{s-1}(\Omega)} \|u\|_{W_2^s(\Omega)}, \qquad 3 \le s \le 4. \tag{19}$$

This estimate is obtained in [21].

Similarly, from (15) and (16) by interpolation we obtain

$$\|\eta_{ij,\bar{x}_i}\|_\omega \le Ch^{s-2}\, \|a_{ij}\|_{W_2^{s-1+\varepsilon(3-s)}(\Omega)} \|u\|_{W_2^s(\Omega)}, \qquad 2 < s < 3,$$

and further

$$\|u-v\|_{W_2^2(\omega)} \leq Ch^{s-2} \max_{i,j} \|a_{ij}\|_{W_2^{s-1+\varepsilon(3-s)}(\Omega)} \|u\|_{W_2^s(\Omega)}, \qquad 2 \leq s \leq 3. \tag{20}$$

Note that the estimate (19) is consistent with the smoothness of data, while the estimate (20) is slightly inconsistent.

In a similar manner we may obtain estimates

$$\|\eta_{ij}\|_{\omega_i} \leq C\,\|a_{ij}\|_{W^{\varepsilon}_{2/\varepsilon+1}(\Omega)} \|u\|_{W_2^{1+\delta}(\Omega)}, \tag{21}$$

$$\|\eta_{ij}\|_{\omega_i} \leq Ch\,\|a_{ij}\|_{W_2^{1+\varepsilon}(\Omega)} \|u\|_{W_2^2(\Omega)}, \tag{22}$$

$$\|\eta_{ij}\|_{\omega_i} \leq Ch^2\,\|a_{ij}\|_{W_2^2(\Omega)} \|u\|_{W_2^3(\Omega)}, \tag{23}$$

where $\varepsilon > 0$ and $\delta > 0$. Applying Lemmma 1 from (22) and (23) we obtain

$$\|\eta_{ij}\|_{\omega_i} \leq Ch^{1+\theta}\,\|a_{ij}\|_{W_2^{1+\theta+\varepsilon(1-\theta)}(\Omega)} \|u\|_{W_2^{2+\theta}(\Omega)}, \qquad 0 < \theta < 1. \tag{24}$$

From (12), (22), (23) and (24), setting $2 + \theta = s$, we obtain the following convergence rate estimate

$$\|u-v\|_{W_2^1(\omega)} \leq Ch^{s-1} \max_{i,j} \|a_{ij}\|_{W_2^{s-1+\varepsilon(3-s)}(\Omega)} \|u\|_{W_2^s(\Omega)}, \qquad 2 \leq s \leq 3. \tag{25}$$

Analogously, from (21) and (22), using (3), (4) and (6), we obtain

$$\|\eta_{ij}\|_{\omega_i} \leq Ch^{\theta}\,\|a_{ij}\|_{W_p^{\theta+\varepsilon}(\Omega)} \|u\|_{W_2^{1+\theta+\delta(1-\theta)}(\Omega)}, \qquad 0 < \theta < 1, \tag{26}$$

where $\theta + \varepsilon \neq$ integer, and

$$\frac{1}{p} = \frac{\varepsilon\,(1-\theta)}{2+\varepsilon} + \frac{\theta}{2}.$$

From (12), (21), (22) and (26), setting $1 + \theta = s$, we obtain

$$\|u-v\|_{W_2^1(\omega)} \leq Ch^{s-1} \max_{i,j} \|a_{ij}\|_{W_p^{s-1+\varepsilon}(\Omega)} \|u\|_{W_2^{s+\delta(2-s)}(\Omega)}, \qquad 1 \leq s \leq 2, \tag{27}$$

where

$$p = p(s) = \frac{2\,(2+\varepsilon)}{2\,(s-1) + \varepsilon\,(3-s)}.$$

Remark. In (21) we can not set $\delta = 0$ because $W_2^1(\Omega) \not\subset C(\overline{\Omega})$ and the error $z = u - v$ may not be defined in the nodes of mesh. If we define the error as

$$z = T_1 T_2 u - v,$$

then relations (11), (12) and (22) are satisfied, having now

$$\eta_{ij} = T_i^+ T_{3-i}^2 (a_{ij} D_j u) - 0.5\left[a_{ij}(T_1T_2u)_{x_j} + a_{ij}^{+i}(T_1T_2u)_{\bar{x}_j}^{+i}\right],$$

while inequality (21) holds true also for $\delta = 0$. Repeating the previous procedure, instead of (27), we obtain

$$\|T_1T_2u - v\|_{W_2^1(\omega)} \leq Ch^{s-1} \max_{i,j} \|a_{ij}\|_{W_p^{s-1+\varepsilon}(\Omega)} \|u\|_{W_2^s(\Omega)}, \qquad 1 \leq s \leq 2.$$

References

1. Bergh, J., Löfström, J.: Interpolation spaces. Springer–Verlag, Berlin – Heidelberg – New York 1976
2. Bramble, J.H. Hilbert, S.R.: Bounds for a class of linear functionals with application to Hermite interpolation. Numer. Math. **16** (1971) 362–369
3. Calderón, A.P.: Intermediate spaces and interpolation, the complex method. Studia Math. **24** (1964) 113–190
4. Dupont, T., Scott, R.: Polynomial approximation of functions in Sobolev spaces. Math. Comput. **34** (1980) 441–463
5. Jovanović, B.S.: On the convergence of finite–difference schemes for parabolic equations with variable coefficients. Numer. Math. **54** (1989) 395–404
6. Jovanović, B.S.: Optimal error estimates for finite–difference schemes with variable coefficients. Z. Angew. Math. Mech. **70** (1990) 640–642
7. Jovanović, B.S.: Convergence of finite–difference schemes for parabolic equations with variable coefficients. Z. Angew. Math. Mech. **71** (1991) 647–650
8. Jovanović, B.S.: Convergence of finite–difference schemes for hyperbolic equations with variable coefficients. Z. Angew. Math. Mech. **72** (1992) 493–496
9. Jovanović, B.S.: The finite–difference method for boundary–value problems with weak solutions. Posebna izdan. Mat. Inst. **16**, Belgrade 1993
10. Jovanović, B.S.: Interpolation of function spaces and the convergence rate estimates for the finite–difference schemes. In: D. Bainov and V. Covachev (eds.), 2nd Int. Coll. on Numerical Analysis held in Plovdiv 1993, VSP, Utrecht 1994, 103–112
11. Jovanović, B.S., Ivanović, L.D., Süli, E.E.: Convergence of a finite–difference scheme for second–order hyperbolic equations with variable coefficients. IMA J. Numer. Anal. **7** (1987) 39–45
12. Jovanović, B.S., Ivanović, L.D., Süli, E.E.: Convergence of finite–difference schemes for elliptic equations with variable coefficients. IMA J. Numer. Anal. **7** (1987) 301–305
13. Lazarov, R.D.: On the question of convergence of finite–difference schemes for generalized solutions of the Poisson equation. Differentsial'nye Uravneniya **17** (1981) 1287–1294 (Russian)
14. Lazarov, R.D., Makarov, V.L., Samarski, A.A.: Application of exact difference schemes for construction and investigation of difference schemes for generalized solutions. Mat. Sbornik **117** (1982) 469–480 (Russian)
15. Lazarov, R.D., Makarov, V.L., Weinelt, W.: On the convergence of difference schemes for the approximation of solutions $u \in W_2^m$ $(m > 0.5)$ of elliptic equations with mixed derivatives. Numer. Math. **44** (1984) 223–232
16. Maz'ya, V.G., Shaposhnikova, T.O.: Theory of multipliers in spaces of differentiable functions. Monographs and Studies in Mathematics **23**. Pitman, Boston, Mass. 1985
17. Samarski, A.A.: Theory of difference schemes. Nauka, Moscow 1983 (Russian)
18. Schwartz, L.: Théorie des distributions. Herman, Paris 1966
19. Süli, E., Jovanović, B., Ivanović, L.: Finite difference approximations of generalized solutions. Math. Comput. **45** (1985) 319–327
20. Triebel, H.: Interpolation theory, function spaces, differential operators. Deutscher Verlag der Wissenschaften, Berlin 1978

21. Živanović, G.: Application of interpolation theory for establishing convergence rate estimates for finite difference schemes. M.Sc. Thesis, University of Belgrade 1994 (Serbian)
22. Zlotnik, A.A.: Convergence rate estimates for projection–difference schemes approximating second order hyperbolic equations Vychisl. Protsessy Sist. **8** (1991) 116–167 (Russian)

Least Squares and Total Least Squares Methods in Image Restoration

Julie Kamm[1] and James G. Nagy[2]

[1] Defense Systems & Electronics, Texas Instruments Inc., Dallas, Texas, USA
[2] Department of Mathematics, Southern Methodist University, Dallas, Texas, USA

Abstract. Image restoration is the process of removing or minimizing degradations (blur) in an image. Mathematically, it can be modeled as a discrete ill-posed problem $H\mathbf{f} = \mathbf{g}$, where H is a matrix of large dimension representing the blurring phenomena, and $\mathbf{g}$ is a vector representing the observed image. Often H is severely ill-conditioned, and both H and $\mathbf{g}$ are corrupted with noise. Regularization is used to reduce the noise sensitivity of the numerical scheme. Most of these methods, however, assume that there is no noise in H, and therefore least squares techniques are used to reconstruct $\mathbf{f}$. In some applications, though, H is also corrupted with noise. In this case a total least squares approach may be more appropriate. These least squares and total least squares methods will be investigated in the context of signal restoration.

1 Introduction

Given a severely ill-conditioned matrix H_t and observation vector $\mathbf{g}_t$, a difficult task is the numerical approximation of the solution to $H_t\mathbf{f}_t = \mathbf{g}_t$. A condition that makes this especially difficult in applications such as signal and image restoration is that H_t and $\mathbf{g}_t$ are not known explicitly [14]. In these applications, observations are used to construct the noise corrupted quantities $H = H_t + E$ and $\mathbf{g} = \mathbf{g}_t + \mathbf{e}$, where the matrix E and vector $\mathbf{e}$ represent noise in H and $\mathbf{g}$, respectively.

Given H and $\mathbf{g}$, the aim is to compute an approximation to $\mathbf{f}_t$. Because H is severely ill-conditioned and H and $\mathbf{g}$ are corrupted with noise, computed solutions are susceptible to large errors [8]. Regularization is often used to reduce this sensitivity to noise. There are many well-known regularization methods, including truncated singular value decomposition [11], Tikhonov regularization [6], and truncated iterations [7, 18]. Most of these schemes assume that there are no errors in H, and therefore are based on least squares (LS) techniques. In many applications, though, H is constructed from measurements and hence is also contaminated with noise. In this case, the total least squares (TLS) method may be more appropriate. Recently, Fierro, Golub, Hansen and O'Leary [2] proposed a truncated TLS regularization method.

In some cases, the matrices H and H_t are structured, and thus so is E. The standard TLS method assumes that the error components in E are unrelated. Alternative formulations that take the structure of E into account, have been proposed [16, 15]. The difficulty with these methods is that they require solving a nonlinear unconstrained minimization problem.

In this paper we focus on the linear LS and TLS methods in the context of signal restoration. Section 2 provides a brief review of signal restoration, and discusses the structure of H_t, H, and E. In Section 3 we discuss a variety of regularized solution methods based on least squares (LS) and total least squares (TLS) techniques. We then compare each of these techniques on a simulated signal restoration example in Section 4.

2 Signal/Image Restoration

Signal restoration [14] is modeled as a first kind integral equation, which is an ill-posed inverse problem [6]. Often this model is not available for analysis purposes. Instead, only a discrete observed signal and a discrete observation of an impulse response, from which the matrix H is constructed, are available. If the imaging system is spatially invariant, then every point is blurred in the same way, and hence only one impulse response is needed to construct H. In this case H will be a Toeplitz matrix for 1-D signals, and block Toeplitz with Toeplitz blocks for 2-D images [14].

In case the blur is spatially varying, then several impulse responses are needed to construct H. For the extreme spatially varying case, H will have little or no structure. This extreme case is not typical, though, and often the assumption that the degradation phenomena is locally spatially invariant is valid [3]. Therefore, we consider only the spatially invariant case in this paper.

It is important to note that often in practice only a noise corrupted column, $\mathbf{h}$, of H can be found. Since we construct H from this one noise corrupted column, it is clear that $H = H_t + E$, where E has the same Toeplitz structure as H_t and H. It should be emphasized, though, that this is true assuming the degradation phenomena is purely spatially invariant.

The following question arises: Is the degradation phenomena purely spatially invariant, or is it *nearly* spatially invariant? In cases such as this, the blur is often assumed to be spatially invariant. Thus the matrix $H = H_t + E$ constructed from one point source will have an error matrix E having little or no structure.

3 Solution Methods

In this section we consider numerical methods for the spatially invariant signal restoration problem discussed above. Our aim is to focus on various regularization methods, and not on specific algorithms for their implementations. A good source for implementations of most regularization schemes discussed below can be found in Hansen's *Regularization Tools* for Matlab [12].

Because the continuous problem is an ill-posed inverse problem, the discrete problem inherits its properties. Specifically, let $H = U\Sigma V^T$ be the singular value decomposition (SVD) of H, where U and V are orthogonal matrices, and

$\Sigma = \text{diag}(\sigma_1, \sigma_2, \cdots, \sigma_n)$ is a diagonal matrix containing the singular values of H. Then the following properties hold:

- the singular values, $\sigma_1 \geq \sigma_2 \geq \cdots \geq \sigma_n$ decay to zero, without a significant gap to indicate numerical rank, and
- the singular vectors associated with small σ_i tend to oscillate significantly.

To see how these properties can cause difficulties when attempting to reconstruct the solution, suppose we use the SVD to naively compute $\mathbf{f}$:

$$\mathbf{f} = V\Sigma^{-1}U^T\mathbf{g} = \sum_{i=1}^{n} \frac{\mathbf{u}_i^T\mathbf{g}}{\sigma_i}\mathbf{v}_i, \tag{1}$$

where $\mathbf{u}_i$ and $\mathbf{v}_i$ are the columns of U and V, respectively. Using the columns of U and V as bases, we can write $\mathbf{f} = \sum_{i=1}^{n} f_i\mathbf{v}_i$, and $\mathbf{e} = \sum_{i=1}^{n} \varepsilon_i\mathbf{u}_i$. Using these relations, (1) can be written as

$$\mathbf{f} = \sum_{i=1}^{n} \left(f_i + \frac{\varepsilon_i}{\sigma_i} \right) \mathbf{v}_i.$$

We see that the small σ_i magnify the noise components ε_i, and, moreover, magnify the high oscillations in $\mathbf{v}_i$. Thus the computed solution is corrupted by noise.

Regularization can be used to alleviate these difficulties by filtering out the effects of the small σ_i. That is, a regularized solution can be written as

$$\mathbf{f}_{reg} = \sum_{i=1}^{n} \phi_i \frac{\mathbf{u}_i^T\mathbf{g}}{\sigma_i}\mathbf{v}_i, \tag{2}$$

where the "filter factors" ϕ_i depend on the regularization scheme and solution method. We now discuss several regularization schemes.

Truncated SVD. Perhaps the most straightforward approach to regularization is truncating the SVD expansion for $\mathbf{f}$ given in (1); this is called the truncated SVD (TSVD) [11]. The filter factors for the TSVD solution are given by

$$\phi_i = \begin{cases} 1 & \text{if } i <= k \\ 0 & \text{otherwise} \end{cases},$$

where $k \leq n$. A disadvantage of using the SVD to compute an approximate solution is that it is expensive, both in computations and storage, and is therefore typically not practical for large problems. Recently, though, Fish, Grochmalicki and Pike [3] proposed a "scanning SVD" method for image restoration. However, this scheme only computes an approximate SVD of H, and not the full decomposition.

Tikhonov regularization. Instead of solving an ill-conditioned least squares problem $\min ||H\mathbf{f} - \mathbf{g}||_2$, Tikhonov regularization [6] reduces noise sensitivity by solving a damped least squares problem:

$$\min \left\| \begin{bmatrix} H \\ \lambda L \end{bmatrix} \mathbf{f} - \begin{bmatrix} \mathbf{g} \\ 0 \end{bmatrix} \right\|_2,$$

where L is a matrix typically chosen to be the identity or a discrete approximation to some derivative operator. The regularization parameter λ needs to be chosen to control the bias introduced by the damping term $||L\mathbf{f}||_2^2$. With this formulation, computational methods can now take advantage of any structure and/or sparsity in H and L. For example, there exist fast direct methods for the case when H is Toeplitz [17], as well as several approaches for efficiently using iterative methods [1].

If $L = I$, then we can use the SVD to determine the filter factors ϕ_i. In particular, it is not difficult to show that

$$\phi_i = \frac{\sigma_i^2}{\sigma_i^2 + \lambda^2}.$$

Thus, there is a smoother transition of filter factors as the singular values range from large to small, than in the TSVD case.

Truncated Iterative Methods. Some iterative methods have the property that if they are applied directly to the (unregularized) least squares problem $\min ||H\mathbf{f} - \mathbf{g}||_2$, then progress can be made in reconstructing the true solution at the beginning of the iteration process. At a certain point, though, the iterations become contaminated with noise, and process begins to diverge. If the iteration is halted at this point where divergence begins, then a reasonable solution can be obtained. Among the iterative methods having this behavior are conjugate gradients applied to the normal equations (CGLS) [18], a symmetric indefinite version of conjugate gradients called MR-II [7], and others [8]. We will focus on the conjugate gradient (CG) type methods.

The regularization properties of CG methods are much more complicated to describe than for TSVD and Tikhonov regularization; details can be found in [8, 18]. In particular, if we write the solution at iteration k as $\mathbf{f}^{(k)}$ then the filter factors $\phi_i^{(k)}$ at iteration k can be written as

$$\phi_i^{(k)} = 1 - r_k(\sigma_i^2).$$

The *Ritz polynomials*, r_k, are the unique sequence of polynomials minimizing

$$F[r_k] = \sum_{i=1}^{n} p_k^2(\sigma_i^2)(\sigma_i f_i + \varepsilon_i)^2$$

over all polynomials p_k of degree k, with $p_k(0) = 1$, $k = 0, 1, 2, \ldots$.

Note that iterative regularization methods can be very efficient for large structured and/or sparse problems. In some cases, preconditioning can be used to improve convergence rates [10, 9].

Truncated Total Least Squares. The regularization methods described above assume that there are no errors in the matrix H. However, as discussed in Section 2, in most practical signal restoration problems the matrix H will also contain noise. In this situation, the method of total least squares (TLS) may be more appropriate than least squares approaches [5, 13]. Recently Fierro, Golub,

Hansen and O'Leary [2] have shown that regularization of TLS for severely ill-conditioned problems can be accomplished by computing a *truncated* TLS solution.

TLS for well-posed problems amounts to minimizing the residual matrix,

$$\min_{[H'\ \mathbf{g}']} \{|||[H\ \mathbf{g}] - [H'\ \mathbf{g}']|||\}, \quad \text{subject to} \quad H'\mathbf{f} = \mathbf{g}'.$$

The TLS solution can be found using the SVD of the augmented matrix $[H\ \mathbf{g}]$ [13]. The truncated TLS approach, as given in [2], is similar to the TSVD (least squares) approach in that the small singular values of the SVD expansion of $[H\ \mathbf{g}]$ are replaced with zeros.

In order to describe the filter factors, let $H = U\Sigma V^T$ and $[H\ \mathbf{g}] = \bar{U}\bar{\Sigma}\bar{V}^T$ be the SVD of H and $[H\ \mathbf{g}]$, respectively. If the singular values of H are distinct, then the filter factors for the truncated TLS solution are [2]

$$\phi_i = \sum_{j=k+1}^{n+1} \frac{\bar{v}_{n+1,j}^2}{||\bar{V}_{22}||_2^2} \left(\frac{\sigma_i^2}{\sigma_i^2 - \bar{\sigma}_j^2} \right), \quad i = 1, 2, \ldots, n,$$

where $\bar{V}_{22} = [\ \bar{v}_{n+1,k+1} \ \cdots \ \bar{v}_{n+1,n+1}\ ]$, and k is the truncation parameter (that is, $\bar{\sigma}_{k+1}, \ldots, \bar{\sigma}_{n+1}$ in the SVD of $[H\ \mathbf{g}]$ are set to zero). If $\bar{\sigma}_j = \sigma_i$ for some j, then the corresponding term does not contribute to ϕ_i.

As with the TSVD method, the SVD approach to compute truncated TLS solutions can be prohibitively expensive for large problems. In this case, a Lanczos-based bidiagonalization algorithm is suggested in [2].

Damped Total Least Squares. The truncated TLS method is one way to regularize discrete ill-posed problems that contain noise in both $\mathbf{g}$ and H. As there are many regularization schemes for LS methods, in this section we investigate a damped total least squares formulation. First we begin by observing that if $\sigma_n(H) > \sigma_{n+1}([H\,\mathbf{g}])$, where $\sigma_k(\cdot)$ denotes the kth singular value of the given matrix, then a closed form representation of the TLS solution is given by [5, 13]

$$(H^T H - \sigma^2 I)\mathbf{f} = H^T\mathbf{g}, \quad \sigma = \sigma_{n+1}([H\,\mathbf{g}]).$$

This closed form expression is typically not recommended to compute TLS solutions, but we note that because $\sigma_n(H) > \sigma_{n+1}([H\,\mathbf{g}])$, the matrix $H^T H - \sigma^2 I$ is symmetric positive definite.

Let $A = H^T H - \sigma^2 I$ and $\mathbf{b} = H^T\mathbf{g}$. Then Franklin [4] shows that a regularization method for $A\mathbf{f} = \mathbf{b}$, where A is symmetric positive definite, can be obtained by solving

$$\min_{\mathbf{f}}\{\mathbf{f}^T A\mathbf{f} - 2\mathbf{f}^T\mathbf{b} + \mu^2\mathbf{f}^T\mathbf{f}\},$$

where μ is a regularization parameter. It is easy to see that the solution to this minimization problem has the form $(A + \mu^2 I)\mathbf{f} = \mathbf{b}$. Using the relations for A and $\mathbf{b}$ given above, we obtain a *damped TLS* solution:

$$(H^T H + (\mu^2 - \sigma^2)I)\,\mathbf{f} = H^T\mathbf{g}. \tag{3}$$

Observe that if the noise in $\mathbf{g}$ satisfies $||\mathbf{e}|| > \sigma$, then in order to dampen the effects of the small singular values, we must have $\mu > \sigma$. Then (3) can be written as

$$\min_{\mathbf{f}} \left\| \begin{bmatrix} H \\ \sqrt{\mu^2 - \sigma^2} I \end{bmatrix} \mathbf{f} - \begin{bmatrix} \mathbf{g} \\ 0 \end{bmatrix} \right\|_2 .$$

Thus, the damped TLS problem reduces to a damped LS problem, with $\lambda = \sqrt{\mu^2 - \sigma^2}$. Therefore, we may use less expensive algorithms developed for solving damped LS problems. Of course there are situations for which the truncated TLS method will provide better solutions (see the numerical experiments in [2]), but there are also situations where the damped TLS method performs better. We illustrate this in the next section.

4 Numerical Tests

In this section we provide some numerical tests that compare the regularization schemes described in the previous sections. All numerical tests were performed on a DEC AlphaStation 200 using Matlab 4.2c. Our tests are set up to simulate the restoration of a gamma ray spectra problem [19], shown in Figure 1a. A Toeplitz matrix H_t is constructed to have entries generated from a Gaussian function: $h_{i,j} = \frac{1}{\sqrt{2\pi\sigma^2}} \exp\left[\frac{-(|i-j|-1)^2}{2\sigma^2}\right]$, where $\sigma = 3.0$. The singular values of H_t are shown in Figure 1b, where it can be seen that in this case H_t has a condition number $\approx 8.65 \times 10^{15}$. A blurred signal can be simulated by computing $\mathbf{g}_t = H_t\mathbf{f}$; this is shown in Figure 1c.

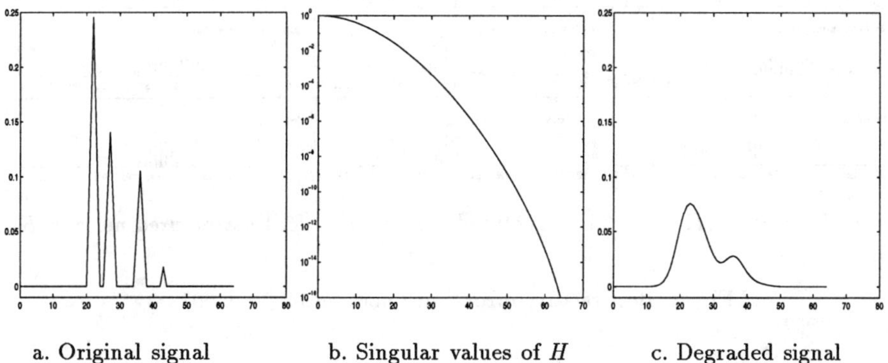

a. Original signal   b. Singular values of H c. Degraded signal

Fig. 1. Simulated signal restoration data.

The noise vector $\mathbf{e}$ and matrix E can be constructed to have entries that are generated randomly from a normal distribution with mean 0.0 and variance 1.0, and scaled so that $||\mathbf{e}||_2/||\mathbf{g}_t||_2 = \gamma$ and $||E||_2/||H_t||_2 = \gamma$. Using this notation, we generated three test problems to simulate the noise situations summarized in Section 2.

Problem 1: In the first test problem, we added noise only to the right hand side vector g, and none to the matrix H.

Problem 2: In addition to adding noise to g, we also added noise to the matrix H. In this problem, the noise was added in a structured fashion. That is, the matrix E has the same Toeplitz pattern as H.

Problem 3: In this case, as well as adding noise to g, we added unstructured noise to H. That is, E is an unstructured matrix containing random entries as described above.

As is done in [2], for each of the test problems we constructed 1000 different noise vectors e and noise matrices E. Each of the methods discussed in Section 3 were used to compute a restoration, with noise level $\gamma = 0.001$. Regularization parameters were chosen in each case so that the resulting restoration produced the smallest relative error. To compare the methods, we tabulated these "best" relative errors for each method, and displayed them using a histogram plot shown in Figure 2. Because damped TLS is equivalent to using Tikhonov regularization, one histogram is used to present results for both methods.

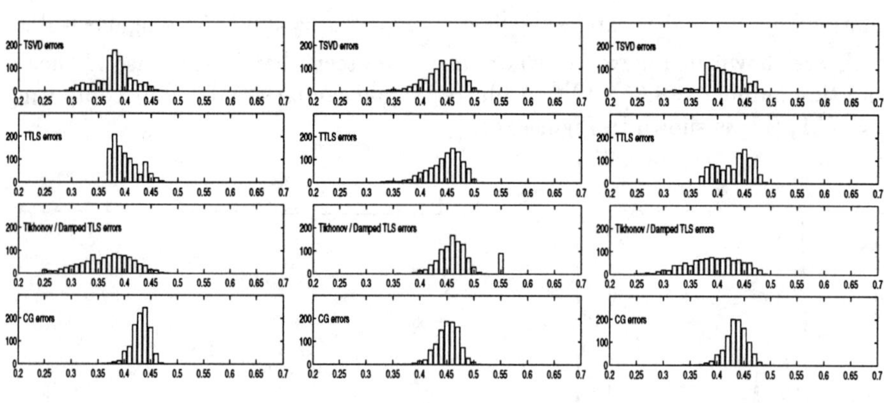

a. No noise in H. b. Structured noise in H. c. Unstructured noise in H.

Fig. 2. Best relative errors with noise level $\gamma = 0.001$.

As seen in Figure 2, Tikhonov regularization performs best for the cases when $E = 0$ and E is unstructured. For structured noise in H, each of the regularization methods performed equally well. The tests were repeated with a higher noise level ($\gamma = 0.05$), with similar results. Notice that the truncated TLS approach did not perform better than the LS approaches, even with errors in the matrix H. This is probably due to the fact that g does not have large components corresponding to the small singular values that are retained [2].

References

1. Chan, R., Nagy, J., Plemmons, R.: Circulant preconditioned Toeplitz least squares iterations. SIAM J. Matrix Anal. Appl. **15** (1994) 80–97.
2. Fierro, R., Golub, G., Hansen, P., O'Leary, D.: Regularization by truncated total least squares. Technical Report UNIC-93-14, UNI•C, Technical University of Denmark, 1993.
3. Fish, D., Grochmalicki, J., Pike, E.: Scanning singular-value-decomposition method for restoration of images with space-variant blur. J. Opt. Soc. Am. A **13** (1996) 1–6.
4. Franklin, J.: Minimum principles for ill-posed problems. SIAM J. Math. Anal. **9** (1978) 638–650.
5. Golub, G., Van Loan, C.: An analysis of the total least squares problem. SIAM J. Numer. Anal. **17** (1980) 883–893.
6. Groetsch, C.: The theory of Tikhonov regularization for Fredholm integral equations of the first kind. Pitman, Boston, 1984.
7. Hanke, M.: Conjugate Gradient Type Methods for Ill-Posed Problems. Pitman Research Notes in Mathematics, Longman Scientific & Technical, Harlow, Essex, 1995.
8. Hanke, M., Hansen, P.: Regularization methods for large-scale problems. Surv. Math. Ind. **3** (1993) 253–315.
9. Hanke, M., Nagy, J.: Restoration of atmospherically blurred images by symmetric indefinite conjugate gradient techniques. Inverse Problems **12** (1996) 157–173.
10. Hanke, M., Nagy, J., Plemmons, R.: Preconditioned iterative regularization. In Numerical Linear Algebra, Reichel, L., Ruttan, A., Varga, R., eds., pages 141–163, de Gruyter, Berlin, 1993.
11. Hansen, P.: The truncated SVD as a method for regularization. BIT **27** (1987) 534–553.
12. Hansen, P.: Regularization tools: A Matlab package for the analysis and solution of discrete ill-posed problems. Numerical Algorithms **6** (1994) 1–35.
13. Van Huffel, S., Vandewalle, J.: The Total Least Squares Problem: Computational Aspects and Analysis. SIAM, Philadelphia, 1991.
14. Jain, A.: Fundamentals of Digital Image Processing. Prentice-Hall, Englewood Cliffs, NJ, 1989.
15. Mesarović, V., Galatsanos, N., Katsaggelos, A.: Regularized constrained total least squares image restoration. IEEE Trans. Image Proc. **4** (1995) 1096–1108.
16. De Moor, B.: Structured total least squares and L_2 approximation problems. Linear Algebra Appl. **188** (1993) 189:163–205.
17. Nagy, J.: Fast inverse QR factorization for Toeplitz matrices. SIAM J. Sci. Comput., **14** (1993) 1174–1193.
18. Plato, R.: Optimal regularization for linear ill-posed problems yield regularization methods. Numer. Funct. Anal. and Optimiz. **11** (1990) 111–118.
19. Trussell, H.: Convergence criteria for iterative restoration methods. IEEE Trans. Acoust., Speech, Signal Processing **31** (1983) 129–136.

Conforming Spectral Domain Decomposition Schemes

Andreas Karageorghis

Department of Mathematics and Statistics
University of Cyprus
P.O. Box 537
1678 Nicosia
Cyprus

Abstract. Spectral domain decomposition schemes are presented for the numerical solution of second and fourth order problems. These schemes, which are formulated in the collocation framework yield spectral approximations which are conforming along the subdomain interfaces for both conforming and non-conforming decompositions. For conforming decompositions the approximations are pointwise C^1 continuous across the interfaces for second order problems and C^3 continuous across the interfaces for fourth order problems. For non-conforming decompositions the corresponding approximations are pointwise C^0 and C^1 continuous for second and fourth order problems, respectively. Efficient direct methods for the solution of the resulting systems are also presented.

1 Introduction

In this paper we discuss various spectral collocation schemes for the solution of second- and fourth-order problems in rectangular domains. We examine two types of domain decompositions, namely, conforming and non-conforming decompositions of the rectangular domains. In the case of conforming domain decompositions the domain is decomposed into an even number of rectangular subdomains and in the case of non-conforming decompositions the domain is decomposed into an odd number of rectangular subdomains. In each case the resulting global matrices have a block structure which may be exploited in the solution of the linear systems. We also present some results of the application of the above schemes to a fluid flow problem.

2 Domain decomposition and spectral approximation

2.1 The problems

We consider the Poisson problem

$$\nabla^2\phi(x,y) = F(x,y) \quad \text{on the rectangle } (\alpha,\beta)\times(a,b) \tag{1}$$

subject to the boundary conditions

$$\phi(\alpha, y) = f_\alpha(y), \tag{2.2a}$$
$$\phi(\beta, y) = f_\beta(y), \tag{2.2b}$$
$$\phi(x, a) = f_a(x), \tag{2.2c}$$
$$\phi(x, b) = f_b(x). \tag{2.2d}$$

We shall assume that the functions $f_\alpha(y), f_\beta(y), f_a(x)$ and $f_b(x)$ are analytic and that they are well approximated by their truncated Taylor series. We shall therefore assume, henceforth, that these functions are polynomials.

We also consider the biharmonic problem

$$\nabla^4\psi(x, y) = F(x, y) \text{ on the rectangle } (\alpha, \beta) \times (a, b), \tag{2.3}$$

subject to the boundary conditions

$\psi(\alpha, y) = f_\alpha(y)$, $\frac{\partial\psi}{\partial x}(\alpha, y) = g_\alpha(y)$ (2.4 a, b)
$\psi(\beta, y) = f_\beta(y)$, $\frac{\partial\psi}{\partial x}(\beta, y) = g_\beta(y)$ (2.4 c, d)
$\psi(x, a) = f_a(x)$, $\frac{\partial\psi}{\partial y}(x, a) = g_a(x)$ (2.4 e, f)
$\psi(x, b) = f_b(x)$, $\frac{\partial\psi}{\partial y}(x, b) = g_b(x)$ (2.4 g, h)

Without loss of generality we shall assume (as before) that the functions $f_\alpha(y)$, $f_\beta(y)$, $f_a(x)$, $f_b(x)$, $g_\alpha(y)$, $g_\beta(y)$, $g_a(x)$ and $g_b(x)$ are polynomials.

2.2 Domain decomposition

For the conforming decomposition of [1], the rectangle $(\alpha, \beta) \times (a, b)$ is decomposed into the subdomains $(\alpha, \alpha_1) \times (a_1, b)$ (Region I), $(\alpha, \alpha_1) \times (a, a_1)$ (Region II) $(\alpha_1, \beta) \times (a, a_1)$ (Region III) and $(\alpha_1, \beta) \times (a_1, b)$ (Region IV), $\alpha < \alpha_1 < \beta, a < a_1 < b,$.

In the non-conforming decomposition of [2], for the partitions $\alpha = \alpha_0 < \alpha_1 < \alpha_2 < ... < \alpha_{N-1} < \alpha_N = \beta$ and $a = a_0 < a_1 < a_2 < ... < a_{N-1} < a_N = b$, $N \in \mathbb{N}$, we consider the decomposition D_{2N-1}: the rectangle $(\alpha, \beta) \times (a, b)$ is decomposed into $2N - 1$ subdomains in the following way: for $k = 1, 2, ..., N - 1$, subdomain $2k - 1$ is the rectangle $(\alpha_{k-1}, \alpha_k) \times (a_{k-1}, a_N)$ and subdomain $2k$ is the rectangle $(\alpha_k, \alpha_N) \times (a_{k-1}, a_k)$. Subdomain $2N - 1$ is the rectangle $(\alpha_{N-1}, \alpha_N) \times (a_{N-1}, a_N)$.

2.3 Spectral approximation

In each of the above cases, in each subdomain the solution is approximated by

$$\phi_s(x, y) = \sum_{m=0}^{M_s} \sum_{n=0}^{N_s} \gamma^s_{mn} \hat{T}^s_m(x) \tilde{T}^s_n(y), \quad s = 1, 2, ..., 2N - 1, \tag{2.5}$$

where the functions $\hat{T}^s_m(x)$ and $\tilde{T}^s_n(y)$ are the shifted Chebyshev polynomials defined on the corresponding intervals of each region and the collocation points on each interval of each region (e.g. $\{x^s_i\}_{i=0}^{M_s}$) are the Gauss-Lobatto points [3,4].

2.4 Continuity

For the conforming case, it can be shown that we can construct approximations which are C^1 continuous across the subdomain interfaces for second order problems and C^3 continuous across the subdomain interfaces for fourth order problems. The details of this construction may be found in [1]. For non-conforming decompositions the corresponding result is weaker. It can be shown that we can construct approximations which are C^0 continuous across the subdomain interfaces for second order problems and C^1 continuous across the subdomain interfaces for fourth order problems. In the case of second order problems we have C^1 continuity only at a finite number of points on the interfaces and in the case of fourth order problems we have C^3 continuity only at a finite number of points on the interfaces. The details of this construction may be found in [2].

2.5 Numerical examples

The techniques described in Section 2 were applied to the following problems:
Example 1

$$\nabla^2\phi(x,y) = (y^2-1)e^x + (x^2-1)e^y + 2e^x + 2e^y \tag{2.6}$$

on the square $(-1,1)^2$ subject to Dirichlet boundary conditions which correspond to the exact solution of this problem, which is $\phi = (y^2-1)e^x + (x^2-1)e^y$

and
Example 2

$$\nabla^4\psi(x,y) = 24(e^x+e^y) + (y^2-1)^2e^x + (x^2-1)^2e^y +$$

$$8((3y^2-1)e^x + (3x^2-1)e^y) \tag{2.7}$$

on the square $(-1,1)^2$, subject to the boundary conditions for ψ and $\frac{\partial\psi}{\partial n}$ corresponding to the exact solution of this problem, which is $\psi = (y^2-1)^2e^x + (x^2-1)^2e^y$.

In Table 1, we present the maximum relative error with m which is the number of degrees of freedom in each direction in each element for a nine-subdomain decomposition.

m	*Total degrees of freedom*	$MRE(Ex.1)$	$MRE(Ex.2)$
4	225	0.87(−3)	0.42(−2)
5	324	0.83(−4)	0.58(−3)
6	441	0.61(−5)	0.25(−4)
7	576	0.39(−6)	0.31(−5)
8	729	0.20(−7)	0.15(−6)
9	900	0.10(−8)	0.98(−8)
10	1089	0.41(−10)	0.47(−9)
11	1296	0.17(−11)	0.21(−10)
12	1521	0.69(−13)	0.84(−12)

Table 1: Maximum Relative Errors (MRE) for test examples.

3 Methods of solution for the global systems

The global matrices for the multidomain conforming and non-conforming domain decompositions are sparse and structured. We next report on the performance of three approaches on a RS 6000 workstation, a SGI Power Challenge and a Cray J-916. A comprehensive analysis of the results may be found in [5], [6] and [11].

3.1 Banded solvers

In the case of the non-conforming decompositions the global matrices have a block diagonal structure. In order to exploit this, some experiments were carried out in [11] with the banded matrix solver pair F07BDF-F07BEF from NAG [8]. This resulted in savings in the CPU times of a factor of two over the dense solvers. It should be pointed out that we are not exploiting the sparsity of the matrix inside the band.

3.2 Capacitance type techniques

In [5], [6] and [11] capacitance-type methods were successfully applied to the solution of the above systems and resulted in considerable savings in CPU times. The savings over the banded solvers varied from a factor of two for the RS 6000 workstation to a factor of three for the SGI Power Challenge.

3.3 Sparse solvers

In [5] and [6] some state-of-the-art sparse solvers, namely the UMFPACK package (versions 1.1 and 2.0, [7]) were applied to the solution of the systems. On the RS 6000 workstation their performance was almost as good as that of the capacitance technique. In the case of the non-conforming decompositions a combination of a capacitance technique with the UMFPACK sparse solver produced the best results. On the Cray J-916 the capacitance technique outperformed the sparse packages in all cases examined.

4 An example from fluid dynamics

In this section we present some preliminary results on the flow through a contracting channel, described in detail in [9]. The flow is steady and laminar and we solve the stream function formulation of the incompressible Navier-Stokes equations. The domain is divided into seven subdomains as follows: Subdomain 1 is the rectangle $(Z_1, Z_2) \times (0, 1)$, subdomain 2 is the rectangle $(Z_2, 0) \times (a_3, 1)$, subdomain 3 is the rectangle $(Z_2, 0) \times (a_2, a_3)$, subdomain 4 is the rectangle $(Z_2, 0) \times (a_1, a_2)$, subdomain 5 is the rectangle $(0, Z_3) \times (a_1, a_2)$, subdomain 6 is the rectangle $(Z_2, Z_3) \times (0, a_1)$, subdomain 7 is the rectangle $(Z_3, Z_4) \times (0, a_2)$. In Figures 1-4 we present the streamlines obtained with thirteen degrees of freedom in each direction in each element for the values of the Reynolds number R = 0, 50, 100 and 150. In Table 2 we present the values of the length of the upstream vortex for Stokes flow for different values of numbers of degrees of freedom in each direction and each element (n) and different values of Z_2. These compare favourably with previously obtained results ([9],[10]).

Z_2	$n = 11$	$n = 13$	$n = 15$
−0.25	0.10	0.32	0.29
−0.30	0.32	0.28	0.30
−0.35	0.22	0.28	0.30
−0.40	0.24	0.28	0.30
−0.50	0.28	0.25	0.28
[9]	0.285 − 0.290		
[10]	0.284		

Table 2: The length of the upstream vortex for Stokes flow.

Acknowledgement
This work was partially supported by a University of Cyprus research grant.

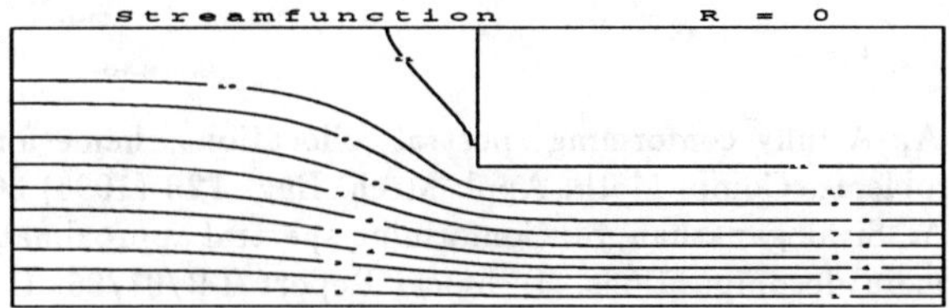

Fig. 1. Streamlines for $R = 0$

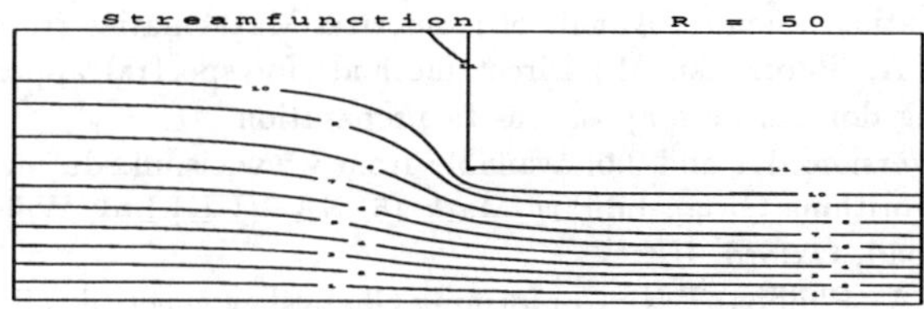

Fig. 2. Streamlines for $R = 50$

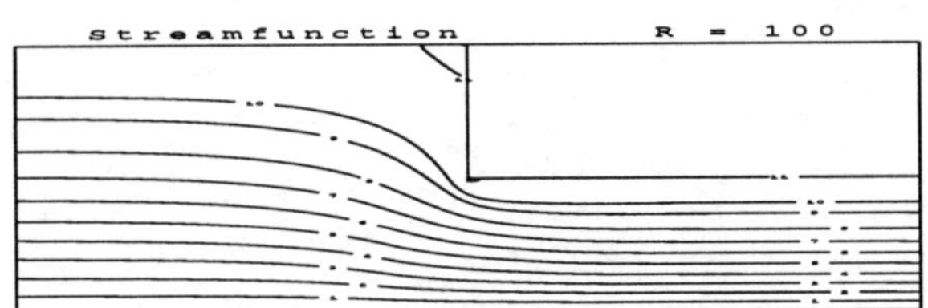

Fig. 3. Streamlines for $R = 100$

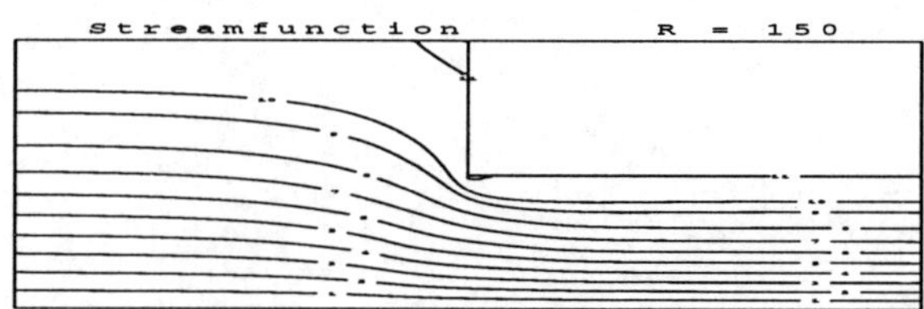

Fig. 4. Streamlines for $R = 150$

References

1. Karageorghis, A.: A fully conforming spectral collocation scheme for second and fourth order problems. Comp. Meth. Appl. Mech. Eng. **126** (1995) 305-314
2. Karageorghis, A., Sivaloganathan, S.: Conforming spectral approximations for non-conforming domain decompositions. Technical Report TR/07/96, Department of Mathematics and Statistics, University of Cyprus (1996)
3. Canuto, C., Hussaini, M., Quarteroni, A., Zang, T.: Spectral Methods in Fluid Dynamics. Springer-Verlag, New York (1988)
4. Boyd, J. P.: Chebyshev and Fourier Spectral Methods. Springer-Verlag, New York (1989)
5. Karageorghis, A., Paprzycki, M.: An efficient direct method for fully conforming spectral collocation schemes. Annals of Numerical Mathematics (to appear)
6. Karageorghis, A., Paprzycki, M.: Direct methods for spectral approximations in non-conforming domain decompositions.In preparation
7. UMFPACK, Versions 1.1 and 2.0, available from www.cis.ufl.edu˜davis
8. Numerical Algorithms Group Library Mark 15, NAG(UK) Ltd, Wilkinson House, Jordan Hill Road, Oxford, UK
9. Karageorghis, A., Phillips, T.N.: Conforming Chebyshev spectral collocation methods for the solution of laminar flow in a constricted channel. IMA Journal of Numer. Anal. **11** (1991) 33-54
10. Dennis, S.C.R., Smith, F.T.: Steady flow through a channel contraction with a symmetrical constriction in the form of a step. Proc. Roy. Soc. London A **372** (1980) 303-414
11. Paprzycki, M., Karageorghis, A.: High performance solution of linear systems arising from conforming spectral approximations for non-conforming domain decompositions. this volume

Newton's Method for Solution of One Complex Eigenvalue Problem

M. Kaschiev and D. Koulova-Nenova
Institute of Mathematics, Acad.G.Bontchev Str., Bl.8, Sofia 1113, Bulgaria
Institute of Mechanics, Acad.G.Bontchev Str., Bl.4, Sofia 1113, Bulgaria

Abstract. An eigenvalue problem with complex coefficient is consider in this paper. The mathematical model described the electrohydrodynamic instability of a layer of insulating liquid bounded by one side by rigid electrode and the other side is liquid-liquid interface and subjected to a electric field. Using the continuous analog of Newton's method, a numerical algorithm for solving this problem is developed. The original problem depend of many physical parameters. The numerical exprements show that there exist such value of parameters for which the imaginary part of the eigenvalue with minimal absolute value vanish. It means that the oscillatory modes of instability can occur.

1. Formulation of the problem

Let us consider the following eigenvalue problem

$$
\begin{aligned}
&L_1(F, W_1) \equiv 4z^2 D(D^2 - k^2)F + 4z(\frac{\omega}{a}\sqrt{z} + 1)(D^2 - k^2)F - DF + W_1 = 0,\\
&b \le z \le b+1,\\
&L_2 W_1 + \lambda M_1 F = 0, \quad b \le z \le b+1,\\
&L_2 W_2 = 0, \quad b - d_2 \le z \le b.
\end{aligned}
\tag{1}
$$

The operators L_2 and M_1 are defined as

$$L_2 W \equiv (D^2 - k^2)(D^2 - k^2 + Re\omega)W, \quad M_1 F \equiv z^{\frac{1}{2}}[(D^2 - k^2) + \frac{1}{4z^2})]F,$$

$$\lambda = k^2 Ta, \quad D = \frac{d}{dz}.$$

The parameters here will be definited below.

The equations (1) describe the following physical phenomena.

Let us consider two layers of immiscible liquids, an upper insulating liquid (the layer thickness is d_1) and a lower conducting layer (the thickness is d_2) bounded by two rigid electrodes and subjected to electric field E (see Fig.1). Free charges of density q can be injected into perfectly insulating liquid from the upper horizontal metal electrode or from the liquid interface.

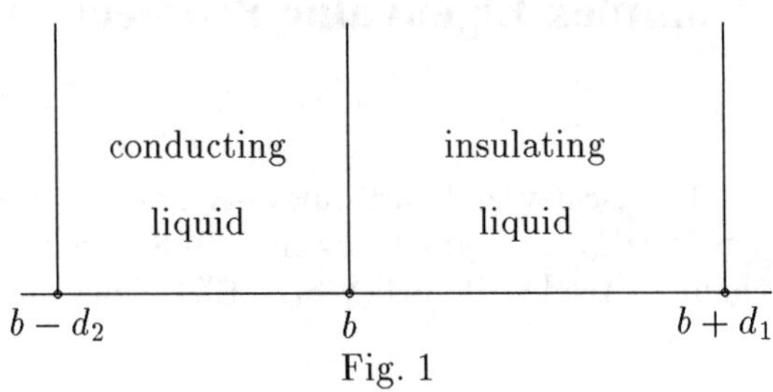

Fig. 1

The liquid is set into motion thus changing the distribution of both the charge density and the electcic field. The lower liquid is assumed to be highly conducting so the elecric field is zero and the electric potential also $\varphi = 0$ (because of the interface will always be equipotential). Thus, the two liquid layer system can be considered as one insulating liquid bounded by rigid and "liquid" parallel electrodes. In this case the rest is a possible solution and it is necessary to examine in what conditions this solution destabilizes.

The behavior of the two liquids is described by the electrodynamic (Poisson's and the charge conservation equation) and hydrodynamic equations (the continuity and Navier-Stokes equation including Coulomb force). Because of $E = 0$ in conducting liquid it remains here the hydrodynamic equations only. Let us consider the case of injection from the liquid interface which is assumed flat and undeformable. The electrical potential is equal to the applied voltage V at the injector and zero at the collector. The fundamental parameters which arises in the instability are: Reynolds number Re, injection parameter C and the ratio of the liquids viscosities $\eta = \dfrac{\eta_1}{\eta_2}$. The equations (1) are normalized with respect to d_1, i.e. the layer thickness of insulating liquid $d_1 = 1$ and of the conducting one is d_2/d_1.

The equations (1) are obtained using the linear instability analysis (see [1]). The pertubations of the electric potential $\varphi(x, y, z, t)$ in the insulating liquid and ones of the liquid velocities $V_j(x, y, z, t),\ j = 1, 2$ in both layers are represented as two-dimensional waves

$$\varphi = F(z)\exp[\omega t + i(k_x x + k_y y)],$$
$$V_j = W_j(z)\exp[\omega t + i(k_x x + k_y y)], \quad j = 1, 2,$$

where $F(z), W_1(z)$ and $W_2(z)$ are amplitudes of the electrical potential and the z-component of the velocities, ω is time constant, k is the wave number, $k^2 = k_x^2 + k_y^2$.

There are two alternative possibilities of the instability: the instability sets in as a stationary convection and the instability setting in as oscillations (i.e. overstability). The the stationary instability ($\omega = 0$) is considered in paper [2], where were worked out the iteration method, discratization of the differential equations (1) and were obtained numerical results, described the instability conditions.

In this paper we study the case ($\omega \neq 0$). Since the neutral state for an oscillatory mode is characterized by $\omega = iR$ with R real number. It is clear that for arbitrary assigned values of the parameters k, Re, η, d_2, C, b and R the value of T will be complex. But the physical meaning of T requires it to be real and positive. Hence, a numerical search must be conducted to find the value of R, which makes the imaginary part of T vanish. If $ImT \leq 0$ for all values of parameters, it means that the neutral state is indeed a stationary one. The onset of overstability in the (k, T) - plane takes place, if $ImT = 0$ for some values of parameters.

The solution F, W_1, W_2 of system (1) have to satisfy the following boundary conditions:

a) on the injector ($z = b$)

$$\begin{array}{l} W_1(b) = W_2(b) = F(b) = 0, \quad DW_2(b-0) = DW_1(b+0), \\ D^2W_2(b-0) = \eta D^2W_1(b+0), \quad D^2F(b) = 0; \end{array} \tag{1.a}$$

b) on the collector ($z = b + 1$)

$$W_1 = DW_1 = F = 0; \tag{1.b}$$

c) on the electrode touching conducting liquid ($z = b - d_2$)

$$W_2 = DW_2 = 0. \tag{1.c}$$

The condition $D^2W_2 = \eta D^2W_1$ expresses the tangential force balance.

The constant b is the root of the equation

$$b^2[(1 + \frac{1}{b})^{1.5} - 1] = \frac{3}{4C}, \quad a = 2C\sqrt{b}. \tag{2}$$

The main problem here consists to find such value of R that the imaginery part of T is vanished in the$k$, Re, η, d_2, C parameter set .

Note that all parameter k, Re, η, d_2, C, b are real numbers.

The aim of the present report is to provide stable numerical methods for solving the eigenvalue problems (1)-(1.c) for non-hermitian operators and to find a such value of R that sutisfiesQQ the conditions of T written up.

The more simple case of one liquid is considered in papers [3], [4].

The aim of the present report is to provide stable numerical methods for solving the eigenvalue problems (1)-(1.c) for non-hermitian operators and to find a such value of R that satisfies the conditions of T written up.

2. Iteration method

There exist many numerical method for solving an hermitian eigenvalue problem such as Lanzosh method [5], Inverse iteration [6], Subspace Iteration Method [7] and e.c. But, in our point of view, to solve a non-hermitian complex eigenvalue problem one can use only the Continuous Analog of Newton Method (CANM), that is proposed by Gavurin [8]. Let us consider the nonlinear equation

$$\varphi(u) = 0. \tag{3}$$

Let u^* is an isolated solution of the equation (3).

The main idea of [8] is to consider the continuous parameter t, $0 < t < \infty$ and to study the evolution equation

$$\frac{d}{dt}\varphi(u(t)) = -\varphi(u(t)), \quad u(0) = u_0, \tag{4}$$

where u_0 is an initial approximation to the solution we are looking for. The function φ is smooth in the vicinity of the solution u^* and $(\varphi')^{-1}$ is bounded, where φ' is the Frechet derivative of φ.

Let $u(t)$ be a solution of (4). It has been shown [8], [9] that the following limit holds:

$$\lim_{t\to\infty} u(t) = u^*.$$

If a discretization of the equation (4) by the Euler method is used, then the iterative method for solving the equation (3) is applied as follows

$$\begin{aligned} &\varphi'(u_n)v_n = -\varphi(u_n), \quad u_{n+1} = u_n + \tau_n v_n, \\ &n = 0, 1, \ldots, \quad 0 < \tau_n \leq 1. \end{aligned}$$

This process coincides with the classical Newton method, when $\tau_n = 1$.

The CAMN for the numerical solution of eigenvalue problem is proposed by I.V.Puzynin [10]. This method for solving the problem (1)-(1.c) when $\omega = 0$ is written in [2]. Following this paper first we will rewrite (1)-(1.c).

Let us consider the equation $L_2 W_2 = 0$. The general solution of this equation can be written in the form

$$W_2(z) = Ae^{kz} + A_1 e^{-kz} + A_2 e^{z\sqrt{k^2 - Re\omega}} + A_3 e^{-z\sqrt{k^2 - Re\omega}}. \tag{5}$$

Here A, A_1, A_2, A_3 are arbitrary complex constants. Using the boundary conditions
$W_2(b-d_2) = W_2'(b-d_2) = W_2(b) = 0$ we find that $A_1 = AB_1$, $A_2 = AB_2$, $A_3 = AB_3$, where B_1, B_2, B_3 depend only on $b - d_2$ and b. So, formula (5) becomes

$$W_2(z) = A(e^{kz} + B_1 e^{kz} + B_2 e^{z\sqrt{k^2 - Re\omega}} + B_3 e^{-z\sqrt{k^2 - Re\omega}}) \equiv AU(z).$$

Now we obtain the following formulation of a problem (1)-(1-c). Find the eigenvalue λ with minimal absolute value, the parameter A and the corresponding eigenfunction $(F(z), W(z))$ as a solution of

$$L_1(F, W) = 0, \quad L_2 W + \lambda M_1 F = 0, \quad b \leq z \leq b + 1, \tag{6}$$

$$A^2 I + \int_b^{b+1} [F^2(z) + W^2(z)]dz - const = 0, \quad I = \int_{b-d_2}^{b} U^2(z)dz, \qquad (7)$$

with boundary conditions

$$\begin{array}{l} W(b) = F(b) = F''(b) = 0, \ \ AU'(b) = W'(b), \ \ AU''(b) = \eta W''(b), \\ W(b+1) = W'(b+1) = F(b+1) = 0. \end{array} \qquad (8)$$

A condition (7) normalizes the eigenfunction the an arbitrary given complex constant *const*.

Now, the iteration method for solving the problem (6)-(8) is derived. Those method is based on the **Algorithm H** (see [2]). Assuming the initial data $F_0,\ W_0,\ \lambda_0,\ A_0$ for problem (6) is given and marking the iteration corrections as $f_n,\ w_n,\ \mu_n,\ \alpha_n$ than the iteration process is written as

Algorithm HΩ

HΩ1. Find (f_n, w_n) as a solution of the system

$$L_1(f_n, w_n) = 0, \qquad L_2 w_n + \lambda_n M_1 f_n = -M_1 F_n$$

with boundary conditions

$$w_n(b) = f_n(b) = f_n''(b) = 0, \ \ A_n U'(b) = w_n'(b),$$

$$w_n(b+1) = w_n'(b+1) = f_n(b+1) = 0.$$

HΩ2. Find α_n from equation $\alpha_n U''(b) = w_n''(b)\eta$;

HΩ3. Using the normalized condition (7) determine the value of μ_n from

$$\mu_n = \frac{1}{2}\{A_n^2 I + \int_b^{b+1} [F_n^2(z) + W_n^2(z)]dz + const\}$$

$$\{A_n \alpha_n I + \int_b^{b+1} [F_n(z) f_n(z) + W_n(z) w_n(z)]dz\}^{-1};$$

HΩ4. Find new approximation

$$\begin{array}{l} F_{n+1} = (1-\tau_n)F_n + \tau_n \mu_n f_n, \ \ W_{n+1} = (1-\tau_n)W_n + \tau_n \mu_n w_n, \\ A_{n+1} = (1-\tau_n)A_n + \tau_n \alpha_n, \\ \lambda_{n+1} = -\{\int_b^{b+1} W_{n+1}(z) L_2 W_{n+1}(z)dz\}\{\int_b^{b+1} W_{n+1}(z) M_1 F_{n+1}(z)dz\}^{-1}. \end{array}$$

HΩ5. If an accuracy criterion is satisfied, then stop the iterations, else go to **HΩ1**.

The numerical schemes for solving of the problems of step **HΩ1** will be presented in next section.

3. Numerical method

In our point of view the finite difference method [11] is most suitable approximating the equations (6). Considering the interval $[b, b+1]$ we construct a regular mesh $S_h = \{z_i = b + ih,\ i = 0, 1, 2, \ldots, N,\ h = 1/N\}$. The value of some mesh function u in the node z_i is noted by u_i. There are no difficulties to approximate the derivatives in the point z_i for $2 \le i \le N-1$. The main problem is to approximate the fourth derivative in the point z_1. We consider an additional point $z_{-1} = b - h$ to solve it and write a formally approximation of D^4W as

$$D^4W(z_1) = \frac{W_{-1} - 4W_0 + 6W_1 - 4W_2 + W_3}{h^4} + O(h^2).$$

On the other hand from the boundary condition $AU'(z_0) = W'(z_0)$ one can obtain
$AU_0' = \dfrac{W_0 - W_{-1}}{h} + O(h)$. So, it follows

$$D^4W(z_1) = \frac{-AU_0'}{h^3} + \frac{-3W_0 + 6W_1 - 4W_2 + W_3}{h^4} + O(h).$$

From boundary condition $AU_0'' = \eta W_0''$ we have

$$AU_0'' = \frac{-5W_1 + 4W_2 - W_3}{h^2} + O(h^2), \quad W_0 = 0,$$

or

$$A = \eta \frac{-5W_1 + 4W_2 - W_3}{U_0'' h^2}.$$

Finally, the following approximation of D^4W is found

$$D^4W_1 = (\frac{6}{h^4} + \eta \frac{5U_0'}{U_0'' h^5})W_1 - (\frac{4}{h^4} + \eta \frac{4U_0'}{U_0'' h^5})W_2 + (\frac{1}{h^4} + \eta \frac{U_0'}{U_0'' h^5})W_3 + O(h).$$

The derivative $F'''(z_i)$, $i = 1, 2, \ldots, N-2$ is approximated by a formula

$$F'''(z_i) = \frac{-F_{i-2} + 2F_{i-1} - 2F_{i+1} + F_{i+2}}{2h^3} + O(h^2),$$

For $i = N$ we use the formula

$$F'''(z_N) = \frac{3F_{N-3} - 6F_{N-2} + 12F_{N-1} - 10F_{N+1}}{2h^3} + O(h^2), \; F_{N+1} = 0.$$

The first and second derivatives of any mesh function are approximated by the central differences. The integral terms are calculated using the trapezoidal formula.

The systems of nonsymmetrical linear algebraic equations, obtained after discretization of step **HΩ1**, are solved by complex Gaussian elimination.

An important issue is how to determine the initial data. We propose the following choice of the functions F_0, W_0 and constant A_0

$$F_0(z) = x(0.5x^2 - 1.5x + 1), \quad W_0(z) = x(x-a)(1-x)^2, \quad x = z - b,$$

where

$$a = -\frac{2U'(b)\eta}{U''(b) + 4\eta U'(b)}, \quad A_0 = \frac{2\eta}{U''(b) + 4\eta U'(b)}.$$

Such choice of initial approximation allows always to use $\tau_n = 1$, that corresponds to the classical Newton method.

4. Numerical results

The numerical method was examined for $R = 0$. The obtained results were the same as those of paper citeDM1. After that we solved the problem in the following way. The values of k, Re, η, d_2, C, b were fixed and we found the eigenvalue T for different values of $R \in [0, 10]$. Our numerical results show that when $Re < 5$ for all values of parameters k, η, d_2, C, b $ImT > 0$. For $5 \leq Re \leq 8$ we find that $ImT = 0$ for some values of parameters. Usually those values do not depend of η, d_2, C,, but they depend very strongly of k. The spectral curves ReT and ImT as a function of R for $C = 5$, $\eta = 2.5$ $d_2 = 2$ and $Re = 6.7$ are shown in Fig.2. The imaginary part of T is shown in Fig.3. It is seen from these figures that there exist the values of k for which $ImT = 0$. In our numerical experiments we use $const = 1 + i$. Note that eigenvalue T of problem does not depend of the value of this constant.

The accuracy of proposed methods is examined by the help of sequences of meshes. Let us construct meshes S_{h_1}, S_{h_2}, S_{h_3}, ..., where $h_{i+1} = 0.5h_i$, $i = 1, 2, \ldots$ and calculate the quantity

$$\alpha = \ln \frac{\lambda^{h_i} - \lambda^{h_{i+1}}}{\lambda^{h_{i+1}} - \lambda^{h_{i+2}}} / \ln 2,$$

where λ^{h_i} is the eigenvalue obtained over the mesh S_{h_i}. But we use this formula only for that values of parameters for which $Im\lambda = 0$. If the accuracy of the method is $O(h^\gamma)$ than we can wait that $\alpha \approx \gamma$ for small h_i. In our cases it is possible to suggest $\gamma = 1$. The values of ReT for problem (1) obtained over different meshes and value of α for parameters $k = 4.5$, $R = 0.55$, $C = 1, d_2 = 2$ are the following: $N = 200$, $Re\ T = 308.8348$, $N = 400$, $Re\ T = 308.9367$, $N = 800$, $Re\ T = 308.9856$, $\alpha = 1.06$. From these results it is seen there is a first order of precision of eigenvalues. So, there is a convergence from below that corresponds to the theoretical results [11]. We note that about 7-8 iterations are needed to satisfy the stop criterion with $\epsilon = 10^{-6}$.

The values of R and k for which $Im\ T = 0$ are shown on Fig.4.

5. Conclusion

The Newton's method for complex eigenvalue problem in electrohydrodinamics is presented in this paper.. Different values of time dependent constant R that the imaginary part of the criteria T is vanished are found. It means that the oscillatory motion takes place for several values of characteristic parameters. In the future this numerical method will be developed for solving of the more complicated problem when the effect of the interface deformation is taken into account.

Acknowledgments. This work is partially supported by Grant No.MM 501/95 with Bulgarian National Foundation "Sientific Investigations".

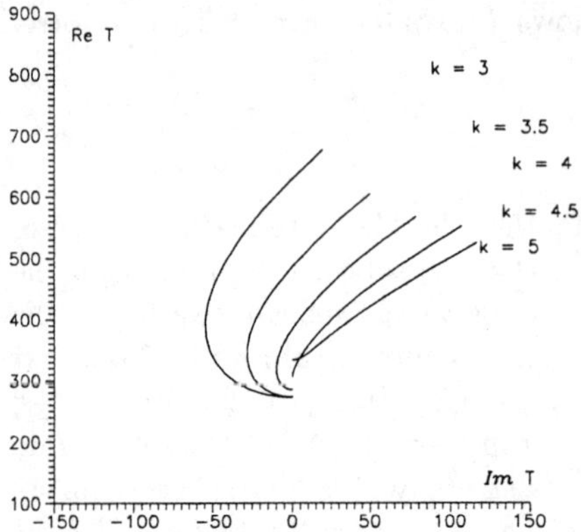

Fig.2. The spectral curves as a function of R for Re=6.7, C=5, η=2.5.

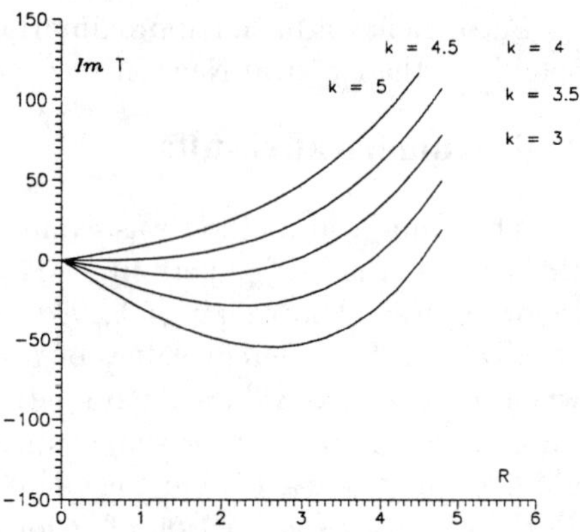

Fig.3. The imaginery part of T as a function of R for Re=6.7, C=5, η=2.5.

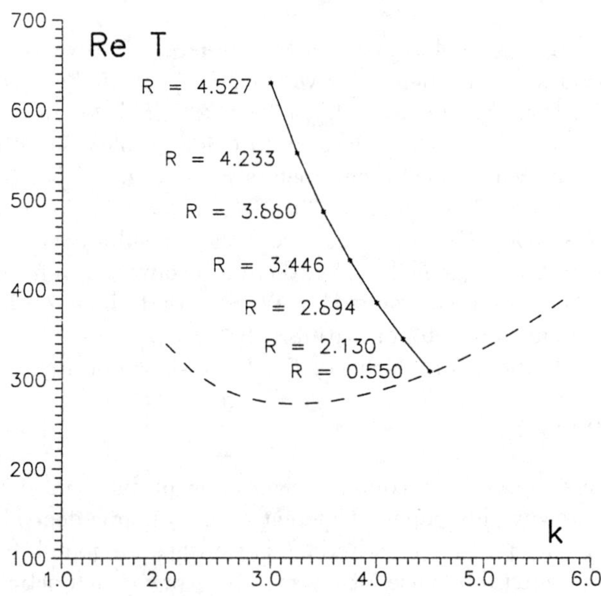

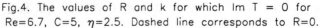

Fig.4. The values of R and k for which Im T = 0 for Re=6.7, C=5, η=2.5. Dashed line corresponds to R=0.

References

1. S.Chandrasekhar, Hydrodynamic and Hydromagnetic Stability, Oxford University Press, ch. 2, **1961**.
2. Iterative Methods In Linear Algebara, II. Proc. of the Second **IMACS** Intern. Symp. on Iter. Meth. in Lin. Alg. Editors Sv.Margenov & P.Vassilevski, **1996**, **IMACS**, pp.390-401.
3. M.Takashima. Journ. of the Phys. Soc. of Japan, v.50, pp.2751-2756, **1981**
4. Zh.Kozhuokharova, S.Slavchev. Journ. of Coll. and Interf. Sci., v.152, pp.473-482, **1992**
5. B.Parlett. The Symmetric Eigenvalue Problem. Moskow, "MIR" , **1983**(*in Russian*).
6. J.Wilkinson. The Algebraic Eigenvalue Problem. Moskow, "MIR" , **1970**(*in Russian*).
7. Kl.-J. Bathe, Ed.Wilson. Numerical Methods in Finite Element Analysis. Prentice-Hall, Cliff. N.J., Englewood, **1976**.
8. M.K.Gavurin. Izv. VUZ, Ser. Mathem. v.5(6), pp 18-31, **1958** (*in Russian*).
9. E.P.Zhidkov, G.I.Makarenko and I.V.Puzynin. EPAN, v.4(1),"Atomizdat",M., pp 127-165, **1973** (*in Russian*).
10. L.I.Ponomarev, I.V.Puzynin, T.P.Puzynina. J.Comput.Phys., v.13, p.1, **1973**.
11. A.A.Samarski. Theory of Difference Schemes. "Nauka",M., **1989** (*in Russian*).

Calculation of Bifurcation States in the Circle Josefsson's Conversion

V.A.Kaschieva
Department of Mathematics, Technical University - Sofia, Bulgaria

1. Formulation of the problem.

Let us consider the steady mathematical model of Josefsson's conversion in the circle with a radius R with ring micro unhomogenious. In the radial - symmetric case this problem is described by the nonlinear boundary value problem

$$-\frac{1}{r}\frac{d}{dr}r\frac{d\varphi}{dr}+[1-\mu\delta(r-r_0)]\sin\varphi=0, \quad \frac{d\varphi(0)}{dr}=0, \quad \frac{d\varphi(R)}{dr}=0. \tag{1}$$

Here $\varphi(r)$ is a difference between the phases of the wave function of superconducting electrons in the upper and down superconductors of the conversion (magnetic flux), $\delta(r-r_0)$ is a Dirac function.

The instability problem here consists the following. Let we know the wave function $\varphi(r)$. This state is stable if the minimal eigenvalue λ_1 of the spectral problem

$$-\frac{1}{r}\frac{d}{dr}r\frac{d\psi}{dr}+[1-\mu\delta(r-r_0)]\cos\varphi\psi=\lambda\psi, \quad \frac{d\psi(0)}{dr}=0, \quad \frac{d\psi(R)}{dr}=0. \tag{2}$$

is positive. If $\lambda_1<0$ the state is unstable. The most important case we have when $\lambda_1=0$. It means that there is an bifurcation point in the parameters μ, r_0, R set. Note that in physical point of view the solution $\varphi(r)$ have to be in the interval $[0,\frac{\pi}{2}]$.

Physical aspects of problems (1)-(2) have been considered in the papers [1], [3]. Numerical methods for solving these problems have been proposed in paper [4]. There were obtained stable and unstable states in the parameter set.

The goal of this paper is to develop the numerical method for finding the bifurcation points of the problem. The bifurcation points are of great interest for applying because in this case the very small variation of parameters the system can jumping to transfer from one stable state to another or to transfer to unstable regime. Knowing the bifurcation points and the corresponding solution allow to find the solution outside of these points by use of the simple bifurcation perturbation theory [3]

Let us put $\lambda = 0$ in equation (2). We assume that the parameters r_0 and R are given and consider the following problem. To find two functions φ, ψ and parameter μ as a solution of

$$L_1(\varphi, \mu) \equiv -\frac{1}{r}\frac{d}{dr}r\frac{d\varphi}{dr} + [1 - \mu\delta(r - r_0)] \sin\varphi = 0, \quad \frac{d\varphi(0)}{dr} = 0, \quad \frac{d\varphi(R)}{dr} = 0,$$

$$L_2(\varphi, \psi, \mu) \equiv -\frac{1}{r}\frac{d}{dr}r\frac{d\psi}{dr} + [1 - \mu\delta(r - r_0)] \cos\varphi\psi = 0,$$

$$\frac{d\psi(0)}{dr} = 0, \quad \frac{d\psi(R)}{dr} = 0. \tag{3}$$

It is clear now that μ is an eigenvalue of the last problem. The eigenfunction $\psi(r)$ is normalized using the condition

$$\int_0^R r\psi^2(r)dr - 1 = 0 \tag{4}$$

The numerical method for solving the the eigenvalue problem (3)-(4) , based on the Newton's method, is worked out in the next section.

2. Iteration method.

An universal method for solving the nonlinear problems is so called Continuous Analog of Newton Method (CANM), that is proposed by Gavurin [5]. Let us consider the nonlinear equation

$$\Phi(u) = 0. \tag{5}$$

Let u^* is an isolated solution of the equation (5).

The main idea of [5] is to consider the continuous parameter t, $0 < t < \infty$ and to study the evolution equation

$$\frac{d}{dt}\Phi(u(t)) = -\Phi(u(t)), \quad u(0) = u_0, \tag{6}$$

where u_0 is an initial approximation to the solution we are looking for. The function Φ is smooth in the vicinity of the solution u^* and $(\Phi')^{-1}$ is bounded, where Φ' is the Frechet derivative of Φ.

Let $u(t)$ be a solution of (6). It has been shown [5], [6] that the following limit holds:

$$\lim_{t\to\infty} u(t) = u^*. \tag{7}$$

Let us discretizate the equation (6) by the Euler's method. Then an iteration method for solving the equation (5) is applied as follows

$$\Phi'(u_n)v_n = -\Phi(u_n), \quad u_{n+1} = u_n + \tau_n v_n,$$
$$n = 0, 1, \ldots, \quad 0 < \tau_n \leq 1.$$

When $\tau_n = 1$ this process coincides with the classical Newton method.

We will used a modification of the CAMN for solving the problem (3)-(4) that have been proposed by Puzynin [7]. Following his idea we will note

$$\nu = \frac{d\mu}{dt}, \quad \frac{d\varphi}{dt} = \varphi_1 + \nu\varphi_2, \quad \frac{d\psi}{dt} = \psi_1 + \nu\psi_2. \tag{8}$$

Proposing that initial functions φ, ψ and number μ are given and using this representation we find the following system for determination the functions $\varphi_1, \varphi_2, \psi_1, \psi_2$:

$$-\frac{1}{r}\frac{d}{dr}r\frac{d\varphi_1}{dr} + [1 - \mu\delta(r - r_0)]\cos\varphi\varphi_1 = -L_1(\varphi, \mu),$$

$$-\frac{1}{r}\frac{d}{dr}r\frac{d\varphi_2}{dr} + [1 - \mu\delta(r - r_0)]\cos\varphi\varphi_2 = \delta(r - r_0)\sin\varphi,$$

$$-\frac{1}{r}\frac{d}{dr}r\frac{d\psi_1}{dr} + [1 - \mu\delta(r - r_0)]\cos\varphi\psi_1 = -[L_2(\varphi, \psi, \mu) - \delta(r - r_0)\sin\varphi\psi\varphi_1],$$

$$-\frac{1}{r}\frac{d}{dr}r\frac{d\psi_2}{dr} + [1 - \mu\delta(r - r_0)]\cos\varphi\psi_2 = \delta(r - r_0)\sin\varphi\psi\varphi_2. \tag{9}$$

The functions $\varphi_1, \varphi_2, \psi_1, \psi_2$ satisfy the same boundary conditions as functions φ and ψ. Using the solution of (7) the number ν is determined as

$$\nu = -0.5\{\int_0^R r\psi(\psi + \psi_1)dr - 1\}\{\int_0^R r\psi\psi_2 dr\}^{-1}. \tag{10}$$

In the special case when the solution $\varphi(r,t)$, $\psi(r,t)$, $\mu(t)$ of continuous problems (9)-(10) can be found, using the initial data and relations (8) one can obtain the solution φ, ψ, μ by (7). In general case allied the Euler's method again we can construct the following iteration process:

Algorithm

Let φ_0, ψ_0, μ_0 is initial approximation of φ, ψ, μ.
For $n = 0, 1, 2, \ldots$ do

A1. Find a solution φ_1^n, φ_2^n of the equations

$$-\frac{1}{r}\frac{d}{dr}r\frac{d\varphi_1^n}{dr} + [1 - \mu\delta(r - r_0)]\cos\varphi_n\varphi_1^n = -L_1(\varphi_n, \mu_n),$$

$$-\frac{1}{r}\frac{d}{dr}r\frac{d\varphi_2^n}{dr} + [1 - \mu\delta(r - r_0)]\cos\varphi_n\varphi_2^n = \delta(r - r_0)\sin\varphi_n. \tag{11}$$

A2. Knowing the functions φ_1^n, φ_2^n find the solution ψ_1^n, ψ_2^n of the equations

$$-\frac{1}{r}\frac{d}{dr}r\frac{d\psi_1^n}{dr}+[1-\mu\delta(r-r_0)]\cos\varphi_n\psi_1^n=-[L_2(\varphi_n,\psi_n,\mu_n)-\delta(r-r_0)\sin\varphi_n\psi_n\varphi_1^n],$$

$$-\frac{1}{r}\frac{d}{dr}r\frac{d\psi_2^n}{dr}+[1-\mu\delta(r-r_0)]\cos\varphi_n\psi_2^n=\delta(r-r_0)\sin\varphi_n\psi_n\varphi_2^n. \quad (12)$$

A3. Find a value of ν_n using the formula

$$\nu_n=-0.5\{\int_0^R r\psi_n(\psi_n+\psi_1^n)dr-1\}\{\int_0^R r\psi_n\psi_2^n dr\}^{-1}. \quad (13)$$

A4. Find the next iteration φ_{n+1}, ψ_{n+1} as

$$\varphi_{n+1}=\varphi_n+\tau_n(\varphi_1^n+\nu_n\varphi_1^n,$$
$$\psi_{n+1}=\psi_n+\tau_n(\psi_1^n+\nu_n\psi_1^n,$$

A5. Find the new value μ_{n+1} as

$$\mu_{n+1}=\mu_n+\tau_n\nu_n.$$

A6. If the criterion of convergence is satisfied then stop the iterations, else go to step A1.

The problems (11)-(13) will be solved numerically. The discretization method is developed in the next section.

3. Numerical method

Numerical method for solving the problems (12)-(13) is based on the finite element method [8], [9]. Let us consider the bilinear form

$$a(u,v;\varphi)=\int_0^R(ru'v'+uv\cos\varphi)dr-\mu r_0u(r_0)v(r_0)\cos\varphi(r_0) \quad (14)$$

and linear forms

$$b_1(v)=-\int_0^R(r\varphi'v'+v\sin\varphi)dr-\mu r_0v(r_0)\sin\varphi(r_0),$$

$$b_2(v)=r_0v(r_0)\sin\varphi(r_0),$$

$$b_3(v)=\int_0^R(r\psi'v'+\psi v\cos\varphi+\psi\varphi_1v)dr-\mu r_0\psi(r_0)\varphi(r_0)v(r_0)\sin\varphi(r_0),$$

$$b_4(v) = \int_0^R r\psi\varphi_2 v \sin\varphi dr - \mu r_0\psi(r_0)\varphi_2(r_0)v(r_0)\sin\varphi(r_0). \quad (15)$$

The weak form of the (3) using (14),(15) can be written:

For given functions $\varphi(r) \in H^1$ and $\psi(r) \in H^1$ find the functions $\varphi_1(r), \varphi_2(r), \psi(r), \psi(r)$ such that the integral identities

$$a(\varphi_1, v; \varphi) = b_1(v), \quad a(\varphi_2, v; \varphi) = b_2(v),$$

$$a(\psi_1, v; \varphi) = b_3(v), \quad a(\psi_2, v; \varphi) = b_4(v) \quad (16)$$

for every function $v \in H^1$.

The discrete schemes are obtained using the second order Lagrangian elements. After discretization of (16) we have to solve four system of linear equations

$$K(\varphi)u_1 = B_1, \quad K(\varphi)u_2 = B_2, \quad K(\varphi)v_1 = B_3, \quad K(\varphi)v_2 = B_4. \quad (17)$$

Here matrix K corresponds to $a(u, v; \varphi$ and vectors B_1, B_2, B_3, B_4 correspond to linear forms b_1, b_2, b_3, b_4. The matrix K is symmetric positive one, so for solving the systems (17) we use the Cholesky decomposition [9].

4. Numerical results

The accuracy of proposed methods is examined by the help of sequences of meshes. Let us construct meshes $\omega_{h_1}, \omega_{h_2}, \omega_{h_3}, \ldots$, where $h_{i+1} = 0.5h_i$, $i = 1, 2, \ldots$ and calculate the quantities

$$\alpha = \ln \frac{\mu^{h_i} - \mu^{h_{i+1}}}{\mu^{h_{i+1}} - \mu^{h_{i+2}}} / \ln 2,$$

where μ^{h_i} is the eigenvalue obtained over the mesh ω_{h_i}. Here the mesh ω_h consists N elements and $h = R/N$. If the accuracy of the method is $O(h^\gamma)$ than we can wait that $\alpha \approx \gamma$ for small h_i. In our cases it is possible to suggest $\gamma = 4$. The values of μ for problem (3),(4) obtained over different meshes and values of α are shown in table 1.

Table 1. Convergence rate for problem (3),(4)

	$R = 6.0, r_0 = 4.0$	$R = 8.0, r_0 = 3.0$
N	μ	μ
40	1.939125606	1.965811220
80	1.939123226	1.965808076
160	1.939123077	1.965807879
	$\alpha = 3.997$	$\alpha = 3.996$

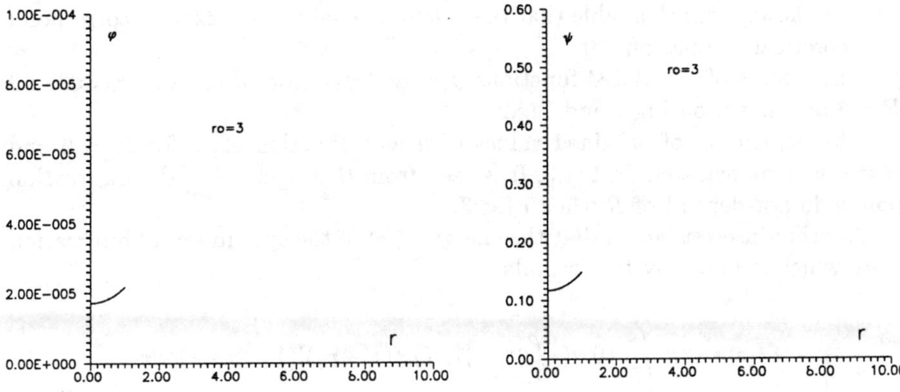

Fig.1. A function φ for ro=3, μ=1.96

Fig.2. A function ψ for ro=3, μ=1.96

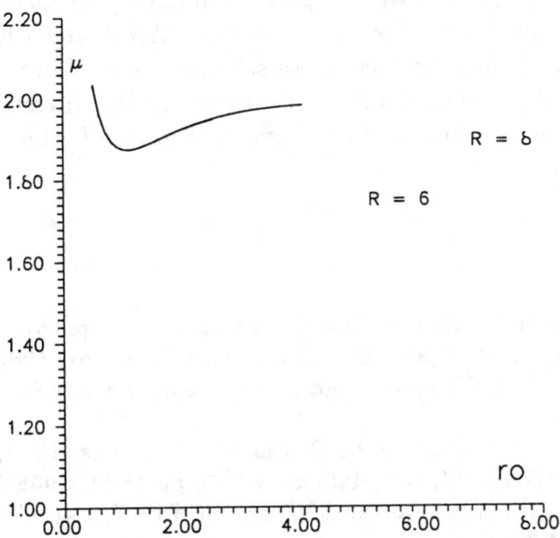

Fig.3. The values of μ a function of ro for R=6 and R=

It is shown from this table that the calculated values of α exactly correspond to theoretical estimation [8].

The graphs of calculated functions $\varphi(r)$ and $\psi(r)$ for parameters $r_0 = 3$ and $R = 8$ are shown on Fig.1 and Fig.2.

The depending of obtained values of μ as a function of r_0 for $R = 6$ and $R = 8$ is demonstrated on Fig.3. It is seen from this figure that the bifurcation points do not depend of R when $r_0 < 3$.

Another interest fact is that the energy $\mathcal{E}(\varphi)$ of the system for all bifurcation state which is found by the formula

$$\mathcal{E}(\varphi) = 2\pi \int_0^R r\{0.5(\frac{d\varphi}{dr})^2 + [1 - \mu\delta(r - r_0)](1 - \cos\varphi)\}dr.$$

is equal to zero. Note that the energy values of the stable and unstable states calculated in paper [4] are between 209 and 248.

5. Conclusions

A new method for finding of the bifurcation points of of Josefsson's conversion in the circle with a radius R with ring micro unhomogenious is developed in this paper. The computational schemes using the CANM and FEM are work out. The numerical analysis show that these schemes have a forth order of accuracy. The depending of the critical values of μ from the R is investigated.

The author has great please th thank prof. Puzynin for him interest and very useful discussions.

References

1. A.T.Filipov & Ju.S.Gal'pern. Solid State Comm. v.48, p. 661, **1983**.
2. A.T.Filipov & Ju.S.Gal'pern. Sov. Jurn. Exp v.48, p. 661, **1983**.
3. Ju.S.Gal'pern & A.T.Filipov. Journ. Exp. Theor. Phys. v.86, p. 1527, **1986** (*in Russian*).
4. V.A.Kaschieva at al. Preprint JINR, P11-84-832, Dubna, **1984** (*in Russian*).
5. M.K.Gavurin. Izv. VUZ, Ser. Mathem. v.5(6), pp 18-31, **1958** (*in Russian*).
6. E.P.Zhidkov, G.I.Makarenko and I.V.Puzynin. EPAN, v.4(1),"Atomizdat",M., pp 127-165, **1973** (*in Russian*).
7. L.I.Ponomarev, I.V.Puzynin, T.P.Puzynina. J.Comput.Phys., v.13, p.1, **1973**.
8. G.Streng, J.Fix Theory of Finite Element Method, "MIR", Moskow, **1977** (*in Russian*).
9. Kl.-J. Bathe, Ed.Wilson. Numerical Methods in Finite Element Analysis. Prentice-Hall, Cliff. N.J., Englewood, **1976**.

Numerical Methods for Computation of the Double Layer Logarithmic Potential

Natalia Kolkovska

Institute of Mathemtics and Informatics,
G. Bonchev str., bl 8, 1113 Sofia, Bulgaria

Abstract. An approximation of the double layer logarithmic potential is obtained as a solution to a discrete Laplacian equation with a new right-hand side. The error is estimated under the assumption that the potential belongs to some Besov spaces. Since the right hand side includes evaluation of line integrals, we use appropriate quadratures for their computation. The well know scheme of A. Mayo, 1984, can be obtained as a partial case.

1 Introduction

Consider the double layer logarithmic potential

$$u(r) = -\frac{1}{2\pi}\int_{\Gamma} g(s)\frac{\partial}{\partial n_s}\ln|r-s|ds, \ r \in R^2. \tag{1}$$

Here Γ is a sufficiently smooth closed curve in R^2, whose properties are specified in the sequel and n is the exterior normal to Γ.

It is well known, that the solution u to the Dirichlet problem for the Laplace equation in the domain D bounded by Γ can be represented as potential (1) with unknown density g. The boundary condition of the differential problem yields a Fredholm integral equation of the second kind for the unknown g. There are fast numerical methods for solving this integral equation. However, once function g has been found, there is still the problem of computing the solution u to the Dirichlet problem for the Laplace equation, which is the potential (1) with already know density. The difficulty particularly occurs near the curve Γ, because the kernel becomes unbounded there. Thus, the computation of potentials with good accuracy needs special procedures (or more computational work).

In 1984 A. Mayo [8] proposed a method for computation of the double layer logarithmic potential. For the method, the domain D is imbedded into a rectangle Ω so that $\mathrm{dist}(\Gamma, \partial\Omega) > 0$. Then, the potential (1) is not evaluated directly, but it is represented as a solution to the boundary value problem

$$\begin{aligned} &\left(\tfrac{\partial^2}{\partial x^2} + \tfrac{\partial^2}{\partial y^2}\right) u(r) = F(r), \ r = (x,y) \in \Omega \\ &u(r) = -\tfrac{1}{2\pi}\textstyle\int_{\Gamma} g(s)\tfrac{\partial}{\partial n_s}\ln|r-s|ds, \ r \in \partial\Omega \end{aligned} \tag{2}$$

where F is a generalized function with support on Γ. At the end the problem (2) is solved numerically, using the usual 5-points discrete Laplacian Λ_h:

$$\begin{aligned} \Lambda_h u_h(r) &= \varphi(r),\ r \in \Omega_h \\ u_h(r) &= -\frac{1}{2\pi}\int_\Gamma g(s)\frac{\partial}{\partial n_s}\ln|r-s|ds, r \in \partial\Omega_h \end{aligned} \tag{3}$$

On $\partial\Omega_h$ quadrature formulas for smooth periodic functions are used to evaluate the potential (1). The solution to the standard discrete problem (3) in the rectangular domain is found effectively by fast solvers.

The main difficulty which occurs here is the numerical treatment of the differential equation, because it always has a generalized function in the right hand side. In the papers of Lazarov, Mokin [7], Yakovlev [11]; Steklov averaging operators are applied in the approximation of the right hand side. In A. Mayo [8], Mokin [10], Drenska [4], the jumps of the derivatives of the potential are included in the approximation of the generalized function. Theoretical investigations of the last idea are carried out by Mayo [8], Mokin [10], Kolkovska [5] in terms of the classical C^k smoothness of the potential. Many numerical experiments are described in Yakovlev V. [11], Golovin G. [3], Gold J., Graham I. [2] Friedman A. [1], Mayo A., Greenbaum A. [9] for the simple and double layer logarithmic potentials, for potentials, arising from the biharmonic equations and for two and three-dimensional domains. The numerical results presented there show good agreement with the convergence results.

The aim of this paper is to investigate the magnitude of the error of the method, used by Mayo, in terms of Besov $B^\alpha_{p,p}$ smoothness of the potential $u, 1 < p < \infty, \alpha \in (2 + \frac{1}{p}, 3 + \frac{1}{p})$. Measuring the smoothness in terms of Besov spaces $B^\alpha_{p,p}$ seems natural, because the order of the error of the discretization (3) to (2) agrees very well with the required $B^\alpha_{p,p}$ smoothness of the potential (1). We establish error estimates using integral representations of the second finite difference of any function, whose behavior is similar to the behavior of the potential (1). The derived integral representations allow us to include the leading terms into the right hand side of the difference scheme and to obtain error estimates for numerical computation of any function with known jumps across Γ (such as the simple and double layer logarithmic potentials, the volume integral, the conjugate function, hence the real and imaginary part of any Cauchy integral and so on). The proofs of all statements in this paper are included in a forthcoming paper of the author.

2 Notations

Let D be a bounded and simply connected domain in R^2. Let the curve Γ, defined with the parametric representation $R(\lambda) = (X(\lambda), Y(\lambda)),\ \lambda \in [0, 2\pi]$, be the boundary of D. We assume diam$(\Omega) < 1$.

Let functions X, Y and X^{-1} (the inverse to X, defined by $X^{-1}(X(\lambda)) = \lambda$) belong to $C^{2,1}[0, 2\pi]$- the space of two times continuously differentiable functions whose second derivatives are Lipschitz continuous.

Let γ be the function: $\gamma(r) = 1$, if $r \in D$, and $\gamma(r) = -1$ otherwise.

We denote by χ_D the characteristic function of the domain $D(\chi_D(r) = 1$ if $r \in D$ and $\chi_D(r) = 0$ otherwise) and by $(x)_+^n$ the truncated power notation $((x)_+^n = x^n$, if $x > 0$ and $(x)_+^n = 0$ otherwise).

Let $1 < p, q < \infty, \alpha > 0$ and let k be a natural number. $L_p(D)$ denotes the space of measurable functions with $\|u\|_{L_p(d)} = \left(\int_D |u(r)|^p dr\right)^{\frac{1}{p}} < \infty$. $W_p^k(D)$ is the Sobolev space

$$W_p^k(D) = \left\{ u \in L_p(D) : \|u\|_{W_p^k(D)} = \|u\|_{L_p(D)} + \sum_{|l|=k} \|D^l u\|_{L_p(D)} < \infty \right\}.$$

For any vector $H = (H_1, H_2)$ in R^2 we denote the k-th finite difference of the function u in H direction at the point z by $\Delta_H^k u$, i.e. $\Delta_H^k u(z) = \sum_{m=0}^{k} (-1)^{m+k} \binom{k}{m} u(z + mH)$, if u is defined on the segment with end points z and $z + kH$, and $\Delta_H^k u(z) = 0$ otherwise.

We introduce the integral modulus

$$\omega_k(u, \delta, D)_p = \sup \left\{ \|\Delta_H^k u(.)\|_{L_p(D)} : |H| \le \delta \right\},$$

which generates Besov space $B_{p,q}^\alpha(D)(k > \alpha)$

$$B_{p,q}^\alpha(D) = \Big\{ u \in L_p(D) : \|u\|_{B_{p,q}^\alpha(D)} = \|u\|_{L_p(D)} + \left\{ \int_0^\infty [\delta^{-\alpha} \omega_k(u, \delta, D)_p]^q \delta^{-1} d\delta \right\}^{1/q} < \infty \Big\}.$$

We shall consider functions u, which have simple jumps across Γ, which means that at every $\overline{r} \in \Gamma$ the limit values $u_i(\overline{r})$ and $u_e(\overline{r})$ from D and from $\Omega \backslash D$ exist and the jump $[u](\overline{r})$ is defined by $[u](\overline{r}) = u_i(\overline{r}) - u_e(\overline{r})$.

Let $\mathcal{B}_{p,p}^\alpha(D, \Omega)$ be the space of all functions, which are defined on $\Omega \backslash \Gamma$; their restrictions on D and on $\Omega \backslash D$ belong to $B_{p,q}^\alpha(D)$ and $B_{p,p}^\alpha(\Omega \backslash D)$ resp. and they can be continuously extended from D to $\overline{D}$ and from $\Omega \backslash D$ to $\overline{\Omega \backslash D}$; the restrictions and their derivatives have simple jumps across Γ.

If $g \in B_{p,q}^\alpha(\Gamma)$, then the double layer logarithmic potential (1) is in $\mathcal{B}_{p,p}^{\alpha + \frac{1}{p}}(D, \Omega)$ and its jumps across Γ can be explicitly written via the density function g and the parametrization of Γ (see e.g. A. Mayo, [8], Kress R., [6]).

We cover the rectangle Ω with an uniform mesh Ω_h with parameters h_1, h_2. Let $h = (h_1, h_2)$ and $|h| = (h_1^2 + h_2^2)^{1/2}$. Define the second finite difference

$$\Lambda_1 u(r) = (h_1)^{-2}(u(x + h_1, y) - 2u(x, y) + u(x - h_1, y)), r = (x, y)$$

in the x direction. For mesh functions, which vanish on $\partial\Omega_h$, we define also the discrete space $W_p^2(\Omega_h)$ with the norm

$$\|u_h\|_{W_p^2(\Omega_h)} = \left(\sum_{r \in \Omega_h} h_1 h_2 |\Lambda_1 u_h(r)|^p \right)^{1/p} + \left(\sum_{r \in \Omega_h} h_1 h_2 |\Lambda_2 u_h(r)|^p \right)^{1/p}.$$

For $r = (x, y) \in \Omega_h$ define E_r to be the rectangle $\{(s,t) : |s-x| \leq h_1, |t-y| \leq h_2\}$ and $\Gamma_r = \Gamma \cap E_r = \{R(\lambda) : \lambda \in [\lambda_1, r, \lambda_2, r]\}$ to be the part of Γ contained in E_r.

Let $P_r, r \in \Omega_h$, be the pattern $P_r = \{(x, \eta) : |y - \eta| \leq h_2\} \cup \{(\xi, y) : |\xi - y| \leq h_1\}$, which is associated with the discrete Laplacian.

Let B be the linear B-spline, defined as a linear function, which is zero at the points $(0, h_2), (h_1, h_2), (h_1, 0), (0, -h_2), (-h_1, -h_2), (-h_1, 0)$ and one at the point $(0, 0)$.

The symbol C denotes positive constant, which depend on α, D, Ω and does not depend on the functions u, g, on the mesh size and so on. The values of C may differ at each occurrence.

3 Integral representations

In this section we derive some local integral representations for the second finite difference of the function $u \in \mathcal{B}^{\alpha}_{p,p}(D, \Omega)$ at the point $r \in \Omega$. In each representation the second finite difference is spread over 4 cells (having the point r as a vertex) as a sum of volume integrals of the derivatives of u and over line integral on the contained in the 4 cells part of Γ. If a finite difference is included in the representation, then all points in its definition are situated at one and the same side of Γ. We use different representations for the second finite difference $\Lambda_1 u(r)$ at the point r, which depend on the location of r and on the smoothness of u. To express the second finite difference in terms of integrals of some derivatives we use several times the Taylor series with remainder in the integral form, modified to functions with jumps.

First we consider a point $r \in \Omega_h$, such that $E_r \cap \Gamma = \emptyset$. Let $u \in B^{\alpha}_{p,p}(E_r)$ and $2 + \frac{1}{p} < \alpha$. Then one has the representation

$$\begin{aligned}\Lambda_1 u(r) &= -\tfrac{1}{2h_1h_2} \int\int_{E_0} B(s,t) \Delta^2_{0,t} \tfrac{\partial^2 u}{\partial x^2}(x+s, y-t) ds dt \\ &+ \tfrac{1}{2h_1h_2} \int\int_{E_0} \tfrac{\partial^2 u}{\partial x^2}(x+s, y+t)(B(s,t) + B(s,-t)) ds dt, r = (x,y).\end{aligned} \tag{4}$$

There are more complicated formulas for the mesh points close to Γ, i.e. this points for which $E_r \cap \Gamma \neq \emptyset$. For simplicity we assume $P_r \cap \Gamma = \{\overline{r}\} =$

$\{(\overline{x}, y)\}, r = (x, y)$. Then the following representation

$$\Lambda_1 u(r).h_1 h_2 = \frac{\gamma(r)(h_1 - |\overline{x} - x|)_+^0 h_2}{h_1}[u(\overline{r})] + \frac{\gamma(r)\text{sgn}(\overline{x} - x)(h_1 - |\overline{x} - x|)_+ h_2}{h_1}[\frac{\partial u}{\partial x}(\overline{r})]$$
$$+ \int\int_{E_0} \frac{\partial^2 u}{\partial x^2}(x + s, y + t)B(s,t)dsdt$$
$$- \int_{-h_1}^{h_1} (1 - \chi_I(x + s)) \int_{-h_2}^{h_2} B(s,t)\Delta_{(0,t)}\frac{\partial^2 u}{\partial x^2}(x + s, y)dtds$$
$$- \int_{-h_1}^{h_1} \chi_I(x + s) \int_{-h_2}^{|Y(\lambda) - y|} B(s, t\text{sgn}(Y(\lambda) - y))\Delta_{(0,t\text{sgn}(Y(\lambda)-y))}\frac{\partial^2 u}{\partial x^2}(x + s, y)dtds$$
$$- \int_{-h_1}^{h_1} \chi_I(x + s) \int_{|Y(\lambda)-y|}^{h_2} B(s, t\text{sgn}(Y(\lambda) - y))\cdot$$
$$\Delta_{(0,y-Y+t\text{sgn}(Y(\lambda)-y))}\frac{\partial^2 u}{\partial x^2}(x + s, Y(\lambda))dtds$$
$$- \int_{-h_1}^{h_1} \chi_I(x + s) \int_{|Y(\lambda)-y|}^{h_2} B(s, t\text{sgn}(Y(\lambda) - y))\Delta_{(0,Y(\lambda)-y)}\frac{\partial^2 u}{\partial x^2}(x + s, y)dtds$$
$$- \int_{\lambda_1}^{\lambda_2} \gamma(X(\lambda), y)\left[\frac{\partial^2 u}{\partial x^2}(R(\lambda))\right] \int_{|Y(\lambda)-y|}^{h_2} B(X(\lambda) - x, t\text{sgn}(Y(\lambda) - y))dtX'(\lambda)d\lambda \tag{5}$$

is true for $2 + \frac{1}{p} < \alpha < 3$.

4 Rate of convergence

Using representations of the type (4) and (5), we include their leading terms into the right-hand side φ of (3). We consider discrete boundary value problem (3) with φ defined by

$$\varphi(r) = 0, \text{ if } r \in \Omega_h : E_r \cap \Gamma = \emptyset,$$

$$\varphi(r) = \frac{\gamma(r)(h_1 - |\overline{x} - x|)_+^0}{h_1^2}[u(\overline{r})] + \frac{\gamma(r)\text{sgn}(\overline{x} - x)(h_1 - |\overline{x} - x|)_+}{h_1^2}[\frac{\partial u}{\partial x}(\overline{r})]$$
$$- \frac{1}{h_1 h_2} \int_{\lambda_{1,r}}^{\lambda_{2,r}} \gamma(X(\lambda), y)[\frac{\partial^2 u}{\partial x^2}(R(\lambda))] \int_{|Y(\lambda)-y|}^{h_2} B(X(\lambda) - x, t\text{sgn}(Y(\lambda) - y))dtX'(\lambda)d\lambda$$
$$- \frac{1}{h_1 h_2} \int_{\lambda_{1,r}}^{\lambda_{2,r}} \gamma(x, Y(\lambda))[\frac{\partial^2 u}{\partial y^2}(R(\lambda))] \int_{|X(\lambda)-x|}^{h_1} B(t\text{sgn}(X(\lambda) - x), Y(\lambda - y)dtY'(\lambda)d\lambda, \tag{6}$$

if $r \in \Omega_h : P_r \cap \Gamma = \{\overline{r}\} = \{(\overline{x}, y)\}$.

At mesh points close to Γ function φ is determined in terms of the known jumps of u and its derivatives across Γ, using the parametrization of Γ and some distances to the curve Γ. We study the accuracy of the proposed equations.

Theorem 1. *Let the potential u be in $B^\alpha_{p,p}(D, \Omega), 2 + \frac{1}{p} < \alpha < 3 + \frac{1}{p}$. If $\alpha = 3$ we assume additionally $p \geq 2$. Then for the solution u_h of (3) with function φ given by (6) the following error estimate holds*

$$\|u - u_h\|_{W^2_p(\Omega_h)} \leq C(\frac{p^2}{p-1})^4 |h|^{\alpha-2} \|u\|_{B^\alpha_{p,p}(D,\Omega)}.$$

An attractive feature of the error estimate in Theorem 1 is its rate of convergence, which is compatible with the smoothness of the potential.

5 Quadrature formulas

The right-hand side φ of the finite difference scheme is given by (6). At points close to Γ computation of φ includes evaluation of two line integrals. If $\alpha > 2+\frac{2}{p}$, then the integrands are continuous functions on Γ. Therefore one can directly approximate any of the integrals in (6) by an one point quadrature formula

$$\int_{\lambda_1}^{\lambda_2} A(\lambda)f(\lambda)d\lambda \approx Cf(\lambda^*), C = \int_{\lambda_1}^{\lambda_2} A(\lambda)d\lambda, \lambda^* \in [\lambda_1, \lambda_2] \tag{7}$$

(exact for any constant). It can be shown, that the total error of the finite difference scheme with the new right hand side, where the integrals are replaced by the quadrature (7), is of the same rate, as the error of the finite difference method (3).

We have the freedom to choose the parameter λ^* in a way to enable an easy evaluation of the coefficients of the quadrature formula. We give two basic cases of relative location of the point r to the curve Γ.

Case A:$r \in \Omega_h, P_r \cap \Gamma = \emptyset$. We set

$$\varphi_1(r) = 0. \tag{8}$$

Case B: $r \in \Omega_h, P_r \cap \Gamma = \{\overline{r}\} = \{(\overline{x}, y)\} = \{R(\overline{\lambda})\}$. We choose λ^* to be $\overline{\lambda}$ and evaluate directly the coefficients of the quadrature formulas. Then we find the following expression φ_1 for φ :

$$\begin{aligned}\varphi_1(r) = &\frac{\gamma(r)(h_1-|x-\overline{x}|)_+^0}{h_1^2}[u(\overline{r})] + \frac{\gamma(r)\mathrm{sgn}(\overline{x}-x)(h_1-|x-\overline{x}|)_+}{h_1^2}[\frac{\partial u}{\partial x}(\overline{r})] \\ &+\frac{\gamma(r)(h_1-|x-\overline{x}|)_+^2}{h_1^2}[\frac{\partial^2 u}{\partial x^2}(\overline{r})]\end{aligned} \tag{9}$$

This right hand side was proposed first by A. Mayo, [8].

We treat the other cases of relative location of the point r to the curve Γ in a similar way.

Theorem 2. *Let the potential u satisfy the assumptions of Theorem 1 and let $\alpha > 2 + \frac{2}{p}$. Let $u_{h,1}$ be the solution to the discrete boundary value problem (3) with the right hand side φ_1, given by (8), (9). Then one has the following error estimate*

$$\|u - u_{h,1}\|_{W_p^2(\Omega_h)} \leq C(\frac{p^2}{p-1})^4|h|^{\alpha-2}\|u\|_{B^\alpha_{p,p}(D,\Omega)}.$$

Acknowledgment

This research was supported by the Bulgarian Ministry of Education, Science and Technology under Grant No. MM 425/94.

References

1. Friedman, A.: Mathematics in industrial problems. Springer (1992)
2. Gold J., Graham I.: Towards automatition of boundary integral methods for Laplace's equation. in "Mathematics of finite elements and applications" Academic Press (1991) 349-360
3. Golovin G., Makarov M., Sablin M., Sukhachev D., Yakovlev V.: Comparison of various methods for solving the Dirichlet problem for the Laplace equation in complicated domains. Zh. Vychisl. Mat. Fiz. **27** (1987) 1662-1679 (in Russian)
4. Drenska, N.: Computation of potentials used in the boundary element method. Colloq. Math. Soc. Janos Bolyai **59** (1990) 157-163
5. Kolkovska, N.: Reconstruction of some potentials used in the boundary element method. J. of Integral Equations and Applications **5** (1993) 345-367
6. Kress, R.: Linear integral equations. Springer (1989) 298
7. Lazarov, R., Mokin Yu. On the computation of the logarithmic potential. Soviet Math. Dokl. **28** (1983) 320-323
8. Mayo A.: The fast solution of Poisson's and the biharmonic equation on irregular regions. SIAM J. Numer. Anal. **21** (1984) 285-299
9. Mayo A., Greenbaum A.: Fast parallel iterative solutions of Poisson's and the biharmonic equations on irregular domains. SIAM J. Sci. Stat. Comp. **13**(1992) 101-118
10. Mokin, Yu.: Methods of calculating a logarithmic potential. Izd. Moskovsk. Gos. Univ., Moscow (1988) 124 (in Russian)
11. Yakovlev V. On the computation of the flaxere of the thin plate with leant edges Vestn. MGU, ser. 15 (1985) 19-22 (in Russian)

Numerical Analysis in Singularly Perturbed Boundary Value Problems Modelling Heat Transfer Processes*

V.L. Kolmogorov[1], G.I. Shishkin[2] and L.P.Shishkina[3]

[1] Institute of Engineering Science, Ural Branch of Russian Academy of Sciences, GSP-207 620219, Ekaterinburg, Russia
[2] Institute of Mathematics and Mechanics, Ural Branch of Russian Academy of Sciences, GSP-384 620219 Ekaterinburg, Russia
[3] Scientific Research Institute of Heavy Machine Building, Ekaterinburg, Russia

Abstract. We construct a finite difference method for boundary value problems modelling heat and mass transfer for fast-running processes. The dimensionless form of the equation in these problems is singularly perturbed, i.e., the highest derivatives are multiplied by a parameter ε^2 which can take any values from the interval (0,1]. The equation involves concentrated sources; the boundary conditions are mixed. As is known, classical numerical methods lead us to large errors that can exceed many times the exact solution for small ε; a similar problem occurs if we are to find the normalized flux, i.e., the gradient multiplied by ε. New special schemes are constructed to converge uniformly with respect to the parameter. The errors in the discrete solution and in the computed fluxes are independent of the parameter. The new schemes can be applied to the analysis of heat exchange in metal working by hot die-forming or for plastic shear.

1 Introduction

Many modern technologies of material working are characterized by fast-running processes. When modelling and analyzing heat and mass transfer, we come to singularly perturbed equations in the dimensionless form. For example, plastic shear in a material can be considered as shifting two parts of the body under tangential stress. As a result, on the slip surface heat is liberated. This process is described by such parameters as the coefficient of temperature conductivity for steel $D^H = 2 \cdot 10^{-5}\, m^2 \cdot sec^{-1}$, the thickness of the shifting parts $L = 1\, m$; the duration of the process is $2\, sec$ when the shift stage $\vartheta = 1\, sec$. Then we come to the problem with concentrated sources for $\varepsilon = \varepsilon_0 = 6.3 \cdot 10^{-3}$ where $\varepsilon^2 = 2D^H L^{-2} \vartheta$. During the process, in a narrow neighbourhood of the slip surface, temperature arises significantly and becomes about some hundreds of Celsium degree. To analyze the process, we are to find both the solution (e.g.,

* This work was supported by the Russian Foundation of Basic Research under Grant N 95-01-00039a

the maximal temperature) and the heat fluxes, which determine the structure of transformations in metal. Thus, singularly perturbed problems are typical for fast-running processes.

For these problems the error in the discrete solution can be large for small values of the parameter ε if we use classical finite difference schemes (see, e.g., [1, 3, 6]). Therefore it is required to develop special methods for which the errors do not depend on the parameter value.

For problems of plastic shear, with the use of the classical scheme (see (6), (10)) for $N = N_0 = 100$, the growth of temperature is 835 ^oC. For the special scheme (see (6), (11)) the growth is 239 ^oC, and the error is 7%.

2 Mathematical Formulation for the Problem

On the set

$$G = D \times (0,T], \quad D = \{\, x: \; d_0 < x < d_1 \,\} \;, \tag{1}$$

we consider the following singularly perturbed parabolic equation with a concentrated source at the point $x = d^*$, $d_0 < d^* < d_1$:

$$\left\{ \varepsilon^2 a(x,t) \frac{\partial^2}{\partial x^2} - p(x,t) \frac{\partial}{\partial t} - c(x,t) \right\} u(x,t) = f(x,t), \;\; (x,t) \in G^* \;, \tag{2a}$$

$$\beta(x) u(x,t) + (1-\beta(x)) \varepsilon \frac{\partial}{\partial n} u(x,t) = \psi(x,t), \;\; (x,t) \in S^L \;, \tag{2b}$$

$$u(x,t) = \varphi(x), \;\; (x,t) \in S_0 \;. \tag{2c}$$

Here $G^* = D^* \times (0,T]$, $D^* = D \setminus \{x = d^*\}$; $S^L = \{x : x = d_0, d_1\} \times (0,T]$ and $S_0 = \overline{D} \times \{t = 0\}$ are the lateral and lower boundary of the set $\overline{G}$. On the set $S^* = \{\, x = d^* \,\} \times (0,T]$ we have the conjugation (interface) conditions

$$[u(x,t)] = 0, \;\; l^*_{(2)} u(x,t) \equiv \varepsilon \left[\frac{\partial}{\partial x} u(x,t) \right] = q(t), \;\; (x,t) \in S^* \;. \tag{2d}$$

By $[\, v(x,t) \,]$ we denote the jump of the function $v(x,t)$ at the point x: $[\, v(x,t) \,] = v(x+0,t) - v(x-0,t)$. All the functions are assumed to be sufficiently smooth, moreover, $a(x,t) \geq a_0 > 0$, $c(x,t) \geq 0$, $p(x,t) \geq p_0 > 0$, $(x,t) \in \overline{G}$.

We shall call the function

$$P(x,t) = \varepsilon \frac{\partial}{\partial x} u(x,t)$$

the normalized diffusion flux. This function is discontinuous on the set S^*. It is convenient to divide the set $\overline{G}$ into two sets

$$\overline{G} = \overline{G}^1 \cup \overline{G}^2, \;\; G^1 = D^1 \times (0,T], \;\; G^2 = D^2 \times (0,T] \;,$$

where $D^1 = \{x : \; d_0 < x < d_*\}$, $D^2 = \{x : \; d^* < x < d_1\}$. Note that, generally speaking, the function $P(x,t)$ is not continuous on each set $\overline{G}^j$, $j = 1,2$. It has discontinuities on the set $S_0^{L*} = \gamma^{L*} \times \{\, t = 0 \,\}$ where γ^{L*} is a set of the points $x = d_0$, d_1 and $x = d^*$. We consider the function $P(x,t)$ on the sets $\overline{G}_0^j$, where $\overline{G}_0^j = \overline{G}^j \setminus S_0^{L*}$, $j = 1,2$. If compatibility conditions on the set S_0^{L*} hold then the function $P(x,t)$ is continuous on each set $\overline{G}^j$, $j = 1,2$.

3 Special Finite Difference Scheme

We give some estimates of the solution and its derivatives for problem (2), (1). Suppose that, for a fixed value of the parameter, the solution of the boundary value problem is sufficiently smooth on each set $\overline{G}^j$ and has the continuous derivatives

$$\frac{\partial^{k+k_0}}{\partial x^k \partial t^{k_0}} u(x,t), \quad (x,t) \in \overline{G}^j, \quad 0 \le k + 2k_0 \le K, \quad j = 1, 2,$$

where $K = 4$. The solution of the boundary value problem on each subset $\overline{G}^j$ can be represented as a sum of the functions

$$u(x,t) = U(x,t) + V(x,t), \quad (x,t) \in \overline{G}^j, \quad j = 1, 2.$$

The functions $U(x,t)$ and $V(x,t)$, i.e. the regular and singular parts of the solution, satisfy the estimates

$$\left| \frac{\partial^{k+k_0}}{\partial x^k \partial t^{k_0}} U(x,t) \right| \le M, \tag{3}$$

$$\left| \frac{\partial^{k+k_0}}{\partial x^k \partial t^{k_0}} V(x,t) \right| \le M\varepsilon^{-k} \exp(-m\varepsilon^{-1} r(x, \Gamma^j)),$$

$$(x,t) \in \overline{G}^j, \quad 0 \le k + 2k_0 \le 4, \quad j = 1, 2,$$

where $r(x, \Gamma^j)$ is a distance from the point x to the set $\Gamma^j = \overline{D}^j \setminus D^j$; m is a sufficiently small arbitrary number.

First, on the segment $[d_1, d_2]$, we design a special mesh condensing in the neighbourhood of the ends. Suppose

$$\overline{\omega}_*(\sigma, N_*, N, d_1, d_2) \tag{4}$$

is a piecewise uniform grid on the segment $[\, d_1, d_2 \,]$. The grid $\overline{\omega}_*$ is uniform on each of the subintervals $[\, d_1, d_1 + \sigma \,]$, $[\, d_1 + \sigma, d_2 - \sigma \,]$, $[\, d_2 - \sigma, d_2 \,]$ with the mesh width $h_{(1)} = \sigma / N_*$ on $[\, d_1, d_1 + \sigma \,]$ and $[\, d_2 - \sigma, d_2 \,]$, and with the mesh width $h_{(2)}$ on $[\, d_1 + \sigma, d_2 - \sigma \,]$; $N+1$ is the overall number of the $\overline{\omega}_*$-grid nodes; $\sigma \le 4^{-1}(d_2 - d_1)$. We assume that $\sigma = \sigma_{(4)}(\varepsilon, N, d_1, d_2, m) = \min\left[4^{-1}(d_2 - d_1), m^{-1}\varepsilon \ln N \right]$, where m is an arbitrary number. Here $N_* = N/4$.

Now we construct a special grid $\overline{G}_h$ which condenses near the boundary and interior layers. Suppose

$$\overline{G}_h = \overline{D}_h \times \overline{\omega}_0 = \overline{\omega}_1^* \times \overline{\omega}_0 \ , \tag{5a}$$

where $\overline{\omega}_0$ is a uniform grid on the interval $[\, 0, T \,]$, $\overline{\omega}_1^*$ is a piecewise uniform grid on $[\, d_0, d_1 \,]$ constructed below; $N+1$ and $N_0 + 1$ are the number of nodes in the grids $\overline{\omega}_1^*$ and $\overline{\omega}_0$. The grid $\overline{D}_h = \overline{\omega}_1^*$ can be defined by the relations

$$\overline{D}^j \cap \overline{D}_h = \overline{D}_h^j, \quad j = 1, 2 \ , \tag{5b}$$

$$\overline{D}_h^1 = \overline{\omega}_{*(4)}(\sigma, N_* = N/8,\ N/2,\ d_0, d^*),\ \text{where } \sigma = \sigma_{(4)}(\varepsilon, N/2, d_0, d^*, m)\ ,$$

$$\overline{D}_h^2 = \overline{\omega}_{*(4)}(\sigma, N_* = N/8,\ N/2,\ d^*, d_1),\ \text{where } \sigma = \sigma_{(4)}(\varepsilon, N/2, d^*, d_1, m)\ .$$

On the grid $\overline{G}_h$ we consider the finite difference scheme [2, 4]

$$\left\{\varepsilon^2 a(x,t)\delta_{\overline{x}\widehat{x}} - p(x,t)\delta_{\overline{t}} - c(x,t)\right\} z(x,t) = f(x,t),\ \ (x,t) \in G_h^*\ , \tag{6}$$

$$\begin{gathered} \varepsilon\{\delta_x - \delta_{\overline{x}}\}\, z(x,t) = q(t),\ \ (x,t) \in S_h^*, \\ \beta(x)z(x,t) + (1-\beta(x))\varepsilon\delta_n\, z(x,t) = \psi(x,t),\ \ (x,t) \in S_h^L\ , \\ z(x,t) = \varphi(x),\ \ (x,t) \in S_h^0\ . \end{gathered}$$

Here $G_h^0 = G^0 \cap \overline{G}_h$, where $G^0 \subseteq \overline{G}$ is one of the sets G^*, S^*, S^L, S^0;

$$\delta_n\, z(x,t) = \begin{cases} -\delta_x\, z(x,t), & x = d_0\ , \\ \delta_{\overline{x}}\, z(x,t), & x = d_1\ . \end{cases}$$

Let us construct the approximation of the function $P(x,t)$. On the sets $\overline{G}_0^j$ we introduce the grid $\overline{G}_{0h}^{j-}$, $j = 1,2$. Here $\overline{G}_{0h}^{j-}$ is a set of such nodes (x,t) from $\overline{G}_{0h}^j = \overline{G}_0^j \cap \overline{G}_h$ for which the operator δ_x is defined. Similarly we define the grids $\overline{G}_h^{j-}$ on the sets $\overline{G}^j$, $j = 1,2$. To approximate the function $P(x,t)$ on $\overline{G}_0^j$, we use the grid function $P^{h+}(x,t) \equiv \varepsilon\delta_x\, z(x,t)$, $(x,t) \in \overline{G}_{0h}^{j-}$, $j = 1,2$. If the function $P(x,t)$ is continuous, then, for the approximation, we apply the function $P^{h+}(x,t)$ on the grid $\overline{G}_h^{j-}$, $j = 1,2$.

We use techniques as in [3, 5, 6]. Estimates (3) imply the ε-uniform convergence of the solution for special scheme (6), (5):

$$\max_{\overline{G}_h} |\, u(x,t) - z(x,t)\,| \le M\left[N^{-1}\ln N + N_0^{-1}\right],\ \ (x,t) \in \overline{G}_{h(5)}\ . \tag{7}$$

If, for a fixed value of the parameter ε, the solution of the problem has the continuous derivatives of required orders, as was said above, the computed flux converges to the real one ε-uniformly on $\overline{G}_h^{j-}$

$$\max_{\overline{G}_h^{j-}} |P(x,t) - P^{h+}(x,t)| \le M\left[N^{-1}\ln N + N_0^{-1}\right],\ \ (x,t) \in \overline{G}_{h(5)}^{j-},\ j = 1,2\ . \tag{8}$$

Theorem 1. *Let the solution of problem (2), (1) be sufficiently smooth for a fixed value of the parameter and satisfy estimates (3). Then for the solution of finite difference scheme (6), (5) and for the computed normalized diffusion flux estimates (7), (8) are valid.*

4 Numerical Study of the Special Scheme

Now we apply the new special finite difference scheme for the solution of a model problem and compare the numerical results with results obtained for a classical discretization scheme on a uniform grid. Assume that the concentrated source acts at the point $x = 0$. On the domain

$$\overline{G} = \overline{G}^1 \cup \overline{G}^2, \quad G^j = D^j \times (0,T], \quad j = 1,2 ,$$

where $T = 1$, $D^1 = (-1,0)$, $D^2 = (0,1)$, we consider the following boundary value problem with the concentrated source:

$$\left\{\varepsilon^2 \frac{\partial^2}{\partial x^2} - \frac{\partial}{\partial t}\right\} u(x,t) = f(x,t), \quad (x,t) \in G^j , \tag{9}$$

$$l^*_{(9)} u(x,t) = q(t), \quad [u(x,t)] = 0, \quad (x,t) \in S^* ,$$

$$l_{(9)} u(x,t) = \psi(x,t), \quad (x,t) \in S^L ,$$

$$u(x,t) = \varphi(x), \quad (x,t) \in S_0 .$$

Here $S^* = \{x = 0\} \times (0,T]$, $S^L = \{x = -1, 1\} \times (0,T]$, $S_0 = \overline{D} \times \{t = 0\}$,

$$f(x,t) = \begin{cases} -1, & (x,t) \in \overline{G}^1 , \\ -t, & (x,t) \in \overline{G}^2 , \end{cases}$$

$$l^*_{(9)} u(0,t) \equiv \varepsilon \left(\frac{\partial}{\partial x} u(+0,t) - \frac{\partial}{\partial x} u(-0,t) \right) =$$

$$= -2 \frac{1}{\sqrt{\pi}} t^{1/2} - \frac{4}{3\sqrt{\pi}} t^{3/2} \equiv q(t), \quad 0 < t \le T ,$$

$$l_{(9)} u(x,t) \equiv \left\{ \begin{array}{ll} u(x,t), & x = -1 , \\ \varepsilon \frac{\partial}{\partial n} u(x,t), & x = 1 \end{array} \right\} =$$

$$= \left\{ \begin{array}{ll} 0, & x = -1 , \\ \frac{2}{3\sqrt{\pi}} t^{3/2}, & x = 1 \end{array} \right\} \equiv \psi(x,t), \quad (x,t) \in S^L ,$$

$$\varphi(x) = 0, \quad x \in \overline{D} .$$

The function $f(x,t)$ has a discontinuity only on the set S^*. On the left end we give the Dirichlet boundary condition, and the Neumann condition on the right end.

The grid equations similar to (6) are considered either on the uniform grids

$$\overline{G}_h = \overline{G}_{h(10)} \tag{10}$$

or on the special grids

$$\overline{G}_h = \overline{G}_{h(11)} \equiv \overline{G}_{h(5)} \text{ for } d_0 = -1,\ d_1 = 1,\ d^* = 0\ . \tag{11}$$

Let us introduce the errors $E_{N^*,N_0^*}(\varepsilon, N, N_0)$ and $\overline{E}_{N^*,N_0^*}(N, N_0)$

$$E_{N^*,N_0^*}(\varepsilon, N, N_0) \equiv \max_{\overline{G}_h} \mid \overline{u}(x,t) - z(x,t) \mid\ ,$$

$$\overline{E}_{N^*,N_0^*}(N, N_0) \equiv \max_{\varepsilon} E_{N^*,N_0^*}(\varepsilon, N, N_0)\ .$$

We set, for $N_0 = N$, that

$$E_{N^*,N_0^*}(\varepsilon, N) = E_{N^*,N_0^*}(\varepsilon, N, N),\quad \overline{E}_{N^*,N_0^*}(N) = \overline{E}_{N^*,N_0^*}(N, N)\ .$$

We introduce also the errrors $Q_{N^*,N_0^*}(\varepsilon, N, N_0)$, $\overline{Q}_{N^*,N_0^*}(N, N_0)$

$$Q_{N^*,N_0^*}(\varepsilon, N, N_0) \equiv \max_{j,\ \overline{G}_h^{j-}} \mid \overline{P}(x,t) - P^{h+}(x,t) \mid\ ,$$

$$\overline{Q}_{N^*,N_0^*}(N, N_0) \equiv \max_{\varepsilon} Q(\varepsilon, N, N_0)\ .$$

Similarly we set for $N_0 = N$

$$Q_{N^*,N_0^*}(\varepsilon, N) = Q_{N^*,N_0^*}(\varepsilon, N, N),\quad \overline{Q}_{N^*,N_0^*}(N) = \overline{Q}_{N^*,N_0^*}(N, N)\ .$$

Here $\overline{u}(x,t) = \overline{u}_{N^*,N_0^*}(x,t)$ and $\overline{P}(x,t) = \overline{P}_{N^*,N_0^*}(x,t)$ are the solution and normalized flux computed by special scheme (6), (11) for sufficiently large $N = N^*$, $N_0 = N_0^*$.

In Tables 1–4 we give the values of the above errors for the case if $N^* = N_0^* = 1024$.

The results in Tables 1–4 agree with the theory. Finite difference schemes on space grids with the arbitrary distribution of nodes, in particular, on uniform grids, can cause errors in the discrete solution which exceed many times the real solution. Besides, for small values of ε, the errors in the computed fluxes tend to some constant exceeding unity if N, $N_0 \to \infty$. We can see that the application of schemes on a special condensing grid allows us to find the approximations of solutions and normalized diffusion fluxes that are ε-uniformly convergent for N, $N_0 \to \infty$.

Table 1. Errors $E_{1024,1024}(\varepsilon, N)$ for classical scheme (6), (10)

$\varepsilon \setminus N$	8	32	128	512
1	2.688e−1	6.182e−2	1.379e−2	1.965e−3
2^{-2}	6.377e−1	1.357e−1	2.944e−2	4.164e−3
2^{-4}	3.169e+0	5.907e−1	1.139e−1	1.554e−2
2^{-6}	1.432e+1	3.104e+0	5.475e−1	7.880e−2
2^{-8}	5.942e+1	1.429e+1	3.095e+0	5.447e−1
2^{-10}	2.400e+2	5.940e+1	1.428e+1	3.093e+0
2^{-12}	9.621e+2	2.399e+2	5.939e+1	1.428e+1
$\overline{E}(N)$	9.621e+02	2.399e+02	5.939e+01	1.428e+01

Table 2. Errors $E_{1024,1024}(\varepsilon, N)$ for special scheme (6), (11)

$\varepsilon \setminus N$	8	32	128	512
1	2.688e−1	6.182e−2	1.379e−2	1.965e−3
2^{-2}	6.377e−1	1.357e−1	2.944e−2	4.164e−3
2^{-4}	1.984e+0	5.907e−1	1.139e−1	1.554e−2
2^{-6}	1.862e+0	8.436e−1	2.342e−1	3.869e−2
2^{-8}	1.846e+0	8.436e−1	2.342e−1	3.869e−2
2^{-10}	1.842e+0	8.436e−1	2.342e−1	3.869e−2
2^{-12}	1.841e+0	8.436e−1	2.342e−1	3.869e−2
$\overline{E}(N)$	1.984e+00	8.435e−01	2.342e−01	3.869e−02

Table 3. Errors of the normalized flux $Q_{1024,1024}(\varepsilon, N)$ for classical scheme (6), (10)

$\varepsilon \setminus N$	8	32	128	512
1	2.633e−1	6.893e−2	1.630e−2	3.622e−3
2^{-2}	6.264e−1	2.193e−1	5.446e−2	7.969e−3
2^{-4}	8.877e−1	6.058e−1	1.962e−1	3.088e−2
2^{-6}	1.018e+0	8.817e−1	5.694e−1	1.013e−1
2^{-8}	1.065e+0	1.018e+0	8.803e−1	5.683e−1
2^{-10}	1.076e+0	1.065e+0	1.018e+0	8.799e−1
2^{-12}	1.079e+0	1.076e+0	1.065e+0	1.018e+0
$\overline{Q}(N)$	1.079e+00	1.076e+00	1.065e+00	1.018e+00

Table 4. Errors of the normalized flux $Q_{1024,1024}(\varepsilon, N)$ for special scheme (6), (11)

$\varepsilon \setminus N$	8	32	128	512
1	2.633e−1	6.893e−2	1.630e−2	3.622e−3
2^{-2}	6.264e−1	2.193e−1	5.446e−2	7.969e−3
2^{-4}	8.703e−1	6.058e−1	1.962e−1	3.088e−2
2^{-6}	8.589e−1	6.938e−1	3.201e−1	4.849e−2
2^{-8}	8.541e−1	6.938e−1	3.201e−1	4.849e−2
2^{-10}	8.527e−1	6.938e−1	3.201e−1	4.849e−2
2^{-12}	8.524e−1	6.938e−1	3.201e−1	4.849e−2
$\overline{Q}(N)$	8.703e−01	6.938e−01	3.201e−01	4.849e−02

Thus, for singularly perturbed diffusion equations with mixed boundary conditions and concentrated sources, we have constructed new finite difference schemes which enable us to find solutions and normalized diffusion fluxes with an ε-uniform accuracy.

References

1. Doolan, E.P., Miller, J.J.H., Schilders, W.H.A.: Uniform Numerical Methods for Problems with Initial and Boundary Layers. Dublin (1980)
2. Marchuk, G.I.: Methods of Numerical Mathematics. Springer, New York (1982)
3. Miller, J.J.H., O'Riordan, E., Shishkin, G.I.: Fitted numerical methods for singular perturbation problems. Errors estimates in the maximum norm for linear problems in one and two dimensions. // World Scientific, Singapore (1996)
4. Samarsky, A.A.: Theory of difference scheme. Nauka, Moscow (1989) (in Russian)
5. Shishkin, G.I.: Grid approximation of a singularly perturbed boundary value problem for quasi-linear parabolic equations in the case of complete degeneracy in spatial variables. Sov. J. Numer. Anal. Math. Modelling **6** (1991) 243–261
6. Shishkin, G.I.: Grid Approximation of Singularly Perturbed Elliptic and Parabolic Equations. Ural Branch of Russ. Acad. Sci., Ekaterinburg (1992) (in Russian)

Improved Perturbation Bounds for the Matrix Exponential

Mihail Konstantinov[1], Petko Petkov[2], Parashkeva Gancheva[1], Vera Angelova[3], Ivan Popchev[3]

[1] Univ. of Arch. & Civil Eng.,1 Hr.Smirnenski Blv.,1421 Sofia, Bulgaria, mmk_fte@bgace5.uacg.acad.bg
[2] Dept. of Automatics, Technical Univ. of Sofia, 1756 Sofia, Bulgaria
[3] IIT, BAS, Akad. G. Bonchev Str., Bl.2, 1113 Sofia, Bulgaria, popchev@bgcict.acad.bg

Abstract. In this paper we give asymptotic series expansions in $\varepsilon = \|E\|$ for the bound of the perturbation $\|\exp(t(A+E)) - \exp(tA)\|$ in the matrix exponential $\exp(tA)$.

1 Introduction

Bounds and perturbation bounds for the matrix exponential $\exp(tA)$, where A is a (complex) $n \times n$ matrix, have been proposed by B. Kagström [5] and C. Van Loan [11] in 1977, see also [4] and [2]. However, as shown in [6] and [10] these bounds give rather pessimistic results for some defective matrices. In particular an overestimation of the real perturbation of hundreds of orders of magnitude was observed for low order and well behaved systems $x' = Ax$.

In this paper we give improved perturbation bounds of the form

$$\Delta(t) = \|\exp(t(A+E)) - \exp(tA)\| \le f(t,\varepsilon),\ \varepsilon = \|E\|$$

where E is a perturbation in A. For this purpose we use bounds for $\|\exp(tA)\|$ based on Schur and Jordan decompositions of A. After that a linear r-th order differential equation for f is derived, where r is the dimension of the dominant Jordan block of A. A study of this equation allows to obtain improved perturbation bounds which are often better than the known in the literature.

Asymptotic series expansions in ε (treated as a small parameter) are also given.

The above results are applicable to the development of condition and error estimates for the solution of linear and nonlinear differential equations.

We denote by $\|.\|$ the matrix 2-norm in $\mathcal{F}^{n.n}$, where $\mathcal{F}$ is the set of real numbers $\mathcal{R}$ or the set of complex numbers $\mathcal{C}$. The unit $n \times n$ matrix is denoted I_n and N_n is the nilpotent $n \times n$ matrix with unit elements at positions $(i, i+1)$ and zeros otherwise.

Throughout the paper A is a fixed $n \times n$ real or complex matrix with spectral abscissa $\alpha = \max\{\mathrm{Re}(\lambda) : \lambda \in \mathrm{spect}(A)\}$, where $\mathrm{spect}(A)$ is the spectrum of A.

2 Problem Statement

The matrix exponential $\exp : \mathcal{F}^{n.n} \to \mathcal{F}^{n.n}$ defined by the power series

$$\exp(A) = \sum_{i=0}^{\infty} \frac{A^i}{i!}$$

appears in the solution of linear differential equations, e.g.

$$\begin{aligned} Y'(t) &= AY(t) + Y(t)B + C(t) \\ Y(0) &= Y_0 \end{aligned}$$

where $Y(t) \in \mathcal{F}^{n.n}$:

$$Y(t) = \exp(tA)Y_0 \exp(tB) + \int_0^t \exp((t-s)A)C \exp((t-s)B)\mathrm{d}s \ .$$

In practice, the mathematical model of a real phenomenon is always contaminated with measurement errors. Also, when solving a numerical problem by a numerically stable method, the computed solution is near to the exact solution of a slightly perturbed problem. In all these situations one has to deal not with the "exact" value $\exp(tA)$ but rather with the perturbed matrix $\exp(t(A+E))$, where $E \in \mathcal{F}^{n.n}$ is the perturbation in the matrix A. Usually the inequality $\|E\|/\|A\| \ll 1$ is fulfilled reflecting the fact that the perturbation is relatively small. Hence the problem arises to estimate the norm of the matrix

$$H(t,E) := \exp(t(A+E)) - \exp(tA)$$

as a function of the current time t and the quantity $\|E\|$. It is easy to show that

$$H(t,E) = \int_0^t \exp((t-s)A)E \exp(s(A+E))\mathrm{d}s \ .$$

Let

$$h(t,\varepsilon) := \max\{\|H(t,E)\| : \|E\| \leq \varepsilon\} \ .$$

Then our aim is to find an asymptotic bound of the form

$$h(t,\varepsilon) \leq \sum_{i=1}^{\infty} \varepsilon^i h_i(t) \ .$$

The expression for $H(t,E)$ may be represented as a sum of terms $H_m(t,E)$ of order m in E:

$$H(t,E) = \sum_{m=1}^{\infty} H_m(t,E)$$

where

$$\|H_m(t,E)\| = \mathrm{O}(\|E\|^m), \ \|E\| \to 0; \ m = 1,2,\ldots$$

We have

$$H_m(t,E) = \int_0^t \exp((t-s)A)G_m(s,E)\mathrm{d}s$$

where

$$G_m(s,E) := \sum_{r=m-1}^{\infty} \frac{s^r}{r!} \sum_{i_1+\cdots+i_m=r-m+1} \prod_{k=1}^{m} (EA^{i_k}) \ .$$

In particular, we have

$$G_1(s,E) = E\sum_{r=0}^{\infty}\frac{s^r}{r!}A^r = E\exp(sA)$$

$$G_2(s,E) = E\sum_{r=1}^{\infty}\frac{s^r}{r!}\sum_{i+j=r-1} A^iEA^j$$

$$G_3(s,E) = E\sum_{r=2}^{\infty}\frac{s^r}{r!}\sum_{i+j+k=r-2} A^iEA^jEA^k \ .$$

Note that $H_1(1,.)$,

$$H_1(1,E) = \int_0^1 \exp((1-s)A)E\exp(sA)\mathrm{d}s,$$

is the Frechet derivative of the function $X \mapsto \exp(X)$ at the point $X = A$, see also [9].

3 Estimates for the Matrix Exponential

When finding perturbation bounds for the matrix exponential, some bounds for the norm of the exponential $\|\exp(tA)\|$ itself are usually used [5, 11, 8].

Several estimates of the form

$$\|\exp(tA)\| \le C(\beta)\exp(t(\alpha+\beta))$$

are known, where β may be chosen arbitrarily from certain interval $(0,b)$, and $C(\beta)$ is a certain expression such that $C(\beta) \to \infty$ as $\beta \to 0$. These estimates lead to immediate perturbation bounds for the matrix exponential; the latter, however, are often too pessimistic if A is defective. That is why we shall use the more sophisticated bounds based on the Schur and Jordan canonical forms of A. To make all the results comparable, we assume that A has a single $n \times n$ Jordan block J with an eigenvalue λ with $\alpha = \mathrm{Re}(\lambda)$. This is not a restrictive assumption since the general case may be reduced to this particular case if we consider only the dominant Jordan block of A corresponding to the eigenvalue λ of A with $\mathrm{Re}(\lambda) = \alpha$.

Denote by c_A the minimum condition number of the transformation matrix $T \in \mathcal{F}^{n.n}$ reducing A into its Jordan normal form $J = T^{-1}AT = \lambda I_n + N_n$:

$$c_A = \min\{\|T\|\,\|T^{-1}\| : T^{-1}AT = J\}$$

(note that such c_A exists, see [7].) Then

$$\|\exp(tA)\| = \|T\exp(tJ)T^{-1}\| \le c_A\|\exp(tJ)\|$$

and since

$$\|\exp(tJ)\| = \left\|\exp(\lambda t)\sum_{k=0}^{n-1} N_n^k \frac{t^k}{k!}\right\| \le \exp(\alpha t)\sum_{k=0}^{n-1}\frac{t^k}{k!}$$

we have

$$\|\exp(tA)\| \le c_A \exp(\alpha t)\sum_{k=0}^{n-1}\frac{t^k}{k!} \ . \tag{1}$$

Consider now the Schur form $S = U^H AU = \lambda I_n + N$ of A, where $U \in \mathcal{F}^{n.n}$ is an unitary matrix and N is a strictly upper triangular matrix. Denote

$$\nu_A = \min\left\{\|N\| : U^H U = I_n,\ U^H AU = \lambda I_n + N\right\} \ .$$

Then we have

$$\begin{aligned}\|\exp(tA)\| &= \|U\exp(tS)U^H\| = \|\exp(tS)\| = \|\exp(\lambda t)\exp(Nt)\| = \\ &= |\exp(\lambda t)|\|\exp(Nt)\| = \exp(\alpha t)\|\exp(Nt)\| \ .\end{aligned}$$

Hence

$$\|\exp(tA)\| = \exp(\alpha t)\left\|\sum_{k=0}^{n-1}\frac{(Nt)^k}{k!}\right\| \le \exp(\alpha t)\sum_{k=0}^{n-1}\frac{(\nu_A t)^k}{k!} \ . \tag{2}$$

Relations (1) and (2) may be written in an unified manner as

$$\|\exp(tA)\| \le e(t) = b\exp(\alpha t)\omega(t),\ \omega(t) := \sum_{k=0}^{n-1}\frac{(\beta t)^k}{k!}$$

where

Table 1.

	b	β
Jordan	c_A	1
Schur	1	ν_A

Denote $F = E/\varepsilon$. Since

$$H'(t,E) = AH(t,E) + \varepsilon F(H(t,E) + \exp(tA))$$
$$H(0,E) = 0$$

we may express H as

$$H(t,E) = \varepsilon \int_0^t \exp((t-s)A)F(H(s,E) + \exp(sA))ds \ .$$

Hence

$$h(t,\varepsilon) \le \varepsilon \int_0^t e(t-s)(h(s,\varepsilon) + e(s))ds \ .$$

Thus

$$h(t,\varepsilon) \le u(t)$$

where u is the solution to the majorant Volterra integral equation

$$u(t) = \varepsilon \int_0^t e(t-s)(u(s) + e(s))ds \ .$$

Setting

$$u(t) = b\exp(\alpha t)z(t) - e(t) = b\exp(\alpha t)(z(t) - \omega(t))$$

we get

$$z(t) = \omega(t) + \mu \int_0^t \omega(t-s)z(s)ds, \ \mu := \varepsilon b \ . \qquad (3)$$

The solution of (3) may be represented as a convergent power series

$$z(t) = \sum_{r=0}^{\infty} \mu^r z_r(t)$$

where

$$z_0(t) = \omega(t)$$
$$z_r(t) = \int_0^t \omega(t-s)z_{r-1}(s)ds, \ r \ge 1 \ .$$

In particular, for $r = 1$ we have

$$z_1(t) = \int_0^t \omega(t-s)\omega(s)ds$$

and the norm of the Frechet derivative is estimated from

$$\|H(1,E)\| \le \mu b \exp(\alpha) \left(\sum_{k=0}^{n-1} \frac{\beta^k}{k!} + \sum_{k=n}^{2n-2} \frac{2n-1-k}{(k+1)!}\beta^k \right) \ .$$

Another way to solve (3) is via a reduction to an n-th order linear differential equation. Indeed, differentiating both sides of (3) n times we get the initial value problem

$$z^{(n)}(t) = \mu \sum_{k=0}^{n-1} \beta^{n-1-i} z^{(i)}(t)$$
$$z^{(k)}(0) = b(\beta + \varepsilon b)^k;\ k = 0, 1, \ldots, n-1 \ .$$

Setting $\tau = \beta t$, $\nu = \mu/\beta$ and $z(\tau/\beta) = y(\tau)$ we obtain (the differentiation is now in τ)

$$y^{(n)}(\tau) = \nu \sum_{k=0}^{n-1} y^{(k)}(\tau) \qquad (4)$$
$$y^{(k)}(0) = (1+\nu)^k;\ k = 0, 1, \ldots, n-1 \ .$$

The solution of the initial value problem (4) may be represented as

$$y = y_n(\tau) = \sum_{s=0}^{\infty} \nu^s y_{n,s}(\tau)$$

where

$$y_{n,s}(\tau) = \sum_{k=s}^{(s+1)n-1} c_k(n,s) \frac{t^k}{k!} \ .$$

Here

$$c_k(n,s) = \binom{s+1}{k-s}_n$$

and $\binom{s}{i}_n$ are the so called n-nomial coefficients defined from

$$\left(1 + x + \cdots + x^{n-1}\right)^s = \sum_{i=0}^{(n-1)s} \binom{s}{i}_n x^i \ .$$

The coefficients $c_k(n,s)$ satisfy the recurrence relation [1, 3]

$$c_k(n,s) = \sum_{i=0}^{n-1} c_{k-i}(n, s-1),\ c_s(n,s) = 1 \ .$$

For $n = 2$ the coefficients $\binom{s}{i}_2$ are equal to the binomial coefficients $\binom{s}{i}$.

4 Examples

Example 1. To illustrate the effectiveness of the estimate proposed, consider the problem of estimating the perturbation in the matrix exponential, where we choose the matrices $A,\ E \in \mathcal{R}^{2.2}$ as

$$A = \begin{bmatrix} -1 & 1 \\ 0 & -1 \end{bmatrix}, E = \begin{bmatrix} 0 & 0 \\ 10^{-4} & 0 \end{bmatrix} .$$

The results are shown at the table bellow, where the second column contains the exact perturbed quantity, est1 is the estimate based on the exact solution of the differential equation for z, est2 is the asymptotic bound using the dominant term in the solution for z and est3 is the bound proposed in [4].

Table 2.

t	$\frac{\\|\exp(t(A+E))-\exp(tA)\\|}{\\|\exp(tA)\\|}$	est1	est2	est3
1	0.763×10^{-4}	$0.606 \times 10^{+2}$	$0.300 \times 10^{+2}$	0.400×10^{-3}
10	0.180×10^{-2}	$0.878 \times 10^{+1}$	$0.438 \times 10^{+1}$	0.122
20	0.681×10^{-2}	$0.406 \times 10^{+1}$	$0.200 \times 10^{+1}$	0.920
30	0.152×10^{-1}	$0.246 \times 10^{+1}$	$0.122 \times 10^{+1}$	$0.316 \times 10^{+1}$
40	0.270×10^{-1}	$0.169 \times 10^{+1}$	0.839	$0.792 \times 10^{+1}$
50	0.423×10^{-1}	$0.125 \times 10^{+1}$	0.628	$0.168 \times 10^{+2}$
100	0.175	0.553	0.349	$0.280 \times 10^{+3}$

Example 2. This example is similar to Example 1 where A is a 3×3 Jordan block with an eigenvalue -1 and E has a single nonzero entry 10^{-4} in position (3,1). The results are as follows.

Table 3.

t	$\frac{\\|\exp(t(A+E))-\exp(tA)\\|}{\\|\exp(tA)\\|}$	est1	est2	est3
1	0.684×10^{-4}	$0.341 \times 10^{+3}$	$0.120 \times 10^{+3}$	0.625×10^{-3}
10	0.204×10^{-2}	$0.135 \times 10^{+2}$	$0.593 \times 10^{+1}$	$0.396 \times 10^{+1}$
20	0.141×10^{-1}	$0.312 \times 10^{+1}$	$0.181 \times 10^{+1}$	$0.152 \times 10^{+3}$
30	0.464×10^{-1}	$0.120 \times 10^{+1}$	$0.100 \times 10^{+1}$	$0.294 \times 10^{+4}$
40	0.110	0.700	0.804	$0.818 \times 10^{+5}$
50	0.218	0.681	0.849	$0.566 \times 10^{+7}$
100	$0.222 \times 10^{+1}$	$0.417 \times 10^{+1}$	$0.379 \times 10^{+1}$	$0.370 \times 10^{+27}$

The proposed estimates are asymptotically better than this from [4]. In fact, we have est3/est1 $\rightarrow \infty$ as $t \rightarrow \infty$. The examples show that our estimates are better even for moderate values of t.

The proposed perturbation bounds require more computational effort compared to those in [4]. However, both approaches involve the preliminary computation (or estimation) of either Jordan or Schur form of A. The extra amount of computations required by our approach is due to the need to find the coefficients $c_k(n, s)$ which, for a given order n of A, may be done in advance.

References

1. Bondarenko, B.: Generalized Pascal's Triangles and Pyramids and their Fractals, Graphs and Applications. FAN. Tashkent (1990)
2. Fong, I., Kuo, T., Kuo, K., Hsu, C., Wu, M.: Sensitivity analysis of linear uncertain systems and its application in the synthesis of an insensitive linear regulator. Int. J. Syst. Sci. **18** (1987) 43–55
3. Freund, J.: Restricted occupancy theory - A generalization of Pascal's triangle. Amer. Math. Monthly **63** 1 (1956) 20–27
4. Golub, G., Van Loan, C.: Matrix Computations. John Hopkins Univ. Press Baltimore 1983
5. Kagström, B.: Bounds and perturbation bounds for the matrix exponential. BIT **17** (1977) 39–57
6. Konstantinov, M., Petkov, P., Christov, N.: Computational methods for linear control systems - some open questions. Proc. 26 IEEE CDC Los Angeles **1** (1987) 818–823
7. Konstantinov, M., Petkov, P., Christov, N.: On best generalized conditioning of matrices from a cone. Proc. 19 Conf. of UBM Bourgas (1990) 105–108
8. Konstantinov, M., Petkov, P., Gu, D., Postlethwaite. L.: Perturbation Techniques for Linear Control Problems. LUED Report 95-7 Dept. of Engineering Leicester UK (1995)
9. Mathias. R.: Evaluating the Frechet derivative of the matrix exponential. Numer. Math. **63** (1992) 213–226
10. Petkov, P., Christov, N., Konstantinov M.: Computational Methods for Linear Control Problems. Prentice Hall. Hemel Hempstead (1991)
11. Van Loan, C.: The sensitivity of the matrix exponential. SIAM J. Numer. Anal. **14** (1977) 971–981

Spectral Portrait of Matrices by Block Diagonalization

P.-F. Lavallée[1], A. Malyshev[2] and M. Sadkane[1]

[1] IRISA-INRIA. Campus Universitaire de Beaulieu. 35042 Rennes Cedex, France.
[2] University of Bergen. Department of Informatics. N-5020 Bergen, Norway.

Abstract. We first describe an algorithm that reduces a matrix A to a block diagonal form using only well conditioned transformations. The spectral properties of A are then carried out from the resulting block diagonal matrix. We show in particular that the spectral portrait of A can be obtained cheaply from that of the block diagonal matrix.

1 Introduction

The spectral portrait or ϵ-pseudospectrum [2, 9] of a matrix A of order n generalizes the notion of eigenvalues of a matrix in the sense that instead of only representing an eigenvalue by its computed approximation, one may consider a neighborhood of it which is defined by some tolerance threshold. It can be applied efficiently in a number of applications in applied sciences such as solid mechanics, fluid mechanics, and more generally, in all applications that involve the characterization of the continuous-time and/or discrete-time stability [10].

Roughly speaking, the spectral portrait of a matrix A is the representation of the function $sp_A(z) = \log_{10}(\sigma_{min}(zI - A))$ by means of level curves in the complex plane, where $\sigma_{min}(zI - A)$ stands for the smallest singular value of the matrix $zI - A$. A classical way for computing sp_A is to use the Singular Value Decomposition algorithm [5] for each z in the complex plane. This approach is acceptable but requires a high computational cost.

We propose another method that first reduces the matrix A to a block diagonal matrix of the form

$$A = S\,\mathrm{diag}(D_1, D_2, \ldots, D_q)\,S^{-1},\ 1 \le q \le n \tag{1}$$

where the condition number $\kappa(S) = ||S|| ||S^{-1}||$ satisfies a tolerance provided by the user. Throughout the paper, the symbol $||\ .\ ||$ denotes the Euclidean norm or its induced matrix norm. The block diagonalization algorithm is similar to the one proposed in [1]. It starts by reducing A to the Schur form $A = QTQ^*$, then the upper triangular matrix T is block diagonalized in such a way that each block contains the eigenvalues corresponding to close (in a sense that will be discussed in Section 2) eigenvectors. The spectral portrait of A is then approximated by those of the matrices D_i, $i = 1, \ldots, q$.

One of the motivations of this approach comes from the definition itself of the spectral portrait : it can be divided into several "spots", each of them contains eigenvalues of the matrix $A+E$ where the norm of E is smaller than $\epsilon\|A\|$, and thus all the points in one "spot" can be grouped in the same diagonal block.

The advantage of this approach over the classical one [2, 9] is its simplicity and rapidity since the spectral portrait of the small matrices D_i, $i = 1,\ldots,q$ can cheaply be obtained and can be done in parallel.

In Section 2 we describe and justify our block diagonalization algorithm. In Section 3 we establish some relationship between the spectral portrait of A and that of the matrices $D_1,\ldots,D_q$ in (1). We also discuss the possibility of using the field of values of these matrices. Section 3 is devoted to cost analysis and numerical experiments.

2 Block diagonalization

The main idea is to use the Schur decomposition to reduce A to upper triangular form

$$Q^*AQ = T = \begin{pmatrix} T_{11} & T_{12} & \ldots & T_{1q} \\ & T_{22} & \ldots & T_{2q} \\ & & \ddots & \vdots \\ & & & T_{qq} \end{pmatrix} \tag{2}$$

and then block diagonalize T thanks to the following theorem (see [5, p.338])

Theorem 1. *Suppose that the decomposition (2) is such that the matrices T_{ii} and T_{jj} have disjoint spectra whenever $i \neq j$, then there exists a nonsingular matrix X such that*

$$(QX)^{-1}A(QX) = diag(T_{11},\ldots,T_{qq}). \tag{3}$$

Our goal is to find reasonable sufficient conditions under which the above theorem gives a stable block diagonalization. In other words, we would like to choose a nonsingular matrix S such that $S^{-1}AS = \text{diag}(D_{11},\ldots,D_{qq})$ with "$\kappa(S)$ small". The following results due to Demmel [3] clarifies the choice of such a matrix S. Let us recall the following results from [3] :

- $S = [S_1|\ldots|S_q]$ is a partition of S in q block columns
- S^i is the subspace spanned by the columns of S_i
- $\mathcal{V}_{ij} = \mathcal{V}(S^i, S^j)$ is the angle between S^i and S^j
- $\mathcal{V}_i = \mathcal{V}(S^i, \text{span}_{j\neq i}\{S^j\})$ is the angle between S^i and the sum of all the other subspaces.

Theorem 2. *Suppose that $S_i^*S_i = I$ for $i = 1,\ldots q$, then*

$$\kappa(S) \leq \sqrt{q}\,\kappa(S_{OPT}) \tag{4}$$

$$\max_i(\csc \mathcal{V}_i + \sqrt{\csc^2 \mathcal{V}_i - 1}) \leq \kappa(S) \leq \sqrt{q}\sqrt{\sum_{i=1}^{q} \csc^2 \mathcal{V}_i} \tag{5}$$

where S_{OPT} *is any matrix* S *whose condition number is as small as possible.*

In the case where $q = 2$*, let* R *be the solution of the Sylvester equation* $T_{11}R - RT_{22} = T_{12}$ *and let* $S = \begin{pmatrix} I & -R \\ 0 & I \end{pmatrix}$*. Then* $S^{-1}TS = diag(T_{11}, T_{22})$*. Moreover*

- $\kappa(S) \geq \cot \mathcal{V}/2$ *with* $S = [S_1|S_2]$ *and* $\mathcal{V} = \mathcal{V}(S_1, S_2)$*.*
- *If* $S_i^* S_i = I$*,* $i = 1, 2$*, then* $\kappa(S) = \kappa(S_{OPT}) = \cot \mathcal{V}/2 = \sqrt{\|R\|^2 + 1} + \|R\|$*.*
- $1/\sin \mathcal{V} \equiv \csc \mathcal{V} = \|S\|$*.*

Theorem 2 can be used to determine a criterion for selecting the different blocks $D_1, \ldots, D_q$. Indeed, let (λ_k, u_k) with $\|u_k\| = 1$, $k = 1, \ldots, n$ denote the eigenelements of A. Assume that the eigenvalues λ_i and λ_j are not in the same block and that

$$\cos \mathcal{V}(u_i, u_j) = |u_i^* u_j| \geq 1 - \eta, \quad 0 \leq \eta \ll 1 \tag{6}$$

then

$$\csc \mathcal{V}(u_i, u_j) \geq (2\eta - \eta^2)^{-\frac{1}{2}}. \tag{7}$$

From (5) and (7) we have

$$\kappa(S) \geq \max_k \csc \mathcal{V}_k \geq \csc \mathcal{V}(u_i, u_j) \approx 1/\sqrt{2\eta}$$

which means that $\kappa(S)$ must be large. We thus have the simple rule for choosing the blocks:

In order that $\kappa(S)$ *remains reasonably small, it is necessary that two eigenvectors* u_i *and* u_j*, corresponding to the eigenvalues* λ_i *and* λ_j *of different blocks, must be such that* $|u_i^* u_j| < 1 - \eta$ *where* η *is a small tolerance fixed by the user.*

This step amounts to applying permutations to the matrix T in (2) in such a way that the resulting diagonal blocks T_{kk}, for $k = 1, \ldots, q$ satisfy

$$\forall \lambda_i \in \lambda(T_{kk}) \ \exists \lambda_j \in \lambda(T_{kk}) \ : \ |u_i^* u_j| \geq 1 - \eta \tag{8}$$

$$\forall \lambda_i \in \lambda(T_{kk}) \ \forall \lambda_j \notin \lambda(T_{kk}) \ : \ |u_i^* u_j| < 1 - \eta \tag{9}$$

where $\lambda(T_{kk})$ denotes the set of eigenvalues of the block T_{kk}.

Because of space limitation, we cannot go into details in the description of the algorithm. Briefly, the idea here is that the eigenvectors of A corresponding to the eigenvalues of two different blocks T_k and T_l must be, in the ideal case, linearly independent. We make this condition less severe by introducing the parameter η. Let us summarize the different steps of the algorithm. The first step is to reduce A into upper triangular form using the Schur decomposition. This Schur decomposition must be carried out in a way such that the block diagonals T_{ii} must observe the rule described above (conditions (8) and (9)). The block diagonalization of T, that is, the elimination of T_{ij}, $i < j$ is carried out using theorem 2. At each step of the elimination, we have to block diagonalize a matrix of the form $E = \begin{pmatrix} E_{11} & E_{12} \\ 0 & E_{22} \end{pmatrix}$ where the eigenvectors of E corresponding to the eigenvalues (of E_{11} and E_{22}) satisfy the conditions (8) and (9). The block diagonalization uses the solution Y of the Sylvester equation

$$E_{11}Y - YE_{22} = E_{12}. \tag{10}$$

It is easy to see that (10) reduces to an upper triangular linear system [3] . If E_{11} and E_{22} have disjoint spectra, then the equation (10) has a unique solution. But the conditions (8) and (9), with small η, are in general sufficient to guarantee the existence (but not necessarily the uniqueness) of Y.

At the end of this step, the matrix T is block diagonalized and the matrix A can be written in the form $A = S\ \mathrm{diag}(D_{11}, \ldots, D_{qq})\ S^{-1}$ with $S = [S_1|\ldots|S_q]$. Now, in order that $\kappa(S)$ be not far from $\kappa(S_{\mathrm{OPT}})$, (see inequality 4 in theorem 2), we orthonormalize each block columns S_i, and therefore a new block diagonalization is obtained. If the user wishes another block diagonalization with smaller $\kappa(S)$, then the number of blocks q is reduced by merging the two blocks T_{i_0,i_0} and T_{j_0,j_0} for which csc $\mathcal{V}_{i_0j_0} = \max_{i\neq j} \mathcal{V}_{ij}$ and a block diagonalization of the new T is accomplished. The choice of T_{i_0,i_0} and T_{j_0,j_0} is not necessarily optimal but it has the advantage of being simple and as shown in the numerical test section, the method gives good decompositions.

Other possible decompositions based on the Jordan canonical form [6, 8] may also be used but the cost involved in these decompositions is high. Our approach is similar to the one proposed in [1] with the following difference: we use the conditions (8) and (9) that allow us to systematically block diagonalize T. In [1], the authors do not impose any conditions on the diagonal blocks T_{kk}, $k = 1, \ldots, q$ obtained from the Schur decomposition. As a consequence, they try to solve, at each iteration, a Sylvester equation that eliminates the upper part (as in (10)). If this solution is ill-conditioned, then the elimination is abandoned and a new larger block is formed.

3 Spectral portrait of matrices

In this section we assume that we have found the block diagonalization (1) whose condition number satisfies a tolerance provided by the user. We want to use it for computing the ϵ-pseudospectrum of A.

A number z in the complex plane is an eigenvalue of the matrix A if and only if $\sigma_{min}(A - zI) = 0$. Similarly, z is near to an eigenvalue of A if and only if $\sigma_{min}(A - zI)$ is near to zero. It is known [5] that, contrary to the smallest eigenvalue, the smallest singular value of a matrix fully quantifies the notion of nearness to singularity. If $z \in \mathbb{C}$, then

$$\sigma_{min}(A - zI) = \min\{||\Delta|| : z \text{ is an eigenvalue of } A + \Delta\}. \tag{11}$$

For each $\epsilon \geq 0$, the ϵ-pseudospectrum of A, denoted hereafter by $\Lambda_\epsilon(A)$, is the set of eigenvalues of all perturbed matrices $A + \Delta$ with $||\Delta|| \leq \epsilon$.

$$\begin{aligned} \Lambda_\epsilon(A) &= \{z \in \mathbb{C} : z \text{ is an eigenvalue of } A + \Delta;\ ||\Delta|| \leq \epsilon\} & (12)\\ &= \{z \in \mathbb{C} : \sigma_{min}(A - zI) \leq \epsilon\}. & (13) \end{aligned}$$

The spectral portrait of A is the "picture" of its ϵ-pseudospectrum. More precisely, it is the representation of the function $sp_A(z) = \log_{10}(\sigma_{min}(zI - A))$ by means of level curves in the complex plane.

Note that $\sigma_{min}(zI - A) = 1/\|R(A,z)\|$ where $R(A,z) = (A - zI)^{-1}$ is the resolvent of the matrix A at the point z. Thus, the spectral portrait shows the behavior of the norm of the resolvent.

Let us now consider the decomposition (1). From the inequalities

$$\frac{1}{\kappa(S)}\,\sigma_{min}(D - zI) \le \sigma_{min}(A - zI) \le \kappa(S)\,\sigma_{min}(D - zI) \qquad (14)$$

we easily obtain the following proposition

Proposition 3. *Let $\epsilon \ge 0$, $\epsilon_1 = \epsilon/\kappa(S)$, and $\epsilon_2 = \epsilon\kappa(S)$, then*

$$\Lambda_{\epsilon_1}(D) \subset \Lambda_\epsilon(A) \subset \Lambda_{\epsilon_2}(D) \qquad (15)$$

$$\Lambda_\epsilon(D) = \cup_{i=1}^{q} \Lambda_\epsilon(D_i). \qquad (16)$$

The inclusions (15) mean that the spectral portrait of A can be approximated by computing $\Lambda_{\epsilon_1}(D)$ and $\Lambda_{\epsilon_2}(D)$. These sets are much easier to compute as (16) shows. The ϵ-pseudospectrum and hence the spectral portrait of D can be obtained from each $\Lambda_\epsilon(D_i)$, $i = 1,\ldots,q$. Because the size of the matrices D_i, $i = 1,\ldots,q$ is small, reliable methods based on the SVD can be used to compute their ϵ-pseudospectrum. Moreover the computations of $\Lambda_\epsilon(D_i)$, $i = 1,\ldots,q$ are fully independent and can thus be done in parallel.

Another concept for studying the spectral properties of a matrix is through the field of values [7]. The field of values of A is defined by

$$F(A) = \{u^*Au : \; u \in \mathbb{C}^n, \; \|u\| = 1\} \qquad (17)$$

The field of values is easy to determine [7, p.34] . It contains the spectrum of A and is related to the ϵ-pseudospectrum as shown by the following proposition

Proposition 4. *For each $\epsilon \ge 0$ we have*

$$\Lambda_\epsilon(A) \subset F(A) + \Delta_\epsilon \qquad (18)$$

$$\Lambda_\epsilon(A) \subset F(D) + \Delta_{\epsilon\kappa(S)} \qquad (19)$$

where $\Delta_\epsilon = \{z \in \mathbb{C} : \; |z| \le \epsilon\}$, and

$$F(D) = Co(\cup_{i=1}^{q} F(D_i)) \qquad (20)$$

where $Co(\cup_{i=1}^{q} F(D_i))$ denotes the convex hull of $\cup_{i=1}^{q} F(D_i)$.

From propositions 3 and 4 we see that

$$\Lambda_\epsilon(D) \subset F(A) + \Delta_{\epsilon_2} \;\text{ with } \epsilon_2 = \epsilon\,\kappa(S). \qquad (21)$$

In a similar way to proposition 3, (20) means that the field of values of the matrix D can be determined by those of the matrices D_i, $i = 1,\ldots,q$.

4 Numerical results and complexity analysis

We have already mentioned that the goal of the block diagonalization is to speed up the computation of $\Lambda_\epsilon(A)$ which we actually have observed in our experiments. Let us compare the complexity of our approach with the standard one. Recall that the computation of $sp_A(z)$ for a matrix A of order n, by the SVD method, is of order $O(n^3)$ for each z on a grid in the complex plane. Suppose that the grid is discretized using $N \times N$ points, then the cost of $\Lambda_\epsilon(A)$ on this grid is of order $O(N^2n^3)$. Now suppose we have q blocs D_i, $i = 1, \ldots, q$, each of which is of order n/q. Then the sequential cost of our approach is of order $O(N^2q(\frac{n}{q})^3)$ plus the cost of one bloc-diagonalization of A. Moreover if we use the fact that the computation of $\Lambda_\epsilon(D_i)$, $i = 1, \ldots, q$, can be done in parallel, then the factor $O(N^2q(\frac{n}{q})^3)$ in the sequential cost becomes $O(N^2(\frac{n}{q})^3)$ and may still be improved. In fact, instead of computing $\Lambda_\epsilon(D_i)$ on the whole grid , we can restrict the domain of computation to the points of a sub-grid. Theoretically, suppose that the domain discretized in $N \times N$ points can be divided into q non-overlapping sub-domains discretized in $\frac{N \times N}{q}$ points, then the parallel complexity for $\Lambda_\epsilon(D_i)$, $i = 1, \ldots, q$ reduces to $O((\frac{N}{q})^2(\frac{n}{q})^3)$.
We have tested our algorithm on several matrices, but because of space limitation, we report the results for only one matrix : GRCAR (see [9]) of order $n = 50$. We give the condition number $\kappa(V)$ of the matrix of its eigenvectors and compare its spectral portrait and that of the block diagonal matrix D in (1) using different block structure.

Example of the GRCAR matrix

Order of $A = 50$
V=matrix of eigenvectors of A
$\kappa(V) = 2.0e+8$

Number of blocks	$\kappa(S)$
13	6188
12	5487
11	3903
10	3409
9	2167
8	1976
7	1684
6	1469
5	1372
4	1359
3	702.8
2	308.5
1	1.0

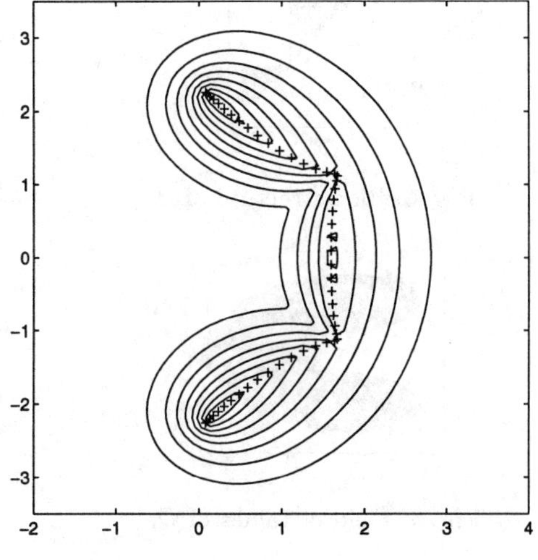

Fig. 1. Spectral portrait of A

- **Decomposition in $q = 3$ blocs : $\kappa(S) = 702.8$**

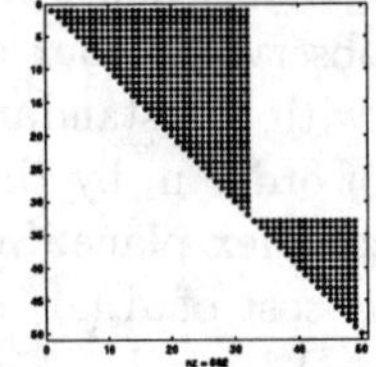

Fig. 2. Bloc-stucture of D

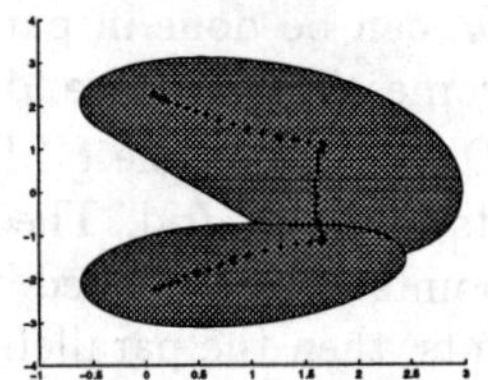

Fig. 3. Field of values of D_i

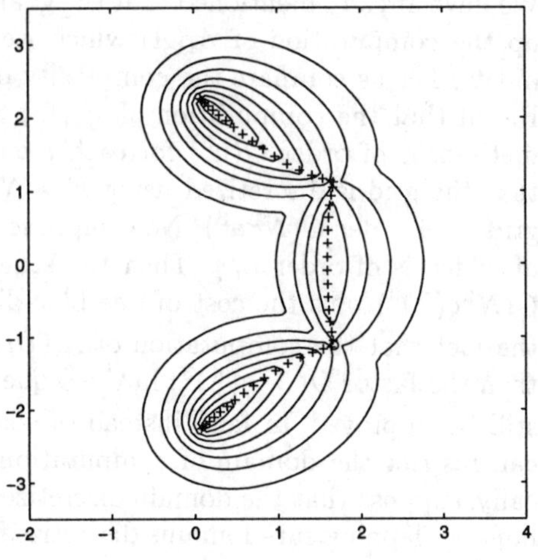

Fig. 4. Spectral portrait of D

- **Decomposition in $q = 6$ blocs : $\kappa(S) = 1469$**

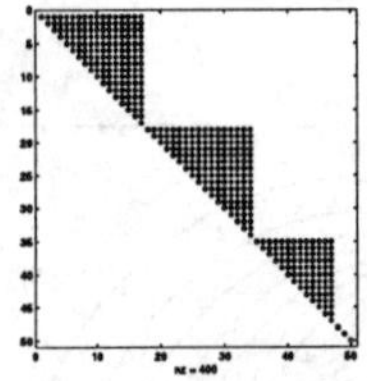

Fig. 5. Bloc-structure of D

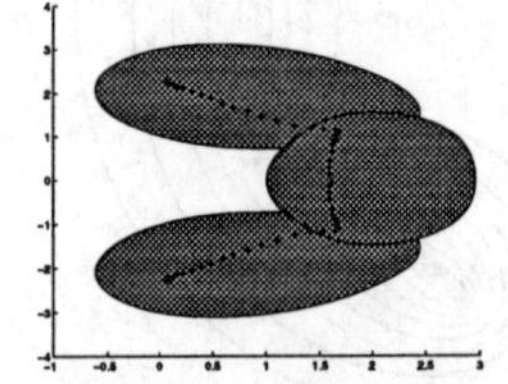

Fig. 6. Field of values of D_i

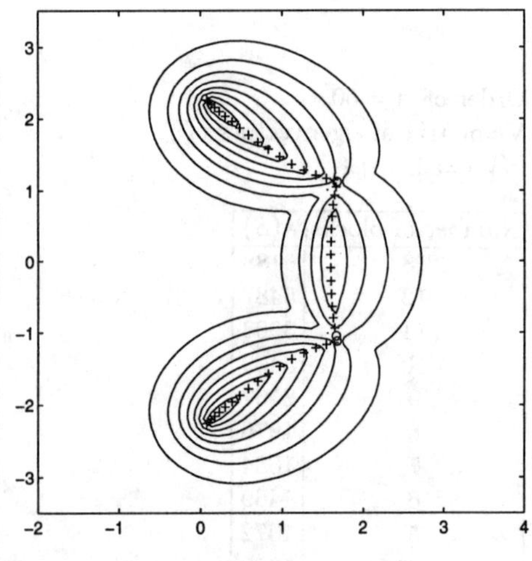

Fig. 7. Spectral portrait of D

- **Decomposition in $q = 13$ blocs : $\kappa(S) = 6188$**

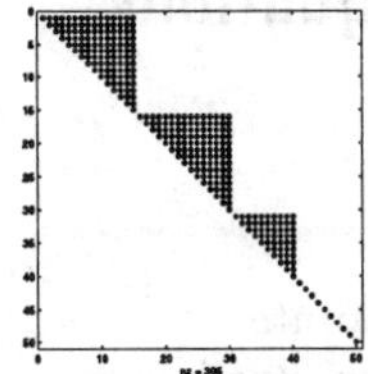

Fig. 8. Bloc-structure of D

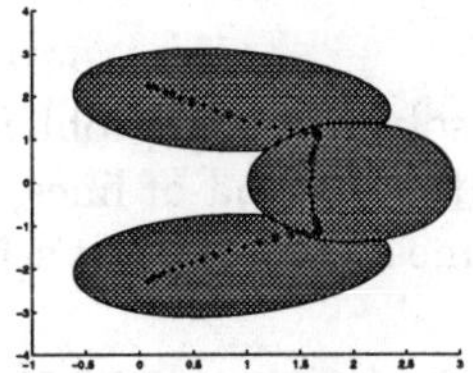

Fig. 9. Field of values of D_i

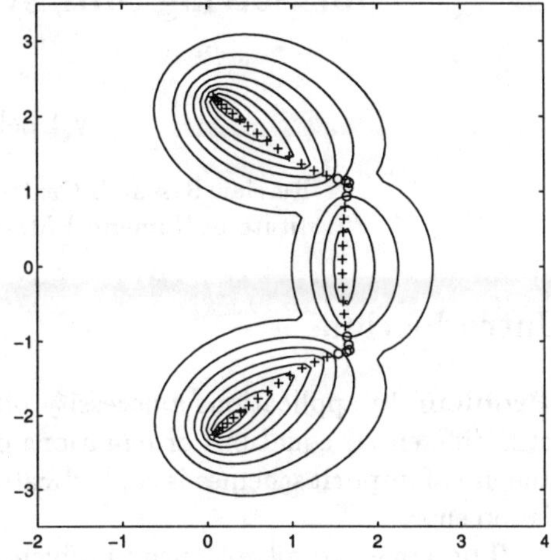

Fig. 10. Spectral portrait of D

References

1. Bavely, C. A., Stewart, G. W.: An algorithm for computing reducing subspaces by block diagonalization. SIAM J. Numer. Anal. Vol. 16, No. 2, (1979) 359–367.
2. Godunov, S. K.: Spectral portrait of matrices and criteria of spectrum dichotomy. In Computer arithmetic and enclosure methods, J. Herzberger and L. Athanassova, eds., Oldenburg, 1991, North-Holland.
3. Demmel, J.: The condition number of equivalence transformations that block diagonalize matrix pencils. SIAM J. Numer. Anal. Vol. 20, No. 3, (1983) 599–610
4. Demmel, J. W.: Computing stable eigendecompositions of matrices. Lin. Alg. Applic, 79 (1986) 163–193.
5. Golub, G. H., Van Loan, C. F.: Matrix computations, 2nd ed., The Johns Hopkins University Press, Baltimore, 1989.
6. Golub, G. H., Wilkinson, J. H.: Ill-conditioned eigensystems and the computation of the Jordan canonical form. SIAM Review, Vol. 18, No. 4, (1976) 578–619.
7. Horn, R. A.,Johnson, C. R.: Topics in matrix analysis. Cambridge University Press, Cambridge, 1991.
8. Kågström, Bo., Ruhe, Axel.: An algorithm for numerical computation of the Jordan normal form of a complex matrix. ACM Trans. Math. Soft. Vol. 6, No. 3, (1980) (1979) 398–419.
9. Trefethen, L. N.: Pseudospectra of matrices. In Numerical analysis 1991, D. F. Griffiths and G. A. Watson, eds., Longman, New York, 1992.
10. Trefethen, L. N.: Pseudospectra of linear operators. In ICIAM'95 : Proceedings of the third international congress on industrial and applied mathematics. Akademie-Verlag, Berlin.

Explicit Difference Schemes with Variable Time Steps for Solving Stiff Systems of Equations

V.I.Lebedev

Russian Research Centre "Kurchatov Institute"
Institute of Numerical Mathematics, Moscow, Russia

Introduction

Problem. In applications a necessity often arises to solve Cauchy problems for stiff differential equations or equations derived from the method of lines, when the use of implicit schemes is sophisticated, and the time step in explicit schemes is too short.

The core. To solve Cauchy problems for stiff differential equations or equations derived from the method of lines, were proposed [1,2,3] explicit stable difference schemes with the steps, varying in time. The conditions of stability of the optimal algorithm of choice of steps were investigated. The algorithm provides a drastic improvement compared with known explicit schemes (up to 30 000 000). A special algorithm is used for interior and boundary layers. Algorithms are based on the properties of T-sequence of roots of Chebyshev polynomials.

The advantages of the proposed method. The methods are convenient to implement on parallel and pipeline computers, is suitable to solve highly multidimensional problems, problems with nonlinear, nonsymmetric, nondefinite operators, for unconstrained optimization. The use of explicit stable difference schemes gives us a possibility to almost absolutely parallelize and vectorize computations.

The code. The method is used in FORTRAN code **DUMKA**. Computations were conducted on CONVEX, CYBER, ELBRUS, BESM and different PCs.

The method is quite flexible. The use in applications is simple, since **DUMKA** user should just to write a subroutine for right-hand side of equations and a subroutine, estimating the greatest in modulus negative eigenvalue of Jacobean.

It seems advantageous to use **DUMKA** for neutron transport problems such as multidimensional nonlinear heat transfer problems and unstationary kinetic problems with delayed neutrons.

1 Explicit stable method for stiff systems

Let's write Cauchy problem for system of ordinary differential equations as

$$\frac{dU}{dt} = f(U,t), \tag{1}$$

$$U|_{t=t_0} = U_0, \tag{2}$$

where $U = (U_1, U_2, \ldots, U_n)$, $f = (f_1, f_2, \ldots, f_n)$.

Let

$$J = \|\frac{df_i}{dU_j}\| \tag{3}$$

be Jacobean matrix, t_k - time-mesh, h_k- steps in time.

Let first the spectrum of J be real and λ be an upper estimate of modulus of negative eigenvalue of Jacobean matrix; the number

$$COU = 2/\lambda \tag{4}$$

we will call courant.

The approximate solution $U_k = U(t_k)$ of the Cauchy problem (1) we will find by the following explicit difference rule

$$\begin{aligned}
&y_{k+1/2} = U_k + h_{k+1} f(U_k, t_k), \\
&t_{k+1/2} = t_k + h_{k+1}, \\
&y_{k+1} = y_{k+1/2} + h_{k+1} f(y_{k+1/2}, t_{k+1/2}), \\
&U_{k+1} = y_{k+1} + \gamma_{k+1} h_{k+1} (f(U_k, t_k) - f(y_{k+1/2}, t_{k+1/2})), \\
&t_{k+1} = t_{k+1/2} + h_{k+1}, \\
&k = 0, 1, \ldots, K-1,
\end{aligned} \tag{5}$$

where h_k, γ_k are parameters chosen in a certain way.

In explicit Euler's rule with the constant step h:

$$U_{k+1} = U_k + hf(U_k, t_k), \tag{6}$$

on the step h by the stability condition the constrain is imposed

$$h \leq COU. \tag{7}$$

The method (4) is implemented in the code **DUMKA**.

The code **DUMKA** from given tolerance EPS derives a series of steps h_{k+1}, $h_{k+2}, \ldots$, h_{k+N}, which guarantee stability and the given tolerance. So the transition operator of the difference rule after N steps forms out the operator Chebyshev polynomial of the first kind [1,2,3] and Zolotarev polynomial of the first kind and of second type [4]. The scheme (1.5) allows us to calculate in the real arithmetic the polynomial transition operator with real and complex roots. It's steps are equal to the inverse values of the halve-sums of the roots of these polynomials.

Let $N = 2K$, $g = h/cou$, $V_N = h^N/cou$,

when

$$h^N = \frac{1}{K}\sum_{i=1}^{K} h_{k+i} \quad (N = 2^k * 3^m, k = 1, \ldots, 14, m = 0, 1, \ldots, 6) \tag{8}$$

be the average step. The values of N, h^N depend on the smoothness of the solution, the tolerance EPS and g.

So are the abilities of **DUMKA** code, that $N \leq 47\,775\,744$, and

$$h^N \leq 3 * 10^7 * COU. \tag{9}$$

The equality in (9) can be reached on the intervals of integration where the solution $U(t)$ is asymptotically linear in t. From the comparison of (7) and (9) follows, that **DUMKA** will solve the Cauchy problem by explicit rule (4) with steps in time $0.4358 * 10^7$ times greater in average then those of explicit Euler's rule (6).

The choice of the mode of implementation procedure of the code due to the spectrum type is determined by ascribing the values of variables: $NP(1 \leq NP \leq 8)$, COU, $SP(SP > -1)$, $SL(SL > -1)$, NK; when NP=1, 2, 3, 4, 6 COU is determinate by formula (4), when NP=5,7,8 by formula:

$$COU = 2/b. \tag{10}$$

Written below is a list of typical problems for ODE and PDE, transformation we recommend to do, modes of implementation (NP, SP, COU, SL, NK), the appearance of the limit (with a unit deviation on spectrum) transition polynomial operator $P_N(z)$ for a cycle of N steps and appearance of the value V_N- the benefit in comparison with the method (6) when $h = COU$.

Let $z = x + iy$, $s = 1 + 2z/\lambda$, and N is big.

The algorithm implemented is able to account on some specific complex spectrum domain configurations of J; stability domain $|P_N(z)| \leq 1$ contains the following sets:

In items 1, 2 NP=1, 2, 3, SL=1,

$$P_N(z) = T_N(s), V_N = N. \tag{11}$$

1.

$$y = 0, \quad -\lambda \leq x \leq 0; \tag{12}$$

SP=0.

2. Inside ellipsis with focus $(0, -\lambda)$ on the real axis with (0,0) belonging to the boundary of the ellipsis and with smaller semiaxis $l = b\frac{\lambda}{2}$:

$$\left(\frac{x}{1 + b^2/4} + \frac{\lambda}{2}\right)^2 + \frac{y^2}{b^2} = \frac{\lambda^2}{4}. \tag{13}$$

$SP = b^2$, $SL = 1$ (see (2.0)).

3. Domain in the neighborhood of the imaginary axis

$$x = 0, \quad |y| \leq b, \quad b > 0. \tag{14}$$

NP=7, 8; $P_4(z) = Q_4(2\sqrt{2}z)$, $Q_4(t) = 1 - t + t^2/2 - t^3/6 + t^4/24$, $V_N = (\sqrt{2}/4)(2/b)$ by NP=7.

In items 4, 5 NP=4, SL=1

$$P_{2N}(z) = T_N(v), \quad v = \frac{2s^2 - 1 + SP}{1 + SP}, \quad V_{2N} = \frac{2N}{1 + SP}. \tag{15}$$

4. The neighborhood of two intervals

$$y = 0, \quad x \in [-\frac{\lambda}{2}(1-b), 0] \bigcup [-\lambda, -\frac{\lambda}{2}(1+b)], \quad 0 \leq b < 1. \tag{16}$$

SP=-b^2.

5. The neighborhood of the "crux"

$$\{y = 0, -\lambda \leq x \leq 0\} \bigcup \{x = -\frac{\lambda}{2}, |y| \leq b\frac{\lambda}{2}\}, \quad 0 \leq b \leq 2. \tag{17}$$

SP=b^2.

In items 6, 7 NP=5,

$$P_{2N}(z) = T_N(v), \quad v = \frac{2t^2 + 1 - d^2}{1 + d^2}, \quad d = c/b, \quad t = (z + c)/b. \tag{18}$$

NP=5, SP=c, SL=1.

6. The neighborhood of the "crux"

$$\{y = 0, |x + c| \leq \delta\} \bigcup \{x = -c, |y| \leq b\}, \quad 0 < \delta \leq c < b. \tag{19}$$

7. Of two vertical intervals

$$\{x = -c, \delta \leq |y| \leq b\}, \quad b > c, \quad 0 \leq \delta. \tag{20}$$

In items 8 – 13 NP=6

$$P_{4N}(z) = T_N(v), \quad v = \frac{2w^2 - 1 + SL}{1 + SL}, \quad w = \frac{2s^2 - 1 + SP}{1 + SP}, \tag{21}$$

$$V_{4N} = \frac{4N}{(1 + SL)(1 + SP)},$$

and $z_j = (s_j - 1)\frac{\lambda}{2}$, when $s_j, j = 1, 2, 3, 4$ are

$$\pm\sqrt{(1 - SP \pm i(1 + SP)\sqrt{SL})/2}.$$

8. The neighborhood of the "cruxes" (17) and piece of hyperbola:

$$\{(x + \frac{\lambda}{2})^2 - y^2 = \frac{(1 - SP)\lambda^2}{8}\}, \tag{22}$$

with ends in the poins $z_j, j = 1, 2, 3, 4$. $SP = b^2$, $SL > 0$.

9. The neighborhood of two the "cruxes"

$$\{y=0; -\lambda \le x \le 0\} \bigcup \{x = -\frac{\lambda}{2}, |y| \le \frac{\lambda}{2}\} \bigcup$$

$$\{-\frac{\lambda}{2}\left(1 \pm \exp{(\pm\frac{i\pi}{4})r}\right), \quad 0 \le r \le c < \sqrt{2}\}; \tag{23}$$

(23) is a partial case (17), (22) by $b = 1$. SP=1, SL=c^4.

10. The neighborhood of two intervals (16) and piece of a hyperbola (22). SP=-b^2, SL=c^4.
11. By SP=1, $SL < 0$, $|SL| < 1$ — the neighborhood of four intervals:

$$\{y=0, \quad x \in [-\frac{\lambda}{2}(1-ss), 0] \bigcup [-\lambda, -\frac{\lambda}{2}(1+ss)]\} \bigcup$$

$$\{x = -\frac{\lambda}{2}, \quad y \in [\frac{\lambda}{2}(1-ss), \frac{\lambda}{2}] \bigcup [-\frac{\lambda}{2}, -\frac{\lambda}{2}(1-ss)]\}$$

when $ss = |SL|^{1/4}$.
In items 12, 13 z_j are real and $-\lambda < z_1 < z_2 < z_3 < z_4 < 0$.

12. $0 < SP < 1$, $-1 < SL < 0$ — the neighborhood of two intervals and of the neighborhood of the "cruxes":

$$\{[-\lambda, z_1] \bigcup [z_4, 0]\} \bigcup$$

$$\{y=0, \quad x \in [z_2, z_3]\} \bigcup \{x = -\frac{\lambda}{2}, |y| \le \sqrt{SP}\frac{\lambda}{2}\}.$$

13. $-1 < SP < 0$, $-1 < SL < 0$ — the neighborhood of four intervals:

$$[-\lambda, z_1] \bigcup [z_2, -\frac{\lambda}{2}(1+\sqrt{|SP|})] \bigcup [-\frac{\lambda}{2}(1-\sqrt{|SP|}), z_3] \bigcup [z_4, 0].$$

The choice of the mode of implementation procedure of the code due to the spectrum type (12) – (23) is determined by ascribing the values of variables: $NP (1 \le NP \le 8)$, COU, SP, SL, NK. Written below is a list of typical problems for ODE and PDE.

14. NP=1, 2, 3 SP=b, where $b = 0$ or small number for solution evolutionary (parabolic) problems. In this problems the spectrum as a rule belongs either to a segment $[-\lambda, 0]$ (see (12)) or to the interior of a narrow ellipse (13) as it is for the problem:

$$\frac{\partial u}{\partial t} = D\Delta u + \sum_i b_i \frac{\partial u}{\partial x_i} + au + f(x,t). \tag{24}$$

15. NP=4, SP=$-b^2$ for evolutionary problems when the spectrum is contained in two rather distant intervals of equal length on the real axis (see (16)). By $NP = 6$, $SP = -b^2$, $SL > 0$ the spectrum is contained in two rather distant intervals of equal length on the real axis (see (16)), and piece of a hyperbola (22).

This is the case for the problems of chemical cinetics or Van der Pols equation.

16. NP=7, 8, SP=0, COU=2/b for the problems with a purely imaginary spectrum (skew symmetric Jacobean) (see (14)) or with the spectrum close to the imaginary axis,say, for the equation

$$\frac{\partial u}{\partial t} = a\frac{\partial u}{\partial x} + bu + f(x,t). \tag{25}$$

17. Problems describing oscillatory processes with dissipation (friction) of the following type

$$\frac{\partial^2 u}{\partial t^2} + 2\nu\frac{\partial u}{\partial t} = F(u,t), \quad \nu > 0, \tag{26}$$

$$u|_{t=t_0} = v_0, \quad \frac{\partial u}{\partial t}|_{t=t_0} = v, \tag{27}$$

one should preliminary transform to the appearance (1), (2) for the variables $u = (w, v)$:

$$\frac{\partial v}{\partial t} = aw,$$

$$\frac{\partial w}{\partial t} = \frac{1}{a}F(v,t) - 2\nu w, \tag{28}$$

$$v|_{t=t_0} = v_0, \quad w|_{t=t_0} = \frac{1}{a}v_1. \tag{29}$$

A value of 'a' is choosen to balance the elements of Jacoby matrix of the system (25). For instance, if Jacoby matrix for $f(w,v)$ has a real spectrum $(\tilde{\lambda}_1, \dots, \tilde{\lambda}_n)$ and $\tilde{\lambda} = max\tilde{\lambda}_i$, then the following numbers will be eigenvalues of matrix of a system (25): $\gamma_i = -\nu \pm \sqrt{\nu^2 + \tilde{\lambda}_i}$. So if $\tilde{\lambda} \gg \nu^2$,we let $a \approx \sqrt{\tilde{\lambda}}$.

18. If for system (27) $\tilde{\lambda} \geq -\nu^2$, then (see (4))

$$\lambda = \nu + \sqrt{\nu^2 + \tilde{\lambda}}, \tag{30}$$

and we let NP=1,2,3 SP=0 (see (12), (11)).

19. NP=4, SP=b^2, if for a system (27) max$|Im\sqrt{\nu^2 + \tilde{\lambda}_i}| \leq b\lambda/2$ with $0 < b < 2$ (see (17)), then

$$\lambda = \nu + max\ Re\sqrt{\nu^2 + \tilde{\lambda}},$$

20. $NP = 5$, $SP = \nu$, $COU = 2/b$ (see (19),(20)), if for a system (27) $\tilde{\lambda} \gg \nu^2$.For example, for wave equation

$$\frac{\partial^2 u}{\partial t^2} + 2\nu\frac{\partial u}{\partial t} = a^2 \Delta u + f(x,t), \tag{31}$$

or for the plate oscillations equation

$$\frac{\partial^2 u}{\partial t^2} + 2\nu\frac{\partial u}{\partial t} + a^2\Delta^2 u = f(x,t). \tag{32}$$

21. $NP = 6$, (see 8.–13.) for the problems with complex spectrum which have no localization of the type (12) – (20).

These possibilities allow us to successfully integrate by explicit rules many of unstationary problems of mathematical physics.

2 Brief description of the code DUMKA

The code is oriented towards the solution of large ODE and PDE multidimensional problems of mathematical physics approximated by the method of lines (that is, with derivatives in time approximated by differences).

To use the code **DUMKA** one should write a subroutine **FUN**, evaluating right - hand side of the system of equations (1), subroutine **COUR**, evaluating **COU**, and main calling sequence (loop 1), the schematic structure of which together with **DUMKA** calling sequence at $T = t_0, \;\; U = U_0, \;\; H = H_0$ is as follows

```
1  CONTINUE
   ...
   CALL DUMKA (FUN, COUR, ST,n,T,U, H,HL,P,CU,EPS,ED,
                       SK,SP,SL,NP,NS,NK,KL, LU,Z1,Z2 )
   ...
   IF (T . LT. ED ) GO TO 1
```

Calling sequences of subroutines **FUN** and **COUR** are the following

```
CALL FUN (n, T, TO, HL, U, Z1)
CALL COUR (NP, n, T, U, SP, COU).
```

The code **DUMKA** does one loop of computations calling once the subroutine COUR and the subroutine **FUN** quite a lot. After the completion **DUMKA** computes new values for **T, U, H**. The codes **DUMKA,FUN, COUR** use three arrays **U(n), Z1(n), Z2(n)**.

On 32-digit computers computations should be performed in double accuracy.

To use **DUMKA** one should know **COU** or some good lower estimate for it, and this is equivalent to good upper estimate $\overline{\lambda}$ for $\max(-\lambda_i)$ where λ_i is an eigenvalue of Jacobean matrix (2) at $\mathbf{T} = t_0$. Sometimes this estimate may be known exactly or from some physical considerations; generally - by the use of some generalizations of Gershgorin circles, Ostrowsky and Fan Ky theorems and so on.

To estimate the predicted step **H** for the next loop **DUMKA** uses by the will of user one of functions: **ST(n,Z), STC(n,Z), STM(n,Z), STMC(n,Z), STQ(n,Z), STQC(n,Z)** (to use that subroutines one should substitute in the call of **DUMKA** the formal argument **ST** with the name of the corresponding subroutine).

On the last step of N steps of the algorithm (5) **DUMKA** estimates the value of **H** - the next step being predicted by the formula

$$H = (q_1 S^r + q_2 (\overline{S})^r)^{1/r}, \tag{33}$$

where $\overline{S}$ is value being predicted at the previous step, $q_1 = 0.92$, $q_2 = 0.08, \;\; r = 0.5$.

By **NP** $= \mathbf{1,4,5,6}$ and **NP** $= \mathbf{2,3}$ if $g > 365$

$$S = HL \cdot (3 \cdot EPS/D), \tag{34}$$

where the value D being computed as follows algorithms. Let

$$\Delta_{iN} = \frac{f_i(U_N, t_N) - f_i(y_{N-1/2}, t_{N-1/2})}{p|U_i(t_N)| + 1 - p + 10^{-13}}, \tag{35}$$

where $0 \leq p \leq 1$ - weight function, characterizing absolute (p =0), relative (p =1) or some other accuracy, let **EPS** is a small parameter, characterizing the accuracy of the local approximation of du_i/dt by the divided difference, Then for subroutines **ST,STC**

$$D = \sum_{i=1}^{n} |\Delta_{iN}| CS_i, \tag{36}$$

for subroutines **STQ,STQC**

$$D = \Big(\sum_{i=1}^{n} \Delta_{iN}^2 CS_i\Big)^{0.5}, \tag{37}$$

$CS_i = 1/n$ for **ST, STQ**; $CS_i \geq 0, \sum_{i=1}^{n} CS_i = 1$ for **STC, STQC**;
for subroutines **STM,STMC**

$$D = \max_i |\Delta_{iN}| CS_i, \tag{38}$$

$CS_i = 1$ for **STM** , $CS_i \geq 0$ for **STMC**. When using **STC** and **STMC, STQC** one should define an array **CS** as

```
COMMON /PCS/ CS(n)
```

and full it before the first call for **DUMKA**.

If **NP** =**2** and $1.393 < g < 365$ and **NP=3**, $324. < g < 365$, the step is predicted as (33) by

$$S = HL \cdot \sqrt{\frac{15EPS}{D}}, \tag{39}$$

where D is defined from (36) — (38) by $t = t_N$ with

$$\Delta_{iN} = \frac{f_i(U(t), t) - 2f_i(U(t - HL), t - HL) + f_i(U(t - 2HL), t - 2HL)}{p|U_i(t)| + 1 - p + 10^{-13}}; \tag{40}$$

by **NP=3** and $g \leq 324$.

$$S = HL \cdot \left(\frac{16EPS}{D}\right)^{1/3}, \tag{41}$$

where Δ_{iN} in (36) – (38) is module 3-s difference quotients by $f(U, t)$.

Let a general Cauchy problem for a system of ODE could be divided into two parts:
relatively ' soft ' with respect to the variables $U = (U_1, \ldots, U_n)$

$$\frac{dU}{dt} = F(U, V, t), \tag{42}$$

$$U|_{t=t_0} = U_0, \tag{43}$$

and relatively stiff with respect to the variables $V = (V_1, \ldots, V_m)$

$$\frac{dV}{dt} = G(U, V, V, t), \tag{44}$$

$$V|_{t=t_0} = V_0, \tag{45}$$

which can be easily resolved with respect to V together with (42) by the use of the following explicit–implicit method

$$\begin{gathered}
y_{k+1/2} = u_k + h_{k+1} F(u_k, v_k, t_k), \\
t_{k+1/2} = t_k + t_{k+1}, \\
w_{k+1/2} = v_k + h_{k+1} G(y_{k+1/2}, w_{k+1/2}, v_k, t_{k+1/2}), \\
y_{k+1} = y_{k+1/2} + h_{k+1} F(y_{k+1/2}, w_{k+1/2}, t_{k+1/2}), \\
u_{k+1} = y_{k+1} + \gamma_{k+1} h_{k+1} (F(u_k, v_k, t_k) - F(y_{k+1/2}, w_{k+1/2}, t_{k+1/2}), \\
t_{k+1} = t_{k+1/2} + t_{k+1}, \\
v_{k+1} = w_{k+1/2} + h_{k+1} G(u_{k+1}, v_{k+1}, w_{k+1/2}, t_{k+1}), \\
k = 0, 1, \ldots, N-1.
\end{gathered} \tag{46}$$

To implement method (46) that allows us to determine (u_k, v_k) – an approximate solution for the problem (42) – (45) – with a help of **DUMKA** code, one should:
1) To define an array V in common block, ascribing initial value (45)
2) Make use of codes **DUMKA** and **COUR** for the solution of the Cauchy problem (42) – (43) only, considering V as a set of parameters.
3) To determine the solution $z = (z_1, \ldots, z_m)$ of a set of equations

$$z - HL \cdot G(U, z, V, T) = V$$

via values U, V, HL, T in the beginning of the subroutine **FUN** .

An example of a system (42) – (43) is a multigroup system of 1D cinetic equations with delayed neutrons. For this particular system a 'soft' part is a set of equations for delayed neutrons

$$\frac{dc_i}{dt} = -\lambda_i c_i + \beta_i \sum_{j=1}^{n} \alpha_j \phi_j, i = 1, \ldots, m,$$

and a stiff part – part equations for instant neutrons

$$\frac{1}{v_i} \frac{\partial \phi_i}{\partial t} = d_i \frac{\partial^2 \phi_i}{\partial x^2} + \sum_{j=1}^{n} \gamma_{ij} \phi_j + \sum_{j=1}^{m} \delta_{ij} c_j + Q_i,$$

$$i = 1, \ldots, n.$$

3 Testing of the code

To test and analyze the code the standard testing set was used [5,6]. The problems were divided into six groups by the sort of spectrum (real, complex), and the sort of equations (linear, nonlinear). 12 stiff test problems from [7] were solved by **DUMKA**. Time of solution and accuracy were compared to those of solvers: **RKC, RADAU5, LSODE, EPISODE, DOPRI8(5), SDIRK4, SEVLEX, SODEX, ROS4, RODAS, STRIDE**. The results of testing confirmed the perspectivity of using **DUMKA** for integration of stiff systems.

Additionally some problems,not included into the sets [5,6,7] were solved, but of interest in investigation of the ability of the code to solve neutron transport problems, and also some problems from [3] (linear heat transfer equation for three - dimensional domain with a large number of nodes, a system of multigroup, one-dimensional kinetic equations with delayed neutrons, Babuska's example, elasticity equation and so on). These results confirmed the assumptions about the method's features and advantages of the code in solution of stiff systems.

References

1. Lebedev, V.I.: Explicit difference schemes with time-variable steps for solution of stiff systems of equations. Preprint DNM AS USSR No.177 1987
2. Lebedev, V.I.: Explicit difference schemes with time-variable steps for solution of stiff systems of equations. Sov. J. Numer. Anal. Math. Modelling. **4** No.2 (1989) 111–135
3. Lebedev, V.I.: How to solve stiff systems of equations by explicit difference schemes. Numerical methods and applications, Ed. G.I.Marchuk, CRC Press, Boca Raton, Ann Arbor, London, Tokyo (1994) 45–80
4. Lebedev, V.I.: Zolotarev polynomials and extremum problems. Russ. J. Numer. Anal. Math. Modelling **9** No.3 (1994) 231–263
5. Enright, W.H., Pryce, J.D.: Two FORTRAN Packages for Assessing Initial Value Methods. ACM Trans. Math. Soft. **13** No.1 (1987) 1–27
6. Byrne, G.D., Hindmarsh, A.C.: Stiff ODE Solvers:A Review of Current and Coming Attractions. J. of Comp. Physics. **70** (1987) 1–62
7. Hairer, E., Wanner, G.: Solving Ordinary Differential Equations, Vol.II. Springer-Verlag 1991

Difference Schemes for the Darboux Problem with Functional Dependence

Henryk Leszczyński[1]

Univ. of Gdańsk, Inst. of Math., ul. Wita Stwosza 57, 80-952 Gdańsk, Poland;
E-mail address: hleszcz@ksinet.univ.gda.pl, mathl@halina.univ.gda.pl

Abstract. We present here some difference schemes for the Darboux problem with a kind of functional dependence which generalizes integrals and deviations of the Volterra type. Convergence results, our main concern, are deduced from certain consistency and stability conditions. Until the right-hand side is independent of any derivatives, one can assume some non-linear comparison estimates instead of the Lipschitz condition, which seem inevitable otherwise.

1 Introduction.

We consider two main Darboux problems with functional dependence

$$D_{xy}z(x,y) = f\big(x, y, z_{(x,y)}\big), \tag{1}$$

$$z(x,y) = \phi(x,y), \qquad (x,y) \in E^0, \tag{2}$$

where $E^0 := [-a_0, a] \times [-b_0, b] \backslash (0, a] \times (0, b]$, the function $z_{(x,y)} : B \to \mathbb{R}$ ($B = [-a_0, 0] \times [-b_0, 0]$) is defined by $z_{(x,y)}(s,t) = z(x+s, y+t)$ for $(s,t) \in B$, $f : \Omega_0 \to \mathbb{R}$, $\phi : E^0 \to \mathbb{R}$ with $\Omega_0 = E \times C(B, \mathbb{R})$; and

$$D_{xy}z(x,y) = f\big(x, y, z_{(x,y)}, (D_x z)_{(x,y)}, (D_y z)_{(x,y)}\big), \tag{3}$$

$$z(x,y) = \phi(x,y), \qquad (x,y) \in E^0, \tag{4}$$

where $f : \Omega_1 \to \mathbb{R}$ with $\Omega_1 = E \times C^{1,1}(B, \mathbb{R}) \times C^{0,1}(B, \mathbb{R}) \times C^{1,0}(B, \mathbb{R})$. The functionals $z_{(x,y)}$, defined above, generalize a multitude of Volterra-type models such as $z(x,y)$, $z(\alpha(x,y), \beta(x,y))$ (so called deviated variables), moreover:

$$\int_{\gamma_1(x,y)}^{\delta_1(x,y)} \int_{\gamma_2(x,y)}^{\delta_2(x,y)} K(x,y,s,t) z(s,t)\, ds\, dt$$

(a double integral over a two-dimensional subset of the set $E_0 \cup E$ with the kernel K), and 'thin' integrals such as:

$$\int_{\gamma_3(x,y)}^{\delta_3(x,y)} K_1(x,y,s) z(s,y)\, ds \quad \text{and} \quad \int_{\gamma_4(x,y)}^{\delta_4(x,y)} K_4(x,y,s) z(x,t)\, dt$$

with the kernels K_1 and K_2, respectively. We assume that there exist classical, sufficiently regular, solutions to problems (1), (2) and (3), (4). Denote by $\mathbf{F}(X, Y)$

the class of all functions from X to Y. We investigate a wide class of difference schemes for these differential-functional problems. Namely, if $F_{hk} : \Omega_0[h,k] := E_{hk} \times \mathbf{F}(D_{h,k}, \mathbb{R}) \to \mathbb{R}$ with a discrete set D_{hk} that is close to the set B for $(h,k) \in I_d$ (the set I_d contains acceptable steps, that is: $hM_0 = a_0$, $kN_0 = N_0$ and $c_0 h \le k \le c_1 h$ for some M_0, $N_0 \in \mathbb{N}$ and $0 < c_0 \le c_1$), then the difference analogue of equation (1) takes the form

$$\delta z^{(i,j)} = F_{hk}\left(x_i, y_j, z_{[i,j]}\right) \qquad i = 0, \ldots, M-1, \; j = 0, \ldots, N-1, \tag{5}$$

where $hM = a$, $kN = b$, $z^{(i,j)} = z(x_i, y_j)$ and

$$\delta z^{(i,j)} = \frac{z^{(i+1,j+1)} - z^{(i,j+1)} - z^{(i+1,j)} + z^{(i,j)}}{hk},$$

$$z_{[i,j]}(s,t) = z(x_i + s, y_j + t) \quad \text{for } (s,t) \in D_{hk}.$$

These schemes include so called explicite and implicite difference equations. Our convergence theorems are proven by means of consistency and stability statements taking into consideration (wherever possible) some nonlinear estimates of the Perron type.

If $\Omega_0[h,k] := E_{hk} \times \mathbf{F}(D_{h,k}, \mathbb{R}^3)$ and $F_{hk} : \Omega_1[h,k] \to \mathbb{R}$, then we write the difference counterpart of problem (3), (4):

$$\delta z^{(i,j)} = F_{hk}\left(x_i, y_j, z_{[i,j]}, (\delta_1 z)_{[i,j]}, (\delta_2 z)_{[i,j]}\right) \tag{6}$$

$$i = 0, \ldots, M-1, \; j = 0, \ldots, N-1,$$

$$z^{(i,j)} = \phi_{hk}^{(i,j)} \qquad \text{for} \quad (x_i, y_j) \in E_{hk}^0. \tag{7}$$

As to the more difficult functional dependence appearing in (3), we consider adequate difference schemes and analyse the question of their stability in a manner which reflect classical methods in existence and uniqueness theory for the Darboux problem. A summation rule becomes a tool playing the same role as an inverse integral formula in the theory of differential-functional problems. In dependence on various sets of assumptions we get convergence theorems in local and global versions.

2 Preliminaries

We start this section with an auxiliary difference equation and an inverse formula to this equation.

$$\delta z^{(i,j)} = H_{hk}^{(i,j)}, \qquad i = 0, \ldots, M-1, \; j = 0, \ldots, N-1, \tag{8}$$

where $H_{hk}^{(i,j)} \in \mathbb{R}$ for all $i = 0, \ldots, M-1, \; j = 0, \ldots, N-1$; and

$$z^{(i,j)} = z^{(i,0)} + z^{(0,j)} - z^{(0,0)} + h\,k \sum_{\mu=0}^{i-1} \sum_{\nu=0}^{j-1} H_{hk}^{(\mu,\nu)} \tag{9}$$

$$\text{for } i = 0, \ldots, M, \; j = 0, \ldots, N,$$

$$\delta_1 z^{(i,j)} = \delta_1 z^{(i,0)} + k \sum_{\nu=0}^{j-1} H_{hk}^{(i,\nu)} \quad \text{for } i = 0, \ldots, M-1, \ j = 0, \ldots, N, \tag{10}$$

$$\delta_2 z^{(i,j)} = \delta_2 z^{(0,j)} + h \sum_{\mu=0}^{i-1} H_{hk}^{(\mu,j)} \quad \text{for } i = 0, \ldots, M, \ j = 0, \ldots, N-1. \tag{11}$$

In particular, we can rewrite scheme (5) with initial conditions given by (7), and scheme (6) with the same initial data, as follows

$$z^{(i,j)} = \phi_{hk}^{(i,0)} + \phi_{hk}^{(0,j)} - \phi_{hk}^{(0,0)} + h\,k \sum_{\mu=0}^{i-1} \sum_{\nu=0}^{j-1} H_{hk}^{(\mu,\nu)}[z] \tag{12}$$
$$\text{for } i = 0, \ldots, M, \ j = 0, \ldots, N,$$

where $H_{hk}^{(i,j)} = H_{hk}^{(i,j)}[z]$ is defined by

$$H_{hk}^{(i,j)} = F_{hk}\left(x_i, y_j, z_{[i,j]}\right) \tag{13}$$

in scheme (5), whereas in scheme (6) it is given by

$$H_{hk}^{(i,j)} = F_{hk}\left(x_i, y_j, z_{[i,j]}, (\delta_1 z)_{[i,j]}, \delta_2 z)_{[i,j]}\right). \tag{14}$$

We introduce some definitions that will be useful in our paper. The first one is an adaptation of an analogous notion in the theory of differential and differential-functional equations, cf. [3].

Definition 1. We will say that $\sigma_{hk} : E_{hk}^{-} \times \mathbf{F}(D_{hk}, \mathbb{R}_{+})$ is a comparison function of the Perron type if $\sigma_{hk}(x_i, y_j, 0) = 0$ and the function $z = 0$ is the only solution to the problem

$$\delta z^{(i,j)} = \sigma_{hk}\left(x_i, y_j, z_{[i,j]}\right), \qquad (x_i, y_j) \in E_{hk}^{-} \tag{15}$$

$$z^{(i,j)} = 0, \qquad (x_i, y_j) \in E_{hk}^{0}. \tag{16}$$

Definition 2. We will say that $\gamma \in \Gamma_0$ if $\gamma : \overline{\mathbb{R}}_{+} \to \overline{\mathbb{R}}_{+}$ and $\gamma(t) \to \gamma(0) = 0$ as $t \to 0$.

Due to the above definition it will be easier to express our consistency statements.

Definition 3. A comparison function $\sigma_{hk} : E_{hk}^{-} \times \mathbf{F}(D_{hk}, \mathbb{R}_{+})$ is said to be stable if for all $\alpha_0, \alpha \in \Gamma_0$ there exists $\beta \in \Gamma_0$ such that $\omega_{hk}^{(i,j)} \leq \beta(h,k)$, where $\omega_{hk} : E_{hk}^{0} \cup E_{hk} \to \overline{\mathbb{R}}_{+}$ is the solution to the problem

$$\delta \omega^{(i,j)} = \sigma_{hk}\left(x_i, y_j, \omega_{[i,j]}\right) + \alpha(h,k), \qquad (x_i, y_j) \in E_{hk}^{-} \tag{17}$$

$$\omega^{(i,j)} = \alpha_0(h,k) \qquad \text{for} \quad (x_i, y_j) \in E_{hk}^{0}. \tag{18}$$

3 Stability of difference schemes without first-order derivatives

In the paper we skip all technical details which concern consistency of our difference schemes with adequate differential equations, however, we remark that applying the Taylor formula is a crucial point in each consistency proof for a sufficiently regular solution of an differential equation, see [3, 5, 6]. Wherever necessary, we will indicate these assumptions which are connected with consistency. We assume that the discrete set D_{hk} consists of all knots which belong to the set B.

If a difference scheme is consistent, then its stability is a sufficient condition for convergence. Once we achieve a theorem on stability, one can regard that the convergence proof is completed.

Assumption 1 *Suppose that there is a stable Perron-type comparison function* $\sigma_{hk} : E_{hk}^{-} \times \mathbf{F}(D_{hk}, \mathbb{R}_+) \to \mathbb{R}_+$ *such that*

$$|F_{hk}(x, y, \omega) - F_{hk}(x, y, \bar{\omega})| \leq \sigma_{hk}(x, y, |\omega - \bar{\omega}|) \tag{19}$$

for $(x, y) \in E_{hk}^{-}$ *and* $w, \bar{w} \in \mathbf{F}(D_{hk}, \mathbb{R}_+)$.

Assumption 2 (Consistency condition) *Suppose that*
1) there is $\alpha_0 \in \Gamma_0$ *such that* $|\phi^{(i,j)} - \phi_{hk}^{(i,j)}| \leq \alpha_0(h, k)$ *for* $(x_i, y_j) \in E_{hk}^0$,
2) $v \in C^1(E^0 \cup E, \mathbb{R})$ *is a solution to problem (1), (2) and* v *is of class* C^3 *on the set* $\overline{E}$,
3) there is $\tilde{\alpha} \in \Gamma_0$ *such that*

$$\left|F_{hk}\left(x_i, y_j, (v_{hk})_{[i,j]}\right) - f\left(x_i, y_j, v_{(x_i,y_j)}\right)\right| \leq \tilde{\alpha}(h, k) \quad \text{for } (x_i, y_j) \in E_{hk}^{-}. \tag{20}$$

Theorem 4. *Suppose that Assumptions 1 and 2 are satisfied. Let* $u_{hk} : E_{hk}^0 \cup E_{hk} \to \mathbb{R}$ *be a solution to (5), (7). Then there is* $\beta \in \Gamma_0$ *such that* $|u_{hk}^{(i,j)} - v_{hk}^{(i,j)}| \leq \beta(h, k)$ *for* $(x_i, y_j) \in E_{hk}$.

Proof. If we denote by w_{hk} the difference $v_{hk} - u_{hk}$, then, regarding the function v_{hk} as a perturbed solution of scheme (5) and applying (12) with $H_{hk}[z]$ defined by (13) for $z = u_{hk}$ and for $z = v_{hk}$ (with an adequate perturbation), we obtain the crucial estimate

$$|w_{hk}^{(i,j)}| \leq 3\alpha_0(h, k) + hk \sum_{\mu=0}^{i-1} \sum_{\nu=0}^{j-1} \left(\sigma_{hk}\left(x_\mu, y_\nu, |w_{hk}|_{hk}\right) + \tilde{\alpha}(h, k)\right) \tag{21}$$

for $(x_i, y_j) \in E_{hk}$. Because the Perron comparison function σ_{hk} is stable, the error function w_{hk} can be estimated by the solution to problem (17), (18), hence it converges uniformly to 0 as $\|(h, k)\| \to 0$.

Notes and Comments. Inequality (21) is definitely explicite. If we put $D_{hk} = \{(x_i, y_j) \mid \mid i = -M_0, \ldots, 1,\ j = -N_0, \ldots, 1\}$, then we face the question of solvability of the implicite difference equation, which touches both: scheme (5) and problem (17). This may result in a presumable necessity of 1) strenthening nonlinear estimates to the Lipschitz condition, 2) putting some constraints on the constants a and b so that the global character of convergence might be lost. On the other hand, implicite schemes include variety of relatively simple examples of much higher accuracy such as:

$$\delta z^{(i,j)} = f\left(x_{i+1/2}, y_{j+1/2}, \frac{1}{4}\sum_{\mu,\nu=0}^{1} z^{(i+\mu,j+\nu)}\right). \tag{22}$$

Note that the above scheme can be easily generalized onto equations with functional dependence, see [3].

4 Stability of difference schemes including first order derivatives

Once again we restrict the discrete set D_{hk} to the set B intersected with the mesh. We start this section with some consistency and Lipschitz condition.

Assumption 3 (Consistency condition) *Suppose that*
1) there are $\alpha_0, \alpha_1, \alpha_2 \in \Gamma_0$ *such that* $|\phi^{(i,j)} - \phi_{hk}^{(i,j)}| \le \alpha_0(h,k)$ *for* $(x_i, y_j) \in E^0_{hk}$, *and*

$$\left|\delta_\kappa \phi^{(i,j)} - \delta_\kappa \phi_{hk}^{(i,j)}\right| \le \alpha_\kappa(h,k) \tag{23}$$

for $\kappa = 1, 2$ *and for all indices* (i, j) *such that the above expressions make sense,*
2) $v \in C^2(E^0 \cup E, \mathbb{R})$ *is a solution to problem (1), (2) and* v *is of class* C^3 *on the set* $\overline{E}$,
3) there is $\tilde{\alpha} \in \Gamma_0$ *such that*

$$\left|F_{hk}\left(x_i, y_j, V_{hk}^{(i,j)}\right) - f\left(x_i, y_j, v_{(x_i,y_j)}\right)\right| \le \tilde{\alpha}(h,k) \quad \text{for } (x_i, y_j) \in E^-_{hk}, \tag{24}$$

where

$$V_{hk}^{(i,j)} = \left((v_{hk})_{[i,j]}, (\delta_1 v_{hk})_{[i,j]}, (\delta_2 v_{hk})_{[i,j]}\right). \tag{25}$$

Assumption 4 (Lipschitz condition) *Suppose that there are* $L_0, L_1, L_2 \in \mathbb{R}$ *such that*

$$|F_{hk}(x_i, y_j, w_0, w_1, w_2) - F_{hk}(x_i, y_j, \bar{w}_0, \bar{w}_1, \bar{w}_2)| \le \sum_{\kappa=0}^{2} L_\kappa \|w_\kappa - \bar{w}_\kappa\| \tag{26}$$

for $(x_i, y_j, w_0, w_1, w_2)$, $(x_i, y_j, \bar{w}_0, \bar{w}_1, \bar{w}_2) \in \Omega_1[h,k]$.

We denote by ξ_{hk} the following residual expression

$$\xi_{hk}^{(i,j)}[z] = \delta z^{(i,j)} - F_{hk}\left(x_i, y_j, z_{[i,j]}, (\delta_1 z)_{[i,j]}, (\delta_2 z)_{[i,j]}\right) \tag{27}$$

for $(x_i, y_j) \in E_{hk}^-$ and $z \in \mathbf{F}(E_{hk}^0 \cup E_{hk}, \mathbb{R})$.

Two parts of the following theorem reflect consistency and stability, respectively.

Theorem 5. *Suppose that Assumptions 3 and 4 are satisfied. Then:*
1) There is $\bar{\alpha} \in \Gamma_0$ *such that* $|\xi_{hk}^{(i,j)}[v_{hk}]| \leq \bar{\alpha}(h,k)$ *for* $i = 0, \ldots, M-1$, $j = 0, \ldots, N-1$.
2) If $|\xi_{hk}^{(i,j)}| \leq \bar{\alpha}(h,k)$, *where* $\bar{\alpha} \in \Gamma_0$, *and* $z, \bar{z} \in \mathbf{F}(E_{hk}^0 \cup E_{hk}, \mathbb{R})$ *are solutions to the problems: (6), (7) and the same perturbed equation with the initial condition* ϕ *instead od* ϕ_{hk}, *then there are* $W_0, W_1, W_2 \in C(E^0 \cup E, \mathbb{R})$ *which tend to* 0 *uniformly as* $\|(h,k)\| \to 0$ *and satisfy (on the whole mesh) the inequalities:*

$$|z^{(i,j)} - \bar{z}^{(i,j)}| \leq W_0(x_i, y_j), \quad |\delta_\kappa z^{(i,j)} - \delta_\kappa \bar{z}^{(i,j)}| \leq W_\kappa(x_i, y_j) \tag{28}$$

for $\kappa = 1, 2$.

Proof. The first part is obtained from Assumption 3 by means the Taylor formula. In order to establish 2), we denote by w the error expression, that is: $w^{(i,j)} = z^{(i,j)} - \bar{z}^{(i,j)}$ on the mesh. If we apply (9) - (11) with $H_{hk}[z]$ and $H_{hk}[\bar{z}]$ defined by (14) in non-perturbed and perturbed versions, respectively, then we get the key inequalities:

$$|w^{(i,j)}| \leq 3\alpha_0(h,k) + x_i y_j \bar{\alpha}(h,k) \tag{29}$$
$$+hk \sum_{\mu=0}^{i-1} \sum_{\nu=0}^{j-1} L_0 \|w_{[\mu,\nu]}\| + L_1 \|(\delta_1 w)_{[\mu,\nu]}\| + L_2 \|(\delta_2 w)_{[\mu,\nu]}\|,$$

$$|\delta_1 w^{(i,j)}| \leq \alpha_1(h,k) + y_j \bar{\alpha}(h,k) \tag{30}$$
$$+k \sum_{\nu=0}^{j-1} L_0 \|w_{[i,\nu]}\| + L_1 \|(\delta_1 w)_{[i,\nu]}\| + L_2 \|(\delta_2 w)_{[i,\nu]}\|,$$

$$|\delta_2 w^{(i,j)}| \leq \alpha_2(h,k) + x_i \bar{\alpha}(h,k) \tag{31}$$
$$+h \sum_{\mu=0}^{i-1} L_0 \|w_{[\mu,j]}\| + L_1 \|(\delta_1 w)_{[\mu,j]}\| + L_2 \|(\delta_2 w)_{[\mu,j]}\|.$$

We define the functions W_0, W_1, W_2 as follows

$$W_0(x,y) = 3\alpha_0(h,k) + xy\bar{\alpha}(h,k) + \int_0^x \int_0^y \tilde{W}(s,t)\, dt\, ds, \tag{32}$$

$$W_1(x,y) = \alpha_1(h,k) + y\bar{\alpha}(h,k) + \int_0^y \tilde{W}(x,t)\, dt, \tag{33}$$

$$W_2(x,y) = \alpha_2(h,k) + x\bar{\alpha}(h,k) + \int_0^x \tilde{W}(s,y)\, ds, \tag{34}$$

on the set E, where the function $\tilde{W} \in C^2(E, \mathbb{R})$ is the solution to the differential problem

$$D_{xy}z(x,y) = L_0 z(x,y) + L_1 D_x z(x,y) + L_2 D_y z(x,y) \tag{35}$$

$$z(0,y) = (3L_0\alpha_0 + L_1\alpha_1 + L_2\alpha_2 + \bar{\alpha}(yL_1+1))\,\frac{e^{yL_1}-1}{L_1}, \tag{36}$$

$$z(x,0) = (3L_0\alpha_0 + L_1\alpha_1 + L_2\alpha_2 + \bar{\alpha}(xL_2+1))\,\frac{e^{xL_2}-1}{L_2}. \tag{37}$$

We write

$$W_\kappa(x,y) = W_\kappa\left(\max\{x,0\}, \max\{y,0\}\right) \tag{38}$$

for $(x,y) \in E^0$ and $\kappa = 0,1,2$. These functions satisfy comparison inequalities with respect to (29) - (31), hence the second assertion of our theorem holds.

Notes and Comments. 1) One can replace in Assumption 4 the Lipschitz condition with respect to w_0 by a non-linear Perron estimate. This way Assumption 4 becomes slighltly weaker, however, we obtain a non-linear differential problem instead of (35) - (37), and its solvability should be guaranteed by an additional assumption. 2) One can consider differential-functional equation (3) with a more subtle version of the Lipschitz condition, which will be reflected by the following modification of condition (26):

$$|F_{hk}(x_i, y_j, w_0, w_1, w_2) - F_{hk}(x_i, y_j, \bar{w}_0, \bar{w}_1, \bar{w}_2)| \le \sum_{\kappa=0}^{2} L_\kappa \|w_\kappa - \bar{w}_\kappa\|_\kappa, \tag{39}$$

where $\|w_0\|_0 = \|w_0\| + \|\delta_1 w_0\| + \|\delta_2 w_0\|$, $\|w_1\|_1 = \|w_1\| + \|\delta_2 w_1\|$ and $\|w_2\|_2 = \|w_2\| + \|\delta_1 w_2\|$. In this case, we should assume that the Lipschitz constants L_1 and L_2 are sufficiently small. Then we get a stability result which can be summarised as follows: if $|\xi_{hk}^{(i,j)}[\bar{z}]| \le \bar{\alpha}$ on E_{hk}^-, then the inequalities

$$|z^{(i,j)} - \bar{z}^{(i,j)}| \le \alpha_0(h,k), \qquad |\delta_1 z^{(i,j)} - \delta_1 \bar{z}^{(i,j)}| \le \alpha_1(h,k), \tag{40}$$

$$|\delta_2 z^{(i,j)} - \delta_2 \bar{z}^{(i,j)}| \le \alpha_2(h,k), |\delta z^{(i,j)} - \delta \bar{z}^{(i,j)}| \le \alpha_3(h,k) \tag{41}$$

assumed on the set E_{hk}^0 with $\alpha_\kappa \in \Gamma_0$ imply analogous estimates on the whole mesh. 3) Note that the tools applied in the theory of difference approximation to the Darboux problem reflect some classical tools out of the theory of differential and differential-functional problems. Especially, this remark concerns so called monotone iterative techniques (see [1, 2, 7]), moreover, our inverse summation rule (9) (also (10) and (11)) is influenced by analogous inverse integral equations. 4) The very fundamental properties of difference schemes for equation (1), contrary to the Darboux problem with partial derivatives, immitate some features of finite difference approximation to certain initial-value problems such as these for ordinary differential equations, or first-order partial differential equations (compare ([5])), or parabolic differential equations (cf. ([6])).

References

1. Byszewski, L.: Existence and uniqueness of solutions of nonlocal problems for hyperbolic equation $u_{xt} = F(x, t, u, u_x)$. Journ. Appl. Math. Stoch. Anal. **3** (1990) 163–168
2. Człapiński, T.: Existence of solutions of the Darboux problem for partial differential-functional equations with infinite delay in a Banach space. Comm. Math. **35** (1995)
3. Kamont, Z., Leszczyński, H.: Numerical solutions to the Darboux problem with the functional dependence. to appear in: Georgian Mathematical Journal
4. Leszczyński, H.: Existence of classical and weak solution the Darboux problem with functional dependence. (unpublished)
5. Leszczyński, H.: Discrete approximations to the Cauchy problem for hyperbolic differential-functional systems in the Schauder canonic form. Comp. Maths Math. Phys. **34** 2 (1994) 151–164 (Zb. Vychisl. Mat. Mat. Fiz. **34** 2 (1994) 185–200)
6. Leszczyński, H.: General finite difference approximation to the Cauchy problem for non-linear parabolic differential-functional equations. Ann. Polon. Math. **53** (1991) 15–28
7. Ladde, G.S., Lakshmikantham, V., Vatsala, A.S.: Monotone Iterative Techniques for Nonlinear Differential Equations. Pitman Advanced Publishing Program, Boston London Melbourne 1985.

Displacement Decomposition Circulant Preconditioners for Almost Incompressible 2D Elasticity Systems*

Ivan Lirkov and Svetozar Margenov

Central Laboratory of Parallel Processing,
Bulgarian Academy of Sciences
Acad. G. Bontchev Str., Bl.25A, 1113 Sofia, Bulgaria

Abstract. The robustness of the recently introduced circulant block-factorization (CBF) preconditioners is studied in the case of finite element matrices arising from the discretization of the 2D Navier equations of elasticity. Conforming triangle finite elements are used for the numerical solution of the differential problem. The proposed preconditioner M_C is constructed by CBF approximation of the block-diagonal part of the stiffness matrix. In other words, we implement in our algorithm the circulant block-factorization into the framework of the displacement decomposition technique. The estimate $\kappa(M_C^{-1}K) = O\left(\sqrt{\frac{N}{1-\nu}}\right)$ is proved asymptotically on N, where N is the size of the discrete problem. Note, that the corresponding known estimate for the widely used incomplete factorization displacement decomposition preconditioner M_{ILU} is $\kappa(M_{ILU}^{-1}K) = O(\frac{\sqrt{N}}{1-\nu})$.
The theoretical estimate as well as the presented numerical tests show some significant advantages of this new approach for a PCG iterative solution of almost incompressible elastic problems, that is when the modified Poisson ratio ν tends to the incompressible limit case $\nu = 1$.

Key words: almost incompressible elasticity, finite elements, circulant preconditioning, displacement decomposition
AMS subject classifications: 65F10, 65F20, 65N30

1 Introduction

This paper is concerned with the numerical solution of the Navier equations of 2D elasticity problem. Using the finite element method, such a problem is reduced to a linear system of the form $K\mathbf{u} = \mathbf{b}$, where K is a sparse matrix. The considered problem is symmetric and positive definite. We assume also, that K is a large scale matrix. It is well known, that in this case the iterative solvers based on the preconditioned conjugate gradient (PCG) method are the best way

* This paper was partially supported by the Bulgarian Ministry of Education, Science and Technology under grant MM 417/94.

to solve the linear algebraic system. The key question is how to construct the preconditioning matrix M.

In this paper we consider an application of the recently introduced circulant block–factorization (CBF) algorithm to the plane strain problem of elasticity. The emphasis is on the robustness of the algorithm in the *almost incompressible* case, i.e., when the modified Poisson ratio $\nu \in (0,1)$ tends to the incompressible limit $\nu = 1$.

There are a lot of works dealing with preconditioning iterative solution methods for the FEM elasticity systems. Here we will briefly comment on some of the used approaches. In an earlier paper, Axelsson and Gustafsson [2] have implemented modified point–ILU factorization for this problem. As the coupled system does not lead to an M–matrix, they construct their preconditioners based on the point–ILU factorization of the displacement decoupled block–diagonal part of the original matrix. This approach is known as *displacement decomposition.* It is based on Korn's inequality, and the convergence deteriorates in the almost incompressible case like $O(\frac{1}{\sqrt{1-\nu}})$. The displacement decomposition remains until now one of the most robust approaches (see also, e.g., [3, 7]). Some new block–ILU factorization preconditioners based on block–size reduction are studied in [5]. This factorization exists for symmetric and positive definite block-tridiagonal matrices that are not necessarily M-matrices. Although the approximate factorization is applied to the original matrix, the dependence on ν of the number of iterations remains the same as above, i.e., $O(\frac{1}{\sqrt{1-\nu}})$.

We study in this paper an implementation of the circulant block–factorization (CBF) algorithm as introduced by Lirkov, Margenov and Vassilevski [11], into the framework of the displacement decomposition. The robustness of the algorithm is based on the efficiency of the CBF preconditioners for strongly anisotropic problems (see for more details in [12]). We prove for the new proposed preconditioner M_C the estimate $\kappa(M_C^{-1}K) = O(\sqrt{\frac{N}{1-\nu}})$ asymptotically on N, where N is the size of the discrete problem. Consequently, the growth with ν of the number of iterations is reduced from $O(\frac{1}{\sqrt{1-\nu}})$ to $O(\frac{1}{\sqrt[4]{1-\nu}})$.

The remainder of the paper is organized as follows. Some background facts about the Navier equations of elasticity, their FEM approximation and the related Korn's inequality are presented in §2. The displacement decomposition CBF (DD CBF) algorithm is described in §3. In §4 we give a model problem analysis of the relative condition number of the studied preconditioner. A set of numerical tests illustrating the performance of the resulting preconditioned conjugate gradient algorithm are presented in the last section §5.

2 FEM 2D elasticity equations

We consider in this paper the Dirichlet boundary value plain strain problem of elasticity in the weak formulation of the Navier system of equations. The unknown displacements $\underline{w}^t = (u, v)$ satisfy the following variational equations:

Find $(u,v) \in H_1^0 \times H_1^0$, such that

$$\begin{aligned} a(u,\tilde{u}) + e_{12}(v,\tilde{u}) &= f_1, \\ & \qquad\qquad \forall(\tilde{u},\tilde{v}) \in H_1^0 \times H_1^0, \qquad (1) \\ e_{21}(u,\tilde{v}) + b(v,\tilde{v}) &= f_2, \end{aligned}$$

where $H_1^0 = \{w \in H_1(\Omega) : w|_{\partial\Omega} = 0\}$, and the related bilinear forms are defined by the formulas:

$$a(\phi,\psi) = \int_\Omega \left(\frac{\partial\phi}{\partial x}\frac{\partial\psi}{\partial x} + \frac{(1-\nu)}{2}\frac{\partial\phi}{\partial y}\frac{\partial\psi}{\partial y} \right) d\Omega,$$

$$b(\phi,\psi) = \int_\Omega \left(\frac{(1-\nu)}{2}\frac{\partial\phi}{\partial x}\frac{\partial\psi}{\partial x} + \frac{\partial\phi}{\partial y}\frac{\partial\psi}{\partial y} \right) d\Omega,$$

$$e_{12}(\psi,\phi) = e_{21}(\phi,\psi) = \frac{1+\nu}{2}\int_\Omega \frac{\partial\phi}{\partial x}\frac{\partial\psi}{\partial y} d\Omega.$$

Here $\nu \in (0,1)$ stands for the modified Poisson ratio. The notion *almost incompressible* is used for the case $\nu = 1-\delta$, where δ is a small positive number. Note that if $\nu = 1$ (the material is incompressible), the problem (1) is ill-posed.

Now, let ω be a square mesh, and let Ω be a polygonal domain, triangulated by right isosceles triangles $T \in \tau$ obtained by a diagonal bisection of the square cells of ω.

Let $W = W_1^0 \times W_1^0$, where $W_1^0 \subset H_1^0$ is the finite element space of conforming piecewise linear functions with nodal Lagrangian basis $\{\phi_i\}_{i=1}^N$ corresponding to the triangulation τ. Then the finite element approximation (u^h, v^h) of the problem (1) is determined as follows:

Find $u^h = \sum_{i=1}^N u_i\phi_i$, $v^h = \sum_{i=1}^N v_i\phi_i$, such that

$$\begin{aligned} a(u^h,\phi_i) + e_{12}(v^h,\phi_i) &= f_{1,i}, \\ & \qquad\qquad \forall i = 1,\ldots,N\cdot \qquad (2) \\ e_{21}(u^h,\phi_i) + b(v^h,\phi_i) &= f_{2,i}. \end{aligned}$$

Equations (2) are equivalent to the linear system

$$K\underline{w}_h = \underline{b},$$

where K is the stiffness matrix, and $\underline{w}_h = \begin{pmatrix} \underline{u}_h \\ \underline{v}_h \end{pmatrix}$ is the vector of the nodal unknowns $\underline{u}_h = \{u_i\}_{i=1}^N$ and $\underline{v}_h = \{v_i\}_{i=1}^N$.

The stiffness matrix K can be written in the following natural block–structure

$$K = \begin{pmatrix} A & E \\ E^T & B \end{pmatrix},$$

where the blocks A and B correspond to the bilinear forms $a(.,.)$ and $b(.,.)$ respectively.

The following theorem plays a key role in the convergence theory of the displacement decomposition methods.

Theorem 1. *The following Korn's inequality holds*

$$\kappa(K_D^{-1}K) \leq \frac{3+\nu}{1-\nu}, \tag{3}$$

where

$$K_D = \begin{pmatrix} A & 0 \\ 0 & B \end{pmatrix}, \tag{4}$$

and $\kappa(.)$ stands for the condition number of the matrix.

Proof. It is easy to see, that $\kappa(K_D^{-1}K) \leq \frac{\lambda_4}{\lambda_1}$, where λ_1 and λ_4 are the minimal and the maximal eigenvalue of the generalized eigenvalue problem

$$\begin{pmatrix} 1 & 0 & 0 & \frac{1+\nu}{2} \\ 0 & \frac{1-\nu}{2} & 0 & 0 \\ 0 & 0 & \frac{1-\nu}{2} & 0 \\ \frac{1+\nu}{2} & 0 & 0 & 1 \end{pmatrix} \underline{w} = \lambda \begin{pmatrix} 1 & 0 & 0 & 0 \\ 0 & \frac{1-\nu}{2} & 0 & 0 \\ 0 & 0 & \frac{1-\nu}{2} & 0 \\ 0 & 0 & 0 & 1 \end{pmatrix} \underline{w} \tag{5}.$$

The eigenvalue problem leads to the characteristic equation

$$(1-\lambda)^2 \left(\frac{1-\nu}{2}\right)^2 \left[(1-\lambda)^2 - \left(\frac{1+\nu}{2}\right)^2\right] = 0. \tag{6}$$

The roots of (6) are as follows $\lambda_1 = \frac{1-\nu}{2}$, $\lambda_2 = \lambda_3 = 1$, and $\lambda_4 = \frac{3+\nu}{2}$, that completes the proof of the theorem.

Remark 1 *The Korn's inequality (3) is proved by a different technique in [2].*

3 DD CBF algorithm

We consider in what follows the problem (1) in the unit square, where $\Omega = (0,1) \times (0,1)$ is covered by a uniform square mesh ω, with a size $h = 1/(n+1)$ for a given integer $n \geq 1$.

To define the displacement decomposition circulant block–factorization (DD CBF) preconditioner M_C of the matrix K we consider the auxiliary problem $-au_{xx} - bu_{yy} = f$ with homogeneous Dirichlet boundary conditions, where a and b are positive constants. This problem is discretized by the same finite elements as the original problem (1). This discretization leads to the stiffness matrix G. We assume that the grid points are ordered along the y-lines if $b < a$, and respectively along the x-lines if $a < b$. Then the matrix G can be written in the following form

$$G = tridiag(-G_{i,i-1}, G_{i,i}, -G_{i,i+1}) \qquad i = 1,2,\ldots,n,$$

where

$$G_{i,i} = tridiag(-g_{j,j-1}, g_{j,j}, -g_{j,j+1}),\ j = (i-1)n+1,\ldots,in,\ i = 1,2,\ldots,n,$$
$$G_{i,i+1} = diag(g_{j,j+n}),\ j = (i-1)n+1,\ldots,in,\ i = 1,\ldots,n-1,$$
$$G_{i,i-1} = diag(g_{j,j-n}),\ j = (i-1)n+1,\ldots,in,\ i = 2,\ldots,n.$$

The coefficients $g_{i,j}$ are positive and $g_{j,j} \geq g_{j,j-1} + g_{j,j+1} + g_{j,j+n} + g_{j,j-n}$, i.e., the matrix G satisfies the maximum principle.

We will use here the CBF preconditioning for the matrix G as introduced by Lirkov, Margenov and Vassilevski in [11] (see also in [4, 6], [12, 13]). The CBF preconditioner G_{CBF} of the matrix G is defined by

$$G_{CBF} = tridiag(-C_{i,i-1}, C_{i,i}, -C_{i,i+1}) \qquad i = 1, 2, \ldots, n, \tag{7}$$

where $C_{i,j} = Circulant(G_{i,j})$ is a circulant approximation of the corresponding block $G_{i,j}$, defined by a diagonal–by–diagonal averaging of the coefficients. Realizing the CBF algorithm we use exact block LU factorization for the preconditioner G_{CBF}. One important property of the CBF preconditioning is, that the solution of systems with the matrix G_{CBF} requires $O(NlnN)$ arithmetic operations, if FFT is used for factorization of the circulant blocks (see for more details in [9, 11]).

Finally, the DD CBF preconditioner for the stiffness matrix K of the problem (1) is defined by

$$M_C = \begin{pmatrix} A_{CBF} & 0 \\ 0 & B_{CBF} \end{pmatrix}. \tag{8}$$

Obviously, the matrix A corresponds to the auxiliary elliptic problem with $a = 1$ and $b = \frac{1-\nu}{2}$, and respectively the matrix B corresponds to the same differential problem with $a = \frac{1-\nu}{2}$ and $b = 1$. This means that the grid points ordering related to the first diagonal block A is along the y-lines, and contrary the ordering related to the diagonal block B is along the x-lines.

Note, that the above ordering of the unknowns is of key importance for the convergence of the DD CBF preconditioner.

4 Model problem condition number analysis

We will estimate in this section the condition number $\kappa(M_C^{-1}K)$ of the preconditioned system by the DD CBF algorithm.

Theorem 2. *The following inequality holds for the relative condition number of the CBF preconditioner (7)*

$$\kappa(G_{CBF}^{-1}G) < \sqrt{2\epsilon}(n+1) + 2, \tag{9}$$

where $\epsilon = \min\{\frac{b}{a}, \frac{a}{b}\}$.

This estimate is based on the exact solution of the corresponding generalized eigenproblem. A detailed proof of the theorem is presented in [12].

The final result of the model problem condition number analysis is given by the next theorem.

Theorem 3. *The preconditioner M_C defined by the DD CBF algorithm satisfies the estimate*

$$\kappa(M_C^{-1}K) < (3+\nu)\left(\frac{n+1}{\sqrt{1-\nu}} + \frac{2}{1-\nu}\right) \tag{10}$$

Proof. The proof follows directly, applying consequently the Korn's inequality from Theorem 1 and the estimate (9) with $\epsilon = \frac{1-\nu}{2}$, i.e.,

$$\begin{aligned}\kappa(M_C^{-1}K) &\le \frac{3+\nu}{1-\nu}\max\left(\kappa(A_{CBF}^{-1}A), \kappa(B_{CBF}^{-1}B)\right)\\ &\le \frac{3+\nu}{1-\nu}\left((n+1)\sqrt{1-\nu}+2\right)\\ &= (3+\nu)\left(\frac{n+1}{\sqrt{1-\nu}}+\frac{2}{1-\nu}\right)\end{aligned}$$

Remark 2 *As a conclusion of the last theorem we get an estimate for the number $n(\varepsilon)$ of the iterations in the PCG algorithm, needed to reduce the relative error with a factor ε, in the form*

$$n(\varepsilon) \le \sqrt{\left(\frac{n+1}{\sqrt{1-\nu}}+\frac{2}{1-\nu}\right)}\ln\frac{2}{\varepsilon}+1.$$

When $N = n^2$ is large enough, the above estimate can be written in the form $n(\varepsilon) = O\left(\sqrt[4]{\frac{N}{1-\nu}}\right)$.

Remark 3 *Although the presented analysis relates to the model problem in a rectangle, the application of circulants is not limited to this case. An efficient circulant based iterative procedure in L-shaped domain is proposed in [10], where the domain Ω is first transformed to the unit square. Another way to treat problems in domains with more complicated geometry is based on the circulant approximation of the Schur complements in the context of the domain decomposition method.*

5 Numerical tests

We analyze in this section the performance rate of our preconditioned iterative method, varying the size parameter n and the modified Poisson ratio ν. The computations are done with double precision on a SUN Sparc Station.

We recall, that the almost incompressible case corresponds to $\nu = 1 - \delta$, where δ is a small positive number.

The Table shows the number of iterations as a measure of the convergence rate of the preconditioners. The iteration stopping criterion is $\|r^{N_{it}}\|/\|r^0\| < 10^{-6}$, where r^j stands for the residual at the jth iteration step of the preconditioned conjugate gradient method.

Asymptotically, the presented data are in a good agreement with the theoretical estimate. The number of iterations has a complex behavior, corresponding to the derived estimate from the last section. One can see how the range, where $n(\varepsilon)$ has a behavior like $n(\varepsilon) = O\left(\sqrt[4]{\frac{N}{1-\nu}}\right)$, grows with n.

Table 1. Number of iterations for the DD CBF preconditioner.

ν	n=32	n=64	n=128	n=256	n=512
0.3	17	21	27	37	49
0.4	18	22	28	37	49
0.5	19	22	29	38	52
0.6	20	25	32	39	54
0.7	22	26	34	41	56
0.8	25	29	37	46	61
0.9	33	37	44	57	71
0.92	36	40	46	60	77
0.94	41	46	52	64	81
0.96	49	53	60	72	91
0.98	67	72	80	91	112
0.99	91	98	106	119	140
0.999	202	266	297	317	344

Remark 4 *Numerical tests for the same problem with pointwise MILU preconditioners are presented in the earlier paper by Axelsson and Gustafsson [2]. Unfortunately these test data are only for coarse grid-sizes with* $n \leq 20$ *that makes the direct comparision not representative. More recent numerical results are presented in [5], where the block-size reduction ILU preconditioner is applied. Both these preconditioners are characterized by* $O(\frac{1}{\sqrt{1-\nu}})$ *growth of the number of iterations in the almost incompressible case.*

References

1. Axelsson, O., Barker, V.A.: Finite Element Solution of Boundary Value Problems: Theory and Computations. Academic Press, Orlando, Fl. (1983).
2. Axelsson, O., Gustafsson, I.: Iterative methods for the solution of the Navier equations of elasticity. Comp.Meth.Appl.Mech, Eng, **15** (1978), 241–258
3. Blaheta, R.: Displacement Decomposition–Incomplete Factorization Preconditioning Techniques for Linear Elasticity Problems. Num. Lin. Alg. Appl. **1** (1994) 107–128
4. Chan, R.H., Chan, T.F.: Circulant preconditioners for elliptic problems. J. Num. Lin. Alg. Appl. **1** (1992) 77–101
5. Chan, T.F., Margenov, S.D., Vassilevski, P.S.: Performance of block–ILU factorization preconditioners based on block–size reduction for 2D elasticity systems. CAM Report 94–33 UCLA, SIAM J.Sci.Comput. (to appear)
6. Huckle, T.: Some aspects of circulant preconditioners. SIAM J. Sci. Comput. **14** (1993) 531–541
7. Jung, M. ,Langer, U., Semmler, U.: Two–level hierarchically preconditioned conjugate gradient methods for solving linear elasticity finite element equations. BIT **29** (1989) 748–768
8. Kocvara, M., Mandel, J.: A multigrid method for three–dimensional elasticity and algebraic convergence estimates. Appl. Math. Comp. **23** (1987) 121–135

9. Van Loan, C.: Computational frameworks for the fast Fourier transform. SIAM, Philadelphia (1992).
10. Lirkov, I., Margenov, S.: On circulant preconditioning of elliptic problems in L-shaped domain. Advances in Numerical Methods and Applications (I.T. Dimov, Bl. Sendov, P.S. Vassilevski eds.), World Scientific (1994) 136–145
11. Lirkov, I., Margenov, S., Vassilevski, P.S.: Circulant block-factorization preconditioners for elliptic problems. Computing **53** 1 (1994) 59–74
12. Lirkov, I., Margenov, S., Zikatanov, L.: Circulant block–factorization preconditioning of anisotropic elliptic problems. Computing (to appear)
13. Strang, G.: A proposal for Toeplitz matrix calculations. Stud. Appl.Math. **74** (1986) 171–176

Numerical Solution of Differential Equations with Trigonometric Coefficients in Mathematical Models of Electric Motors

Adam Marlewski[1] and Stanislaw Rawicki[2]

[1] Institute of Mathematics, Poznan University of Technology, 60-296 Poznań, Poland
[2] Institut of Industrial Electrotechnics, Poznan University of Technology, 60-296 Poznań, Poland

Abstract. Mathematical models of many phenomena in electric motors contain differential equations with trigonometric coefficients. Existing methods for solving such equations characterize by a complicate structure which is introduced by a great number of auxiliary unknowns. This disvantage is avoided in the method proposed in this paper. The primary equation system is replaced by the new equivalent infinite system of equations with constant coefficients. This system is reduced to the finite one and the Sylvester method is applied. Obtained solution determines the analytic form of the approximate solution of primary equation. As an example, the transient zero-sequence components of electric currents and the torque in the three–phase slip–ring induction motor are calculated. The result is compared to the solution produced by the Runge–Kutta method.

1 Introduction

Effects which occur in electric motors are frequently described (see e.g. [1], [2], [5], [6]) by means of differential equations with trigonometrical coefficients (DETC). The determination of the solutions of these equations is discussed by numerous authors. The results do not satisfy engineer's expectations for their coherence or the complexity. For example, in [4] it is given the method based on the Laplace operational calculus. Within the solving procedure a great number of unknowns appears and the final formulae have the structure of the infinite continued fractions. In this paper the new method of solution of DETC is presented. The primary equation system is replaced by the new equivalent infinite system of equations with constant coefficients. This new system is reduced to the finite one (and it obviously implies that the accuracy of the produced solution has to be examined). The standard Sylvester approach (in the theory of control system it is known as the matrix state–space method) is applied for calculations of the transient components of electrical currents. By way of example, the asymmetrical state of the three–phase slip–ring induction motor is analysed. In particular, within the symmetrical–component method, the zero–sequence system of the motor is described by DETC. For some number of the equations forming the reduced system with constant coefficients, convergence of solutions for the stator and rotor currents as well as for for the electromagnetic torque is presented.

2 Description of the method

The differential equation with trigonometric coefficients, we deal with in this paper, has the form

$$[a + b\cos(\omega t + \alpha)]\frac{dy}{dt} = c\cos(\omega_1 t + \gamma) + [f + g\cos(\omega t + \beta)]\,y\ . \tag{1}$$

Here t stands for the independent variable (the time),
y is the function depending on the time t,
$a, b, c, f, g, \omega, \omega_1, \alpha, \beta, \gamma$ are constants.

Moreover, there is given the initial value $y_0 = y(0)$.

In the formula (1) we replace all cosine functions with the exponential functions and we obtain

$$a\frac{dy}{dt} + b'e^{j\omega t}\frac{dy}{dt} + b'^{*}e^{-j\omega t}\frac{dy}{dt} = c_{(0)} + fy + g'e^{j\omega t}y + g'^{*}e^{-j\omega t}y\ , \tag{2}$$

where

$c_{(0)} = \frac{c}{2}\left[e^{j(\omega_1 t+\gamma)} + e^{-j(\omega_1 t+\gamma)}\right]$,

$b' = \frac{b}{2}e^{j\alpha}$,

$g' = \frac{g}{2}e^{j\beta}$,

$j = \sqrt{-1}$ stands for the imaginary unit,

s^{*} is the complex conjugate of the value s.

Now we introduce the following denotations:

$$y_{(x)}(t) = y(t)e^{jx\omega t}\ , \tag{3}$$

$$c_{(x)} = c_{(0)}e^{jx\omega t} = \frac{c}{2}\left\{e^{j[(\omega_1+x\omega)t+\gamma]} + e^{-j[(\omega_1-x\omega)t+\gamma]}\right\}\ , \tag{4}$$

where the parametr x is an arbitrary integer number.

It is easy to see that

$$e^{j\omega t}\frac{d}{dt}y_{(0)} = \frac{d}{dt}y_{(1)} - j\omega y_{(1)}\ ,$$

$$e^{-j\omega t}\frac{d}{dt}y_{(0)} = \frac{d}{dt}y_{(1)} + j\omega y_{(1)}\ .$$

In consequence, the equation (2) can be written in the form

$$\begin{aligned} b'^{*}\tfrac{d}{dt}y_{(-1)} + a\tfrac{d}{dt}y_{(0)} + b'\tfrac{d}{dt}y_{(1)} &= \\ = c_{(0)} + \left(g'^{*} - j\omega b'^{*}\right)y_{(-1)} + fy_{(0)} &+ \left(g' + j\omega b'\right)y_{(1)}\ . \end{aligned} \tag{5}$$

Let us successively multiply both sides of this equation by the factors

$$e^{\pm j\omega t}\ ,\ e^{\pm j2\omega t}\ ,\ e^{\pm j3\omega t}\ ,\ldots$$

In the resulting equations we take into account the relation

$$e^{jm\omega t}\frac{d}{dt}y_{(x)} = \frac{d}{dt}y_{(x+m)} - jm\omega y_{(x+m)}\ ,$$

which holds true for arbitrary integer m.

This way we come to the following infinite system

$$\mathcal{F}\frac{d}{dt}y = y + \mathcal{Z}c\ . \tag{6}$$

Here we have the infinite vectors

$$y = \left[\ldots, y_{(-3)}, y_{(-2)}, y_{(-1)}, y_{(0)}, y_{(1)}, y_{(2)}, y_{(3)}, \ldots\right]^T\ , \tag{7}$$

$$c = \left[\ldots, c_{(-3)}, c_{(-2)}, c_{(-1)}, c_{(0)}, c_{(1)}, c_{(2)}, c_{(3)}, \ldots\right]^T \tag{8}$$

and the matrices

$$\mathcal{F} = \begin{bmatrix} \ddots & & & & & & & \\ & b'^* & a & b' & & & & \\ & & b'^* & a & b' & & & \\ & & & b'^* & a & b' & & \\ & & & & b'^* & a & b' & \\ & & & & & b'^* & a & b' \\ & & & & & & & \ddots \end{bmatrix}\ , \tag{9}$$

$$\mathcal{Z} = \begin{bmatrix} \ddots & & & & & & & \\ & n_{(-3)} & d_{(-2)} & w_{(-1)} & & & & \\ & & n_{(-2)} & d_{(-1)} & w_{(0)} & & & \\ & & & n_{(-1)} & d_{(0)} & w_{(1)} & & \\ & & & & n_{(0)} & d_{(1)} & w_{(2)} & \\ & & & & & n_{(1)} & d_{(2)} & w_{(3)} \\ & & & & & & & \ddots \end{bmatrix}\ , \tag{10}$$

where

$$n_{(x)} = g'^* + jx\omega b'^*\ , \qquad d_{(x)} = f + jx\omega a\ , \qquad w_{(x)} = g' + jx\omega b'\ . \tag{11}$$

Let us note that the infinite system (6) is equivalent to the equation (1) and it has constant coefficients.

For an arbitrary natural number r we set $y_{(x)} = c_{(x)} = 0$ if $|x| > r$. This way we reduce the system (6). If the parameters a and b generate the nonsingular matrix

$$\mathcal{F}_r = \begin{bmatrix} a & b' & & & \\ b'^* & a & b' & & \\ & b'^* & a & b' & \\ & & & \ddots & \\ & & & b'^* & a \end{bmatrix}$$

corresponding to the matrix $\mathcal{F}$, we obtain the system

$$\frac{d}{dt}y_r = \mathcal{A}_r y_r + \mathcal{B}_r c_r \ , \tag{12}$$

where

$$y_r = \left[y_{(-r)}, y_{(-r+1)}, \ldots, y_{(r)}\right]^T \ ,$$

$$c_r = \left[c_{(-r)}, c_{(-r+1)}, \ldots, c_{(r)}\right]^T \ ,$$

the matrix $\mathcal{A}_r$ is simply the inverse $\mathcal{F}_r^{-1}$ of the matrix $\mathcal{F}_r$, the matrix $\mathcal{B}_r = \mathcal{F}_r^{-1}\mathcal{Z}_r$ and $\mathcal{Z}_r$ corresponds to the matrix $\mathcal{Z}$.

So we came to the system (12) which is well-known in the control theory: y_r, c_r, $\mathcal{A}_r$ and $\mathcal{B}_r$ are called the state vector, the input vector, the circuit matrix and the input matrix, respectively. Now we can solve it via the Sylvester method (see e.g. [3]).

3 Example of application

Within the framework of a practice, the three–phase induction motors runs often in asymmetrical states. The examples of asymmetries, owing to motor demages, are the following: break in the stator or rotor phase, the short–circuit of turns of the winding, line–to–line fault, earth fault. Within the symmetrical–component method (the zero–, the positive– and negative–sequence systems are superposed), the primary asymmetrical system of currents and voltages is transformed to the new system of symmetrical components. At the constant rotor speed ω, equations of the mathematical model of the zero–sequence system of the three–phase slip–ring induction motor, which contain the trigonometric coefficients, are written as follows [5]

$$\begin{aligned}
&\tfrac{d}{dt}\psi_S = U_m \cos(\omega_1 t + \gamma) - R_S i_S \ , \\
&\tfrac{d}{dt}\psi_r = -R_r i_r \ , \\
&\psi_S = L_S i_S + M \cos 3(\omega t + \varphi_0)\, i_r \ , \\
&\psi_r = L_r i_r + M \cos 3(\omega t + \varphi_0)\, i_S \ ,
\end{aligned} \tag{13}$$

where

ψ_S , ψ_r and i_s , i_r are the linkage fluxes and the currents of the stator (S) and the rotor (r), respectively,

U_m , ω_1 , γ are the maximum value, the angular frequency and the initial phase of the supply voltage, respectively,

R_S , R_r are the resistances of the stator and rotor phase winding,

L_S , L_r are the inductances of the stator and rotor phase winding,

M is the maximum value of the mutual inductance between the stator and rotor phase winding,

φ_0 is the angular position of the rotor in relation to the stator for the initial instant $t = 0$.

All the time in accordance with [5], the electromagnetic torque m is expressed by the formula

$$m = -3pMi_S i_r \sin 3\,(\omega t + \varphi_0) \quad ,$$

where p is the pole–pair number of an induction motor.

Exemplary calculations based on the new method have been done for a zero–sequence system of a four–pole three–phase slip–ring induction motor. There were assumed the following rated values: power: 5.5 kW, voltage: 380 V, frequency: 50 Hz. The resistances and the inductances of the induction motor were: $R_S = 0.79\,\Omega$, $R_r = 0.11\,\Omega$, $L_S = 0.0217\,\mathrm{H}$, $L_r = 0.00228\,\mathrm{H}$, $M = 0.00568\,\mathrm{H}$. The system (16) of the DETC was solved by means of the method presented in Chapter 2. The motor currents were denoted as follows

$$i_{S(x)} = i_S e^{jx\omega t} \quad , \qquad i_{r(x)} = i_r e^{jx\omega t} \quad .$$

So there was defined the following infinite state vector:

$$i = \left[\ldots, i_{S(-6)}, i_{r(-6)}, i_{S(-3)}, i_{r(-3)}, i_S, i_r, i_{S(3)}, i_{r(3)}, i_{S(6)}, i_{r(6)}, \ldots\right]^T \quad .$$

It is interesting that the considered system consists of two independent subsystems. One of these subsystems contain the unknown vector

$$i^1 = \left[\ldots, i_{S(-6)}, i_{r(-3)}, i_S, i_{r(3)}, i_{S(3)}, \ldots\right]^T \quad ,$$

the second subsystems is in the unknown vector

$$i^2 = \left[\ldots, i_{r(-6)}, i_{S(-3)}, i_r, i_{S(3)}, i_{r(3)}, \ldots\right]^T \quad .$$

These two infinite systems, already having constant coefficients, were reduced (up to 5, 7 9 and 11 equations) by symmetrical cancelling of the rows and columns of the matrices. The eigenvalues of truncated matrices were numerically determined and the Sylvester formula was applied to produce the final results. These results appeared acceptably coherent with experimental observations and also with solutions produced by the standard Runge–Kutta method (for example, the differences between values of the transient zero-sequence component of the stator do not exceed 15, 10, 6 and 3 percentages). For r big enough the coherence is almost ideal in case of the stator current (for 11 equations the differenced do not exceed 3%), while in case of the electromagnetic torque the differences are still significant (and they decrease slowly as the parameter r determining the number of equations grows up).

4 Conclusions

The method presented in this paper yields the analytical form of the approximate solution (and the more equations are forming the truncated system, the higher is the accuracy of the approximation). Thanks to this analytical form the proposed method has the advantage of strictly numerical methods such as the Runge–Kutta one. Now an engineer can observe, in much more easy way, the values of amplitudes and frequencies of current and torque components. In consequence, he can improve the design parameters of the systems of the electric drive.

The method does not apply the Laplace transformations and, thanks to its convergence observed in many practical cases, lets to consider relatively small systems of linear equations (having the tridiagonal structure). These features are essential while comparing it to the methods given in [4] and [7].

References

1. Brown, I., Butler, O.: The zero–sequence parameters and performance of three–phase induction motors. Proc. IEE, part IV, vol.101, 1954, 219–224
2. Heller, V., Hamata, V.: Harmonic field effects in induction machines. Elsevier, Amsterdam, 1977
3. Kaczorek, T.: Theory of automatic control systems (in Polish). WNT, Warsaw, 1977
4. Levinstein, M.L.: Application of operational calculus to determination of transient states of three–phase circuits (in Polish). WNT, Warsaw, 1967
5. Rawicki, S.: Mathematical model of zero–sequence component of three–phase induction machine (in Polish). Archiwum Elektrotechniki, 1984, No.1/2, 147–168
6. Rusek, I.: Drehmoment und Strom einer über stromeinprägenden Wechselrichter gespeisten Asynchronmaschine in stationären Betriebszustand. Archiv für Elektrotechnik, vol.67, 1984, 151–160
7. Sobczyk, T.: Infinitely-dimensional linear and quadratic forms of electric machines. Rozprawy Elektrotechniczne, No.29, 1983, 101–113

Stability of Difference Schemes with Variable Weights

P.P. Matus[1]

Institute of Mathematics, Academy of Science of Belarus, Minsk, Belarus

Abstract. The paper considers stability and convergence of difference schemes with variable weights. Such schemes appears in the theory of adaptive grids, hybrid computational methods and others. Various explanatory examples are introduced.

Introduction

Difference schemes with variable weights are frequently used in numerical modelling of non–stationary problems with singularities [1]. As classical examples may serve hybrid methods, explicit–implicit schemes [2], computational algorithms on dynamic locally refined grids [3]–[9], difference schemes of increased order of approximation on irregular grids [10, 11].

Investigations on development of general stability theory of operator–difference schemes [12] for the case of difference schemes with operator–weight factors are began in present time. In connection with this in first order it should be pointed out papers [13, 14], and also my joint papers with A.A.Samarskii, P.N.Vabishchevich, V.S.Shcheglik, I.A.Mickhiliouk [3, 4], [9, 11], [15]–[17].

Obtaining of stability estimates with respect to initial data and right sides are of certain interest in such norm from which convergence of the grid method in strong norms will follow in the case of nonsmooth weight functions. Such stability estimates for two–level and three–level operator–difference are presented in this paper.

Theoretical results are illustrated by numerous examples, among which it is necessary to point out computational algorithms with adaptation on time [1], [5]–[7] and difference schemes of increased order of approximation on irregular grids [10].

1 Subsidiary results

Let us assign real finite–dimensional Hilbert space H and time grid

$$\bar{\omega}_\tau = \{t_n = n\tau,\ n = 0, 1, ..., n_0;\ \tau n_0 = t_0\} = \omega_\tau \cup \{t_0\}. \qquad (1)$$

Denote by $A, B : H \to H$ – linear operators in H that are depend of τ and t_n. Let us consider initial value problem for operator–difference equation

$$B\frac{y_{n+1} - y_n}{\tau} + Ay_n = \varphi_n, \qquad t_n \in \omega_\tau, \qquad (2)$$

$$y(0) = u_0, \tag{3}$$

where $y_n = y(t_n) \in H$ – is required function and $\varphi_n, u_0 \in H$ – are given. We will use following notation

$$y = y_n, \quad \hat{y} = y_{n+1}, \quad \check{y} = y_{n-1}, \quad y_t = \frac{\hat{y} - y}{\tau}, \quad y_{\bar{t}} = \frac{y - \check{y}}{\tau}.$$

Under two–level scheme it is meant the set of initial value problems (2), (3) that are depend of parameter τ and notation (2), (3) is called as canonical form of two–level schemes [12].

Example 1. Here and further for simplicity we will consider initial–boundary problem for heat–transfer equation of the form

$$\frac{\partial u}{\partial t} = \frac{\partial^2 u}{\partial x^2} + f(x,t), \qquad 0 < x < l, \quad 0 < t \leq t_0, \tag{4}$$

$$u(0,t) = u(l,t) = 0, \qquad u(x,0) = u_0(x). \tag{5}$$

On uniform grid $\omega = \omega_h \times \omega_\tau$, $\omega_h = \{x_i = ih, \ i = \overline{0,N}, \ hN = l\}$ we will approximate differential problem by the finite–difference scheme with weights

$$y_t = y_{\bar{x}x}^{(\sigma)} + \varphi, \qquad (x,t) \in \omega, \tag{6}$$

$$\hat{y}_0 = \hat{y}_N = 0, \qquad y_i^0 = u_0(x_i). \tag{7}$$

Here $y = y_i^n = y(x_i, t_n)$, $y_{\bar{x}x} = (y_{i+1}^n - 2y_i^n + y_{i-1}^n)/h^2$, $\varphi = f^{(\sigma)}$, $v^{(\sigma)} = \sigma\hat{v} + (1-\sigma)v$.

To present scheme (6) (7) in canonical form of two–level operator–difference scheme (2), (3) assume that $H = \Omega_h$, where Ω_h – is the set of grid functions defined on ω_h, $y_n = (y_1^n, y_2^n, ..., y_{N-1}^n)$, $u_0 = (u_1, ..., u_{N-1})$, $\varphi_n = \{\varphi_1^n, \varphi_2^n, ..., \varphi_{N-1}^n\}$,

$$(Ay)_i = \begin{cases} (2y_1 - y_2)/h^2 & , \ i = 1, \\ -y_{\bar{x}x,i} & , \ i = 2, ..., N-2, \\ (-y_{N-2} + 2y_{N-1})/h^2 & , \ i = N-1. \end{cases} \tag{8}$$

Let us use identity $v^{(\sigma)} = v + \tau\sigma v_t$. Then in operator–difference scheme (2), (3) for constant σ: $B = E + \sigma\tau A, \quad A = A^* > 0, \quad B = B^* > 0$.

Let us introduce in H scalar product $(\cdot,\cdot)$ and define self–adjoint positive operator D that is for arbitrary elements $y, v \in H$ under $y \neq 0$ following relations $(Dy, v) = (y, Dv)$, $(Dy, y) > 0$ take the place. Let us denote by H_D Hilbert space that is consist of elements of the space H and equipped with scalar product $(y,v)_D = (Dy, v)$ and norm $\|y\|_D = \sqrt{(Dy,y)}$.

Definition 1 [12]. Difference scheme (2), (3) is said to be stable on initial data and right sides if for it solution in sufficiently small $\tau < \tau_0$ and arbitrary $n = 0, 1, ..., n_0 - 1$ following estimate take place:

$$\|y_{n+1}\|_D \leq M_1\|u_0\|_D + M_2 \max_{0 \leq t' \leq t_n} \|\varphi(t')\|_*,$$

where positive constants M_1, M_2 are independent of τ, $\|\cdot\|_*$ – certain norm in H.

To use general theory of stability of operator–difference schemes of form (2), (3) it is necessary for operators A and B to satisfy to the following conditions $B(t) > 0$, $A(t) = A^*(t) > 0$, $t \in \omega_\tau$ and operator $A(t)$ is Lipschitz-continuous on t:

$$|((A(t) - A(t-\tau))v, v)| \leq \tau c_0 (A(t-\tau)v, v)$$

for all $v \in H$, $t \in \omega_\tau$, where $c_0 = \text{const} > 0$.

Let us consider one of classical results obtained in [12].

Theorem 2. *If $A = A^* > 0$ – is constant operator and*

$$B(t) \geq 0.5\tau A. \tag{9}$$

Then the scheme (2), (3) is stable with respect to initial data and right sides and following estimate is true for it solution

$$||y_{n+1}||_A \leq ||y_0||_A + ||\varphi_0||_{A^{-1}} + \sum_{k=1}^{n} \tau ||\varphi_{\bar{t},k}||_{A^{-1}}. \tag{10}$$

For the mentioned above difference scheme with weights (6), (7) under constant σ, condition (9) is fulfilled for $\sigma \geq 0,5$.

According to (10) and difference analog of embedding theorem [12]

$$||y||_C \leq \frac{\sqrt{l}}{2} ||y||_A, \qquad ||y||_C = \max_{x \in \omega_h} |y(x)|,$$

solution of difference problem is stable in grid seminorm W_2^1 and in uniform metric too.

2 Difference schemes with variable weights

Let us consider initial value problem for evolutionary equation of the first order

$$\frac{du}{dt} + Au = \varphi, \qquad 0 < t \leq t_0, \quad u(0) = u_0, \tag{11}$$

where $A : H \to H$ – linear finite–difference operator. Let $y^{(\Sigma)} = \Sigma\hat{y} + (E - \Sigma)y$, where $\Sigma = \Sigma(t)$ – linear operator and E – identical operator.

In what follows we will consider that operator A and Σ are not pairwise commutative

$$\Sigma A \neq A\Sigma.$$

We will put in correspondence two–level scheme with operator–weight factors

$$y_t + Ay^{(\Sigma)} = \varphi, \qquad y(0) = u_0 \tag{12}$$

for initial value problem (11).

Two–level operator–difference scheme (12) is reduced to canonical form (2), (3) with $B = E + \tau A\Sigma$. Because operators A and Σ are not pairwise commutative nothing can be said about sign of operator B so it is impossible to use directly results of general theory of stability [12].

Let us show some results on investigation of stability of the difference scheme (12), that were obtained in [3], [16].

Theorem 3. *Let in difference scheme (12) $A = A(t) > 0$, then for*

$$\Sigma(t) \geq 0.5E, \qquad t \in \omega_\tau \tag{13}$$

following a priory stability estimate of the difference solution with respect to initial data and right side is true

$$\|y_{n+1}\| \leq \|y_0\| + \|(A^{-1}\varphi)_0\| + \|(A^{-1}\varphi)_n\| + \sum_{k=0}^{n-1} \tau\|(A^{-1}\varphi)_{\bar{t},k}\|. \tag{14}$$

Theorem 4. *Let $A(t) = A^*(t) > 0$ – is self-adjoint operator and*

$$\Sigma(t) \geq \sigma_0 E, \quad \sigma_0 = \frac{1}{2} - \frac{1}{\tau\|A\|}, \quad t \in \omega_\tau, \tag{15}$$

then a priory estimate (14) take place for the scheme (12).

Theorem 5. *Let in difference scheme (12) $A(t) > 0$, $\varphi = \varphi_1 + \tau^{1/2}A\varphi_2$ and condition*

$$\Sigma(t) \geq (0.5 + \varepsilon)E, \qquad t \in \omega_\tau \tag{16}$$

is fulfilled, where $\varepsilon > 0$ – is arbitrary constant. Then a priory stability estimate for the difference solution of the problem (12) on initial data and right side is true

$$\|y_{n+1}\| \leq \|y_0\| + \|(A^{-1}\varphi_1)_0\| + \|(A^{-1}\varphi_1)_n\| +$$

$$+ \sum_{k=0}^{n-1} \tau\|(A^{-1}\varphi_1)_{\bar{t},k}\| + M_1 \max_{0 \leq k \leq n} \|\varphi_2(t_k)\|,$$

where $M_1 = \sqrt{t_0/(2\varepsilon)}$.

Let us show how to use obtained results for investigation of stability and convergence of the difference schemes with variable weights.

Example 2. We will consider boundary problem for transfer equation

$$\frac{\partial u}{\partial t} + \frac{\partial u}{\partial x} = f, \qquad u(x,0) = u_0(x), \quad 0 \leq x \leq l; \qquad u(0,t) = 0.$$

Differential problem is approximated by difference one

$$y_t + (y^{(\sigma)})_{\bar{x}} = \varphi, \qquad y(x_i, 0) = u_0(x_i), \qquad y(0, t_{j+1}) = 0. \tag{17}$$

Here $\sigma = \sigma(x,t) > 0$ – bounded grid function, $(x,t) \in \omega$, φ – some pattern functional from f.

Let us introduce for the problem (17) H – space of grid functions defined on $\omega_h^+ = \omega_h \cup \{x_N = l\}$ with scalar product $(y,v) = \sum_{i=1}^{N} y_i v_i h$ and norm $\|y\| = \sqrt{(y,y)}$.

We will define operator $A : H \to H$ by the following way

$$(Ay)_i = \begin{cases} y_1/h, & i = 1, \\ y_{\bar{x},i}, & i = \overline{2,N}. \end{cases}$$

It is obvious, that $A > 0$, $A \neq A^*$ – is not self–adjoint positive operator. By representing the scheme (17) in operator form (12) we find $\Sigma = \mathrm{diag}(\sigma_1^n, \sigma_2^n, ..., \sigma_N^n)$, $\Sigma : H \to H$ – linear operator. Operator inequality (13) is fulfilled for $\sigma(x,t) \geq 0.5$ and on the base of theorem 3 for solution of the difference scheme (17) take place a priory estimate (14).

Let us consider convergence of the scheme under assumptions that $\sigma(x,t)$ is limited function only. Problem for the error $z = y - u$ of the scheme (17) take the form

$$z_t + (z^{(\sigma)})_{\bar{x}} = \psi, \qquad z(x_i, 0) = 0, \quad z(0, t_{j+1}) = 0. \tag{18}$$

Obviously, that in grid nodes where function $\sigma(x,t)$ is discontinuous on variable x that is approximation $\psi = O(h + \tau^2 + \tau/h)$ is conditional even for smooth solutions. Besides this if $\sigma_t = O(\tau^{-1})$ (there is no smoothness on variable t) then $\|(A^{-1}\psi)_t\| = O(1)$. So that obtained a priory estimate on right side (14) can not be used for investigation of convergence. Let us use theorem 5 to analyze precision. By the use of identity $(u^{(\sigma)})_{\bar{x}} = u_{\bar{x}}^{(0.5)} + \tau((\sigma - 0.5)u_t)_{\bar{x}}$ we represent approximation error ψ in the form $\psi = \psi_1 + \tau^{1/2} A\psi_2$, where

$$\psi_1 = -(u_t + u_{\bar{x}}^{(0.5)}) = O(h + \tau^2), \qquad \psi_2 = -\tau^{1/2}(\sigma - 0.5)u_t = O(\tau^{1/2}).$$

According to theorem 5 for $\sigma(x,t) \geq 0.5 + \varepsilon$ the following energetic inequality take place for the problem (18)

$$\|z_{n+1}\| \leq \|(A^{-1}\psi_1)_0\| + \|(A^{-1}\psi_1)_n\| + \sum_{k=0}^{n-1} \tau\|(A^{-1}\psi_1)_{t,k}\| + M_1 \max_{0 \leq k \leq n} \|\psi_2(t_k)\|.$$

Let us show that $\|A^{-1}\psi_1\| \leq l\|\psi_1\|$. By the use of identity [2] $(A^{-1}\psi_1)_i = \sum\limits_{k=1}^{i} h(\psi_1)_k$ according to definition of scalar product in H we have

$$\|A^{-1}\psi_1\|^2 = \sum_{i=1}^{N} h\Big(\sum_{k=1}^{i} h(\psi_1)_k\Big)^2 \leq \sum_{i=1}^{N} hNh\Big(\sum_{k=1}^{N} h(\psi_1)_k^2\Big) \leq l^2\|\psi_1\|^2.$$

On the basis of carried out computations we conclude that solution converges in norm L_2 with the rate $O(h + \tau^{1/2})$ without any restrictions on relations between grid steps τ and h.

3 Sufficient stability condition for the schemes of nondivergent type

Let us consider following class of difference schemes with operator–weight factors for solution of initial value problem (11)

$$y_t + (Ay)^{(\Sigma)} = \varphi, \qquad y(0) = u_0, \tag{19}$$

that is not solution y is weighting but vector Ay. As a characteristic example we may consider explicit–implicit scheme ("classic" scheme)

$$y_{t,i} = \begin{cases} y_{\bar{x}x,i} + f_i\,,\ i - \text{even}, \\ \hat{y}_{\bar{x}x,i} + \hat{f}_i\,,\ i - \text{odd}, \end{cases}$$

that is deduced to the scheme (6) with variable weight

$$y_{t,i} = (y_{\bar{x}x,i})^{(\sigma_i)} + f^{(\sigma_i)}, \qquad \sigma_i = \begin{cases} 0,\ i - \text{even}, \\ 1,\ i - \text{odd}. \end{cases} \tag{20}$$

For $H = \omega_h$, $y = (y_1, y_2, ..., y_{N-1})$ and for $A = A^* > 0$ – constant operator of form (8), $\varphi = f^{(\Sigma)}$, $f = (f_1, f_2, ..., f_{N-1})$, $\Sigma = \text{diag}\{\sigma_1, \sigma_2, ..., \sigma_{N-1}\}$ scheme (20) is reduced to two–level operator–difference scheme (19) or to canonical form (2) with $B = E + \tau\Sigma A$, $\Sigma A \neq A\Sigma$. Let us show some results on investigation of stability of the difference schemes with variable weight factors of the form (19). It is necessary to point out that if Σ is independent from t and $\varphi = 0$ then correspondent stability estimates on initial data are obtained in [13, 14].

Theorem 6. *If in the difference scheme (19) operator $A > 0$ is constant. Then under fulfillment of operator inequality (13) following estimate take place*

$$\|Ay_{n+1}\| \le \|Ay_0\| + \|\varphi_0\| + \|\varphi_n\| + \sum_{k=0}^{n-1} \tau\|\varphi_{t,k}\|. \tag{21}$$

Remark. If $A = A^* > 0$ — constant operator then theorem 6 is true under less hard restrictions

$$\Sigma(t) \ge \sigma_0 E, \qquad \sigma_0 = 0.5 - (\tau\|A\|)^{-1}. \tag{22}$$

But obtained estimate can not be used for investigation of convergence rate of the difference schemes with discontinuous with respect to time–weights because even for smooth solutions approximation error in norm $\|\psi_t\| = O(1)$. So that we show stability estimate with respect to right side in more weak norm L_2.

Theorem 7. *Let $A = A^* > 0$ — is constant operator in the scheme(19), and $\Sigma(t) = \Sigma^*(t)$ comply with one of inequalities:* $0.5(1+\varepsilon)E \le \Sigma(t) \le E$; $E \le \Sigma(t) \le 0.5(3-\varepsilon)E$ *with* $0 < \varepsilon \le 1$. *Then for difference solution take place estimate of the form*

$$\|y_{n+1}\|_*^2 \le \|y_0\|_*^2 + \frac{1}{\varepsilon}\sum_{k=0}^{n} \tau\|\varphi_k\|^2; \qquad \|v\|_*^2 = \|v\|_A^2 + \tau\|Av\|_{\Sigma_{n-1}}^2.$$

Example 3. Schemes with adaptive grid on time [1], [5]–[7]. Examine the first boundary value problem for parabolic equation (4), (5). Let introduce uniform grid ω_h, $\omega_{\tau_0} = \{t_{n+\alpha/p} = (n+\alpha/p)\tau, \alpha = \overline{0,p}, n = \overline{0, n_0 - 1}\}$, $\bar{\omega}_1 = \{(x_i, t_{n+\alpha/p}),$ $0 < i \le m_{1n}$, $m_{2n} \le i \le N-1$, $\alpha = \overline{0,p}\}$, $\omega_2 = \{(x_i, t_{n+\alpha/p}),\ m_{1n} < i < m_{2n},$ $\alpha = \overline{0,p}\}$. Suppose that when $t = t_n$ the values of approximate solution y_i^n

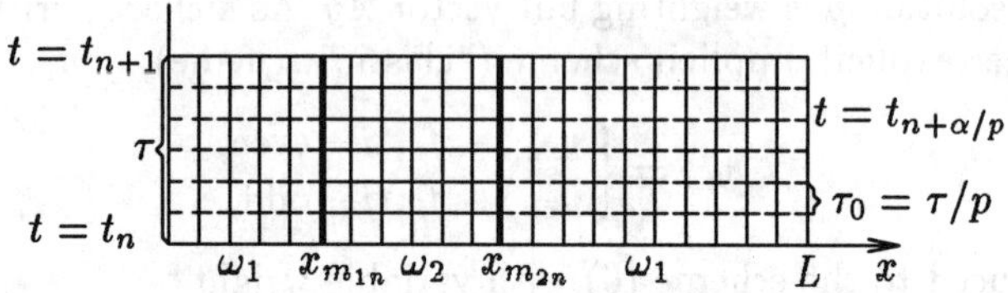

Fig. 1. Adaptive grid.

are found in the x_i $(i = \overline{0,N})$ nodes of the grid. Assume that in region ω_1 the solution is relatively smooth and naturally in this region one could use a coarse step τ to find approximate solution. Suppose it is also known that in region $\bar{\omega}_2$ solution has a singularity which leads to the need of using a rather small step $\tau_0 = \tau/p$ ($p \ge 1$ is an integer). The implicit difference scheme enabling to find rigorously the grid solution in region ω_1 with step τ and in $\bar{\omega}_2$ – with step τ_0 has a form [1]

$$\begin{aligned} \frac{y_{(\alpha+1)} - y^n}{(\alpha+1)\tau_0} &= (\Lambda y + f)_{(\alpha+1)}, \qquad (x,t) \in \omega_1; \\ \frac{y_{(\alpha+1)} - y_{(\alpha)}}{\tau_0} &= (\Lambda y + f)^{(\sigma_c)}, \qquad (x,t) \in \omega_2 \end{aligned} \tag{23}$$

in which $\Lambda y = y_{\bar{x}x}$, $\sigma_c = \text{const} > 0$, $y_{(\alpha)} = y_i^{n+\alpha/p} = y(x_i, t_{n+\alpha/p})$. Difference scheme (23) is algebraically equivalent to the scheme

$$y_{t,\alpha} = (\Lambda y + f)^{(\sigma_{\alpha+1})}, \quad (x,t) \in \omega_h \times \omega_{\tau_0}, \quad y_{t,\alpha} = (y_{(\alpha+1)} - y_{(\alpha)})/\tau_0, \tag{24}$$

$$\sigma_{(\alpha+1)} = \sigma(x, t_{n+(\alpha+1)/p}) = \begin{cases} \alpha+1, & (x,t) \in \omega_1, \\ \sigma_c, & (x,t) \in \omega_2. \end{cases}$$

Now write scheme (24) in operator form (19)

$$y_{t,\alpha} + (A y_{(\alpha)})^{(\Sigma)} = \varphi(t), \qquad t = t_{n+\alpha/p},$$

where $\varphi = f^{(\Sigma(t))}$, $\Sigma(t) = \text{diag}\{\sigma_1^{n+(\alpha+1)/p}, \ldots, \sigma_{N-1}^{n+(\alpha+1)/p}\}$. It is difficult to make sure that operators Σ and A are not commutative. Let us check the performance of the sufficient stability condition (22). It is obvious that operator inequality $\Sigma(t) \geq \sigma_0 E$, $\sigma_0 \geq 0.5 - (\tau||A||)^{-1}$ is satisfied if only $\sigma \geq \sigma_0$. Hence the difference scheme on the time–adaptive grid (23), (7) when $\sigma_c \geq \sigma_0$ is absolutely stable with respect to initial data right side and a priory estimate (21) is valid for it.

Example 4. Adaptation on space. Above we have considered difference schemes on time–adaptive grids. But adaptation on space play important rule too. For nonstationary problems this question is connected with problem of interpolation of approximate solution on level t_n if it is necessary to condense (add new nodes) grid on spatial variable for t_{n+1} – level. As input problem we will consider heat transfer problem (4), (5). Then solution y_{n+1} on new level one may find by natural algorithm

$$y_{t,i} = \hat{y}_{\bar{x}x,i} + \varphi_i, \quad \text{if } y_i^n \text{ is known,} \tag{25}$$

$$\frac{\hat{y}_i - 0.5(y_{i-1} + y_{i+1})}{\tau} = \hat{y}_{\bar{x}x} + \varphi, \quad \text{if } y_i^n \text{ is unknown.} \tag{26}$$

Lax scheme (26) is absolutely stable in norm C and may be reduced to the form

$$y_t = \hat{y}_{\bar{x}x} + \frac{h^2}{2\tau} y_{\bar{x}x} + \varphi,$$

so that local approximation take place only for $\tau > h^2$. Danger of such scheme is that it may converge (because it is absolutely stable) not to exact solution that is in fact mean divergence for $\tau \ll h^2$. Let eliminate this shortage. For this instead of scheme (26) we consider scheme

$$\frac{\hat{y}_i - 0.5(y_{i-1} + y_{i+1})}{\tau} = \left(1 - \frac{h^2}{2\tau}\right)\hat{y}_{\bar{x}x} + \varphi$$

that is equivalent to the scheme with unconditional approximation

$$y_t = \hat{y}_{\bar{x}x} + \frac{h^2}{2} y_{t\bar{x}x} + \varphi$$

and beside this it may be written in the form

$$y_t = \left(1 - \frac{h^2}{2\tau}\right)\hat{y}_{\bar{x}x} + \frac{h^2}{2\tau} y_{\bar{x}x} + \varphi, \quad \text{if } y_i^n \text{ is unknown.} \tag{27}$$

Now we may write the scheme (25), (27) as

$$y_{t,i} = (y_{\bar{x}x})^{(\sigma_i)} + \varphi_i, \quad \sigma_i = \begin{cases} 1 \quad , & \text{if } y_i^n \text{ is known,} \\ 1 - \dfrac{h^2}{2\tau}, & \text{if } y_i^n \text{ is unknown.} \end{cases} \tag{28}$$

Test of sufficient condition of stability (13) $\Sigma(t) \geq 0.5E$ of operator–difference scheme (19) lead to necessity of inequality $\sigma_i \geq 0.5$ fulfillment. It is follows from (28) that the scheme is stable in norm (21) (according to theorem 6) for $\tau \geq h^2$. Although stability conditions of the scheme (28) are similar to demands of approximation of Lax scheme (26) but for $\tau = h^{3/2}$ order of approximation of the scheme (27) $\psi = \mathrm{O}(h^2 + \tau) = \mathrm{O}(h^{3/2})$ is higher than approximation of the scheme (26): $\psi = \mathrm{O}(h^{1/2})$.

4 Operator–difference schemes of divergent type

Violation of conservative property of the scheme (27) in the case of variable weight $\sigma = \sigma(x)$, $x \in \omega_h$

$$\left((y_{\bar{x}})^{(\sigma_i)}\right)_{x,i} \neq \left(y_{\bar{x}x,i}\right)^{(\sigma_i)}.$$

is the shortage of considered above difference schemes with variable weight factors of the form (27).

In this paragraph we select the one more class of difference schemes with variable weights, that is used for equation with divergent operators. In solution of heat transfer equation such class of difference schemes correspond to the weighting not of solution but stream. Let us consider two–level operator difference scheme of form

$$y_t + T^*(STy)^{(\Sigma)} = \varphi, \quad t \in \omega_\tau, \quad y(0) = u_0, \tag{29}$$

where, as above $v^{(\Sigma)} = \Sigma\hat{v} + (E - \Sigma)v$, $\Sigma(t)$, $S(t)$, T, T^* — linear limited operators. Schemes of such type may occur in approximation of parabolic equations with operator $Lu = \operatorname{div}(k \operatorname{grad} u)$. We will define as H^* Euclidean space with scalar product $(y, v]$ and norm $\|y]\| = \sqrt{(y, y]}$. Let operator T act from H to H^*, operators S, Σ are given in H^* and operator T^* act from H^* to H. Then operator $A = T^*ST$ transform H to H, that is $A : H \to H$. We will consider that:

1. Operators T and T^* are constant and conjunctive in following mean $(Ty, v] = [y, T^*v)$ for all $y \in H$, $v \in H^*$;
2. $S = S(t)$, $\Sigma = \Sigma(t)$ — self–adjoint commutative operators, that satisfy for all $t \in \bar{\omega}_\tau$ to inequalities: $k_1 E \leq S(t) \leq k_2 E$, $0.5(1 + \varepsilon)E \leq \Sigma(t) \leq k_3 E$, where k_1, k_2, k_3 — positive constants, that are independent from grid steps, $0 < \varepsilon \leq 1$ — arbitrary real number;
3. $S(t)$ — Lipschitz-continuous on t with constant c_0: $\|S(t + \tau) - S(t)]\| \leq \tau c_0$.

Theorem 8 [15]. *Let conditions 1.—3. take place. Then operator-difference scheme (29) is absolutely stable with respect to initial data and right side and following estimate is true for it solution*

$$\|y_n\|^2_{A_n} \leq M^2\Big(\|y(0)\|^2_{A_0} + 0.5\sum_{k=0}^{n-1} \tau\|\varphi_k\|^2\Big), \tag{30}$$

where $M = \exp(0.5\, c\, t_0)$, $c = 2c_0(1 + k_3)/(k_1\sqrt{\varepsilon})$.

Remark. If operator S is constant, then theorem 8 take place for $\Sigma(t) \geq 0.5E$ with $M = 1$.

Remark. For investigation of convergence of conservative schemes with discontinuous weight function [6] we will represent right side φ in the form $\varphi = \xi + \tau^{1/2}T^*\eta$. Then estimate (30) may be written in the form

$$\|y_n\|^2_{A_n} \le M^2\Big(\|y(0)\|^2_{A_0} + 0.5\sum_{k=0}^{n-1}\tau\big(\|\xi_k\|^2 + \frac{4}{\varepsilon k_1}\|\eta_k]|^2\big)\Big). \tag{31}$$

Example 5. Let us use conservative scheme with variable weight factors

$$y_{t,i} = \big((y_{\bar{x},i})^{(\sigma_i^n)}\big)_{x,i} + \varphi_i, \qquad (x,t) \in \omega. \tag{32}$$

to approximate the problem (4), (5).

We write scheme (32) (7) in operator form (29). For this define $y_n = (y_1^n, y_2^n, ..., y_{N-1}^n)$, $H = \Omega_h$ — is the space of grid functions given on the grid ω_h with scalar product $(y,v) = \sum\limits_{x\in\omega_h} y(x)v(x)h$; $y, v \in H$ and $H^* = \Omega_h^+$ — is space of functions defined on the set of nodes $\omega_h^+ = \omega_h \cup \{x_N = l\}$. Scalar product in H^* is $(y,v] = \sum\limits_{x\in\omega_h^+} y(x)v(x)h$; $y, v \in H^*$. Then operator $A = T^*T : H \to H$ take the form (8) where $(Ty)_1 = y_1/h$, $(Ty)_i = y_{\bar{x},i} = (y_i - y_{i-1})/h$ for $i = 2,3,...,N-1$, $(Ty)_N = -y_{N-1}/h$ so that $v = Ty \in H^*$ if $y \in H$; $(T^*v)_i = -v_{x,i} = -(v_{i+1} - v_i)/h$ for $i = 1,2,...,N-1$ so that $T^*v \in H$ if $v \in H^*$; $(\Sigma v)_i^n = \sigma_i^n v_i^n$ $i = 1,2,..,N$ that is $\Sigma v \in H^*$ if $v \in H^*$.

So operator S is constant, then according to the first remark to the theorem 8, for $\sigma_i^n \ge 0.5$ following estimate of the scheme (32), (7) stability with respect to initial data and right side take place:

$$\|y_{n\bar{x}}]|^2 \le \big(\|u_0\|^2_A + 0.5\sum_{k=0}^{n-1}\tau\|\varphi(t_k)\|^2\big).$$

Let us consider convergence. It is necessary to point out that because of discontinuity of weight function σ approximation error as in norm C as in norm L_2 have the conditional character [6]. We will use estimate (31) to obtain estimates of error $z = y - u$ without restrictions on steps τ and h. To meet this aim by the use of identity $(v_{\bar{x}})^{(\sigma_i)} = (v_{\bar{x}})^{(0.5)} + 0.5\tau(\sigma_i - 0.5)v_{\bar{x}t}$ we will represent residual of the scheme in form (for simplicity we will consider that $f = 0$) $\psi = \xi + \tau^{1/2}T^*\eta$, $\xi = -u_t + (u_{\bar{x}}^{(0.5)})_x = O(h^2 + \tau^2)$, $T^*\eta = -\eta_x$, $\eta = -\tau^{1/2}(\sigma - 0.5)u_{\bar{x}t} = O(\tau^{1/2})$.

¿From theorem 8 for $\sigma \ge 0.5(1+\varepsilon)$

$$\|z\|_C \le 0.5\sqrt{l}\|z_{\bar{x}}]| = 0.5\sqrt{l}\|z\|_A \le c(h^2 + \sqrt{\tau}).$$

Example 6. Difference scheme of increased order of approximation on irregular grids [10, 11]. By the use of fact that on arbitrary irregular grid

$$\hat{\bar{\omega}}_h = \{x_i = x_{i-1} + h_i,\ i = 1,2,...,N,\ x_0 = 0,\ x_N = l\}$$

$$u_{\bar{x}\hat{x},i} - u''(\bar{x}_i) = O(\hbar_i^2),$$

where $u_{\bar{x}\hat{x},i} = ((u_{i+1} - u_i)/h_{i+1} - (u_i - u_{i-1})/h_i)/\hbar_i$, $\hbar_i = 0.5(h_i + h_{i+1})$, $\bar{x}_i = x_i + (h_{i+1} - h_i)/3 = (x_{i-1} + x_i + x_{i+1})/3$ conservative difference schemes of the second order of approximation $O(\hbar_i^2 + \tau)$ on irregular grid $\omega = \hat{\omega}_h \times \omega_\tau$ of the form

$$y_{t,i} + \left((h^2/6)y_{t\bar{x}}\right)_{\hat{x},i} = y^{(\sigma_0)}_{\bar{x}\hat{x},i} + f^{(\sigma_0)}(\bar{x}_i, t_n) \tag{33}$$

are constructed in the paper [10] for parabolic equation (4), where $\sigma_0 > 0$ — is constant. We may write scheme (33) as the scheme with variable weight (32), where $\sigma_i = \sigma_0 - h_i^2/(6\tau)$, $\varphi_i^n = f^{(\sigma_0)}(\bar{x}_i, t_n)$. This scheme is stable with respect to initial data and right side in energetic norm H_A for $\sigma_i \geq 0.5$ or $\sigma_0 \geq 0.5 + h_i^2/(6\tau)$. Note that in paper [11] difference schemes of the second order of approximation on irregular grids are constructed for elliptic equations of arbitrary dimension.

5 Three–level operator–difference schemes with variable weight factors

Similar investigation may be fulfilled for evolutionary equation of the second order

$$\frac{d^2u}{dt^2} + Au = \varphi, \; 0 < t \leq t_0, \qquad u(0) = u_0, \quad \frac{du}{dt}(0) = \bar{u}_0. \tag{34}$$

We will put in correspondence for the initial value problem (34) three classes of operator–difference schemes:

$$y_{\bar{t}t} + Ay^{(\Sigma_1,\Sigma_2)} = \varphi(t), \; t \in \omega_\tau, \qquad y(0) = y_0, \quad y_t(0) = \bar{y}_0. \tag{35}$$

$$y_{\bar{t}t} + (Ay)^{(\Sigma_1,\Sigma_2)} = \varphi(t), \; t \in \omega_\tau, \qquad y(0) = y_0, \quad y_t(0) = \bar{y}_0. \tag{36}$$

$$y_{\bar{t}t} + T^*(STy)^{(\Sigma_1,\Sigma_2)} = \varphi(t), \; t \in \omega_\tau, \qquad y(0) = y_0, \quad y_t(0) = \bar{y}_0. \tag{37}$$

in which $v^{(\Sigma_1,\Sigma_2)} = \Sigma_1\hat{v} + (E - \Sigma_1 - \Sigma_2)v + \Sigma_2\check{v}$ and $\Sigma_k(t) : H \to H$ — are uniformly limited operators, where $\Sigma_k A \neq A\Sigma_k$, and $\Sigma_k(t)$ are not Lipschitz-continuous operator with respect to t. Results obtained in [4], [15], [17] under such strong restrictions on operators make possible investigations of various classes of difference schemes with variable and discontinuous weight factors for hyperbolic equations of the second order. Note, for example, that schemes (35)–(36) may be written in canonical form [12]

$$Dy_{\bar{t}t} + By_{\overset{\circ}{t}} + Ay = \varphi(t) \tag{38}$$

with operator D and B respectively: $D = E + 0.5\tau^2 A(\Sigma_1 + \Sigma_2)$, $B = \tau A(\Sigma_1 - \Sigma_2)$ for the scheme (35); and $D = E + 0.5\tau^2(\Sigma_1 + \Sigma_2)A$, $B = \tau(\Sigma_1 - \Sigma_2)A$ for the scheme (36).

Requirements $D = D^* > 0$, $D(t)$ — Lipschitz-continuous on t and $B(t) \geq 0$ are not fulfilled because of imposed above conditions on operators A and Σ_k. So that schemes (35)–(37) are not belong to "original" class of three–level difference schemes, for which general stability theory take the force.

Because of restriction on volume of this paper we will consider only result from [17] and example from [10].

Theorem 9. *Assume that in initial value problem (35) $A = A^* > 0$ — is constant operator and*

$$\Sigma_2(t) = \Sigma_2^*(t) \geq -\frac{1}{\tau^2\|A\|}E, \qquad \Sigma_1(t) \geq \Sigma_2^{(0.5)} + 0.5E.$$

Then the following estimate take place

$$\|y_{n+1}\| \leq M_1\left(\|y(0)\| + \|y_t(0)\|_{R(\tau)} + \|A^{-1}\varphi(0)\|\right) + \\ + \|A^{-1}\varphi(t)\| + M_2 \max_{0\leq t'<t} \|A^{-1}\varphi_t(t')\|,$$

where $R(\tau) = A^{-1} + \tau^2(E + \Sigma_2)$; $M_1, M_2 = \mathrm{const} > 0$.

Example 7. Constructed in [14] conservative difference scheme of the second order of approximation $O(\hbar^2 + \tau)$ on arbitrary irregular grid $\hat{\omega}_h \times \omega_\tau$ for wave equation belong to the class of schemes with variable weight factors and have the form

$$y_{\bar{t}t,i} = \left(y_{\bar{x}}^{(\sigma_{1i},\sigma_{1i})}\right)_{\hat{x},i} + \varphi_i, \qquad \sigma_{1i} = \sigma - h_i^2/(6\tau^2). \qquad (39)$$

This scheme may be reduced to the equation (37) or to canonical form

$$Dy_{\bar{t}t} + Ay = \varphi$$

for $A = T^*T$, $A = A^* > 0$, $D = E + \tau^2 T^* \Sigma_1 T$, $\Sigma_1 = \mathrm{diag}\{\sigma_{11}, \sigma_{12}, ..., \sigma_{1N}\}$. On the base of results obtained in [10], [17] we conclude that scheme (39) is stable with respect to initial data and right side in energetic norm H_A for $\sigma \geq h_1^2/(6\tau^2) + (1+\varepsilon)/4$ where $\varepsilon > 0$ — is arbitrary real number.

References

1. Matus, P.P.: On one class of difference schemes on composite grids for nonstationary problems of mathematical physics. Differents. Urav. **26** (1990) 1241–1254
2. Samarskii, A.A., Gulin, A.V.: Theory of Stability of Difference Schemes. Moscow. Nauka (1973)
3. Matus, P.P., Mickhiliouk, I.A.: Difference schemes with variable weights for the systems of hyperbolic equations. Mathematical Modelling **5** (1993) 35–60
4. Matus, P.P.: Difference Schemes on Time-Adaptive Grids for Boundary Problems of Mathematical Physics. Author's summary of dissertation of doctor of physics and mathematics. Moscow (1995)
5. Matus, P.P.: On the problem of constructing difference schemes for multidimentional parabolic equations on time-adaptive grids. Differents. Urav. **27** (1991) 1961–1971
6. Matus, P.P.: Conservative difference schemes for parabolic and hyperbolic second order equations in subdomains. Differents. Urav. **29** (1993) 700–711
7. Matus, P.P.: Conservative difference schemes for quasilinear parabolic equations in subdomains. Differents. Urav. **29** (1993) 1222–1231
8. Matus, P.P.: About difference schemes on composite grids for hyperbolic equations. Zh. Vychisl. Mat. Mat. Fiz. **34** (1994) 870–885

9. Vabishchevich, P.N., Matus, P.P., Rychagov, V.G.: About one class difference schemes on dynamic locally–condensed grids. Differents. Urav. **31** (1995) 849–857
10. Samarskii, A.A., Vabishchevich, P.N., Matus P.P.: Difference schemes of increased order of precision on nonuniform grids. Differents. Urav. **32** (1996) 313–322
11. Samarskii, A.A., Vabishchevich, P.N., Matus, P.P.: Difference schemes of increased order of approximation on nonuniform grids for elliptic equations. Dokl. Akad. Nauk Belarusi (to appear)
12. Samarskii, A.A.: Theory of Difference Schemes. Moscow. Nauka (1989)
13. Samarskii, A.A., Gulin, A.V.: Criteria of stability of the class of difference schemes. Dokl. Ross. Akad. Nauk **330** (1993) 694–695
14. Samarskii, A.A., Gulin, A.V.: On the stability of one class of difference schemes. Differents. Urav. **29** (1993) 1163–1174
15. Vabishchevich, P.N., Matus, P.P., Shcheglik, V.S.: Operator–difference equations of divergent type. Differents. Urav. **30** (1994) 1175–1186
16. Vabishchevich, P.N., Matus, P.P.; Two–level difference schemes with variable weights. Dokl. Akad. Nauk Belarusi **37** (1993) 15–17
17. Vabishchevich, P.N., Matus, P.P., Shcheglik, V.S. Difference schemes with variable weights for evolutionary equations of second order. Dokl. Akad. Nauk Belarusi **38** (1994) 13–15

Relative Precision in the Inductive Assertion Method

W. D. Maurer

Department of Electrical Engineering and Computer Science
The George Washington University
Washington, DC 20052, USA
and
Naval Research Laboratory, Code 5580
4555 Overlook Avenue SW
Washington, DC 20375, USA

Abstract

The inductive assertion method of Floyd is here applied to programs involving floating point numbers, using a new verification condition generator for C programs known as ProveIt. The exit assertions of such programs need to state that the answers are correct to within some tolerance. We define this notion of tolerance, and show that it is equivalent to Olver's notion of relative precision. As an example, we present an O(*ln n*) program which takes the *n*th power of *a*, and show that the speed of the program does not improve the relative precision, which remains 2*n* rather than the expected 2 *ln n*.

1 Introduction

We describe here some numerical aspects of our ongoing program to develop computing tools to aid in the proofs of correctness of programs. For this purpose we use the inductive assertion method, described in §2.1 below. A commonly encountered variant of this method is the back substitution method, developed by Hoare [1]. We do not use this, for reasons which we describe in §2.2 below.

The proof of correctness of a numerical algorithm requires special techniques to take account of the fact that the results are often not exact. It is possible to perform floating point calculations on quantities which remain exact throughout the program, and we describe this in §3.1 below. Normally, however, we can prove only that the calculated answers are correct within a certain tolerance, which may be absolute or relative, as described in §3.2 and §3.3 below.

2 Proving the Correctness of Programs

Floyd [2] introduced the inductive assertion method of proving programs correct. Over the years we have developed numerous improvements to this method (see, for example, [3, 4, 5, 6]). We here improve on what is often called forward error analysis (see, e. g., [7]) by applying it to specific programs, running on specific computers. (The present system does not perform backward error analysis in the sense of [7].) We are also not concerned here with approximations of differential equations by difference equations, and the attendant errors (even when minimized by predictor-corrector or Runge-Kutta methods).

2.1 The Basic Inductive Assertion Method

What we seek to prove about a program P is that if some assertion A (called the *entry assertion*) is true when P starts, then:

(*a*) P will not loop endlessly;

(*b*) P will not encounter any undefined statements;

(*c*) when P ends, some other assertion B (called the *exit assertion*) will be true when it ends. (It is possible for a program to end abnormally in more than one place, in which case there may be a separate exit assertion associated with each of these.)

By an undefined statement we mean one whose effect is undefined when it is executed because some requirement for its proper execution is not satisfied. Examples are division by zero and the execution of a command involving a subscripted variable at a time when some subscript is out of range. A function call may also be undefined and will always be so if its own entry assertion is not satisfied.

To prove statements like this, we introduce *intermediate assertions* at certain key points in our program. These must be such that any closed loop in the program contains at least one such key point. One way to insure this is to choose, as our key points, the *join points*; that is, those points with indegree greater than 1, in the directed graph of the program. Using join points as key points minimizes the size of the proof in certain ways which we have described in [4].

The *assertion points* in a program are now the entry, exit, and intermediate assertion points. A *verification path* is a path in the directed graph of the program from one assertion point to another, without any assertion points in between. The *initial* (*final*) *assertion* of such a path is that assertion which is associated with its first (last) statement. An *execution* of the path is an execution, in sequential order, of all of its statements except the last. (The last statement in a path is always executed as part of the next path after this one, unless it is a halt statement, in which case its execution is undefined and meaningless.)

We then associate with each verification path a *verification condition*. There are three levels of these, depending on what you want to prove about your program. At the first level, the condition says that any execution of the path which starts with its initial assertion valid will end with its final assertion valid. If all first-level verification conditions hold, the program is partially correct; we have shown statement (*c*) above, but not (*a*) or (*b*).

At the second level, we introduce *statement guards*; these are such that a statement will never be undefined if its guard is satisfied. (They need not be necessary and sufficient conditions, and indeed such conditions are occasionally impossible to formulate.) The verification condition now says that, if a statement is executed within a path (starting with its initial assertion valid), its guard will be true at that time. This must hold for all statements in the path. If all second-level verification conditions hold, the program is clean; we have shown statements (*b*) and (*c*) above, but not (*a*).

At the third level, we introduce *termination expressions*. Each of these is an integer expression associated with some intermediate assertion, which must imply that it is non-negative. The third level affects only verification conditions which start and end at *intermediate* assertion points. In an execution which starts at point L1, with associated termination expression T1 having value V1 at the time, and ends at point L2 with associated expression T2 having value V2 at the time, the third-level expression must imply V2 < V1. If all third-level conditions hold, the program is correct; we have shown statements (*a*), (*b*), and (*c*) above. All this is shown rigorously in [3].

2.2 Back Substitution and Modification Indices

The back substitution method was developed by Hoare [1] and widely used by many researchers. In [5] and [6] we introduced the modification index method as an alternative to back substitution. We give here a brief review of these two methods.

Given a verification path containing K+1 statements, an execution of this path is an execution of the first K statements (call them $P_1, P_2, ..., P_K$). Suppose now that P_J is $V_J := E_J$, for $1 \leq J \leq N$, and we have some assertion A_{K+1} at the end of the path. We can now generate assertions $A_1, ..., A_K$ to be associated with the statements $P_1, ..., P_K$. We generate A_K by substituting E_K for V_K throughout A_{K+1}; then A_{K-1} by substituting E_{K-1} for V_{K-1} throughout A_K; and so on, moving backward through the path, which is why this method is called *back substitution*. The actual assertion at the start of the path must then imply A_1 as thus derived, and this becomes the (first-level) verification condition of the path.

In [5] and [6] we showed that the back substitution method can lead, in the worst case, to verification conditions whose size is exponential in the path length. This size dependence is reduced to polynomial (and usually linear) size by using the *modification index method* introduced there, in which a new variable is generated every time an existing variable changes its value. The successive modifications of J are called J1, J2, and so on, unless (for example) there is already a variable J2 in this path, in which case J3 is the next modification after J1. In turn, the successive modifications of J2 are J2A, J2B, and so on, unless there is already a J2B in the program, in which case we skip to J2C.

The modification index method also has advantages in proofs of correctness in which the human programmer is constructing the proof. Classic unsolvability results imply that there is no way of using a universal system as a "magic box" to prove correctness in the way that a compiler, for example, compiles a program without human intervention. Our expectation has therefore always been that the programmer constructs the proof and debugs the proof, with the system acting as a proof checker. Modification indices provide a natural way of talking about the values which a variable can take on as a program is executed.

2.3 Verification Condition Generators

Computer systems which apply the inductive assertion method have been written many times, and for many source languages, since the pioneering work of King [8]. Such a program is generally known as a verification condition generator, or VCG. Our particular VCG is called ProveIt, and it acts on programs written in the source language C. (It is itself written in C++.) ProveIt differs from other VCGs in the source language (although see [9] for another C verifier); in the great variety of C statements handled; and in many other ways, one of which is of primary concern to us here: it is the first VCG to treat floating point computation in relative precision (see §3.3 below).

3 Floating Point Computation

Suppose now that we are proving the correctness of a program which manipulates real numbers. In what sense is such a program correct, if it incurs floating point approximation and roundoff errors? Clearly we cannot prove that the answers are exact, if in fact they are not exact. We can, however, prove that the answers are within a certain tolerance. The question then is how to define this tolerance, and how to use it in the formulation of intermediate and exit assertions.

3.1 Exact Arithmetic and Overflow

Before discussing tolerance, we briefly take up those cases in which it is not needed because the computed quantities, including all intermediate results, are exact. The techniques employed in such a case are extensions of those employed for integer overflow. In both cases we have two basic problems: formulating the proper statement guards and formulating the proper entry assertion.

Given a statement which changes the value of a variable in such a way that the computed result might overflow, there is a statement guard which says that the result has not in fact overflowed. The form of this guard depends on the size of the variable in bits, and whether it is a signed integer, an unsigned integer, or an exact floating point quantity. Once these guards are formulated, the derivation of second and third level verification conditions proceeds as before. The entry assertion may now include the condition that all variables are initially within their ranges.

For d-bit integers, the ranges are 0 through 2^d-1 (unsigned) or -2^{d-1} through $2^{d-1}-1$ (signed twos' complement arithmetic). For floating point numbers, we have found it useful to employ three kinds of range. Most exact floating arithmetic is done with floating point numbers which in fact are integers within the contiguous range of -2^m through 2^m where m is the number of bits in the mantissa. Note that 2^m+2 and -2^m-2 are also representable exactly, but 2^m+1 and -2^m-1 are not. The guards $-2^m \leq v$ and $v \leq 2^m$ may therefore be included for each modified exact floating point variable v. These are to be included both as hypotheses in the initial assertion and as statement guards to be proved in the usual way.

Two further levels of exact floating point range are useful. One of these encompasses all exactly representable integers, including 2^m+2 and the like, as above. The other one encompasses all exactly representable numbers whatsoever, most of these being fractions of the form $i/2^m$ where i is some exactly representable integer. In either of these cases, the guards which say that v is representable are treated as before.

A more difficult problem concerns the formulation of an entry assertion which will hold if and only if the computation will not overflow. For a computation involving several variables, the formulation of all intermediate results, together with the statement that none of these overflow, is actually the shortest way to state this assertion! Of course this is not meaningful if what we need are guidelines within which we can easily determine usable values of our input quantities.

In our work we have found a simple but radical solution to this problem: we ask users to specify the normal ranges in which their variables lie, and then show that, if all variables are initially within these ranges, overflow will never occur. If the normal ranges increase, we redo these calculations. Normally, of course, mathematicians seek necessary and sufficient conditions, and so we use this method only where the determination of such conditions is intractable.

3.2 Absolute Precision

We now proceed to those cases in which the results are known only within some tolerance. Suppose that we have a variable v which is supposed to be equal to v', but which in fact is only known to be close to v'. It is not too hard to see that the obvious definition of tolerance, namely that v is in the interval from $v'-\delta$ to $v'+\delta$ for some δ, is usable only in a small number of cases, in most of which the absolute value of v is bounded both above and below (say, $0.5 \leq v \leq 2$). In fact, if v' is very small, absolute precision is meaningless; for example, if $v' = \delta/2$, then v is in the interval from $-\delta/2$ to $3\delta/2$. In other words, we do not even know, in this case, whether v is positive, negative, or zero.

There are a few very simple cases, involving addition and subtraction only, in which this is actually acceptable. In any event, we can define $v \approx v'$, with absolute precision δ, if $v'-\delta \leq v \leq v'+\delta$. This is called absolute precision by contrast with relative precision (see §3.3 below), and not to imply that the results are "absolutely precise" (that is, no tolerance whatsoever).

From now on we take ε to be that number (normally 2^{-m} for an m-bit mantissa) such that $1-\varepsilon$ is the largest exactly representable number less than 1. The representable numbers between $\frac{1}{2}$ and 1 are $\frac{1}{2}$, $\frac{1}{2}+\varepsilon$, $\frac{1}{2}+2\varepsilon$, ..., $1-\varepsilon$, 1. Thus $v-\varepsilon$ is the largest exactly representable number less than v, for $\frac{1}{2} < v \leq 1$. By adding k to all exponents, we can see, therefore, that $v-2^k\varepsilon$ is the largest exactly representable number less than v, for $2^{k-1} < v \leq 2^k$. (For example, the largest representable number smaller than 21 is $21-32\varepsilon$.)

For every positive number v, let p_v be the largest power of 2 which is not greater than v. If p_v is in the floating point range, then there is a representable number v' such that $v \approx v'$ with absolute precision $p_v\varepsilon$. (For example, all numbers from 21 down to $21-16\varepsilon$ can be represented by 21, and those from there down to $21-48\varepsilon$ can be represented by $21-32\varepsilon$.)

Suppose now that we have N positive numbers $x_1, ..., x_n$, which are represented in some computer by the floating point numbers $x'_1, ..., x'_n$. If these numbers are all bounded from above, in absolute value, by some quantity M, then $x_i \approx x'_i$ with absolute precision $p_M\varepsilon$. From this we should see the need to keep M as small as possible, as well as one of the disadvantages of absolute precision. (Relative precision, as we shall see, does not depend on how large a number is.)

Consider now the addition of two such numbers. If $v_1'-\delta_1 \le v_1 \le v_1'+\delta_1$ and $v_2'-\delta_2 \le v_2 \le v_2'+\delta_2$, then $(v'_1+v'_2)-(\delta_1+\delta_2) \le v_1+v_2 \le (v'_1+v'_2)+(\delta_1+\delta_2)$. That is, if $v_3 = v_1+v_2$, $v'_3 = v'_1+v'_2$, and $\delta_3 = \delta_1+\delta_2$, then $v_3'-\delta_3 \le v_3 \le v_3'+\delta_3$. Addition, in other words, adds the tolerances. (Subtraction does also.) If we wish to add together all the numbers $x'_1, ..., x'_n$, the result will be equal to the sum of $x_1, ..., x_n$ with absolute precision $2np_M\varepsilon$. The factor of 2 may seem surprising, but for each x_i there are two possible sources of error: we may have $x_i \neq x'_i$, and also the addition itself may have an inexact result, which has to be rounded.

3.3 Relative Precision

The idea of relative precision was introduced by Olver [10] as an improvement in mathematical elegance over former methods of determining the accuracy of floating point calculations. Independently of Olver, we have developed a notion of relative precision (see [3]) which we will show to be equivalent to Olver's.

Following [10], we first redefine $v \approx v'$, with absolute precision δ, if $-\delta \le v'-v \le \delta$; and we now define $v \approx v'$, with relative precision δ, if $e^{-\delta} \le v'/v \le e^{\delta}$. For Olver, the fundamental property of $v \approx v'$ with relative precision δ is that it is equivalent to $ln\ v \approx ln\ v'$ with absolute precision δ (this clearly holds because the first definition above is obtain by taking natural logarithms in the second). Proved properties of absolute precision therefore correspond to proved properties of relative precision.

In [3] we developed our own notion of floating point approximation, based on the quantity ε introduced in §3.2 above. We first define *flap*(x), the floating approximation of x, as follows. If $x = 0$, then *flap*(x) = 0. If $x > 0$, and if k is the smallest integer such that $2^k > x$, then *flap*(x) = $int(\frac{1}{2}+x\cdot 2^{t-k})2^{k-t}$, where *int*($e$) is the greatest integer in e. If $x < 0$, then *flap*(x) = $-$*flap*($-x$). We now define x to be exactly representable, or *fexact*(x), if either $x = 0$ or there exist e and a with $|x| = a\cdot 2^{e-t}$, $-2^{s-1} \le e \le 2^{s-1}-1$, and $-2^{t-1} \le e \le 2^{t-1}-1$ (here e is the s-bit exponent and a is the t-bit mantissa, both as integers).

It is then not difficult to show two basic properties of *flap*. First, *fexact*(*flap*(x)) if *flap*(x) is in the floating point exponent range, that is, $2^{-m} \le |flap(x)| \le 2^{m-1}$ where $m = 2^{s-1}$. Furthermore, $1-\varepsilon \le v/flap(v) \le 1/(1-\varepsilon)$, or *fapproxr*($v$, *flap*($v$), 1) in our notation. Note that $1/(1-\varepsilon) = 1+\varepsilon+\varepsilon^2+...$, and if ε^2 and higher order terms could be neglected, we would have $1-\varepsilon \le v/flap(v) \le 1+\varepsilon$, but this definition is inferior because it is not symmetric. (If $1-\varepsilon \le v/v' \le 1+\varepsilon$, then $1/(1-\varepsilon) \ge v'/v \ge 1/(1+\varepsilon)$ and thus $1/(1+\varepsilon) \le v'/v \le 1/(1-\varepsilon)$; but $1/(1-\varepsilon) > 1+\varepsilon$ and thus $v'/v \le 1/(1-\varepsilon)$ does not imply $v'/v \le 1+\varepsilon$.)

In general, we define *fapproxr*(v, v', k) as $(1-\varepsilon)^k \le v/v' \le (1-\varepsilon)^{-k}$. This is equivalent to Olver's relative precision $k\ ln\ (1-\varepsilon)$, since $(1-\varepsilon)^{\pm k} = e^{\pm k\ ln(1-\varepsilon)}$, and so the basic properties of Olver's notion apply to ours. In particular, multiplication and division add the relative precisions, while the sum of positive numbers with relative precision i and j has relative precision *max*(i, j). Also, $u \approx v$ and $v \approx w$ imply $u \approx w$, and the relative precisions add in this case as well.

4 Floating Point Application of the Assertion Method

Here is an example of the use of relative precision in verification conditions. Consider the following program:

{z=1.0; j=1; *goto loop*2; *loop*: j=j+j; *loop*2: *if* ($j \leq n$) *goto loop*; k=n; *goto loop*4; *loop*3: z=z*z; *if* ($k < j$) *goto loop*4; k=k–j; z=z*a; *loop*4: j=j/2; *if* ($j > 0$) *goto loop*3;}

If you don't know what this program does, it is not obvious, even after a great deal of perusal. Actually, it calculates $z = a^n$, using repeated squaring. Thus for $n = 23$, instead of multiplying a by itself 23 times, it calculates sequentially a^2, a^4, a^5, a^{10}, a^{11}, a^{22}, and a^{23}. The total number of multiplication steps here is not greater than 2 *ln n*, and we would like to investigate the precision of the result. (This is exactly the kind of program that ought to be proved correct: it is short, fast, and hard to understand.)

Let us first assume exact arithmetic. The entry assertion is then $n \geq 0$, and the exit assertion is $z = a^n$. We have two join points, *loop*2 and *loop*4; we can get to each of these either in normal sequence or by a *goto* (or, for *loop*4, by either of two *goto*s). There are six verification paths. One goes from the start to *loop*2; one goes from *loop*2 around the loop, to *loop*2 again; one goes from *loop*2 to *loop*4; two different paths go from *loop*4 around the loop, to *loop*4 again (one for $k < j$ and the other for $k \geq j$); and one path goes from *loop*4 to the end of the program.

The intermediate assertion at *loop*2 is that $j/2 \leq n$ and $j = 2^p$ for some $p > 0$. As we go around this loop, we add 2^p to itself, producing 2^{p+1}. If $j \leq n$ fails, then $j > n$ but we still have $j/2 \leq n$, so that, if $j1$ is the new value of j, then $j1/2 \leq n < j1$. We then set k equal to n, so that now $j/2 \leq k < j$ for the new value of j.

The intermediate assertion at *loop*4, if the arithmetic were exact, would be that j is still 2^p for some $p > 0$; that $j/2 \leq k < j$, as above; and that, if q and r are the quotient and remainder, respectively, upon dividing n by j (so that $n = qj+r$, $q \geq 0$, $r \geq 0$, and $r < j$), then $z = a^q$. Note that if ($j > 0$) fails at the end, then we must have had $j = 1$ before dividing j by 2, and thus $r = 0$ (since $r \geq 0$ and $r < 1$) and $n = qj+r = q$, implying $z = a^n$, which is what we want.

This proof for exact arithmetic is now modified so as to produce a proof in the general case. This time we have $z \approx a^n$ at the end, and $z \approx a^q$ at *loop*4, with some relative precision. The total number of multiplication steps here is not greater than 2 *ln n*, and one might, therefore, expect that the relative precision would be 2 *ln n* at the end, and 2 *ln q* at *loop*4. What we will show, in fact, is that this is not the case; indeed, although the given program is fast, it is no more precise than simply multiplying a by itself n times.

This rather surprising result is obtained as follows. Consider the path going around the second loop here with $k < j$. Its (first-level) verification condition, in the exact arithmetic case, is that if $j = 2^p$, $j/2 \leq k < j$, $n = qj+r$, $q \geq 0$, $r \geq 0$, $r < j$, $z =$

a^q, $j1 = j/2$, $j1 > 0$, $z1 = z \cdot z$, and $k < j1$, then (for some $p1$, and for $q1$ and $r1$ defined as q and r were) $j1 = 2^{p1}$, $k/2 \leq j1 < k$, $z1 = q1 \cdot j1 + r1$, $q1 \geq 0$, $r1 \geq 0$, $r1 < j1$, and $z1 = a^{q1}$. In fact, $p1 = p-1$ (and $j1 = j/2 = 2^p/2 = 2^{p-1} = 2^{p1}$); $q1 = 2q$; and $r1 = r$. Thus $n = qj+r = 2q \cdot (j/2)+r$ (this holds since $j = 2^p$) $= q1 \cdot j1+r1$; $q1 = 2q \geq 0$; $r1 = r \geq 0$; and $z1 = z \cdot z = a^q \cdot a^q = a^{2q} = a^{q1}$. Finally, $j1/2 < j/2 \leq k < j1$, completing the proof of validity of this verification condition.

The problem arises when we pass to the general case. In calculating the new value of z, we are assuming that the relative precision would increase by 2. In fact, however, the relative precision doubles. We saw in §3.3 above that multiplication adds the relative precisions. Whatever the relative precision d of z is, when we multiply z by z, the resulting relative precision is therefore $d+d$, or $2d$. Thus as the exponent of a doubles, so does the relative precision, which therefore remains linear in the exponent. What the inductive assertion method has shown us here is therefore that, if we want to improve the precision of calculating a^n, we need a different idea.

References

1. Hoare, C. A. R., *An axiomatic basis for computer programming*, Communications of the ACM 12, 10 (1969), pp. 576-580, 583.

2. Floyd, R. W., *Assigning meanings to programs*, Proc. Symp. Applied Math. 19 (Mathematical Aspects of Computer Science), American Mathematical Society, Providence, R. I., 1967, pp. 19-32.

3. Maurer, W. D., The *correctness of computer programs, Part I: preliminary version,* Memorandum GWU-IIST-88-23, Institute for Information Science and Technology, George Washington University, August 1988.

4. Maurer, W. D., *Some minimization theorems for verification conditions*, 8th International Conference on Computing and Information (ICCI '96), Waterloo, Ontario, Canada, June 1996.

5. Maurer, W. D., *A new method of generating verification conditions,* Proc. 1977 Conf. on Information Sciences and Systems, Johns Hopkins Univ., Baltimore, March 1977, pp. 128-133.

6. Maurer, W. D., *The modification index method of generating verification conditions,* Proc. 15th Annual Southeast Regional Conf., Biloxi, Miss., April 1977, pp. 426-440.

7. Ralston, A., and P. Rabinowitz, *A First Course In Numerical Analysis* (2nd ed.), McGraw-Hill, 1978.

8. King, J. C., *A program verifier*, Ph. D. Thesis, Department of Computer Science, Carnegie-Mellon University, 1969.

9. Subramanian, S., and J. V. Cook, *Mechanical verification of C programs*, Trusted Information Systems, 444 Castro St., Suite 800, Mountain View, CA.

10. Olver, F. W. J., *A new approach to error arithmetic*, SIAM J. Numer. Anal. 15, 2 (Apr. 1978), pp. 368-393.

Third Order Explicit Method for the Stiff Ordinary Differential Equations

Medovikov*. Alexei. A.

Institute Numerical Mathematics.
117334 Russia. Leninski prospect 32-a, Moscow, e-mail: nucrect@inm.ras.ru

Abstract. The time-step in integration process has two restrictions. The first one is the time-step restriction due to accuracy requirement τ_{ac} and the second one is the time-step restriction due to stability requirement τ_{st}. The stability property of the Runge-Kutta method depend on stability region of the method. The stability function of the explicit methods is the polynomial. The stability regions of the polynomials are relatively small. The most of explicit methods have small stability regions and consequently small τ_{st}. It obliges us to solve the ODE with the small step size $\tau_{st} \ll \tau_{ac}$. The goal of our article is to construct the third order explicit methods with enlarged stability region (with the big τ_{st}: $\tau_{st} \geq \tau_{ac}$). To achieve this aim we construct the third order polynomials: $1 - z + z^2/2 - z^3/6 + \sum_{i=4}^{n} d_i z^i$ with the enlarge stability regions. Then we derive the formula for the embedded Runge-Kutta third order accuracy methods with the stability functions equal to above polynomials. The methods can use only three arrays of the storage. It gives us opportunity to solve large systems of differential equations.

1 Introduction.

Let us consider the system of ordinary differential equations:

$$\frac{du}{dt} = f(u,t), \qquad u|_{t=t_0} = u_0, \tag{1}$$

with sufficiently smooth function $f(u,t)$ and Runge-Kutta explicit method:

$$\begin{aligned} Y_i &= y_0 + h \sum_{j=1}^{s} a_{ij} f(t_0 + c_j h, Y_j), \\ y_1 &= y_0 + h \sum_{j=1}^{s} b_j f(t_0 + c_j h, Y_j). \end{aligned} \tag{2}$$

The table of the method (2) is the following:

* This article was written with the kind help of professors Lebedev V.I., Wanner G., Hairer E. and Russian Fund Fundamental Researches.

$$\begin{array}{c|ccccc} 0 & 0 & 0 & \dots & 0 \\ c_2 & a_{21} & 0 & \dots & 0 \\ c_3 & a_{31} & a_{32} & \dots & 0 \\ \vdots & \vdots & \vdots & \ddots & \vdots \\ c_s & a_{s1} & \dots & a_{s,s-1} & 0 \\ \hline 1 & b_1 & b_2 & \dots & b_s \end{array}, or \quad \begin{array}{c|c} \mathbf{c} & A \\ \hline 1 & \mathbf{b}^t \end{array}, \tag{3}$$

where $\mathbf{c^t} = (0, c_2, c_3, \dots, c_s)$, $\mathbf{b^t} = (b_1, \dots, b_s)$, $\mathbf{e^t} = (1, \dots, 1)$. Let us apply method (3) for the simple test problem:

$$u' = -\lambda u, \qquad u\Big|_{t=t_0} = u_0 = y_0. \tag{4}$$

The numerical result is expressed in term of stability function:

$$y_1 = R_s(\lambda h)y_0 = \Big(1 + \sum_{i=0}^{s-1} \mathbf{b}^t A^i \mathbf{e}(-\lambda h)^{i+1}\Big)y_0. \tag{5}$$

Stability function $R_s(z), z = \lambda h$ is the polynomial of degree s.

Definition1.1: *We will call a region* $U = \{z : |R_s(z)| \le 1\}$ *stability region.*

Definition1.2: *We will call an interval* $I = U \bigcap [0, \infty[$ *real stability interval.*

We will construct a third order explicit methods with enlarged real stability interval. This means that order conditions must satisfy:

$$\begin{aligned} \textstyle\sum_i^s b_i = 1, \\ \textstyle\sum_{i,j=1}^s b_i a_{ij} = \sum_i^s b_i c_i = 1/2, \\ \textstyle\sum_{i,j,k=1}^s b_i a_{ij} a_{jk} = \mathbf{b}^t A^2 \mathbf{e} = 1/6, \\ \textstyle\sum_{i,j,k=1}^s b_i a_{ij} a_{ik} = \sum_i^s b_i c_i^2 = 1/3. \end{aligned} \tag{6}$$

From the order conditions (6) it follows that:

$$R_s(z) = 1 - z + z^2/2 - z^3/6 + \sum_{i=4}^s d_i z^i. \tag{7}$$

Absolute value of stability function $|R_s(z)|$ decreases in the small vicinity of the point $z = +0$. It means that stability region U and real stability interval are not empty set. The aim of our speculations is to construct function (7) with the maximum or 'nearly' real stability interval.

Theorem 1.3([2]): *If* $s > 3$ *and*

$$\max_{0 \le t \le M} |R_s(t)| \le \eta = 1. \tag{8}$$

The polynomial $R_s(t)$ *has a maximal value* M *if there exist* $n + 1 - k$ *(k=3 for the third order methods) points* $t_i : 0 < t_1 < t_2 \dots < t_{n+1-k} \le M$ *such that:*

$$R_s(t_1) = -\eta = -1, R_s(t_2) = +\eta = 1, \dots, R_s(z_{n+1-k}) = (-\eta)^{n+1-k}, \tag{9}$$

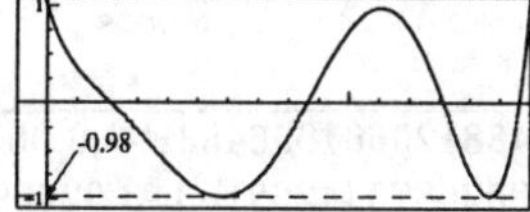

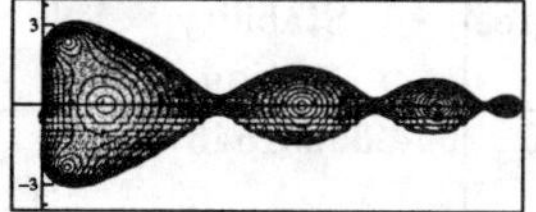

Fig. 1. Third order polynomial and it's stability region.

For practical calculations we will take η not equals to 1 but 'near' optimal value i.e. $\eta \approx 0,98 < 1$. It gives us more stability for many practical problems but it decreases real stability interval. The algorithm of construction of the optimal second and third order polynomials was described in the works of Kovalenko, Medovikov and Lebedev [2, 4]. Here we reproduce only roots of the some third order polynomials. These roots we used for the calculation of the parameters of the third order methods and used it in the program DUMKA3. In the **table 1.1** we represent the roots of the polynomials reduced to the interval [0,1]. To calculate the roots $\{\gamma_i\}_1^s$ of stability function (7) you have to multiply roots from the table and the parameter of stability region from the second column of the table.

TABLE 1.1 The roots of the third order stability polynomials.

Degree	Stability region	Roots
3	2.5005127005	(0.638297752962491,0.0000000000E+0), (0.280728100628313,0.722787568361731), (0.280728100628313,-0.722787568361731)
6	15.96769685542662	(1.316188704042163E-001,0.00000000E+0), (5.217998085157960E-2,-1.472133692919474E-1), (5.217998085157960E-2,1.472133692919474E-1), 5.397127885347366E-1,8.210181090527608E-1, 9.792807844727616E-1
9	38.31795251315424	(5.707036703430203E-2,0.000000000E+0), (2.307842599268251E-2,-6.407179746204085E-2), (2.307842599268251E-2,6.407179746204085E-2), 2.650900447972151E-1,4.564443606877882E-1, 6.434022749551114E-1,8.066819069334241E-1, 9.275476063065802E-1,9.917867224786107E-1
15	109.9635751502718	(2.027487087133956E-2,0.00000000E+0), (8.316021861212946E-3,-2.280465621150311E-2), (8.316021861212946E-3,2.280465621150311E-2), 1.002074585617464E-1,1.825798689818222E-1, 2.765670440070977E-1,3.791595999288834E-1, 4.861890912273565E-1,5.931003215440658E-1 6.952794913650315E-1,7.882921191244471E-1, 8.680911462040328E-1,9.311998388255081E-1, 9.748666481030083E-1,9.971869309844605E-1

Degree	Stability region	Roots
36	644.3020154572322	(3.488129601956453E-3,0.000000000E+0) (1.441852687344269E-3,-3.926697828110118E-3) (1.441852687344269E-3,3.926697828110118E-3) 1.778795197054982E-2,3.322288535643486E-2, 5.192760724673927E-2,7.393156860341134E-2, 9.912096771275233E-2,1.273224815410135E-1, 1.583307101392706E-1,1.919142061529804E-1, 2.278200310551857E-1,2.657765020564892E-1, 3.054957626118224E-1,3.466761985247533E-1, 3.890048660640773E-1,4.321599477783770E-1, 4.758132470553982E-1,5.196327141245989E-1, 5.632849914369575E-1,6.064379628604224E-1, 6.487632895565875E-1,6.899389145951866E-1, 7.296515180870848E-1,7.675989046864292E-1, 8.034923056404166E-1,8.370585780984603E-1, 8.680422850997063E-1,8.962076405169030E-1, 9.213403042299864E-1,9.432490139204415E-1, 9.617670411057193E-1,9.767534603597429E-1, 9.880942220801697E-1,9.957030206519658E-1, 9.995219514104168E-1
48	1145.804705468596	(1.963379226522905E-3,0.000000000E+0), (8.122094719300525E-4,-2.210430853325917E-3), (8.122094719300525E-4,2.210430853325917E-3), 1.006092366834490E-2,1.874663175494967E-2, 2.938902171273918E-2,4.199468711545039E-2, 5.653456570661632E-2,7.295652733345687E-2, 9.119500031334919E-2,1.111743678974924E-1, 1.328104842674921E-1,1.560115518572836E-1, 1.806787615356976E-1,2.067068432197892E-1, 2.339845866405663E-1,2.623953579846579E-1, 2.918176237070712E-1,3.221254861934987E-1, 3.531892327054925E-1,3.848758973520228E-1, 4.170498349018754E-1,4.495733047149258E-1, 4.823070627473591E-1,5.151109593853533E-1, 5.478445407355358E-1,5.803676509224324E-1, 6.125410328983705E-1,6.442269252512830E-1, 6.752896524954869E-1,7.055962063465749E-1, 7.350168155120516E-1,7.634255015728661E-1, 7.907006185865761E-1,8.167253741097289E-1, 8.413883294145599E-1,8.645838767626988E-1, 8.862126916957106E-1,9.061821584084480E-1, 9.244067663858447E-1,9.408084766063493E-1, 9.553170557451642E-1,9.678703769472033E-1, 9.784146858826100E-1,9.869048309461775E-1, 9.933044566154082E-1,9.975861591395987E-1, 9.997316038935454E-1

2 Third order Runge-Kutta method with enlarged stability region

Let us consider method (2) with $s = 3k$. Stability function can be represented in the form:

$$R_s(z) = \prod_{i=1}^{k}(1 - d_1^i z + d_2^i z^2 - d_3^i z^3) = \prod_{i=1}^{k} R_{s_i}^i(z). \quad (10)$$

We will consider composition method [1] where method (2) is comprises k submethods:

$$\begin{aligned}
v_0 &= y_0 \\
Y_2^i &= v_{i-1} + h a_{21}^i f(t_{i-1}, v_{i-1}) \\
Y_3^i &= v_{i-1} + h\left(a_{31}^i f(t_{i-1}, v_{i-1}) + a_{32}^i f(t_{i-1} + hc_2^i, Y_2^i)\right) \\
v_i &= v_{i-1} + h\left(b_1^i f(t_{i-1}, v_{i-1}) + b_2^i f(t_{i-1} + hc_2^i, Y_2^i) + b_3^i f(t_{i-1} + hc_3^i, Y_3^i)\right) \\
t_i &= t_{i-1} + h * (b_1^i + b_2^i + b_3^i) = t_{i-1} + h_i, \\
i &= 1, \ldots, k = s/3, \\
y_1 &= v_k.
\end{aligned}$$

Stability function of each submethods equals to:

$$R_{s_i}^i = 1 - d_1^i z + d_2^i z^2 - d_3^i z^3 \quad (11)$$

To satisfy order conditions (6) we have to set:

$$\begin{cases}
b_1^i + b_2^i + b_3^i = d_1^i \\
b_2^i c_2^i + b_3^i c_3^i = d_2^i \\
b_3^i a_{32}^i c_2^i = d_3^i \\
b_2^i (c_2^i)^2 + b_3^i (c_3^i)^2 = (d_3^i)^3/3 + ((d_1^i)^2 - 2d_2^i)t_{i-1} = B
\end{cases} \quad (12)$$

The fourth equality of (12) follows from the fourth equality of (6). We want every submethod to satisfy cubature formulae:

$$\int_{t_{i-1}}^{t_i} \tau^2 d\tau = b_1^i (t_{i-1})^2 + b_2^i (hc_2^i + t_{i-1})^2 + b_3^i (hc_3^i + t_{i-1})^2 = (t_i^3 - t_{i-1}^3)/3 \quad (13)$$

Another equalities of the system (12) follow from the stability function of the submethod (11). The system (12) has four equations and six variables. There are two free parameters and one can choose it to achieve some additional purpose. We consider only one case here. Let us take:

$$b_1^i = a_{31}^i, b_2^i = a_{32}^i.$$

In this case formulas for parameters of the methods are the following:

$$\begin{aligned}
b_3^i &= \rho_1^i,\\
c_3^i &= \rho_2^i + \rho_3^i,\\
c_2^i &= \frac{B - b_3^i (c_3^i)^2}{\rho_2^i \rho_3^i},\\
a_{32}^i &= \frac{\rho_2^i \rho_3^i d_3^i}{B - b_3^i (c_3^i)^2},\\
b_2^i &= a_{32}^i,\\
b_1^i &= \rho_2^i + \rho_3^i - b_2^i,\\
a_{31}^i &= b_1^i,\\
i &= 1, \ldots, k = s/3,
\end{aligned} \tag{14}$$

where ρ_1^i,ρ_2^i,ρ_3^i are the inverse values for the roots of the stability function $R_{s_i}^i$. Now to construct the method we take the roots of the third order polynomial, separate it on the groups : $1/\gamma_{j_{i_1}} = \rho_1^i, 1/\gamma_{j_{i_2}} = \rho_2^i, 1/\gamma_{j_{i_3}} = \rho_3^i$ (ρ_1^i is chosen as a real value), construct stability function R_3^i and solve systems (12) for $i = 1, \ldots, k = s/3$. This method require only three arrays to store and it is sufficient also for step-size control procedure. Consider the algorithm used in program DUMKA3. Let us h be a step-size of the method, $\{h_i\}_{i=1}^{k=s/3}$ -step sizes of each submethod $\sum_i^{k=s/3} h_i = h$. We remind that value h is chosen so that the spectrum of the problem lie inside the stability region: $h \leq M/\lambda_{max}$.

$$\begin{aligned}
&v_0 = y_0\\
&Y_2 = v_{i-1} + h a_{21}^i f(t_{i-1}, v_{i-1})\\
&Y_3 = Y_2 + h\left((a_{31}^i - a_{21}^i) f(t_{i-1}, v_{i-1}) + a_{32}^i f(t_{i-1} + h c_2^i, Y_2)\right)\\
&v_i = Y_3 + h b_3^i f(t_{i-1} + h c_3^i, Y_3)\\
&\quad i = 1, \ldots, k = s/3,\\
&y_1 = v_k.
\end{aligned} \tag{15}$$

Let us check the order conditions (6). Submethod (15) is Runge-Kutta method with stability function $R_3^i = (1 - d_1^i z + d_2^i z^2 - d_3^i z^3)$ for every i. Stability function of the method is the product $R_s = \prod_{i=1}^{k=s/3} R_3^i$. We use polynomials with the roots from the table 1.1 so we have automatically that $R_s = 1 - z + z^2/2 - z^3/6 + \sum_{i=4}^{s} d_i z^i$. Hence the first three equality of the order conditions (6) are satisfied. The fourth one follows from (13):

$$\int_{t_0}^{t_s} \tau^2 d\tau = \sum_{i=1}^{k=s/3} \left(\int_{t_{i-1}}^{t_i} \tau^2 d\tau \right) = \sum_{i=1}^{k=s/3} \frac{t_i^3 - t_{i-1}^3}{3} = \frac{t_{k=s/3}^3 - t_0^3}{3}.$$

Consequently the last of (6) equalities of the order conditions is satisfied as well:

$$\sum_{i=1}^{s} b_i c_i^2 = \frac{1}{3}.$$

Finally we consider the time-step control procedure. We took this idea from the book by Hairer E., Wanner G.[1]. We use embedded formula to step-size control procedure. For this aim let us derive second order method with solution $\hat{y}_1$. With the help of the same speculation as in [1] we derive the formula:

$$\begin{aligned} err &= \|\hat{y}_1 - y_1\|, \\ h_{new} &= h \min(fac_{max}, \max(fac_{min}, fac(tol/err)^{1/(p+1)})), \end{aligned} \tag{16}$$

where $\|.\|$ is some norm in R^n, h- step-size of the previous step, fac_{min} and fac_{max} - some factors of step-size diminishing and increasing, tol is the tolerance which we require from our calculations and p -accuracy of the embedded method ($p = 2$ in our case). To calculate $\hat{y}_1$ consider the last step in a method (15). Solution $y_1 = v_k$ is the third order solution, so we can calculate second order solution $\hat{v}_{k-1}$ in the point $t_{k-1} = t_k - hc_4^k = t_0 + h$:

$$\hat{v}_{k-1} = v_k - h(c_4^k)\frac{f(t_k, v_k) + f(t_{k-1}, v_{k-1})}{2},$$

and recalculate solution with the second order $\hat{v}_k$ in the point t_k with the help of the formulas:

$$\hat{Y}_2 = \hat{v}_{k-1} + h(c_2^k)\frac{f(t_{k-1}, v_{k-1}) + f(t_{k-1} + hc_2^k, Y_2)}{2},$$

$$\hat{Y}_3 = \hat{Y}_2 + h(c_3^k - c_2^k)\frac{f(t_{k-1} + hc_2^k, Y_2) + f(t_{k-1} + hc_3^k, Y_3)}{2},$$

$$\hat{v}_k = \hat{Y}_3 + h(c_4^k - c_3^k)\frac{f(t_{k-1} + hc_3^k, Y_3) + f(t_k, v_k)}{2}.$$

We can derive:

$$\begin{aligned} \hat{y}_1 - y_1 = \hat{v}_k - v_k = -h(c_4^k)\tfrac{f(t_k,v_k)+f(t_{k-1},v_{k-1})}{2} &+ \\ h(c_2^k)\tfrac{f(t_{k-1},v_{k-1})+f(t_{k-1}+hc_2^k,Y_2)}{2} &+ \\ h(c_3^k - c_2^k)\tfrac{f(t_{k-1}+hc_2^k,Y_2)+f(t_{k-1}+hc_3^k,Y_3)}{2} &+ \\ h(c_4^k - c_3^k)\tfrac{f(t_{k-1}+hc_3^k,Y_3)+f(t_k,v_k)}{2} & \end{aligned} \tag{17}$$

We will use this difference to calculate next step size by formula (16). The algorithm (15) require 3 arrays to store, moreover the last step in (15) use only 2 arrays. This gives us opportunity to calculate the difference (17) without any help of additional array. We calculate the sum:

$$Z2 = h\frac{c_2^k - c_4^k}{2}f(t_{k-1}, v_{k-1}) + h\frac{c_3^k}{2}f(t_{k-1} + hc_2^k, Y_2),$$

after the second step in the formula(15), then calculate $f(t_{k-1} + hc_3^k, Y_3)$ and add to the sum:

$$Z2 = Z2 + h\frac{c_4^k - c_2^k}{2}f(t_{k-1} + hc_3^k, Y_3),$$

after calculation of the solution $y_1 = v_k$ we calculate $f(t_k, v_k)$ which we need for the next step and add this term to the sum:

$$Z2 = Z2 - h\frac{c_3^k}{2} f(t_k, v_k).$$

The sum $Z2$ we use as a difference $\hat{y}_1 - y_1$ and substitute it to the formula (16) calculate new step size. If $||Z2||$ exceeds tolerance: $||\hat{y}_1 - y_1|| > tol$ we can reject this step. We have written the program DUMKA3 to test this algorithm and next chapter is dedicated to numerical results.

3 Numerical results.

To test the program DUMKA3 we use the test problems from the book of Hairer E. and Wanner G.[1]. We take three programs RKC(Sommeijer[3]), DUMKA ([4]) and DUMKA3 and solve these problems. The results are represented in the figure 2 in axes: y- the time of calculations, x - accuracy. To use the program

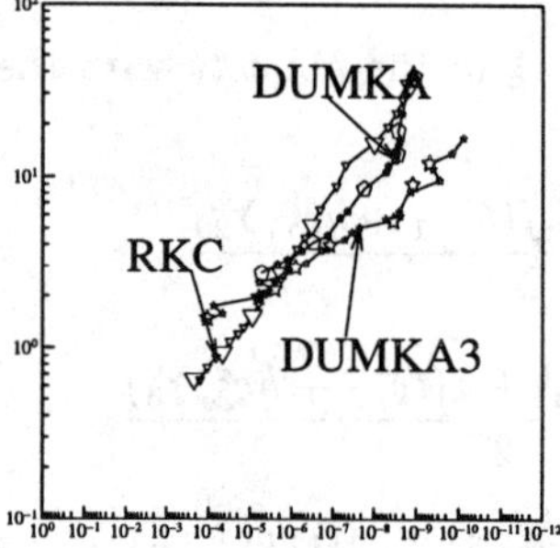

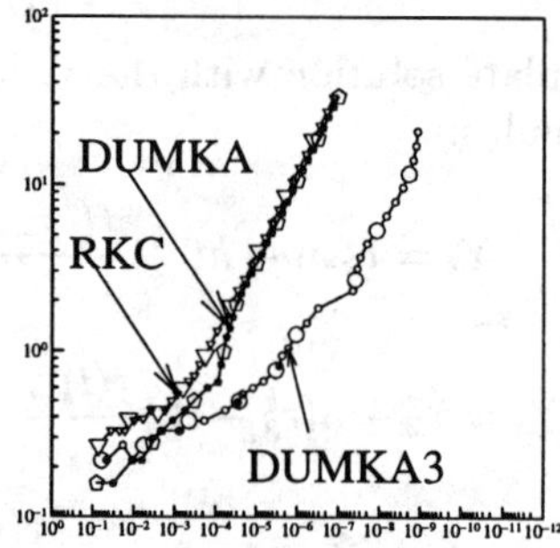

Fig. 2. The result of comparison of the programs for problems: 1-CUSP, 2-BURGERS

DUMKA3 one needs to write two SUBROUTINEs: subroutine of calculation of the right hand of equation and program for evaluation maximal eigen value λ_{max} and the value of $COU = 2/\lambda_{max}$. It is convenient to use and you don't need calculations of Jacoby matrix and linear algebra procedures.

References

1. Hairer E., Wanner G. (1991) Solving ordinary differential equations 2. Stiff problems. Springer-Verlag Berlin Heidelbers New York London Paris Tokyo.
2. Lebedev V.I., Medovikov A.A. (1995) The methods of the second order accuracy with variable time steps. Izvestiya Vuzov. Matematika. N10. Russia.
3. P.J. van der Houwen, B.P. Sommeijer (1980) On the internal stability of explicit m-stage Runge-Kutta methods for large m-values. Z.Angew. Math. Mech. 60.
4. Lebedev V.I. (1987) Explicit difference scheams with time-variable steps for solution of stiff system of equations. Preprint DNM AS USSR N177.

Justification of Difference Schemes for Derivative Nonlinear Evolution Equations

Tadas Meškauskas and Feliksas Ivanauskas

Department of Differential Equations and Numerical Analysis, Vilnius University, Naugarduko 24, 2006 Vilnius, Lithuania

Abstract. We consider the difference schemes applied to a derivative nonlinear system of evolution equations. For the boundary-value problem with initial conditions

$$\frac{\partial u}{\partial t} = A\frac{\partial^2 u}{\partial x^2} + B\frac{\partial u}{\partial x} + f(x,u) + g(x,u)\frac{\partial u}{\partial x}, \quad (t,x) \in (0,T] \times (0,1),$$
$$u(t,0) = u(t,1) = 0, \quad t \in [0,T],$$
$$u(0,x) = u^{(0)}(x), \quad x \in (0,1)$$

we use the Crank-Nicolson discretizations. A is complex and B – real diagonal matrixes; u, f and g are complex vector-functions. The analysis shows that proposed schemes are uniquely solvable, convergent and stable in a grid norm W_2^2 if all (diagonal) elements in $Re(A)$ are positive. This is true without any restrictions on the ratio of time and space grid steps.

1 Introduction

In recent years there has been a growing interest in nonlinear evolution equations. Such well-known equations (as well as their systems) as nonlinear Schrödinger equation (NLS), nonlinear reaction-diffusion equation (NLRD) and the nonlinear Kuramoto-Tsuzuki equation (NLKT) appear in many models of mathematical physics. For example, one often finds NLS in nonlinear optics [1], plasma physics [2]. NLRD systems are used in investigating a wide class of nonlinear processes [3, 5]. NLKT describes the behavior of two-component systems in a neighborhood of a bifurcation point [3, 4]. In some models there is necessary to study the effects born by higher order perturbations and the derivative nonlinear (DN) equations appear.

We deal with the difference schemes applied to a derivative nonlinear system of evolution equations. For the boundary-value problem with initial conditions

$$\frac{\partial u}{\partial t} = A\frac{\partial^2 u}{\partial x^2} + B\frac{\partial u}{\partial x} + f(x,u) + g(x,u)\frac{\partial u}{\partial x}, \qquad (t,x) \in Q, \tag{1}$$
$$u(t,0) = u(t,1) = 0, \qquad t \in [0,T], \tag{2}$$
$$u(0,x) = u^{(0)}(x), \qquad x \in \Omega \tag{3}$$

we use the Crank-Nicolson discretization. Here $u = (u_1, u_2, \ldots, u_n)$ is a complex vector-function, $\Omega = (0,1)$ and $Q = (0,T] \times \Omega$. Diagonal matrices A and B contain complex and real coefficients, respectively.

We consider the "truncation" of corresponding Cauchy problem to a bounded domain. Such approach is often used solving the problem numerically.

2 Difference scheme

Denote the diagonal elements of matrices A and B by a_{jj} and b_{jj}, respectively. By introducing new functions $y_j = u_j \exp(b_{jj}x/2a_{jj})$ one may neglect the first order partial derivatives $\partial u_j/\partial x$ in the linear part of system (1). Also note that since the matrix A is diagonal, there is no essential difference between the study of system (1) and the study of one equation. For simplicity we assume that the nonlinear functions $f(x,u)$ and $g(x,u)$ do not depend on x. Therefore we further consider one equation

$$\frac{\partial u}{\partial t} = a\frac{\partial^2 u}{\partial x^2} + f(u) + g(u)\frac{\partial u}{\partial x}, \qquad (t,x) \in Q. \tag{4}$$

(4) is a DN Schrödinger equation if $Re(a) = 0$. In the case $Im(a) = 0$ (4) represents a DN reaction-diffusion equation and, finally, when both $Re(a), Im(a) \neq 0$, it stands for a DN Kuramoto-Tsuzuki type equation.

We assume that:

a) the partial derivatives of f and g with respect to u are continuous, and

$$|f(u)|, \quad |g(u)|, \quad \left|\frac{\partial f(u)}{\partial u}\right|, \quad \left|\frac{\partial g(u)}{\partial u}\right| \le \varphi(|u|),$$

where φ is a continuous nondecreasing function,

b) $f(0) = g(0) = 0$,

c) $Re(a) \ge \delta > 0$.

Conditions a) and b) are satisfied for the models [1–6]. The condition c) means the positivity of the heat conduction coefficient.

Using the notation of [6], we introduce the uniform grids ω_τ and Ω_h with steps τ and h for the variables t and x, respectively. We relate the problem (4),(2), (3) with the following Crank–Nicolson type symmetric difference scheme:

$$v_t = a\,\mathring{v}_{\bar{x}x} + f\left(\mathring{v}\right) + g\left(\mathring{v}\right)\mathring{v}_{\overset{\circ}{x}}, \qquad (t,x) \in \omega_\tau \times \Omega_h, \tag{5}$$

$$v(t,0) = v(t,1) = 0, \qquad t \in \bar{\omega}_\tau, \tag{6}$$

$$v(0,x) = u^{(0)}(x), \qquad x \in \bar{\Omega}_h, \tag{7}$$

where $\hat{v} = v(t+\tau, x)$, $\mathring{v} = (\hat{v} + v)/2$, $v_{\overset{\circ}{x}} = (v(t,x+h) - v(t,x-h))/2h$, $v_t = (\hat{v} - v)/\tau$, $v_{\bar{x}x} = (v(t,x+h) - 2v(t,x) + v(t,x-h))/h^2$.

The scheme (5) – (7) is implicit and nonlinear. To calculate a solution on the upper layer $t = t_{j+1}$ one can apply the iterative method:

$$\frac{v^{s+1} - v}{\tau} - a\frac{v^{s+1}_{\bar{x}x} + v_{\bar{x}x}}{2} = f\left(\frac{v^s + v}{2}\right) + g\left(\frac{v^s + v}{2}\right)\frac{v^s_{\overset{\circ}{x}} + v_{\overset{\circ}{x}}}{2}, \qquad x \in \Omega_h,$$

$$v^{s+1}(0) = v^{s+1}(1) = 0,$$

$$v^0 = v.$$

The next iteration v^{s+1} can be found by the matrix sweep method, for example.

Using the grid analogues of a new type *a priori* estimates we justify (5) – (7) difference scheme. It appears to be convergent and stable without any restrictions on the ratio of time and space grid steps. Note only, that proving the boundedness of numerical problem solution, the usual estimate for transition to the upper layer doesn't work here, as the mathematical induction, based on this estimate, fails (see further). The main difficulty concerns the treatment of the DN terms in the equation. For justification of difference schemes without gradient-dependent nonlinearity see [7]. To overcome this, we modified the above method, estimating a sum of vanishing geometric progression (see proof of Lemma 1). For details, consider the auxiliary linear difference scheme

$$v_t - a \overset{\circ}{v}_{\bar{x}x} = r, \qquad (t,x) \in \omega_\tau \times \Omega_h, \tag{8}$$

$$v(t,0) = v(t,1) = 0, \qquad t \in \bar{\omega}_\tau, \tag{9}$$

$$v(0,x) = v^{(0)}(x), \qquad x \in \bar{\Omega}_h, \tag{10}$$

with the right-hand side $r(t) \in \overset{\circ}{W}{}^1_2(\Omega_h)$, $t \in \omega_\tau$, and an initial data $v^{(0)}(x) \in \overset{\circ}{W}{}^2_2(\Omega_h)$.

Lemma 1. *Suppose the hypothesis* c) *is satisfied,* $r(t) \in \overset{\circ}{W}{}^1_2(\Omega_h)$, $t \in \omega_\tau$ *and* $v^{(0)}(x) \in \overset{\circ}{W}{}^2_2(\Omega_h)$. *Then with all* $t_j \in \bar{\omega}_\tau$, j *– the layer number, for the solution of the problem* (8) – (10) *the following estimates hold:*

$$\|v(t_j)\|_{L_2(\Omega_h)} \le \|v(0)\|_{L_2(\Omega_h)} + \tau \sum_{s=0}^{j-1} \|r(t_s)\|_{L_2(\Omega_h)}, \tag{11}$$

$$\|v_{\bar{x}}(t_j)\|_{L_2(\Omega_h^+)} \le \|v_{\bar{x}}(0)\|_{L_2(\Omega_h^+)} + \frac{1}{\sqrt{2\delta}} \sqrt{\tau \sum_{s=0}^{j-1} \|r(t_s)\|^2_{L_2(\Omega_h)}}, \tag{12}$$

$$\|v_{\bar{x}x}(t_j)\|_{L_2(\Omega_h)} \le \|v_{\bar{x}x}(0)\|_{L_2(\Omega_h)} + \frac{1}{\sqrt{2\delta}} \sqrt{\tau \sum_{s=0}^{j-1} \|r_{\bar{x}}(t_s)\|^2_{L_2(\Omega_h^+)}}, \tag{13}$$

where $\| \ \|$ *are grid analogues of the corresponding continuous spaces norms (see* [6]).

Proof. We apply the Fourier separation of variables method. As a basis, consider the functions

$$\mu_k(x) = \sqrt{2}\sin(k\pi x), \qquad k = 1,2,\ldots,N-1, \qquad h = \frac{1}{N}.$$

It is known [8, p. 285, 286] that $\{\mu_k(x)\}$ is an orthonormal and complete function system in $L_2(\Omega_h)$; the systems of difference derivatives

$$\mu_{k_{\bar{x}}}(x) = \sqrt{\lambda_k}\sqrt{2}\cos(k\pi[x - h/2]), \qquad k = 1,2,\ldots,N-1,$$

and

$$\mu_{k_{\bar{x}x}}(x) = \lambda_k \mu_k(x), \qquad \lambda_k = \frac{4}{h^2}\sin^2\left(\frac{k\pi h}{2}\right), \qquad k = 1, 2, \ldots, N-1,$$

appear to be orthogonal in $L_2(\Omega_h^+)$ and $L_2(\Omega_h)$, respectively.

For grid functions introduce the inner product

$$(y,\, v)_{\Omega_h} = h \sum_{x \in \Omega_h} u(x) v^*(x),$$

where v^* denotes the complex conjugate of v.

We look for the solution of problem (8)–(10) of the form

$$v(t,x) = \sum_{k=1}^{N-1} v_k(t)\mu_k(x), \tag{14}$$

where $v_k(t) = (v(t,x),\, \mu_k(t))_{\Omega_h}$ are the Fourier coefficients. By Eq. (8) we have the relations

$$\hat{v} = \alpha_k v_k + \tau \beta_k r_k, \tag{15}$$

where $r_k = (r,\, \mu_k)_{\Omega_h}$, $\beta_k = \left(1 + \frac{\tau a \lambda_k}{2}\right)^{-1}$, $\alpha_k = \left(1 - \frac{\tau a \lambda_k}{2}\right)\beta_k$. By the condition c) it follows that

$$|\alpha_k|, |\beta_k| \le 1, \qquad k = 1, 2, \ldots, N-1. \tag{16}$$

From (14), (15), using the Parseval identity, estimate (16), and Minkowski inequality we obtain

$$\|\hat{v}\|_{L_2(\Omega_h)} \le \|v\|_{L_2(\Omega_h)} + \tau \|r\|_{L_2(\Omega_h)}.$$

This completes the proof of estimate (11).

We now pass to the estimate of the first order difference derivatives $v_{\bar{x}}$. Since $\{\mu_{k_{\bar{x}}}(x)\}$ is an orthogonal system, by (15) it follows that

$$\|v_{\bar{x}}(t_j)\|_{L_2(\Omega_h^+)} = \sqrt{\sum_{k=1}^{N-1} \lambda_k \left|v_k\left(t_j\right)\right|^2} =$$

$$\sqrt{\sum_{k=1}^{N-1} \lambda_k \left|\alpha_k^j v_k(0) + \tau\beta_k \left(\alpha_k^{j-1} r_k(0) + \alpha_k^{j-2} r_k\left(t_1\right) + \cdots + r_k\left(t_{j-1}\right)\right)\right|^2}.$$

Apply Minkowski and Cauchy inequalities. We get

$$\|v_{\bar{x}}(t_j)\|_{L_2(\Omega_h^+)} \le \|v_{\bar{x}}(0)\|_{L_2(\Omega_h^+)} + \sqrt{\tau \sum_{k=1}^{N-1}\sum_{s=0}^{j-1} \left|r_k\left(t_s\right)\right|^2 \tau\lambda_k \left|\beta_k\right|^2 \sum_{s=0}^{j-1} \left|\alpha_k\right|^{2s}}. \tag{17}$$

A simple estimate of the sum of a geometrical progression shows that

$$\tau\lambda_k \left|\beta_k\right|^2 \sum_{s=0}^{j-1} \left|\alpha_k\right|^{2s} \leq \frac{\tau\lambda_k \left|\beta_k\right|^2}{1-\left|\alpha_k\right|^2} \leq \frac{1}{2\delta}.$$

Now estimate (12) can be obtained from (17). In a similar way we can deduce (13). Lemma is proved.

Remark. Note that appearance of derivative type nonlinearities in (4) requires a lower norm on the right-hand side r in the estimates (12) and (13). This enable us further to treat the problem in a way similar to the usual nonlinear evolution equation (see [7]). Also, taking $j = 1$ in (12), one could get the estimate for transition to the upper layer

$$\|\hat{v}_{\bar{x}}\|_{L_2(\Omega_h^+)} \leq \|v_{\bar{x}}\|_{L_2(\Omega_h^+)} + c\sqrt{\tau}\|r\,(t_s)\,\|_{L_2(\Omega_h)},$$

which is not sufficient to apply mathematical induction principle.

Now we state the results of our analysis:

Theorem 2 (Convergence). *Suppose the hypotheses* a) – c) *are held, and the solution of the problem* (4),(2),(3) *is smooth enough. Then there exist constants* $\tau_0, h_0 > 0$ *such that, for* $\tau \leq \tau_0$, $h \leq h_0$, *there exists a unique solution of the problem* (5) – (7), *converging to the solution of the problem* (4),(2),(3) *and the following estimate holds:*

$$\|u-v\|_{W_2^1(\Omega_h)} \leq c\left(\tau^2+h^2\right), \qquad t \in \bar{\omega}_\tau.$$

Remark. By the imbedding theorem $W_2^1 \to C$ the convergence in C follows.

If $f(u)$ and $g(u)$ are polynomials then one can prove the convergence of the difference method in W_2^2:

$$\|u-v\|_{W_2^2(\Omega_h)} \leq c\left(\tau^2+h\right), \qquad t \in \bar{\omega}_\tau.$$

In this case the scheme is convergent in C^1, too.

Theorem 3 (Stability). *Let* v_1 *and* v_2 *be the solutions of the problem* (5) – (7) *with initial data* $u_1^{(0)}$ *and* $u_2^{(0)}$. *Suppose the hypotheses of* Theorem 1 *are satisfied. Then there exist constants* $\tau_0, h_0 > 0$ *such that, for* $\tau \leq \tau_0$, $h \leq h_0$, *the following estimate holds:*

$$\|v_1(t)-v_2(t)\|_{W_2^1(\Omega_h)} \leq \|v_1(0)-v_2(0)\|_{W_2^1(\Omega_h)}, \qquad t \in \bar{\omega}_\tau.$$

Remark. If $f(u)$ and $g(u)$ are polynomials then the difference scheme (5) – (7) is stable in W_2^2.

Remark. In a similar way one can prove the convergence and stability of the difference schemes of the form

$$
\begin{aligned}
&v_t = a v^{(\sigma)}_{\bar{x}x} + F(v,\hat{v}) + G(v,\hat{v})D_x(v,\hat{v}), \qquad (t,x)\in\omega_\tau\times\Omega_h,\\
&v(t,0) = v(t,1) = 0, \qquad t\in\bar{\omega}_\tau,\\
&v(0,x) = u^{(0)}(x), \qquad x\in\bar{\Omega}_h,
\end{aligned}
$$

where F, G, D_x approximate f, g, $\frac{\partial u}{\partial x}$, respectively, $v^{(\sigma)} = \sigma\hat{v} + (1-\sigma)v$, $0.5 \le \sigma \le 1$.

Notes and Comments. We have justified Crank–Nicolson type finite difference schemes for evolution equation systems. There is no any restriction on the ratio of time and space grid steps τ and h. Note that for $Re(a) = 0$ (as in (4)), i.e. derivative nonlinear Schrödinger equations these schemes remain questionable. Parameter $Re(a) = \delta > 0$ can be viewed as artificial viscosity in this case.

(5) – (7) scheme was practically examined on a computer for the problem

$$\frac{\partial u}{\partial t} = (\delta + i\alpha)\frac{\partial^2 u}{\partial x^2} + i\sigma_1|u|^2u + \sigma_2|u|^2\frac{\partial u}{\partial x}, \qquad (t,x)\in Q. \tag{18}$$

$$u(t,0) = u(t,1) = 0, \qquad t\in[0,T], \tag{19}$$

$$u(0,x) = u^{(0)}(x), \qquad x\in\Omega. \tag{20}$$

No negative signs were observed, even if $\delta = 0$. Note that for $\delta = 0$ (18) – (20) problem has an energy conservation law in L_2 norm and (5) – (7) scheme is conservative one.

References

1. Akhmanov, S. A., Vysloukh, V. A., Chirkin, A. S.: Wave Packets Self-action in a Nonlinear Medium and Femto-second Laser Pulses Generation [in Russian]. Uspekhi Fiz. Nauk. **149** (1986) 449–509
2. Mio, W., Ogino, T., Minami, K., Takeda, S.: Modified nonlinear Schrödinger equation for Alfven waves propagating along the magnetic field in cold plasmas. J. Phys. Soc. Japan **41** (1976) 265–271
3. Akhromeeva, T. S., Kurdyumov, S. P., Malinetskii, G. G., Samarskii, A. A.: On Classification of Solutions of Nonlinear Diffusion Equations in a Neighborhood of a Bifurcation Point [in Russian]. Itogi Nauki i Tekhniki. Sovrem. Probl. Mat. Noveishye Dostizheniya, VINITI AN SSSR, Moscow **28** (1986) 207–313
4. Kuramoto, Y., Tsuzuki, T.: On the Formation of Dissipative Structures in Reaction-diffusion Systems. Progr. Theor. Phys. **54** (1975) 687–699
5. Nikolis, G., Prigozhin, I.: Self-organization in Nonequilibrium Systems [in Russian]. Mir, Moscow (1979)
6. Samarskii, A. A.: Difference Scheme Theory [in Russian]. Nauka, Moscow (1989)
7. Ivanauskas, F. F.: Convergence and Stability of Difference Schemes for Nonlinear Schrödinger Equations, the Kuramoto-Tsuzuki Equation, and Systems of Reaction-Diffusion Type. Russian Acad. Sci. Dokl. Math. **50** (1995) 122–126
8. Samarskii, A. A., Andreev, V. B.: Difference Methods for Elliptic Equations [in Russian]. Nauka, Moscow (1976)

On the Growth Problem for *D*-Optimal Designs

M. Mitrouli[1] and C. Koukouvinos[2]

[1] Department of Mathematics, University of Athens, Panepistimiopolis 15784, Athens, Greece.
[2] Department of Mathematics, National Technical University of Athens, Zografou 15773, Athens, Greece.

Abstract. When Gaussian elimination with complete pivoting (GECP) is applied to a real $n \times n$ matrix A, we will call $g(n, A)$ the associated growth of the matrix. The problem of determining the largest growth $g(n)$ for various values of n is called the growth problem. It seems quite difficult to establish a value or close bounds for $g(n)$. For specific values of n $(n = 1, 2, 3, 4)$ and for a special category of matrices, such as Hadamard matrices, $g(n)$ has been evaluated exactly. In the present paper, we discuss the maximum determinant and the growth problem of $n \times n$ matrices with elements ± 1, which are called D-optimal designs. Specific examples of $n \times n$ weighing matrices W attaining $g(n, W) = n-1$ are exhibited.

1 Introduction

Let $A = [a_{ij}] \in \mathcal{R}^{n\times n}$. We reduce A to upper triangular form by using Gaussian elimination with complete pivoting (GECP) [11]. Let $A^{(k)} = [a_{ij}^{(k)}]$ denote the matrix obtained after the first k pivoting operations, so $A^{(n-1)}$ is the final upper triangular matrix. A diagonal entry of that final matrix will be called a pivot. Matrices with the property that no exchanges are actually needed during GECP are called completely pivoted (CP). The following problem arises during the elimination process.

The growth problem
Let $g(n, A) = \max_{i,j,k} |a_{ij}^{(k)}|/|a_{11}^{(0)}|$ denote the growth associated with GECP on A and $g(n) = \sup\{\, g(n, A)/A \in \mathcal{R}^{n\times n}\}$. The problem of determining $g(n)$ for various values of n is called the growth problem.

The determination of $g(n)$ remains a mystery. Wilkinson [11] conjectured that $g(n, A) \leq n$, $\forall\ A \in \mathcal{R}^{n\times n}$. This conjecture is now known to be false [7]. A best general bound for $g(n)$ known so far is $g(n) \leq [n\, 2\, 3^{1/2} \ldots n^{1/n-1}]^{1/2}$. This bound is also due to Wilkinson [11] and is much larger than his conjecture n. While it is easy to see that $g(1) = 1$ and $g(2) = 2$ for all $n > 2$, even $n = 3$ or 4, it seems quite difficult to establish a value or close bound. By using algebraic methods with many cases, it was proved [4] that $g(3) = 2.25$, $g(4) = 4$ and $g(5) \leq 4\ \ 17/18$ [4]. For a special category of orthogonal matrices $H \in \mathcal{R}^{n\times n}$

with elements ± 1 and $HH^T = nI$, has been observed that $g(n, H) = n$. This equality has been proved for a certain class of $n \times n$ Hadamard matrices [4].

One of the curious frustrations of the growth problem is that it is quite difficult to construct any examples of $n \times n$ matrices A ,other than Hadamard, for which $g(n, A)$ is very close to n. Wilkinson has remarked that in real–world problems, $g(n, A)$ has never been observed to be very large [11]. In [3] Cryer did numerical experiments in which he computed $g(n, A)$, doing complete pivoting on $n \times n$ matrices A with entries chosen randomly from the interval $[-1, 1]$ and for sizes up to $n = 8$. He had to generate over 50000 3×3 examples before finding one with $g(3, A) > 2$. Also the largest $g(n, A)$ he obtained by testing 10000 random matrices for sizes up to $n = 8$ was 2.8348.

Thus, in order to obtain matrices with large growth sophisticated numerical optimization techniques must be applied [7]. By using such methods matrices with growth larger than $n = 13, 14, 15, 16, 18, 20, 25$ were specified. The matrices that give rise to larger than n growth factors are often extremely sensitive to small perturbations in their entries in that tiny perturbations to a complete elimination matrix rarely results in another such matrix. This makes it rather difficult to specify matrices which give rise to large growth.

Wilkinson's initial conjecture seems to be connected with Hadamard matrices. Since Hadamard matrices are D–optimal designs of order $n \equiv 0 \pmod 4$, in the present paper we study the growth problem for D–optimal designs. Especially orthogonal designs (Hadamard and weighing matrices) of order n can achieve large growth equal or very close to n.

2 The growth for D–optimal designs of order $n \equiv 1 \pmod 4$

In this section we discuss the problem of the maximum determinant of all $n \times n$ matrices with elements ± 1 when $n \equiv 1 \pmod 4$, and we study the growth of these matrices. A D–optimal, n–observation, 2^{n-1} design of resolution III, defined for all positive integers n, is a square matrix of size n with entries ± 1 having maximum determinant.

Let d_n denote the maximum determinant of all $n \times n$ matrices with elements ± 1. It follows from Hadamard's inequality that $d_n \leq n^{n/2}$ and it is easily shown that equality can only hold if $n = 1$ or 2 or if $n \equiv 0 (\mathrm{mod} 4)$. If $n \equiv 1 (\mathrm{mod} 4)$, $n \neq 1$, it is well known (see [9]) that $d_n \leq (n-1)^{(n-1)/2}\sqrt{2n-1}$ and equality can hold only if $n = 2s^2 + 2s + 1$, $s = 1, 2, 3, \ldots$. For more details and the construction of the corresponding D-optimal designs see [1, 9].

Comments for growth

By detecting the pivot structure of D–optimal designs when $n \equiv 1 (\mathrm{mod} 4)$ the following table was computed.

n	growth	Pivot Pattern
5	4	$(1, 2, 2, 4, 3)$
9	7	$(1, 2, 2, 4, 3, 2.6667, 4, 4, 7)$
13	10	$(1, 2, 2, 4, 3, 3.3333, 3.6, 4.4, 4.5, 6, 6, 10)$
17	13.3333	$(1, 2, 2, 4, 3, 3.3333, 3.6, 4, 4.8889, 4.3636, 5, 4.8, 5.3333, 4, 8, 8, 13.3333)$
21	18.125	$(1, 2, 2, 4, 3, 3.3333, 3.6, 4, 4.2222, 4.5789, 4.9655, 4.6667, 5.2381,$ $4.6212, 5.6557, 6.9565, 6.6667, 5.9375, 8.4211, 10, 18.125)$
25	21	$(1, 2, 2, 4, 3, 3.3333, 3.6, 4, 4, 4.5, 5.1825, 5.4857, 4.75, 5.4737, 5.1923,$ $5.3333, 7.2, 5.3333, 7.5, 7.2, 8, 7.5, 9.6, 12, 21)$

Table 1

The following proposition is derived.

Proposition 1. *Let* $n = 5, 9, 13, 17, 21, 25$ *and* A *be the* $n \times n$ *D–optimal desugn. Reduce* A *by GECP then the magnitude of the first four pivots are* $1, 2, 2$ *and* 4*; The magnitude of* $|a_{55}^{(4)}|$ *is* ≤ 3.

Proof. Due to the structure of these matrices, we can always have in the upper 4×4 part of A a CP Hadamard matrix. Thus, the first four pivots will be $1, 2, 2$ and 4 [4]. Because every entry in $A^{(3)}$ is of magnitude $0, 2$ or 4, pivoting on $a_{44}^{(3)}$ will only involve adding ± 1 or $\pm 1/2$ times the fourth row of $A^{(3)}$ to the rows below, and this will create only integer entries in $A^{(4)}$. Thus $|a_{55}^{(4)}|$ must be an integer satisfying the relation

$$A(1\,2\,3\,4\,5) = 16|a_{55}^{(4)}| \leq 4^{4/2}\sqrt{10-1} \Rightarrow |a_{55}^{(4)}| \leq 3.$$

where $A(1\,2\,3\,4\,5)$ denotes the determinant of the 5×5 principal submatrix of A.

Remark. We notice that when $n = 5, 9, 13, 17, 21, 25$ and A is the $n \times n$ D–optimal design the growth factor satisfies the relation $n - 4 \leq g(n, A) < n$.

3 The growth for D–optimal designs of order $n \equiv 2 \pmod 4$

If $n \equiv 2 \pmod 4$, $n \neq 2$, it is well known (see [9]) that $d_n \leq (2n-2)(n-2)^{\frac{n}{2}-1}$ and equality can hold only if $2n - 2 = x^2 + y^2$, where x and y are integers. For more details and a survey for the construction of the corresponding D-optimal designs see [8].

Comments for growth

By applying $GECP$ to specific designs of this category we produced the following table.

n	growth	Pivot Pattern
6	4	(1, 2, 2, 4, 3, 10/3)
10	8	(1, 2, 2, 4, 3, 2, 4, 4, 8, 6)
14	8.6667	(1, 2, 2, 4, 3, 10/3, 18/5, 4, 4, 5, 4.8, 7.5, 5.4, 26/3)
18	12.3636	(1, 2, 2, 4, 3, 10/3, 18/5, 4, 4, 4.6667, 5.1429, 5.3333, 4.8889, 7.2727, 6, 5.8667, 8, 12.3636)
26	18.75	(1, 2, 2, 4, 3, 10/3, 18/5, 4, 4.3333, 5.0909, 5.0714, 5.1362, 4.44, 4.8652, 5.7766, 5.8059, 7.7093, 6.5158, 7, 8.4, 8.3265, 7.0588, 10.2, 11.2941, 18.75)
30	22.5556	(1, 2, 2, 4, 3, 3.3333, 3.6, 4, 4, 5, 5.0667, 5.2632, 5.28, 4.8977, 5.0812, 4.9029, 6.5436, 5.7465, 6.1981, 6.7749, 8.0926, 7.1220, 6.8082, 9.1690, 8.5806, 10.3158, 8.3333, 10.08, 14, 22.5556)

Table 2

Remark. (i) We notice that the first five pivots always equal to $1, 2, 2, 4$ and 3. The three first pivots will equal to $1, 2$ and 2 due to the structure of the constructed matrices. Since we start with a matrix A of ± 1's clearly $|a_{11}^{(0)}| = 1$. If we suppose that A is CP we can apply $GCEP$ to A. Every entry of $A^{(1)}$ is 0 or ± 2. Thus $a_{22}^{(1)} = 2$. Every entry of $A^{(2)}$ must be $0, \pm 2$, thus $a_{33}^{(2)} = 2$.

(ii) The growth factor satisfies the relation $n - 8 \leq g(n, A) < n$.

4 On Hadamard and weighing matrix growth conjecture

On Hadamard matrices

A Hadamard matrix, H, of order n has elements ± 1 and satisfies $HH^T = nI_n$. If $H = I + S$ where $S^T = -S$ then H is called skew–Hadamard.

It is well known [10] that if H is a Hadamard matrix of order $n > 2$, then n is a multiple of 4. There is a large and growing body of literature on existence and applications of Hadamard matrices, and they have been constructed for many infinite families of value n [10].

We give here the simplest method for constructing Hadamard matrices of order $n = 2^k$, which we will use below:
If we say

$$H_1 = \begin{bmatrix} 1 & 1 \\ 1 & -1 \end{bmatrix}, \; H_2 = \begin{bmatrix} H_1 & H_1 \\ H_1 & -H_1 \end{bmatrix}$$

then we obtain, in general

$$H_{k+1} = \begin{bmatrix} H_k & H_k \\ H_k & -H_k \end{bmatrix}$$

Two Hadamard matrices are called equivalent (or Hadamard equivalent) if one can be obtained from the other by a sequence of row exchanges, column exchanges, row negations, and column negations. The number of equivalence classes of Hadamard matrices of order n has been determined for some values of n. For

$n = 2, 4, 8, 12$ there is only one class. For $n = 16$ there are 5 classes, for $n = 20$ there are 3 classes, for $n = 24$ there are 60 classes, and for $n = 28$ there are 486 classes. Also we know that for $n = 32$ there are over 15 classes and for $n = 36$ there are over 109 classes ([10], p.440).

The growth conjecture for Hadamard matrices

When Gaussian elimination is done on an $n \times n$ Hadamard matrix A the following properties hold true

Proposition 2. (see [4]) Let A be an $n \times n$ Hadamard matrix. The six last pivots for GECP are at most $n, n, n/2, n/2, n/2$, and n respectively. □

Proposition 3. (see [4]) Let $n \geq 4$ and A be an $n \times n$ Hadamard matrix. Reduce A be GECP. Then the magnitude of the first four pivots are $1, 2, 2$ and 4; and if $n > 4$ the fifth pivot is 2 or 3. □

Proposition 4. (see [5]) Let A be a 12×12 Hadamard matrix. Reduce A by GECP. Then the fifth pivot equals to 3. □

Proposition 5. (see [5]) Let A be a 12×12 Hadamard matrix. Then the growth factor $g(12, A) = 12$. □

Proposition 6. Let H_{k+1} be an $n \times n$ CP Hadamard matrix, $n = 2^{k+1}$. If we reduce H_{k+1} by GECP and denote by p_{k+1} the computed pivots, the pivoting structure $p_{k+1} = (p_k, 2p_k)$ will always appear.

Proof. Let H_k be a $2^k \times 2^k$ Hadamard matrix for some k. If GECP is applied to H_k, then let us denote by HG_k the upper triangular matrix and by p_k the attained pivot structure. Then H_{k+1} will be constructed by using the formula $H_{k+1} = \begin{bmatrix} H_k & H_k \\ H_k & -H_k \end{bmatrix}$ which is row equivalent with the form $H_{k+1} = \begin{bmatrix} H_k & H_k \\ \bigcirc & -2H_k \end{bmatrix}$. Since the last matrix is block diagonal with a specific structure the application of GECP to it will end up by giving $HG_{k+1} = \begin{bmatrix} HG_k & HG_k' \\ \bigcirc & -2HG_k \end{bmatrix}$ and thus the pivot structure p_{k+1} of HG_{k+1} will equal to $(p_k, 2p_k)$. □

Corollary 7. Let H_k be an $n \times n$ CP Hadamard matrix, $n = 2^k$. If we reduce H_k by GECP then $g(n, H_k) = n$. □

By applying GECP to Hadamard matrices we produced the following table:

n	growth	Pivot Pattern	existence
2	2	$p_2 = (1, 2)$	unique
$4 = 2^2$	4	$p_4 = (1, 2, 2, 4) = (p_2, 2p_2)$	unique
$8 = 2^3$	8	$p_8 = (1, 2, 2, 4, 2, 4, 4, 8) = (p_4, 2p_4)$	unique
12	12	$p_{12} = (1, 2, 2, 4, 3, 10/3, 18/5, 4, 3, 6, 6, 12)$	unique
$16 = 2^4$	16	$p_{16} = (1, 2, 2, 4, 2, 4, 4, 8, 2, 4, 4, 8, 4, 8, 8, 16)$ $= (p_8, 2p_8)$	Class I one representative
20	20	$p_{20} = (1, 2, 2, 4, 3, 10/3, 18/5, 4, 4, 4.5556, 4.878,$ $4.5, 5, 5.5556, 6, 6.6667, 5, 10, 10, 20)$	Class I one representative
20	20	$p_{20} = (1, 2, 2, 4, 3, 10/3, 3.2, 5, 4, 4.5, 4.4444, 6,$ $3.3333, 5, 7.5, 6.6667, 5, 10, 10, 20)$	Class II one representative
20	20	$p_{20} = (1, 2, 2, 4, 3, 10/3, 18/5, 4, 4.6667, 4.2857,$ $4.6667, 4.7619, 4, 6.25, 6, 6.6667, 5, 10, 10, 20)$	Class III one representative
24	24	$p_{24} = (p_{12}, 2p_{12})$	one representative
28	28	$p_{28} = (1, 2, 2, 4, 3, 10/3, 3.6, 3.7778, 4.4706, 5.0526, 4.9583,$ $4.9076, 4.6849, 5.1711, 5.771, 6.2222, 5.1204, 6.2025,$ $5.6, 6.6667, 5.6, 8.75, 8.4, 9.3333, 7, 14, 14, 28$	Class I one reperesentative
$32 = 2^5$	32	$p_{32} = (p_{16}, 2p_{16})$	one representative

Table 3

Remark. For $n = 16$ they have been computed many different pivot patterns for the existing 5 equivalent classes. See [4] for more examples.

For $n = 20$ after applying GECP to many CP representatives of the existing 3 equivalent classes, we discovered at least 13 possible pivot patterns. For all these patterns the fifth and sixth pivots were 3 and 10/3.

For $n = 24$ the following structure was used: $\begin{bmatrix} H_{12} & H_{12} \\ H_{12} & -H_{12} \end{bmatrix}$,where H_{12} is a Hadamard matrix of order 12.

From the above study the following conjecture still remains unsolved and forms open questions:

The Hadamard conjecture

Let A be an $n \times n$ Hadamard matrix. Reduce A by GECP. Then

(i) $g(n, A) = n$.
(ii) The four last pivots equal to $n/4, n/2, n/2, n$.
(iii) Every pivot before the last will have magnitude at most $n/2$.

On weighing matrices

A conference matrix C of order $n \equiv 2 \pmod 4$ has zero diagonal, other elements ± 1 and satisfies $CC^T = (n-1)I_n$. Conference matrices cannot exist unless $n - 1$ is the sum of two squares: thus they cannot exist for orders $22, 34, 58, 70, 78, 94$.

A $(0,1,-1)$ matrix $W = W(n,k)$ of order n satisfying $WW^T = kI_n$ is called a weighing matrix of order n and weight k or simply a weighing matrix. A $W(n,n)$, $n \equiv 0 \pmod 4$, is a Hadamard matrix of order n. A $W = W(n,k)$ for which $W^T = -W$ is called a skew–weighing matrix.

A $W = W(n, n-1)$ satisfying $W^T = W$, $n \equiv 2 \pmod 4$, is called a symmetric conference matrix.

We say that the weighing matrix $W = W(2n, k)$ is constructed from two circulant matrices M, N of order n if $W = \begin{bmatrix} M & N \\ N^T & -M^T \end{bmatrix}$

Comments for growth No investigation has been performed on weighing matrices. By detecting the pivot structure of several weighing matrices the following table was computed.

n, k	growth	Pivot Pattern
4,3	3	$(1, 2, 1.5, 3)$
6,5	5	$(1, 2, 2, 2.5, 2.5, 5)$
10,9	9	$(1, 2, 2, 3, 3, 4, 2.25, 4.5, 4.5, 9)$
12,10	10	$(1, 2, 2, 3, 3, 3.1111, 3.2143, 4.4444, 2.5, 5, 5, 10)$
14,13	13	$(1, 2, 2, 3, 3.3333, 3.6, 4, 3.25, 4.3333, 4.3333, 3.25, 6.5, 6.5, 13)$
15,9	9	$(1, 2, 2, 3, 2.6667, 3, 3, 3, 3, 3, 4.5, 2.25, 4.5, 4.5, 9)$
18,17	17	$(1, 2, 2, 4, 3, 10/3, 3.6, 4, 4, 4.25, 4.25, 4.7222, 5.1, 5.6667, 4.25, 8.5, 8.5, 17)$
20,15	15	$(1, 2, 2, 3, 3, 10/3, 10/3, 3.84, 4.25, 4.1569, 4.4575, 4.2857, 3.75, 4.1667, 4.5, 5, 3.75, 7.5, 7.5, 15)$

Table 4

From the examples we have tested, we think that the following conjecture holds.

The weighing conjecture

Let $W = W(n,k)$ be a weighing matrix. Reduce W by GECP. Then

(i) $g(n, W) = k$.
(ii) The pivot before the last equals to $k/2$.
(iii) Every pivot before the last will have magnitude at most $k/2$.

We can construct $n \times n$ conference matrices by two circulants which can have growth factor equal to $n-1$. We tested several such examples and in the following table we summarize some orders of such matrices and the attained growth. For these orders Hadamard matrices do not exist, thus the corresponding conference matrices are the most simple $n \times n$ known matrices (since their elements are only $0, \pm 1$) attaining growth factor equal to $n-1$, except for $n = 22, 34$ where the growth is 20 and 32 respectively.

order	18	22	26	30	34	38	42	50	54
growth	17	20	25	29	32	37	41	49	53

Table 5

References

1. Beth, T., Jungnickel, D., Lenz, H.: Design Theory. Cambridge University Press, Cambridge, Engalnd, 1986
2. Cohen, A. M.: A note on pivot size in Gaussian elimination. Lin. Alg. Appl. **8** (1974) 361–368
3. Cryer, C. W.: Pivot size in Gaussian elimination. Numer. Math. **12** (1968) 335–345
4. Day, J., Peterson, B.: Growth in Gaussian elimination. Amer. Math. Monthly **95** (1988) 489–513
5. Edelman, E., Mascarenhas, W.: On the complete pivoting conjecture for a Hadamard matrix of order 12. Linear and Multilinear Algebra **38** (1995) 181–187
6. Geramita, A. V., Seberry, J.: Orthogonal designs: Quadratic forms and Hadamard matrices. Marcel Dekker, New York-Basel, 1979
7. Gould, N.: On growth in Gaussian elimination with pivoting. SIAM J. Matrix Anal. Appl. **12** (1991) 354–361
8. Koukouvinos, C.: Linear models and D-optimal designs for $n \equiv 2 \bmod 4$. Statistics and Probability Letters **26** (1996) 329–332
9. Raghavarao, D.: Constructions and Combinatorial Problems in Design of Experiments. J. Wiley and Sons, New York, 1971
10. Seberry, J., Yamada, M.: Hadamard matrices, sequences and block designs. Contemporary Design Theory: A collection of surveys. Edited by J. Dinitz and D. R. Stinson, J. Wiley and Sons, New York, (1992) 431–560
11. Wilkinson, J. H.: The Algebraic Eigenvalue Problem. Oxford University Press, London, 1988

A Cray T3D Performance Study

Asha Nallana[1] and David R. Kincaid[2]

[1] Interact, Inc., 9390 Research Blvd., Austin, TX 78758, USA
[2] Center for Numerical Analysis, The University of Texas at Austin, Austin, TX 78713–8510, USA kincaid@cs.utexas.edu

Abstract. We carry out a performance study using the Cray T3D parallel supercomputer to illustrate some important features of this machine. Timing experiments show the speed of various basic operations while more complicated operations give some measure of its parallel performance.

1 Introduction

Recently, high-performance computers have become an important tool for obtaining the solution of complex scientific problems. In spite of the enormous advances in performance of machines and method, they fall short of providing computational solutions to many important applications. To successfully solve these problems, one needs an increase in computational power of several orders of magnitude. Since the speed of the fastest processor already approaches the limits set by the laws of physics, such an increase will only be feasible through the integration of hundreds or thousands of powerful processors into a massively parallel computer. In principle, there is no limit to the aggregate speed of parallel computers, although the growing communication requirements limit the useful size for practical computer systems. Parallel computers are also superior to conventional systems if one considers their cost-effectiveness—a parallel machine employing off-the-shelf processors is usually much less expensive than a sophisticated serial computer.

The basic strategy for programming a massively parallel computer system is to assign the work to the appropriate data locations while keeping all the processors busy with the overall goal of solving the problem in the shortest possible time. Thus, the algorithm chosen must be highly local and parallel. The Cray T3D parallel supercomputer system is based upon the multiple instruction multiple data (MIMD) multiprocessor computational model. It also supports the single program multiple data (SPMD) and the single instruction multiple data (SIMD) computational models.

The objective of this research is to become acquainted with the Cray T3D computer and some of the modes of parallel programming available on it. To do this, we perform some numerical experiments and analyze their performance. In the report by Nallana [2], additional details concerning the Cray T3D are discussed.

2 Results

Following the procedures outline in [1], we present the results of some example programs run on the Cray T3D. While this computer supports several styles of parallel programming, we are mainly interested in *data sharing* and *work sharing.* Data sharing distributes data, such as an array, over the memories of the processing elements (PEs) using mostly implicit communication. The goal is to let as many PEs as possible perform operations on their own data rather than going off to another PE's memory to get the data since operations on local data are faster than those on remote data. Work sharing distributes the statements of the application program among the computer's PEs with the goal of executing them in parallel. For instance, iterations of a do-loop can be distributed among the processors. Work sharing provides a combination of implicit and explicit communication. We consider examples divided into two classes—timing and numerical.

2.1 Method of Timing an Operation

We begin with a discussion of a procedure for timing elementary operations. The basic idea is to measure the time `t` that the computer takes to do a large number `n` of the same operation so that the individual operation time is given by `t/n`. First, we set the initial time `t1` before doing any operation, then we perform the operation a large number (`r1`) of times. We measure the new time `t2` after this computation. These two function calls return the floating-point value of the real-time clock (in clock ticks). The difference between the two timings (`t2` − `t1`) is the elapsed time. This elapsed time includes the overhead for the loop control arithmetic which should not be included in the operation time since it is just an artifact of the technique we use to measure the time. To remove the loop overhead, we time the operation once again. This time we start with time `t3`, perform the computation again with a different number (`r2`) of repetitions, and measure the new time `t4`. Since the overhead for both do-loops is the same, the time for the loop control arithmetic is cancelled out by the expression ((`t2` − `t1`) − (`t4` − `t3`)) and we store it in `dtime`. Hence, the variable `dtime` represents the elapsed time for `nrep*dup` executions of the statement containing the operation. Here `nrep` is the number of repetitions in the timing loops. The variable `dup` is the difference between `r2` and `r1` which is the effective number of operations timed. We use `(r1, r2)` as either (16, 8) or (2, 1). We repeat this procedure `nsamp` times and collect `dtime` into `sumtime`. Expression `sumtime` contains the time for performing (`nsamp` ∗ `nrep` ∗ `dup`) repetitions of the operation. Thus, the average time for one execution of the operation is given by `sumtime` divided by `(nsamp*nrep*dup)`. Apart from calculating the average operation time, the code also calculates the average rate of executing the statement in millions of floating-point operations (`mflops`) which is (number of floating-point operations) divided by [(average operation time) $*$ 10^6]. In the following tables, all timing results are given in seconds.

2.2 Timing Examples

Arithmetic Operations. Using the procedure outline above, we begin by timing the basic arithmetic operations of addition, multiplication, and division. Values of the scalar quantities x and y used are 0.0 and 0.1, respectively. Here `(r1, r2) = (16, 2)`. We present the results in Table 1.

Table 1. Arithmetic operations: $z = x\,\texttt{op}\,y$

	nrep	nsamp	dup
	16384	10	8
op	avg. op. time	mflops	z
+	3.9707E–08	25.18	1.0000E–01
*	3.9816E–08	25.12	0.0000E+00
/	4.1677E–07	2.40	0.0000E+00

Serial Dot Product. This example calculates the dot product of two vectors $\left(\texttt{dp} = \mathbf{x}^T\mathbf{y} = \sum_{i=1}^{n} x_i * y_i\right)$ using the routine SDOT from the Basic Linear Algebra Subprograms (BLAS). In the code, each of the vector elements x_i and y_i are assigned the value 1.1. Here `(r1, r2) = (16, 2)`. The average operation time is for a single dot-product operation and we count $2n$ floating-point operations per dot-product operation. The results are given in Table 2.

Table 2. Serial Dot Product

n	nrep	nsamp
16384	50	10
avg. op. time	mflops	dp
1.2247E–03	26.8	1.9825E+04

Parallel Dot Product. This example calculates the dot product of two vectors but the difference between it and the routine discussed above is that in this one the computation is distributed over p nodes—each doing approximately $1/p$ of the computation. Hence, each node computes a portion of the dot product

$$x_i * y_i + x_{i+1} * y_{i+1} + \cdots + x_{i+k} * y_{i+k} \tag{1}$$

where $k \approx n/p$ and n is the length of the vector. We use the Cray MPP Fortran programming model knows as CRAFT and use some compiler directives such

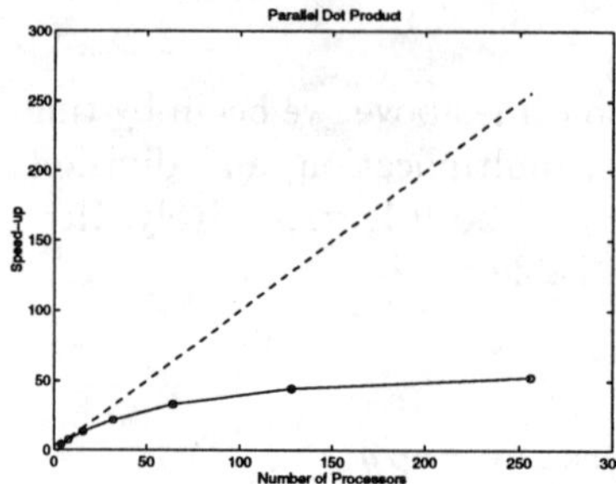

Fig. 1. Speed-up (Table 3)

Table 3. Parallel Dot Product

	n	nrep	nsamp
	16384	50	10
p	avg. op. time	mflops	dp
1	1.2276E–03	26.7	1.9825E+04
2	6.2260E–04	52.6	1.9825E+04
4	3.2108E–04	102.1	1.9825E+04
8	1.6921E–04	193.6	1.9825E+04
16	9.3824E–05	349.2	1.9825E+04
32	5.6204E–05	583.0	1.9825E+04
64	3.7174E–05	881.5	1.9825E+04
128	2.7604E–05	1187.1	1.9825E+04
256	2.3287E–05	1407.1	1.9825E+04

as `CDIR$ SHARED V(:BLOCK)`, `CDIR$ DOSHARED (K) ON V(K)`, `CDIR$ MASTER`, and `CDIR$ BARRIER`. The first one relates to the *data sharing* and it distributes the data among the memories of the various PEs ensuring that each processor works on its own data. The second one relates to *work sharing* and it causes the execution of different iterations of the loops to be distributed over different PEs with the goal of executing then in parallel. Communication among the PEs is mostly implicit. A subroutine call is used to force shared-to-private coercion which allows one to call the BLAS routine `SDOT` on the *local* data. Here `(r1, r2) = (2, 1)`. The results are given in Table 3.

Next in Table 4, we not only double the number of processors used but also double the problem size to give some indication of the *scalability*.

Rather than calling the routine `SDOT`, if we had written the dot-product calculation in Fortran, then various programming tricks would be necessary to obtain optimal performance on the Cray T3D; e.g., a four-way unrolled loop plus read-ahead would improve the number of cache hits. However, this version runs at approximately 15 `mflops` per node while the `SDOT` version goes at approximately 26 `mflops` per node.

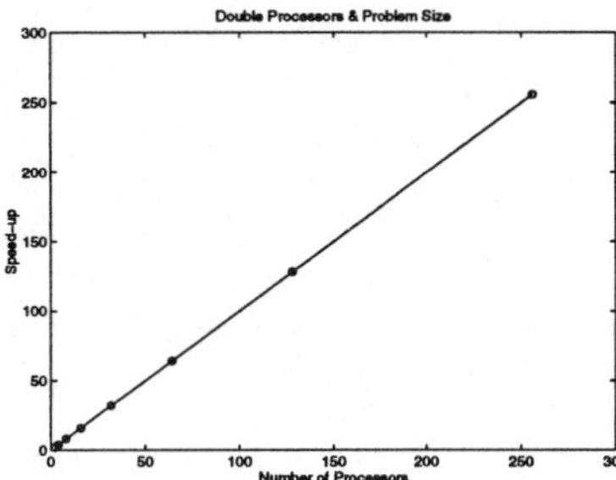

Fig. 2. Speed-up (Table 4)

Table 4. Parallel Dot Product: Scalability

		nrep	nsamp	
		50	10	
p	n	avg. op. time	mflops	dp
1	16384	1.2276E–03	26.7	1.9825E+04
2	32768	1.2281E–03	53.4	3.9649E+04
4	65536	1.2288E–03	106.7	7.9299E+04
8	131072	1.2284E–03	213.4	1.5860E+05
16	262144	1.2281E–03	426.9	3.1719E+05
32	524288	1.2283E–03	853.7	6.3439E+05
64	1048576	1.2281E–03	1707.7	1.2688E+06
128	2097152	1.2284E–03	3414.5	2.5376E+06
256	4194304	1.2283E–03	6829.4	5.0751E+06

Alternatively, one could use the Parallel BLAS (PBLAS) dot-product routine `PDDOT` from ScaLAPACK. The PBLAS are written as an internal component of this library so little effort was made to simplify their use. Also, the PBLAS is not a stand alone library and they require the use of an additional set of routines (BLACS) to handle the data distribution and communication. Consequently, their arguments are a bit difficult to understand if viewed only in the context of the PBLAS. Fortunately, an example of a dot-product program is available at the URL site: `http://www.netlib.org/blacs/BLACS/Examples.html`

Now we present an interesting numerical result. When timing a Fortran parallel code for computing the distributed dot product, we move the global sum to the end so that it is outside the timing loops. Hence, the code does just the multiplications and additions on p processors and, consequently, it is ideally parallelizable since the global sum communications are not timed. (Assuming an efficient global sum, the multiplications and additions should be the most time consuming part of the calculations.) The results are given in Table 5. We note that as the number of processors increases by powers of 2 the relative speed-up

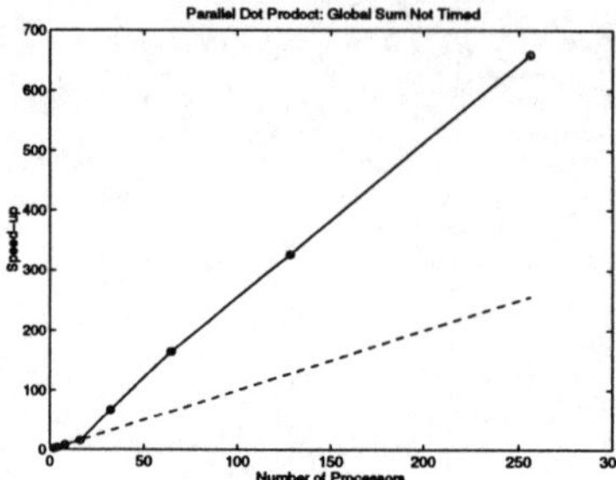

Fig. 3. Speed-up (Table 5)

Table 5. Parallel Dot Product: Golbal Sum Not Timed

	n	`nrep`	`nsamp`
	16000	50	10
p	avg. op. time	`mflops`	`dp`
1	3.2760E–03	9.8	1.9360E+04
2	1.6335E–03	19.6	1.9360E+04
4	8.1685E–04	39.2	1.9360E+04
8	4.0875E–04	78.3	1.9360E+04
16	2.1192E–04	151.0	1.9360E+04
32	4.9309E–05	649.0	1.9360E+04
64	1.9939E–05	1604.9	1.9360E+04
128	1.0034E–05	3189.3	1.9360E+04
256	4.9701E–06	6438.5	1.9360E+04

from $p-1$ to p processors, T_p/T_{p-1}, is approximately 2 for all cases except for 32 and 64 processors which are 4.3 and 2.43, respectively. Consequently, we obtain *superlinear* speed-up as shown in Fig. 3. The reason for this is that as the number of processors increases the amount of work per node decreases until at 32 processors the data just fits into high-speed cache. The data size is 2×16000 and on 32 PEs the data fits in the 8 Kbyte = 1 Kword cache: $32000/32 = 1000$. So for 32 PEs and above, the code is running from cache at approximately 20 `mflops` while for least than 32 PEs it is running from memory at approximately 10 `mflops`.

2.3 Numerical Examples

Next, we discuss the results of some parallel numerical examples; namely, polynomial evaluation and numerical integration.

Polynomial Evaluation. The first numerical code computes a short table of the values of the polynomial $x^3 + 2x^2 + 3x + 4$ for equally spaced argument values on the interval [0,1]. Here all the processors share the computation for each argument. In this program, the argument values are dependent on the number of processors being used. Here we exploit both the data-sharing and the do-shared loops of the work-sharing method of programming. Special compiler directives **CDIR$ SHARED P_VALUE**, **CDIR$ DOSHARED**, and **CDIR$ MASTER** are used to assign the shared data and direct the parallel do-loops utilizing this data. From a sample output of this program with eight processors, we obtain the results in Table 6.

Table 6. Parallel Polynomial Evaluation

x Value	Polynomial Value	Processor
0.	4.	(PE 6)
0.14285714285714285	4.4723032069970845	(PE 4)
0.2857142857142857	5.0437317784256557	(PE 5)
0.42857142857142855	5.7317784256559765	(PE 7)
0.5714285714285714	6.5539358600583082	(PE 2)
0.71428571428571419	7.5276967930029146	(PE 3)
0.8571428571428571	8.6705539358600578	(PE 1)
1.	10.	(PE 0)

Numerical Integration. This example computes the value of the integral

$$\int_0^{\pi/2} \sin^3(x)dx \qquad (2)$$

using Simpson's rule with a fixed number of partition points. The exact value of the integral is $\int_0^{\pi/2} \sin^3 x\, dx = -\frac{1}{3}\cos x(\sin^2 x + 2)\big|_0^{\pi/2} = \frac{2}{3}$. In this example, the work is shared by all the nodes with each doing the same amount of work. In particular, each node is assigned to work on a different subinterval of the interval of integration obtaining a portion of the integral value. In the code, the outer do-loop uses the number of processors assigned to determine the limits for the inner do-loops which are executed in parallel. Contributions from each node are added as soon as they become available. The compiler directive **CDIR$ MASTER** is used to add all the contributions on a single processing element. In the results in Table 7, the eight processors finish in order 1, 6, 3, 4, 7, 5, 2, 0.

3 Summary

In doing performance tests, one has to think about the behavior of the cache. For example when putting code inside a do-loop for timing purposes, sometimes the vector lengths may exceed the cache size and result in a *cache miss* while other times they may be totally within the cache. If there is a cache misalignment

Table 7. Parallel Simpson's Rule

```
Number of processors = 8
Number of panels = 80
Lower limit = 0, Upper limit = 1.5707963267949001
Value of the Integral = 0.6666666635697946
```

there could be a *thrashing* of the cache. Since any of these situations may effect the timing results, one has to be very careful to determine what actually is going on before trying to analyze the results.

Our experience with CRAFT is that there are definite tricks that need to be used; e.g., shared-to-private coercion. The processors used in the T3D have limited memory bandwidth so the results can be disappointing using CRAFT. On the other hand, the new Cray T3E preserves the macroarchitecture and programming environment of the Cray T3D as well as having faster processing and communication speeds which are coupled with a larger memory. With the Cray T3E, the memory bandwidth is increased and there is a secondary cache with a three-way associativity. Moreover, the CRAFT model is simplified to improve its performance.

Comparing the results from the serial and the parallel dot-product routines, we come to the conclusion that parallel computation gives much improved performance. This is clearly evident from the speed-up graphs. This reiterated the effectiveness of parallelism and of the Cray T3D supercomputer, in particular.

Acknowledgments

This work was supported, in part, by the National Science Foundation grant CCR–9504954, Cray Research, Inc., grant LTR DTD, and the Texas Advanced Research Program grant TARP–266 with The University of Texas at Austin. The work was accomplished with the aid of the Cray T3D at the National Energy Research Supercomputer Center. We thank Alex Kluge and Robert Harkness of the Computation Center at The University of Texas at Austin. Also, Bob Numrich of Cray Research, Inc., was helpful in explaining some of our parallel timing results. In addition, we thank the University of Colorado High Performance Scientific Computing Group for the use of their excellent material [1].

References

1. Fosdick, Lloyd D., Jessup, Elizabeth R., Schauble, Carolyn S. C., Domick, Gitt: *An Introduction to High Performance Scientific Computing.* MIT Press, Boston, 1995. (URL: `http://www.cs.colorado.edu/95-96/courses/materials/hpsc.html`)
2. Nallana, Asha: *Cray-T3D Performance Study*, Report CNA-281, Center for Numerical Analysis, University of Texas at Austin, December 1995.

Using Dense Matrix Computations in the Solution of Sparse Problems

Tz. Ostromsky[1] and Z. Zlatev[2]

[1] Central Laboratory for Parallel Information Processing, Bulgarian Academy of Sciences, Acad. G.Bonchev str., bl. 25-A, 1113 Sofia, Bulgaria; e-mail: ceco@iscbg.acad.bg

[2] National Environmental Research Institute, Frederiksborgvej 399, P. O. Box 358, DK-4000 Roskilde, Denmark; e-mail: luzz@sun2.dmu.dk

Abstract. On many high-speed computers the dense matrix technique is preferable to sparse matrix technique when the matrices are not very large, because the high computational speed compensates fully the disadvantages of using more arithmetic operations and more storage. Dense matrix techniques can still used if the computations are successively carried out in a sequence of large dense blocks. A method based on this idea will be discussed.

1 Statement of the problem

Consider: $Ax = b - r$ with $A^T r = 0$, where $A \in \mathbf{R}^{m \times n}$, $m \geq n$, $rank(A) = n$, $b \in \mathbf{R}^{m \times 1}$, $r \in \mathbf{R}^{m \times 1}$ and $x \in \mathbf{R}^{n \times 1}$. It is difficult to use efficiently high-speed computers in for general sparse matrices. For dense matrices all difficulties disappear, and the speed of the dense matrix computations is normally close to the peak performance of the computer used; see Table 1.

matrix gemat1 (10595 × 4929)	
Computing time (in seconds)	511
Speed of computations in MFLOPS	878
Peak performance of the computer	902
Efficiency (in percent)	97%

Table 1. Results on CRAY C92A by using LAPACK dense subroutines.

A new method will be described. The method is based on a reordering algorithm LORA, [3], [4], which allows us to form easily a sequence of relatively large blocks. The blocks are handled as dense matrices. This is **a trade off procedure: it is accepted to perform more computations, but with higher**

speed. All computations are performed by dense kernels (LAPACK subroutines, [1] are used at present, but other dense subroutines may also be applied). The dense kernels perform the most time-consuming part of the work; Table 3. Therefore, it should be expected to obtain good results on any computer for which high quality software for dense matrices is available.

The new method is described in Section 2. Numerical results, obtained on CRAY C92A and POWER CHALLENGE (Silicon Graphics) are given in Section 3. Plans for future work on the algorithm are outlined in Section 4.

2 Description of the algorithm

The main objective is to compute the QR-factorization of A by applying dense orthogonal transformations to a sequence of large blocks. The algorithm consists of five steps: (i) reordering by LORA, (ii) scatter, (iii) compute, (iv) gather and (v) deal with the last block. The actions performed during the five steps are discussed in §2.1 - §2.5)

2.1 Using LORA for rectangular matrices

LORA ("locally optimized reordering algorithm"; [3, 4]) reorders the matrix to a block upper triangular form (Fig. 1) with an important additional requirement to put as many zeros as possible under the diagonal blocks ("the separator").

The complexity of LORA is $O(NZ \ \log \ n)$ assuming that A has NZ non-zeros and n columns. The complexity can be reduced to $O(NZ)$. The more expensive version allows us to introduce additional criteria, [3], by which the quality of the ordering is improved. A criterion that puts more non-zeros close to the diagonal blocks is applied here.

2.2 Scattering the non-zeros in a two dimensional array

The next task is to form large dense block-rows. It is easier to explain the main ideas by taking the simplest (but not the best) case: each block-row is formed by taking a dense block of the separator (Fig. 1). If the first dense block of the separator contains r_1 rows and q_1 columns, then the non-zeros in the first block-row are stored in a two-dimensional array with r_1 rows and c_1 columns, where c_1 is the union of the sparsity patterns of its r_1 rows. The number of orthogonal stages is q_1 ($rank(A) = n$ implies $q_1 \leq r_1$).

Assume that the second dense block of the separator contains r_2 rows and q_2 columns. The second block-row can be formed by taking this dense block and adding the relevant "unfinished rows" from the first block-row. The total number of unfinished rows is $r_1 - q_1$ but there may be some with no non-zero in the first q_2 columns. Therefore the number of relevant (when the second block is treated) unfinished rows is $r_1^* \leq r_1 - q_1$. The remaining $r_1 - q_1 - r_1^*$ unfinished rows are moved to the end of the second block-row. They will be treated in some of the following block-rows. The non-zeros of the relevant rows of the second

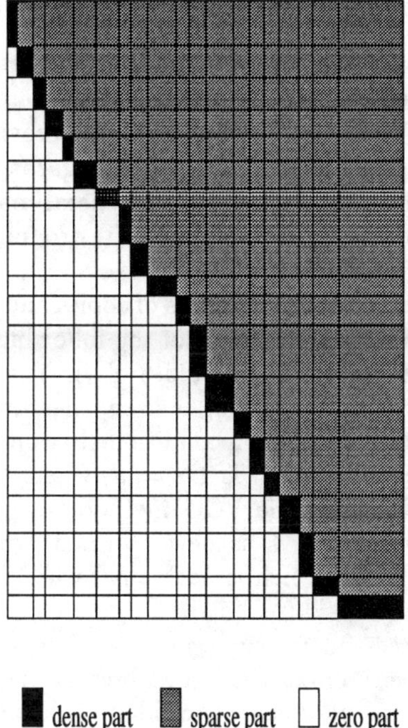

Fig. 1. Sparsity pattern of a rectangular matrix, reordered by LORA .

block-row (the first $r_2 + r_1^*$ rows) are stored in a two-dimensional array with $r_2+r_1^*$ rows and c_2 columns, where c_2 is the union of the sparsity patterns of its $r_2+r_1^*$ relevant rows. The number of orthogonal stages is q_2 (again, $rank(A) = n$ implies $q_2 \leq min(r_2 + r_1^*, c_2)$).

Continuing in this way, one can factorize the whole matrix. This may be inefficient, because the blocks of the separator are in general too small. One can combine several blocks of the separator into a larger diagonal block and to apply the same technique, as above. It is appropriate to use two parameters: LROWS and LSTAGE as lower limits respectively for the number of rows and the number of stages in a composite block.

2.3 Calling subroutines for dense matrices

The i'th dense block contains r_i rows and c_i columns, while q_i stages are to be performed. The LAPACK subroutine SGEQRF is called to produce zeros under the diagonal elements of first q_i columns of the $(r_i \times c_i)$ dense matrix by Householder reflections. After that the LAPACK subroutine SORMQR is called to modify the last $c_i - q_i$ columns of the i'th dense block-row by using the orthogonal matrix Q_i obtained during the call of SGEQRF. The complexity of SGEQRF is $O(r_i q_i^2 - q_i^3/3)$. The complexity of SORMQR is $O(r_i q_i (c_i - q_i) - q_i^2 (c_i - q_i)/2)$. The cost of Step 3 is the dominant part of the work; Table 3.

2.4 Gathering the non-zeros in one-dimensional arrays

After Step 3 for block-row i, the non-zeros must be gathered in the sparse arrays. This can cause difficulties, because the number of non-zeros per row is in general changed in the Step 3. Therefore some operations (moving rows to the end of the sparse arrays and even performing occasionally garbage collections; [5]) that are traditionally used in sparse techniques for general matrices must be carried in Step 4. This extra work can be reduced by (i) dropping small elements and (ii) avoiding the storage of Q_i .

Dropping has two effects: (i) the numbers of copies and/or garbage collections is often reduced and (ii) the sizes of some of the following blocks are sometimes reduced (when c_i is reduced for some values of i). The second effect is more important than the first one. If dropping is used, then one should try to regain the accuracy lost by using R in a preconditioned conjugate gradients (PCG) method. The preconditioned system is $Cz = d$, where $C = (R^T)^{-1}A^TAR^{-1}$, $z = Rx$ and $d = (R^T)^{-1}A^Tb$. The PCG method is applicable, because C is symmetric and positive definite. C is never formed explicitly; one works the whole time with A and R. Q, which is normally rather dense, is neither stored nor used in the iterative process (see [5]).

Dropping is very successful for some matrices (see §3.2), but it should not be used if the matrix is very ill-conditioned. Direct methods may work better in the latter case. The storage of Q can be avoided by calculating $c = Q^Tb$ during the decomposition and then x can be found from $Rx = c$.

2.5 Final switch to dense matrix technique

If the density of the non-zeros in the active submatrix (in percent) becomes greater than some parameter (30% is normally a good choice), then the algorithm switches to Step 5. The same work as in Step 2 - Step 4 has to be done, with two simplifications: (i) no need to determine the union of the sparsity patterns of the rows in the last block; (ii) no need of gather step after the dense orthogonal decomposition (the part of R calculated during Step 5 can be used in the solution process directly from the dense array).

3 Numerical results

Most of the results have been obtained on CRAY C92A. A few results on POWER CHALLENGE from Silicon Graphics will also be presented.

3.1 Experiments with rectangular Harwell-Boeing matrices

Results obtained when the two largest rectangular matrices from [2] have been run, are given in Table 2. The new algorithm has been compared with three other algorithms: (i) **Totally dense.** Matrix A is stored in a full-size two-dimensional array (the empty locations are filled with zeros). LAPACK is directly used to solve the problem. (ii) **Partially sparse**. The large block-rows are created in the same way (as explained in §2.2). Givens rotations are used to produce zero elements. A Givens rotation is performed only if both leading elements of the two rows involved are non-zeros. (iii) **Pure sparse**. No dense matrix is used (the sparse algorithm used is discussed in [5]).

Matrix	Characteristics measured	Totally dense	New algorithm	Partially sparse	Pure sparse
gemat1	Computing time	511	28	10	34
	MFLOPS	878	662	196	4
	Efficiency	97%	73%	22%	0.4%
	Number of blocks	1	47	49	-
bellmedt	Computing time	14	10	16	1490
	MFLOPS	870	685	458	6
	Efficiency	96%	76%	51%	0.7%
	Number of blocks	1	8	8	-

Table 2. Results of comparing the four algorithms on CRAY C92A (one processor, peak performance 902 MFLOPS).

If the matrix is relatively small, then the direct use of LAPACK ("Totally dense") is quite competitive with the new algorithm. The situation changes when the matrix becomes larger. The general conclusion is that the use of a sequence of large dense blocks together with dense kernels should be preferred when a standard package is needed.

In Table 3 the computing times spent in the different steps of the new algorithm is given. Most of the time is spent in the LAPACK subroutines.

Step	Action	gemat1	bellmedt
Step 1	LORA	0.67	0.25
Step2 and the beginning of Step 5	Scatter	1.30	0.07
Step 3 and the second part of Step 5	Compute	22.55	9.38
Step 4	Gather	3.22	0.14
	Solve	0.09	0.03
	Total time	**27.79**	**9.91**

Table 3. Computing times (in seconds) spent in the different steps. "Solve" refers to the solution part carried out after the factorization (the direct solver is used in this example).

3.2 Running larger problems with matrices of class F2

LAPACK can be used directly for solving problems with Harwell-Boeing matrices (only for gematl the results are inefficient; Table 2). Therefore some very large matrices, must also be tested. Matrices of class F2 ([5]) are used. Five parameters determine a matrix of that class: (i) *the number of rows* M, (ii) *the number of columns* N, (iii) *the sparsity pattern* C, (iv) *the average number of non-zeros per row* NR, (v) *the condition parameter* $ALPHA$. The results in Table 4 show that there is no difficulty with the computing times (they are small even when $M = 500000$).

Matrix identifiers			Computing times		
M	N	C	$NR = 5$	$NR = 10$	$NR = 15$
50000	25000	15000	10	18	35
100000	50000	30000	18	34	60
150000	75000	45000	27	40	76
200000	100000	60000	36	54	74
250000	125000	75000	46	73	109
300000	150000	90000	59	81	107
350000	175000	105000	70	98	141
400000	200000	120000	82	117	167
450000	225000	135000	88	125	162
500000	250000	150000	100	142	179

Table 4. Computing times (on CRAY C92A) spent for the factorization of 30 matrices of class F2 ($ALPHA = 1$, $NZ = NR * M + 110$) with the new algorithm.

3.3 Dropping small elements and using PCG

Dropping small elements and computing an approximate QR-factorization is discussed in [5]. It can be used (often with a great positive effect) in our new algorithm. The approximate R (obtained by dropping small non-zeros in Step 4) is used as a preconditioner; §2.4 . Sometimes this leads to considerable reductions of the computing time and/or the storage; see Tables 5, 6. Large drop tolerance may be inefficient for very ill-conditioned matrices, For such matrices it is better to apply the direct method (this is true for gemat1).

3.4 Preliminary results obtained on POWER CHALLENGE

Table 7 contain some results for the performance of our algorithm on a POWER CHALLENGE computer from Silicon Graphics. The results are obtained by just taking the CRAY version of the sparse code and calling the LAPACK routines

Density	Direct	Iterative	Eval. error	Exact error
$R = 10$	10	6.9 (0.9, 5)	1.1E-08	4.5E-10
$R = 20$	15	8.4 (1.2, 7)	4.9E-08	1.2E-09
$R = 30$	20	10.1 (1.3, 7)	1.3E-08	2.4E-09
$R = 40$	30	11.1 (1.3, 7)	7.9E-08	1.5E-10
$R = 50$	44	12.6 (1.4, 7)	2.9E-08	6.1E-10
$R = 60$	64	14.4 (2.1,10)	2.1E-08	8.6E-10
$R = 70$	74	16.0 (2.9,13)	2.4E-08	1.3E-09
$R = 80$	95	18.3 (3.0,13)	2.3E-08	1.4E-09
$R = 90$	105	20.1 (3.1,13)	2.4E-08	1.4E-09
$R = 100$	124	24.2 (5.5,23)	1.1E-08	4.0E-11

Table 5. Results of running 10 matrices from class F2 on CRAY C92A with drop-tolerance TOL= 0.0625. LSTAGE= 75, the accuracy required: 10^{-7}. The total time for the solution of the problem by the iterative method is given under "Iterative" (the time spent in the iterative procedure and the number of iterations being given in brackets). The two-norms of the evaluated by the code error and the exact error are given in the last two columns.

TOL	Fact.	Sol.	Iters	Eval. error	Exact error
2^{-1}	1.6	1.39	113	9.5E-08	3.7E-09
2^{-2}	5.0	0.49	36	8.4E-08	9.0E-09
2^{-4}	7.9	0.26	18	5.9E-08	1.6E-08
2^{-8}	10.1	0.10	5	1.8E-08	4.9E-10
2^{-12}	10.1	0.06	3	6.0E-09	3.2E-12
2^{-16}	10.1	0.04	2	7.3E-10	2.4E-13

Table 6. Results of running bellmedt on CRAY C92A with different values of the drop-tolerance TOL and the accuracy required 10^{-7}. The two-norms of the evaluated by the code error vector and the exact error vector are given in the last two columns.

available on POWER CHALLENGE. Therefore, the results are quite satisfactory: the main idea (to obtain standard modules that perform reasonably well on different high-speed computers on which the LAPACK modules perform well) seems to work (we get reasonably good time, speed-up, MFLOPS and efficiency). The results could be improved.

4 Conclusions and plans for the future work

Only the computations in Step 3 and Step 5 have been optimized. These two steps are the most time-consuming parts of the computational work (which has

Processors	Comp. time	Speed-up	MFLOPS	Efficiency
1	416	-	167	64%
2	266	1.7	262	50%
4	157	3.3	443	43%
8	94	4.4	741	36%

Table 7. Results on a POWER CHALLENGE computer from Silicon Graphics for a problem with coefficient matrix gemat1.

been illustrated by two examples in Table 3), but it is nevertheless necessary to optimize also the other steps of the computational work.

The main conclusion is that although the new method will not always give best results it is **a standard tool for solving efficiently large sparse linear least squares problems**. It will produce good results on any high-speed computer on which the LAPACK modules perform well.

Acknowledgements

This research was partially supported by the BRA III Esprit project APPARC (# 6634), the Bulgarian Ministry of Education (grant I-505/95), NATO (OUTR.CGR 960312) and the Danish Natural Sciences Research Councill (SNF). Dr. Guodong Zhang from the Application Group at Silicon Graphics sent us the last updated version of the LAPACK routines tunned on POWER CHALLENGE.

References

1. Anderson, E., Bai, Z., Bischof C., Demmel J., Dongarra, J., Du Croz, J., Greenbaum, A., Hammarling, S., McKenney, A., Ostrouchov, S. and Sorensen, D, "LAPACK: Users' guide", SIAM, Philadelphia, 1992.
2. Duff, I. S., Grimes, R. G. and Lewis, J. G., "Sparse matrix test problems", ACM Trans. Math. Software, 15 (1989), 1-14.
3. Duin, A. C. N. van, Hansen, P. C., Ostromsky, Tz., Wijsoff, H. and Zlatev, Z., "Improving the numerical stability and the performance of a parallel sparse solver", Comput. Math. Applics., Vol. 30, No. 12 (1995), 81-96.
4. Gallivan, K., Hansen, P. C., Ostromsky, Tz. and Zlatev, Z., "A locally optimized reordering algorithm and its application to a parallel sparse linear system solver", Computing, 54 (1995), 39-67.
5. Zlatev, Z., "Computational methods for general sparse matrices", Kluwer Academic Publishers, Dordrecht-Toronto-London, 1991.

Integration of Some Constitutive Relations of Plain Strain Elastoplasticity Using Modified Runge-Kutta Methods

G. Papakaliatakis[1] and **T.E. Simos**[2]

[1] Section of Mechanics, Department of Civil Engineering, University of Thrace, GR-67 100 Xanthi, GREECE.
[2] Laboratory of Applied Mathematics and Computers, Department of Sciences, Technical University of Crete, Kounoupidiana, GR-73100 Chania, Crete, GREECE.

Abstract. In this paper we solve via high order Runge-Kutta methods a set of coupled ordinary differential equations in which the constitutive relations of Tresca and Mohr-Coulomb materials in plain strate elastoplasticity has been transformed. In our comparisons we have included a low order modified Runge-Kutta method with phase-lag of order twelve. Conclusions on efficiency of the used methods have been made, based on extensive numerical tests.

1 Introduction

In the mathematical theory of plasticity two types of theories have been advanced. The one which is called **deformation theory** or **total deformation theory** is based on the hypothesis that the plastic strain is a solely function of the stress. In the other type, which is referred as **flow** or **incremental theory** it is stated that the plastic strain increments are related to the stress state and stress increment. In general these relations are not integrable and the integral depends on the loading path. Using finite element method and incremental techniques, t he flow theory can easily be applied to solve plasticity problems.

In incremental theory the constitutive behaviour of an elastoplastic material may be expressed with an incremental stress-stain relation of the form

$$\dot{\sigma} = \mathbf{D_{ep}} \dot{\epsilon} \tag{1}$$

where $\dot{\sigma}$ denotes the stress increment vector, $\dot{\epsilon}$ is the strain increment vector and the superior dot represents a derivative with respect to time. Also $\mathbf{D_{ep}}$ denotes the elastoplastic constitutive matrix, that for the perfectly plastic Tresca and Mohr-Coulomb structural material to be considered in this work depends only on the components of the stress within the element.

Using a finite element computer program based on flow theory of plastisity the nodal displacement increments and the strain increments are computed that corresponds to the nodal forces. Once these are known, the above equation can be used to find the stress increments. This equation constitutes a set of ordinary differential equations and the integration of this set presents difficultes because the matrix $\mathbf{D_{ep}}$ is a highty non-linear function of the stresses.

Assuming that the strain rates are constant, equation (1) can be analysed to a of coupled ordinary differential equations of the form

$$\dot{\sigma} = \mathbf{D_{ep}} \frac{\Delta\epsilon}{\Delta \mathbf{t}} \tag{2}$$

where $\Delta \mathbf{t}$ is the time interval and $\mathbf{t_0} \leq \mathbf{t} \leq \mathbf{t_0} + \Delta \mathbf{t}$. This set of equations defines a standard initial-value problem since $\dot{\sigma}_\mathbf{0} = \dot{\sigma}(\mathbf{t_0})$ and $\Delta\epsilon$ are known and the matrix $\mathbf{D_{ep}}$ is the function of stresses σ. Solving the above system of ordinary differential equations we find an approximation for the stresses.

Sloan and Booker [1] has compared the well known embedded Runge-Kutta method of Fehlberg [2], the embedded Runge-Kutta method of Dormand and Prince [3] and the embedded Runge-Kutta method of England [4]. All these methods are of order four and five. The conclusion of the work of Sloan was that for this class of problems the embedded Runge-Kutta Dormand and Prince method was the most efficient.

The purpose of this paper is to compare some high order embedded Runge-Kutta methods for the numerical solution of initial-value problems of the form (2). We use for comparison the embedded Runge-Kutta methods of order five and six of Dormand and Prince [5], of Prince and Dormad [6], of Sharp and Verner [7], of Verner [8]-[11]. We use, also, the embedded Runge-Kutta method of order six and seven of Verner [9], [11] and of Sharp and Verner [7] and the embedded Runge-Kutta methods of order seven and eight of Prince and Dormand [6] and of Verner [9]. Finally we use the modified Runge-Kutta Dormand-Prince method of order four and five with phase-lag of order twelve produced by Papakaliatakis and Simos in [12]. This method is based on the well known Runge-Kutta Dormand-Prince method RKDP5(4)8M.

2 Increments stress-strain relations for perfectly plastic materials

During an infinitesimal increment of stress, changes of strain are assumed to be divisible into elastic and plastic parts. Thus

$$\dot{\epsilon} = \dot{\epsilon}^\mathbf{e} + \dot{\epsilon}^\mathbf{p} \tag{3}$$

where $\dot{\epsilon}^\mathbf{l}$ denotes the elastic and $\dot{\epsilon}^\mathbf{p}$ the plastic strain increment.

The elastic strain inrements are related to the stress increments by the Hooke's law as usual, by a symmetric matrix of constants $\mathbf{D}$, so that

$$\dot{\epsilon}^\mathbf{e} = \mathbf{D}^{-1}\dot{\sigma} \tag{4}$$

Using the normality principle of the Von-Mises flow rule, the plastic strain increment $\dot{\epsilon}^\mathbf{p}$ can be written as

$$\dot{\epsilon}^\mathbf{p} = \lambda \frac{\partial \mathbf{F}}{\partial \sigma} \tag{5}$$

where $\mathbf{F}$ denotes the yield surface which for a perfectly plastic material may be written as a function of σ (yielding of material can occur only if the stress σ satisties the yied criterion $\mathbf{F}(\sigma) = \mathbf{0}$. Also *lambda* denotes a proportionality constant.

A reduction of the restriction of the above rule can be obtained by specifying a plastic potential

$$\mathbf{Q} = \mathbf{Q}(\sigma) \tag{6}$$

which defines the plastic strain increment as

$$\dot{\epsilon}^{\mathbf{P}} = \lambda \frac{\partial \mathbf{Q}}{\partial \sigma} \tag{7}$$

The particular case of $\mathbf{Q} = \mathbf{F}$ is known as associated plasticity flow rule and the matrix $\mathbf{D_{ep}}$ is symmetric. Otherwise the material is said to have an non associated flow rule and the matrix $\mathbf{D_{ep}}$ is in general unsymetric.

Comparing equations (3), (4) and (7) we can write

$$\dot{\epsilon} = \mathbf{D}^{-1}\dot{\sigma} + \lambda \frac{\partial \mathbf{Q}}{\partial \sigma} \tag{8}$$

When plastic yield is occuring the stress are on the yield surface. Differentiating we can write therefore

$$\left\{\frac{\partial \mathbf{F}}{\partial \sigma}\right\}^T \dot{\sigma} = 0 \tag{9}$$

The indeterminate constant *lambda* can be eliminated. This results in an explicit expansion which determins the stress changes in terms of imposed strain changes with

$$\dot{\sigma} = \mathbf{D_{ep}}\dot{\epsilon}$$

$$\mathbf{D_{ep}} = \mathbf{D} - \mathbf{D}\left\{\frac{\partial \mathbf{Q}}{\partial \sigma}\right\}\left\{\frac{\partial \mathbf{F}}{\partial \sigma}\right\}^{\mathbf{T}} \mathbf{D}\left[\left\{\frac{\partial \mathbf{F}}{\partial \sigma}\right\}^{\mathbf{T}} \mathbf{D}\left\{\frac{\partial \mathbf{Q}}{\partial \sigma}\right\}\right]^{-1} \tag{10}$$

The elasto-plastic matrix $\mathbf{D_{ep}}$ takes the place of the elasticity matrix $\mathbf{D_T}$ in incremental analysis.

For the yield surfaces we have the following relations

Tresca relation

$$\mathbf{F} = 2\bar{\sigma} cos\theta - \mathbf{Y}(\mathbf{k}) = \mathbf{0} \tag{11}$$

where $\mathbf{Y}(\mathbf{k})$ is the yield stress from uniaxial tests.

Mohr-Coulomb relation

$$\mathbf{F} = \sigma_m sin\phi + \bar{\sigma} cos\theta - \frac{\bar{\sigma}}{\sqrt{3}} sin\phi sin\theta - c cos\phi = \mathbf{0} \tag{12}$$

where $c(k)$ and $\phi(k)$ are the cohesion and angle of friction, respectively, which could depend on some strain hardening parameter k and

$$\sigma_m = \frac{J_1}{3} = \frac{\sigma_x + \sigma_y + \sigma_z}{3}$$
$$\bar{\sigma} = J_2^{\frac{1}{2}} = \left[\frac{1}{2}\left(\mathbf{s}_x^2 + \mathbf{s}_y^2 + \mathbf{s}_z^2\right) + \tau_{xy}^2 + \tau_{yz}^2 + \tau_{xz}^2\right]^{\frac{1}{2}}$$
$$\theta = \frac{1}{3} sin^{-1}\left[-\frac{3\sqrt{3}J_3}{2\bar{\sigma}^3}\right] \tag{13}$$

with $-\frac{\pi}{6} < \theta < \frac{\pi}{6}$, where

$$\mathbf{J_3} = \mathbf{s}_x\mathbf{s}_y\mathbf{s}_z + 2\tau_{xy}\tau_{yz}\tau_{xz} - \mathbf{s}_x\tau_{yz}^2 - \mathbf{s}_y\tau_{xz}^2 - \mathbf{s}_z\tau_{xy}^2$$
$$\mathbf{s}_x = \sigma_x - \sigma_m,\ \mathbf{s}_y = \sigma_y - \sigma_m,\ \mathbf{s}_z = \sigma_z - \sigma_m \tag{14}$$

3 Embedded Runge-Kutta methods

To solve the system of coupled ordinary differential equations (2) via embedded Runge-Kutta methods we must estimated at each step size the local truncation error. To do this we must have one method of lower order q and a method of higher order $q+1$. The lower order method produces an approximation of the stress σ, while the higher order method produces an approximation of the stress $\hat{\sigma}$ given by

$$\sigma = \sigma_0 + [\sum_{i=1}^{m} b_i \Delta\sigma_\mathbf{i}]$$
$$\hat{\sigma} = \sigma_0 + [\sum_{i=1}^{m} \hat{b}_i \Delta\sigma_\mathbf{i}] \tag{15}$$

where

$$\Delta\sigma_\mathbf{1} = \mathbf{D_{ep}}(\sigma_\mathbf{0}) h_i \Delta\epsilon$$
$$\Delta\sigma_\mathbf{j} = \mathbf{D_{ep}}\left(\sigma_\mathbf{0} + [\sum_{k=1}^{j-1} a_{j,k}\Delta\sigma_\mathbf{k}]\right) h_n \Delta\epsilon,\ j = 1(1)m \tag{16}$$

where b_i, $\hat{b}_i$ and $a_{j,k}$, $j = 2(1)m$, $k = 1(1)j-1$, $i = 1(1)m$ are the coefficients of the appropriate Runge-Kutta methods, h_n is the steplength of the current subinterval.

The estimate of the relative error in **sigma** in the current subinterval is given by

$$E_n = \frac{||\hat{\sigma} - \sigma||}{||\hat{\sigma}||} \tag{17}$$

The current steplength is accepted if the relative error is less than a tolerance TOL specified by the user. If the relative error is greater or equal to the specified tolerance then the steplength is rejected. In both cases, the steplength of the next subinterval h_{n+1} is given by

$$h_{n+1} = wh_n, \; w > 0 \tag{18}$$

where w is given by

$$w = 0.9 \left(\frac{TOL}{E_n} \right)^{1/q} \tag{19}$$

where q is the order of the Runge-Kutta method used.

To start the integration we apply the methods (15) with $h_n = 1$. If $E_n \geq TOL$ then this steplength will be rejected and the procedure is repeated without updating the stresses. It is obvious that in this case only an additional evaluation of the matrix $\mathbf{D_{ep}}$ is required. If the steplength is accepted, i.e. if $E_n < TOL$ then the stresses are calculated via methods (15) and also the new steplength of the next subinterval is computed. In our tests the relative error tole rances, TOL, are equal to 10^{-s}, $s = 3(1)6$. But, in this paper we present the results for 10^{-5}. We have similar conclusion for the other tolerances.

4 Numerical tests

In the above section we have described the general embedded Runge-Kutta method for integrating Tresca and Mohr-Coulomb constitutive relations.

The numerical procedure for examing the efficiency of the proposed methods is fully described in [1].

The accuracy of the Runge-Kutta methods is defined from the errors in the elastoplastic stresses which are computed from the formula

$$E = \frac{||\sigma - \sigma_{\mathrm{e}}||_\infty}{||\sigma_{\mathrm{e}}||_\infty} \tag{20}$$

where $||.||_\infty$ is the maximum norm and σ_{e} are the stresses calculated from the exact solution which for the Tresca constitutive law is fully defined in [1]. For the Mohr-Coulomb constitutive law (in which there isn't analytical solution) the matrix σ_{e} is calculated using a high order Runge-Kutta method (RKDP87 defined in [6]) with constant stepsizes equal to 0.01.

In our numerical tests we measure the average error and the maximum error given by

$$E_{av} = \frac{1}{N} \sum_{n=1}^{N} E_n \tag{21}$$

$$E_{max} = max_{1 \leq n \leq N} (E_n) \tag{22}$$

where E_n is the error calculated from (17) for each of the N subintervals.

4.1 Tresca constitutive relation

The various methods are used to integrate the Tresca constitutive law in order to approximate the elastic stress in the plane $Z = 0$. The initial stresses are chosen to be equal to $\sigma_x = \sigma_y = 0$ and $\tau_{xy} = c$. The material paramneters are $\nu = 0.49$, $G = 100$ and $c = 1$. We note here that the material parameters are appeared in the elastoplastic constitutive matrix $\mathbf{D_{ep}}$.

a/a	Method	a/a	Method
MI	RKDP6(5)8M of [6]	MVIII	RKVE6(5)9a of [10]
MII	RKVE5(6) of [9]	MIX	CIRK65 of [7]
MIII	RKDP6(5)8C of [5]	MX	RKVE6(7) of [9]
MIV	RKDP6(5)8S of [5]	MXI	RKVE7(6) of [11]
MV	RKVE6(5)9c of [8]	MXII	RKVE7(8) of [9]
MVI	RKVE6(5)8M of [8]	MXIII	RKDP8(7) of [6]
MVII	RKVE5(6) of [11]	MXIV	CIRK76 of [7]
		MXV	RKMDP5(4)of [12]

Table 1. Runge-Kutta methods used for the integration of the system of coupled ordinary differential equations

In Table 2 we present the average error, the maximum error and the real time of computation for the integration of the Tresca constitutive law using the Runge-Kutta methods presented in Table 1. The relative error tolerance, TOL, is equal to 10^{-5}.

4.2 Mohr-Coulomb constitutive relation

The various methods are used, also, to integrate the Mohr-Coulomb constitutive law in order to approximate the elastic stress in the plane $Z = 0$. The initial stresses are chosen to be equal to $\sigma_x = \sigma_y = 0$ and $\tau_{xy} = c cos\phi$. The material paramneters are $\nu = 0.3$, $G = 100$, $c = 1$ and $\phi = 30^o$.

In Table 3 we present the average error, the maximum error and the real time of computation for the integration of the Mohr-Coulomb constitutive law using the Runge-Kutta methods presented in Table 1. The relative error tolerance, TOL, is equal to 10^{-5}.

From the results presented we have that the most rapid methods are the method RKDP6(5)8S of [5] for the Tresca criterion and the methods RKVE7(6) of [11] and RKDP8(7) of [6] for the Mohr-Coulomb criterion. The most accurate methods are the RKVE7(6) of [11] and the RKMDP5(4)of [12] for the Tresca criterion. We note that for the Mohr-Coulomb criterion the method RKVE7(6) of [11] is also very accurate but the most accurate methods are the methods

Method	E_{av}	E_{max}	Real time of computation (secnds)
MI	1.410^{-6}	9.710^{-6}	55
MII	3.210^{-6}	1.510^{-5}	65
MIII	7.010^{-7}	4.110^{-6}	85
MIV	4.110^{-7}	2.810^{-6}	50
MV	1.610^{-7}	8.610^{-7}	135
MVI	2.010^{-7}	9.710^{-7}	78
MVII	3.010^{-6}	9.710^{-6}	96
MVIII	7.010^{-8}	2.810^{-7}	81
MIX	1.810^{-7}	1.210^{-6}	108
MX	1.510^{-6}	9.510^{-6}	93
MXI	3.010^{-8}	1.510^{-7}	113
MXII	4.510^{-8}	4.710^{-7}	83
MXIII	6.610^{-8}	1.010^{-6}	58
MXIV	7.510^{-7}	8.710^{-6}	100
MXV	1.110^{-9}	4.010^{-9}	165

Table 2. Average error, maximum error and real time of computation for the integration of the Tresca constitutive law using the Runge-Kutta methods presented in Table 1

Method	E_{av}	E_{max}	Real time of computation (secnds)
MI	7.710^{-7}	8.110^{-6}	31
MII	3.110^{-6}	1.310^{-5}	73
MIII	4.510^{-7}	3.610^{-6}	26
MIV	6.410^{-7}	8.810^{-6}	26
MV	1.510^{-7}	9.110^{-7}	66
MVI	1.310^{-7}	9.510^{-7}	47
MVII	3.010^{-6}	1.210^{-5}	49
MVIII	6.010^{-8}	7.210^{-7}	71
MIX	9.010^{-8}	1.110^{-6}	84
MX	1.810^{-6}	2.110^{-5}	53
MXI	1.710^{-7}	2.410^{-6}	25
MXII	6.010^{-8}	9.510^{-7}	42
MXIII	7.010^{-8}	1.810^{-6}	25
MXIV	5.710^{-7}	1.110^{-6}	48
MXV	2.110^{-9}	1.810^{-8}	93

Table 3. Average error, maximum error and real time of computation for the integration of the Mohr-Coulomb constitutive law using the Runge-Kutta methods presented in Table 1

RKVE6(5)9a of [10], RKVE7(8) of [9] and RKMDP5(4)of [12]. Because of the significant interst for the accuracy of the used methods we propose the methods RKVE6(5)9a of [10], RKVE7(8) of [9], RKMDP5(4)of [12] and RKVE7(6) of [11] as the most efficient methods for these problems.

References

1. Sloan, S.W., Booker, J.R.: Integration of Tresca and Mohr-Coulomb constitutive relations in plane strain elastoplasticity. Inter. J. Numer. Meth. Engrg. **33** (1992) 163–196
2. Fehlberg, E.: Klassische Runge-Kutta Formeln Vierter und Niedrigerer Ordnung mit Schrittweiten-kontrolle und Ihre Anwendung auf Wärmeleitungs-probleme. Computing **6** (1970) 61–71
3. Dormand,J.R., Prince, P.J.: A family of embedded Runge-Kutta formulae. J. Comput. Appl. Math. **6** (1980) 19–26
4. Enlgand, R.: Error estimates for Runge-Kutta type solutions to systems of ordinary differential equations. Comput. J **12** (1969) 166–170
5. Dormand, J.R., Prince, P.J.: A reconsideration of some embedded Runge-Kutta formulae. J. Comput. Appl. Math. **15** (1986) 203–211
6. Prince, P.J., Dormand, J.R.: High order embedded Runge-Kutta formulae. J. Comput. Appl. Math. **7** (1981) 67–75
7. Sharp, P.W., Verner, J.H.: Completely imbedded Runge-Kutta pairs. SIAM J. Numer. Anal. **31** (1994) 1169–1190
8. Verner, J.H.: The comparison of some Runge-Kutta formula pairs using DETEST. Dept. Mathematics and Statistics, Queen's University, Kingston, Ontario, Canada, 1990
9. Verner, J.H.: Explicit Runge-Kutta methods with estimates of the local truncation error. SIAM J. Numer. Anal. **15** (1978) 772–790
10. Verner, J.H.: Some Runge-Kutta formula pairs. SIAM J. Numer. Anal. **28** (1991) 496–511
11. Verner, J.H.: Strategies for deriving new explicit Runge-Kutta pairs. Annals Numer. Math. **1** (1994) 225–244
12. Papakaliatakis, G., Simos, T.E.: Numerical methods for the integration of Tresca and Mohr-Coulomb relations in plane strain elastoplasticity. Section of Mechanics, Department of Civil Engineering, University of Thrace, 1996

High Performance Solution of Linear Systems Arising from Conforming Spectral Approximations for Non-Conforming Domain Decompositions

Marcin Paprzycki[1] and Andreas Karageorghis[2]

Department of Mathematics and Statistics, University of Cyprus
1678 Nicosia, Cyprus
andreask@pythagoras.mas.ucy.ac.cy

Abstract. We apply a conforming spectral collocation technique to non-conforming domain decompositions. The resulting global matrices have a particular block structure. We study the performance of various direct methods of solution of the resulting linear system on a RS6000 workstation, a SGI Power Challenge and a Cray J-916 supercomputer.

1 Introduction

In this paper we study the performance of three direct methods for the solution of the global systems resulting from spectral approximations for a certain class of domain decompositions. In particular, we examine the linear systems arising from conforming spectral approximations in non-conforming decompositions in rectangular domains, developed in [7]. The spectral approximations which are used are conforming, that is, the solution is C^0 continuous at all points across the subdomain interfaces for second order problems and C^1 continuous at all points across the subdomain interfaces for fourth order problems. The matrices resulting from these approximations possess a specific block diagonal structure which can be exploited by applying a banded system solver [8] or a capacitance-type technique [5, 6]. The performance of these two approaches is compared to the performance of two full matrix solvers from the NAG library [8].

2 Domain decomposition and spectral approximation

We consider the problem

$$\nabla^2\phi(x,y) = F(x,y) \tag{1}$$

on the rectangle $(\alpha,\beta)\times(a,b)$ subject to Dirichlet boundary conditions. We shall assume that the boundary conditions can be expressed as polynomials. As in [5, 7], for the partitions $\alpha=\alpha_0<\alpha_1<\alpha_2<...<\alpha_{N-1}<\alpha_N=\beta$ and $a=a_0<a_1<a_2<...<a_{N-1}<a_N=b$, $N\in\mathbb{N}$, we consider the following decomposition: the rectangle $(\alpha,\beta)\times(a,b)$ is decomposed into $2N-1$ subdomains in the following way: for $k=1,2,...,N-1$, subdomain $2k-1$ is the rectangle

$(\alpha_{k-1}, \alpha_k) \times (a_{k-1}, a_N)$ and subdomain $2k$ is the rectangle $(\alpha_k, \alpha_N) \times (a_{k-1}, a_k)$. Subdomain $2N-1$ is the rectangle $(\alpha_{N-1}, \alpha_N) \times (a_{N-1}, a_N)$. In each subdomain the solution is approximated by

$$\phi_s(x,y) = \sum_{m=0}^{M_s} \sum_{n=0}^{N_s} \gamma_{mn}^s \hat{T}_m^s(x) \tilde{T}_n^s(y), \quad s = 1, 2, ..., 2N-1, \tag{2.2}$$

where the functions $\hat{T}_m^s(x)$ and $\tilde{T}_n^s(y)$ are the shifted Chebyshev polynomials defined on the corresponding intervals of each region and the collocation points on each interval of each region (e.g. $\{x_i^s\}_{i=0}^{M_s}$) are the Gauss-Lobatto points [1, 2]. We shall assume that $M_{2k} \leq \min\{M_{2k+1}, M_{2k+2}\}$ and that $N_{2k-1} \leq \min\{N_{2k}, N_{2k+1}\}$, $k = 1, 2, ..., \mathrm{N}-1$

For for the above problem and domain decomposition it can be shown that the spectral approximation (2.2) yields C^0 conforming approximations on all the subdomain interfaces [5, 7].

3 Methods of solution

3.1 Capacitance technique

The structure of the global system for the five subdomain decomposition is of the form given in Figure 1.

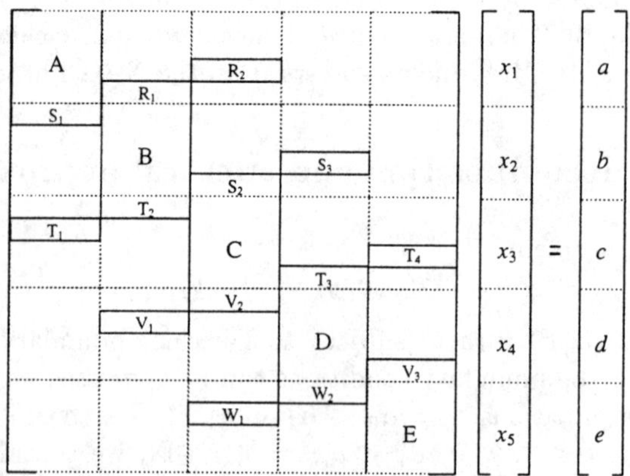

Fig. 1. Matrix for the 5-subdomain decomposition.

From the above system (Figure 1) we may express $\underline{x}_1$ and $\underline{x}_5$ in terms of $\underline{x}_2$, $\underline{x}_3$ and $\underline{x}_4$. We then substitute these expressions into the original system to yield a reduced system of the form:

$$\begin{aligned}\bar{A}_2\underline{x}_2 + \bar{S}_2\underline{x}_3 + \bar{S}_3\underline{x}_4 &= \underline{\bar{\alpha}}_2\\ \bar{T}_2\underline{x}_2 + \bar{A}_3\underline{x}_3 + \bar{T}_3\underline{x}_4 &= \underline{\bar{\alpha}}_3\\ \bar{V}_1\underline{x}_2 + \bar{V}_2\underline{x}_3 + \bar{A}_4\underline{\bar{x}}_4 &= \underline{\bar{\alpha}}_4\end{aligned}$$

The above system is then solved for $\underline{x}_2, \underline{x}_3$ and $\underline{x}_4$. The unknown vectors $\underline{x}_1$ and $\underline{x}_5$ may be obtained by back subsitution. This process can be easily generalized for any non-conforming multidomain decomposition [6].

3.2 Banded solvers

As can be seen from Figure 1 the linear system can be viewed as banded with bandwith equal to $\max\{(\sum_{n=0}^{2}(M_{S-2+K} - 1)(N_{S-2+K} + 1)), (\sum_{n=0}^{2}(M_{S+K} + 1)(N_{S+K} + 1))\}$ (the bandwidth is independent of the number of subdomains in the decomposition). It can be also observed that the banded approach may not be the most efficient as there is a substantial number of zeros inside the band that the solver cannot exploit.

4 Experimental results

4.1 Numerical example

The performance of the various techniques described in Section 3 was tested on the following test problem:

$$\nabla^2\phi(x,y) = (y^2 - 1)e^x + (x^2 - 1)e^y + 2e^x + 2e^y \quad \text{on} \quad (-1,1)^2$$

subject to Dirichlet boundary conditions which correspond to the exact solution of this problem $\phi(x,y) = (y^2-1)e^x + (x^2-1)e^y$. We used the decomposition (in the notation of Section 2), $\alpha_i = \alpha_{i-1} + (1/2)(\alpha_N - \alpha_{i-1}), i = 1,2,...,N-1, \alpha_0 = \alpha = -1, \alpha_N = \beta = 1$ and $a_i = a_{i-1} + (1/2)(a_N - a_{i-1}), i = 1,2,...,N-1, a_0 = a = -1, a_N = b = 1$. We also took (in Equation (2.2)) $M_s = N_s = n, s = 1,2,...,L$. The total number of unknowns (u_T) is therefore $u_T = L(n+1)^2$.

4.2 Implementation

The experiments were performed on a RS6000-550 workstation, a SGI Power Challenge and a Cray J-916 supercomputer. Timings on the RS6000 were obtained on an empty machine using the *time* function, on the Power Challenge the timings were obtained using the system timer *getrusage* and on the Cray using the *perftrace* utility. Each result presented here is an average of multiple runs. We experimented with two dense solvers. The first is the pre-packaged general solver routine F04ATF from the NAG library [8], which performs an LU decomposition, solves the linear system and if necessary improves the solution by iterative refinement(s). The second is a LAPACK [4] based pair of routines F07ADF (LAPACK routine _GETRF) performing the LU decomposition and F07AEF (LAPACK routine _GETRS) solving the factorized system. We used both these approaches inside the capacitance technique. For the banded solver we used the F07BDF/F07BEF pair (LAPACK pair _GBTRF/ _GBTRS).

4.3 Dense Solver Performance

In Table 1 the timings in seconds for $n = 4, 8, 12$ are presented for the dense solvers for the three computers we have experimented with for the three, five, seven and nine element decompositions.

		*RS*6000		*SGI*		*Cray*	
n	u_T	*F*04	*F*07	*F*04	*F*07	*F*04	*F*07
			Three elements				
4	75	0.08	0.06	0.02	0.02	0.04	0.06
8	243	0.80	0.53	0.19	0.08	0.33	0.24
12	507	7.47	4.83	0.94	0.42	1.40	0.87
			Five elements				
4	125	0.15	0.11	0.05	0.03	0.11	0.11
8	405	3.11	1.63	0.53	0.24	0.89	0.56
12	845	22.83	15.20	3.43	1.19	4.48	2.81
			Seven elements				
4	175	0.30	0.18	0.09	0.05	0.19	0.16
8	567	7.31	3.38	1.18	0.55	1.79	1.11
12	1183	49.35	31.09	8.94	5.51	10.20	6.76
			Nine elements				
4	225	0.51	0.23	0.16	0.08	0.30	0.23
8	729	11.41	5.93	2.25	1.10	3.18	1.96
12	1521	77.75	51.64	18.72	12.50	19.30	13.60

Table 1. Timings for the full matrix solution

It can be observed that the LAPACK pair outperforms the F04ATF solver for all machines and all decompsition sizes. This can be related to the iterative refinements which when performed are a series of level 2 BLAS operations which are characterized by a much poorer performance than the level 3 BLAS based LAPACK kernels [4, 9].

It can also be observed that as the matrix size u_T increases the Cray's performance reaches that of the Power Challenge. This fact is rather peculiar as one processor of the Cray J-916 reaches only 195 MFlops in comparison with 270 MFlops of the practical peak performance of the Power Challenge. There are two points that need to be raised: Firstly, the performance of the SGI has been observed to decrease as the matrix size increases, whereas the performance of the Cray remains unchanged. Secondly, the BLAS/LAPACK kernels are much better optimized by the Cray then the same routines in the SGI's scientific computing library. The crossing point where the Cray outperforms the Power Challenge occurs at $n = 14$ (for the nine element decomposition).

Finally, for the largest systems the Cray's performance reaches 187 MFlops, which is approximately 95% of the practical peak.

4.4 Banded Solver Performance

It is clear that for the three element decomposition the banded structure is non-existent. This is why we did not applied the banded solver in that case. In Table 2 the results for the banded technique are summarized for the five and seven element decompositions for $n = 4, 7, 10, 13, 16$ (the results of the nine element decompostion are presented in section 4.5).

n	*Bandwidth*	*RS6000*	*SGI*	*Cray*
		Five	*elements*	
4	75	0.09	0.03	0.11
7	192	0.94	0.12	0.32
10	363	4.96	0.52	0.95
13	507	19.06	2.08	2.83
16	867		7.92	7.85
		Seven	*elements*	
4	75	0.16	0.04	0.15
7	192	1.81	0.25	0.56
10	363	10.15	1.24	2.03
13	507	38.84	5.72	6.78
16	867		20.06	19.10

Table 2. Timings for the banded solver

The missing result for the RS6000 means that the system of this size did not fit into the memory. When compared with the results from Table 1 it can be observed that the performance of the banded solver is approximately 1.5 times better. This is most apparent for the RS6000 and least visible for the SGI. As above, the increase in matrix size improves the performance of the Cray relatively to that of the SGI.

4.5 Capacitance Technique Performance

As was the case in the banded system solver approach, the capacitance technique as described in Section 3 cannot be applied to the three-domain decomposition. Based on the results of Section 4.3 we applied the capacitance technique with the F07ADF/F07AEF pair. Further experiments with F04ATF confirmed the validity of this choice. The performance gain in the largest cases was about 10%. In Table 3 the results for the banded technique are summarized for the five and seven element decompositions for $n = 4, 7, 10, 13, 16$ (the nine element decompostion results are presented below).

n	u_T	*RS6000*	*SGI*	*Cray*
		Five	*elements*	
4	125	0.07	0.02	0.07
7	320	0.46	0.09	0.21
10	605	2.35	0.30	0.55
13	980	9.67	0.99	1.49
16	2601		2.94	3.78
		Seven	*elements*	
4	125	0.13	0.03	0.08
7	320	0.69	0.13	0.26
10	605	3.95	0.64	0.75
13	980	17.17	2.19	2.24
16	2601		6.86	5.98

Table 3. Timings for the capacitance technique

Finally, in Table 4 we summarize the nine element performance for the banded solver and the capacitance technique for all three computers and for $n = 4, 5, \ldots, 16$.

	RS6000		*SGI*		*Cray*	
n	*banded*	*capacit*	*banded*	*capacit*	*banded*	*capacit*
4	0.28	0.10	0.07	0.05	0.20	0.11
5	0.70	0.23	0.13	0.08	0.33	0.18
6	1.49	0.53	0.24	0.15	0.53	0.31
7	3.09	1.19	0.45	0.28	0.86	0.55
8	5.31	2.18	0.90	0.51	1.38	0.88
9	9,79	3.95	1.56	0.88	2.27	1.33
10	16.64	6.93	2.61	1.48	3.61	1.93
11	26.45	11.54	4.57	2.43	5.70	2.82
12	42.68	18.08	8.20	3.81	8.57	3.63
13	64.09	29.83	13.24	5.82	12.80	5.91
14			20.82	8.58	18.90	7.23
15			31.41	12.67	29.00	26.00
16			55.71	18.18	38.93	13.10

Table 4. Timings for the capacitance and banded solvers

Comparing results in Tables 2, 3 and 4 one can observe that the capacitance technique outperforms the banded solver in all cases. In the largest case (nine subdomain decomposition) the performance gain is about 2.15 for the RS6000, 3.06 for the SGI Power Challenge and 2.97 for the Cray. The results reconfirm the fact that as the matrix size increases the Cray performs better than the

SGI Power Challenge. For the largest reported case the performance gain is 1.43 for the banded solver and 1.39 for the capacitance technique. In this case the Cray's banded solver reaches 163 MFlops and the capacitance technique runs at 112 MFlops (83% and 57% of the practical peak perfomance respectively). The Cray's performance drop for $n = 15$ is related to the memory bank conflicts (the blocks are of sizes that are multiples of 16; see [4, 9]).

5 Conclusions

In this study we present efficient direct methods of solution for the global systems resulting from conforming approximations for a class of non-conforming domain decompositions. A comparison of these methods is carried out for several decompositions and the results indicate that a capacitance-type technique which expoits the block structure is best suited for the solution of such systems. Our results also indicate that a substantial improvement is necessary in the SGI's scientific computing library for it to be able to compete with the Cray in terms of performance. The next step of our research is to study the performance characteristics of the general sparse multifrontal solver UMFPACK [3] when applied to the problem in question.

6 Acknowledgement

This work was partially supported by a grant from the University of Cyprus. The computer time grant from the NCSA at Urbana-Champain is kindly acknowledged.

References

1. Boyd J. P.: Chebyshev and Fourier Spectral Methods. Springer-Verlag, New York (1989)
2. Canuto C., Hussaini M., Quarteroni A., Zang T.: Spectral Methods in Fluid Dynamics. Springer-Verlag, New York (1988)
3. Davis T. A., Duff I. S.: A combined unifrontal/multifrontal method for unsymmetric sparse matrices. Technical Report TR-95-020, CISE Department, University of Florida (1995)
4. Dongarra J. J., Duff J. S., Sorensen D. C., van der Vorst H. A.: Solving Linear Systems on Vector and Shared Memory Computers. SIAM, Philadelphia (1991)
5. Karageorghis A.: Conforming Spectral Domain Decomposition Schemes. (this volume)
6. Karageorghis A., Paprzycki M.: An efficient direct method for fully conforming spectral collocation schemes. Annals of Numerical Mathematics. (to appear)
7. Karageorghis A., Sivaloganathan S.: Conforming spectral approximations for non-conforming domain decompositions. Technical Report TR/07/96, Department of Mathematics and Statistics, University of Cyprus (1996)
8. Numerical Algorithms Group Library Mark 15. NAG(UK) Ltd, Wilkinson House, Jordan Hill Road, Oxford, UK
9. Paprzycki M., Cyphers C.: Multiplying matrices on the Cray – Practical Considerations, CHPC Newsletter **6** (1991) 77-82

Stabilization and Experience with the Partitioning Method for Tridiagonal Systems *

Velisar Pavlov and Daniela Todorova

Center of Applied Mathematics and Informatics,
University of Rousse, 7017 Rousse, Bulgaria

Abstract. The partitioning algorithm which is a modification of Wang's method for tridiagonal equations is stabilized to the case of arbitrary well conditioned matrix. A realization on Parallel Virtual Machine (PVM) is presented. The parallel solution is analysed under different loads: system dimension, variable numbers of virtual machines and different kind of bandwidth local area networks.

1 Introduction

Large tridiagonal systems appear in many applications, such as finite element technology, difference approximation to differential equations, power distribution systems, etc. If we want to solve such systems in parallel we can use the partitioning method which is a modification of the algorithm proposed by H. H. Wang in [13]. The aim of this work is to apply a new approach which stabilizes this partitioning method significantly and analyzes its parallel implementation on PVM.

The generalization for banded systems of this modification is given in [5]. The difference between [5] and this paper is that here we propose a stabilization procedure and analyze the tridiagonal case on PVM. Other types of parallel algorithms for tridiagonal systems are presented in [1], [10], [11], [12].

The parallel paradigm applied is single program multiple data (SPMD) and the master program has to coordinate the concurrent work of all started slaves (see [6]). The obtained data could be used to predict the optimal number of started processes which minimize the total time for the algorithm's execution. By a process we mean a piece of the code which is implemented sequentially and executed on one slave. We have studied the influence of two important factors on the performance of the parallel partitioning algorithm, namely system dimension and number of used host machines.

Let the linear system under consideration be denoted by

$$Ax = d, \tag{1}$$

where $A = tridiag(a, b, c)$ is a well conditioned tridiagonal matrix of order n, here a, b, c, d are vectors of size n, where $a = (a_2, a_3, \ldots, a_n)^T, b = (b_1, b_2, \ldots, b_n)^T$, $c = (c_1, c_2, \ldots, c_{n-1})^T, d = (d_1, d_2, \ldots, d_n)^T$.

* This work supported in part by Grant MM-434/94 from the National Scientific Research Fund of the Bulgarian Ministry of Education and Science.

The partitioning method for solving (1) can break down when it is necessary to divide by numbers which are less than a certain limit δ. In such cases we improve the algorithm perturbing the inputs or intermediate data. But the result which we get is perturbed. In order to make the solution more accurate we use iterative refinement (see [8]). Hence, it is necessary to solve (1) several times with different right hand sides. A similar perturbation approach is used in [4] for a Strassen-type matrix inversion algorithm. Then such an approach is presented in [9] applied to a fast Toeplitz solver, and the convergence of the iterative refinement is analysed in [16].

Let us note that similar algorithms and their stability analysis is presented in [15] (the partitioning algorithm applied to bidiagonal sytems), and in [2], [3], [14] and [17] (analysis of another parallel algorithm for tridiagonal and block tridiagonal systems).

The outline of the paper is as follows. Section 2 presents the partitioning algorithm. In Section 3 we consider perturbations and iterative improvement of the solution. In the next section we present some numerical experiments in MATLAB. Finaly in Section 5 we present an implementation of the partitioning algorithm on PVM.

2 The partitioning algorithm

Let us assume that $n = ks$. We partition matrix A and the right hand side d of the system (1) into the block-tridiagonal form shown below

$$\begin{pmatrix} B_1 & C_1 & & & \\ A_2 & B_2 & C_2 & & \\ & \ddots & \ddots & \ddots & \\ & & \ddots & \ddots & C_{s-1} \\ & & & A_s & B_s \end{pmatrix} \begin{pmatrix} X_1 \\ X_2 \\ \vdots \\ X_{s-1} \\ X_s \end{pmatrix} = \begin{pmatrix} D_1 \\ D_2 \\ \vdots \\ D_{s-1} \\ D_s \end{pmatrix}, \tag{2}$$

where $B_i, 1 \le i \le s$, is a tridiagonal matrix of order k

$$B_i = \begin{pmatrix} b_{(i-1)k+1} & c_{(i-1)k+1} & & & \\ a_{(i-1)k+2} & b_{(i-1)k+2} & c_{(i-1)k+2} & & \\ & \ddots & \ddots & \ddots & \\ & & \ddots & \ddots & c_{ik-1} \\ & & & a_{ik} & b_{ik} \end{pmatrix},$$

A_i, C_i are matrices of order k of the following kind

$$A_i = \begin{pmatrix} 0 & \cdots & 0 & a_{(i-1)k+1} \\ \vdots & & \vdots & \vdots \\ 0 & \cdots & 0 & 0 \end{pmatrix}, \quad C_i = \begin{pmatrix} 0 & 0 & \cdots & 0 \\ \vdots & \vdots & & \vdots \\ c_{ik} & 0 & \cdots & 0 \end{pmatrix},$$

$$2 \le i \le s, \qquad\qquad 1 \le i \le s-1,$$

and $X_i, D_i, i = 1,\ldots,s$ are vectors of size k of the following form

$$X_i = (x_{(i-1)k+1}, x_{(i-1)k+2},\ldots,x_{ik})^T, D_i = (d_{(i-1)k+1}, d_{(i-1)k+2},\ldots,d_{ik})^T.$$

The algorithm consists of three stages. At the first one s systems of equations with the matrices B_i and right hand sides $A_i, C_i, D_i, 1 \leq i \leq s$, are solved (this can be done in parallel). Here, for simplicity of notation, we introduce $A_1 = C_s = 0$.

Stage 1. Obtain the LU-factorization

$$B_i = L_i U_i, \quad i = 1,\ldots,s,$$

where L_i is unit lower triangular and U_i is a upper triangular matrix with diagonal elements $u_1^{(i)}, u_2^{(i)},\ldots,u_k^{(i)}$, using Gaussian elimination (with pivoting, if necessary) one process per factorization.

If we premultiply both sides of (2) by $diag(B_1^{-1}, B_2^{-1},\ldots,B_s^{-1})$ we obtain a system of the form $Fx = g$, where

$$F = \begin{pmatrix} I_k & P_1 & & & \\ Q_2 & I_k & P_2 & & \\ & \ddots & \ddots & \ddots & \\ & & \ddots & \ddots & P_{s-1} \\ & & & Q_s & I_k \end{pmatrix}, \quad g = \begin{pmatrix} G_1 \\ G_2 \\ \vdots \\ G_{s-1} \\ G_s \end{pmatrix},$$

here for $i = 1, 2,\ldots,s$, we denote

$$\begin{aligned} P_i &= (p^{(i)}, 0, 0,\ldots,0), \quad \text{where} \quad p^{(i)} = (p_{(i-1)k+1}, p_{(i-1)k+2},\ldots,p_{ik})^T, \\ Q_i &= (0, 0,\ldots,0, q^{(i)}), \quad \text{where} \quad q^{(i)} = (q_{(i-1)k+1}, q_{(i-1)k+2},\ldots,q_{ik})^T, \\ G_i &= (g_{(i-1)k+1}, g_{(i-1)k+2},\ldots,g_{ik})^T. \end{aligned}$$

In other words P_i, Q_i and G_i are obtained by solving the linear systems

$$L_i U_i [P_i, Q_i, G_i] = [C_i, A_i, D_i] \quad \text{for} \quad 1 \leq i \leq s, \tag{3}$$

Each process $2 \leq i \leq s-1$ handles three linear systems of the form $L_i U_i y = f$, while process 1 and s each handles two linear systems of the same structure.

Now let us notice that when we solve (3) it is necessary to divide by $u_j^{(i)}$ for $i = 1,\ldots,s$, and $j = 1,\ldots,k$. In this case it is sufficient that at least one of the quantities $u_j^{(i)}$ becomes very small or zero and the algorithm can break down, or behave poorly. In order to avoid this dangerous situation we propose to perturb them with δ, where δ is sufficiently small. The realization of this idea is presented in the next section.

Stage 2. Let us take into account equations with numbers $k, k+1, 2k, 2k+1,\ldots,(s-1)k,(s-1)k+1$. In this way we obtain the reduced system

$$\hat{F}\hat{x} = \hat{g}, \tag{4}$$

where

$$\hat{F} = \begin{pmatrix} 1 & p_k & & & & & & & \\ q_{k+1} & 1 & 0 & p_{k+1} & & & & & \\ q_{2k} & 0 & 1 & p_{2k} & & & & & \\ & & q_{2k+1} & 1 & 0 & p_{2k+1} & & & \\ & & q_{3k} & 0 & 1 & p_{3k} & & & \\ & & & & & \ddots & & & \\ & & & & & q_{(s-2)k+1} & 1 & 0 & p_{(s-2)k+1} \\ & & & & & q_{(s-1)k} & 0 & 1 & p_{(s-1)k} \\ & & & & & & & q_{(s-1)k+1} & 1 \end{pmatrix}, \quad \hat{g} = \begin{pmatrix} g_k \\ g_{k+1} \\ g_{2k} \\ g_{2k+1} \\ g_{3k} \\ \vdots \\ g_{(s-2)k+1} \\ g_{(s-1)k} \\ g_{(s-1)k+1} \end{pmatrix}.$$

At the second stage the solution of the system (4) $\hat{x}$ is computed by Gaussian elimination using the specific sparse structure of the matrix $\hat{F}$. If we know the components of the solution
$\hat{x} = (x_k, x_{k+1}, x_{2k}, x_{2k+1}, \ldots, x_{(s-1)k}, x_{(s-1)k+1})^T$ the rest of the components of the solution x are defined straightforward.

Stage 3. Compute the rest of the components of x by simple substitution:

$$x_i = d_i - p_i x_{k+1}, \quad \text{for } 1 \le i \le k-1,$$
$$x_i = d_i - q_i x_{n-k}, \quad \text{for } n-k+2 \le i \le n,$$
$$x_{ik+j} = d_{ik+j} - q_{ik+j} x_{ik} - p_{ik+j} x_{(i+1)k+1}, \quad \text{for } 1 \le i \le s-2, \text{ and } 2 \le j \le k-1.$$

3 The Stabilized Algorithm

As was noticed in the previous section the algorithm can break down, or behave poorly, when $u_i^{(j)}$, for $i = 1, \ldots, k$ and for $j = 1, \ldots, s$, are zero or small. So, we can perturb them in such a way that it would be away from zero. The stabilization step can be summarized as follows:

if $(|u_j^{(i)}| < \delta)$
  if $(|u_j^{(i)}| = 0)$
    $u_j^{(i)} = \delta;$
  else
    $u_j^{(i)} = u_j^{(i)} + sign(u_j^{(i)})\delta;$
  end
end

In this way we shift $u_j^{(i)}$ away from zero. Hence, the algorithm ensures that we do not divide by a small number.

The iterative refinement procedure is a standard one (see [8]), and we shall omit it here. We note that, when $\delta = 10^{-5}$, in practice the perturbed solution is very close to the exact one and we need usually only one or two steps of iterative refinement, depending what accuracy we require.

The theoretical parallel time for the unstabilized algorithm (without the communication time) is $22k + 18s + \mathcal{O}(1)$. In this way the theoretical speedup is approximately $(4/11)s$. For the stabilized algorithm the parallel time is $22k + 18s + N_{it}(13k + 11s) + \mathcal{O}(1)$, where N_{it} is the number of iterations. For the most spread case $N_{it} = 2$ we have a theoretical speedup of $(1/6)s$ which is of the same order as in the unstabilized case. So we do not add too much computation.

4 Numerical Experiments

Numerical experiments in this section are done in MATLAB, where the round-off unit is $\rho_0 \approx 2.22 \times 10^{-16}$. The exact solution in the our example is $x = (1, 1, \ldots, 1)^T$, and we measure two types of errors:

1. The relative forward error
$$FE = \frac{||x - \tilde{x}||_\infty}{||x||_\infty},$$
where $\tilde{x}$ is the computed solution;
2. The componentwise backward error (see [7])
$$BE = \max_{1 \le i \le n} \frac{|(A\tilde{x} - d)_i|}{(|A||\tilde{x}| + |d|)_i},$$

We stop the iterations when at least one of the following conditions is fulfilled:

1. $||Ax - d||_\infty / ||d||_\infty \le 1000\rho_0$;
2. Number of iterations is > 10;

	$\delta = 0$	$\delta = 10^{-3}$	$\delta = 10^{-4}$	$\delta = 10^{-5}$	$\delta = 10^{-6}$	$\delta = 10^{-7}$
BE	∞	1.63×10^{-14}	1.50×10^{-12}	1.44×10^{-15}	6.79×10^{-14}	6.83×10^{-13}
FE	∞	1.13×10^{-13}	9.01×10^{-12}	1.02×10^{-14}	2.27×10^{-13}	2.73×10^{-12}
N_{it}	0	4	2	2	2	2

Table 1. The forward and backward error and the number of iterations for the matrix
$A = tridiag(1, 0, 1)$ when $k = 101, s = 10$

Let us consider examples with $A = tridiag(1, eps, 1)$. In our tests we give to eps the values 0 and 10^{-12}. Let us note that when n is odd the matrix A is singular, and when n is even the matrix A is very well conditioned. To show the effectiveness of our stabilization approach we take n to be even and k to be odd, i.e. the matrix A is very well conditioned. Nevertheless the original algorithm can break down (when $eps = 0$) on account of dividing by zero, or behave poorly (when $eps = 10^{-12}$). At the same time the stabilized algorithm gives much better results. The three tables which we present in this section show the results with different values of δ. When $\delta = 0$ we obtain the original algorithm without stabilization.

Let us note that at most two steps of iterative refinement are enough when $\delta = 10^{-5}$, and that only a few perturbations are necessary in all the examples.

	$\delta = 0$	$\delta = 10^{-3}$	$\delta = 10^{-4}$	$\delta = 10^{-5}$	$\delta = 10^{-6}$	$\delta = 10^{-7}$
BE	0.33	1.63×10^{-14}	1.50×10^{-12}	1.55×10^{-15}	2.10×10^{-15}	5.55×10^{-17}
FE	1	1.34×10^{-13}	9.02×10^{-12}	2.23×10^{-14}	1.87×10^{-14}	7.32×10^{-15}
N_{it}	0	4	2	2	2	3

Table 2. The forward and backward error and the number of iterations for the matrix
$A = tridiag(1, 10^{-12}, 1)$ when $k = 101, s = 10$

5 Parallelization

5.1 Parallel Virtual Machine

The PVM is a software system that enables a collection of heterogeneous computers to be used as a coherent and flexible parallel computational resource. It is especially suited for the development and execution of large concurrent or parallel applications. PVM was developed initially at the Emory University and Oak Ridge National Laboratory. The individual computers may be shared or local memory multiprocessors, vector supercomputers or scalar workstations, that can be interconnected by various networks (Ethernet, Fiber Distributed Data Interface (FDDI), etc.). User programs written in C or Fortran access to PVM through the use of calls to PVM library routines for functions such as process initiation, message transmission and reception. The PVM system handles message routing, data conversion for incompatibile architectures and tasks that are necessary for operations in a heterogeneous network environment.

5.2 Results

All tests were performed on Local Area Network with Ethernet protocol. The virtual machine consists of 8 workstations IBM RISC 6000/250.

We are mainly interested in two questions: would it be possible to achieve good speedup when using a rather small matrix size and what would be the influence of the task size on the performance using different numbers of slave. All measurements are taken in single-user mode.

The size of the task has to be specified by the user. We note the fact that the matrix dimension is importante. From onc side it seems that small tasks (small dimension) cause less idle time at the end of the computation (measured on master processor). On the other side the biggest dimension will increase the effectiveness of the parallel implementation. The speedup obtained on 2, 4, 6, and 8 slaves we report in Table 3. We could notice that:

1. The best results are obtained with 2 and 4 slaves.
2. Some of the possible ways to increase the performance of the proposed parallel algorithm is:
   - to apply it on FDDI network of workstations;
   - to use cluster organization of all machines connected under PVM;
   - to develop the partitioning algorithm for another virtual parallel machine such as Message-Passing Interface.

dimension	2 slaves	4 slaves	6 slaves	8 slaves
10000	0.74	1.41	3.76	4.97
20000	1.02	1.97	4.78	5.87
40000	1.30	2.69	4.98	6.01
60000	1.41	3.06	5.02	6.31
80000	1.48	3.39	5.23	6.54
100000	1.52	3.45	5.34	6.70
150000	1.55	3.56	5.45	6.82
200000	1.61	3.67	5.54	6.95
250000	1.85	3.71	5.60	7.02
300000	1.91	3.79	5.70	7.23

Table 3. Parallel speedup of the partitioning algorithm

References

1. Amodio, P., Brugnano, L., Politi, T.: Parallel Factorizations for tridiagonal matrices. SIAM J. Numer. Anal. **30** (1993) 813–823

2. Amodio, P., Mazzia, F.: Backward error analysis of cyclic reduction for the solution of tridiagonal systems. Mathematics of Computation **62** (1994) 601–617
3. Amodio, P., Trigiante, D.: A parallel direct method for solving initial value problems for ordinary differential equations. Applied Numerical Mathematics **11** (1993) 85–93
4. Balle, S., Hansen, P., Higham, N.: A Strassen-Type Matrix Inversion Algorithm for the Connection Machine. APPARC PaA2 Deliverable, Esprit BRA Contract # 6634; Report UNIC-93-11, UNI•C (1993)
5. Dongarra, J., Sameh, A.: On some parallel banded system solvers. Parallel Computing **1** (1984) 223–235
6. Geist, A., Beguelin, A., Dongarra, J., Jiang, W., Manchenk, R., Sunderam, V.: PVM: Parallel Virtual Machine. A Users' Guide and Tutorial for Networked Parallel Computing. The MIT Press (1994)
7. Higham, D., Higham, N.: Backward error and condition of structured linear systems. SIAM J. Matrix Anal. Appl. **13** (1992) 162–175
8. Golub, G., Van Loan, C.: Matrix Computations. The John Hopkins University Press (1989)
9. Hansen, P., Yalamov, P.: Stabilization by Perturbation of a $4n^2$ Toeplitz Solver. Preprint N25, Technical University of Rousse (1995)
10. Stone, H.: An efficient parallel algorithm for the solution of a tridiagonal linear system of equations. J. Assoc. Comput. Mach. **20** (1973) 27–38
11. Stone, H.: Parallel tridiagonal solvers. ACM Trans. Math. Software **1** (1975) 289–307
12. Van Der Vorst, H.: Large tridiagonal and block tridiagonal linear systems on vector and parallel computers. Parallel Comput. **5** (1987) 45–54
13. Wang, H.: A parallel method for tridiagonal equations. ACM Trans. Math. Software **7** (1981) 303–311
14. Yalamov, P.: On the Stability of the Cyclic Reduction without Back Substitution for Tridiagonal Systems. BIT **34** (1994) 428–447
15. Yalamov, P.: Stability of a partitioning algorithm for bidiagonal systems. (submitted to SIAM J. Matrix Anal. Appl.)
16. Yalamov, P.: Convergence of the Iterative Refinement Procedure Applied to Stabilization of a Fast Toeplitz Solver. Preprint N26, Technical University of Rousse (1995)
17. Yalamov, P., Pavlov, V.: Stability of the Block Cyclic Reduction. Linear Algebra and Its Applications (to appear)

Numerical Solution of High Order Matrix Riccati Equations

P.Hr. Petkov[1], N.D. Christov[1] and M.M. Konstantinov[2]

[1] Dept. of Automatics, Technical Univ. of Sofia, 1756 Sofia, Bulgaria
[2] Univ. of Arch. & Civil Eng., 1 Hr. Smirnenski Blv., 1421 Sofia, Bulgaria

Abstract. The numerical solution of matrix algebraic Riccati equations with condition and forward error estimates is considered. A comparison of the accuracy of Schur and matrix sign function methods in the solution of high order Riccati equations is done.

1 Introduction

In this paper we consider the implementation of Schur and matrix sign function methods to the numerical solution of continuous–time matrix algebraic Riccati equations. The numerical solution involves computation of condition and forward error estimates permitting evaluation of equation sensitivity and solution accuracy. A comparison of the Schur and the matrix sign function methods in the solution of well conditioned and ill conditioned high order Riccati equations is done.

The following abbreviations are used: $\mathcal{R}$ denotes the field of real numbers; $\mathcal{R}^{m\times n}$ – the space of $m \times n$ matrices $A = [a_{ij}]$ over $\mathcal{R}$; A^T – the transposed matrix A; $\|A\|$ – the norm of the matrix A (we use the 1-norm $\|A\|_1$, the spectral norm $\|A\|_2$ and the Frobenius norm $\|A\|_F = (\sum |a_{ij}|^2)^{1/2}$); I_n – the $n \times n$ unit matrix and $A \otimes B$ – the Kronecker product of A and B.

Consider the continuous–time algebraic Riccati equation

$$A^T X + XA + C - XDX = 0 \tag{1}$$

where $A \in \mathcal{R}^{n\times n}$ and the matrices $C \in \mathcal{R}^{n\times n}$, $D \in \mathcal{R}^{n\times n}$ are symmetric. We assume that C and D are non–negative definite matrices with the pair (A, D) stabilizable and the pair (C, A) detectable. It is well known that under these conditions there exist unique non–negative definite solutions of (1) and the matrix $A - DX$ is stable.

The Schur method [6] for solving (1) involves computation of an orthonormal basis of the invariant subspace associated with the stable eigenvalues of the Hamiltonian matrix

$$H = \begin{bmatrix} A & -D \\ -C & -A^T \end{bmatrix}.$$

Let $U \in \mathcal{R}^{2n\times 2n}$ be an orthogonal transformation matrix which reduces H into real Schur form

$$T = U^T H U = \begin{bmatrix} T_{11} & T_{12} \\ 0 & T_{22} \end{bmatrix}$$

where $T_{ij} \in \mathcal{R}^{n\times n}$ and T_{11}, T_{22} are upper quasitriangular with all eigenvalues of T_{11} having negative real parts. Partitioning U into four $n \times n$ blocks

$$U = \begin{bmatrix} U_{11} & U_{12} \\ U_{21} & U_{22} \end{bmatrix}$$

one obtains the unique non–negative definite solution of the Riccati equation as

$$X = U_{21}U_{11}^{-1}.$$

Let

$$H = V \begin{bmatrix} J_- & 0 \\ 0 & J_+ \end{bmatrix} V^{-1}$$

be the Jordan decomposition of H where the eigenvalues of the $n \times n$ submatrix J_- are the eigenvalues of H having negative real parts. The matrix sign function of H is defined as [8]

$$\text{sign}(H) = V \begin{bmatrix} -I_n & 0 \\ 0 & I_n \end{bmatrix} V^{-1}.$$

The matrix

$$P_- = \frac{1}{2}(I_{2n} - \text{sign}(H))$$

is the spectral projector corresponding to the stable eigenvalues of H. If

$$Q \equiv \begin{bmatrix} Q_{11} & Q_{12} \\ Q_{21} & Q_{22} \end{bmatrix}$$

is the orthogonal matrix in the rank revealing QR decomposition of P_-,

$$QR\Pi = P_-,$$

then the non–negative solution of the Riccati equation is given by

$$X = Q_{21}Q_{11}^{-1}.$$

The matrix sign function of H may be computed efficiently using a simple iteration which involves inversion of a Hamiltonian matrix at each step.

The matrix sign function method has the advantage that it is simply constructed from a small set of highly parallelizable matrix building blocks, including matrix multiplication, QR decomposition and matrix inversion. This makes it preferable to the solution of high order Riccati equations.

2 Condition Estimates

Let $A_C = A - DX$. Using a first-order sensitivity analysis it is possible to show that the variation ΔX in the solution X of the Riccati equation, due to small perturbations $\Delta A, \Delta C, \Delta D$ in the matrices A, C, D, satisfies [2, 4]

$$\Delta X = -\Omega^{-1}(\Delta C) - \Theta(\Delta A) + \Pi(\Delta D) \tag{2}$$

where

$$\begin{aligned} \Omega(Z) &= A_C^T Z + Z A_C \\ \Theta(Z) &= \Omega^{-1}(Z^T X + XZ) \\ \Pi(Z) &= \Omega^{-1}(XZX) \end{aligned}$$

are linear operators in the space of $n \times n$ matrices.

Based on (2) Byers [2] introduced the approximate condition number

$$K_B = \frac{\|\Omega^{-1}\|\|C\| + \|\Theta\|\|A\| + \|\Pi\|\|D\|}{\|X\|} \tag{3}$$

where

$$\|\Omega^{-1}\| = \max_{Z \neq 0} \frac{\|\Omega^{-1}(Z)\|}{\|Z\|}$$

$$\|\Theta\| = \max_{Z \neq 0} \frac{\|\Theta(Z)\|}{\|Z\|}$$

$$\|\Pi\| = \max_{Z \neq 0} \frac{\|\Pi(Z)\|}{\|Z\|}$$

are the corresponding induced operator norms.

The computation of K_B may be done in the following way.

The quantities $\|\Omega^{-1}\|_1, \|\Theta\|_1, \|\Pi\|_1$ can be efficiently estimated by using the condition estimator, proposed in [3], which estimates the norm $\|T\|_1$ of a linear operator T, given the ability to compute Tv and $T^T w$ quickly for arbitrary v and w. This esimator is implemented by the subroutine xLACON from LAPACK [1] which is called via a reverse communication interface, providing the products Tv and $T^T w$. In respect to the determination of

$$\|\Omega^{-1}\|_F = \|P^{-1}\|_2$$

where

$$P = A_C^T \otimes I_n + I_n \otimes A_C^T$$

the using of xLACON means to solve a few linear equations

$$Py = q, \; P^T z = q,$$

q being determined by xLACON, which is equivalent to solve a few Lyapunov equations.

Estimates of $\|\Theta\|_1$ and $\|\Pi\|_1$ are found in a similar way.

3 Forward Error Estimate for the Schur Method

Let $\hat{X}$ be the solution of the Riccati equation computed by the Schur method in machine arithmetic with relative precision ε. In our implementation of the Schur approach we use an estimate of the forward error

$$\frac{\|\hat{X} - X\|}{\|X\|}.$$

based on the numerical analysis presented in [7]. According to this analysis the solution of the Riccati equation computed by the Schur method satisfies

$$\frac{\|\hat{X} - X\|_2}{\|X\|_2} \leq \frac{c(n)\varepsilon}{\delta}(1 + \frac{1}{\|X\|_2})\|H\|_2\|\hat{U}_{11}^{-1}\|_2 + \frac{\|\hat{X} - X^0\|_2}{\|X\|_2} \tag{4}$$

where

$$X^0 = \hat{U}_{21}\hat{U}_{11}^{-1}$$

is the exact result found from the computed orthonormal basis of the stable invariant subspace of H and $c(n)$ is a low order polynomial in n.

The bound (4) is used to obtain an a posteriori estimate of the relative error in $\hat{P}$.

An estimate of the forward error for the matrix sign function method is not available up to the moment but it is known that this method may be numerically unstable in some cases. In our experiments the matrix sign function method performed stably if the Hamiltonian matrix was scaled as proposed in the next section.

4 Scaling

Consider a similarity transformation of the Hamiltonian matrix

$$H = \begin{bmatrix} A & -D \\ -C & -A^T \end{bmatrix}$$

with a matrix

$$T = \begin{bmatrix} pI_n & 0 \\ 0 & qI_n \end{bmatrix}; p \neq 0, q \neq 0.$$

As a result one obtaines

$$\bar{H} = \begin{bmatrix} A & -\rho D \\ -C/\rho & -A^T \end{bmatrix}$$

where $\rho = q/p$.

Depending on the choice of ρ we may scale the solution of (1) in a different way. If, for instance, $\rho = \|X\|$, then $\|\bar{X}\| = 1$ and this is the "optimal scaling" which ensures backward stability of the Schur method. A disadvantage of this scaling is the necessity to know $\|X\|$ which requires to solve first the unscaled equation. Here we propose to use the scaling

$$\rho = 1 \text{ if } \|C\| \leq \|D\|,$$

$$\rho = \frac{\|C\|}{\|D\|} \text{ if } \|C\| > \|D\|.$$

This scaling guarantees that $\|C/\rho\| \leq \|\rho D\|$ which improves the numerical properties both of the Schur method and matrix sign function method. This scaling, of course, may give worse results than the optimal scaling, but in most cases it gives satisfactory results with much less computational work.

5 Numerical Examples

In this section we present some examples which illustrate the numerical properties of the unscaled and scaled Schur and matrix sign function methods.

In order to have a closed form solution the matrices in the Riccati equation are chosen as

$$A = TA_0T^{-1}, \quad C = T^{-T}C_0T^{-1}, \quad D = TD_0T^T$$

where A_0, C_0, D_0 are diagonal matrices and T is a nonsingular transformation matrix. The solution of (1) is then given by

$$X = T^{-T}X_0T^{-1}$$

where X_0 is a diagonal matrix whose elements are determined simply from the elements of A_0, C_0, D_0. To avoid large rounding errors in constructing and inverting T this matrix is chosen as

$$T = H_2SH_1$$

where

$$H_1 = I_n - 2ee^T/n, e = [1, 1, ..., 1]^T$$

$$H_2 = I_n - 2ff^T/n, f = [1, -1, 1, ..., (-1)^{n-1}]^T,$$

$$S = \text{diag}(1, s, s^2, ..., s^{n-1}), s > 1.$$

Example 1. Consider 15-th order well conditioned Riccati equations constructed as described above for

$$A_0 = \text{diag}(A_1, A_1, A_1, A_1, A_1),$$

$$C_0 = \text{diag}(C_1, C_1, C_1, C_1, C_1),$$

$$D_0 = \text{diag}(D_1, D_1, D_1, D_1, D_1).$$

where

$$A_1 = \mathrm{diag}(1 \times 10^k, 2 \times 10^k, 3 \times 10^k),$$

$$C_1 = \mathrm{diag}(1 \times 10^{-k}, 1, 1 \times 10^k),$$

$$D_1 = \mathrm{diag}(10^{-k}, 10^{-k}, 10^{-k}).$$

This equation is very well conditioned (K_B is of order 1) but in the unscaled version of the Schur method the quantity $\|U_{11}^{-1}\|$ increases quickly with k which introduces large errors in the solution.

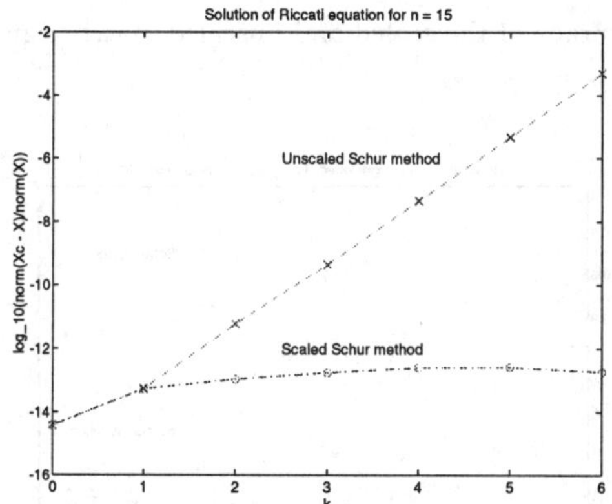

Fig. 1. Solution accuracy of the scaled and unscaled Schur method

The solution accuracy for $s = 1.2$ and different values of k is shown in Figure 1 for the unscaled and scaled Schur method and in Figure 2 for the unscaled and scaled matrix sign function method. All computations were done using precision $\varepsilon \approx 2.22 \times 10^{-16}$. The computation of the unscaled and scaled matrix sign functions was done using MATLAB. The unscaled matrix sign function was computed by the m-files presented in [5]. For $k = 6$ the scaled versions of both methods produce solutions which are about 10^{10} times more accurate than the solutions obtained by the unscaled versions.

Example 2. Consider well conditioned Riccati equations of order $n = 450$ constructed by using the matrices A_1, C_1, D_1 given in Example 1. The comparison of the accuracy of scaled Schur and matrix sign function methods for $s = 1.01$ and different values of k is shown in Figure 3.

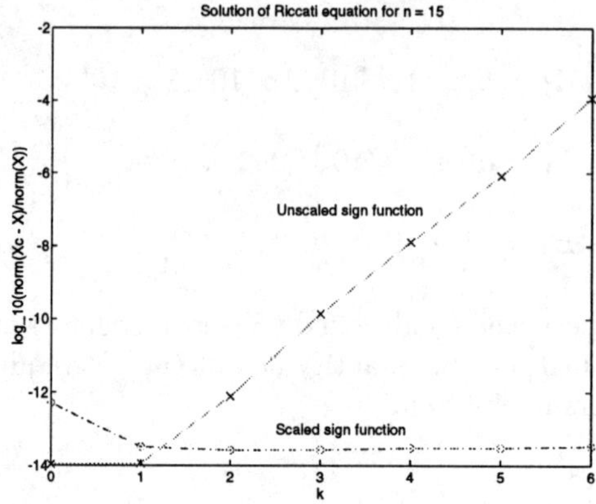

Fig. 2. Solution accuracy of the scaled and unscaled matrix sign function method

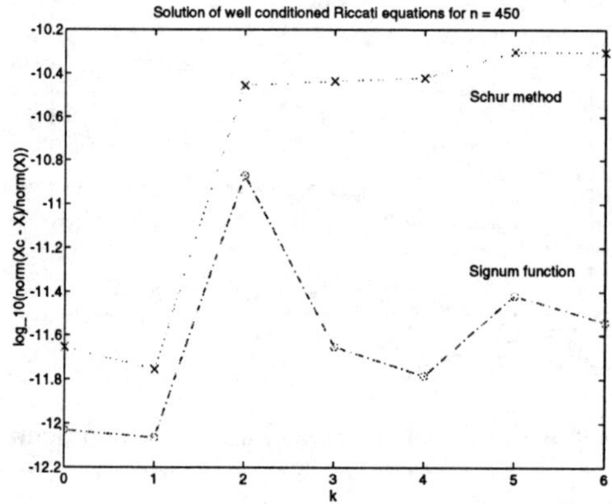

Fig. 3. Solution of well conditioned Riccati equations, n = 450

Example 3. In this case ill conditioned Riccati equations of order $n = 450$ are constructed by using the matrices

$$A_1 = \mathrm{diag}(-1 \times 10^{-k}, -2, -3 \times 10^{k}),$$

$$C_1 = \mathrm{diag}(3 \times 10^{-k}, 5, 7 \times 10^{k}),$$

$$D_1 = \mathrm{diag}(10^{-k}, 1, 10^{k}).$$

The condition number of these equations grows quickly with k.

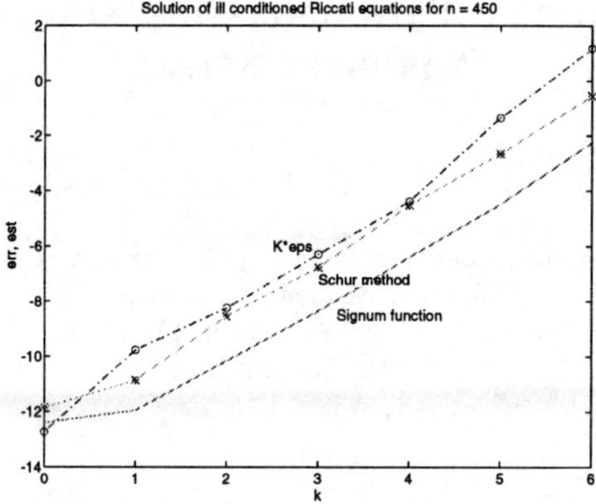

Fig. 4. Solution of ill conditioned Riccati equations, n = 450

In Figure 4 we show the accuracy of both methods together with the error estimate $est = \tilde{K}\varepsilon$, computed by using the estimate $\tilde{K}$ of the condition number K_B. It is seen that both methods produce solutions whose accuracy is close to the accuracy predicted by the sensitivity analyzis.

References

1. Anderson E., Bai Z., Bischof C.H., Demmel J.M., Dongarra J.J., DuCroz J.J., Greenbaum A., Hammarling S.J., McKenney A., Ostrouchov S., Sorensen D.C.: LAPACK Users's Guide, 2nd ed., SIAM, Philadelphia (1995)
2. Byers R.: Numerical condition of the algebraic Riccati equation. Contemporary Math., **47**, (1984), 35–49
3. Higham N.J.: Algorithm 674: FORTRAN codes for estimating the one-norm of a real or complex matrix, with application to condition estimation. ACM Trans. Math. Software, **14**, (1988), 381–396
4. Kenney C., Hewer G.: The sensitivity of the algebraic and differential Riccati equations. SIAM J. Control Optim., **28**, (1990), 50–69
5. Kenney C., Laub A.J.: On scaling Newton's method for polar decomposition and the matrix sign function. SIAM J. Matrix Anal. Appl., **13**, (1992), 688–706
6. Laub A.J.: A Schur method for solving algebraic Riccati equations. IEEE Trans. Autom. Control, **24**, (1979), 913–921
7. Petkov P.Hr., Christov N.D., Konstantinov M.M.: Computational Methods for Linear Control Systems. Prentice Hall, Hemel Hempstead (1991)
8. Roberts J.: Linear model reduction and solution of the algebraic Riccati equation. Int. J. Control, **32**, (1980), 677–687

A Variational Parameters-to-Estimate-Free Nonlinear Solver

Svetozara I. Petrova and Panayot S. Vassilevski *

Central Laboratory of Parallel Processing,
Bulgarian Academy of Sciences,
"Acad. G. Bontchev" street, block 25A,
1113 Sofia, Bulgaria

Abstract. The paper introduces and analyzes an extension of the simple steepest descent variational iterative method for solving nonlinear equations defined from non–differentiable nonlinear mappings. This method is originally proposed for solving a class of nonlinear equations that arise in inner-outer iterative methods for solving linear equations with matrices of a two–by–two block form, see Axelsson and Vassilevski [2]. With minor modifications it turns out to be applicable to a more general class of nonlinear equations defined from non–differentiable mappings that are however sufficiently close to differentiable mappings in a neighborhood of the solution. An extension of the preconditioned steepest descent variational (PSD) method for solving semi–linear elliptic PDEs is illustrated with numerical experiments for a class of finite element discretization techniques based on two subspaces. The convergence of the PSD versus versions of Newton's method is compared and discussed.

1 Introduction

This paper deals with an extension of the method studied in Axelsson and Vassilevski [2] which originally was proposed for solving non–linear equations that are defined from a mapping that is globally sufficiently close to a linear mapping. The method from [2] is modified for solving nonlinear equations defined from generally a non–differentiable nonlinear mapping. Research in the area of solving nonlinear equations defined from non–differentiable mappings is found in Brown and Saad [7], Kaporin and Axelsson [14] and Axelsson and Kaporin [3] and earlier in Bank and Rose [5] and Dembo, Eisenstat and Steihaug [8]. A general approach of solving nonlinear PDEs by multigrid methods is the so–called FAS (*Full Approximation Scheme*, cf., Hackbusch [12]) which has been analyzed in a rather general situation in Reusken [15]. The influence of the discretization on the convergence of certain iterative methods for solving nonlinear equations has been studied in Allgower, Böhmer, Potra and Rheinboldt [1]. We do not, however, pay special attention to this topic in the present paper.

* This research has been supported in part by the Bulgarian Ministry of Education, Science and Technology under Grants I–504/1995 and MM–415/1994.

The purpose of the present paper is to extend a variational type method proposed originally for solving a special type nonlinear equations that arise in inner–outer iterative methods for solving certain class of linear equations. As it will be demonstrated in the following section, this method with minor modification, turns out to be applicable to a more general class of nonlinear equations that include nonlinear equations defined from nondifferentiable mappings that are, however, sufficiently close to differentiable mappings in a neighborhood of the solution. The method, is also shown to be parameters to estimate free and hence can offer an alternative to other known, based, e.g., on fictitious time stepping procedures, or to the so-called inexact Newton methods (cf., Dembo, Eisenstat, and Steihaug [8], Brown and Saad [7], Kaporin and Axelsson [14]) for differentiable and nondifferentiable mappings.

The remainder of the paper is organized as follows. In §2 the method is described and a general convergence result, including the case of nondifferentiable mappings, is proven. The numerical results are presented in §3.

2 The method

Consider the nonlinear equation

$$F(u) = f, \tag{1}$$

where $F(.)$ is a nonlinear mapping defined in a Hilbert space H, i.e., $F(.)$: $H \to H$. We next describe the method to be studied.

Consider a current approximation u^0 to the solution of (1) and define the residual equation:

$$A[c] = d, \tag{2}$$

where $A[.] = A(u^0)[.] \equiv F(u^0 + \cdot) - F(u^0)$ and $d = f - F(u^0)$.

The goal is to find an approximate correction c and then the next iterate is $u = u^0 + c$. We now describe the assumptions on the nonlinear mapping $A[.]$. Note that it depends on the current approximation u^0 but we assume below uniform constants, i.e., independent of u^0.

(i) Coercivity assumption:

$$(A[r], r) \geq \delta_1 \|r\|^2 \quad \text{for all } r \ : \ \|r\| < \Delta,$$

for some constants $\Delta > 0$ and $\delta_1 > 0$. It is then clear that (i) implies,

$$\frac{\|r\|}{\|A[r]\|} \leq \delta_1^{-1}. \tag{3}$$

Note that (i) is actually a monotonicity property of the mapping F, i.e.,

$$(F(v) - F(w), v - w) \geq \delta_1 \|v - w\|^2, \tag{4}$$

with $v = u^0 + r$ and $w = u^0$.

(ii) Boundedness assumption:

$$\|A[r]\| \leq \delta_2\|r\| \quad \text{for all } r \;:\; \|r\| < \Delta,$$

for some constant $\delta_2 > 0$. Note that (ii) is actually a Lipschitz continuity property of the mapping $F(.)$.

(iii) Approximate linearity:

$$\|A[c+\alpha r] - A[c] - \alpha A[r]\| \leq \delta\left\{|\alpha|\|r\| + \|c\|\right\},$$

for an accuracy constant $\delta > 0$ for all $\|c\| < \Delta$, $|\alpha| < R$ and $\|c+\alpha r\| < \Delta$.

We will next show that a variational (steepest descent like) method can give better approximation $u = u^0 + c$ to the solution of (1) in the sense that $\|f - F(u^0 + c)\| \leq \gamma\|f - F(u^0)\|$ for a constant $\gamma \in (0,1)$.

Consider the following algorithm:

Algorithm 1 *(Steepest Descent.) Choose* $c^0 = 0$ *and let* $r^0 = d - A[c^0] = d$. *For* $k = 1, 2, \ldots,$ *compute:*

$$\begin{aligned}
\tilde{r}^{k-1} &= A[r^{k-1}];\\
\alpha &= \tfrac{(r^{k-1},\tilde{r}^{k-1})}{(\tilde{r}^{(k-1)},\tilde{r}^{(k-1)})};\\
c^k &= c^{k-1} + \alpha r^{k-1};\\
r^k &= d - A[c^k]\cdot
\end{aligned}$$

To analyze the contraction property of the above algorithm we consider the residual $r^+ = d - A[c+\alpha r]$. Here $r = r^{k-1}$, $\alpha = \frac{(r,A[r])}{(A[r],A[r])}$, and $r^+ = r^k$. We first note that (see (3))

$$\alpha = \frac{(r, A[r])}{(A[r], A[r])} \leq \frac{\|r\|}{\|A[r]\|} \leq \delta_1^{-1}.$$

Then we have:

$$\begin{aligned}
\|r^+\| &= \|d - A[c+\alpha r]\|\\
&\leq \|d - A[c] - \alpha A[r]\| + \|A[c+\alpha r] - A[c] - \alpha A[r]\|\\
&= \|r - \alpha A[r]\| + \|A[c+\alpha r] - A[c] - \alpha A[r]\|\\
&\leq \sqrt{1 - \left(\tfrac{\delta_1}{\delta_2}\right)^2}\,\|r\| + \delta\left\{|\alpha|\|r\| + \|c\|\right\}\\
&\leq \left[\delta_1^{-1}\delta + \sqrt{1 - \left(\tfrac{\delta_1}{\delta_2}\right)^2}\right]\|r\| + \delta\|c\|\cdot
\end{aligned}$$

Noting that from Algorithm 1 $A[c] = d - r$, using (i) and the triangle inequality, one has:

$$\|c\| \leq \delta_1^{-1}\|A[c]\| \leq \delta_1^{-1}(\|d\| + \|r\|)\cdot$$

Therefore, the last estimate for r^+ becomes:

$$\|r^+\| \leq \delta\delta_1^{-1}\|d\| + \left\{2\delta\delta_1^{-1} + \sqrt{1 - \left(\frac{\delta_1}{\delta_2}\right)^2}\right\}\|r\|\cdot$$

Hence, assuming that δ is sufficiently small such that $q \equiv 2\delta\delta_1^{-1} + \sqrt{1 - \left(\frac{\delta_1}{\delta_2}\right)^2} < 1$, one then gets

$$\begin{aligned}\|r^+\| &\leq (1 + q + \cdots + q^{k-1})\delta\delta_1^{-1}\|d\| + q^k\|r^0\| \\ &\leq q^k\|r^0\| + \tfrac{1}{1-q}\delta\delta_1^{-1}\|d\| = \left[q^k + \tfrac{1}{1-q}\delta\delta_1^{-1}\right]\|d\|.\end{aligned}$$

Again, if δ is sufficiently small, we get

$$\|f - F(u^0 + c^+)\| = \|r^+\| \leq \gamma\|f - F(u^0)\|, \tag{5}$$

where $\gamma = q^k + \frac{1}{1-q}\delta\delta_1^{-1} \leq \gamma_1 \equiv q + \frac{1}{1-q}\delta\delta_1^{-1} < 1$ (if δ is sufficiently small).

We proved also that the iterates c remain in a ball with radius $\Delta = \delta_1^{-1}(\|d\| + \|r\|) \leq (1 + \gamma_1)\delta_1^{-1}\|d\| \leq 2\delta_1^{-1}\|d\|$. This radius ($\Delta$) can be made sufficiently small if the current iterate u^0 is sufficiently close to the solution of (1) (in this case $d = f - F(u^0) = F(u) - F(u^0)$ will then have a small norm). Also, since $\|r\| < \|r_0\| = \|d\| < \Delta$ we see that the approximate linearity assumption (iii) is applicable if $R \geq \delta_1^{-1}$.

We next verify the assumption on approximate linearity of the mapping $A(u^0)[.]$ assuming that $F(.)$ is differentiable, i.e. let us assume

$$\|F(u^0 + h) - F(u^0) - \nabla F(u^0)h\| \leq C\|h\|^{1+\tau} \quad \text{for } \|h\| < \delta_0, \tag{6}$$

for some constant $\tau > 0$ and $\delta_0 > 0$ small enough constant. Here $\nabla F(u^0)$ is a linear operator, the gradient of F at the point u^0. Estimate (6) implies that

$$\|A(u^0)[h] - \nabla F(u^0)h\| \leq \frac{1}{2}\delta\|h\| \quad \text{for } \|h\| < \delta_0, \tag{7}$$

where $\delta = 2C\delta_0^\tau$; i.e., $A(u^0)[.]$ is close to a linear operator in a ball about the origin with a small radius δ_0.

Note that estimate (7) can be satisfied for a mapping $\nabla K(u^0)$, the gradient of a differentiable mapping $K(.)$ that is sufficiently close to F in the following sense (cf., Axelsson and Kaporin [3] and earlier in Zinćello [19]):

$$\|F(v) - K(v) - F(w) + K(w)\| \leq \frac{1}{4}\delta\|v - w\|, \tag{8}$$

for a sufficiently small $\delta > 0$ and for all v and w contained in a sufficiently large ball around u^0. We also assume that K is differentiable in the sense (6); i.e., the following estimate holds for a linear mapping $\nabla K(u^0)$:

$$\|K(u^0 + h) - K(u^0) - \nabla K(u^0)h\| \leq C\|h\|^{1+\tau} \quad \text{for all } h : \|h\| < \delta_0,$$

for $\delta_0 > 0$ sufficiently small and for some positive constant $\tau > 0$.

One then easily gets from the last estimate and estimate (8) with $v := u^0 + h$ and $w := u^0$,

$$\begin{aligned}\|A(u^0)[h] - \nabla K(u^0)h\| &\leq \|K(u^0 + h) - K(u^0) - \nabla K(u^0)h\| \\ &\quad + \|F(u^0 + h) - F(u^0) - K(u^0 + h) + K(u^0)\| \\ &\leq \left(\tfrac{1}{4}\delta + C\|h\|^\tau\right)\|h\| \\ &\leq \tfrac{1}{2}\delta\|h\|.\end{aligned}$$

That is, an estimate of the form (7) holds with $\nabla F(u^0)$ replaced by $\nabla K(u^0)$. This allows us to relax the assumption on the differentiability of F and to have F to be close enough to a differentiable mapping K in the sense (8) instead. Here we will use the notation ∇F even in the case when F is not differentiable but satisfies (8) for a differentiable mapping K and then ∇F will denote the gradient of K.

Estimate (7) used for $h := c + \alpha r$, c and r, assuming that $\Delta < \delta_0$, implies:

$$\begin{aligned} \|A(u^0)[c+\alpha r] - \nabla F(u^0)(c+\alpha r)\| &\le \tfrac{1}{2}\delta\|c+\alpha r\|, \\ \|A(u^0)[c] - \nabla F(u^0)c\| &\le \tfrac{1}{2}\delta\|c\|, \\ \|\alpha A(u^0)[r] - \alpha\nabla F(u^0)r\| &\le \tfrac{1}{2}\delta|\alpha|\|r\|. \end{aligned}$$

Therefore,

$$\begin{aligned} \|A(u^0)[c+\alpha r] - A(u^0)[c] - \alpha A(u^0)[r]\| &\le \tfrac{1}{2}\delta(\|c+\alpha r\| + \|c\| + |\alpha|\|r\|) \\ &\le \delta\left\{\|c\| + |\alpha|\|r\|\right\}. \end{aligned}$$

This represents the verification of the assumption (iii) on the approximate linearity of the mapping $A(u^0)[.]$ in the case of differentiable nonlinear mapping $F(.)$ or close to a differentiable mapping in the sense (8). Note that here we do not have any additional restriction on α except that $\|c + \alpha r\| < \Delta$.

The assumptions (i–ii) on the coercivity and boundedness of $A(u^0)[.]$ depend on the particular application which will reflect the choice of an appropriate norm $\|.\|$ defined from a proper preconditioner to $\nabla F(u^0)$. However, if the gradient of F, $\nabla F(u^0)$, is coercive and bounded in a given norm, then $A(u^0)[.]$ is coercive and bounded as well. This is easily seen (cf., [2]). Assume that the following estimates hold:

$$(\nabla F(u^0)r, r) \ge \tilde{\delta}_1\|r\|^2 \quad \text{for all } r, \tag{9}$$

for a positive constant $\tilde{\delta}_1 > 0$;

$$\|\nabla F(u^0)r\| \le \tilde{\delta}_2\|r\| \quad \text{for all } r, \tag{10}$$

for a positive constant $\tilde{\delta}_2$.

Then, using (7), one easily gets,

$$\begin{aligned} (A(u^0)[r], r) &= (A(u^0)[r] - \nabla F(u^0)r, r) + (\nabla F(u^0)r, r) \\ &\ge (\tilde{\delta}_1 - \tfrac{1}{2}\delta)\|r\|^2. \end{aligned}$$

This shows the coercivity estimate (i) with $\delta_1 = \tilde{\delta}_1 - \frac{1}{2}\delta$ (for sufficiently small δ). Similarly,

$$\|A(u^0)[r]\| \le \|A(u^0)[r] - \nabla F(u^0)r\| + \|\nabla F(u^0)r\| \le (\frac{1}{2}\delta + \tilde{\delta}_2)\|r\|.$$

That is, with $\delta_2 = \tilde{\delta}_2 + \frac{1}{2}\delta$, the boundedness property (ii) of $A(u^0)[\cdot]$ is verified.

In summary, we can state the following result.

Theorem 1. *Consider the residual equation (2) and assume that $A[.]$ satisfies the properties (i–iii). Then if $\delta > 0$ is sufficiently small the presented variational (steepest descent like) algorithm provides an approximate correction c such that the new iterate $u := u^0 + c$ satisfies the estimate (5). In particular, the presented method is convergent if $F(.)$ is differentiable in the sense (6) or close to a differentiable mapping K in the sense (8) and its gradient $\nabla F(u^0)$ (either of F or of K) satisfies (9) and (10) for an accurate enough initial approximation u^0. Then $\|d\| = \|f - F(u^0)\|$ is small enough such that $\Delta = 2\delta_1^{-1}\|d\| < \delta_0$ needed for (6) to hold. The method has a geometric rate of convergence specified by estimate (5) with a convergence factor at least $\gamma \in (0,1)$.*

We remark, that we do not explicitly need in our algorithm neither ∇F nor ∇K (cf., in contrast Axelsson and Kaporin [3]). We instead need however a good preconditioner G (a symmetric positive definite linear mapping) to $A[\cdot]$ such that $G^{-1}A(u^0)[.]$ is coercive and bounded in terms of G, i.e., we need the estimates:

(i') coercivity:

$$(A(u^0)[r], r) \geq \delta_1(Gr, r) \quad \text{for all } r \;:\; \|r\| < \Delta;$$

(ii') boundedness:

$$\|G^{-1}A(u^0)[r]\|_G \leq \delta_2\|r\|_G \quad \text{for all } r \;:\; \|r\| < \Delta\cdot$$

Here (in (i') and (ii')), δ_1 and δ_2 are positive constants and Δ specifies the radius of a sufficiently large ball around the origin.

Then Algorithm 1 (*Steepest Descent*) takes the following preconditioned form. We first transform the residual equation (2) to

$$\hat{A}[c] = \hat{d}, \tag{11}$$

where $\hat{A} = G^{-1}A[c]$ and $\hat{d} = G^{-1}d$. Then by rewriting Algorithm 1 (*Steepest Descent*) applied to (11), i.e., applied for $\hat{A}[.]$ and $\hat{d}$, and inner product $(.,.)_G = (G\cdot,\cdot)$, one ends up with the following:

Algorithm 2 *(Preconditioned Steepest Descent.)*

0. Initiate:
Set $c^0 = 0$, $r^0 = d - A[c^0] = d$.

1. For $k = 1, 2, \ldots$, iterate:

$$\begin{aligned} \hat{r} &= G^{-1}r^{k-1}\,; \\ g &= A[\hat{r}]\,; \\ h &= G^{-1}g\,; \\ \alpha &= \frac{(\hat{r}, A[\hat{r}])}{(h, A[\hat{r}])} = \frac{(\hat{r}, g)}{(h, g)}\,; \\ c^k &= c^{k-1} + \alpha\hat{r}\,; \\ r^k &= d - A[c^k]\cdot \end{aligned}$$

This algorithm requires to invert G. Since we intend to use this algorithm for discretized versions of (1), in practice one is interested in a symmetric positive definite linear mapping (matrix) G for which the constants δ_1 and δ_2 in (i') and (ii') are mesh independent (i.e., independent of the discretization parameter when (1), and hence (2), is discretized.)

3 Numerical experiments

In this section we present some numerical results to demonstrate the efficiency of the preconditioned steepest descent (PSD) method described in Algorithm 2 of §2 for solving nonlinear differential equations for a class of finite element discretization techniques based on two subspaces in a framework close to that presented by Xu [17]. Our model problem is

$$F(u) = -div(a(x,y) \bigtriangledown u) + u^3 = f \qquad \text{in } \Omega, \tag{12}$$

where Ω is the unit square $(0,1) \times (0,1)$ partitioned into two families T_H and T_h of isosceles right triangulations. Let $V_H \subset V_h \subset H_0^1(\Omega)$ be two finite element subspaces consisting of piecewise linear functions on T_H and T_h, respectively. On the boundaries $x = 0$ and $y = 0$ we have Dirichlet boundary conditions and on $x = 1$ and $y = 1$ boundary conditions of Neumann type are imposed.

We denote by n the grid size in the fine grid and by ω the set of all points in the fine grid with $h = 1/n$. The coarse grid step is $H = mh$ for $m \geq 1$. We tested a model problem of the type (12) with the coefficient $a(x,y) = 1$ and exact solution $u(x,y) = \sin \pi x \sin \pi y$.

The computations were executed on the SunSparc Station with Fortran double precision. On the coarse grid level we solve the nonlinear problem by PSD method considering the number of outer and inner iterations. We vary the grid size on the fine grid $n = 1/h = 80, 160$, the discretization ratio $m = H/h = 2, 4, 8$ and the number of inner iterations *inner* $= 1, 2, 4, 8$. As an initial guess on the coarse grid we choose $u^0 = 0$. The inner iterative procedure in Algorithm 2 of §2 is to compute a correction c and then to update the solution by $u := u^0 + c$. Solving the problem on the coarse and fine grids we get the number of outer iterations to achieve the tolerance, denoted in the tables below by *outer coarse* and *outer fine*, respectively.

The stopping criterion for outer iterations is

$$\frac{r^T r}{(r^0)^T r^0} \leq 10^{-12}, \tag{13}$$

where $r^0 = d - A[c^0]$ is the initial residual for solving the residual problem $\hat{A}[c] = \hat{d}$ and $r = d - A[c]$ is the current one. Here we have $\hat{A} = G^{-1}A[c]$, $\hat{d} = G^{-1}d$ and as a preconditioner G we choose a block-incomplete LU factorization of A.

The so obtained solution on the coarse grid is linearly interpolated and is used as an initial guess solving the problem on the finer discretization level. At this level Xu [17] proposed to apply one or two Newton's iterations which lead to linear solvers on the fine grid.

We describe briefly how to implement the Newton's method for the equation (1). Let u^k be a current approximation of $u(x,y)$. The next iterate is sought as

$$u^{k+1} := u^k - v, \quad \text{where } v = \bigtriangledown F(u^k)^{-1}(F(u^k) - f).$$

Here $\bigtriangledown F(u^k)$ is a linear operator defined in (6), i.e., the gradient of nonlinear mapping F at the point u^k.

In the numerical experiments we consider the following algorithm:

Algorithm 3 *(Newton's method for $F(u) = f$)*

0. Initiate: Let u^0 be an initial guess;
1. For $k = 0, 1, 2, \ldots$, iterate:

$$g = F(u^k) - f\,;$$
$$v = \nabla F(u^k)^{-1} g\,;$$
$$u^{k+1} = u^k - v.$$

To find the correction v we need to solve the linear system $\nabla F(u^k)v = g$. It requires, in general, a solver with a nonsymmetric matrix. To avoid it, we use a lumping of the mass matrix. Then the preconditioned conjugate gradient (PCG) method with the same stopping criterion (13) and an incomplete LU factorization as a preconditioner is applied to solve the fine linearized equation.

Another approach is to use PSD method on the fine grid. The results on the accuracy for both (Newton's and PSD) methods are compared in Tables 1-2 for our model problem. For Newton's method we have only *outer coarse* and for PSD we have in addition *outer fine* when apply this method on the fine grid.

The following error estimators are used:

(1) the discrete L^2–error: $\varepsilon_0 = \left\{\sum_{(x,y)\in\omega} h^2[\tilde{u}(x,y) - u(x,y)]^2\right\}^{1/2}$;

(2) the discrete H^1–error: $\varepsilon_1 = \left\{(\tilde{u} - u)^T A(\tilde{u} - u)\right\}^{1/2}$.

The global CPU-time solving the problem is reported in seconds. Our conclusion is that PSD gives better results in both L^2- and H^1- norms but it is more expensive in CPU-time due to the nonlinear solver on the fine grid. It is worth mentioning that the results from two Newton's iterations can not be significantly improved by performing more Newton's iterations.

References

1. E.L. Allgower, K. Böhmer, F.A. Potra and W.C. Rheinboldt, A mesh–independence principle for operator equations and their discretizations, *SIAM J. Numer. Anal.*, **23**, 1986, 160-169.
2. O. Axelsson and P.S. Vassilevski, Construction of variable–step preconditioners for inner–outer iteration methods, *Iterative Methods in Linear Algebra (eds. R. Beauwens and P. de Groen), Elsevier Science Publishers B.V. (North–Holland)*, Amsterdam, 1992, 1-14.
3. O. Axelsson and I.E. Kaporin, On the solution of nonlinear equations for non-differentiable mappings, In fast solvers for flow problems (W. Hackbusch, ed.), Vieweg-Verlag, Braunschweig/Wiesbaden, 1994, 38-51.
4. O. Axelsson, On mesh independence and Newton–type methods, *Applications of Mathematics*, **38**, 1993, 249–265.
5. R.E. Bank and D.J. Rose, Global approximate Newton methods, *Numer. Math.*, **37**, 1981, 279–295.
6. M.O. Bristeau, R. Glowinski, J. Periaux, P. Perrier, and O. Pironneau, On the numerical solution of nonlinear problems in fluid dynamics by least squares and finite element methods (I) Least squares formulations and conjugate gradient solution of the continuous problems, *Comp. Meth. Appl. Mech. Eng.*, **17/18**, 1979, 619–657.

Table 1. h=1/80, number of outer and inner iterations, CPU-time in seconds; L^2- and H^1- norm of the error.

H	inner	outer coarse	outer fine	CPU
$H = 2h$	1	121	136	114
	2	61	68	102
	4	31	34	96
	8	15	17	92
coarse	$\varepsilon_0 = 4.397 \times 10^{-4}$, $\varepsilon_1 = 5.092 \times 10^{-3}$			
one Newton's	$\varepsilon_0 = 1.427 \times 10^{-4}$, $\varepsilon_1 = 1.383 \times 10^{-3}$			
two Newton's	$\varepsilon_0 = 9.846 \times 10^{-5}$, $\varepsilon_1 = 1.373 \times 10^{-3}$			
PSD	$\varepsilon_0 = 1.004 \times 10^{-4}$, $\varepsilon_1 = 1.372 \times 10^{-3}$			
$H = 4h$	1	34	204	141
	2	17	102	126
	4	9	51	117
	8	4	26	115
coarse	$\varepsilon_0 = 2.003 \times 10^{-3}$, $\varepsilon_1 = 1.861 \times 10^{-2}$			
one Newton's	$\varepsilon_0 = 4.119 \times 10^{-4}$, $\varepsilon_1 = 1.663 \times 10^{-3}$			
two Newton's	$\varepsilon_0 = 1.042 \times 10^{-4}$, $\varepsilon_1 = 1.381 \times 10^{-3}$			
PSD	$\varepsilon_0 = 1.004 \times 10^{-4}$, $\varepsilon_1 = 1.372 \times 10^{-3}$			
$H = 8h$	1	11	252	171
	2	5	126	152
	4	4	63	142
	8	3	32	138
coarse	$\varepsilon_0 = 9.534 \times 10^{-3}$, $\varepsilon_1 = 6.635 \times 10^{-2}$			
one Newton's	$\varepsilon_0 = 1.554 \times 10^{-3}$, $\varepsilon_1 = 4.168 \times 10^{-3}$			
two Newton's	$\varepsilon_0 = 2.734 \times 10^{-4}$, $\varepsilon_1 = 1.530 \times 10^{-3}$			
PSD	$\varepsilon_0 = 1.004 \times 10^{-4}$, $\varepsilon_1 = 1.372 \times 10^{-3}$			

7. P.N. Brown and Y. Saad, Hybrid Krylov methods for nonlinear systems of equations, *SIAM J. Sci. Stat. Comput.*, **11**, 1990, 450–481.
8. R.S. Dembo, S.C. Eisenstat and T. Steihaug, Inexact Newton methods, *SIAM J. Numer. Anal.*, **19**, 1982, 400–408.
9. V. Giarault and P.-A. Raviart, *Finite Element Methods for Navier-Stokes equations, Springer–Verlag*, Berlin–Heidelberg–New York–Tokyo, 1986.
10. R. Glowinski, H.B. Heller and L. Reinhart, Continuation–conjugate gradient methods for the least squares solution of nonlinear boundary value problems, *SIAM J. Sci. Stat. Comput.*, **6**, 1985, 793–832.
11. R. Glowinski, J. Periaux, and Q.V. Dihn, Domain decomposition methods for nonlinear problems in fluid dynamics, Rapports de Recherche, N° 147, 1982, INRIA, Domaine de Voluceau, Rocquencourt, France.
12. W. Hackbusch, *Multigrid Methods and Applications, Springer–Verlag*, Berlin–Heidelberg–New York–Tokyo, 1985.
13. L.V. Kantorovich and G.P. Akilov, *Functional Analysis (Russian), Nauka*, Moscow, 1977.

Table 2. h=1/160, number of outer and inner iterations, CPU-time in seconds; L^2- and H^1- norm of the error.

H	inner	outer coarse	outer fine	CPU
$H = 2h$	1	449	266	1032
	2	228	132	915
	4	114	66	863
	8	57	33	833
coarse	$\varepsilon_0 = 1.004 \times 10^{-4}$, $\varepsilon_1 = 1.372 \times 10^{-3}$			
one Newton's	$\varepsilon_0 = 3.535 \times 10^{-5}$, $\varepsilon_1 = 3.690 \times 10^{-4}$			
two Newton's	$\varepsilon_0 = 2.350 \times 10^{-5}$, $\varepsilon_1 = 3.664 \times 10^{-4}$			
PSD	$\varepsilon_0 = 2.297 \times 10^{-5}$, $\varepsilon_1 = 3.665 \times 10^{-4}$			
$H = 4h$	1	121	510	1414
	2	61	255	1242
	4	31	127	1168
	8	15	64	1128
coarse	$\varepsilon_0 = 4.397 \times 10^{-4}$, $\varepsilon_1 = 5.092 \times 10^{-3}$			
one Newton's	$\varepsilon_0 = 1.033 \times 10^{-4}$, $\varepsilon_1 = 4.366 \times 10^{-4}$			
two Newton's	$\varepsilon_0 = 2.507 \times 10^{-5}$, $\varepsilon_1 = 3.684 \times 10^{-4}$			
PSD	$\varepsilon_0 = 2.297 \times 10^{-5}$, $\varepsilon_1 = 3.665 \times 10^{-4}$			
$H = 8h$	1	34	655	1964
	2	17	361	1736
	4	9	181	1643
	8	4	91	1579
coarse	$\varepsilon_0 = 2.003 \times 10^{-3}$, $\varepsilon_1 = 1.861 \times 10^{-2}$			
one Newton's	$\varepsilon_0 = 3.962 \times 10^{-4}$, $\varepsilon_1 = 1.072 \times 10^{-3}$			
two Newton's	$\varepsilon_0 = 6.916 \times 10^{-5}$, $\varepsilon_1 = 4.052 \times 10^{-4}$			
PSD	$\varepsilon_0 = 2.297 \times 10^{-5}$, $\varepsilon_1 = 3.665 \times 10^{-4}$			

14. I.E. Kaporin and O. Axelsson, On a class of nonlinear equation solvers based on the residual norm reduction over a sequence of affine subspaces, *SIAM J. Sci. Comput.*, **16**, 1994, 228-249.
15. A.A. Reusken, Convergence of the multilevel full approximation scheme (FAS) including the V–cycle, *Preprint # 492, Department of Mathematics, University of Utrecht, The Netherlands*, 1987.
16. G. Vainikko, Galerkin's perturbation method and the general theory of approximation for nonlinear equations, *USSR Comp. Math. Phys.*, **7**, 1967, 1–41.
17. J. Xu, Two–grid finite element discretization for nonlinear elliptic equations, Report No. AM 105, Department of Mathematics, Penn State University, 1992, (SINUM, to appear).
18. T.J. Ypma, Local convergence of inexact Newton methods, *SIAM J. Numer. Anal.*, **21**, 1984, 583–590.
19. D. Zinćello, A class of approximate methods for solving operator equations with nondifferentiable operators, *Dopovidi Akad. Nauk Ukrain SSR*, 1963, 852–856.

Computation Model for Internal Chemical-Reacting Flows Through Curved Channels

B.V.Rogov[1] and I.A. Sokolova[2]

[1] Institute for High Temperatures
Izhorskaya , 13/19, Moscow, 127412, Russia,
[2] Institute for Mathematical Modelling of Rus. Ac. of Science
Miusskaya pl. 4a, Moscow, 125047, Russia

Abstract. A new model for numerical computation chemical-reacting viscous flows through variable cross section channel with wall curvature is proposed. It is based on the full system of Navier-Stokes equations, which has been written in adaptive system of orthogonal curvilinear coordinates attached to channel. The simplified equations have been deduced. New terms due to nonzero curvature of wall appear in equations. Thus, the model improves the well-known slender channel approximation, particular in case of curved channel. Numerical combined marching method providing calculations of supersonic and critical flows was worked out. Critical flow through supersonic nozzle with controlled accuracy was calculated.

1 Introduction

One of the important problem of computational hydrodynamic is the problem of numerical calculation of viscous, chemical-reacting, heat conducting internal flows. Numerical methods for such flows are very complicated and consume a large amount of computer time, especially in case of 3D and 2D geometry of channel. For efficient solving of these problems it is necessary to derive an asymptotically accurate simplified hydrodynamic equations, and to work out a fast algorithm with controlled accuracy. The well-known slender channel approximation (SlC) ordinary used for calculation of chemical-reacting viscous flows is not suitable to curved channel. Direct marching method gives no way of finding accurate level for critical mass flow and no effect of errors on flow parameters.

The objects of this report are: (1) simplification of the full system of Navier-Stokes (NS) equations for viscous flows through channels with variable cross sections and curved walls to parabolic equations; (2) short description of algorithm and numerical methods for calculation both subsonic and supersonic flows, providing controlled accuracy of critical mass flow; (3) numerical investigations of different characteristics of hydrodynamic flows.

2 Simplified Hydrodynamic Equations

2.1 Adaptive Curvilinear Orthogonal Coordinates

We will consider laminar, viscous axisymmetric flows of premixed, heat conducting gas through smooth channels with variable cross section and nonzero wall curvature. Simplification of the full system of NS equations is based on the new curvilinear orthogonal coordinates (ξ, η, ζ) adapted to the wall shape $y_w(x)$:

$$\xi = \xi(x, y), \quad \eta = y/y_w(x), \quad \zeta = \begin{cases} z, \text{for plane geometry} \\ \varphi, \text{for cylindric geometry,} \end{cases}$$

where (x, y, z) are the Cartesian coordinates. The typical plane channel and coordinates lines (ξ, η) and (x, y) are shown in Fig.1.

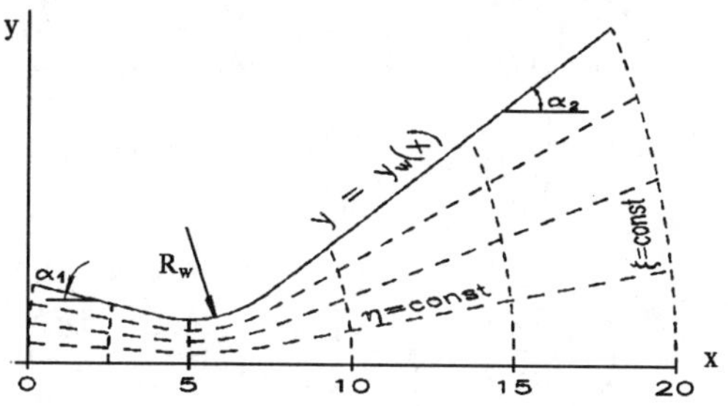

Fig. 1. Typical Laval's nozzle. Notation of geometry

For $2D$ flows the function $\xi(x, y)$ is derived from differential equation of the first order

$$\frac{y_w'(x)}{y_w(x)}\frac{\partial \xi}{\partial x} - \frac{1}{y}\frac{\partial \xi}{\partial y} = 0, \quad \text{given that} \quad \xi = x \quad \text{at} \quad y = 0, \tag{1}$$

and can be found from formula (1) in the following explicit form

$$\Phi(\xi) = \Phi(x) + y^2/2, \quad \text{where} \quad \Phi(x) = \int [y_w(x)/y_w'(x)]dx. \tag{2}$$

2.2 Smooth Channel Approximation

Assume, that $y_w(x)$ and its first's derivatives are continuous functions, curvature of axis (symmetry line of channel) is small, $y_w/R_{sl} \ll 1$, and $y_w/R_w \sim 1$ at nonzero curvature of channel wall. Here R_{sl} is the curvature radius of axis (index 'sl' refers to symmetry line), R_w is the curvature radius of the wall. The full system of NS equations written in the new curvilinear orthogonal coordinates is transformed to parabolic equations provided that the ratio $r^*/L^* \ll 1$, and neglecting small terms of second order, and provided that $\mathrm{Re}_L = \rho^* u^* L^*/\mu^* \gg 1$, where Re_L is the longitudinal Reynolds number, and neglecting small terms of order $1/\mathrm{Re}_L$. The resulting system of equations has been named 'smooth channel approximation' (SmC) [1]. In case of small wall curvature, $1/R_w \to 0$, the equations of SmC approximation are similar to SlC equations [2]. The difference is in the coordinate system used. As result, the pressure equation of SmC approximation gives pressure distribution over cross section of channel owing to relation $\xi = \xi(x, y)$.

In case of steady-state plane flows the equations for dimensionless variables u, p, T (longitudinal velocity, pressure and temperature, correspondingly) are transformed accounting stream function Ψ

$$\frac{\partial}{\partial \xi}\left(u\frac{\partial \Psi}{\partial \eta}\right) - \frac{\partial}{\partial \eta}\left(u\frac{\partial \Psi}{\partial \xi}\right) - \frac{1}{\mathrm{Re}_r}\left\{\frac{\partial}{\partial \eta}\left[\mu\frac{H_\xi}{H_\eta}\left(\frac{\partial u}{\partial \eta} + \varepsilon u\right)\right] - \right.$$
$$\left. -\varepsilon\mu\frac{H_\xi}{H_\eta}\left(\frac{\partial u}{\partial \eta} + \varepsilon u\right)\right\} + \varepsilon u\frac{\partial \Psi}{\partial \xi} + \frac{1}{\gamma}H_\eta\frac{\partial p}{\partial \xi} = 0, \tag{3}$$

$$\frac{\partial \ln p}{\partial \eta} = -\varepsilon\gamma\frac{u^2}{T}, \tag{4}$$

$$\frac{\partial}{\partial \xi}\left(H\frac{\partial \Psi}{\partial \eta}\right) - \frac{\partial}{\partial \eta}\left(H\frac{\partial \Psi}{\partial \xi}\right) - \frac{1}{\mathrm{Re}_r}\left\{\frac{1}{\mathrm{Pr}}\frac{\partial}{\partial \eta}\left(\lambda\frac{H_\xi}{H_\eta}\frac{\partial T}{\partial \eta}\right) + \right.$$
$$\left. +(\gamma - 1)\frac{\partial}{\partial \eta}\left[\mu u\frac{H_\xi}{H_\eta}\left(\frac{\partial u}{\partial \eta} + \varepsilon u\right)\right]\right\} = 0. \tag{5}$$

The stream function Ψ has been introduced to satisfy the equation of mass continuity

$$\Psi = \int_0^\eta \rho u H_\eta d\eta, \quad \frac{\partial \Psi}{\partial \eta} = \rho u H_\eta, \quad \frac{\partial \Psi}{\partial \xi} = -\rho \nu H_\xi. \tag{6}$$

In the above equations H_ξ, H_η are the Lame parameters, $\varepsilon = H_\eta/R_\eta$ is the function of (ξ, η), R_η is the curvature radius of coordinate line $\eta = \mathrm{const}$, $H = T + (\gamma - 1)u^2/2$ is the total enthalpy, ν is the lateral velocity, ρ is the mass density, μ and λ are the coefficients of dynamic viscosity and heat conductivity, $\mathrm{Re}_r = \rho_0 u_0 r^*/\mu_0$ is the Reynolds number, $\mathrm{Pr} = \mu_0 c_{p0}/\lambda_0$ is the Prandtl number, $\gamma = c_{p0}/c_{\nu 0}$ is the ratio of heat capacities. Reference values of density, pressure and temperature ρ_0, p_0, T_o are taken as stagnation parameters of flows at the entrance of channel, u_o is the sonic velocity, $u_o = a_0 = \sqrt{\gamma R T_0}$, R is the gas constant, r^* is the half height of the critical cross section. Equation (4) has been derived from reduced equation of momentum on axis ξ taking into account the equation of state $\rho = p/T$.

The system of equations (3-(5) is calculated under conditions that on the wall $u = 0$ and $T = T_w$; the relative velocity and temperature profiles u/u_{sl} and $T/T_{sl.}$ are given at $\xi = 0$ and the specified stagnation parameters $p = p_0$, $T = T_0$.

Conditions on the symmetry line $\eta = 0$ are $\partial u/\partial \eta = \partial T/\partial \eta = 0$. The equation of mass flow

$$\frac{\partial \Psi(\xi, 1)}{\partial \xi} = 0 \quad \text{or} \quad \Psi_w = Q = \text{const}, \tag{7}$$

closes the system of the SmC equations. Q is the dimensionless mass flow through channel. The equation (7) is derived from equation (6) provided that lateral velocity $v = 0$ at the wall.

3 Computational Method

As well-known, marching method ordinary used to computing real steady-state flows equations in the slender channel approximation, has principal difficulties. This is a singularity in transonic area and uncontrolled accuracy while calculating the critical flow through nozzle [3]. We will overcome these difficulties with successive modification of standard marching method.

3.1 Marching Method I

At first, the longitudinal velocity, temperature, and pressure gradient are calculated simultaneously at the every characteristic ξ = const of parabolic system of the differential equations. This scheme is the modification of the finite difference scheme [4] with the fourth order of approximation over lateral coordinate η, and second one over coordinate ξ. The dissimilarity of this scheme from the ordinary used scheme is in the another approximation of derivative with respect to longitudinal coordinate

$$\left(\frac{\partial a}{\partial \xi}\right)^+ = 2\frac{a^+ - \tilde{a}^-}{\Delta\xi}, \quad \tilde{a}^- = a^- + \frac{1}{2}\left(\frac{\partial a}{\partial \xi}\right)^- \Delta\xi \tag{8}$$

where $a = a(\xi, \eta)$, index " + " and " − " refers to values at $\xi = \xi^+$ and $\xi = \xi^-$, correspondingly, $\Delta\xi = \xi^+ - \xi^-$ is the mesh width. Thus, the solution at the next characteristic $\xi = \xi^+$ is obtained by calculating finite difference equations at the same characteristic, contrary to the method [4], where the difference equations are calculated at the intermediate characteristic $\hat{\xi} = (\xi^+ + \xi^-)/2$.

The finite difference equations for u and T are calculated by vector sweep method. The equation for pressure gradient has been derived specially, as follows.

Upon integrating equation (4) with respect to the η from 0 to η we obtain:

$$p = p_{sl}(\xi)\alpha(\xi, \eta), \quad \alpha = \exp(-\gamma \int_0^\eta \varepsilon \frac{u^2}{T} d\eta). \tag{9}$$

So, if profiles of u and T are known, the pressure distribution within SmC approximation is defined by axial pressure p_{sl}. Using equation (9), the equation

(3) for longitudinal velocity u may be written as follows

$$\frac{\partial}{\partial\xi}\left(u\frac{\partial\Psi}{\partial\eta}\right)-\frac{\partial}{\partial\eta}\left(u\frac{\partial\Psi}{\partial\xi}\right)-\frac{1}{\mathrm{Re}_r}\left\{\frac{\partial}{\partial\eta}\left[\mu\frac{H_\xi}{H_\eta}\left(\frac{\partial u}{\partial\eta}+\varepsilon u\right)\right]-\right.$$
$$\left.-\varepsilon\mu\frac{H_\xi}{H_\eta}\left(\frac{\partial u}{\partial\eta}+\varepsilon u\right)\right\}+\varepsilon u\frac{\partial\Psi}{\partial\xi}-\alpha p_{sl}H_\eta\frac{\partial}{\partial\xi}\int\limits_0^\eta\varepsilon\frac{u^2}{T}d\eta+\frac{1}{\gamma}\alpha H_\eta\frac{dp_{sl}}{d\xi}=0. \quad (10)$$

Substitution of equation (9) into equation (7) and use of the equation of state and the symmetry conditions give the equation for axial pressure

$$p_{sl}=Q\bigg/\int\limits_0^1\alpha\frac{u}{T}H_\eta d\eta. \quad (11)$$

It was found, if equation (10) and equation (5) for u and T and equation (11) for axial pressure were iterated one after another on the next characteristic, then numerical process was divergent. For overcoming this difficulty, the equation for pressure gradient, providing computational convergence, is deduced.

Upon integrating equation (10) with respect to the η from 0 to 1 in view of (6), (7) and boundary conditions for u on the wall and symmetry line we obtain

$$\frac{1}{\gamma}\int\limits_0^1\alpha H_\eta d\eta\frac{dp_{sl}}{d\xi}=\frac{1}{2}H_{\xi w}c_f-\frac{d}{d\xi}\left(p_{sl}\int\limits_0^1\alpha\frac{u^2}{T}H_\eta d\eta\right)-\int\limits_0^1\varepsilon u\frac{\partial\Psi}{\partial\xi}d\eta+$$
$$+p_{sl}\int\limits_0^1\left(\frac{\partial}{\partial\xi}\int\limits_0^\eta\varepsilon\frac{u^2}{T}d\eta\right)\alpha H_\eta d\eta-\frac{1}{\mathrm{Re}_r}\int\limits_0^1\varepsilon\mu\frac{H_\xi}{H_\eta}\left(\frac{\partial u}{\partial\eta}+\varepsilon u\right)d\eta, \quad (12)$$

where value p_{sl} in the right side of equation (12) is defined by equation (11), and c_f is the wall friction coefficient

$$c_f=-\frac{2}{\mathrm{Re}_r}\left(\frac{\mu}{H_\eta}\frac{\partial u}{\partial\eta}\right)_w. \quad (13)$$

The pressure gradient $(dp_{sl}/d\xi)^+$ at the next characteristic $\xi=\xi^+$ is defined using the known profiles of u^+ and T^+ from equation (12). And axial pressure p_{sl}^+ is defined in accordance with equation (8). Substitution of axial pressure and its gradient into equation (5) and equation (10) gives the new profiles for longitudinal velocity and temperature. This process of successive iterations is convergent. The above described direct marching method gives possibility to compute the flows at given mass flow Q in the transonic section of nozzle and at the origin of expansion. In transonic area the pressure and wall friction coefficient are very sensitive to the value of critical mass flow. The distributions of axial pressure and wall friction coefficient, calculated in frame of this method, show branching over corresponding limiting distributions in variation of mass flow (see Fig.2) near critical mass flow Q_{cr}. This branching allows us to find the value of critical mass flow with controlled accuracy when the corresponding difference grid is taken.

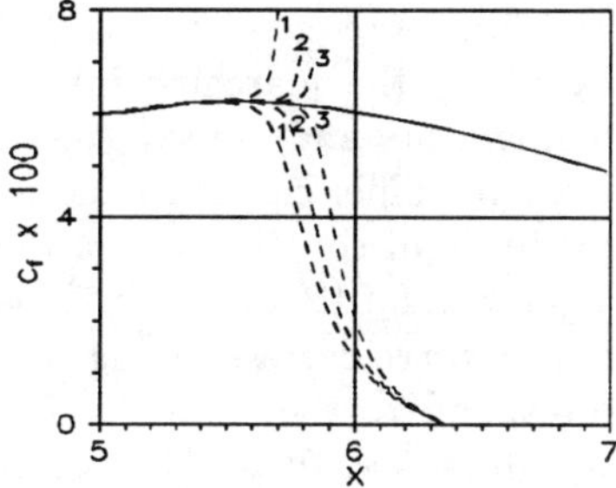

Fig. 2. Friction coefficient versus Cartesian coordinate x. The dashed curves are the results of the first marching method. The solid curve is the result of the second one

3.2 Marching Method II

For calculating the flow after the critical cross section of nozzle the known critical mass flow and preceding profiles of u^- and T^- are taken. The direct marching method is added by global iterations over pressure, and the new difference equation for axial pressure .

At first, the distribution of axial pressure is taken. Then equation (5) and (10) for velocity and temperature are integrated by means of marching method and the new axial pressure distribution $p_{sl}(\xi)$ is derived from equation (11). The iteration process is repeated. Regularization procedure of numerical method reduces to another calculation of pressure gradient as compared to marching method I. The pressure gradient on the next coordinate $\xi = \xi^+$ is defined by following difference equation at constant value of mesh width $\Delta\xi$

$$\left(\frac{\partial p_{sl}}{\partial \xi}\right)^+ = \frac{p_{sl}(\xi^+ + \Delta\xi) - p_{sl}(\xi^+ - \Delta\xi)}{2\Delta\xi}. \tag{14}$$

4 Numerical Results

The test calculation of viscous flows in the nozzle has been performed at $\mathrm{Re}_r = 100$. Contraction and extension of sections has been set equal to $\alpha_1 = 10°$, and $\alpha_2 = 30°$, correspondingly, curvature radius of wall in transonic area is $R_w = 5.0$, $r^* = 1$ (see Fig.1). Viscosity is taken in terms of power of temperature, $\mu = T^n$, $n = 0.647$, Prandtl number is constant, $\mathrm{Pr} = 0.71$, $\gamma = 1.4$, $T_w = 0.3$. The velocity and temperature profiles at input channel section are taken uniform in main part of the flow with smooth parabolic approach to the wall. The grid with variable mesh width over lateral coordinate η has 41 points, thickening to the wall . The uniform grid in respect to longitudinal coordinate has 41 points per length unit. Such grid provides convergent solution in the whole computational domain. The results of calculations of viscous friction coefficient c_f in the transonic region of nozzle after the critical section $x = 5.0$ are presented in Fig.2. The dashed curves show the results of the calculation by the marching method I with a simultaneous finding of the velocity, temperature and pressure distributions on every step along the marching coordinate for different values of the mass flow Q. In the process of calculation the condition of conservation of the mass flow along the nozzle were met with an accuracy of 0.01% or better. It is seen, that the dashed curves branch out from limiting curve corresponding to the limiting mass flow $Q_{\lim}$. The upper curves correspond to the values of the mass flow $Q > Q_{\lim}$, and the lower curves to the values $Q < Q_{\lim}$. The curves 1,2 and 3 correspond to the values of the mass flow differing from the value of the limiting mass flow in sixth, seventh and eighth significant figure respectively. In turn, the value of the limiting mass flow tends to that of the critical mass flow Q_{cr} , as step of finite-difference grid tends to zero. The finite-difference grid described above allows the critical mass flow to be found with a precision of 5 significant figures.

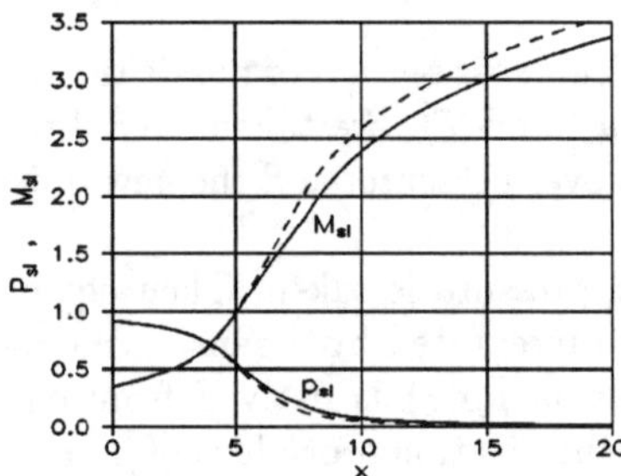

Fig. 3. Axial distribution of pressure p_{sl} and Mach number M_{sl} . The solid curves are the results of the SmC model; the dashed curves are the results of the SlC model

Using the calculated value of the critical mass flow, there were computed characteristics of the critical regime of flow from the critical section to the output one by the marching method II. The solid curve in Fig.2 shows results of the friction coefficient corresponding to the critical flow calculated by the marching method II (with global iteration of pressure). To obtain the values of the friction coefficient with an accuracy of 1% in critical cross section it was necessary to meet the condition of conservation of mass flow along the nozzle with an accuracy of 0.01%.

The results of axial pressure p_{sl} and Mach number M_{sl} are shown in Fig.3. The solid curves correspond to the results of the SmC approximation. The dashed curves show SlC results. The difference between the curves after critical cross section is seen. The cause is that the SmC model yields the pressure distribution over cross section of channel with the pressure gradient, which is proportional to the local curvature of coordinate line $\eta = \text{const}$. This is illustrated in Fig.4 by

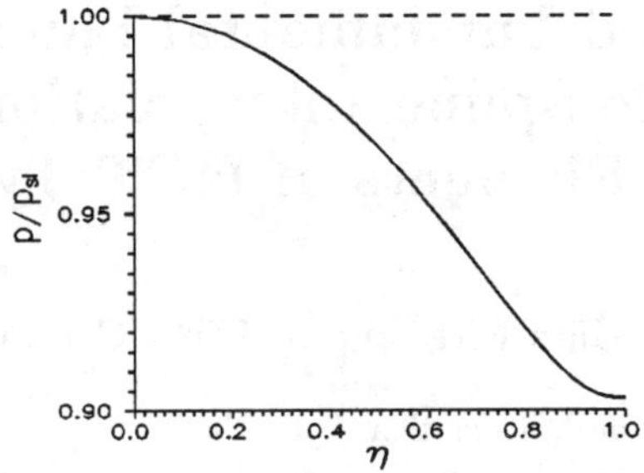

Fig. 4. Lateral profile of pressure at the critical section of the nozzle. The solid curve is the profile, calculated by the SmC model; the dashed straight line is the results of the SlC model

solid line. The SlC model gives the constant pressure, which is shown by dashed straight line in Fig.4. The profiles of relative pressure over lateral coordinate η are given in the critical cross section of the nozzle. In conclusion we must say, that the new computation model and speed algorithm give efficient tool for study chemical-reactivity internal viscous flows through curved channel.

Acknowledgments

Authors are thankful to Prof. N.N.Kalitkin for interest to work and useful discussions.

The work was supported by GRAND RFFI, code N 94-01-01618

References

1. Rogov, B.V., Sokolova, I.A.: Viscous flows in variable cross section channel // Doklady of Russian Academy of Science **345** 5 (1995) 615-618
2. Lapin, Yu.V., Streletz, M.H.: Internal Flows of gas mixtures. Moscow: Nauka, (1989) 368
3. Lapin, Yu.V., Nehamkina, O.A., Pospelov, V.A. et.al.: Numerical modeling of internal flow of viscous gas mixtures WINITI. Science and Technique ser. Mechanic of Fluids and Gas Moscow **19** (1985) 86-185
4. Petuhov, I.B.: Numerical analysis of two-dimension flow in boundary layer // Numerical methods of solution of differential and integral equations and quadrature formulae. Moscow Nauka (1964) 304-325

Compactly Supported Fundamental Functions for Bivariate Hermite Spline Interpolation and Triangular Finite Elements of HCT Type

D. Barrera Rosillo, M. J. Ibáñez Pérez and A. López Carmona

Departamento de Matemática Aplicada
Facultad de Ciencias, Universidad de Granada
Campus Universitario de Fuentenueva s/n
18071 Granada, ESPAÑA

Abstract. Let τ be the triangulation of the plane generated by a uniform three direction mesh of the plane. If τ_3 is the subtriangulation of τ obtained joining the center of gravity of each triangle $T \in \tau$ to its vertices, we construct certain bivariate Hermite spline interpolation operator which preserves a chosen space of polynomials, making use of HCT finite elements.

1 Introduction

Let τ be the triangulation of the plane $\mathbb{R}^2$ induced by $x = k$, $y = k$ and $x - y = k$, $k \in \mathbb{Z}$. Let τ_3 be the subtriangulation of τ obtained joining the center of gravity of each triangle $T \in \tau$ to its vertices.

The problem we are going to deal with appears motivated by the problem considered in [4], in which it is undertaken a Lagrange spline interpolation particular problem with data in $\mathbb{Z}^2$, and that is solved constructing its fundamental function (of compact support), L, by assembling Hsieh-Clough-Tocher finite elements (HCT, [3]), which are C^1-cubic splines defined on the subtriangulation τ_3 de T.

In [5, 6] P. Sablonnière and M. Laghchim-Lahlou carry out a wide study of the HCT type finite elements of class C^r, $r \geq 1$, for an arbitrary triangulation of the plane. Those generalized splines permit them to define class C^ρ global interpolants over a bounded polygonal domain, Ω, of $\mathbb{R}^2$ for a function sufficiently

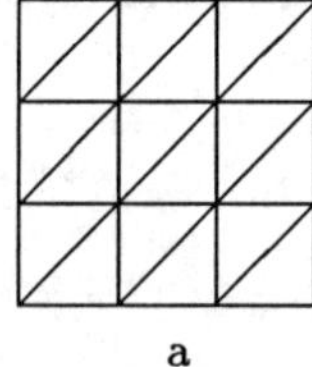

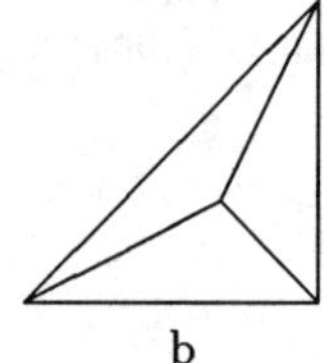

Fig. 1. The τ triangulation of a subset of the plane and the τ_3 the subtriangulation of one of its triangles.

regular. They need to know the values of the interpolated function and the ones of its partial derivatives up to order $\rho + \left[\frac{\rho}{2}\right]$ at the vertices of every triangle included in Ω, apart from certain directional derivatives at distinct points inside its edges ($[x]$ is the integer part of x).

Our objective is to construct Hermite interpolation operators of prefixed order $r \geq 1$, whose fundamental function be of compact support, which provide interpolants of global class C^s, which determines the degree of HCT finite element that has to be used. If we do not have at our disposal the values at the triangulation vertices of all the needed partial derivatives nor the additional directional derivatives, we need to impose some condition that let us define the required interpolant. Concretely, we will demand that the interpolation operator be exact on any predetermined polynomial space, $\mathbb{P}_m$, that is to say, that the restriction of the operator to $\mathbb{P}_m$ be the identity. This requeriment will be translated in that the interpolation error be of order $O\left(h^{m+1}\right)$ if $h\tau$ and $h\tau_3$ and data in $h\mathbb{Z}^2$, $h > 0$, are considered.

Henceforth, if S is a finite subset of $\mathbb{Z}^2$, $[S]$ and $card\,(S)$ will denote its convex hull and its cardinality, respectively. In the same way, $supp\,(f)$ will represent the support of f and will use the directions $e_1 = (1,0)$, $e_2 = (0,1)$ and $e_3 = (1,1)$ as well as their directional derivatives $\overline{e}_1 = (0,1)$, $\overline{e}_2 = (1,0)$ and $\overline{e}_3 = (1,-1)$, respectively. To each $\alpha \in Z^2$ we associate the hexagon $H(\alpha) = [\{\alpha + e_1, \alpha + e_3, \alpha + e_2, \alpha - e_1, \alpha - e_3, \alpha - e_2\}]$ and $H(S) = \bigcup_{\alpha \in S} H(\alpha)$ to $S \subset \mathbb{Z}^2$. If $S \subset \mathbb{R}^2$ and $\gamma \in Z^2$ we define $t_\gamma(S) = \{s + \gamma : s \in S\}$ and $\sigma_\gamma(S) = \{2\gamma - s : s \in S\}$. Given $r \in \mathbb{N}$, for $0 \leq i \leq r$ we make $\Gamma_i = \{\beta \in \Delta_i : |\beta| := i\}$ and $\Delta = \bigcup_{i=0}^{r} \Delta_i$, where $\Delta_j = \left\{\beta = (\beta_1, \beta_2) \in (\mathbb{N} \cup \{0\})^2 : |\beta| := \beta_1 + \beta_2 \leq j\right\}$, $j \in \mathbb{N} \cup \{0\}$. In like manner, we use the notation $\partial_\alpha^\beta(f) = \frac{\partial^{|\beta|} f}{\partial x^{\beta_1} \partial y^{\beta_2}}(\alpha)$, $\beta \in (\mathbb{N} \cup \{0\})^2$, $\alpha \in \mathbb{R}^2$. $D_v^s f(\alpha)$ designates the directional derivative of f in the direction v at the point α.

If Ω is a bounded polygonal domain of $\mathbb{R}^2$ endowed of τ triangulation, $\mathbb{S}_{2k}(\Omega)$ will denote the set composed by the $C^{2k}(\Omega)$ polynomial functions of total degree at most $6k+1$ on each $t \in \tau_3$, of class C^{3k} at the vertices of $T \in \tau$ and C^{3k+2} at its gravity center and whose restrictions to T be of class $C^{2k+1}(T)$. Also $\mathbb{P}_{6k+1}^0(T, \tau_3)$ will designate the space of polynomial functions of degree at most $6k+1$ on the τ_3 triangles of T. We make $\mathbb{P}_{6k+1}^{2k+1}(T, \tau_3) = \mathbb{P}_{6k+1}^0(T, \tau_3) \cap C^{2k+1}(T)$ and $\mathbb{S}_{2k}(T)$ will be the space constituted by the functions of $\mathbb{P}_{6k+1}^{2k+1}(T, \tau_3)$ which are of C^{3k} class at the vertices of T.

2 A First Example

We consider $P = \{(0,0), (1,0), (1,1), (0,1), (-1,0), (-1,-1), (0,-1)\}$, $P_1 = \{(0,0), (1,0), (1,1), (0,1)\}$ and $P_0 = \{(-1,0), (-1,-1), (0,-1)\}$.

There exists a unique family $\left\{q_{2\gamma-\alpha}^{(0,0)}\right\}_{\alpha \in P_0 \cup P_1} \cup \left\{q_{2\gamma-\alpha}^{(1,0)}, q_{2\gamma-\alpha}^{(0,1)}\right\}_{\alpha \in P_1}$ in $\mathbb{P}_4$

a			b			c		
	•	•		1	1		0	0
•	•	•	0	1	1	1	1	0
•	•		0	0		1	1	

Fig. 2. a) The points system P. b) The subsets P_1 and P_0. c) The symmetric ones.

for each $\gamma \in \mathbb{Z}^2$such that every polynomial $p \in \mathbb{P}_4$ can be expressed as

$$p = \sum_{\alpha \in P_0 \cup P_1} \partial_\alpha^{(0,0)}(p)\, q_{2\gamma-\alpha}^{(0,0)} + \sum_{\alpha \in P_1} \left(\partial_\alpha^{(1,0)}(p)\, q_{2\gamma-\alpha}^{(1,0)} + \partial_\alpha^{(0,1)}(p)\, q_{2\gamma-\alpha}^{(0,1)} \right)$$

We can construct ([5]) $L_\gamma^{(0,0)} \in \mathbb{S}_2\left(t_\gamma\left(H\left(P_0 \cup P_1\right)\right)\right)$ that satisfies the following properties:

- $\partial_\alpha^{(0,0)}\left(L_\gamma^{(0,0)}\right) = 1\!\!1_{\alpha\gamma} = \begin{cases} 1, & \alpha = \gamma \\ 0, & \alpha \neq \gamma \end{cases}, \partial_\alpha^\beta\left(L_\gamma^{(0,0)}\right) = 0, \beta \in \Gamma_1, \alpha \in t_\gamma\left(P_0 \cup P_1\right).$
- $\partial_\alpha^\beta\left(L_\gamma^{(0,0)}\right) = \partial_\alpha^\beta\left(q_{2\gamma-\alpha}^{(0,0)}\right)$, $\beta \in \Gamma_2 \cup \Gamma_3$, $\alpha \in t_\gamma\left(P_0 \cup P_1\right)$.
- For $1 \le k \le 3$ and $\alpha \in t_\gamma\left(P_0 \cup P_1\right)$
  - $D_{\overline{e}_k}^r L_\gamma^{(0,0)}\left(\alpha \pm \frac{1}{2} e_k\right) = D_{\overline{e}_k}^r q_{2\gamma-\alpha}^{(0,0)}\left(\alpha \pm \frac{1}{2} e_k\right)$, $r = 1, 3$.
  - $D_{\overline{e}_k}^2 L_\gamma^{(0,0)}\left(\alpha \pm \frac{j}{3} e_k\right) = D_{\overline{e}_k}^2 q_{2\gamma-\alpha}^{(0,0)}\left(\alpha \pm \frac{j}{3} e_k\right)$, $1 \le j \le 2$.
- At the τ vertices which are on the boundary of $t_\gamma\left(H\left(P_0 \cup P_1\right)\right)$ the values of $L_\gamma^{(0,0)}$ and the ones of its partial derivatives up to order three are zero, as well as the normal derivatives of orders at most three at similar points to the considered ones in the previous condition which are interior to the edges that are part of the boundary of $t_\gamma\left(H\left(P_0 \cup P_1\right)\right)$.

We extend $L_\gamma^{(0,0)}$ by zero to all $\mathbb{R}^2$. The resulting function, which we will keep on naming $L_\gamma^{(0,0)}$, is of $C^2\left(\mathbb{R}^2\right)$ class and belongs to $\mathbb{S}_2(T)$ for every $T \in \tau$.

In an analogous manner, but by means of the points of $t_\gamma\left(P_1\right)$, we can define functions $L_\gamma^{(1,0)}$ and $L_\gamma^{(0,1)}$ with support $t_\gamma\left(H\left(P_1\right)\right)$ and similar properties. We claim that the interpolant defined by

$$H_1(f) = \sum_{\alpha \in \mathbb{Z}^2} \left(\partial_\alpha^{(0,0)}(f)\, L_\alpha^{(0,0)} + \partial_\alpha^{(1,0)}(f)\, L_\alpha^{(1,0)} + \partial_\alpha^{(0,1)}(f)\, L_\alpha^{(0,1)} \right)$$

is exact on $\mathbb{P}_4$, that is to say, $H_1(p) = p$ for $p \in \mathbb{P}_4$.

3 The Interpolation Problem

The objective of this section is to define Hermite interpolants of even global class, what covers the above section example, making use of the proper HCT finite elements. The remaining case is treated in a similar way.

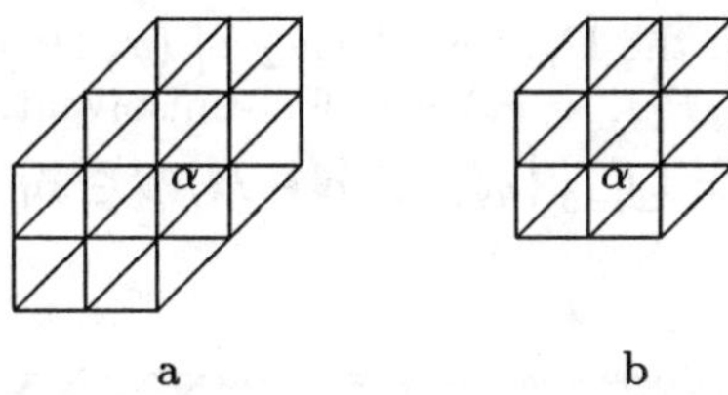

Fig. 3. The supports of the fundamental functions $L_\alpha^{(0,0)}$ (a) and $L_\alpha^{(1,0)}$ and $L_\alpha^{(0,1)}$ (b).

Definition 1. Let P be a finite subset of $\mathbb{Z}^2$ and r and m be natural numbers. We say that $(P_0, P_1, \ldots, P_r)$ is a $\mathbb{P}_m$-admissible configuration if $P_r \subset P$, $(0,0) \in P_r$, $P_{r-i} \subset P - \bigcup_{j=r-i+1}^{r} P_j$, $1 \leq i \leq r$, and $\sum_{i=0}^{r} \frac{(i+1)(i+2)}{2} card\,(P_i) = \frac{(m+1)(m+2)}{2}$.

Definition 2. If for every subset $\bigcup_{i=0}^{r} \left\{p_\alpha^\beta : \alpha \in P_i,\ \beta \in \Delta_i\right\}$ of real numbers the problem consisting in finding $p \in \mathbb{P}_m$ such that $\partial_\alpha^\beta(p) = p_\alpha^\beta$, $\alpha \in P_i$, $\beta \in \Delta_i$, $0 \leq i \leq r$, admits a unique solution, we say that the $\mathbb{P}_m$-admissible configuration $(P_0, P_1, \ldots, P_r)$ is $\mathbb{P}_m$-unisolvent.

¿From $H(P_i)$, $0 \leq i \leq r$, let $H_i = \bigcup_{j=i}^{r} H(P_j)$ be, $0 \leq i \leq r$. The results in ([5, 6]) permit us to enunciate the following result:

Proposition 3. *For $r, k \geq 1$, $1 \leq m \leq 6k+3$, let $(P_0, P_1, \ldots, P_r)$ be a $\mathbb{P}_m$-unisolvent configuration. Then there exists a unique family $\left\{L_\alpha^\beta : \beta \in \Delta\right\}_{\alpha \in \mathbb{Z}^2}$ of functions that satisfy the following properties:*

1. $L_\alpha^\beta \in C^{2k}(\mathbb{R}^2)$, *supp* $(L_\alpha^\beta) = t_\alpha(H_{|\beta|})$, $L^\beta_{\alpha|t_\alpha(H_{|\beta|})} \in \mathbb{S}_{2k}(t_\alpha(H_{|\beta|}))$ *and* $L_{\alpha|T}^\beta \in \mathbb{S}_{2k}(T)$ *for every τ triangle T.*
2. $\partial_\gamma^\delta(L_\alpha^\beta) = 1\!\!1_{\alpha\gamma} 1\!\!1_{\beta\delta}$, $\delta \in \Delta_r$, $\gamma \in t_\alpha(P_{|\beta|} \cup \ldots \cup P_r)$.
3. $\partial_\gamma^\delta(L_\alpha^\beta) = \partial_\gamma^\delta\left(q_{2\alpha-\gamma}^\beta\right)$, $\gamma \in t_\alpha(P_{|\beta|} \cup \ldots \cup P_r)$, $\delta \in \Gamma_{r+1} \cup \ldots \cup \Gamma_{3k}$.
4. *For each $1 \leq k \leq 3$ and $\gamma \in t_\alpha(P_{|\beta|} \cup \ldots \cup P_r)$,*

$$D_{\bar{e}_k}^s L_\alpha^\beta\left(\gamma \pm \tfrac{j}{s+1} e_k\right) = D_{\bar{e}_k}^s q_{2\alpha-\gamma}^\beta\left(\gamma \pm \tfrac{j}{s+1} e_k\right), \quad \begin{array}{c} 1 \leq s \leq 2k \\ 1 \leq j \leq s \end{array}$$

$$D_{\bar{e}_k}^s L_\alpha^\beta\left(\gamma \pm \tfrac{j}{4k-s+1} e_k\right) = D_{\bar{e}_k}^s q_{2\alpha-\gamma}^\beta\left(\gamma \pm \tfrac{j}{4k-s+1} e_k\right), \quad \begin{array}{c} 2k+1 \leq s \leq 4k-1 \\ 1 \leq j \leq 4k-s \end{array}$$

5. *If $\gamma \in \mathbb{Z}^2$ is in the boundary of L_α^β then $\partial_\gamma^\delta(L_\alpha^\beta) = 0$ for $\beta \in \Delta_{3k}$.*
6. *The normal derivatives analogous to the considered above corresponding to the edges of the L_α^β boundary are zero.*

Proposition 4. *In the conditions of the Prop. 3, the Hermite operator of order r defined by*

$$\mathrm{H}_r(f) = \sum_{\alpha \in \mathbb{Z}^2} \sum_{i=0}^{r} \sum_{\beta \in \Delta_i} \partial_\alpha^\beta(f)\, L_\alpha^\beta, \quad f \in C^{4k-1}(\mathbb{R}^2)$$

is exact on $\mathbb{P}_m$.

The proof is based on the $\mathbb{P}_m$-unisolvency of $(\sigma_\gamma(P_0), \sigma_\gamma(P_1), \ldots, \sigma_\gamma(P_r))$ for each $\gamma \in \mathbb{Z}^2$, as $(P_0, P_1, \ldots, P_r)$ was $\mathbb{P}_m$-unisolvent. Hence, there exists a unique $\mathbb{P}_m$ subset $B_m = \bigcup_{i=0}^r \left\{ q_{2\gamma-\alpha}^\beta : \alpha \in P_i,\ \beta \in \Delta_i \right\}$ such that, for every $p \in \mathbb{P}_m$,

$$p = \sum_{i=0}^r \sum_{\alpha \in P_i} \sum_{\beta \in \Delta_i} \partial_{2\gamma-\alpha}^\beta (p)\, q_{2\gamma-\alpha}^\beta = \sum_{i=0}^r \sum_{\alpha \in P_i \cup \ldots \cup P_r} \sum_{\beta \in \Gamma_i} \partial_{2\gamma-\alpha}^\beta (p)\, q_{2\gamma-\alpha}^\beta$$

This representation and the conditions about L_α^β summarized in Prop. 3 lead us to the exactness indicated above of the interpolation operator (cf. [1] for further details).

4 Some Considerations about the Configurations

This section is intended to describe some generic facts about the classification of admissible configurations ([10]).

The problem of determining the $\mathbb{P}_m$-admissible $(P_0, P_1, \ldots, P_r)$ configurations constructed with points in P for the polynomial Hermite interpolation of order r is equivalent to find the forms in which the symbols s, l and h_i, $1 \le i \le r$, for instance, can be assigned to the P points so that the condition which defines the admissibility is fulfilled (cf. Sect. 1). We will make use of the $r+2$-tuple $sym = (s, l, h_1, \ldots, h_r)$ and its components will be named as sym_j, $0 \le j \le r+1$.

Definition 5. We say that the $r+2$-tuple $t = (t_0, t_1, \ldots, t_{r+1})$, where $t_0 = card\,(P) - \sum_{i=0}^r card\,(P_i)$ and $t_i = card\,(P_{i-i})$, $1 \le i \le r+1$, is the type of the $\mathbb{P}_m$-admissible configuration $(P_0, P_1, \ldots, P_r)$.

The number of $\mathbb{P}_m$-admissible configurations is the coefficient of $x^{\frac{(m+1)(m+2)}{2}}$ in the development of $\left(1 + \sum_{i=0}^r x^{\frac{(i+1)(i+2)}{2}}\right)^{card(P)}$.

As the plane triangulation is regular, for studying the $\mathbb{P}_m$-unisolvent configurations it is reasonable to classify the $\mathbb{P}_m$-admissible ones of a given type taking into account the $\mathbb{P}_m$-unisolvency preservation by means of the isometries of the P points system or of some associated ones. Concretely, to P we can make correspond the $\widetilde{P}$ system formed in the natural way by the vertices of the equilateral triangulation, τ^*, relative to τ and we note P^* to the points system of the considered ones with bigger number of isometries.

To each configuration $\mathbb{P}_m$-admissible $(P_0, P_1, \ldots, P_r)$ it corresponds a $r+1$-tuple $(P_0^*, P_1^*, \ldots, P_r^*)$, with P_i^* subset of P^*, $0 \le i \le r$.

Definition 6. We say that two $\mathbb{P}_m$-admissible configurations $(P_0, P_1, \ldots, P_r)$ and $(Q_0, Q_1, \ldots, Q_r)$ of type c are equivalent if there exists an isometry σ de P^* such that $\sigma\,(P_i^*) = Q_i^*$, $0 \le i \le r$.

The defined relation produces a partition in the set of $\mathbb{P}_m$-admissible configurations of the type c. The $\mathbb{P}_m$-unisolvency of a configuration representative of a particular equivalency class implies the one of the remaining members of this class, what justifies our interest in the equivalency classes and in the number of equivalency classes of this prefixed type.

Let $G(P^*)$ be the isometries group of P^*. Every σ isometry of P^* produces a permutation, $\pi(\sigma)$, of the P points, which is split as a product of cycles, whose orders form a set which we note $J_{\pi(\sigma)}$. Let $z_i(\pi(\sigma))$ denote the number of cycles of $\pi(\sigma)$ having length $i \in J_{\pi(\sigma)}$. The polynomial $\prod_{i \in J_{\pi(\sigma)}} x_i^{z_i(\pi(\sigma))}$ is named representation of the structure of $\pi(\sigma)$ cycles.

Definition 7. For $n = \max \bigcup_{\sigma \in G(P^*)} J_{\pi(\sigma)}$, we define the cycle index of $G(P^*)$ as $I_{G(P^*)}(x_1, x_2, \cdots, x_n) = \frac{1}{card(G(P^*))} \sum_{\sigma \in G(P^*)} \prod_{i \in J_{\pi(\sigma)}} x_i^{z_i(\pi(\sigma))}$.

Proposition 8. *The number of equivalency classes of type t is the coefficient of* $\prod_{i=0}^{r+1} sym_i^{t_i}$ *in* $I_{G(P^*)}\left(\sum_{i=0}^{r+1} sym_i, \sum_{i=0}^{r+1} sym_i^2, \ldots, \sum_{i=0}^{r+1} sym_i^n\right)$.

The unisolvency of a multivariate Hermite polynomial interpolation problem has been widely studied and is the object of an intense research ([2, 9] show the state of the subject in a precise way). The interpolation system theory developed by M. Gasca, J. I. Maeztu, V. Ramírez and J. Lorente is a very interesting tool in the determination of the unisolvency of a given problem and in the construction of the solution ([7, 8] and included references).

We finish giving in the following tables the configurations $\mathbb{P}_m$-admissible with points in a particular set P for low values of m, and some $\mathbb{P}_m$-unisolvent classes.

Table 1. $P = \{(0,0), (1,0), (1,1), (0,1), (-1,0), (-1,-1), (0,-1)\}$ and configurations for $\mathbb{P}_m$, $m = 2, 3, 4$.

Space of polynomials	Types of configurations	Number of configurations	Number of classes
$\mathbb{P}_2$	$(1,6,0)$	7	2
	$(3,3,1)$	140	15
	$(5,0,2)$	21	4
$\mathbb{P}_3$	$(1,4,2)$	105	12
	$(3,1,3)$	140	15
$\mathbb{P}_4$	$(0,3,4)$	35	6
	$(2,0,5)$	21	4

Acknowledgment

This work was partially supported by Junta de Andalucía, Grupo de Investigación 1107.

Table 2. $P = [\{(-1,1), (1,1), (1,-1), (-1,-1)\}] \cap \mathbb{Z}^2$. Results about types of configurations $\mathbb{P}_5$-admissible with points in P and their number and classes.

Types of configurations	Number of configurations	Number of classes
$(0,3,6)$	84	16
$(2,0,7)$	36	8

Table 3. Representatives of equivalency classes of $\mathbb{P}_5$-unisolvent types.

Types of configurations	Representantive configurations of $\mathbb{P}_5$-unisolvent classes
(2.0.7)	1 1 111 111 11 11 1 1
$(0,3,6)$	001 010 011 101 101 101 100 001 111 111 111 111 001 001 001 001 110 111 111 111 111 011 101 110 010 010 011 101 111 111 110 010 011 101 101 111

Table 4. $P = [\{(-2,0), (-2,-2), (0,-2), (2,0), (2,2), (0,2)\}] \cap \mathbb{Z}^2$ and configurations for $\mathbb{P}_m$, $m = 6,7,8$.

Space of polynomials	Types of configurations	Number of configurations	Number of classes
$\mathbb{P}_6$	$(0,13,5)$	162792	13748
	$(3,10,6)$	7759752	648203
	$(5,7,7)$	39907296	3329248
	$(7,4,8)$	24942060	2081515
	$(9,1,9)$	923780	77440
$\mathbb{P}_7$	$(1,9,9)$	923780	77440
	$(3,6,10)$	7759752	648203
	$(5,3,11)$	4232592	353808
	$(7,0,12)$	50388	4356
$\mathbb{P}_8$	$(6,0,13)$	27132	2376
	$(2,3,14)$	116280	9871
	$(4,0,15)$	3876	364

Table 5. Some $\mathbb{P}_m$-unisolvent configurations, $m = 6, 7, 8$.

$m=6$	$m=7$	$m=8$
0	1 1 1	1 1
1 1 0	1 0	1 1 1 1
0 1 1	1 0 1 1 0	1 1 0 0 1
1 1 1 0	1 1 0 1	1 1 0
1	0 0	1 1 1

References

1. Barrera Rosillo, D.: Esquemas locales de interpolación de Lagrange y de Hermite. Su extensión a varias variables. Tesis doctoral, Universidad de Granada (in preparation)
2. Bojanov, B. D., Hakopian, H. A., Sahakian, A. A.: Spline Functions and Multivariate Interpolations. Kluwer Academic Publishers (1993)
3. Ciarlet, P. G.: Interpolation error estimates for the reduced Hsieh-Clough-Tocher triangles. Math. Comp. **32** (1978) 335–344
4. Dahmen, W., Goodman, T. N. T., Micchelli, C. A.: Local Spline Interpolation Schemes in One and Several Variables. Approximation & Optimization, Proceedings Havana 1987, ed. A. Gómez, F. Guerra, M.A. Jiménez, G. López, LNM **1354**. Springer Verlag (1987)
5. Laghchim-Lahlou, M.: Elements Finis Composites de Classe C^k dans $\mathbb{R}^2$. Thèse, I.N.S.A. de Rennes (1991)
6. Laghchim-Lahlou, M., Sablonnière, P.: Triangular Finite Elements of HCT type and class C^ρ. Advances in Mathematics **2** (1994) 101–122
7. Lorente Pardo, J.: Funciones Splines: Interpolación Conservativa. Interpolación Multivariada. Tesis doctoral, Universidad de Granada (1987)
8. Lorente Pardo, J., Ramírez, V.: On Interpolation Systems and H-Reducible Interpolation Problems. Topics in Multivariate Approximation, ed. C. K. Chui, L. L. Schumaker, F. I. Utreras. Academic Press (1987)
9. Lorentz, R. A.: Multivariate Birkhoff Interpolation. LNM **1516**. Springer Verlag (1992)
10. Van Lint , J. H., Wilson, R. M.: A Course in Combinatorics. Cambridge University Press (1992)

The Effectiveness of Band-Toeplitz Preconditioners: A Survey

Stefano Serra

Dipartimento di Energetica, Via Lombroso 6/17, 50100 Firenze, Italy

Abstract. In many applications such as signal processing [26], differential equations [40], image restoration [20] and statistics [22] we have to solve $n \times n$ Hermitian Toeplitz linear systems $A_n(f)\mathbf{x} = \mathbf{b}$ where the symbol f, *the generating function*, is an L^1 function and the entries of $A_n(f)$ along the k-th diagonal coincide with the k-th Fourier coefficient a_k of f. When the generating function has (essential) zeros of even orders and is nonnegative, several very satisfactory band-Toeplitz preconditioners have been proposed [9, 18, 12, 32] leading to optimal and superoptimal preconditioned conjugate gradient (PCG) methods with a total cost of $O(n \log n)$ ops, even in the ill-conditioned case where other celebrated techniques fail [17]. More recently these preconditioners have been successfully extended to the block [27] and nondefinite [29] cases as well as to the case of zeros of any (also noninteger) orders [34]. The latter extension is very important because it concerns with a lot of practical situations not considered before. Therefore, the only relevant criticism to this approach continues to be the assumption that, at least, the sign of f, the position of the zeros and their order are known. In fact, in some applications, it is possible to assume the availability of this information but, in other fields, e.g. image restoration, this just results an ideal hypothesis. In the latter case, very recently, we have introduced [38] a practical and economical criterion in order to discover all the needed analytical properties of f by using as data input the coefficients a_k. In such way, we are convinced that the set of all the band-Toeplitz preconditioners based strategies can become much more attractive from an applicative point of view.

1 Introduction

The aim of this paper is to present and analyse new strategies for the solution by PCG method of $n \times n$ symmetric Toeplitz systems [22] $A_n(f)\mathbf{x} = \mathbf{b}$. Toeplitz matrices are assumed to be generated by 2π-periodic integrable real-valued functions f defined on the fundamental interval $[-\pi, \pi]$, in the sense that the entry of $A_n(f)$ along the k-th diagonal is given by the k-th Fourier coefficient a_k of f:

$$a_k = \frac{1}{2\pi} \int_{-\pi}^{\pi} f(x) e^{-\mathbf{i}kx} dx, \quad \mathbf{i}^2 = -1, \quad k \in \mathbb{Z} \ .$$

In some applications the function f is explicitly given or can be easily obtained as in the case of Finite Difference discretization of constant coefficient elliptic operators on rectangular domains. In other ones, we know that a "generating" function exists but we have not its analytic expression and/or we have not information about important properties such as the presence of zeros, their localisation, their orders etc. This is, for instance, the case of applications like signal processing, image restoration, linear prediction of a stochastic stationary process.

Since the dimension of these applicative problems are often very large it is not convenient to use general purpose solution techniques. Actually, in the relevant literature, several specialised direct methods have been proposed. For instance, for the positive definite case, the so called *superfast methods* require $O(n \log^2 n)$ arithmetic operations (ops) [1, 15]; for the nondefinite case the technique of T.Chan and P.Hansen [14] is a *fast method* and needs $O(n^2)$ ops. On the other hand, the iterative procedures have two main advantages: they are easily parallelizable and are, for their intrinsical nature, more flexible in the sense that they can be finely adapted to specific problems (we are thinking to the prolongation operators in the multigrid techniques or to the sophisticated idea of the preconditioning).

When the generating function f is (essentially) strictly positive, then the best iterative methods are the preconditioned conjugate gradient (PCG) ones based on matrix algebra preconditioners [11, 8, 13, 6, 7]: they lead to superoptimal methods having a total cost of $O(n \log n)$ ops. For completeness, we have to say that this positive case is the easiest and, actually, even a relaxed Jacobi method costs asymptotically only $O(n \log n)$ ops. A more complicated problem is encountered if f is essentially nonnegative with the essential infimum m_f equal to zero. Under this assumption F.Di Benedetto has proved [17] that circulant and Hartley preconditioners do not work and, therefore, the only optimal PCG methods are those based on band-Toeplitz (BT) preconditioners [9, 18] and τ preconditioners [17, 18, 33]. Anyway, we recall that the proof of the good behaviour of the τ PCG method is strongly depending on some BT preconditioners. In addition, the band Toeplitz preconditioning idea has been successfully applied to the nondefinite case [29, 34, 37] and to the block case [27]. We observe that the technique introduced in [29] is the only PCG method for nondefinite Toeplitz systems with nondefinite generating functions having zeros (see also [25, 10]), while the (double) BT PCG method for block Toeplitz systems devised in [27] is the only optimal iterative algorithm for the ill-conditioned positive definite case. Moreover, by means of approximation theory tools, the basic method has been substantially enriched and, under regularity assumption on f, some very fast [12, 32] and superlinear [32] PCG methods based on BT preconditioners have been proposed. It is useful to recall that the total cost remains of $O(n \log n)$ ops that is the asymptotic cost of a fast Fourier transform (FFT) [39]. Anyway, even if the BT preconditioning seems to be the most versatile idea, it must be stressed that, in all the cited papers, two somehow heavy hypotheses have been implicitly or explicitly assumed:

H1 The generating function is known.
H2 f has only zeros of "even orders".

More recently, these restrictions have been substantially overcome. In the paper [34], by exploiting new properties about the extreme eigenvalues of BT preconditioned matrices, the BT PCG technique is successfully extended to the case where the generating function has zeros of any (also noninteger) order. The resulting iterative procedure is fast even if it is not optimal in the sense stated, for instance, in [3].

Finally, in [38], an economical technique is discussed in order to obtain information about f, when only the Fourier coefficients that is the entries of $A_n(f)$ are assumed to be known. It is shown how to discover the sign of f, the position of the possible zeros and their approximated orders. We recall that this last information is crucial to understand the order of the asymptotic ill-conditioning of $A_n(f)$ [24, 30, 31] and, above all, in order to practically construct the BT preconditioner [9, 18, 12, 32, 34].

The paper is organised as follows. In section 2 we first discuss the main strategies to handle the nonnegative case with zeros of even orders, then we relax this assumption by considering the case of one zero of any order. Section 2 is also accompanied by some explanatory numerical experiments.

2 The Fast Preconditioning

Here, by exploiting the correspondence between Toeplitz operator theory and function theory, we consider the *ambitious* aim of devising optimal and superoptimal PCG methods for the ill-conditioned case. Actually we recall that if f is nonnegative and has zeros of orders α_i then the Euclidean condition number of $A_n(f)$ grows asymptotically as n^α with $\alpha = \max \alpha_i$ (see for instance [24, 9, 30, 31]). As said before, we will use a correspondence between Toeplitz operators and related generating functions: this idea is well resumed in the following theorems which relates the spectrum of Toeplitz matrices, the spectrum of a special class of preconditioned Toeplitz matrices and the analytical properties of the generating functions.

Theorem 1. *[22, 9] Let m_f and M_f be the essinf and the esssup of f in $[-\pi, \pi]$. If $m_f < M_f$ then $\forall n > 0$ we have*

$$m_f < \lambda_i(A_n(f)) < M_f$$

where $\lambda_i(X)$ is the i–th eigenvalue of X arranged in nondecreasing order. If $m_f \geq 0$ then $A_n(f)$ is positive definite.

Theorem 2. *[18, 28] Let $f, g \in L^1[-\pi, \pi]$ functions essentially nonnegative, i.e., $m_f, m_g \geq 0$. The matrices $A_n(f), A_n(g)$ are positive definite (see Theorem (1)) and the eigenvalues λ_i^n of $P_n(f; g) = A_n^{-1}(g)A_n(f)$ arranged in nondecreasing order are such that:*

1. *$\lambda_i^n \in (r, R)$, r, R being the essinf and the esssup of f/g, respectively.*

2. $\bigcup_{n\in\mathbf{N}}\bigcup_{i\leq n}\lambda_i^n$ *is dense in the "essential range"* $\mathcal{ER}(f/g)$ *of* f/g *(the essential range of an integrable function* h *defined on* I *is the set of all* y *real numbers for which,* $\forall\epsilon>0$ *the Lebesgue measure of* $\{x\in I : h(x)\in(y-\epsilon,y+\epsilon)\}$ *is positive [28]).*
3. $\lim_{n\to\infty}\lambda_1^n=r,\quad \lim_{n\to\infty}\lambda_n^n=R.$

In order to have an optimal preconditioning we have to satisfy the following conditions:

1. The systems related to the preconditioning matrix $A_n(g)$ must be "easy" to solve.
2. The preconditioned matrix should have the spectrum contained in a positive interval well separated from zero.

In order to meet the first requirement, we choose g in the class of trigonometric polynomials and therefore $A_n(g)$ is a BT preconditioner [9, 18]: the associated systems can be solved in a sequential model of computation in $O(n)$ ops [21] and, in a parallel model, in $O(\log n)$ ops [4, 19].

The second requirement is just a simple function theory problem: we have to find a trigonometric polynomial g which matches with f the same zeros with their orders. In this way, we directly obtain that $0<r\leq R<\infty$ and therefore, in the light of theorem 2, we fulfil the condition 2.

The strong limitation of this theoretical tool is that g must be, at the same time, a nonnegative polynomial and a function with the same zeros of f. By exploiting the regularity features of g, we deduce that g can have only zeros of *even* orders (!). Consequently, we can directly handle only the case where f is nonnegative and has zeros $x_1,\ldots,x_q$ of even orders $2k_1,\ldots,2k_q$. However, even with this limitation, this is a nontrivial result since the proposed technique is the most economical by requiring only $O(n\log n)$ ops and $O(\log n)$ parallel steps. The best direct methods use $O(n\log^2 n)$ ops and are structurally sequential while the circulant and Hartley PCG methods [11, 7] are, in this case, not optimal [17].

Within this framework, the technique has been developed very much. The main tool is the use of some theorems derived by the approximation theory [23]: this approach is very natural since the original linear algebra problem has been transformed into a function theory problem. By simplifying a little, the main approximation strategies used in the literature can be reduced to the following ones:

S1 the basic BT preconditioner $A_n(g_1)$ [9, 18];
S2 the optimal BT preconditioner $A_n(g_2)$ [12];
S3 the quasi optimal BT preconditioner $A_n(g_3)$ [32];
S4 the superlinear BT preconditioner $A_n(g_4)$ [32].

We briefly recall that $g_1=\prod_{j=1}^{q}(2-2cos(x-x_j))^{2s_j}$ where $s_j=k_j$ if $x_j\neq 0$ and $s_j=k_j/2$ otherwise. In 1994, R.Chan and P.Tang [12] have proposed to increase the bandwidth of $A_n(g_1)$ to get extra degrees of freedom. They calculate g_2 by means of the Remez algorithm by minimising $h=\|(f-g_2)/f\|_\infty$ over all the polynomials g of fixed degree l. We observe that this minimisation property enables one not only to match the zeros of f but also to minimise R/r obtaining (by using Theorem 3.1 in [18]) the best BT preconditioner in the class of all the BT matrices of fixed bandwidth $2l+1$.

Since the Remez algorithm can be heavy from a computational point of view, in [32] we proposed new techniques to minimise "in a certain sense" $(f-g)/f$. For instance, g_3 is easily calculated as $g_1 * g$ where g interpolates the function f/g_1 at the Chebychev zeros. This calculation can be done in $O(l \log l)$ ops and $O(\log l)$ parallel steps, by using only few fast Fourier transforms (FFT) [39] of order $l-k$ where l is the degree of g_3. The low cost of the computation of the coefficients of the third preconditioner suggest us, very naturally, to increase the bandwidth l as function of n. In particular, in [32], it is shown that the choice of $l = \log n$, which leads to the preconditioner $A_n(g_4)$, obtains two distinct aims: the total cost of the procedure is of $O(n \log n)$ ops, but the convergence speed of the PCG method is superlinear (superoptimal iterative technique!).

The next section gives a very impressive numerical evidence of the discussed ideas.

3 Numerical Results

In this section, we compare the convergence rate of the basic BT preconditioner [9, 18] with the quasi optimal BT preconditioner [32], the optimal BT preconditioner [12] and with the optimal circulant preconditioner [13] on two different generating functions having zeros. They are $(x-1)^2(x+1)^2$ and $1-e^{-x^2}$ and are associated to ill-conditioned matrices A_n having Euclidean condition numbers equal to $O(n^2)$ (see for instance [30, 31]). The matrices A_n are formed by evaluating the Fourier coefficients of the generating functions by using FFTs (see [12]). In the considered tests, the vector of all ones is the right-hand side vector, the zero vector is the initial guess and the stopping criterion is $\|r_q\|_2/\|r_0\|_2 \leq 10^{-7}$, where r_q is the residual vector after q iterations. All computations are done by using Matlab.

In the subsequent tables, I denotes that no preconditioning is used, C is the T.Chan optimal circulant preconditioner [13], $P^*_{n,l}$ is the optimal BT preconditioner [12] and $P_{n,l}$ is the BT preconditioner defined according to S3; here l denotes the half-bandwidth of the band preconditioners. We observe that the "optimal" S2 and the S3 BT PCG methods perform, substantially, in the same way, but the second one is much more economical with respect to the computation of the related generating function. This fact is not so considerable when the bandwidth is fixed, but it becomes crucial in order to increase l, say, as $\log n$. Actually, in the case S4, for any dimension n, it is not expensive to calculate a different preconditioner, since the related cost $O(\log n \log(\log n))$ is strongly dominated by the cost $O(n \log n)$ of each PCG iteration.

Finally, the reduction of the number of required iterations, as the dimension increases, shown in table 2 gives a numerical evidence of the superlinear convergence claimed in [32]. We stress that the exceptional convergence behaviour of the PCG algorithm related to $P_{n,6}$ (table 2, last column) is explained by the good approximation properties of the first-kind Chebyshev interpolation: to have a practical measure of this, it is sufficient to notice that the reduction of the condition number from A_n to $(P_{n,6})^{-1}A_n$, for $n = 512$, is from $2.7 * 10^4$ to $1 + 5 * 10^{-4}$.

If we want to solve a linear system of the form $A_n(f)\mathbf{x} = \mathbf{b}$ where f has zeros of any order ρ, then, in the light of theorem 2, it follows that is is impossible to find optimal BT preconditioners because, if there exists a zero of noneven order, then we have $r = 0$ or $R = \infty$. Surprisingly enough, the BT preconditioners result effective even in this case as strictly proved in [34]. Here we show only a numerical test (table 3) which confirms the foresight of the theoretical analysis given in [34].

Table 1. $f(x) = (x^2 - 1)^2$

n	I	C	$A_n(g_1)$	$P^*_{n,4}$ $P_{n,4}$	$P^*_{n,5}$ $P_{n,5}$	$P^*_{n,6}$ $P_{n,6}$
128	193	22	18	11 11	8 8	7 7
256	465	28	19	11 11	8 9	7 7
512	> 1000	34	19	11 11	8 8	7 7

Table 2. $f(x) = 1 - e^{-x^2}$, superlinear PCG, Prec= $P_{n,l(n)} = A_n(g_4)$, $l(n) = \log(n) - 2$

n	16	32	64	128	256	512
$l(n)$	2	3	4	5	6	7
Iter	9	7	5	3	2	2

Table 3. $f(x) = x^2|x|^{\frac{1}{10}}$, $\rho = 2.1$, $2k = 2$, Prec $= A_n(g_1)$, $g_1(x) = 2 - 2\cos(x)$

n	128	256	512
I_n	86	187	397
$A_n(g_1)$	20	22	22

References

1. Ammar, G., Gragg, W.: Superfast solution of real positive definite Toeplitz systems. SIAM J. Matr. Anal. Appl. **9** (1988) 61–76
2. Axelsson, O., Lindskög, G.: The rate of convergence of the preconditioned conjugate gradient method. Num. Math. **52** (1986) 499–523
3. Axelsson, O.,Neytcheva, M.: The algebraic multilevel iteration methods - theory and applications. Proc. of the 2nd Int. Coll. on Numerical Analysis, D. Bainov Ed., Plovdiv (Bulgaria), august 1993, 13–23
4. Bini, D.: Matrix structure in parallel matrix computation. Calcolo **25** (1988) 37–51
5. Bini, D., Capovani, M.: Spectral and computational properties of band symmetric Toeplitz matrices. Lin. Alg. Appl. **52/53** (1983) 99–126
6. Bini, D., Di Benedetto, F.: A new preconditioner for the parallel solution of positive definite Toeplitz linear systems. Proc. 2nd SPAA conf., Crete (Greece), july 1990, 220–223
7. Bini, D., Favati, P.: On a matrix algebra related to the discrete Hartley transform. SIAM J. Matr. Anal. Appl. **14** (1993) 500–507
8. Chan, R.H.: Circulant preconditioners for Hermitian Toeplitz systems. SIAM J. Matr. Anal. Appl. **10** (1989) 542–550
9. Chan, R.H.: Toeplitz preconditioners for Toeplitz systems with nonnegative generating functions. IMA J. Numer. Anal. **11** (1991) 333–345
10. Chan, R.H., Ching, W.: Toeplitz–circulant preconditioners for Toeplitz systems and their applications to queueing network with batch arrivals. SIAM J. Sci. Comp. (to appear)
11. Chan, R.H., Strang, G.: Toeplitz equations by conjugate gradients with circulant preconditioner. SIAM J. Sci. Stat. Comp. **10** (1989) 104–119
12. Chan, R.H., Tang, P.: Fast band–Toeplitz preconditioners for Hermitian Toeplitz systems. SIAM J. Sci. Comp **15** (1994) 164–171
13. Chan, T.F.: An optimal circulant preconditioner for Toeplitz systems. SIAM J. Sci. Stat. Comp. **9** (1988) 766–771
14. Chan, T.F., Hansen P.C.: A look–ahead Levinson algorithm for indefinite Toeplitz systems. SIAM J. Matr. Anal. Appl. **13** (1992) 491–506
15. De Hoog, F.: On the solution of Toeplitz systems. Lin. Alg. Appl. **88** (1987) 123–138
16. Di Benedetto, F.: Iterative solution of Toeplitz systems by preconditioning with discrete sine transform. Proc. in Advanced Signal Processing Algorithms, Architectures, and Implementations - SPIE conference, F. Luk Ed., San Diego (CA), july 1995 302–312
17. Di Benedetto, F.: Analysis of preconditioning techniques for ill–conditioned Toeplitz matrices. SIAM J. Sci. Comp. **16** (1995) 682–697
18. Di Benedetto, F., Fiorentino, G., Serra, S.: C.G. Preconditioning for Toeplitz Matrices. Comp. Math. Applic **6** (1993) 35–45
19. Fiorentino, G., Serra, S.: Multigrid methods for Toeplitz matrices. Calcolo **28** (1991) 283–305
20. Gohberg,I., I. Feldman, I.: Convolution Equations and Projection Methods for Their Solution. Trans. of Math. Mon. **41** American Mathematical Society, Providence, Rhode Island 1974
21. Golub, G.H., Van Loan, C.F.: Matrix Computations. The Johns Hopkins University Press, Baltimore 1983

22. Grenander, U., Szegö, G.: Toeplitz Forms and Their Applications. Second Edition, Chelsea, New York 1984
23. Jackson, D.: The Theory of Approximation. American Mathematical Society, New York 1930
24. Kac, M., Murdoch, W., Szegö, G.: On the extreme eigenvalues of certain Hermitian forms. J. Rat. Mech. Anal. **13** (1953) 767–800
25. Ku, T., Kuo, C.: Spectral properties of preconditioned rational Toeplitz matrices: the nonsymmetric case. SIAM J. Matr. Anal. Appl. **14** (1993) 521–542
26. Oppenheim, A.: Applications of Digital Signal Processing. Prentice–Hall, Englewood Cliffs 1978
27. Serra, S.: Preconditioning strategies for asymptotically ill–conditioned block Toeplitz systems. BIT **34** (1994) 579–594
28. Serra, S.: Conditioning and solution, by means of preconditioned conjugate gradient methods, of Hermitian (block) Toeplitz systems. Proc. in Advanced Signal Processing Algorithms, Architectures, and Implementations - SPIE conference, F. Luk Ed., San Diego (CA), july 1995 326–337
29. Serra, S.: Preconditioning strategies for Hermitian Toeplitz systems with nondefinite generating functions. SIAM J. Matr. Anal. Appl. **17-4** (1996) (in press)
30. Serra, S.: On the extreme eigenvalues of Hermitian (block) Toeplitz matrices. Lin. Alg. Appl. (to appear)
31. Serra, S.: On the extreme spectral properties of Toeplitz matrices generated by L^1 functions with several minima (maxima). BIT **36** (1996) 135–142
32. Serra, S.: Optimal, quasi-optimal and superlinear band-Toeplitz preconditioners for asymptotically ill-conditioned positive definite Toeplitz systems. Math. Comp. (1997) (in press)
33. Serra, S.: Superlinear PCG methods for symmetric Toeplitz systems. (Submitted)
34. Serra, S.: New PCG based algorithms for the solution of Hermitian Toeplitz systems. Calcolo (to appear)
35. Serra, S.: On the conditioning and solution, by means of multigrid methods, of symmetric (block) Toeplitz systems. Proc. 5th Int. Coll. on Differential Equations, D. Bainov Ed., Plovdiv (Bulgaria), august 1995 249–256
36. Serra, S.: Sulle proprietà spettrali di matrici precondizionate di Toeplitz. Boll. Un. Mat. Ital. (in press)
37. Serra, S.: The extension of the concept of *generating* function to a class of preconditioned Toeplitz matrices. (Submitted)
38. Serra, S.: A practical algorithm to design fast and optimal band-Toeplitz preconditioners for Hermitian Toeplitz systems. (In preparation)
39. Van Loan, C.: Computational Frameworks for the Fast Fourier Transform. SIAM, Philadelphia 1992
40. Varga, R.S.: Matrix Iterative Analysis. Prentice Hall, Englewood Cliffs 1962

Grid Approximations of the Solution and Diffusion Flux for Singularly Perturbed Equations with Neumann Boundary Conditions *

Grigorii I. Shishkin

Institute of Mathematics and Mechanics, Ural Branch of Russian Acad. Sci., 16 S.Kovalevskoi Street, 620219 Ekaterinburg, Russia, sgi@eqmph.imm.intec.ru

Abstract. Neumann problems for singularly perturbed parabolic equations are considered on a segment and on a rectangle. The second-order derivatives are multiplyed by a small parameter ε^2. When $\varepsilon = 0$, the parabolic equation degenerates, and only the time derivative remains. The normalized diffusion flux, i.e., the product of ε and the derivative in the direction of normal, is given on the boundary. The solution of a classical discretization method on a uniform grid does not converge ε-uniformly. Moreover, we show with numerical examples that, in the case of a Neumann problem, the approximate solution and, thereupon, the discretization error may increase without bound for a vanishing ε. The error can exceed the real solution many times for small ε. For the solution of the boundary value problems new special finite difference schemes are constructed. These schemes allow us to approximate the solution and the normalized diffusion fluxes ε-uniformly.

1 Introduction

Let the diffusion flux of substance in a solid material (sample) be described by Fik's law, and the process of substance diffusion satisfy the diffusion equation. Suppose, the substance concentration in the material at an initial instant as well as the diffusion flux on the boundary at the current moment are known. It is required to find the distribution of substance and the diffusion fluxes in the material at the current moment.

Note that the diffusion Fourier number, which is proportional to the diffusion coefficient of substance, can be sufficiently small that cause the large variation of concentration with depth of the sample. For small values of Fourier number, the diffusion boundary layer appear in the narrow neighbourhood of the sample boundary. These problems are said to be singularly perturbed. For such problems the perturbation parameter is a small coefficient multiplying the highest derivatives of the differential equation.

* This work was supported by the Russian Foundation of Basic Research under Grant N 95-01-00039a

Difficulties arise when it is required to find the solution of the Dirichlet boundary value problem (see, e.g., [1–5]). More significant difficulties arise for the Neuman boundary value problem when it is required to find the space derivatives (or the diffusion fluxes) of the solution as well as the solution of the boundary value problem itself. Using a simple example, we shall illustrate the problems that appear with numerical solving.

Let it be required to find the solution of the singularly perturbed diffusion equation[1]

$$L_{(1)}u(x,t) \equiv \varepsilon^2 \frac{\partial^2}{\partial x^2}u(x,t) - \frac{\partial}{\partial t}u(x,t) = 0, \quad (x,t) \in G\ , \tag{1}$$

$$\varepsilon\frac{\partial}{\partial n}u(x,t) = \psi(t),\ x = 0, \quad \varepsilon\frac{\partial}{\partial n}u(x,t) = 0,\ x = 1, \quad (x,t) \in S^L\ ,$$

$$u(x,t) = 0, \quad (x,t) \in S^0\ .$$

Here $G = D \times (0,T]$, $D = (0,1)$, $T = 1$, $S^0 = \overline{D} \times \{t = 0\}$, $S = \overline{G} \setminus G$, $S = S^0 \cup S^L$, $\partial/\partial n$ is the derivative in the direction of the outward normal to the set $\overline{G}$; $\psi(t) = -8/3\pi^{-1/2}t^{3/2}$; the parameter ε takes arbitrary values in the interval $(0,1]$. The solution of this problem is bounded ε-uniformly. Note that, if the parameter ε does not enter into the boundary condition, no boundary layer appears in practice.

To solve problem (1) numerically, we use a classical finite difference scheme. In the set $\overline{G}$, the grid $\overline{G}_h = \overline{D}_h \times \overline{\omega}_0$ is introduced. Here $\overline{D}_h$ and $\overline{\omega}_0$ are uniform grids, respectively, on the segment $\overline{D}$ with a grid step $h = N^{-1}$ and on the segment $[0,T]$ with a grid step $h_0 = TN_0^{-1}$; $N+1$ and N_0+1 are the number of nodes in the grids $\overline{D}_h$ and $\overline{\omega}_0$ respectively. For problem (1) we apply the following difference scheme

$$\Lambda_{(2)}z(x,t) \equiv \varepsilon^2\delta_{x\overline{x}}z(x,t) - \delta_{\overline{t}}z(x,t) = 0, \quad (x,t) \in G_h\ , \tag{2}$$

$$\varepsilon\delta_x z(0,t) = \psi(t), \quad \varepsilon\delta_{\overline{x}}z(1,t) = 0,\ t \in \omega_0, \quad z(x,0) = 0,\ x \in \overline{D}_h\ ,$$

Here $\delta_{x\overline{x}}z(x,t)$ and $\delta_x z(x,t)$, $\delta_{\overline{x}}z(x,t)$, $\delta_{\overline{t}}z(x,t)$ are respectively the second central difference derivative and the first one-sided difference derivatives (straightforwarded and backforwarded).

The solution of problem (1) has the form $u(x,t) = W(x,t)+v(x,t)$, $(x,t) \in \overline{G}$, where

$$W(x,t) = \operatorname{erfc}\left(\frac{x}{2\varepsilon\sqrt{t}}\right)\left(\frac{x^4}{12\varepsilon^4} + \frac{x^2}{\varepsilon^2}t + t^2\right) - \\ -\frac{1}{\sqrt{\pi}}\exp\left(-\frac{x^2}{4\varepsilon^2 t}\right)\left(\frac{x^3}{6\varepsilon^3}t^{1/2} + \frac{5x}{3\varepsilon}t^{3/2}\right)\ ,$$

is the principal term in the singular part of the solution, $v(x,t)$ is the remainder term, $|W(x,t)| \le M$, $v(x,t)| \le M\varepsilon^n$, $(x,t) \in \overline{G}$, n is an arbitrary sufficiently

[1] The subscript number (within brackets) for a symbol denotes the equation in which the symbol is defined.

large number. The function $P(x,t) = \varepsilon(\partial/\partial x)u(x,t)$, $(x,t) \in \overline{G}$ (see (5)) called the normalized diffusion flux is bounded ε-uniformly: $|P(x,t)| \le M$, $(x,t) \in \overline{G}$. By M (m) we denote sufficiently large (small) positive constants which are independent of the parameter ε and the distribution of nodes in the used grids.

Using the solutions of the difference scheme, we found the values

$$E(\varepsilon, N) = \max_{\overline{G}_h} |\, u(x,t) - z(x,t)\, |, \quad Q(\varepsilon, N) = \max_{\overline{G}_h^-} |\, P(x,t) - P^{h+}(x,t)\, | \ ,$$

which are respectively errors in the approximate solution and the computed normalized flux for various values of ε and N, with $N_0 = N$ and $T = 1$, where $P^{h+}(x,t) \equiv \varepsilon \delta_x\, z(x,t)$, $\overline{G}_h^- = \{\, D_h \cup \{x = 0\}\,\} \times \overline{\omega}_0$.

In Tables 1 and 2 we give the values of $E(\varepsilon, N)$, $\overline{E}(N)$ and $Q(\varepsilon, N)$, $\overline{Q}(N)$ computed with using the uniform grid $\overline{G}_h$ for the given values of ε and $N = N_0$, $\varepsilon = 2^{-12} - 1.0$, $N = 4 - 1024$. Here

$$\overline{E}(N) = \max_{\varepsilon} E(\varepsilon, N), \quad \overline{Q}(N) = \max_{\varepsilon} Q(\varepsilon, N), \quad \varepsilon = 2^{-12} - 1.0 \ .$$

Table 1. Errors of the solution $E(\varepsilon, N)$, $\overline{E}(N)$ for classical scheme (2)

$\varepsilon \setminus N$	4	16	64	256	1024
1.0	4.736e−01	9.723e−02	2.332e−02	5.771e−03	1.439e−03
2^{-2}	1.098e+00	2.347e−01	5.581e−02	1.377e−02	3.431e−03
2^{-4}	5.213e+00	1.014e+00	2.114e−01	4.994e−02	1.230e−02
2^{-6}	2.312e+01	5.177e+00	9.935e−01	2.056e−01	4.848e−02
2^{-8}	9.530e+01	2.311e+01	5.169e+00	9.884e−01	2.041e−01
2^{-10}	3.842e+02	9.530e+01	2.311e+01	5.167e+00	9.871e−01
2^{-12}	1.540e+03	3.842e+02	9.530e+01	2.311e+01	5.166e+00
$\overline{E}(N)$	1.540e+03	3.842e+02	9.530e+01	2.311e+01	5.166e+00

From Tables 1, 2 we can see that the solution of difference scheme (2) for $N_0 = N$ and also the normalized flux converge for a fixed value of the parameter. However, these do not converge uniformly with respect to the parameter ε or, in short, ε-uniformly. The ε-uniform convergence means that the error should be independent of ε. Here the error in the approximate solution unboundedly increases for $N \to \infty$ when $\varepsilon = \varepsilon(N)$, $\varepsilon(N)N \to 0$.

So, the solutions of classical discretization methods do not approximate the solution of the boundary value problem ε-uniformly, even roughly in qualitative terms. Thus, for singularly perturbed problems with Neumann boundary conditions the development of special numerical methods is necessary.

Table 2. Errors of the normalized flux $Q(\varepsilon, N)$, $\overline{Q}(N)$ for classical scheme (2)

$\varepsilon \setminus N$	4	16	64	256	1024
1.0	1.675e−01	3.658e−02	8.844e−03	2.193e−03	5.470e−04
2^{-2}	8.878e−02	1.728e−02	3.903e−03	9.482e−04	2.353e−04
2^{-4}	4.679e−02	5.094e−02	6.211e−03	1.095e−03	2.444e−04
2^{-6}	3.116e−03	3.844e−02	4.071e−02	3.421e−03	3.910e−04
2^{-8}	1.954e−04	2.531e−03	3.642e−02	3.813e−02	2.722e−03
2^{-10}	1.221e−05	1.585e−04	2.391e−03	3.591e−02	3.749e−02
2^{-12}	7.633e−07	9.911e−06	1.498e−04	2.357e−03	3.579e−02
$\overline{Q}(N)$	1.675e−01	5.094e−02	4.071e−02	3.813e−02	3.749e−02

2 Boundary Value Problem on a Segment. Classical Scheme

On a segment $D = \{x : \ 0 < x < d\}$ we consider the boundary value problem for the parabolic equation

$$L_{(3)}u(x,t) \equiv \left\{\varepsilon^2 a(x,t)\frac{\partial^2}{\partial x^2} - c(x,t) - p(x,t)\frac{\partial}{\partial t}\right\} u(x,t) = f(x,t), \quad (x,t) \in G \ ,$$

$$l_{(3)}u(x,t) \equiv \varepsilon\frac{\partial}{\partial n}u(x,t) = \psi(x,t), \quad (x,t) \in S^L \ , \tag{3}$$

$$u(x,t) = \varphi(x), \quad (x,t) \in S^0 \ .$$

Here

$$G = D \times (0,T], \quad S = S(G) = \overline{G} \setminus G, \quad S = S^0 \cup S^L \ , \tag{4}$$

S^0 and S^L are respectively lower and lateral sides of the boundary S; the functions $a(x,t)$, $c(x,t)$, $p(x,t)$, $f(x,t)$ and also the functions $\psi(x,t)$ and $\varphi(x)$ are sufficiently smooth functions on the sets $\overline{G}$ and S^1, S^0 respectively, moreover $a(x,t) \geq a_0$, $c(x,t) \geq 0$, $p(x,t) \geq p_0$, $(x,t) \in \overline{G}$, a_0, $p_0 > 0$; $\varepsilon \in (0,1]$.

The solution of the boundary value problem is a function $u \in C^{2,1}(G) \cap C^{1,0}(\overline{G})$ which satisfies the equation on G and the boundary condition on S.

Suppose that on the set $S^\star = \{(x,t) : \ x = 0, d, \ t = 0\}$ the compatibility conditions [6], which ensure the required smoothness of the solution of the problem for a fixed value of the parameter, are satisfied.

We wish to find the solution $u(x,t)$ and its gradient $(\partial/\partial x)u(x,t)$, $(x,t) \in \overline{G}$.

When the parameter tends to zero, the parabolic boundary layer appear in the neighbourhood of the set S^L. Herewith, the derivative $(\partial/\partial x)u(x,t)$ increases unboundedly in the neighbourhood of the boundary layer. Therefore it is convenient to consider, instead of the gradient $(\partial/\partial x)u(x,t)$, the product $\varepsilon(\partial/\partial x)u(x,t)$, $(x,t) \in \overline{G}$ which is bounded ε-uniformly. The value

$$P(x,t) = \varepsilon\frac{\partial}{\partial x}u(x,t), \quad (x,t) \in \overline{G} \tag{5}$$

shall be called the normalized diffusion flux (or shorter, the normalized flux or the flux). So, in the case of problem (3) the normalized diffusion flux is given on the lateral boundary S^L and it is required to find the functions $u(x,t)$, $P(x,t)$, $(x,t) \in \overline{G}$.

On the set $\overline{G}$ we introduce the grid

$$\overline{G}_h \equiv \overline{D}_h \times \overline{\omega}_0 = \overline{\omega}_1 \times \overline{\omega}_0 \ , \tag{6}$$

where $\overline{\omega}_1$ is a grid on the interval $[0,d]$ on the axis x, generally speaking, a nonuniform grid, $\overline{\omega}_0$ is a uniform grid on the interval $[0,T]$. Suppose $h^i = x^{i+1} - x^i$, x^i, $x^{i+1} \in \overline{\omega}_1$, $h = \max_i h^i$, $h_0 = TN_0^{-1}$. By $N+1$ and N_0+1 we denote the number of nodes in the grids $\overline{\omega}_1$ and $\overline{\omega}_0$ respectively; let $h \le MN^{-1}$. For problem (3) we use the difference scheme

$$\Lambda_{(7)} z(x,t) = f(x,t), \quad (x,t) \in G_h \ , \tag{7}$$

$$\lambda_{(7)} z(x,t) = \psi_{(7)}(x,t),\ (x,t) \in S_h^L, \quad z(x,t) = \varphi(x),\ (x,t) \in S_h^0 \ .$$

Here $G_h = G \cap \overline{G}_h, \quad S_h = S \cap \overline{G}_h,$

$$\Lambda_{(7)} z(x,t) \equiv \left\{ \varepsilon^2 a(x,t) \delta_{\overline{x}\widehat{x}} - c(x,t) - p(x,t)\delta_{\overline{t}} \right\} z(x,t) \ ,$$

$$\lambda_{(7)} z(x,t) \equiv \begin{cases} -\varepsilon \delta_x z(x,t) + 2^{-1}\varepsilon^{-1}(x^{i+1} - x^i) a^{-1}(x,t) \times \\ \qquad \times \left[p(x,t)\delta_{\overline{t}} z(x,t) + c(x,t) z(x,t) \right], \quad x = 0 \ , \\ \varepsilon \delta_{\overline{x}} z(x,t) + 2^{-1}\varepsilon^{-1}(x^i - x^{i-1}) a^{-1}(x,t) \times \\ \qquad \times \left[p(x,t)\delta_{\overline{t}} z(x,t) + c(x,t) z(x,t) \right], \quad x = d, \end{cases} \qquad x = x^i \ ,$$

$$\psi_{(7)}(x,t) = \psi_{(3)}(x) - \begin{cases} 2^{-1}\varepsilon^{-1}(x^{i+1} - x^i) a^{-1}(x,t) f(x,t),\ x = 0 \ , \\ 2^{-1}\varepsilon^{-1}(x^i - x^{i-1}) a^{-1}(x,t) f(x,t),\ x = d, \end{cases} \qquad x = x^i \ .$$

The finite difference scheme (7), (6) on a uniform grid approximates boundary value problem (3) with the second-order accuracy in the space variable and with the first order in the time variable.

Also we shall apply the difference scheme

$$\Lambda_{(7)} z(x,t) = f(x,t), \quad (x,t) \in G_h \ , \tag{8}$$

$$\lambda_{(8)} z(x,t) = \psi(x,t),\ (x,t) \in S_h^1, \quad z(x,t) = \varphi(x),\ (x,t) \in S_h^0 \ ,$$

where

$$\lambda_{(8)} z(x,t) \equiv \begin{cases} -\varepsilon \delta_x\, z(x,t),\ x = 0 \ , \\ \varepsilon \delta_{\overline{x}}\, z(x,t),\ \ x = d \ . \end{cases}$$

The difference operator $\lambda_{(8)}$ approximates the differential operator $l_{(3)}$ only with the first-order accuracy with respect to the space variable.

To approximate the function $P(x,t) = P_{(5)}(x,t)$, we use the grid function

$$P^{h+}(x,t) \equiv \varepsilon \delta_x z(x,t), \quad (x,t) \in \overline{G}_h^- \ , \tag{9}$$

where $\overline{G}_h^- = \overline{\omega}_1^- \times \overline{\omega}_0$ is a grid on $\overline{G}$, $\overline{\omega}_1^- = \omega_1 \cup \{x{=}0\}$.

The difference schemes (7), (6) and (8), (6) are monotonic [7] for arbitrary distribution of nodes in the grid $\overline{\omega}_1$ generating the grid $\overline{G}_{h(6)}$.

Using the maximum principle we establish the convergence of the difference scheme for a fixed value of the parameter

$$| u(x,t) - z(x,t) | \leq M \left[\varepsilon^{-1} N^{-1} + N_0^{-1} \right], \quad (x,t) \in \overline{G}_h \,. \tag{10}$$

In the case of the grid

$$\overline{G}_h = \left\{ \overline{G}_{h(6)}, \text{ where } \quad \overline{D}_h = \overline{\omega}_1 \quad \text{is an uniform grid} \right\} \tag{11}$$

for the solution of difference scheme (7), (6) we have the estimate

$$| u(x,t) - z(x,t) | \leq M \left[\varepsilon^{-2} N^{-2} + N_0^{-1} \right], \quad (x,t) \in \overline{G}_{h(11)} \,. \tag{12}$$

Using estimate (12) and a priori estimates similar to these given in [5,8,9], we attain the convergence of the computed diffusion flux to the exact one for a fixed value of the parameter. For difference scheme (7), (11) and difference schemes (7), (6) and (8), (6) we have the following estimates respectively

$$| P(x + 2^{-1}h, t) - P^{h+}(x,t) | \leq M \left[\varepsilon^{-2} N^{-2} + N_0^{-1} \right], \; (x,t) \in \overline{G}^{-}_{h(11)} \,, \tag{13}$$

$$| P(x,t) - P^{h+}(x,t) | \leq M \left[\varepsilon^{-1} N^{-1} + N_0^{-1} \right], \quad (x,t) \in \overline{G}^{-}_{h(6)} \,. \tag{14}$$

Theorem 1. *Let the solution of boundary value problem (3), (4) be sufficinetly smooth for a fixed value of the parameter, and, to solve the boundary value problem, classical finite difference schemes (7), (6); (8), (6) and (7), (11) are used. Then the solutions of the difference schemes and the computed diffusion fluxes converge to the solution of problem (3), (4) and the exact diffusion fluxes for a fixed value of the parameter. For the solutions of difference schemes (7), (6) and (8), (6) (scheme (7), (11)) estimates (10), (14) (estimates (12), (13)) are valid.*

3 The Motivation of Necessity to Construct Special Schemes

Let us describe difficulties that appear when problem (3) is solved by classical finite difference schemes for small values of the parameter.

For example, we consider boundary value problem (3) where

$$a(x,t) = c(x,t) = p(x,t) \equiv 1, \quad f(x,t) \equiv 0, \quad (x,t) \in \overline{G} \;; \tag{15}$$

$$\psi(x,t) = 1, \quad (x,t) \in S^1, \; t \geq t_1; \qquad \varphi(x) \equiv 0, \quad x \in \overline{D} \;,$$

when the related difference scheme (7), (11) is used. Here $\psi(x,t)$ is a nondecreasing function, t_1 is a sufficiently large number.

Taking into account the asymptotic behaviour (for $T \to \infty$ and $\varepsilon = \varepsilon(h)$, $\varepsilon h^{-1} \to O(1)$, $h \to 0$) of the solutions $u(x,t)$ and $z(x,t)$, and also the fluxes $P(x,t) = P_{(5)}(x,t)$ and $P^{h+}(x,t) = P^{h+}_{(9)}(x,t)$ we come to the relations

$$| u(0,T) - z(0,T) | \geq m, \quad | P(h,T) - P^{h+}(h,T) | \geq m \tag{16}$$

for any values $h \le h_0$, $\tau \le \tau_0$, $\varepsilon = O(h)$, i.e. the solution of difference scheme (7), (11) and the computed flux do not converge ε-uniformly for $h,\ \tau \to 0$. Here $u(x,t)$ and $z(x,t)$ are the solutions of problems (3), (15) and (7), (11), (15), $P(x,t)$ and $P^{h+}(x,t)$ are the normalized diffusion fluxes corresponding to these solutions, h_0, τ_0 are sufficiently small numbers. Moreover, we have

$$\max_{\overline{G}_h} |\, z(x,t)\,| \left[\max_{\overline{G}} |\, u(x,t)\,|\right]^{-1} \to 0 \ , \tag{17}$$

$$\max_{\overline{G}_h} |\, u(x,t) - z(x,t)\,| \to \max_{\overline{G}} |\, u(x,t)\,| = \max_{\overline{D}} |\, u(x,T)\,| \tag{18}$$

$$\text{for } h,\ \tau \to 0, \quad \varepsilon = o(h) \ .$$

So, for the functions $u(x,t)$ and $P(x,t)$ estimates (16)–(18) are fulfilled.

If difference scheme (8), (11) is used, we have the relations

$$|\, u(0,T) - z(0,T)\,| \ge m, \ |\, P(h,T) - P^{h+}(h,T)\,| \ge m \ \text{ for } \ \varepsilon = O(h) \ , \tag{19}$$

$$\max_{\overline{G}_h} |\, z(x,t)\,| \left[\max_{\overline{G}} |\, u(x,t)\,|\right]^{-1} \to \infty \ , \tag{20}$$

$$\max_{\overline{G}_h} |\, u(x,t) - z(x,t)\,| \ge |\, u(0,T) - z(0,T)\,| \to \infty \quad \text{for } \ \varepsilon = o(h) \tag{21}$$

as $h,\ \tau \to 0$. Thus, for the functions $u(x,t)$ and $P(x,t)$ estimates (19)–(21) are fulfilled. So we come to the following conclusions.

Theorem 2. *Let the functions $u(x,t)$ and $P(x,t)$, $(x,t) \in \overline{G}$ be the solution of boundary value problem (3) and the normalized diffusion flux. Then for N, $N_0 \to \infty$ the functions $z(x,t)$, $(x,t) \in \overline{G}_h$ and $P^{h+}(x,t)$, $(x,t) \in \overline{G}_h^-$, which are the solution of the classical difference scheme on the grid $\overline{G}_{h(6)}$ and the computed normalized diffusion flux, do not converge ε-uniformly to the functions $u(x,t)$ and $P(x,t)$.*

Theorem 3. *Let for the solution of boundary value problem (3), (15) the finite difference schemes (7), (11) and (8), (11) are used. Then for N, $N_0 \to \infty$ estimates (16)–(18) and (19)–(21) for the solutions of difference schemes (7), (11) and (8), (11) respectively and their computed normalized diffusion fluxes are fulfilled. For ratious of the values $\max_{\overline{G}_h} |\, z(x,t)\,|$ and $\max_{\overline{G}} |\, u(x,t)\,|$ estimate (17) for scheme (7), (11) and estimate (20) for scheme (8), (11) are valid.*

Thus, in the case of the singularly perturbed boundary value problem (3), (4) we arrive at the problem of working out special finite difference schemes which allow us to approximate both the solution of the boundary value problem and the normalized diffusion flux ε-uniformly.

4 Special ε-Uniformly Convergent Schemes for Problem (3)

On the set $\overline{G}$, we introduce a special grid condensing in the boundary layer similar to the grid introduced in [4,8,9]

$$\overline{G}_{h(22)} = \overline{G}_{h(22)}(\sigma) \equiv \overline{D}_h \times \overline{\omega}_0 = \overline{\omega}_1^* \times \overline{\omega}_0 \ , \tag{22}$$

where $\overline{\omega}_1^* = \overline{\omega}_1^*(\sigma)$ is a piecewise uniform grid on $[0, d]$, σ is a parameter depending on ε and N. Mesh steps of the grid $\overline{\omega}_1^*$ on the intervals $[0, \sigma]$, $[d - \sigma, d]$ and $[\sigma, d - \sigma]$ are constant and equal to $h^{(1)} = 4\sigma N^{-1}$ and $h^{(2)} = 2(d - 2\sigma)N^{-1}$ respectively, $\sigma \leq 4^{-1}d$. The value σ is taken to satisfy the condition $\sigma = \sigma(\varepsilon, N) = \min\left[4^{-1}d, m^{-1}\varepsilon \ln N\right]$, where m is an arbitrary positive number.

Taking into account the a priori estimates, similar to [10] we establish the ε-uniform convergence of scheme (8), (22)

$$|\, u(x,t) - z(x,t)\,| \leq M\left[N^{-1}\ln N + N_0^{-1}\right], \quad (x,t) \in \overline{G}_h \ . \tag{23}$$

For scheme (7), (22) we have

$$|\, u(x,t) - z(x,t)\,| \leq M\left[N^{-2}\ln^2 N + N_0^{-1}\right], \quad (x,t) \in \overline{G}_h \ . \tag{24}$$

To prove estimates (23), (24), we use the maximum principle and the ε-uniform approximation of the differential operators $L_{(3)}$ and $l_{(3)}$ by the difference operators $\Lambda_{(7)}$ and $\lambda_{(7)}$, $\lambda_{(8)}$ on the solution of boundary value problem (3).

According to the a priori estimates we can see that the differentiation with respect to the time variable, both for the solution $u(x,t)$ and for its derivatives, does not increase the singularity. This argument implies the ε-uniform convergence of the difference derivative (with respect to the time variable) of the solution of difference schemes (7), (22) and (8), (22)

$$\left|\frac{\partial}{\partial t}\,u(x,t) - \delta_{\overline{t}}\,z(x,t)\right| \leq M\left[N^{-1}\ln N + N_0^{-1}\right], \quad (x,t) \in \overline{G}_h,\ t > 0 \ . \tag{25}$$

In virtue of estimates (23)–(25) we have that the product $\varepsilon^2\delta_{\overline{x}\widehat{x}}\,z(x,t)$ converges to the product $\varepsilon^2(\partial^2/\partial x^2)u(x,t)$ ε-uniformly

$$\varepsilon^2\left|\frac{\partial^2}{\partial x^2}\,u(x,t) - \delta_{\overline{x}\widehat{x}}\,z(x,t)\right| \leq M\left[N^{-1}\ln N + N_0^{-1}\right], \quad (x,t) \in G_h \tag{26}$$

for difference schemes (7), (22) and (8), (22).

Taking into account estimates (25), (26) and the a priori estimates of the function $u(x,t)$ and its derivatives, we derive the estimate

$$|\, P(x,t) - P^{h+}(x,t)\,| \leq M\left[N^{-1}\ln N + N_0^{-1}\right], \quad (x,t) \in \overline{G}_h^{\,-} \ . \tag{27}$$

Here $P^{h+}(x,t)$ are the fluxes computed from the solutions of problems (7), (22) and (8), (22).

Theorem 4. *Let the solution of boundary value problem (3), (4) be sufficinetly smooth for a fixed value of the parameter. Then the solutions of difference schemes (7), (22) and (8), (22), and also the functions $P^{h+}(x,t)$, $(x,t) \in \overline{G}_h^-$ converge ε-uniformly. For the solutions of the difference schemes and the computed fluxes $P^{h+}(x,t)$ estimates (23)–(26), (27) are valid.*

5 Numerical Study of the Special Scheme

Now we can examine the special difference scheme (2), (22) (similar to scheme (8), (22)) for problem (1). In Tables 3, 4 the results computed are given.

Table 3. Errors of the solution $E(\varepsilon, N)$, $\overline{E}(N)$ for special scheme (2), (22)

$\varepsilon \setminus N$	4	16	64	256	1024
1.0	4.736e−01	9.723e−02	2.332e−02	5.771e−03	1.439e−03
2^{-2}	1.098e+00	2.347e−01	5.581e−02	1.377e−02	3.431e−03
2^{-4}	3.367e+00	1.014e+00	2.114e−01	4.994e−02	1.230e−02
2^{-6}	3.221e+00	1.492e+00	4.650e−01	1.398e−01	4.196e−02
2^{-8}	3.184e+00	1.492e+00	4.650e−01	1.398e−01	4.196e−02
2^{-10}	3.175e+00	1.492e+00	4.650e−01	1.398e−01	4.196e−02
2^{-12}	3.172e+00	1.492e+00	4.650e−01	1.398e−01	4.196e−02
$\overline{E}(N)$	3.367e+00	1.492e+00	4.650e−01	1.398e−01	4.196e−02

Table 4. Errors of the normalized flux $Q(\varepsilon, N)$, $\overline{Q}(N)$ for special scheme (2), (22)

$\varepsilon \setminus N$	4	16	64	256	1024
1.0	1.675e−01	3.658e−02	8.844e−03	2.193e−03	5.470e−04
2^{-2}	8.878e−02	1.728e−02	3.903e−03	9.482e−04	2.353e−04
2^{-4}	2.524e−02	5.094e−02	6.211e−03	1.095e−03	2.444e−04
2^{-6}	9.922e−03	7.228e−02	1.431e−02	2.137e−03	3.520e−04
2^{-8}	1.153e−02	7.228e−02	1.431e−02	2.137e−03	3.520e−04
2^{-10}	1.163e−02	7.228e−02	1.431e−02	2.137e−03	3.520e−04
2^{-12}	1.163e−02	7.228e−02	1.431e−02	2.137e−03	3.520e−04
$\overline{Q}(N)$	1.675e−01	7.228e−02	1.431e−02	2.137e−03	3.520e−04

From Tables 3, 4 we have the conclusion that the approximate solution and the computed normalized flux converge (with increasing N, $N_0 = N$) ε-uniformly.

Thus, the numerical results illustrate the theoretical results and show the efficiency of the special scheme.

6 Generalizations

The techniques developed for boundary value problem (3), (4) can be generalized to multidimensional boundary value problems. For example, on the rectangle $D = \{x : 0 < x_s < d_s, s = 1,2\}$ we consider the boundary value problem

$$L_{(28)}u(x,t) = f(x,t), \quad (x,t) \in G , \tag{28}$$

$$\varepsilon \frac{\partial}{\partial n} u(x,t) = \psi(x,t), \ (x,t) \in S^L, \quad u(x,t) = \varphi(x), \ (x,t) \in S^0 .$$

Here

$$G = D \times (0,T], \quad S = \overline{G} \setminus G, \quad S = S^0 \cup S^L , \tag{29}$$

$$L_{(28)} \equiv \varepsilon^2 L^2_{(28)} - c(x,t) - p(x,t)\frac{\partial}{\partial t} ,$$

$$L^2_{(28)} \equiv \sum_{s=1,2} a_s(x,t)\frac{\partial^2}{\partial x_s^2} + \sum_{s=1,2} b_s(x,t)\frac{\partial}{\partial x_s} - c^0(x,t) ,$$

S^0 and S^L are lower and lateral sides of the boundary S, the functions $a_s(x,t)$, $b_s(x,t)$, $c^0(x,t)$, $c(x,t)$, $p(x,t)$, $f(x,t)$, $s = 1,2$ and also the function $\varphi(x)$ are sufficiently smooth functions on the sets $\overline{G}$ and $\overline{D}$ respectively, moreover $a_0 \le a_1(x,t), a_2(x,t) \le a^0$, $c(x,t)$, $c^0(x,t) \ge 0$, $p(x,t) \ge p_0$, $(x,t) \in \overline{G}$, a_0, $p_0 > 0$; $\varepsilon \in (0,1]$. The function $\psi(x,t)$ is smooth only on each sides of the set G forming the lateral boundary S^L; $\psi \notin C(S^L)$. Suppose that on the set $S^* = S^{0*} \cup S^{L*}$, $S^{0*} = \{(x,t) : x \in \Gamma, t = 0\}$, $S^{L*} = \{(x,t) : x \in \Gamma^*, 0 < t \le T\}$, where $\Gamma = \overline{D} \setminus D$, Γ^* is a set of corner points of the rectangle D, the compatibility conditions are satisfied so that the required smoothness of the solution is ensured for each fixed value of the parameter.

The solution of boundary value problem (28), (29) is a function $u \in C^{2,1}(G) \cap C^{1.0}(\overline{G})$, which satisfies the equation on G and the boundary condition on S. It is required to find the solution of the boundary value problem and also the functions $P_1(x,t)$, $P_2(x,t)$ where $P_s(x,t) = \varepsilon(\partial/\partial x_s)u(x,t)$, $(x,t) \in \overline{G}$, $s = 1, 2$ are the normalized diffusion fluxes along the axes x_1 and x_2.

The classical difference scheme on special grids

$$\overline{G}_{h(30)} = \overline{G}_{h(30)}(\sigma) = \overline{D}_h \times \overline{\omega}_0 = \overline{\omega}_1^* \times \overline{\omega}_2^* \times \overline{\omega}_0 , \tag{30}$$

where $\overline{\omega}_1^*$, $\overline{\omega}_2^*$ are piecewise uniform grids similar to the grid $\overline{D}_{h(22)}$, allows us to approximate both the solution $u(x,t)$ and the normalized diffusion fluxes $P_s(x,t)$, $s = 1,2$ ε-uniformly.

Let $z(x,t)$, $(x,t) \in \overline{G}_h$ be a solution of the finite difference scheme

$$\Lambda_{(31)}z(x,t) = f(x,t), \quad (x,t) \in G_h , \tag{31}$$

$$\lambda_{(31)} z(x,t) = \psi(x,t),\ (x,t) \in S_h^L \setminus S_h^{L*},\quad z(x,t) = \varphi(x),\ (x,t) \in S_h^0 \ .$$

Here $G_h = G \cap \overline{G}_h, \quad S_h = S \cap \overline{G}_h,$

$$\Lambda_{(31)} z(x,t) \equiv \left\{ \varepsilon^2 \Lambda^2_{(31)} - c(x,t) - p(x,t)\delta_{\overline{t}} \right\} z(x,t) \ ,$$

$$\Lambda^2_{(31)} \equiv \sum_{s=1,2} a_s(x,t)\delta_{\overline{xs}\,\widehat{xs}} + \sum_{s=1,2} \left[b_s^+(x,t)\delta_{xs} + b_s^-(x,t)\delta_{\overline{xs}} \right] - c^0(x,t) \ ,$$

$$\lambda_{(31)} z(x,t) \equiv \begin{cases} -\varepsilon\delta_{xs}\, z(x,t),\ x_s = 0, & x_{3-s} \in \omega_{3-s} \ , \\ \varepsilon\delta_{\overline{xs}}\, z(x,t),\ x_s = d_s, & x_{3-s} \in \omega_{3-s},\ s = 1,2 \ . \end{cases}$$

Then the following estimate is valid:

$$|\, u(x,t) - z(x,t) \,| \le M \left[N^{-1} \ln N + N_0^{-1} \right], \quad (x,t) \in \overline{G}_{h(31)} \ . \tag{32}$$

Now we construct the grid approximation of the functions $P_s(x,t)$ on the grid $\overline{G}_h$ with arbitrary distribution of nodes:

$$\overline{G}_h = \overline{D}_h \times \overline{\omega}_0 = \overline{\omega}_1 \times \overline{\omega}_2 \times \overline{\omega}_0 \ . \tag{33}$$

With this aim, we use the modified difference derivatives. Let the solution of the difference scheme satisfy the estimate

$$|\, u(x,t) - z(x,t) \,| \le \beta(N, N_0), \quad (x,t) \in \overline{G}_{h(33)} \ .$$

We define the computational parameter h_s^* by the relation $h_s^* = h_s^*(\varepsilon, \beta(N,N_0)) = \min\left[4^{-1} d_s, M\varepsilon\beta^{1/2}(N,N_0)\right]$, where M is an arbitrary number. We introduce the grid sets $\overline{G}_h^{*1-}$, $\overline{G}_h^{*2-}$

$$\overline{G}^{*s-}_{h(33)} = \overline{G}_{h(33)} \cap \{(x,t):\ x_s \le d_s - h_s^*\}, \quad s = 1,\, 2 \ .$$

Then we form the modified difference derivatives

$$\delta^*_{x1}\, z(x,t) = (h_1^*)^{-1} \left[\widetilde{z}^1(x_1 + h_1^*, x_2, t) - z(x,t) \right], \quad (x,t) \in \overline{G}_h^{*1-} \ ,$$
$$\delta^*_{x2}\, z(x,t) = (h_2^*)^{-1} \left[\widetilde{z}^2(x_1, x_2 + h_2^*, t) - z(x,t) \right], \quad (x,t) \in \overline{G}_h^{*2-} \ ,$$

where $\widetilde{z}^s(x,t)$ is a continuous function along x_s and a discrete one along the variables x_{3-s}, t, $s = 1,2$; $\widetilde{z}^s(x,t)$ is the linear interpolation along x_s from the grid function $z(x,t)$; the function $\beta(N,N_0)$ is the right-hand side of (32).

The normalized fluxes $P_1(x,t)$, $P_2(x,t)$ are approximated by the grid functions $P_1^{*h+}(x,t)$, $P_2^{*h+}(x,t)$, where $P_s^{*h+}(x,t) = \varepsilon\delta^*_{xs}\, z(x,t)$, $(x,t) \in \overline{G}_h^{*s-}$, $s = 1,2$.

Using estimate (32) we come to the ε-uniform convergence of the functions $P_1^{*h+}(x,t)$, $P_2^{*h+}(x,t)$

$$|P_s(x,t) - P_s^{*h+}(x,t)| \le M \left[N^{-1} \ln N + N_0^{-1} \right]^{1/2},\ (x,t) \in \overline{G}_h^{*s-},\ s = 1,2 \tag{34}$$

for scheme (31), (30) and to the convergence for a fixed value of the parameter

$$|P_s(x,t) - P_s^{*h+}(x,t)| \le M \left[\varepsilon^{-1} N^{-1} + N_0^{-1} \right]^{1/2},\ (x,t) \in \overline{G}_h^{*s-},\ s = 1,2 \tag{35}$$

for scheme (31), (33).

Let the following condition be valid for problem (28):

$$b_s(x,t) \equiv 0, \quad (x,t) \in \overline{G}, \quad s = 1,2\ . \tag{36}$$

Then, similar to the derivation of estimates (27) we obtain

$$|\, P_s(x,t) - P_s^{h+}(x,t)\, | \le M\left[N^{-1}\ln N + N_0^{-1}\right], \quad (x,t) \in \overline{G}{}_{h(30)}^{s-}\ . \tag{37}$$

Here $\overline{G}{}_h^{s-} = \overline{D}{}_h^{s-} \times \overline{\omega}_0$, $s = 1,2$, $\overline{D}{}_h^{1-} = \overline{\omega}_1^- \times \overline{\omega}_2$, $\overline{D}{}_h^{2-} = \overline{\omega}_1 \times \overline{\omega}_2^-$, $\overline{\omega}_s^- = \omega_s \cup \{x_s = 0\}$, $s = 1,2$; $P_s^{h+}(x,t) = \varepsilon\delta_{xs}\, z(x,t)$, $(x,t) \in \overline{G}{}_h^{s-}$, $s = 1,2$.

Theorem 5. *Let the solution of boundary value problem (28), (29) and its regular and singular parts, including a corner boundary layer [6], be sufficiently smooth for a fixed value of the parameter. Then the solution of difference scheme (31), (30) and also the function $P_s^{*h+}(x,t)$, $(x,t) \in \overline{G}{}_h^{*s-}$, $s = 1,2$ converge ε-uniformly. For the solution of difference scheme (31), (30) and the computed fluxes $P_1^{*h+}(x,t)$, $P_2^{*h+}(x,t)$ estimates (32), (34), (35) are valid. Under condition (36), for the computed fluxes $P_1^{h+}(x,t)$, $P_2^{h+}(x,t)$ estimate (37) is fulfilled.*

References

1. Bakhvalov N.S.: On optimization of methods to solve boundary value problems with boundary layers. Zh. Vychisl. Mat. i Mat. Fiz. **9** (1969) 841–859 (in Russian)
2. Doolan, E.P., Miller, J.J.H., Schilders, W.H.A.: Uniform Numerical Methods for Problems with Initial and Boundary Layers. Dublin (1980)
3. Il'in, A.M.: Difference scheme for a differential equation with a small parameter at the highest order derivative. Mat. Zametki **6** (1969) 237–248 (in Russian)
4. Miller, J.J.H., O'Riordan, E., Shishkin, G.I.: Fitted Numerical Methods for Singular Perturbation Problems. Errors Estimates in the Maximum Norm for Linear Problems in One and Two Dimensions. World Scientific, Singapore (1996)
5. Shishkin, G.I.: Grid Approximation of Singularly Perturbed Elliptic and Parabolic Equations. Ural Branch of Russ. Acad. Sci., Ekaterinburg (1992) (in Russian)
6. Ladyzhenskaya, O.A., Solonnikov, V.A., Ural'tseva, N.N.: Linear and Quasi-linear Equations of Parabolic Type. Nauka, Moscow (1967) (in Russian)
7. Samarsky, A.A.: Theory of Difference Schemes. Nauka, Moscow (1989) (in Russian)
8. Shishkin, G.I.: Grid approximation of singularly perturbed boundary value problem for quasi-linear parabolic equations in the case of complete degeneracy in spatial variables. Sov. J. Numer. Anal. Math. Modelling **6** (1991) 243–261
9. Shishkin, G.I.: Grid approximation of singularly perturbed boundary value problem for quasi-linear parabolic equation in the case of complete degeneracy. Zh. Vychisl. Mat. i Mat. Fiz. **31** (1991) 1808–1826 (in Russian)
10. Shishkin, G.I.: Difference scheme for singularly perturbed parabolic equation with discontinuous coefficients and lumped factors. Zh. Vychisl. Mat. i Mat. Fiz. **29** (1989) 1277–1290 (in Russian)

Bessel and Neumann Fitted Methods for the Numerical Solution of the Schrödinger Equation

T.E. Simos[1] and **P.S. Williams**[2]

[1] Laboratory of Applied Mathematics and Computers, Technical University of Crete, Kounoupidiana, 73100 Hania, Greece.

[2] Department CISM, London Guildhall University, London EC3N 1JY, U.K.

Abstract. Two 2-step methods for the numerical solution of some problems of the Schrödinger equation are developed in this paper. One is of the Numerov- type and of algebraic order 4 and the other is of the Runge-Kutta type and of algebraic order 5. Each of these methods have free parameters which will be defined such that the methods are *fitted* to spherical Bessel and Neumann functions. Based on these methods we have obtained a variable-step method. The results produced based on the phase-shift problem of the radial Schrödinger equation indicate that this new approach is more efficient than other well known methods.

1 Introduction

Numerous numerical techniques exist in the literature for solving the Schrödinger equation (see [2], [3], [7]).

The one dimensional Schrödinger equation is a boundary value problem which has the form:

$$y''(x) = f(x, y) = [l(l+1)/x^2 + V(x) - k^2]\ y(x). \tag{1}$$

with one boundary condition given by:

$$y(0) = 0 \tag{2}$$

and the other boundary condition, for large values of x, is determined by physical considerations. Equations of this type occur very frequently in theoretical physics (see [5]) and there is a real need to be able to solve them both efficiently and reliably by numerical methods. In (1) the function $W(x) = l(l+1)/x^2 + V(x)$ denotes *the effective potential*, which satisfies $W(x) \to 0$ as $x \to \infty$, k is a constant which may be complex but in this paper we will work exclusively with the case where k is a real number related to the energy $E = k^2$, l is a given integer and V is a given function which denotes the potential.

The precise form of the second boundary condition depends crucially on the sign of E. Here we will investigate the case $E = k^2 > 0$. In this case, in general, the potential function $V(x)$ dies away faster than the term $l(l+1)/x^2$; and then equation (1) then effectively reduces to

$$y''(x) + (E - \frac{l(l+1)}{x^2})y(x) = 0 \tag{3}$$

for large x. The above equation has linearly independent solutions $kxj_l(kx)$ and $kxn_l(kx)$, where $j_l(kx)$ and $n_l(kx)$ are the spherical Bessel and Neumann functions respectively. Thus the solution of equation (1) has the asymptotic form:

$$y(x) \cong_{x\to\infty} Akxj_l(kx) - Bkxn_l(kx)$$
$$\cong_{x\to\infty} A\left[\sin\left(kx - \frac{l\pi}{2}\right) + \tan\delta_l \cos\left(kx - \frac{l\pi}{2}\right)\right],$$

where δ_l is the **phase shift** which may be calculated from the formula:

$$\tan\delta_l = \frac{y(x_2)S(x_1) - y(x_1)S(x_2)}{y(x_1)C(x_2) - y(x_2)C(x_1)} \tag{4}$$

for x_1 and x_2 distinct points on the asymptotic region with $S(x) = kxj_l(kx)$ and $C(x) = -kxn_l(kx)$.

For equation (1) there are two categories of numerical solution : methods with constant coefficients and methods with coefficients which depend on the frequencies involved in the problem. The construction of these methods has been the subject of great activity the last few years (see the references in [3]).

The iterative Numerov method of Allison [1] is one of the most widely used techniques for solving molecular scattering problems associated with Schrödinger's equation (see [14]). Recently Raptis and Cash [8] have developed second order Bessel and Neumann fitted techniques which are more efficient than the Numerov method or the iterative Numerov method of Allison [1].

Exponential fitting is another approach for developing efficient methods for the solution of (1). This approach is appropriate because for large x the solution of (1) is *periodic*. A Numerov-type exponentially fitted method has derived by Raptis and Allison [6]. Numerical results presented there indicate that these fitted methods are much more efficient than Numerov's method for the solution of (1). Many authors have investigated the idea of exponential fitting, since Raptis and Allison. The most recent and interesting contribution to these methods is that proposed by Simos [11] (see Ixaru et. al. [4]).

In section 2 we will develop the fourth and fifth algebraic order Bessel and Neumann fitted methods. Based on these methods we will develop in section 3 a variable-step method. An application of this variable-step method to the one-dimensional Schrödinger equation is given in section 4.

2 Bessel and Neumann fitted methods of order four and six

Consider the two parameter family of two-step fourth order methods $M_{2,4}(a,b)$

$$\overline{y}_{n+1} = y_{n+1} - ah^2(f_{n+1} - f_n)$$

$$\overline{y}_n = y_n - bh^2(\overline{f}_{n+1} - 2f_n + f_{n-1}) \tag{5}$$
$$y_{n+1} - 2y_n + y_{n-1} = \frac{h^2}{12}\ (f_{n+1} + 10\overline{f}_n + f_{n-1})$$

where $f_p = f(x_p, y_p)$, $p = n-1(1)n+1$, $\overline{f}_q = f(x_q, \overline{y}_q)$, $q = n,\ n+1$.

The corresponding local truncation error is given by

$$LTE = \frac{h^6\ [(200\ b - 1)\ y_n^{(4)}\, F_n + y_n^{(6)}]}{240} + O(h^8) \tag{6}$$

where $F_n = \left(\frac{\partial f}{\partial y}\right)_n$.

Demanding now that the family of methods $M_{2,4}(a,b)$ integrate exactly the functions $kxj_l(kx)$ and $kxn_l(kx)$, where $j_l(kx)$ and $n_l(kx)$ are the spherical Bessel and Neumann functions respectively, we have the following equations

$$\begin{aligned}\overline{J}_{n+1} &= J_{n+1} - a(F_{n+1}J_{n+1} - F_nJ_n)\\ \overline{J}_n &= J_n - b(F_{n+1}\overline{J}_{n+1} - 2F_nJ_n + F_{n-1}J_{n-1})\\ J_{n+1} - 2J_n + J_{n-1} &= \frac{1}{12}\ (F_{n+1}J_{n+1} + 10F_n\overline{J}_n + F_{n-1}J_{n-1})\end{aligned} \tag{7}$$

and

$$\begin{aligned}\overline{Y}_{n+1} &= Y_{n+1} - a(F_{n+1}Y_{n+1} - F_nY_n)\\ \overline{Y}_n &= Y_n - b(F_{n+1}\overline{Y}_{n+1} - 2F_nY_n + F_{n-1}Y_{n-1})\\ Y_{n+1} - 2Y_n + Y_{n-1} &= \frac{1}{12}\ (F_{n+1}Y_{n+1} + 10F_n\overline{Y}_n + F_{n-1}Y_{n-1})\end{aligned} \tag{8}$$

where

$$F_q = \left[\frac{l(l+1)}{x_q^2} - k^2\right]\ h^2$$
$$J_q = kx_qj_l(kx_q),\ Y_q = kx_qy_l(kx_q),\ q = n-1(1)n+1$$

In Appendix A we present the values of the parameters a and b obtained by solving the above system of equations. It can be seen that the parameters a and b must computed at each step.

Consider the two parameter family of two-step fifth order methods $M_{2,5}(a,b)$

$$\begin{aligned}\overline{y}_{n+1} &= y_{n+1} - ah^2(f_{n+1} - f_n)\\ \overline{y}_{n+\frac{1}{2}} &= \frac{1}{2}(y_n + y_{n+1}) - \frac{h^2}{384}(34f_n + 19\overline{f}_{n+1} - 5f_{n-1})\\ \overline{y}_{n-\frac{1}{2}} &= \frac{1}{2}(y_n + y_{n-1}) - \frac{h^2}{384}(34f_n + 19f_{n-1} - 5\overline{f}_{n+1})\\ \overline{y}_n &= y_n - bh^2[(f_{n+1} - 2f_n + f_{n-1}) - 4(\overline{f}_{n+\frac{1}{2}} - 2f_n + \overline{f}_{n-\frac{1}{2}})]\\ y_{n+1} - 2y_n + y_{n-1} &= \frac{h^2}{60}\left[f_{n+1} + 26\overline{f}_n + f_{n-1} + 16(\overline{f}_{n+\frac{1}{2}} + \overline{f}_{n-\frac{1}{2}})\right]\end{aligned} \tag{9}$$

where $f_p = f(x_p, y_p)$, $p = n-1(1)n+1$, $\overline{f}_q = f(x_q, \overline{y}_q)$, $q = n,\ n+1$, $\overline{f}_{n\pm\frac{1}{2}} = f(x_{n\pm\frac{1}{2}}, \overline{y}_{n\pm\frac{1}{2}})$.

The corresponding local truncation error is given by

$$LTE = -\frac{7\ a\ h^7\ y_n^{(3)}}{720} + O(h^8) \tag{10}$$

Demanding that the family of methods $M_{2,5}(a,b)$ integrate exactly the functions $kxj_l(kx)$ and $kxn_l(kx)$, we have the following equations

$$\begin{aligned}
\overline{J}_{n+1} &= J_{n+1} - a(F_{n+1}J_{n+1} - F_nJ_n)\\
\overline{J}_{n+\frac{1}{2}} &= \frac{1}{2}(J_n + J_{n+1}) - \frac{1}{384}(34F_nJ_n + 19F_{n+1}\overline{J}_{n+1} - 5F_{n-1}J_{n-1})\\
\overline{J}_{n-\frac{1}{2}} &= \frac{1}{2}(J_n + J_{n-1}) - \frac{1}{384}(34F_nJ_n + 19F_{n-1}J_{n-1} - 5F_{n+1}\overline{J}_{n+1})\\
\overline{J}_n &= J_n - b[(F_{n+1}J_{n+1} - 2F_nJ_n + F_{n-1}J_{n-1}) -\\
&\quad -4(F_{n+\frac{1}{2}}\overline{J}_{n+\frac{1}{2}} - 2F_nJ_n + F_{n-\frac{1}{2}}\overline{J}_{n-\frac{1}{2}})]\\
J_{n+1} - 2J_n + J_{n-1} &= \frac{1}{60}[F_{n+1}J_{n+1} + 26F_n\overline{J}_n + F_{n-1}J_{n-1} +\\
&\quad +16(F_{n+\frac{1}{2}}\overline{J}_{n+\frac{1}{2}} + F_{n-\frac{1}{2}}\overline{J}_{n-\frac{1}{2}})]
\end{aligned} \tag{11}$$

and

$$\begin{aligned}
\overline{Y}_{n+1} &= Y_{n+1} - a(F_{n+1}Y_{n+1} - F_nY_n)\\
\overline{Y}_{n+\frac{1}{2}} &= \frac{1}{2}(Y_n + Y_{n+1}) - \frac{1}{384}(34F_nY_n + 19F_{n+1}\overline{Y}_{n+1} - 5F_{n-1}Y_{n-1})\\
\overline{Y}_{n-\frac{1}{2}} &= \frac{1}{2}(Y_n + Y_{n-1}) - \frac{1}{384}(34F_nY_n + 19F_{n-1}Y_{n-1} - 5F_{n+1}\overline{Y}_{n+1})\\
\overline{Y}_n &= Y_n - b[(F_{n+1}Y_{n+1} - 2F_nY_n + F_{n-1}Y_{n-1}) -\\
&\quad -4(F_{n+\frac{1}{2}}\overline{Y}_{n+\frac{1}{2}} - 2F_nY_n + F_{n-\frac{1}{2}}\overline{Y}_{n-\frac{1}{2}})]\\
Y_{n+1} - 2Y_n + Y_{n-1} &= \frac{1}{60}[F_{n+1}Y_{n+1} + 26F_n\overline{Y}_n + F_{n-1}Y_{n-1} +\\
&\quad +16(F_{n+\frac{1}{2}}\overline{Y}_{n+\frac{1}{2}} + F_{n-\frac{1}{2}}\overline{Y}_{n-\frac{1}{2}})]
\end{aligned} \tag{12}$$

where

$$F_q = \left[\frac{l(l+1)}{x_q^2} - k^2\right] h^2$$

$$J_q = kx_qj_l(kx_q),\ Y_q = kx_qy_l(kx_q),\ q = n-1(1)n+1$$

In Appendix B we present the values of the parameters a and b obtained by solving the above system of equations. It can be seen that the parameters a and b must computed at each step.

3 Error Estimation

For the integration of systems of initial-value problems, several methods have been proposed for the estimation of the local truncation error (LTE) (see for example [8], [9], [10]).

In this paper we base our local error estimation technique on an embedded pair of integration methods. For oscillatory solutions the best approximation to the solution occurs by selecting the member of the pair with maximal order.

The new variable-step procedure consists the following:

(1) We divide the integration range into two distinct parts.

(2) The first part of the integration runs from $r = 0$ to r_c, where r_c is a point such that $\mid V(r) \mid$ is small. In practice we consider that $V(r)$ is small when $\mid V(r) \mid < \frac{1}{8}$. In the range $0 \leq r < r_c$ the variable-step method is exactly as described in [8].

(3) For $r \geq r_c$ we use a variable-step procedure:

For the purpose of local error estimation, we use as a lower order solution y_{n+1}^L, the fourth order Bessel and Neumann fitted method developed above. As a higher order solution y_{n+1}^H we use the fifth order Bessel and Neumann fitted method obtained above. The local truncation error in y_{n+1}^L is estimated by

$$LTE = \mid y_{n+1}^H - y_{n+1}^L \mid \tag{13}$$

We use the step length control mechanism fully described in [8], [9], [12].

We illustrate the new variable-step procedure described above by applying it to the solution of (1), where $V(x)$ is the *Lennard-Jones potential* which has been widely discussed in the literature. For this problem the potential $V(x)$ is given by:

$$V(x) = m(1/x^{12} - 1/x^6), \text{ where } m = 500. \tag{14}$$

We solve this problem as an initial value one and, in order to be able to use a two-step method we need an extra initial condition to be specified, e.g. $y_1 (= y(h))$. It is well known that, for values of x close to the origin, the solution of (1) behaves like

$$y(x) \simeq Cx^{l+1} \text{ as } x \to 0. \tag{15}$$

In view of this we use $y_1 = h^{l+1}$ as our extra initial condition.

The problem we consider is the computation of the relevant phase shifts correct to 4 decimal places for energies $k = 1$, $k = 5$ and $k = 10$ and for $l = 0(1)10$. We will consider four approaches:

Method MI: The iterative Numerov method developed by Allison [1]

Method MII: based on the variable step method of Raptis and Cash [8],

Method MIII: based on the variable step method of Raptis and Cash [9],

Method MIV: based on the variable step procedure developed by Simos [12],

Method MV: based on the variable step procedure developed by Simos and Mousadis [13],

Method MVI: based on the variable step procedure developed by Avdelas and Simos [3],

Method	Average time of computation (in seconds)
MI	2.750
MII	2.144
MIII	1.716
MIV	1.623
MV	1.543
MVI	1.204
MVII	0.967

Table 1. Phase shift problem. Average time of computation for the calculation of the phase shifts correct to 4 decimal places for energies $k = 1$, $k = 5$ and $k = 10$ and for $l = 0(1)10$.

Method MVII: based on the variable-step method developed above.

In Table 1 we present the average time of computation of the phase shifts correct to four decimal places.

It can be seen from the theoretical and numerical results that the new method is more efficient than the other most recent and well-known published variable-step methods.

All computations were carried out on a PC i486 using double precision arithmetic (16 significant digits accuracy).

Appendix A

$$a = \frac{12(t_3 + 2S_1 - t_5 + t_{18} - t_8 + t_{10} - t_{11} + t_{14} + t_{17})}{F_{n+1}[12S_2 + t_{17} + F_{n-1}t_5 - 11(t_8 - t_3) - f_{n+1}t_{11} + t_{14}]} \tag{16}$$

where $t_1 = F_{n+1}Y_{n+1}$, $t_2 = J_n t_1$, $t_3 = F_n t_2$, $t_4 = F_n J_n Y_{n+1}$, $t_5 = J_{n-1}t_1$, $t_6 = F_n Y_n$, $t_7 = J_{n+1}t_6$, $t_8 = F_{n+1}t_7$, $t_9 = F_{n+1}J_{n+1}Y_n$, $t_{10} = F_{n+1}J_{n+1}Y_{n-1}$, $t_{11} = F_{n-1}J_{n+1}Y_{n-1}$, $t_{12} = Y_{n-1}J_n$, $t_{13} = F_{n-1}t_{12}$, $t_{14} = F_n t_{13}$, $t_{15} = J_{n-1}t_6$, $t_{16} = F_n t_{12}$, $t_{17} = -F_{n-1}t_{15}$, $t_{18} = F_{n-1}Y_{n+1}J_{n-1}$, $t_{19} = F_{n-1}J_{n-1}Y_n$, $S_1 = t_2 - t_4 - t_9 + t_{15} - t_{19} - t_{16} + t_7 + t_{13}$, $S_2 = t_{15} - t_4 + 2t_2 + t_7 + t_{10} - t_{16} - 2t_9 - t_5$.

$$b = -\frac{12W_1 - t_8 + t_9 - 11(t_{10} - t_{11}) - t_{12} + t_{13}}{10F_n(-t_{10} + t_{11} + t_{12} - t_{13} - t_9 + t_8)} \tag{17}$$

where $t_1 = F_n Y_n$, $t_2 = J_{n-1}t_1$, $t_3 = F_{n+1}Y_{n+1}$, $t_4 = J_n t_3$, $t_5 = J_{n+1}t_1$, $t_6 = Y_{n-1}J_n$, $t_7 = J_{n-1}t_3$, $t_8 = F_{n-1}t_2$, $t_9 = F_{n-1}t_7$, $t_{10} = F_{n+1}t_5$, $t_{11} = F_n t_4$, $t_{12} = F_{n+1}J_{n+1}F_{n-1}Y_{n-1}$, $t_{13} = F_n F_{n-1}t_6$, $W_1 = t_2 - F_n J_n Y_{n+1} + 2t_4 + t_5 + F_{n+1}J_{n+1}Y_{n-1} - F_n t_6 - 2F_{n+1}J_{n+1}Y_n - t_7$.

Appendix B

$$a = \frac{24t_{16} + 408t_4 + 252t_5 + 156t_6 + 84t_7 + 228t_{12}}{F_{n+1}(5F_{n-\frac{1}{2}} - 19F_{n+\frac{1}{2}})(12t_{14} - t_{17} + 11t_{15} + t_{18})} \tag{18}$$

where $t_1 = J_{n-1}F_{n+1}Y_{n+1} - F_{n+1}J_{n+1}Y_{n-1} + J_{n+1}F_{n-1}Y_{n-1} - Y_{n+1}F_{n-1}J_{n-1}$, $t_2 = J_{n+1}F_nY_n - F_nJ_nY_{n+1} - F_nJ_nY_{n-1} + J_{n-1}F_nY_n + J_nF_{n+\frac{1}{2}}Y_{n+1} - F_{n+\frac{1}{2}}J_{n+1}Y_n - Y_nJ_{n-1}F_{n-\frac{1}{2}} + Y_{n-1}J_nF_{n-\frac{1}{2}}$, $t_3 = F_{n+1}J_{n+1}Y_n + F_{n+\frac{1}{2}}J_{n+1}Y_{n-1} - J_nF_{n+1}Y_{n+1} + Y_{n+1}F_{n-\frac{1}{2}}J_n - F_{n-\frac{1}{2}}J_{n+1}Y_n - Y_{n+1}F_{n+\frac{1}{2}}J_{n-1} - F_{n-1}Y_{n-1}J_n + Y_nF_{n-1}J_{n-1} - F_{n+\frac{1}{2}}Y_nJ_{n-1} + Y_{n+1}J_{n-1}F_{n-\frac{1}{2}} - F_{n-\frac{1}{2}}J_{n+1}Y_{n-1} + Y_{n-1}J_nF_{n+\frac{1}{2}}$, $t_4 = F_{n-\frac{1}{2}}J_{n+1}F_nY_n + J_{n-1}F_{n+\frac{1}{2}}F_nY_n - F_{n+\frac{1}{2}}F_nJ_nY_{n-1} - Y_{n+1}F_{n-\frac{1}{2}}F_nJ_n$, $t_5 = -F_{n+\frac{1}{2}}J_{n+1}F_{n-1}Y_{n-1} + F_{n-\frac{1}{2}}F_{n+1}J_{n+1}Y_{n-1} - J_{n-1}F_{n-\frac{1}{2}}F_{n+1}Y_{n+1} + Y_{n+1}F_{n-1}J_{n-1}F_{n+\frac{1}{2}}$, $t_6 = F_{n+1}J_{n+1}F_{n+\frac{1}{2}}F_nY_n - F_{n+\frac{1}{2}}F_nJ_nF_{n+1}Y_{n+1} - F_{n-\frac{1}{2}}F_nJ_nF_{n-1}Y_{n-1} + F_{n-1}J_{n-1}F_{n-\frac{1}{2}}F_nY_n$, $t_7 = -F_{n+1}J_{n+1}F_{n-\frac{1}{2}}F_nY_n + F_{n+\frac{1}{2}}F_nJ_nF_{n-1}Y_{n-1} + F_{n-\frac{1}{2}}F_nJ_nF_{n+1}Y_{n+1} - F_{n-1}J_{n-1}F_{n+\frac{1}{2}}F_nY_n$, $t_8 = -F_{n-\frac{1}{2}}F_{n+1}J_{n+1}F_{n-1}Y_{n-1} + F_{n+\frac{1}{2}}F_{n+1}J_{n+1}F_{n-1}Y_{n-1} - F_{n-1}J_{n-1}F_{n+\frac{1}{2}}F_{n+1}Y_{n+1} + F_{n-1}J_{n-1}F_{n-\frac{1}{2}}F_{n+1}Y_{n+1}$, $t_9 = F_{n+\frac{1}{2}}F_nJ_nY_{n+1} + F_{n-\frac{1}{2}}F_nJ_nY_{n-1} - J_{n+1}F_{n+\frac{1}{2}}F_nY_n - J_{n-1}F_{n-\frac{1}{2}}F_nY_n$, $t_{10} = F_{n-\frac{1}{2}}F_{n+1}J_{n+1}Y_n - J_nF_{n-\frac{1}{2}}F_{n+1}Y_{n+1} + Y_nF_{n-1}J_{n-1}F_{n+\frac{1}{2}} - F_{n-1}Y_{n-1}J_nF_{n+\frac{1}{2}}$, $t_{11} = F_{n+\frac{1}{2}}F_{n+1}J_{n+1}Y_n - J_nF_{n+\frac{1}{2}}F_{n+1}Y_{n+1} - F_{n-1}Y_{n-1}J_nF_{n-\frac{1}{2}} + Y_nF_{n-1}J_{n-1}F_{n-\frac{1}{2}}$, $t_{12} = -F_{n+\frac{1}{2}}F_{n+1}J_{n+1}Y_{n-1} + J_{n-1}F_{n+\frac{1}{2}}F_{n+1}Y_{n+1} + F_{n-\frac{1}{2}}J_{n+1}F_{n-1}Y_{n-1} - Y_{n+1}F_{n-1}J_{n-1}F_{n-\frac{1}{2}}$, $t_{13} = F_{n+1}J_{n+1}F_nY_n - F_nJ_nF_{n-1}Y_{n-1} - F_nJ_nF_{n+1}Y_{n+1} + F_{n-1}J_{n-1}F_nY_n$, $t_{14} = J_{n+1}F_nY_n - J_{n-1}F_{n+1}Y_{n+1} - F_nJ_nY_{n+1} - F_nJ_nY_{n-1} + J_{n-1}F_nY_n + F_{n+1}J_{n+1}Y_{n-1} - 2F_{n+1}J_{n+1}Y_n + 2J_nF_{n+1}Y_{n+1}$, $t_{15} = -F_{n+1}J_{n+1}F_nY_n + F_nJ_nF_{n+1}Y_{n+1}$, $t_{16} = 48t_1 + 288t_2 + 96t_3 + t_8 + 63t_9 + 3t_{10} + 27t_{11} + 16t_{13}$, $t_{17} = F_{n+1}J_{n+1}F_{n-1}Y_{n-1}$, $t_{18} = F_nJ_nF_{n-1}Y_{n-1} + F_{n-1}J_{n-1}F_{n+1}Y_{n+1} - F_{n-1}J_{n-1}F_nY_n$.

$$b = -\frac{2880t_1 + 5760t_2 + 106t_3 + 58t_4 + 10t_5 + 38t_6 + 384t_7 + 48t_8 + 1296t_9}{13F_n(53t_{10} + 29t_{11} + 5t_{12} + 19t_{13} + 192t_{14} + 96t_{15} + 672t_{16})} \tag{19}$$

where $t_1 = J_{n-1}F_{n+1}Y_{n+1} - J_{n+1}F_nY_n + F_nJ_nY_{n+1} + F_nJ_nY_{n-1} - J_{n-1}F_nY_n - F_{n+1}J_{n+1}Y_{n-1}$, $t_2 = F_{n+1}J_{n+1}Y_n - J_nF_{n+1}Y_{n+1}$, $t_3 = -F_{n+1}J_{n+1}F_{n+\frac{1}{2}}F_nY_n + F_{n+\frac{1}{2}}F_nJ_nF_{n+1}Y_{n+1}$, $t_4 = -F_{n+1}J_{n+1}F_{n-\frac{1}{2}}F_nY_n + F_{n-\frac{1}{2}}F_nJ_nF_{n+1}Y_{n+1}$, $t_5 = -F_{n+\frac{1}{2}}F_nJ_nF_{n-1}Y_{n-1} + F_{n+\frac{1}{2}}F_{n+1}J_{n+1}F_{n-1}Y_{n-1} - F_{n-1}J_{n-1}F_{n+\frac{1}{2}}F_{n+1}Y_{n+1} + F_{n-1}J_{n-1}F_{n+\frac{1}{2}}F_nY_n$, $t_6 = -F_{n-\frac{1}{2}}F_{n+1}J_{n+1}F_{n-1}Y_{n-1} + F_{n-\frac{1}{2}}F_nJ_nF_{n-1}Y_{n-1} + F_{n-1}J_{n-1}F_{n-\frac{1}{2}}F_{n+1}Y_{n+1} - F_{n-1}J_{n-1}F_{n-\frac{1}{2}}F_nY_n$, $t_7 = -F_{n+\frac{1}{2}}F_nJ_nY_{n+1} + F_{n-\frac{1}{2}}F_{n+1}J_{n+1}Y_n + F_{n-\frac{1}{2}}F_{n+1}J_{n+1}Y_{n-1} - F_{n-\frac{1}{2}}F_nJ_nY_{n-1} + F_{n+\frac{1}{2}}F_{n+1}J_{n+1}Y_n + J_{n+1}F_{n+\frac{1}{2}}F_nY_n - J_{n-1}F_{n-\frac{1}{2}}F_{n+1}Y_{n+1} + J_{n-1}F_{n-\frac{1}{2}}F_nY_n - J_nF_{n+\frac{1}{2}}F_{n+1}Y_{n+1} - J_nF_{n-\frac{1}{2}}F_{n+1}Y_{n+1}$, $t_8 = F_{n+1}J_{n+1}F_{n-1}Y_{n-1} - F_nJ_nF_{n-1}Y_{n-1} - F_{n-1}J_{n-1}F_{n+1}Y_{n+1} + F_{n-1}J_{n-1}F_nY_n$, $t_9 = F_{n+1}J_{n+1}F_nY_n - F_nJ_nF_{n+1}Y_{n+1}$, $t_{10} = -F_{n+1}J_{n+1}F_{n+\frac{1}{2}}F_nY_n + F_{n+\frac{1}{2}}F_nJ_nF_{n+1}Y_{n+1}$, $t_{11} = -F_{n+1}J_{n+1}F_{n-\frac{1}{2}}F_nY_n + F_{n-\frac{1}{2}}F_nJ_nF_{n+1}Y_{n+1}$, $t_{12} = -F_{n+\frac{1}{2}}F_nJ_nF_{n-1}Y_{n-1} + F_{n+\frac{1}{2}}F_{n+1}J_{n+1}F_{n-1}Y_{n-1} - F_{n-1}J_{n-1}F_{n+\frac{1}{2}}F_{n+1}Y_{n+1} + F_{n-1}J_{n-1}F_{n+\frac{1}{2}}F_nY_n$, $t_{13} = -F_{n-\frac{1}{2}}F_{n+1}J_{n+1}F_{n-1}Y_{n-1} + F_{n-\frac{1}{2}}F_nJ_nF_{n-1}Y_{n-1} + F_{n-1}J_{n-1}F_{n-\frac{1}{2}}F_{n+1}Y_{n+1} - F_{n-1}J_{n-1}F_{n-\frac{1}{2}}F_nY_n$, $t_{14} = -F_{n+\frac{1}{2}}F_nJ_nY_{n+1} + F_{n-\frac{1}{2}}F_{n+1}J_{n+1}Y_n + F_{n-\frac{1}{2}}F_{n+1}J_{n+1}Y_{n-1} - F_{n-\frac{1}{2}}F_nJ_nY_{n-1} + F_{n+\frac{1}{2}}F_{n+1}J_{n+1}Y_n + J_{n+1}F_{n+\frac{1}{2}}F_nY_n - J_{n-1}F_{n-\frac{1}{2}}F_{n+1}Y_{n+1} + J_{n-1}F_{n-\frac{1}{2}}F_nY_n - J_nF_{n+\frac{1}{2}}F_{n+1}Y_{n+1} -$

$J_n F_{n-\frac{1}{2}} F_{n+1} Y_{n+1}$, $t_{15} = -F_{n+1} J_{n+1} F_{n-1} Y_{n-1} + F_n J_n F_{n-1} Y_{n-1} + F_{n-1} J_{n-1} F_{n+1} Y_{n+1} - F_{n-1} J_{n-1} F_n Y_n$, $t_{16} = -F_{n+1} J_{n+1} F_n Y_n + F_n J_n F_{n+1} Y_{n+1}$.

References

1. Allison, A.C.: The numerical solution of coupled differential equations arising from the Schrödinger equation. J. Comput. Phys. **6** (1970) 378–391
2. Avdelas, G., Simos, T.E.: A generator of high-order embedded P-stable methods for the numerical solution of the Schrödinger equation. J. Comput. Appl. Math., in press.
3. Avdelas, G., Simos, T.E.: Embedded methods for the numerical solution of the Schrödinger equation. Comput. Math. Appl. **31** (1996) 85–102
4. Ixaru, L.Gr., Rizea, M., Vertse, T.: Piecewise perturbation methods for calculating eigensolutions of a complex optical potential. Comput. Phys. Commun. **85** (1995) 217–230
5. Landau, L.D., Lifshitz, F.M.: Quantum Mechanics. Pergamon, New York, 1965
6. Raptis, A.D., Allison, A.C.: Exponential-fitting methods for the numerical solution of the Schrödinger equation. Comput. Phys. Commun. **14** (1978) 1–5
7. Raptis, A.D.: Numerical solution of coupled differential equations. Ph.D. Thesis, Glasgow University, Glasgow, 1977
8. Raptis, A.D., Cash, J.R.: A variable step method for the numerical integration of the one-dimensional Schrödinger equation. Comput. Phys. Commun. **36** (1985) 113–119
9. Raptis, A.D., Cash, J.R.: Exponential and Bessel fitting methods for the numerical solution of the Schrödinger equation. Comput. Phys. Commun. **44** (1987) 95–103
10. Shampine, L.F., Watts, H.A., Davenport, S.M.: Solving nonstiff ordinary differential equations - The state of the art. SIAM Rev. **18** (1976) 376–411
11. Simos, T.E.: Exponential fitted methods for the numerical integration of the Schrödinger equation. Comput. Phys. Commun. **71** (1992) 32–38
12. Simos, T.E.: New variable-step procedure for the numerical integration of the one-dimensional Schrödinger equation. J. Comput. Phys. **108** (1993) 175–179
13. Simos, T.E., Mousadis, G.: A two-step method for the numerical solution of the radial Schrödinger equation. Comput. Math. Appl. **83** (1994) 1145–1153
14. Thomas, L.D., Alexander, M.H., Johmson, B.R., Lester Jr., W.A., Light, J.C., McLenithan, K.D., Parker, G.A., Redmon, M.J., Schmaltz, T.G., Secrest, D., Walker, R.B.: Comparison of numerical methods for solving second order differential equations of molecular scattering theory. J. Comput. Phys. **41** (1981) 401–426

On The Data Re-Processing-Free Algorithm In The Initial State Estimation Problem For Parabolic Systems *

Irina Sivergina

Institute for Mathematics and Mechanics, Russian Academy of Sciences, Ekaterinburg, 620219, GSP-384, Russia

Abstract. The paper deals with the problem of initial state estimation for parabolic systems on the basis of observations generated by sensors. The observations are assumed to be taken in presence of unknown but bounded disturbances. The model to these latter leads to a procedure of guaranteed estimating that involves calculating an informational domain of the initial states being consistent with measurement data and a priori restrictions on the disturbances. This problem is ill-posed and need to be regularized to provide a numerically stable solution. In the paper a regularizing procedure is proposed being data re-processing-free (with observations on a more long-run time interval) and admitting an on-line solution. This procedure also offers a way of approximating the informational domain as a whole.

1 The Guaranteed Estimation Problem

Let $(\mathbb{H}, |\cdot|_{\mathbb{H}})$ be a Hilbert space, $\mathbb{H} = \mathbb{H}^*$, $(\cdot,\cdot)_{\mathbb{H}}$ stand for the scalar product in $\mathbb{H}$, $(V, \|\cdot\|)$ be a separable reflexive Banach space, and V be imbedded into $\mathbb{H}$ densely and continuously. Consider a parabolic system

$$\dot{u}(t) + Au(t) = Bf(t), \ t \in [0,\theta] = T, \qquad u(0) = u_0 \in \mathbb{H} \ . \tag{1}$$

Here $A : V \mapsto V^*$ is a linear symmetrical continuous operator satisfying the condition

$$< Ay, y > + \lambda \mid y \mid^2 \geq c \, \|y\|^2 \quad \forall y \in V$$

with some $c > 0$ and $\lambda \in R$. Here $< \cdot, \cdot >$ stands for a bilinear form on $V^* \times V$. $u(t)$ is said to be the state of the system (1) at the time instant t, and $f(t)$ to be an input to that one, $f(\cdot) \in L_2(T; U)$. U is supposed to be a real Hilbert space; $B \in \mathcal{L}(U; V)$.

Following [8], a function $u(\cdot)$ is called a solution to (1) on T if

* This research was supported by the Russian Foundation of Basic Researches, grants 94-01-803 and 96-01-50. The author wishes to express appreciation to Academician A.B.Kurzhanskii for useful discussions.

1. $u(\cdot) \in W(T;V) = \{u(\cdot) \in L_2(T;V),\ \dot{u}(\cdot) \in L_2(T;V^*)\}$;
2. the equality (1) is fulfilled in that

$$< \dot{u}(t), v > + < Au(t), v > = < Bf(t), v > \quad for\ a.a.\ t \in T\ \forall v \in V\ .$$

For every $u_0 \in \mathbb{H}$, $f(\cdot) \in L_2(T;U)$, as it is known, there exists an unique solution to (1).

It is further assumed that the input $f(t)$, $t \in T$, is given; contrastingly, the initial state u_0 is unknown in advance. As this takes place, the dynamic information about the solution $u(t)$ is available through a finite-dimensional measurement equation

$$y(t) = G(t)u(t),\ t \in [\tau, \theta] = T_\tau\ .$$

Here $y(t)$ is a measurement data, $y(t) \in \mathbb{R}^m$; $G(t)$ is a linear observation operator (a "sensor") with its range in $\mathbb{R}^m$. A parameter $\tau \geq 0$ defines an interval of observation. We will consider $y(\cdot)$ as an element of the space $L_2(T_\tau; \mathbb{R}^m)$. The problem is to reconstruct the initial state u_0 on the basis of observations $y(\cdot)$. It is a deterministic inverse ill-posed problem [7, 9]. It is reasonable, however, to expect that the measurements are inexact. Let an uncertainty be of a sort

$$y(t) = G(t)u(t) + \xi(t),\ t \in T_\tau \tag{2}$$

with $\xi(t)$ be a measurement "noise". We will suppose a restriction on $\xi(t)$ to be imposed

$$\| \xi(\cdot) \|_{L_2(T_\tau;\mathbb{R}^m)} \leq \mu\ . \tag{3}$$

For this latter the reconstruction problem has not, in general, an unique solution.

Definition 1 [2, 3, 4]. The informational domain $\mathcal{U}_0[y_\theta(\cdot)]$ of initial states of system (1) that are consistent with the measurement data $y_\theta(\cdot) = y(t)$, $t \in [\tau, \theta]$ of form (2) and the restriction (3) is the set of all those elements $u_0 \in \mathbb{H}$ such that conditions (2) and (3) are valid for a solution of the equation (1) corresponding to u_0 and some $\xi(\cdot)$.

The informational domain is not empty set and in any case there are two possibilities: either $\mathcal{U}_0[y_\theta(\cdot)]$ consist of only one element or it is unbounded. The first might happen only if the system were observable.

Definition 2 [4, 5]. The system (1)-(3) is said to be observable if the measurements $y(t) = 0$, $t \in T_\tau$, provided that $\xi(t) = 0$, $f(t) = 0$ for $t \in T_\tau$, yield $u_0 = 0$.

Calculating the informational domains and their spacial elements are the issues of the guaranteed estimation problem. The goal of this paper is to propose some stable, data re-processing free schemes of doing that. (A direct calculation of those elements may obviously lead to unstable numerical procedures.)

2 Regularizing procedure

An approach used in this Section was announced in [5]. Consider an auxiliary evolutionary equation in a backward time

$$\dot{w}(t) + Aw(t) + \epsilon\frac{d}{dt}Aw(t) = Bf(t),\ t \in T,\ w(\theta) = w_0 \tag{4}$$

with a positive parameter ϵ and an initial condition w_0 given for $t = \theta$. According to [1] for $w_0 \in \mathcal{D}(A) = \{w \in V,\ Aw \in \mathbb{H}\}$ there exists one solution $w(t)$ to the equation (4) being $w(t) \in \mathcal{D}(A)$, $t \in T_\tau$. For $w_0 \in \mathbb{H}$ a function $w(t)$, $t \in T$ means the solution to (4) iff for any sequence w_{0k} being convergent to w_0 in $\mathbb{H}$ with $k \to \infty$, $w_{0k} \in \mathcal{D}(A)$, the corresponding solutions $w_k(t)$ to (4) offer the property $w_k(t) \to w(t)$ in $\mathbb{H}$, too. In such a sense of a solution the Cuachy problem (4) is well-posed for $w_0 \in \mathbb{H}$.

Let w_0 be unknown and there be given an observation equation

$$z(t) = G(t)w(t) + \eta(t),\ t \in T_\tau \tag{5}$$

where $z(t)$ is a measurement data, $\eta(t)$ denotes a "noise" of observations. Here A, B, $G(t)$, $f(t)$ are supposed to be the same ones as in the assumption problem. Assume w_0 and $\eta(t)$ to be satisfying a joint quadratic constraint

$$(w_0, Nw_0)_{\mathbb{H}} + \int_\tau^\theta \eta'(t)M(t)\eta(t)dt \le \mu^2 + \gamma^2\ . \tag{6}$$

N and $M(t)$ are the linear non-negative self-adjoint operators in $\mathbb{H}$ and $\mathbb{R}^m$ accordingly, N is invertible, $M(t)$ is continuous on t. Let $\mathcal{W}_0[z(\cdot)] \subset \mathbb{H}$ stand for an informational domain of the states $w(0)$ of the system (4) at the time instant $t = 0$ that are consistent with the measurement data $z(t)$, $t \in T_\tau$ in (5) and the restriction (6). The requirements to N and $M(t)$ achieve $\mathcal{W}_0[z(\cdot)]$ to be a bounded closed set in $\mathbb{H}$, whatever $z(\cdot)$ is.

Lemma 3. *The domain $\mathcal{W}_0[z(\cdot)]$ is an ellipsiod in $\mathbb{H}$ in the sense that its support function is*

$$\begin{aligned}\rho(l \mid \mathcal{W}_0[z(\cdot)]) &\equiv \sup\{(l,w)_{\mathbb{H}} \mid w \in \mathcal{W}_0[z(\cdot)]\} \\ &= (l, w^*)_{\mathbb{H}} + (\mu^2 + \gamma^2 - h^2)^{\frac{1}{2}}(l, Pl)_{\mathbb{H}}^{\frac{1}{2}}\ .\end{aligned} \tag{7}$$

The center w^* may be found out as a value of a function $w^*(t)$ as $t = 0$ solving the following equation

$$\begin{gathered}\dot{w}^*(t) + Aw^*(t) + \epsilon\tfrac{d}{dt}Aw^*(t) = Bf(t) + \chi_\tau(t)P(t)G^*(t)M(t)(z(t) - G(t)w^*(t)), \\ t \in T,\ w^*(\theta) = 0\ .\end{gathered} \tag{8}$$

This equation need to be solved in parallel with the operator differential Riccati equation

$$\begin{gathered}\dot{P}(t) + AP(t) + \epsilon\tfrac{d}{dt}AP(t) + P(t)A^* + \epsilon\tfrac{d}{dt}P(t)A^* = -P(t)G^*(t)M(t)G(t)P(t), \\ t \in T,\ P(\theta) = N^{-1}\ .\end{gathered} \tag{9}$$

The operator P in the formulae for the support function is equal to $P(0)$.

Here $\chi(t)$ stands for an indicator function of the interval $[\tau, \theta]$

$$\chi_\tau(t) = \begin{cases} 1 & \text{if } \tau \le t \le \theta \, , \\ 0 & \text{otherwise.} \end{cases}$$

To complete we accompany this lemma by an differential equation, which the parameter h^2 in (3) as a value of a function $h^2(t)$ at the zero time instant may be found out from:

$$\dot{h}^2(t) = \chi_\tau(t)(z(t) - G(t)w(t))' M(t)(z(t) - G(t)w(t)),$$
$$t \in T, \; h^2(\theta) = 0 \; .$$

Denote by E the identical operator in $\mathbb{R}^m$. Let φ_k, $k = 1, 2, \ldots$ be the eigenfunctions in the spectral problem

$$A\varphi_k = -\lambda_k \varphi_k, \; \phi_k \in V$$

making up an orthonormal basis in $\mathbb{H}$. λ_k are the corresponding eigenvalues, $\lambda_k \to \infty$ with $k \to \infty$.

Theorem 4. *Let $N \equiv N_{\epsilon\beta}$ where*

$$N_{\epsilon\beta} w = \beta \sum_{k=1}^{\infty} e^{\frac{2\lambda_k \theta}{1+\epsilon\lambda_k}} w_k \varphi_k, \; w \in \mathbb{H} \; ,$$
$$w_k = < w, \varphi_k > \; ;$$

$$M(t) \equiv E \; .$$

E symbolizes the identical operator in $\mathbb{R}^m$. Assume that $z(\cdot)$ in (5) is equal to $y_\theta(\cdot)$. If the system (1)-(3) is observable, then

$$w^* \to u_0 \textit{ in norm } \mathbb{H}$$
$$\textit{when } \mu \to 0, \; \epsilon \to 0, \; \beta \to 0, \; (\mu^2 + \epsilon)/\beta \to 0 \; .$$

Corollary 5. *If the system (1)-(3) were not be observable, the reconstruction problem of u_0 would not have an unique solution even with the exact observations, i.e. with $\xi(t) \equiv 0$. The requirements of the theorem 4 in this situation ensure w^* does convergent to an element u_0 having a minimal norm among those that are consistent with the measurement data $y_\theta(\cdot)$.*

Remark. For sufficiently large γ^2 the set $\mathcal{W}_0[y_\theta(\cdot)]$ in fact should not be empty and w^* might be really interpreted as the center of the ellipsoid (3).

Remark. The regularizing algorithm described in the theorem has the grave practical disadvantage. Namely, assume the measurement data are given not on the interval $[\tau, \theta]$ but on $[\tau, \theta + \delta]$ with $\delta > 0$. In this case the solutions $w^*(t)$ and $P(t)$ to the equations (8) and (9) on $[0, \theta]$ may not work for calculating a new estimate w^*. This is caused by a dependence of the operator $N_{\epsilon\beta}$ on θ. That is why the regularizing algorithms of an evolutionary kind are of prime interest.

3 Data Re-Processing-Free Procedure

For the moment let us restrict the problem to sensors of a special sort. Suppose m functions $g_i(t)$, $g_i(\cdot) \in L_2(T;\mathbb{H})$ $(i = 1, 2, \ldots, m)$ are given and the sensor $G(t)$ in (2) is defined through the equality

$$G(t)u(t) = \mathrm{col}[(g_1(t), u(t))_{\mathbb{H}}, (g_2(t), u(t))_{\mathbb{H}}, \ldots, (g_m(t), u(t))_{\mathbb{H}}],\ t \in T_\tau \ .$$

Specify a new sensor $\tilde{G}(t)$ on the same time interval T_τ

$$\tilde{G}(t)w(t) = \mathrm{col}[(\tilde{g}_1(t), w(t))_{\mathbb{H}}, (\tilde{g}_2(t), w(t))_{\mathbb{H}}, \ldots, (\tilde{g}_m(t), w(t))_{\mathbb{H}}] \tag{10}$$

where $\tilde{g}_i(t)$ $(i = 1, 2, \ldots, m)$ is the value at the time instant t of the solution to the Cauchy problem $\dot{\tilde{g}}(s) + A\tilde{g}(s) = 0,\ 0 \le s \le t,\ \tilde{g}(0) = g_i(t)$. Consider a problem of estimating of an initial state w_0 in the system

$$\begin{cases} \dot{w}(t) = 0,\ t \in T,\ w(0) = w_0 \ , \\ z(t) = \tilde{G}(t)w(t) + \eta(t),\ t \in T_\tau \ , \\ \epsilon(w_0, w_0)_{\mathbb{H}} + \int_\tau^\theta \eta'(t)M(t)\eta(t)dt \le \mu^2 + \gamma^2 \ . \end{cases} \tag{11}$$

Here $M(t)$ is of the same kind as in Sect. 2. Resoning similar to the previous section shows that an informational domain $\mathcal{W}_0[z(\cdot)]$ of the states w_0 of the system (11), that are consistent with the measurement data $z(t)$, $t \in T_\tau$ and the quadratic restriction, is a bounded closed ellipsoid in $\mathbb{H}$. Its center w^* is equal to $w^*(\theta)$ subject to

$$\dot{w}^*(t) = Bf(t) + \chi(t)_\tau P(t)\tilde{G}^*(t)M(t)(z(t) - \tilde{G}(t)w^*(t))\ ,\ t \ge 0,\ w^*(0) = 0\ ;$$
$$\dot{P}(t) = -P(t)\tilde{G}^*(t)M(t)\tilde{G}(t)P(t)\ ,\ t \ge 0,\ P(0) = \epsilon^{-1}\mathcal{E}\ .$$

Here $\mathcal{E}$ is identical operator in $\mathbb{H}$.

Theorem 6. *Assume that $z(\cdot)$ in (11) to be equal to $y_\theta(\cdot)$. If the system (1)-(3) is observable, then for $M(t) \equiv E$ we will have*

$$w^* \to u_0 \ \textit{in norm}\ \mathbb{H}$$
$$\textit{when}\ \mu \to 0,\ \epsilon \to 0,\ \mu^2/\epsilon \to 0\ .$$

Corollary 7. *Let $G(t) \equiv G$, $t \in T_\tau$. Then $\tilde{g}_i(t)$ $(i = 1, 2, \ldots, m)$ in (10) is the solution to the Cauchy problem*

$$\dot{\tilde{g}}(t) + A\tilde{g}(t) = 0,\ t \in T,\ \tilde{g}(0) = g_i\ . \tag{12}$$

Corollary 8. *Let $\Omega = (0, \pi) \subset \mathbb{R}^1$, $\mathbb{H} = L_2(\Omega)$, $V = \overset{0}{\mathbb{H}}{}^1(\Omega)$, and A be the Laplace operator on Ω. Hence $u(t)$ is a function $u(t, x)$, $x \in \Omega$, and $u(t, \cdot) \in C(\bar{\Omega})$ for a.a. $t > 0$. Suppose that $G(t) \equiv G$, $t \in T_\tau$ and G is of pointwise type:*

$$G(t)u(t) = col[u(t, \hat{x}_1), u(t, \hat{x}_2), \ldots, u(t, \hat{x}_m)],\ 0 < \tau \le t \le \theta$$

with some fixed points $\hat{x}_1, \hat{x}_2, \ldots, \hat{x}_m \in \Omega$. Now the conclusion in the Theorem 6 is provided by the sensor $\tilde{G}(t)$ of a form (10) where $\tilde{g}_i(t) = \tilde{g}_i(t, x)$ and the last one adheres to the equation (12) but with the initial condition $\tilde{g}_i(0, x) = \delta(x - \hat{x}_i)$.

A proposition akin to Corollary 5 has a place for the regularizing procedure described in Theorem 6. This will be illustrated in Sect.5 by some examples.

Theorem 9. *Let $M(t)$ and the sensor $G(t)$ defining on $[\tau,\infty)$ be such that $P(t)$ tends to zero in an operator norm when $\theta \to \infty$. Suppose also that $\xi(t) \equiv 0$ in (2) and $z(\cdot)$ in (11) is equal to $y_\theta(\cdot)$. If the system (1)-(3) is observable, then for any $\epsilon > 0$*

$$w^* \to u_0 \text{ in norm } \mathbb{H} \text{ when } \theta \to \infty \ .$$

Theorem 10. *Suppose $M(t) \equiv E$; nevertheless the assumption related to $P(t)$ in Theorem 9 is fulfilled. Let there exist $\bar{\mu}^2 < \infty$ such that $||\xi(\cdot)||_{L_2((\tau,\infty);\mathbb{R}^m)} < \bar{\mu}$. If the system (1)-(3) is observable, then the conclusion of Theorem 9 is true.*

4 Approximations for $\mathcal{U}_0[y_\theta(\cdot)]$

In this Section there will be given some approximations to the informational domain $\mathcal{U}_0[y_\theta(\cdot)]$ that are pertaining to the regularizing procedure described in Theorem 6. But this needs some additonal assumptions about the system considered. Firstly, suppose that a constant $\alpha > 0$ is given such that the inequality

$$||u_0||_{\mathbb{H}} \leq \alpha \tag{13}$$

is valid for the initial state u_0 in the system (1).Then an ("outer") inclusion $\mathcal{U}_0[y_\theta(\cdot)] \subseteq \mathcal{W}_0^+[y_\theta(\cdot)]$ is true for all $\epsilon < \epsilon_0$, $\epsilon_0 > 0$. Secondly, let the set $\mathcal{U}_0[y_\theta(\cdot)]$ be not one-point one. Then it is not difficult to show the ("inner") inclusion $\mathcal{W}_0^-[y_\theta(\cdot)] \subseteq \mathcal{U}_0[y_\theta(\cdot)]$ for $\epsilon < \epsilon_0$. Here $\mathcal{W}_0^-[y_\theta(\cdot)]$ is the informational domain of the initial states for the system (11) with $\gamma = 0$, and $\mathcal{W}_0^+[y_\theta(\cdot)]$ is the one with $\gamma = \epsilon M$. In much the same way as in [3, 6] one can prove the following

Theorem 11. *Let (13) be valid and $\mathcal{U}_0[y_\theta(\cdot)]$ be not a one-point set. Then*

$$\mathcal{U}_0[y_\theta(\cdot)] = \mathrm{cl} \bigcup_{\epsilon \leq \epsilon_0} \mathcal{W}_0^-[y_\theta(\cdot)] = \bigcap_{\epsilon \leq \epsilon_0} \mathcal{W}_0^+[y_\theta(\cdot)] \ .$$

In the last formulae the closure is taken in $\mathbb{H}$.

5 Examples

Let us take the environment of Corollary 5 and consult the regularizing procedure described in Theorem 6.

Example 1. Let $\epsilon = 10^{-6}$, $\mu = 15$, $M = 6$, $G(t)u(t) = (\varphi_1, u(t))_{\mathbb{H}}$. Fig.1 shows the projection of $w^*(\theta)$ and of the boundary of the informational domain $\mathcal{W}_0^+[y_\theta(\cdot)]$ onto $\mathrm{span}\{\varphi_1\} \subset \mathbb{H}$ for the observation data $y_\theta(t)$, given by $\xi(t) = 10\mathrm{e}^{-10t}\cos(100t)$, and $u_0(x) = 5\varphi_1(x)$, for $0 \leq \theta \leq 0.5$.

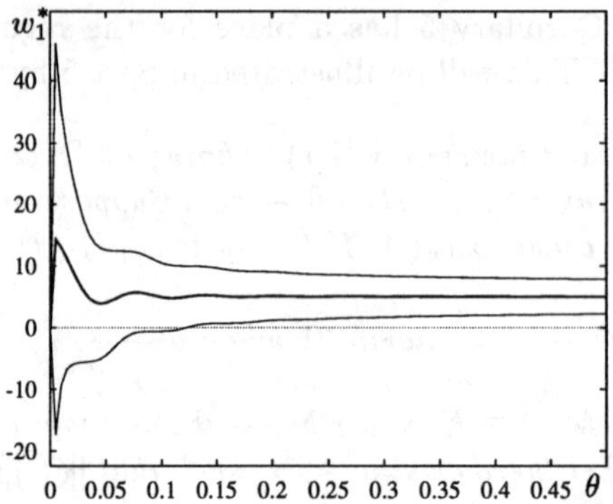

Fig. 1. The projections of w^* and of the boundaries of $\mathcal{W}_0^+[y_\theta(\cdot)]$ onto span$\{\varphi_1\}$

Example 2. Here all the parameters are identical to those of Example 1 but $G(t)u(t) = (\varphi_1, u(t))_{\mathbb{H}} + (\varphi_2, u(t))_{\mathbb{H}}$. Fig.2 shows the projection of the informational domains $\mathcal{W}_0^+[y_\theta(\cdot)]$ onto the subspace span$\{\varphi_1, \varphi_2\} \subset \mathbb{H}$ for the $y_\theta(t)$, given by the same $\xi(t)$ but by $u_0(x) = 5\varphi_1(x)+\varphi_2$. The graphs of $(\varphi_1, w^*)_{\mathbb{H}}$ and $(\varphi_2, w^*)_{\mathbb{H}}$ are outlined in their own right on Fig(s).3a, 3b correspondingly.

References

1. Gajewski, J., Zacharias, K.: Zur Regularisierung einer Klasse nichtkorrekter Probleme bei Evolutionsgleichungen. J. Math. Anal. and Appl. **38** (1972) 784-789
2. Kurzhanski, A.B.: Control and observation under uncertainty conditions. Nauka Moscow (1977)
3. Kurzhanski, A.B., Khapalov, A.Yu.: On the state estimation problem for distributed systems. Lect. Notes in Control and Inform. Sci. **83** (1986) 102-113
4. Kurzhanski, A.B., Khapalov, A.Yu.: An observation theory for distributed-parameter systems. J. Math. Sys., Estimat., and Control. **1** (1991) 389-440
5. Kurzhanski, A.B., Sivergina, I.F.: On the inverse problems for evolutionary systems: guaranteed estimates and the regularized solutions. Lect. Notes in Control and Inform. Sci. **154** (1990) 93-101
6. Kurzhanski, A.B., Valyi, I.: Ellipsoidal Calculus to Estimation and Control. Birkhauser Boston (to appear)
7. Lavrentyev, M.M., Romanov, V.G., Shishatskii, S.P.: Ill-posed problems in analysis and mathematical physics. Nauka Moscow (1980)
8. Lions, J.L.: Optimal control of systems governed by partial differential equations. Springer NY (1971)
9. Tikhonov, A.N., Arsenin, V,Ya.: Solution methods for ill-posed problems. Nauka Moscow (1979)

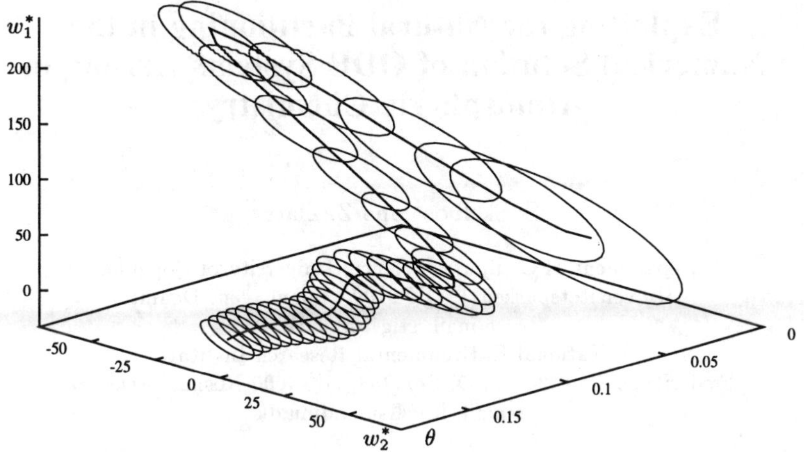

Fig. 2. The projections of w^* and of the boundaries of $\mathcal{W}_0^+[y_\theta(\cdot)]$ onto span$\{\varphi_1\ ,\varphi_2\}$

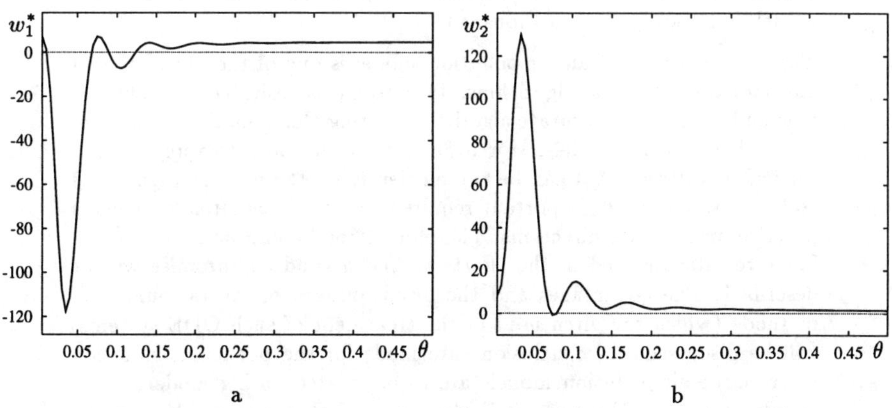

Fig. 3. The graphs of $(\varphi_1, w^*)_{\mathbb{H}}$ and $(\varphi_2, w^*)_{\mathbb{H}}$

Exploiting the Natural Partitioning in the Numerical Solution of ODE Systems Arising in Atmospheric Chemistry

S. Skelboe[1] and Z. Zlatev[2]

[1] Department of Computer Science, University of Copenhagen, Universitetsparken 1, DK-2100 Copenhagen, Denmark
e-mail: stig@diku.dk
[2] National Environmental Research Institute, Frederiksborgvej 399, P. O. Box 358, DK-4000 Roskilde, Denmark
e-mail: luzz@sun2.dmu.dk

Abstract. Large air pollution models are commonly used to study transboundary transport of air pollutants. Such models are described mathematically by systems of partial differential equations (the number of equations being equal to the number of pollutants involved in the model). The use of appropriate splitting procedures leads to several sub-models. If the model is discretized on a (96x96x10) grid and if the number of pollutants is 35, then a system of ODE's containing 3225600 equations is to be treated in each sub-model. The ODE system of the chemical sub-model can be decoupled to (96x96x10) small systems, each of them containing 35 equations. The number of time-steps, needed in each sub-model, is typically several thousand.

The chemical part of an air pollution model is one of the most difficult parts for the numerical algorithms. Therefore it is desirable to apply reliable and sufficiently accurate algorithms during the numerical treatment of the chemical sub-models. Moreover, it is also desirable to apply fast numerical algorithms that can be run efficiently on the modern high-speed computers. These two important requirements work, as often happens in practice, in opposite directions. Therefore a good compromise is needed. Some results achieved in the efforts to find a good compromise will be described. The advantages and the disadvantages of several numerical methods (which are often used in the treatment of such ODE systems) will be discussed. All conclusions are made for the particular situation where large air pollution models are to be treated on big modern high-speed computers. Moreover, it is also assumed that a particular air pollution model, the **Danish Eulerian Model**, is used. However, the ODE systems that arise in the chemical sub-models have at least three rather common properties, which appear again and again when large scientific and engineering problems are studied. These systems are large, stiff and badly scaled. Therefore some of the conclusions are also valid in a much more general context, i.e. in all cases where large, stiff and badly scaled systems of ODEs are to be handled numerically.

1 What is typical for the ODE systems in atmospheric models?

1. Large number of relatively small ODE systems are to be solved. The chemical sub-models are an important part in the models for studying long-range transport of air pollutants. These models are described mathematically by PDE systems. The use of any discretization followed by a splitting procedure leads to the solution of relatively small but numerous ODE systems in the chemical sub-model. For example, if the space domain of the model is discretized by using a (96x96x10) grid and if the number of chemical species involved in the model is 35, then (96x96x10) ODE systems, each of them containing 35 equations, are to be treated during many time-steps (in conjunction with the other physical processes).

2. The chemical ODE systems are normally very stiff. Some of the chemical reactions are very fast, others are rather slow. This fact implies stiffness. There are periods (sun-rises and sun-sets) when certain chemical processes (the so-called photochemical reactions) are activated or disactivated. This also causes quick changes of some species and, thus, introduces extra difficulties. Therefore one should be very careful in the selection of numerical methods for the chemical sub-models.

3. The chemical ODE systems are badly scaled. The concentrations of the chemical species, which are often measured in $molecules/cm^3$, vary in different ranges. An example is given in Table 1. It is seen that (i) the different species have different magnitudes and (ii) some species vary very slowly (as CO), while other species vary in very large intervals (compare the minimal concentrations with the maximal concentrations for NO and $O^{(1)}D$ in Table 1).

Species	Starting concentration	Minimal concentration	Maximal concentration
NO	10^{+10}	$3.6 * 10^{+03}$	$1.8 * 10^{+10}$
NO_2	10^{+11}	$8.8 * 10^{+08}$	10^{+11}
CO	$3.8 * 10^{+14}$	$3.7 * 10^{+14}$	$3.8 * 10^{+14}$
$O^{(1)}D$	10^{+03}	$2.1 * 10^{-43}$	$1.3 * 10^{+03}$

Table 1
Variations of some chemical species.

4. Coupling with the other atmospheric processes. The use of a splitting procedure allows us to treat the chemical sub-model with separate numerical methods. However, the chemical sub-model has to be coupled with the other physical processes: horizontal transport (advection), horizontal diffusion, deposition, emissions and vertical exchange between the different layers. This implies that the chemical ODEs should be integrated over a series of very short intervals. After the integration over a short interval, the other physical processes are handled (which leads to considerable changes of the concentrations) and the modified concentrations are used as starting values in the chemical sub-model for the next short interval. This means that the problem of starting the integration process occurs very often for the chemical ODEs.

5. The use of efficient integration techniques is crucial. The difficulties discussed above show that the choice of numerical methods for the solution of the ODE systems in the chemical sub-model is important. In many atmospheric models, the treatment of the chemical sub-models is the most time-consuming part. The problem of finding the optimal numerical method for chemical ODEs is still open. A choice where the chemical species are partitioned into several blocks will be discussed in the following section.

2 How can natural partitioning be exploited?

The particular chemical scheme used in this paper is the condensed CBM IV scheme, but the method described below can also be applied in the numerical treatment of other chemical schemes; as, for examples, those discussed in [5]. The condensed CBM IV scheme has been proposed in [1] and has been used in several air pollution models. It contains 35 species and 71 chemical reactions. In this paper the chemical scheme will be considered as a box model. This is relevant, because in large air pollution models the splitting procedure leads to several sub-models which can be treated separately over short time-intervals. The chemical sub-model could be viewed as a set of box-models (the number of the box-models being equal to the number of grid-points used in the discretization of the space domain).

The different species involved in the CBM IV chemical scheme are coupled in a different way. Some species are strongly coupled, other species are weakly coupled. One can try to detect this and decompose the system into different loosely connected sub-systems. One can also investigate the rate coefficients of the different chemical reactions and try to separate the species into two groups: species involved in fast chemical reactions and species involved in slow chemical reactions. Chemists could help substantially in the division process. We shall assume that some division into species involved in fast chemical reactions and species involved in slow chemical reactions has somehow been obtained and shall discuss the treatment of the partitioned system. It should also be noted that the first partitioning discussed when a linear analysis of the method is carried out (Section 4) is very similar to the partitioning used in [2].

Assume that a suitable splitting procedure has been applied in the air pollution model under consideration and consider one of the box-models in the chemical part. It can be defined by a system of ordinary differential equations,

$$y' = f(t, y), \; y(t_0) = y_0 \text{ and } t \geq t_0 \tag{1}$$

where the concentrations $y: \; R \to R^S$, f is Lipschitz continuous in y and $f: R \times R^S \to R^S$. Stable systems of differential equations are considered stiff when the stepsize of the discretization by an *explicit* integration method is limited by stability of the discretization and not by accuracy. Efficient numerical integration of the chemical sub-model therefore requires *implicit* integration methods.

Let the original problem (1) be partitioned as follows,

$$\begin{pmatrix} y_1' \\ y_2' \\ \vdots \\ y_q' \end{pmatrix} = \begin{pmatrix} f_1(t,y) \\ f_2(t,y) \\ \vdots \\ f_q(t,y) \end{pmatrix}, \quad y = \begin{pmatrix} y_1 \\ y_2 \\ \vdots \\ y_q \end{pmatrix}, \quad y(t_0) = \begin{pmatrix} y_{1,0} \\ y_{2,0} \\ \vdots \\ y_{q,0} \end{pmatrix} \tag{2}$$

where $y_r : R \to R^{s_r}$, $f_r : R \times R^S \to R^{s_r}$ and $\sum_{i=1}^{q} s_i = S$.

Each sub-system corresponds to a group of strongly interacting species (or just one single equation) which on the other hand interact weakly with the remaining species. This property is formalized by the monotonic max-norm stability condition introduced in [4].

3 Implementation of partitioned integration

The decoupled implicit Euler method is defined in [7] as follows, where the r'th sub-system is discretized by the backward Euler formula:

$$y_{r,n} = y_{r,n-1} + h_n f_r(t_n, \tilde{y}_{1,n}, \ldots, \tilde{y}_{r-1,n}, y_{r,n}, \tilde{y}_{r+1,n}, \ldots, \tilde{y}_{q,n}), \tag{3}$$

where $n = 1, 2, \ldots$, and the variables $\tilde{y}_{i,n}$ are convex combinations of values in $\{y_{i,k} \mid k \geq 0\}$ for $i \neq r$. The time step is denoted by $h_n = t_n - t_{n-1}$ such that $t_n = t_0 + \sum_{j=1}^{n} h_j$. The method exploits the partitioning in (2) by limiting the implicitness to the individual sub-systems thus reducing the necessary nonlinear and linear algebra required for solving the corrector equations.

The convexity of $\tilde{y}_{i,n}$ is necessary for the proof of stability of the decoupled implicit Euler method in [4]. This paper also discusses a multirate formulation where the stepsize, $h_{r,n}$, may be different for each sub-system. In practice, improved accuracy at the expense of weaker stability properties is obtained by using extrapolation,

$$\tilde{y}_{i,n} = y_{i,n-1} + h_n/h_{n-1}(y_{i,n-1} - y_{i,n-2}) \tag{4}$$

The decoupled implicit Euler equations (3) can be solved in parallel on a parallel computer, but if the sequence of sub-systems $r = 1, 2, \ldots$ is solved on a sequential computer, it is advantageous to use a "Gauss-Seidel" type formulation,

$$y_{r,n} = y_{r,n-1} + h_n f_r(t_n, y_{1,n}, \ldots, y_{r-1,n}, y_{r,n}, \tilde{y}_{r+1,n}, \ldots, \tilde{y}_{q,n}) \tag{5}$$

The decoupled implicit Euler formula (3) can be iterated in either a block Jacobi or the following block Gauss-Seidel scheme,

$$y_{r,n}^{(m+1)} = y_{r,n-1} + h_n f_r(t_n, y_{1,n}^{(m+1)}, \ldots, y_{r-1,n}^{(m+1)}, y_{r,n}^{(m+1)}, y_{r+1,n}^{(m)}, \ldots, y_{q,n}^{(m)}), \tag{6}$$

with $y_{i,n}^{(0)} = \tilde{y}_{i,n}$. The inner loop of the iteration involves the solution of nonlinear systems of dimension s_r (where s_r can be 1) for $r = 1, \ldots, q$. When the partitioning is chosen appropriately, the Gauss-Seidel iteration will converge. If the iteration (6) is carried to convergence, the discretization is the classical backward Euler formula where the discretization equations are solved using block Gauss-Seidel instead of a Newton type iteration applied to the full system.

4 Linear analysis of the method

The paper [4] gives a sufficient condition for stability of the discretization (3) called monotonic max-norm stability. It is easy to verify that this stability condition is *not* fulfilled for the partitioned chemical sub-model described in the previous section (this condition is also violated for the chemical schemes discussed in [5]). The analysis is based on linearizations of the system, and this linearization also makes it possible to explain the observed behaviour of the implementation.

Consider the linear system of ODEs, $y' = Ay$ where $A = \partial f / \partial y$ for some y along the solution trajectory. Two different partitionings are specified by the block diagonal matrices D_1 and D_2. The first five diagonal blocks of D_1 have dimensions 4, 2, 3, 2, 2, the dimension of the remaining diagonal blocks being 1. D_2 has one single diagonal block of dimension 13 corresponding to the same 13 rows and columns as the blocks of D_1. The remaining diagonal blocks have dimension 1 and they are identical to the corresponding part of D_1. The non-zero elements in the diagonal blocks of D_1 and D_2 are identical to the non-zero elements with the same row and column indices in A. The off-diagonal parts of A are defined as $B_1 = A - D_1$ and $B_2 = A - D_2$, respectively.

The decoupled implicit Euler formula is now expressed as follows (the "Jacobi" formulation equivalent to (3)),

$$y_n = y_{n-1} + h_n(D_1 y_n + B_1 \tilde{y}_n) \tag{7}$$

Let the diagonal square block r of D_1 be denoted d_r and the corresponding row blocks of B_1 be denoted by b_{rs} where $b_{rr} = 0$. Construct the matrix S with off-diagonal elements $\|b_{rs}\|$ and diagonal elements $\mu(d_r)$ where $\mu(\cdot)$ is the logarithmic norm corresponding to the norm $\|\cdot\|$ used for the off-diagonal elements [4]. Then $y' = Ay$ is monotonically max-norm stable if the eigenvalues of S are all in the left-hand half plane. For a typical Jacobian A of the chemical sub-model, the spectral abscissa of S is computed as 0.3052 thus violating the monotonic max-norm stability condition.

With $\tilde{y}_n = y_{n-1}$, (7) can be expressed as $y_n = M_1 y_{n-1}$ where $M_1 = (I - h_n D_1)^{-1}(I + h_n B_1)$. The discretization is stable if all the eigenvalues of M_1 are inside the unit circle. This would have been guaranteed by monotonic max-norm stability. While the spectral abscissa of D_1 is close to zero leading to eigenvalues of S in the right-hand half plane, the norm of D_1 is quite large, namely $7.5 * 10^4$ for the same example compared with the norm of B_1 which is 93.6. Therefore the eigenvalues of M_1 remain on (since some of the eigenvalues of A are zero) and inside the unit circle for all relevant values of h_n.

With $\tilde{y}_n$ being expressed by (4), the decoupled implicit Euler formula is

$$y_n = y_{n-1} + h_n(D_1 y_n + B_1(y_{n-1} + h_n/h_{n-1}(y_{n-1} - y_{n-2}))$$

This can be reformulated as $(y_n, y_{n-1})^T = M_a \ (y_{n-1}, y_{n-2})^T$ where

$$M_a = \begin{pmatrix} (M_1 + h_n^2/h_{n-1}(I - h_n D_1)^{-1} B_1) & -h_n^2/h_{n-1}(I - h_n D_1)^{-1} B_1 \\ I & 0 \end{pmatrix}$$

The eigenvalues of M_a are also on and inside the unit circle just as the eigenvalues of M_1 although the use of extrapolation for $\tilde{y}_n$ weakens the stability properties. However, this only shows up in the small eigenvalues where the eigenvalues of M_a stay larger than the eigenvalues of M_1 and $M = (I - h_n A)^{-1}$ for the classical implicit Euler formula. The large eigenvalues determine the accuracy since they approximate $\exp(h_n \lambda_r)$ where λ_r for $|h_n \lambda_r| \ll 1$ are the eigenvalues of A corresponding to the smooth (non-stiff) solution components. The large eigenvalues of M_a and M are very similar and more accurate than the large eigenvalues of M_1.

Numerical experiments show that $||M_a|| > 1$ (instability) for very large values of h_n while M_b, the analogue of M_a based on D_2 and B_2, remains stable. Both schemes give unstable discretizations for $h_n/h_{n-1} \gg 1$. Notice that M_1 does *not* depend on this stepsize ratio.

5 Some numerical results

The CBM IV scheme with the second partitioning from the previous section (specified by the block diagonal matrix D_2, one big block containing 13 species and 22 small blocks) has been used in the experiments.The scheme has been run over a time period of 42 hours (starting at 6 o'clock in the morning and finishing at 24 o'clock in the next day). This period contains several changes from day to night or from night to day. A diurnal variation of the temperature has been simulating by using a cosine function. The method based on partitioning has been compared with the QSSA, the quasi-steady-state-approximation (proposed for air pollution models in [3]; an improved version, similar to that discussed in [5], has been used here) as well as with two classical methods: the Backward Euler Method and the Trapezoidal Rule (these methods have also been used in [8]). The experiments have been carried out on a SUN work-station. Both the computing times and the accuracy of the methods involved have been compared. The computing times are measured in seconds. A reference solution obtained with a very small stepsize ($h = 0.001$) has been used to check the accuracy; the reference solution is practically the same as the solution obtained by the variable stepsize variable order code from [6] with an accuracy requirement $EPS = 10^{-7}$.

The specific variant of the decoupled implicit Euler formulas which is used to obtain the numerical results is (5) with constant stepsize and extrapolation (4). The sub-systems are solved with a Newton type iteration where the derivatives (or Jacobian and it's LU-factorization) are kept constant as long as the convergence rate is satisfactory.

The computing times are given in Table 2. The accuracy obtained in the calculation of some important chemical species is shown in Table 3. The accuracy of the concentration of the i'th species at a given step n is measured by the quantity:

$$\epsilon_{i,n} = \frac{\mid y_{i,n}^{calc} - y_{i,n}^{ref} \mid}{max(y_{i,n}^{ref}, 10^{-6} y_{i,n}^{background})} \tag{8}$$

where $y_{i,n}^{calc}$ is the value calculated by the numerical method under consideration, $y_{i,n}^{ref}$ is the corresponding value of the reference solution and $y_{i,n}^{background}$ is some small background value (this factor is only used to prevent division by zero when the concentrations become very small and could be set by the computer to zero when underflows take place). The maximal values (taken over the whole time-integration interval of 42 hours) of the relative errors $\epsilon_{i,n}$ for the four selected species are given in Table 3.

Method	$h = 30s$	$h = 1s$
QSSA	1.17	37.71
Backward Euler	1.11	28.61
Trapezoidal Rule	1.38	44.50
Partitioning	0.71	18.16

Table 2
Computing times obtained in the integration of the test-problem by four integration methods

Method	NO_2	O_3	OH	SO_2
QSSA	2.01E-2	3.20E-3	3.46E-1	5.35E-5
Backward Euler	8.65E-3	5.81E-4	1.10E-2	1.76E-5
Trapezoidal Rule	1.11E-3	5.81E-4	1.10E-2	1.76E-5
Partitioning	9.17E-3	5.74E-4	1.21E-2	1.04E-5

Table 3
The accuracy obtained in the integration of the test-problem by four integration methods with a stepsize $h = 30s$.

It is seen that the method based on partitioning is quicker than the other three algorithms (see the computing times in Table 2).

It is also seen that the last three algorithms give approximately the same accuracy, which is higher than the accuracy achieved by the QSSA. QSSA gives a rather poor accuracy for some radicals (see, for example, the results for the hydroxyl radical OH in Table 3). For the species which vary slowly (see the results for SO_2) all algorithms produce rather accurate results.

6 Future plans

Several tasks must be solved in the near future. The implementation of the method based on partitioning could be improved (for example, the use of some sparse matrix algorithm may be appropriate). Experiments with some other partitionings, as the partitioning discussed in Section 4, will be carried out. Experiments with other series of starting values and with some emission scenarios are also needed. Finally, the best among the partitionings tested has to be implemented in the Danish Eulerian Model ([8]) and the performance of the new model on high-speed computers have to be compared with previous results (in which the QSSA has been used in the chemical sub-model; see [8] and [9].)

References

1. Gery, M. W., Whitten, G. Z., Killus, J. P. and Dodge, M. C., "A photochemical kinetics mechanism for urban and regional computer modeling". J. Geophys. Res., 94 (1989), 12925-12956.
2. Hertel, O., Berkowicz, R., Christensen, J. and Hov, Ø., "Test of two numerical schemes for use in atmospheric transport-chemistry models". Atmospheric Environment, 27A (1993), 2591-2611.
3. Hestvedt, E., Hov, Ø. and Isaksen, I. A., "Quasi-steady-state approximation in air pollution modelling: comparison of two numerical schemes for oxidant prediction". Internat. J. Chem. Kinetics, 10i (1979), 971-994.
4. Sand, J. and Skelboe, S., "Stability of backward Euler multirate methods and convergence of waveform relaxation", BIT 32 (1992), pp. 350-366.
5. Sandu, A., Verwer, J. G., van Loon, M., Carmichael, G. R., Potra, F. A., Dabdub, D. and Seinfeld, J. H., "Benchmarking stiff ODE solvers for atmospheric chemistry problems I: Implicit versus explicit" Submitted for publication in Atmospheric Environment.
6. Skelboe, S., "INTGR for the integration og stiff systems of ordinary differential equations", Report IT 9, March 1977, Institute of Circuit Theory and Telecommunication, Technical University of Denmark.
7. Skelboe, S., "Parallel integration of stiff systems of ODEs", BIT 32 (1992), pp. 689-701
8. Zlatev, Z., "Computer treatment of large air pollution models". Kluwer Academic Publishers, Dordrecht-Boston-London, 1995.
9. Zlatev, Z., Dimov, I and Georgiev, K., "Modeling the long-range transport of air pollutants". Computational Science and Engineering, Vol. 1, No. 3 (1994), 45-51.

Existence and Stability of Traveling Discrete Shocks

Yiorgos Sokratis Smyrlis[1] and Shih Hsien Yu[2]

[1] Department of Mathematics and Statistics,
University of Cyprus,
Nicosia 1678, CYPRUS
[2] Department of Mathematics,
University of California at Los Angeles,
Los Angeles, CA 90024

Abstract. We are concerned with existence and stability questions for finite difference schemes approximating solutions of scalar conservation laws with shocks. A suitable model for the study of the artifacts created by these schemes near the shocks are the traveling discrete shock profiles; these are discrete shock profiles $v = (v_k)_{k\in\mathbb{Z}}$ which reappear shifted, when the scheme is applied on them, according to the speed of the shock. Existence of such profiles connecting entropy admissible shocks is already established for monotone schemes, first and third order accurate schemes and the Lax-Wendroff scheme. Jennings showed existence and l^1-stability of these profiles for conservative monotone schemes. Smyrlis showed existence and parametrization by the amount of excess mass and stability for stationary profiles of the Lax-Wendroff scheme. Shih Hsien Yu showed existence of traveling profiles of mild strength for the Lax-Wendroff scheme using inertial manifolds theory. Here we study traveling discrete shock profiles for Lax-Wendroff, Engquist-Osher and monotone schemes. We show existence of such profiles with small shock speed. We also show that these profiles are stable with respect to suitably weighted l^2-norms.

1 Introduction

A scalar conservation law is an initial value problem of the form

$$u_t + f(u)_x = 0\ ,\ \ u(x,0) = u_0(x) \tag{1}$$

where $(x,t)\in\mathbb{R}\times\mathbb{R}_0^+$, u, f scalar, u the conserved quantity and f the flux function, nonlinear in general. The above is one of the simplest quasi-linear partial differential equations and as it is well known, regardless how smooth or small the initial data are, their solutions may very well fail to be smooth for all times. Nevertheless, since conservation laws, derive from integral relations, we could define solutions in a weaker sense, the sense of distributions:

Definition 1. A function $u\in L^\infty(\mathbb{R}\times\mathbb{R}_0^+)$ is called a weak solution of (1) if it satisfies :

$$\int_{\mathbb{R}} dx \int_{\mathbb{R}_0^+} dt\, (u\phi_t + f(u)\phi_x) + \int_{\mathbb{R}} u_0(x)\phi(x)dx = 0 \tag{2}$$

for all test functions $\phi \in C_0^\infty(\mathbb{R}\times\mathbb{R}_0^+)$.

Clearly if u is smooth then (1) imply (2). However, if u has a jump discontinuity across the curve $x = \xi(t)$ and

$$\lim_{x\to\xi(t)+} u(x,t) = u_l\ ,\quad \lim_{x\to\xi(t)-} u(x,t) = u_r \tag{3}$$

then the definition of the weak solution implies that

$$s = \frac{d\xi}{dt} = \frac{f(u_r) - f(u_l)}{u_r - u_l} \tag{4}$$

and s is the speed of propagation of the discontinuity. Equation (4) is known as the *Rankine-Hugoniot shock condition.*

Formulation (2) does not guarantee uniqueness. More specifically, the same discontinuous initial data may produce weak solutions with shocks, as well as, weak solutions with rarefactions. Nevertheless only one of them has physical relevance. Criteria selecting the right one are called *Entropy conditions.* P.D. Lax in [7] proposed one such criterion which, in the scalar case, reduces to the following formulation :

The shock (u_l, u_r) is admissible whenever

$$f'(u_r) < s < f'(u_l) \tag{5}$$

provided that f is strictly convex.

The simplest admissible discontinuous weak solution, is a *shock*

$$u(x,t) = \begin{cases} u_l \text{ if } & x < st \\ u_r \text{ if } & x > st \end{cases} \tag{6}$$

with u_l, u_r and s related by (4).

A variety of finite difference schemes have been employed for the approximation of the solution of (1). Such schemes not only provide numerical answers but one anticipates that they could reveal patterns and information about the solutions and the theory to be developed. When a finite difference scheme approximates the solution of a partial differential equation, there is a truncation error expressed by introducing higher order terms in the equation. The form and effects of these artificially introduced terms are related to the order of accuracy of the scheme. In the case of *monotone schemes,*[3] which are first order accurate, the new terms create artificial viscosity and thus smeared shocks. See Figure 1. There is a continuous analogue. The traveling wave solutions of viscous conservation laws converge to a sharp shock as viscosity parameter tends to zero. Extensive study of existence and stability of traveling wave solutions of

[3] A finite difference scheme

$$u_k^{n+1} = G(u_{k-r}^n, u_{k-r+1}^n, \dots, u_{k+r}^n)$$

is called monotone if the partial derivatives $G_{v_\rho}(v_{-r}, v_{-r+1}, \dots, v_r)$ are positive for all $\rho = -r, \dots, r$. One such scheme is the Lax-Friedrichs.

viscous conservation laws could be found in Il'in & Oleinik [4], S. Kawashima & A. Matsumura [10], Tai-Ping Liu [9] to mention a few.

For second-order accurate schemes the prevalent higher order terms are dispersive. One such example is the Lax-Wendroff scheme [8] :

$$u_k^{n+1} = u_k^n - \frac{\Delta t}{\Delta x}\Big(F(u_k^n, u_{k+1}^n) - F(u_{k-1}^n, u_k^n)\Big) \tag{7}$$

where the numerical flux function is given by

$$F(v,w) = \frac{1}{2}\Big(f(v) + f(w)\Big) - \frac{1}{2}\frac{\Delta t}{\Delta x} f'\left(\frac{v+w}{2}\right)\Big(f(w) - f(v)\Big)$$

and u_k^n approximates the value of $u(n\Delta t, k\Delta x)$. Lax-Wendroff is a nonmonotone scheme in *conservation form*[4]

These numerical artifacts, which persist in approximate solutions of conservation laws near the shocks, were investigated by Gray Jennings [5], who studied discrete shocks. Let $u^{n+1} = \mathcal{L}_\lambda(u^n)$ be a finite difference operator where $u^n = (u_k^n)_{k\in\mathbb{Z}}$, $n\in\mathbb{N}$,

$$\mathcal{L}(u)_k = G_\lambda(u_{k-r}, \dots, u_{k+r})$$

and $\lambda = \Delta t/\Delta x$.

Definition 2. A profile $v = (v_k)_{k\in\mathbb{Z}}$ is said to be traveling with speed s with respect to the finite difference operator $\mathcal{L}$ if p iterations of $\mathcal{L}$ shift v by q positions, i.e.

$$\mathcal{L}^p(v) = S_-^q v \tag{8}$$

[4] A finite difference scheme $u_k^{n+1} = G(u_{k-r}^n, u_{k-r+1}^n, \dots, u_{k+r}^n)$, $k\in\mathbb{Z}$, $n\in\mathbb{N}$, is said to be in conservation form and thus a *conservative scheme* if

$$G(v_{k-r}^n, v_{k-r+1}^n, \dots, v_{k+r}^n) = v_k^n - \frac{\Delta t}{\Delta x}(F(v_{k-r+1}^n, v_{k-r+1}^n, \dots, v_{k+r}^n) - F(v_{k-r}^n, v_{k-r+1}^n, \dots, v_{k+r-1}^n))$$

Remark. (i) A conservative scheme is *consistent*, i.e. at least first order accurate, if it satisfies

$$F(u, u, \dots, u) = f(u) \qquad \text{[Consistency condition]}$$

(ii) Thus if $(v_k)_{k\in\mathbb{Z}}$ is a profile connecting u_l and u_r then consistency condition and conservation form implies that

$$\begin{aligned}\sum_{k\in\mathbb{Z}}(v_k^{n+1} - v_k^n) &= -\frac{\Delta t}{\Delta x}\sum_{k\in\mathbb{Z}}(F(v_{k-r+1}^n, v_{k-r+1}^n, \dots, v_{k+r}^n) - F(v_{k-r}^n, v_{k-r+1}^n, \dots, v_{k+r-1}^n)) \\ &= -\frac{\Delta t}{\Delta x}(f(u_r) - f(u_l)) .\end{aligned}$$

Thus the *total mass* increases according to the conservation law. in which the artificially introduced dispersive higher order terms create oscillations near the shocks. See Figure 2.

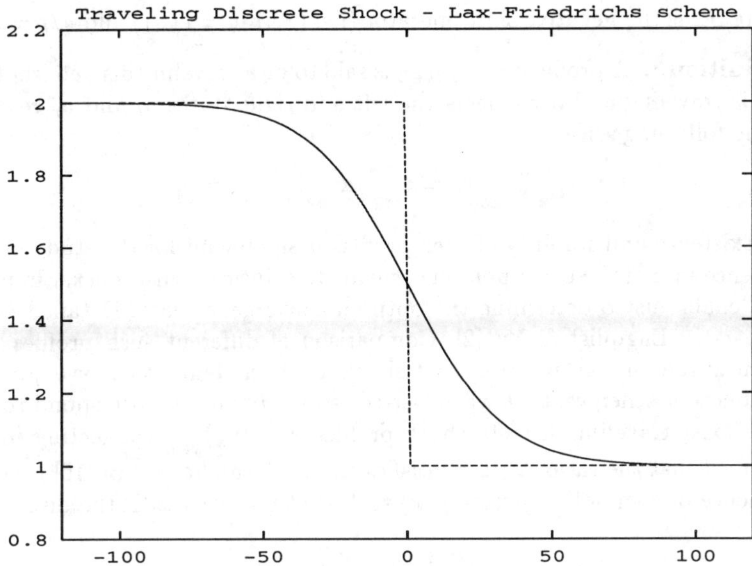

Fig. 1. Effects of Numerical viscosity.

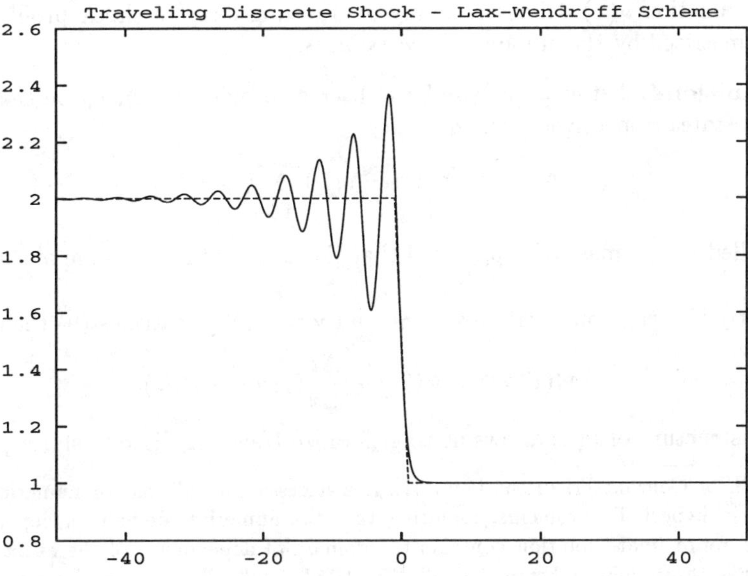

Fig. 2. Effects of Numerical dispersion.

where S_- is the negative shift operator (i.e. $(S_-v)_k = v_{k-1}$) and $q/p = s\Delta t/\Delta x$.

Definition 3. A profile $v = (v_k)_{k\in\mathbb{Z}}$ is said to be a traveling discrete shock profile if it is traveling and it connects the left and right states u_l and u_r respectively, in the following sense :

(9) $$\lim_{k\to-\infty} v_k = u_l \textit{ and } \lim_{k\to+\infty} v_k = u_r$$

Existence and stability of such profiles is significant for the study of the convergence rate of the corresponding scheme to solutions with shocks. In particular they enable us to determine the optimal convergence rate. Detailed discussion appears in Engquist & Yu [2]. Comparison of different such profiles plays an essential role in tracking the exact shock location. Jennings showed, in the case of monotone schemes, that whenever the states u_l and u_r correspond to a shock, there exist traveling discrete shock profiles $v = \left(v_k\right)_{k\in\mathbb{Z}}$ connecting u_l and u_r, provided that the ratio $\Delta t/\Delta x$ satisfies the *CFL condition.*[5] In [16] we prove the existence of a smooth function $\nu(x)$ such that for every $x\in\mathbb{R}$ the discrete profile

(11) $$\boldsymbol{\nu}^x = \Big(\nu(x+k)\Big)_{k\in\mathbb{Z}}$$

is a traveling discrete shock profile and

(12) $$\sum_{k\leq 0}\Big(\nu(x+k) - u_l\Big) + \sum_{k>0}\Big(\nu(x+k) - u_r\Big) = (u_r - u_l)x$$

This function $\nu(x)$ defines a family of traveling discrete shock profiles is thus parametrized by the amount of *excess mass.*

Definition 4. Let $\mathbf{v} = (v_k)_{k\in\mathbb{Z}}$ be a discrete profile and u_l, u_r be the left and right states respectively. The quantity

(13) $$\mathcal{M}(\mathbf{v}) = \sum_{k\leq 0}(v_k - u_l) + \sum_{k>0}(v_k - u_r)$$

is called excess mass of $\mathbf{v}$ provided that the above sums converge absolutely.

Clearly if $\mathcal{L}$ is a conservative scheme and $\mathbf{v} = (v_k)_{k\in\mathbb{Z}}$ satisfies (9) then

(14) $$\mathcal{M}(\mathcal{L}(\mathbf{v})) = \mathcal{M}(\mathbf{v}) + \frac{\Delta t}{\Delta x}(f(u_l) - f(u_r))$$

This structure of $\nu(x)$ allows us to generalize traveling discrete shock profiles:

[5] CFL or Courant-Friedrichs-Lewy [1], is a necessary condition for numerical stability in hyperbolic problems, requiring that the numerical domain of dependence of the approximate solution contains the domain of dependence of the exact solution. For a three point scheme, i.e. $u_k^{n+1} = G(u_{k-1}^n, u_k^n, u_{k+1}^n)$, approximating a scalar conservation law, the CFL condition reduces to the inequality

(10) $$\frac{\Delta t}{\Delta x}\left|f'\left(u_k^0\right)\right| \leq 1 \text{ for all } k \in \mathbb{Z}$$

Definition 5. A continuous function $\nu(x)$, $x \in \mathbb{R}$, is said to be a parametrized traveling discrete shock profile of the finite difference scheme $\mathcal{L}$ connecting the states u_l and u_r, i.e.

$$\lim_{x \to -\infty} \nu(x) = u_l \ , \ \lim_{x \to +\infty} \nu(x) = u_r \ , \tag{15}$$

if it satisfies (12) and $\boldsymbol{\nu}^{x - s\Delta t/\Delta x} = \mathcal{L}(\boldsymbol{\nu}^x)$ for all $x \in \mathbb{R}$ where $\boldsymbol{\nu}^x$ is defined by (11) and u_l, u_r and s satisfy (4).

This parametrization follows a linear relation with the amount of excess mass. Indeed, if $\boldsymbol{\nu}^x$ is the discrete profile defined by (11), then using (13) and (14) we obtain $\mathcal{M}(\boldsymbol{\nu}^x) = (u_r - u_l)x$. Thus, if the traveling discrete shock profiles are *stable*, the limit of the iterations of a perturbed one such profile is not necessarily the same unperturbed one, but instead, a member of the resulting one-parameter family of profiles. Particularly the one with the right amount of excess mass.

Parametrized traveling discrete shock profiles are more convenient since they requires only one application of our operator $\mathcal{L}$ and our problem is reduced to studying the fixed points of $\mathcal{L}$ instead of $\mathcal{L}^p$. Furthermore it allows the speeds of propagation of the discrete shocks to be even *irrational numbers.*

Jennings has also shown l^1-stability of such parametrized profiles for conservative monotone schemes. Michelson [11] and Majda & Ralston [12] showed existence of discrete such profiles for first and third order schemes. Harten, Hyman & Lax [3] observed numerically that such discrete profiles exist and persist even in the case of Lax-Wendroff scheme, which does not belong in the previous categories and its shock profiles are oscillatory. They also realized that these profiles are not l^2-stable. Smyrlis [15] showed existence, parametrization by the amount of excess mass and stability of stationary profiles, i.e. $s = 0$. Stability was obtainable with respect to a weighted l^2-norm. Shih Hsien Yu [17] showed existence of traveling discrete profiles of the Lax-Wendroff scheme in the case of shocks with mild strength. His proof relied on the theory of inertial manifolds.

In [16] we prove existence and stability of traveling discrete shock profiles for Lax-Wendroff and Engquist-Osher [13] schemes.

The stability is with respect to a suitably weighted l^2-norm. Weights are necessary since our linearized finite difference operators are not l^2-stable. These weights depend on the states u_l and u_r. Their introduction deforms the continuous spectrum of the linearized operator in a way that it lies strictly inside the unit circle of the complex plane. We follow the methods of Jones et al [6], Sattinger [14] and Smyrlis [15] in the construction of our weighted norm.

More specifically we prove the following:

Theorem 6. *Let $\nu(x)$ be a stable parametrized traveling discrete shock profile of the Lax-Wendroff, Engquist-Osher or of a monotone scheme connecting the states u_l are u_r which correspond to an admissible shock. If $\bar{u}_r$ is sufficiently close to u_r then there exists a parametrized traveling discrete shock profile $\bar{\nu}(x)$ connecting u_l and $\bar{u}_r$. Furthermore this profile is stable with respect to iterations* *of the finite difference operator in the norm of the weighted space $C_{\alpha,\beta}$ where*

$$C_{\alpha,\beta} = \left\{ u \in C(\mathbb{R}) \;\middle|\; \|u\|_{\alpha,\beta} < +\infty \text{ and } u \text{ satisfies } (12) \right\}$$

where

$$\|u\|_{\alpha,\beta} < +\infty \max \left\{ \sup_{x<0} \alpha^{-x} |u(x)|, \; \sup_{x \geq 0} \beta^{x} |u(x)| \right\}$$

with the weights $\alpha, \beta > 1$ depending on u_l and u_r .

Remark. (i) The existence proof is obtainable using a suitable form of the implicit function theorem. We need that the initial profile $\nu(x)$ is *stable*, which means, stable with respect to the norm of the Banach space $C_{\alpha,\beta}$. Special cases of parametrized traveling discrete shock profiles are obtained in [15] and their stability with respect to norm of $C_{\alpha,\beta}$ is also obtainable. (ii) The weights α, β in the definition of $C_{\alpha,\beta}$ are necessary for stability. Such profiles are known to be unstable with respect to the standard L^2 norm in the case of the Lax-Wendroff scheme. See [3]. The presence of these weights deforms the continuous spectrum of the linearized finite difference operator around the profile in a way that the whole continuous spectrum lies strictly inside the unit circle of the complex plane. See Figure 3.

References

1. Courant, R., Friedrichs, K. O., Lewy, H., *Über die Partiellen Differenzialgleichungen der Mathematischen Physik,* Math. Ann., **100** (1928), pp. 32-74.
2. Björn Engquist, Shih Hsien Yu, *Convergence of finite difference schemes for piecewise smooth solutions with shocks,* to appear.
3. A. Harten, J.M. Hyman, P.D. Lax, *On finite-difference approximations and entropy conditions for shocks,* Comm. Pure Appl. Math., **29** (1976), pp. 297-322.
4. A.M. Il'in, O.A. Oleinik, *Behavior of the solution of the Riemann problem for certain quasilinear equations for unbounded increase of the time,* Amer. Math. Soc. Transl. Ser. 2, **42** (1964), pp. 19-23.
5. G. Jennings, *Discrete shocks,* Comm. Pure Appl. Math., **27** (1974), pp. 25-37.
6. C.K.R.T. Jones, R. Gardner, T. Kapitula, *Stability of traveling waves for non-convex scalar viscous conservation laws,* Comm. Pure Appl. Math., **46** (1993), pp. 505-526.
7. P.D. Lax, *Hyperbolic systems of conservation laws II,* Comm. Pure Appl. Math., **10** (1957), pp. 537-566.
8. P.D. Lax, B. Wendroff, *On Stability of finite difference schemes,* Comm. Pure Appl. Math., **15** (1962), pp. 363-371.
9. Tai-Ping Liu, *Nonlinear stability of shock waves for viscous conservation laws,* Memoirs of the Amer. Math. Soc., Number 328, July 1985.
10. S. Kawashima, A. Matsumura, *Asymptotic stability of traveling wave solutions of systems for one-dimensional gas motion,* Comm. Math. Phys., **101** (1985), pp. 97-127.

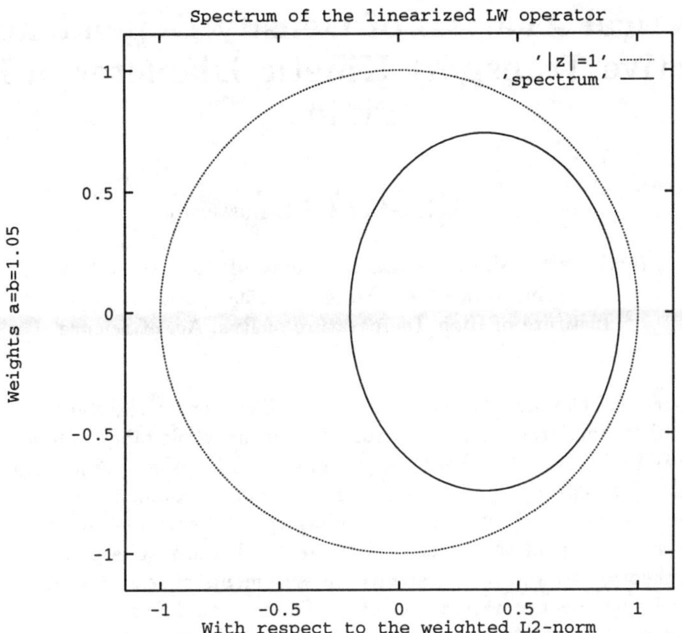

Fig. 3. The deformed continuous spectrum of the linearized LW operator.

11. D. Michelson, *Discrete shocks for difference approximations to system of conservation laws,* Adv. Appl. Math., **5** (1984), pp. 433-469.
12. A. Majda, J. Ralston, *Discrete shock profiles for systems of conservation laws,* Comm. Pure Appl. Math., **32** (1979), pp. 445-482.
13. B. Engquist, S. Osher, *One-sided difference approximations for nonlinear conservation laws,* Math. Comp., **36** (1981), pp. 321-351.
14. D.H. Sattinger, *On the stability of waves of nonlinear parabolic systems,* Adv. Math., **22** (1976), pp. 312-355.
15. Y.S. Smyrlis, *Existence and Stability of Stationary Profiles of the LW Scheme,* Comm. Pure. Appl. Math., **42** (1990), pp. 509-545.
16. Y.S. Smyrlis, Shih Hsien Yu, *On the stability of traveling discrete shocks of conservation laws,* to appear.
17. Shih Hsien Yu, *Existence of Discrete Shock Profiles for the Lax-Wendroff Scheme,* to appear.

Numerical Analysis of Density Dependence for Effective Transport Kinetic Diameter of Real Fluids

I.A. Sokolova[1] and B.E.Lusternik[2]

[1] Institute for Mathematical Modelling of Rus. Ac. of Science
Miusskaya pl. 4a, Moscow, 125047, Russia
[2] Institute of High Temperature of Rus. Ac. of Science

Abstract. The model of effective kinetic diameter for real fluids is considered to simulating transport properties in the whole range of temperature and pressure on the base of equations of Enskog theory. The model is relies on the effective potential model of pair interaction, which has been taken in terms of density due to shielding effect of surrounding particles. The comparison of the effective diameter with the results of perturbation theory, MD and from viscosity measurements of real matters shows a good agreement at various densities. The new model of effective diameter makes possible to economize computer's resources while computing transport processes and solving applied tasks.

1 Introduction

The viscous flow of fluids are of great interest nowadays. The behavior of viscosity in wide ranges of temperature and pressure is a subject of many investigations. But direct measurements as well as generalization of viscosity into universal relations enclose the bounded domain of data, and some area on the map of knowledge of properties remains empty. A mathematical simulation alone is the best way to investigate the transport properties in various matter states.

The most descriptions of transport processes in liquids are in some way to statistical mechanic, molecular dynamic, perturbation theory. But, the 'pure gas' model is not enough accurate representation of the fluid state, as expansion away from its ideal state. The many more terms would be required to describe the liquid state adequately. In addition, convergence of the expansion into a liquid state is not assured, if all the terms could be calculated. From the other hand, equilibrium molecular dynamic requires a very large system, if one wishes to reproduce the viscosity or thermal conductivity to within an acceptable accuracy. And although the state of the art of computer simulation has been broadened recently by the introduction of nonequilibrium molecular dynamic, they are not acceptable for the multiple calculations owing to high resource-consume.

Another way to implement computer simulation is to use the kinetic theory foundation and Enskog method of generalization for dense gases and fluids [1]. But the Enskog theory deals with hard sphere model of particles. It is unsound for the real fluids states because the real interactions between particles are not taken into consideration. But if the model of effective diameter of sphere is taken properly as well as the effective potential of particle interaction, the real fluids transport properties may be successfully simulated. The formalism of hard-sphere theory remains just the same. In principle, selection of a proper effective diameter σ_{eff} gives the possibility to describe viscosity and other properties of real substance in a wide range of state parameters with the help of well-developed and simple methods of the hard spheres theory.

2 Models of Effective Diameters

2.1 Collisions Diameter

According to collision mechanism of impulse transfer, the collision diameter σ_{coll} is determined as an average distance for impulse transfer from one molecule to another and is formed under governing effect of repulsive forces. The collision diameter can be found from different properties of substances, as well as from the statistic mechanic theory, using the pair interaction potential $U(r)$. The collision diameter due to perturbation theory of Barker-Henderson and Andersen-Chandler-Weeks is defined according the to effective volume of hard spheres

$$b = 2\pi \int_0^{r_{\min}} \left\{1\left[1 + \frac{U_{rep}(r)}{kT}\right] \exp\left(\frac{U_{rep}(r)}{kT}\right)\right\} r^2 dr\,, \tag{1}$$

with the help of the following relation $b = 2/3\pi N\sigma_{col}^3$, where $U_{rep}(r)$ is the repulsive part of real potential function, $r_{\min}$ is the position of the minimum of interaction energy and N is the Avogadro's number.

As one can see from (Eq.1), σ_{coll} is a function of temperature alone. From the first order perturbation theory, if the repulsive potential is taken in form of inverse power of distance $U_{rep}(r) = (\sigma_0/r)^n$, the dependence $\sigma_{col} = \sigma_{col}(T)$ has the following form [3]

$$\frac{\sigma_{col}}{\sigma_0} = (\beta^*)^{1/n}(1 + \gamma/n)\,, \quad \beta^* = 1/kT\,, \quad \gamma = 0.5772\,.$$

Another group of methods for determination of the effective collisions diameter is based on possibility of molecules because of strong mutual repulsion to form a firm space structure in solid phase or closed order structure in liquids. Measurements of structure factor solid or liquid by means of diffraction methods allow to find the radial distribution function (RDF) $g(r)$ in with the position of the first maximum r_1, which accords to a collision diameter of the molecule $\sigma_{coll} = r_1$.

Collision diameter σ_{coll} also determines the stable distance between particles in a solid state and can be determined from mole density of the solid phase of the substance ρ_s according to the expression

$$\sigma_{coll} = \left(\frac{6}{\pi N}\frac{\xi_S}{\rho_S}\right)^{1/3}$$

where ξ_S is the geometrical factor of close packing spheres in a solid phase. The lowest value of collision diameter σ_{coll} can be defined indirectly from the closest approach r_m of scattering particles at the zero of impact parameter in the spherical-symmetric field $U(r)$,

$$\sigma_{coll} \geq \lim_{\beta \to 0} r_m .$$

Comparison of perturbation theory calculations [2] σ_{coll} with experiments on P-V-T data for argon are good agreement and gathered into a group of data with the same temperature dependence. The values of σ_{coll}, while obtained by such methods, do not depend on density.

2.2 Kinetic Diameter

Kinetic diameter allows to modeling impulse transfer along the flight trajectory. It must be taken so that the deflection angles of scattering particles coincide with deflection angles of scattering particles of effective diameters. According to rare gas theory [1] this condition guarantees the coincidence of the viscosity η_0 of real gas and the viscosity of model system of hard spheres with the diameter of σ^0_{kin}.

$$\eta_0[\mu P \cdot s] = \frac{2.669\sqrt{MT}}{(\sigma^0_{kin})^2} = \frac{2.669\sqrt{MT}}{\sigma_0^2 \Omega^{(2.2)^*}(T)} , \tag{2}$$

where $\Omega^{(2.2)^*}$ is the collision integral calculated by using potential function $U(r)$ of particles interaction. From the formula (2) follows

$$\sigma^0_{kin}(T^*) = \left(\frac{2.669\sqrt{MT}}{\eta_0}\right)^{1/2} , \quad \text{or} \quad \sigma^0_{kin}(T^*) = \sigma_0[\Omega^{(2.2)^*}(T^*)]^{1/2} , \tag{3}$$

where $T^* = kT/\varepsilon$, σ_0 and ε are the parameters of the potential model, k is the Boltzman's constant. The kinetic diameters derived from formula (3) at different potential models as well as from viscosity data of direct measurements exhibit the identical function of temperature. But the kinetic diameter behavior is quite different from the collision diameter. There is a systematic difference between data: $\sigma^0_{kin}(T) > \sigma_{coll}(T)$ at all temperatures. The diffusion, thermal conductivity show the same behavior.

2.3 Effective Kinetic Diameter in Terms of Density

From computer simulation by MD methods [2] it is evident, that effective pair potential in dense media is a truncated function of distance owing to screening effect of neighbor particles. That is why the molecular dynamic calculations were carried out on the base of the Leonard-Jones model with the cut-off at some distance: $r^{\sim}(2.5 - 3)\sigma_0$, or using other modifications, for example (exp-6-8) truncated model. But such truncation at constant value of distance is not true for interactions of particles in real media [2].

We will regard the interaction of particles taking into account the overlap of distance fields due to close molecular packing in terms of density taking the effective potential in form of reference potential model with screening function. The screening function is an exponential to the power $m = 2$:

$$U(r) = [U_{rep}(r) + U_{att}(r)] \exp\left[-\left(\frac{r}{b_{cut}}\right)^2\right], \tag{4}$$

where b_{cut} is the screening parameter depending on density. We will use the average distance between the particles $s = n^{-1/3}$ as a geometrical screening parameter. Thus, $b_{cut} = s$ or $b_{cut}/\sigma_0 = b^*_{cut} = (n\sigma_0^3)^{-1/3} = (n^*)^{-1/3}$, where n is the number density. In case of the low density the model rearranges to the usual form of interaction potential. In case of the dense liquids the influence of attractive forces will be neglected.

Computer-calculated cross sections and collision integrals in terms of density make possible to define the effective kinetic diameter of real fluids.

3 Numerical Calculations

Cross section of scattering particles is defined as follows

$$Q^{(l)}(E) = 2\pi \int_0^\infty [1 \cos^{(l)}(\chi, E)]\beta d\beta \tag{5}$$

where β is the impact parameter, E is the relative energy of collision particles, χ is the deflection angle,

$$\chi = \pi - 2\beta \int_{r_m}^\infty \frac{dr/r^2}{\left[1 \frac{U(r)}{E}\frac{\beta^2}{r^2}\right]^{1/2}} \tag{6}$$

r_m, the classical turning point, is the outermost zero of

$$F(r) = 1 - \frac{U(r)}{E} - \frac{\beta^2}{r^2} = 0.$$

Because of the fact that the interaction of particles is restricted in condensed media along the trajectory by large impact parameters, the scattering on the tails of potential is not essential (direct numerical calculations of cross sections proved that). So, we will consider the collisions of particles only with impact parameter less than b_{cut}, $\beta_{\max} = b_{cut}$:

$$Q^{(l)}(E) = 2\pi \int_0^\infty (1 - \cos^{(l)}(\chi))\beta d\beta =$$

$$= 2\pi \int_0^{b_{cut}} (1 \cos^{(l)}(\chi))\beta d\beta + 2\pi \int_{b_{cut}}^\infty (1 \cos^{(l)}(\chi))\beta d\beta\,. \tag{7}$$

$$Q^{(l)} \approx 2\pi \int_0^{b_{cut}} (1 \cos^{(l)}(\chi))\beta d\beta \,.$$

Or in dinemsionless form

$$Q^{(l)^*}(E) = \int_0^{b^*_{cut}} (1 - \cos^{(l)}(\chi))\beta^* d\beta^* \tag{8}$$

where ,$Q^{(l)^*} = Q^{(l)}/(2\pi\sigma_0^2)$, $b^*_{cut} = b_{cut}/\sigma_0$ $\beta^*_{cut} = \beta/\sigma_0$.

And effective kinetic diameter is derived from the reduced collision integrals

$$\Omega^{(l.s)^*} = \frac{4(l+1)}{\{(s+1)![2l+1-(1)^l]\}\pi\sigma_0^2} \int_0^{\infty} \exp(\gamma^2)\gamma^{(2s+1)} Q^{(l)}(E)\gamma d\gamma \,. \tag{9}$$

The numerical calculation of the collision integrals reduces to the evaluation of a triple integral represented by formulae (6) , (8) and (9). The numerical integration of triple integral is laborious due to the singularities in the integral expressions for deflection angle formula (6) and rapid oscillation of the integrant in formula (8) corresponding to the spiraling of the particles about one another at certain impact parameters - an orbiting phenomena. These difficulties complicate and lengthen the calculations, and usual methods as, for example, the Gaussian quadratures, Gauss-Mahler, Gauss-Cristoffle methods are not available to build a computer program to deal with any shape of interaction potential with automatic control accuracy. The new numerical method for calculation of the classical deflection angles for any potential function to a prescribe accuracy have been used. The algorithm is based on the Runge-Romberg's rule and evaluation of remainder term which have been derived using the Taylor- series expansion [3]

$$R = C_1 h^{0.5} + C_2 h^{1.5} + C_3 h^2 + C_4 h^{2.5} + C_5 h^{3.5} + C_6 h^4 + \ldots \,. \tag{10}$$

where $h = 1/n$, n is the number of points. In order to evaluate the integral formula (6) we make use of the substitution $t = r/r_m$ thus the deflection angle is given by

$$\chi = \pi - 2\beta \int_0^1 \frac{dt}{[1 - U(r)/E\beta^2 t^2/r_m^2]^{1/2}} \,. \tag{11}$$

The calculation of χ was carried out by the mean scheme, corrected with the triple embedded grids and exclusion of approximation errors of form $C_1 h^{0.5} + C_2 h^{1.5}$. The calculations of cross sections and collision integrals were performed using embedded grids.

The cross sections by formula (8) are shown in Fig.1 at different b^*_{cut}. The corresponding collision integrals are shown in Fig.2. The upper curves (line 7) in Fig.1 and Fig.2, correspondingly, are at the biggest value of b^*_{cut} ($b^*_{cut} = 20$). They really coincide with the cross sections and with the collision integrals for

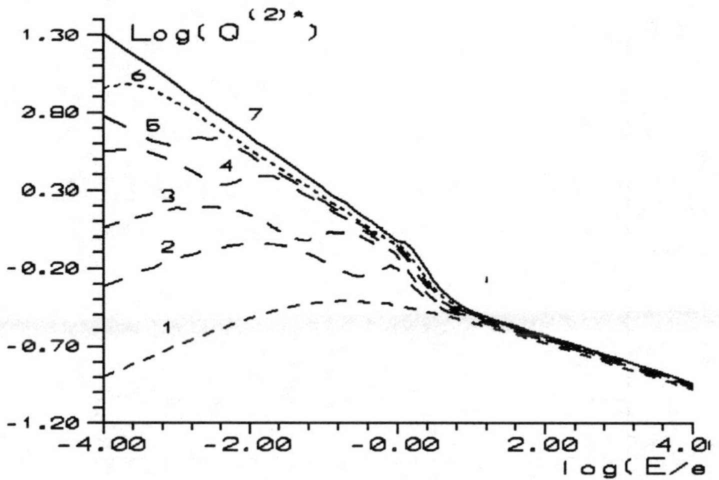

Fig. 1. Truncated cross section $\log[Q^{*(2)}(E^*)]$
1: $b^*_{cut} = 1.0$; 2: $b^*_{cut} = 1.5$; 3: $b^*_{cut} = 2.0$; 4: $b^*_{cut} = 3.0$; 5: $b^*_{cut} = 1.0$; 6: $b^*_{cut} = 6.0$; 7: $b^*_{cut} = 20.0$.

usual Lennard-Jones potential [1]. The effect of density is the more when the cut-off parameter is the less.

As one can see from Fig.1, the damping of the attractive forces with the increase of density ($b^*_{cut} \to 1$) changes the character of cross section. The bend on the curve (7) corresponds to the energy range where the transition from the dominating role of repulsive forces (on the right) to the dominating role of attractive forces (on the left) is in the energy scale. When the values of b^*_{cut} decrease, that is, when the density increases, the attractive forces at low energy range are shielded, the cross section decreases.

For given density, the effective kinetic diameter have been derived from relation

$$\sigma_{kin}(T^*, b\rho) = [\Omega^{2.2)}(T, b_{cut})]^{1/2} = \sigma_0[\Omega^{2.2)^*}(T^*, b^*_{cut})]^{1/2}, \qquad (12)$$

where $b\rho = 2/3\pi n^*, \rho$ is the mass density.

The comparison of the effective kinetic diameter with the results of viscosity in terms of density is shown in Fig.3. For correct consideration of viscosity in Enskog's theory the kinetic part and the collision part of viscosity have been calculated separately [4],

$$\eta = \eta_{kin} + \eta_{coll} \,. \qquad (13)$$

For verification of the model of shielding by formula (4) the test calculations have been performed at different power of shielding function, $m = 0, 1, 2, ...\infty$. The best value $m = 2$ had been confirmed.

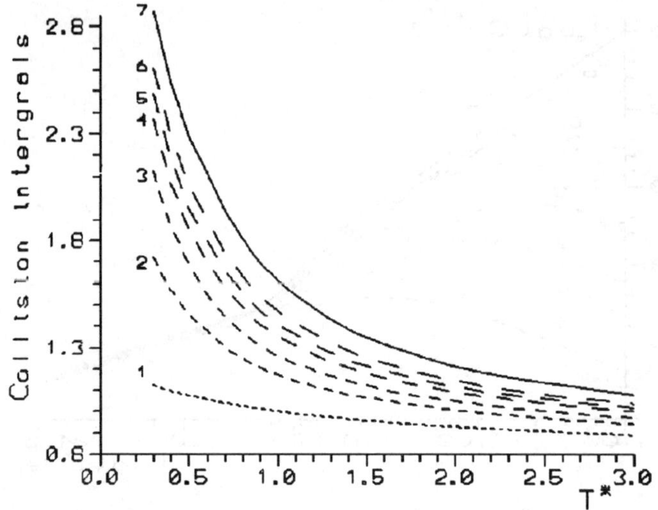

Fig. 2. Reduced collision integrals, $\Omega^{*(2.2)}(T^*)$

1: $b^*_{cut} = 1.0$; 2: $b^*_{cut} = 1.5$; 3: $b^*_{cut} = 2.0$; 4: $b^*_{cut} = 3.0$; 5: $b^*_{cut} = 1.0$; 7: $b^*_{cut} = 20.0$.

Our calculations at $m = 2$ (line 2) are in good agreement with the results derived from viscosity data of Ar, N_2, H_2 from different measurements: points (4)-(8) (see References in [4]) on Fig.3. The line (1) on Fig.3 demonstrates the calculation at shielding function with $m = 1$. The limiting case $m = 0$ corresponds to the calculation without shielding potential function. Another limiting case at $m = 20$ is shown on Fig.3 by line (3).

The line (9) on the Fig.3 demonstrates the calculations [5]. It somewhat differs from the true results (curve 2). We suggest, however, that the reason of such disagreement in disregard of the density dependence of kinetic diameter while calculating viscosity. The calculations [5] had been performed on the base of generalization without considering two different mechanism of impulse transfer separately. And although disagreement is not dramatic, the slope of curve (9) is not true, therefore approach [5] may lead to the wrong asymptotic behavior. In the limiting case of high density $\rho b \to 2$ the effective kinetic diameter coincides with the collision one (Fig.3).

In conclusion it may be said, that the new model has been tested for different matters, and different states of matters. It makes possible to describe the viscosity of matters from liquids to vapor in wide ranges of temperature, density and pressure. Numerical calculation proves, that the new model is essentially less recourse-consuming. It reduces computer time more than once, in particular as compared with the MD and perturbation theory, which requires many hours for calculations in some individual points.

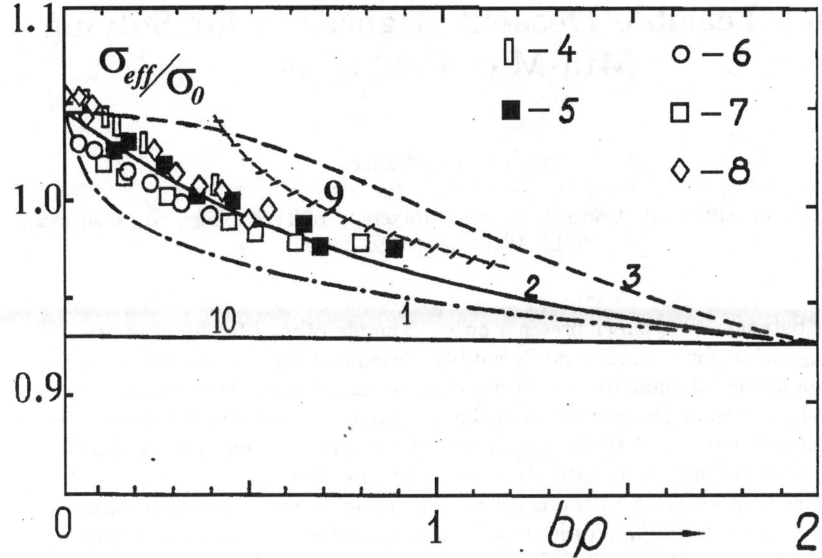

Fig. 3. Effective kinetic diameter

Our calculations at isotherm $T^* = 2.51 : m = 1; 2 : m = 2; 3 : m = 20$. From viscosity experiments 4: Ar; 5: H_2; 6: N_2; 7: H_2; 8: Ar. 9: calculations [5]. 10: collision diameter.

Thus, the new numerical model of simulation of liquids transport properties may be useful for practical applications and liquids mixtures.

Acknowledgments

We would like to thank Professor N.N. Kalitkin for many useful discussions.

The work was suported by the Grand of RFFI, code N 96-01-00305

References

1. Chapman, S., Cowling, T.G.: The mathematical theory of non-inform gases. Cambridge University Press (1952)
2. Hansen, J.P., McDonald, I.R.: Theory Simple Fluids. New York (1986)
3. Repin, I.V., Sokolova, I.A.: Algorithms for calculating deflection angle and the phase shifts. Mathematical Modeling **3** 4 (1993) 76-88
4. Lusternic, V.E., Voronin, M.P.: About the equation of state of viscosity of liquids and gases. Thermophysica of High Temperature. **21** 3 (1983) 464-471
5. Rabinovich, V.A., Wasserman, A.A., Nedostup, B.I., Veksler, L.D.: Thermophysical properties of Neon, Argon, Kripton and Xenon. Moscow Standards (1976) 636-648

On a Feasible Descent Algorithm for Solving Min-Max Problems*

Andrzej Stachurski

Institute of Automatic Control, Warsaw University of Technology, Nowowiejska 16/19, Warszawa, POLAND

Abstract. The paper presents on an example of max functions a new algorithm for solving nondifferentible optimization problems when full knowledge of subderivative is available at each point. We generate an easy to handle representation of the improvement cone with the computational cost equal to the calculation of the inverse of one matrix. Due to that we find at each iteration a descent direction.
This is in contrary to the usual nondifferentiable methods which make use of only one subgradient at each point and solve a QP direction search problem at each step. This is much more computationally expensive and does not ensure that the direction found is descent.
The results of numerical experiments on the sequential machine are presented. They have proved that the current algorithm works quite good on convex min-max problems. The case of nonconvex problems requires careful elaboration of a new minimization algorithm.
We present also an idea of a parallel method of bundle type making use of our mechanism of feasible directions generation.

1 Introduction

The paper is devoted to a new solution method for solving min-max problems of the following structure:

$$f(x) = \max_{i \in I} f_i(x), \quad \text{where } I = \{1, 2, \ldots, m\} \tag{1}$$

We restrict ourselves to minmax problems, anyhow our considerations are valid for any problems involving minimization of a subdifferentiable function with a readily available full knowledge of its subdifferential at each feasible point.

This is the main difference comparing our approach with the standard minimization algorithms for nondifferentiable optimization, where one usually assumes the knowledge of one subgradient at each trial point, cf. Kiwiel (1985, 1988), Lemarechal (1977), Mifflin (1982) and others.

The second important difference between our approach and the traditional way of generating is that we generate exclusively feasible directions. For that purpose we use a similar mechanism as in the generalized elimination method for

* sponsored by KBN, grant No. 3 P40301806 "Parallel computations in linear and nonlinear optimization"

solving Quadratic Programming QP problems with equality constraints (Fletcher 1987). The traditional nondifferentiable optimization algorithms finds the search direction by solving a special QP problems with constraints. This approach does not guarantee feasibility. Furthermore it should be stressed that our way is definitely computationally cheaper. We generate the search direction with the cost of one matrix inversion and a few matrix multiplications.

The paper is organized as follows. In section (2) some notation and basic facts from the subdifferential calculus are introduced. Section (3) contains the description of the algorithm. Section (4) is devoted to the problem of finding feasible directions and section (5) to the directional minimization method. Section (7) presents discussion on the parallelization of our method.

2 Preliminaries

Let us start with the definition of the generalized gradient for locally Lipschitz functions.

Definition 1. Subderivative (generalized gradient) of $f : R^n \to R^1$ at any point x is the set

$$\partial f(x) = co\{y \in R^n : \nabla f(y^j) \to g \text{ for some sequence } y^j \to x \\ \text{with } f \text{ differentiable at points } y^j\}$$

Elements of the subderivative are called subgradients of f at point x.

Nonempty, compact set $\partial f(x)$ reduces to $\{\nabla f(x)\}$ when gradient ∇f is continuous at x.

One of the simplest cases in the subdifferential calculus represents the finite max functions - ($f(x) = \max_{i \in I} f_i(x)$ then

$$\partial f(x) = co\{\frac{\partial f_i}{\partial x}(x);\ i \in I,\ f_i(x) = f(x)\} \tag{2}$$

i.e. $\partial f(x)$ is the convex hull of the gradients of the member functions with values equal to $f(x)$.

Surveys of subgradient calculus, which generalizes rules like $\partial(F_1+F_2)(x) = \partial F_1(x) + \partial F_2(x)$, may be found in Clarke (1983) and Kiwiel (1985, 1988).

The following condition is necessary for $x^* \in R^n$ to be optimal

$$0 \in \partial f(x^*).$$

For convex functions this condition is sufficient.

3 Formulation of the algorithm

The algorithm shall be presented in a similar way as the traditional linearization algorithms for solving nondifferentiable problems. To simplify the presentation we assume now that the minimized function is convex. Furthermore we assume the full knowledge of the subderivative at each point x^k. As it is easily seen from formula (2) the subderivative $\partial f(x^k)$ is the convex hull of the gradients of the "active" member functions, i.e.

$$\partial f(x^k) = \{d \in R^n;\, d = \sum_{j \in I(x^k)} \alpha_i \frac{d}{dx} f_i(x^k)\}, \tag{3}$$

where $I(x^k) = \{i \in I;\, f_i(x^k) = f(x^k)\}$.

The general structure of the algorithm is as follows:

Step 0: (Initialization) Choose the starting point x^1, tolerance of final accuracy $\epsilon_s \geq 0$. Compute $f^1 = f(y^1)$ and gradients $\nabla f_i(x^1)$ for $i \in I^1 = \{i \in I;\, f_i(x^1) = f(x^1)\}$ of functions active at point x^1.

Step 1: (Direction search) Find d^k - any point satisfying the following inequalities:

$$\left[\frac{\partial}{\partial x} f_i(x^k)\right]^T d \leq 0 \qquad \text{for all} \qquad i \in I(x^k). \tag{4}$$

Step 2: (Stopping criterion) If $d^k = 0$ is the only solution of (4) then STOP.

Step 3: (Directional minimization) Find the optimum stepsize τ_k such that

$$f(x^k + \tau_k d^k) = min_{\tau \geq 0} f(x^k + \tau d^k) \text{ and set } x^{k+1} = x^k + \tau_k d^k.$$

Step 4: (Generation of the new subderivative) Compute

$$f^{k+1} = f(x^{k+1}) = \max_{i \in I} f_i(x^{k+1}), \quad I^{k+1} = \{i \in I; \quad f_i(x^{k+1}) = f(x^{k+1})\}$$

and

$$\nabla f_i(x^{k+1}) \text{ for } i \in I^{k+1}$$

Increase iteration number k by 1 and return to Step 1.

4 Generation of Feasible Directions

For finite max functions subderivative is expressed as:

$$\partial f(x^k) = co\{\nabla f_i(x^k);\, i \in I(x^k)\}, \text{ where } I(x^k) = \{i \in I;\, f_i(x^k) = f(x^k)\}$$

For differentiable functions descent directions are those which forms an angle not greater than $\frac{\pi}{2}$ with the minus gradient directions. In our nondifferentiable case a descent direction should form an angle not greater than $\frac{\pi}{2}$ with all vectors opposite to that belonging partial derivative $\partial f(x)$.

4.1 Improvement Cone

Let us denote by A the rectangular matrix with columns equal to the gradients of the functions "active" at x^k.

The improvement cone K at x^k may be represented as follows:

$$K = \{d; a_i^T d \leq 0 \,\mathrm{for}\, i \in I^k\}. \tag{5}$$

Now let us consider two separate cases.

At the beginning let A has got the full column rank and S, Z be the rectangular matrices such that

$$A^T S = I, \qquad A^T Z = 0, \qquad [S|Z] \quad \text{is a nonsingular square matrix}$$

Let's introduce the following transformation of variables $d = S\alpha + Zy$. Then $A^T d = \alpha \leq 0$ and

$$K = \{d = S\alpha + Zy; \alpha \leq 0,\ y \,\text{is free}\} \tag{6}$$

Transformation of such type is typical for generalized elimination method used for solving QP problems with linear equality constraints, cf. Fletcher, 1987. Matrix Z is a basis of the null space of the constraint matrix A while S is a basis of the image space of A respectively.

If A hasn't got the full column rank let $\bar{A}$ be a submatrix of A formed with the linearly independent columns of A and $rank\bar{A} = rankA$ and S, Z be such that

$$\bar{A}^T S = I, \qquad \bar{A}^T Z = 0, \qquad [S|Z] \quad \text{is a nonsingular square matrix}$$

Improvement cone K is represented as previously by (5).

Let $d = S\alpha + Zy$. Then $\bar{A}^T d = \alpha \leq 0$ and $B^T\alpha \leq 0$ and

$$K = \{d = S\alpha + Zy; \alpha \leq 0,\ B^T\alpha \leq 0,\ y \,\text{is free}\} \tag{7}$$

Usually submatrix B will be composed of one column vector and therefore taking it into account does not present any special difficulty.

4.2 Generation of matrices S and Z

Basis matrices S and Z may be generated in many ways. Perhaps the simplest idea is to augment the rectangular matrix A to full square matrix, i.e. take V such that square matrix $[A|V]$ is nonsingular. Then

$$[A|V]^{-T} = [S|Z],$$

The next possibility is to use QR factorization of A. Let Q, R be a result of an orthogonal factorization of A, i.e.

$$QA = \begin{bmatrix} R \\ - \\ 0 \end{bmatrix} \quad \text{and let's denote} \quad Q^T = [Q_1|Q_2]$$

Then one may take $S = Q_1 R^{-T}$ and $Z = Q_2$.

5 Directional Minimization

Below we shall formulate an algorithm for finding parameter τ minimizing $\phi(\tau) = f(x^k + \tau d^k)$. Below $\phi'(p; d)$ - means the directional derivative of function ϕ at point τ in direction d, i.e.

$$\phi'(\tau; d) = \lim_{t \to 0+} \frac{\phi(\tau + t * d) - \phi(\tau)}{t}$$

Symbol $\phi'(\tau; 1)$ denotes the directional derivative at point τ into direction 1 (towards increasing numbers) ($\phi'(\tau; -1)$ towards decreasing numbers).

Step 0. Choose $a > 0$. Find τ_1 and τ_2 such that $\tau_1 < \tau_2$ and $\phi'(\tau_1; 1) < 0$ and $\phi'(\tau_2; -1) < 0$.

Step 1. (Calculation of a new value of parameter τ) Compute t_1 and t_2 such that

$$\phi(\tau_1) + t_1\phi'(\tau_1; 1) = \phi(\tau_2) + t_2\phi'(\tau_2; -1); \quad \tau_1 + t_1 = \tau_2 - t_2$$

and set $\tau := \tau_1 + t_1 = \tau_2 - t_2$.

Step 2. Compute $\phi(\tau)$ and $\phi'(\tau, 1)$. If $\phi'(\tau, 1) < 0$ then set $\tau_1 := p$, go to Step 4 else go to Step 3.

Step 3. Compute $\phi'(\tau, -1)$. If $\phi'(\tau, -1) > 0$ then STOP ($p := \tau_1$ - the optimal solution) else set $p := \tau_2$, go to Step 4.

Step 4. If $||\tau_1 - \tau_2|| < a$ then STOP else return to Step 1.

6 Numerical Experiments with the Algorithm

Our algorithm has been implemented in the standard ANSI C language. We have used the Borland C++ compiler v. 3.1 to compile the code. The program has been run on several small size problems. Some of them are listed below. We have used the Schittkowski constrained test problems (Schittkowski 1987) converted to our min-max form via the l_1-exact penalty function.

Example 1.

$$\min_{x \in R^n} \max\, [1 - x_1,\, 1 - x_2,\, x_1, x_2]$$

Starting point: $(x^0)^T = (10, 10)$.
End point: $(\hat{x})^T = (0.500000, 0.500000)$.

Example 2.

$$\begin{aligned}\min_{x \in R^n} \max\ [&x_1,\, x_2,\, x_3,\\ &-0.49x_1 - 0.497x_2 - 0.167x_3 + 2.152,\\ &-0.467x_1 - 0.145x_2 - 0.283x_3 + 1.895,\\ &0.036x_1 - 0.305x_2 + 0.2x_3 + 1.069,\\ &0.449x_1 - 0.226x_2 - 0.056x_3 + 0.833,\\ &-0.392x_1 + 0.198x_2 + 0.064x_3 + 1.13,\\ &-0.459x_1 - 0.335x_2 + 0.315x_3 + 1.479,\\ &0.185x_1 + 0.264x_2 + 0.327x_3 + 0.224]\end{aligned}$$

Starting point: $(x^0)^T = (10, 10, 10)$.
End point: $(\hat{x})^T = (1.000000, 1.000000, 1.000000)$.

Example 3.

$$\min_{x \in R^n} \max \left[x_2,\ x_2 + (x_1^2 - x_2)\right]$$

Starting point: $(x^0)^T = (1, 1)$.
End point: $(\hat{x})^T = (0.000210, 0.000000)$.

Example 4.

$$\begin{aligned}\min_{x \in R^n} \max\ \big[&100(x_1^2 - x_2)^2 + (x_1 - 1)^2 + x_1(x_1 - 4) - 2x_2 + 12,\\ &100(x_1^2 - x_2)^2 + (x_1 - 1)^2 - x_1(x_1 - 4) + 2x_2 - 12\big]\end{aligned}$$

Starting point: $(x^0)^T = (1, 1)$.
End point: $(\hat{x})^T = (1.999422, 4.000000)$.

Example 5.

$$\begin{aligned}\min_{x \in R^n} \max\ \big[&(x_1 - 2)^2 + (x_2 - 1)^2,\\ &(x_1 - 2)^2 + (x_2 - 1)^2 + x_1^2 - x_2,\\ &(x_1 - 2)^2 + (x_2 - 1)^2 - x_1 + x_2^2\big]\end{aligned}$$

Starting point: $(x^0)^T = (0.5, 0.5)$.
End point: $(\hat{x})^T = (1.166641, 1.166641)$.

Example 6.

$$\min_{x \in R^n} \max \left[-x_2,\ -x_2 - 1 + 2x_2 - x_1,\ -x_2 - x_1^2 - x_2^2,\ -x_2 - 1 + x_1^2 + x_2^2\right]$$

Starting point: $(x^0)^T = (-0.1, -0.9)$.
End point: $(\hat{x})^T = (0.600000, 0.800000)$.

Example 7.

$$\begin{aligned}\min_{x \in R^n} \max\ \big[&x_1^2 + x_2 + x_1 + x_2 - 1,\\ &x_1^2 + x_2 + x_1 + x_2^2 - 1,\\ &x_1^2 + x_2 - x_1^2 - x_2^2 + 9\big]\end{aligned}$$

Starting point: $(x^0)^T = (-3, 0)$.
End point: $(\hat{x})^T = (-0.499548, -2.263846)$.

Example 8.

$$\begin{aligned}\min_{x \in R^n} \max\ \big[&(x_1 - 10)^3 + (x_2 - 20)^3,\\ &(x_1 - 10)^3 + (x_2 - 20)^3 - (x_1 - 5)^2 - (x_2 - 5)^2 + 100,\\ &(x_1 - 10)^3 + (x_2 - 20)^3(x_1 - 6)^2 - (x_1 - 5)^2,\\ &(x_1 - 10)^3 + (x_2 - 20)^3 - 82.81 + (x_1 - 6)^2 + (x_2 - 5)^2\big]\end{aligned}$$

Starting point: $(x^0)^T = (14.35, 8.6)$.
The problem tested has appeared to be unconstrained. The algorithm has found a point and direction on which the minimized max function is unbounded from below.

Table 1. Test results for Examples (1-8)

Example	ITER	F0	F	NF
1	23	10.000000	0.500000	85
2	70	10.000000	1.000000	295
3	5	1.000000	0.000000	86
4	50	7.000000	0.999379	1034
5	10	2.500000	0.916667	167
6	54	0.900000	-0.800000	277
7	22	9.000000	1.611154	385
8	2	-1399.231125	unbounded	

The numerical results are summarized in the table given below.

Numerical tests proved that the algorithm may be a reliable and useful tool for solving convex max problems, although fast convergence could not be rather expected. It is natural since the algorithm corresponds to the steepest descent method for differentiable optimization. We merely ensure the descent property. For nonconvex problems another directional minimization procedure should be used.

There are still some open questions. First of all how to use the second order information of the max member functions. Furthermore even this preliminary computational experience indicates that the accuracy of generating bases matrices Z and S is crucial for the behavior of the algorithm. Another crucial point is the choice value of the parameter governing the ϵ - subgradient calculation. Its introduction has appeared to be necessary.

7 Parallel Version of the Algorithm

In this section we shall discuss the possibilities of using our algorithm in a parallel way.

The algorithm itself does not offer many direct possibilities for parallelization. Only the operation of finding matrices S and Z may be carried out faster on machine with parallel processors. The details on finding the QR factorization using parallelism may be found for instance in Cosnard at al. (1986). For informations concerning calculation of an inverse of a matrix on parallel machines we refer to Bertsekas and Tsitsiklis (1989).

However it is also possible to use parallel processors to carry out directional minimization in Step 3 into several directions simultaneously. The basic idea is to use at the current point approximating the optimal solution x^k a bundle of feasible directions, i.e. directions belonging to the improvement cone K.

Afterwards the synchronization follows at which the "best" point found in the parallel directional searches is selected, the stopping criterion is verified and the whole process is either repeated or stopped.

Our representation of the improvement cone via formula (6 & 7) is very convenient for such purpose. We suggest to select the bundle of search directions randomly applying an even distribution of parameters α over a box $[0,1]\times[0,1]$ (of course other strategies are possible).

The parallel version of the algorithm is now under preparation.

8 Conclusions

The main novelty lies in the generation strategy of the improvement directions. We assumed the full knowledge at each point x of the whole subdifferential of the minimized max function f. This is advantageous compared to general nondifferentiable optimization problems where we know usually only one subgradient from the subdifferential. The use of the generalized elimination idea known in literature within the context of QP problems with linear equality constraints allows to obtain an easy to handle representation of the cone of improvement directions. The net result is that we are able to generate a set of descent directions with the cost practically equal to the cost of computing the inverse to a certain square matrix of dimension $n \times n$. The case when the Haar condition is violated is also considered and a possible solution to overcome this difficulty is proposed.

References

1. Bertsekas, D.P., Tsitsiklis, J.N.: Parallel and distributed computation. Numerical methods. Prentice-Hall, Englewood Cliffs, New Jersey (1989).
2. Clarke, F.H.: Optimization and nonsmooth analysis. Wiley-Interscience, New York (1983).
3. Cosnard, M., Muller, J.-M., Robert, Y.: Parallel QR Decomposition of a Rectangular Matrix. Numerische Mathematik **48** (1986) 239-249.
4. Fletcher, R.: Practical Methods of Optimization. Second ed., John Wiley & Sons, New York (1987).
5. Kiwiel, K.C.: Methods of descent for nondifferentiable optimization. Springer Verlag, Berlin, Heidelberg, New York (1985).
6. Kiwiel, K.C.: Some Computational Methods of Nondifferentiable Optimization. (in polish), Ossolineum, Wrocław (1988).
7. Lemarechal, C.: Nonsmooth Optimization and Descent Methods. Technical Report RR-78-4, International Institute for Applied Systems Analysis, Laxenburg, Austria (1977).
8. Mifflin, R.: A Modification and an Extension of Lemarechal's Algorithm for Nonsmooth Minimization, in: Nondifferentiable and Variational Techniques in Optimization (eds. Sorensen D.C. and Wets R.J.-B.), Mathematical Programming Study **17**, North Holland, Amsterdam (1982) 77-90.
9. Schittkowski, K.: More Test Examples for Nonlinear Programming Codes, Lecture Notes in Economics and Mathematical Systems **282**, Springer Verlag, Berlin, Heidelberg (1987).

Rate of Convergence of a Numerical Procedure for Impulsive Control Problems

Mabel TIDBALL

Department of Mathematics, Universidad de Rosario,
Pellegrini, 250 - 2000 Rosario, Argentina.

Abstract. We consider a deterministic impulsive control problem. We discretize the Hamilton-Jacobi-Bellman equation satisfied by the optimal cost function and we obtain discrete solutions of the problem. We give an explicit rate of convergence of the approximate solutions to the solution of the original problem. We consider the optimal switching problem as a special case of impulsive control problem and we apply the same structure of discretization to obtain also a rate of convergence in this case. We present a numerical example.

1 Introduction

Optimization problems of dynamical systems lead to the treatment of non linear partial differential equations: the Hamilton-Jacobi-Bellman (HJB) equations or Isaacs equations, associated to optimal control problems or differential game problems respectively. Except in special cases, the analytical solutions of these equations are unknown. Therefore, it is interesting to find characterizations of the existence, unicity, regularity of the solutions of these equations in order to develop numerical methods to obtain approximate solutions. With these solutions it is possible to obtain suboptimal feedback controls, equilibrium strategies, etc.

After the development of viscosity solutions (see [6]), the approximation of HJB equations was deeply studied. The techniques to obtain numerical approximations are based on time-space discretizations. The time discretization procedure was introduced in [5] for example, for deterministic control problems. The space discretization procedure uses finite elements techniques; see [3] and [10] for control problems and [1] for game problems.

Another type of discretization, which involves only space discretization, has also been studied, in [9] and [13] for example.

In this paper we consider a deterministic impulsive control problem (see [2]). We apply the time-space discretization procedure to obtain the discrete solutions.

It is organized as follows. In §1 we introduce the continuous problem and we give the properties of the optimal cost function. In §2 we introduce a time discretization of the HJB equation, we also give a convergence rate. The results obtained here are extensions of the results presented in [5] for impulsive control problems. The time discretization scheme involves a delay between impulsions.

We can prove the existence and unicity of the discrete time solution because it is the fixed point of a contractive operator. We also define a non contractive scheme of time discretization. This scheme corresponds to a process where instantaneous impulsions are allowed. We prove that both schemes are equivalent. In §3 we also discretize in the spatial variable in order to obtain an approximate problem whose solution can be found numerically. We give a rate of convergence of these solutions to the solution of the original problem. In §4 we present a numerical example and finally, in §5, we consider the optimal switching problem (see [4]). This type of problem is a special case of impulsive control problems and we can apply the previously developed theory. Due to the special structure of the optimal switching problem, we obtain a better rate of convergence than the one obtained for general impulsive control problems.

2 Description of the continuous problem.

We consider a deterministic impulsive control problem, where the state of the system $y(.)$ is given by the following ordinary differential equation:

$$\frac{dy}{ds}(s) = g(y(s)) \quad s > 0, \qquad y(0) = x \tag{1}$$

where $x \in \Omega$, Ω an open set of $I\!R^n$.

Equation (1) is valid $\forall s > 0$ except at times θ_ν where an impulsive control is applied. The impulsions are given by $z(\theta_\nu) \in Z$, Z a compact set of $I\!R^j$, with $\theta_\nu < \theta_{\nu+1}$. The impulsive controls, that we denote by $\mathbf{z}$, are determined by the sequence of values $\mathbf{z}(.) = \{(\theta_\nu, z(\theta_\nu)), \nu = 1, 2...\}$. We call $\mathcal{Z}$ the set of impulsive controls. The impulsions $z(\theta_\nu)$ produce a jump given by $\bar{g}(y(\theta_\nu^-), z(\theta_\nu))$ that makes the system change instantaneously from position $y(\theta_\nu^-)$ to $y(\theta_\nu^+)$; i.e.

$$y(\theta_\nu^+) = y(\theta_\nu^-) + \bar{g}(y(\theta_\nu^-), z(\theta_\nu)) \tag{2}$$

We assume that $y(s) \in \Omega \;\; \forall s > 0$. The problem consists on finding the optimal cost function u, defined $\forall x \in I\!R^n$ by:

$$u(x) = \inf_{\mathbf{z}(.)\in\mathcal{Z}} J(x, \mathbf{z}(.)) = \inf_{\mathbf{z}(.)\in\mathcal{Z}} \left\{ \int_0^\infty f(y(s)) e^{-\lambda s} ds + \sum_{\nu=1}^{\infty} q(y(\theta_\nu^-, z(\theta_\nu)) e^{-\lambda\theta_\nu} \right\}. \tag{3}$$

Here, f is the instantaneous cost, $\lambda > 0$ is the discount rate and $q(x, z)$ is the cost of applying each impulsion.

Properties of the optimal cost function u. We assume that $\forall x, \tilde{x} \in \Omega$, $\forall z \in Z$:

$$\bar{g}(x, .), q(x, .), \; \textit{continuous} \; \forall \; x \in \Omega. \tag{4}$$

$$g, f, \bar{g}, q, \; \textit{bounded functions, Lipschitz continuous in } x. \tag{5}$$

$$q_0 = \inf_{x,z} q(x, z) > 0 \; . \tag{6}$$

We call $L_g, L_f, L_{\bar{g}}$ and L_q, the Lipschitz constants of g, f, $\bar{g}$ and q respectively. Likewise, let M_f be the bound on f. By virtue of (4), (5) and (6) we can prove (see [8]) that:

Lemma 2.1. *If a control* $\mathbf{z}(.)$ *has more than* $\mu_0 = \dfrac{2eM_f}{q_0\lambda}$ *impulsions in* $[t, t+\delta)$, *with* $\delta = 1/\lambda$ *then there exists another control* $\bar{\mathbf{z}}$ *with cost strictly lower, i.e:* $J(x, \bar{\mathbf{z}}) < J(x, \mathbf{z})$.

Remark 2.2. By Lemma 2.1 we can deduce that for optimization purpose, we can consider only controls with at most μ_0 impulsions in $[t, t+\delta)$. In consequence hereafter we shall assume that the controls verify this condition.

If we denote by $\mu(s) = \mu(s, \mathbf{z})$ the number of impulsions in $[0, s)$, with $\mu(\infty)$ the total number of impulsions of a generic control $\mathbf{z}(.)$, we have: $\mu(s) \leq \mu_0(1+s/\delta)$. We call $\mu_\delta = \mu_0/\delta = \lambda\mu_0$ and

$$\lambda_{\bar{g}} = \sup\left\{\frac{\| x + \bar{g}(x,z) - x' - \bar{g}(x',z) \|}{\| x - x' \|} \;:\; z \in Z,\; x, x' \in \Omega,\; x \neq x'\right\}. \quad (7)$$

By (5) we obtain that $\lambda_{\bar{g}} \leq 1 + L_{\bar{g}}$. We can prove that:

Lemma 2.3. *Under hypotheses* (4), (5) *and* (6), *u satisfies:* $|u(x)| \leq M_f/\lambda$, $|u(x) - u(x')| \leq C \| x - x' \|^\gamma$, $\forall\, x, x' \in \mathbb{R}^n$ *and* $\gamma = 1$ *if* $\lambda > L$, $\gamma \in (0,1)$ *if* $\lambda = L$, $\gamma = \lambda/L$ *if* $\lambda < L$, *with* $L = (\mu_\delta \; ln\lambda_{\bar{g}})^+ + L_g$.

Theorem 2.4. *The optimal cost function u is the unique solution, in the viscosity sense, of the following Hamilton-Jacobi-Bellman equation.*

$$\min\left\{\frac{\partial u}{\partial x}\cdot g + f - \lambda u,\; Mu - u\right\} = 0, \quad Mu(x) = \min_z \{u(x + \bar{g}(x,z)) + q(x,z)\}. \quad (8)$$

3 Time discretization of the HJB equation

Time discretization scheme. Let h be suitably small. To find a discrete time approximation of (3) we consider the solution of

$$u = Tu \quad (9)$$

where $T : X \longrightarrow X$, $X = C^{0,\gamma}$ with $\| u \|_\gamma = \sup_x |u(x)| + \sup_{x_1 \neq x_2} \frac{|u(x_1) - u(x_2)|}{|x_1 - x_2|^\gamma}$ and $T = \min(P_0, P_1)$ where

$$P_0 u(x) = (1 - \lambda h)\, u(x + hg(x)) + h\, f(x) \quad (10)$$

$$P_1 u(x) = \min_z \{q(x,z) + h\, f(x + \bar{g}(x,z)) + (1 - \lambda h) u(x + \bar{g}(x,z) + hg(x + \bar{g}(x,z)))\}. \quad (11)$$

P_0 and P_1 are time discretizations of (8). We can easily prove, following the same ideas of [5] that P_0 and P_1 are contractive operators, and so is T. Then we have the following Lemma:

Lemma 3.1. *There exists a unique solution of* (9) *that we call* u^h.

Interpretation of u^h. For all $x \in \mathbb{R}^n$ the following representation for u^h is valid:

$$u^h(x) = \min_{\mathbf{z}(.)\in \mathcal{Z}^h} J^h(x, \mathbf{z}(.)) \tag{12}$$

where: $\mathcal{Z}^h = \{\mathbf{z}(.) \in \mathcal{Z} : \theta_i = n_i h, \ n_i \in \mathbb{N}, n_i < n_{i+1}, i = 1, ...\}$

$$J^h(x, \mathbf{z}(.)) = h \sum_{j=0}^{\infty} f(y_h^+(j))(1 - \lambda h)^j + \sum_{j=1}^{\infty} q(y_h^-(n_j), z_j)(1 - \lambda h)^{n_j} . \tag{13}$$

Let $I(\mathbf{z}(.)) = \{j : \exists i \,/\, j = n_i\}$. We define the sequence $y_h(j)$ by the following recurrence formulae:

$$\begin{cases} y_h^-(j+1) = y_h^+(j) + hg(y_h^+(j)) \\ y_h^+(n_j) = y_h^-(n_j) + \bar{g}(y_h^-(n_j), z_j) \text{ if } j \in I(\mathbf{z}(.)) \\ y_h^+(j) = y_h^-(j) \qquad\qquad\qquad \text{if } j \notin I(\mathbf{z}(.)) \\ y_h^+(0) = x \end{cases} \tag{14}$$

Remark 3.2. With the same arguments used in [8], we can prove that there exists an optimal discrete policy that realizes u^h, i.e., there exists $\tilde{\mathbf{z}}(.) \in \mathcal{Z}^h$ such that $u^h(x) = J^h(x, \tilde{\mathbf{z}}(.))$.
Moreover, we can prove (as in the continuous case), that the optimal policy has at most μ_0 impulsions in each interval of length $1/\lambda$.

Lemma 3.3. *Under hypotheses* (4), *(5) and* (6), u^h *satisfies:* $|u^h(x)| \leq M_f/\lambda$, $|u^h(x) - u^h(x')| \leq C \parallel x - x' \parallel^\gamma$, $\forall x, x' \in \mathbb{R}^n$ *and* $\gamma = 1$ *if* $\lambda > L$, $\gamma \in (0,1)$ *if* $\lambda = L$, $\gamma = \lambda/L$ *if* $\lambda < L$.

Rate of convergence of the h-approximate solution

Theorem 3.4. *If* (4), (5) *and* (6) *hold, then:*

$$|u(x) - u^h(x)| \leq Ch^\gamma,$$

where $\gamma = 1$ *if* $\lambda > L$, $\gamma \in (0,1)$ *if* $\lambda = L$, $\gamma = \dfrac{\lambda}{L}$ *if* $\lambda < L$.

Remark 3.5. It is possible to work with a non contractive scheme of discretization. The time discretization of (8) can be understood as a problem where we allow simultaneous impulsions. We can consider the solution of: $u = \bar{T}u$, where $\bar{T} : X \longrightarrow X$ is defined as $\bar{T} = \min(P_0, \bar{P}_1)$, P_0 is defined in (10) and

$$\bar{P}_1 u(X) = \min_z (q(x,z) + u(x + \bar{g}(x,z))).$$

By using mainly the hypotheses (6) and the theory of B-L (Bensoussan-Lions) algorithm, introduced in [11] we obtain that there exists a unique solution of $u = \overline{T}u$ that we call $\bar{u}^h$. Moreover we have that this problem and problem (9) are equivalent in the following sense:

$$0 \leq u^h - \bar{u}^h \leq Ch^\gamma.$$

4 The fully discrete solution of HJB equation

Description of the fully discrete problem. The above introduced time discretization remains a theoretical one. To obtain computational results it is also necessary to perform a space discretization. We will use the same discretization as the one introduced in [9], and [12]. Let Ω be an open set of $\mathbb{R}^n$ and S_j^k a family of regular triangulations of Ω such that $k = \max_j(diam S_j^k)$. Let $\Omega_k = \cup_j S_j^k$.

Let $W_k : \Omega_k \longrightarrow \mathbb{R}$ be the set of finite linear elements. Then, the fully discrete problem is:

Problem P_k: Find the fixed point of operator T in W_k

We understand operator T (see the definitions of P_0 and P_1) as an operator $T : W^k \to W^k$. This means, for example, that $u(x+hg(x)) = \sum_j \lambda_{ij} u(x_j)$ where $x_j,\ j = 1, ..., \bar{N}$ is the set of nodes of the triangulation and λ_{ij} the barycentric coordinates such that $x + hg(x)$ belongs to the simplex S_j. Therefore, to obtain u_k^h it is only necessary to compute $u_k^h(x_i),\ i = 1, ..., \bar{N}$, where $\{x_i : i = 1, ..., \bar{N}\}$ is the set of nodes of the triangulation.

Theorem 4.1. *There exists a unique solution of problem P_k that we call u_k^h.*

Rate of convergence of the fully discrete solution

Theorem 4.2. *If (4), (H) and (6) hold, then:*

$$|u(x) - u_k^h(x)| \leq C(h + \frac{k}{\sqrt{h}})^\gamma \tag{15}$$

where $\gamma = 1$ if $\lambda > L$, $\gamma \in (0,1)$ if $\lambda = L$, $\gamma = \dfrac{\lambda}{L}$ if $\lambda < L$.

Remark 4.3. In the usual case, i.e. when h is of the same order than k (which means when there exists m_1 and m_2 such that $m_1 h \leq k \leq m_2 h$), formula (15) becomes: $|u(x) - u_k^h(x)| \leq M k^{\frac{\gamma}{2}}$. However, optimizing (15) with respect to h, we obtain the optimal value for $h = k^{\frac{2}{3}}$ and consequently: $|u(x) - u_k^h(x)| \leq M k^{\frac{2\gamma}{3}}$.

5 A numerical example

Consider the case where $\Omega = (0,2)$, $f(x) = -x$, $q(x,z) = q_0$, and the dynamics of the system given by: $\frac{dy}{ds} = -by(s)$, $y(0) = x$, $b > 0$ while no impulsive control is applied. When an impulsion is applied the jump is given by: $\bar{g}(y) = 1$ if $0 < y < 1$, $\bar{g}(y) = 2 - y$ if $1 \leq y < 2$.

After a few calculations, it can be shown that the cost function is given by:

$$u(x) = \begin{cases} u(2)\left(\frac{2}{x}\right)^{\frac{\lambda}{b}} - \frac{x-2\left(\frac{2}{x}\right)^{\frac{\lambda}{b}}}{\lambda+b} & \text{if } \xi < x \leq 2 \\ u(2) + q_0 & \text{if } 1 < x \leq \xi \\ u(x+1) + q_0 & \text{if } 0 < x \leq 1 \end{cases} \tag{16}$$

The value of ξ is given by the largest point where the following equality is valid:

$$u(2) = u2(\xi) = \frac{-\frac{\xi-(2/\xi)^{\lambda/b}}{\lambda+b} - q_0}{1-(2/\xi)^{\lambda/b}}.$$

Remark 5.1. The structure of an optimal feedback policy is the following: $(\xi, 2)$ is the set of continuation; $(0, \xi)$ is the set of application of impulsive control.

In this example, with $\lambda = 0.5$, $q_0 = 0.5$, $b = 0.125$, we obtain $\xi = 1.5$ and $u(2) = -3.125$.

Table 1 gives us the maximum error between the real and the approximate solution for different values of $\bar{N} = 1/k$ using space discretization $h = k^{\frac{2}{3}}$. Table 2 shows the same measure using $h = k$.

Table 1

$\bar{N}$	error
80	0.24473
160	0.01215
320	0.0060813
640	0.0029928
1280	0.0014999

Table 2

$\bar{N}$	error
80	0.076639
160	0.04788
320	0.030016
640	0.018822
1280	0.011816

6 Optimal switching problems

Description of the problem. We consider in this section a deterministic optimal switching problem, (see [4]), of a system described by an ordinary differential equation, which dynamics can be modified, at the price of a positive switching cost, into anyone of a different setting.

The problem consists in finding the optimal way to modify the dynamics with the purpose of minimizing an associated cost.

More precisely, let us define an admissible control $\alpha(.) = \{(\theta_\nu, d_\nu) : \nu = 1, 2, ...\}$ to be a sequence of switching times θ_ν and control setting or switching decisions

$d_{\nu-1} \longrightarrow d_\nu$, where $\theta_\nu \in [0,+\infty)$, $\theta_\nu < \theta_{\nu+1}$ and $d_\nu \in D = \{1,...,m\}$. Then, the control $d(.)$ remains constant in each interval $[\theta_\nu, \theta_{\nu+1})$ being:

$$d(t) = \begin{cases} d \in D & \text{if } 0 \le t < \theta_1 \\ d_\nu \in D & \text{if } \theta_\nu \le t < \theta_{\nu+1}, \quad \nu \ge 1 \end{cases}$$

For each $d \in D$ we also define A^d, the set of all admissible controls with initial setting d. For a given $x \in \mathbb{R}^n$, $d \in D$, $\alpha \in A^d$, the response of the system to the control $\alpha(.)$ is given by the following ordinary differential equation: $\frac{dy}{ds} = g(y(s), d_\nu)$, $\theta_\nu \le t < \theta_{\nu+1}$ with $y(0) = x$.
Our goal is to design for each $x \in \mathbb{R}^n$, $d \in D$, an optimal control $\bar{\alpha}$, such that: $u^d(x) = \inf_{\alpha \in A^d} J^d(x,\alpha) = J^d(x,\bar{\alpha})$ where

$$J^d(x,\alpha) = \sum_{\nu=1}^{\infty} \left\{ \int_{\theta_{\nu-1}}^{\theta_\nu} f(y(s), d_{\nu-1}) e^{-\lambda s} ds + q(d_{\nu-1}, d_\nu) e^{-\lambda\theta_\nu} \right\}$$

$f : \mathbb{R}^n \times D \longrightarrow \mathbb{R}$ is the instantaneous cost and $q(d,\tilde{d})$ is the transition cost to replace d by $\tilde{d}$. Besides the usual assumptions of regularity and boundedness for f, g and q, we assume that: $q(d,\tilde{d}) \ge q_0 > 0$, $q(d,\tilde{d}) \le q(d,\hat{d}) + q(\hat{d},\tilde{d})$.

Equivalent impulsive problem. To obtain a fully discrete solution of the optimal switching problem and to estimate the rate of convergence towards the real solution, we show that this type of problem is equivalent to a special impulsive control problem.

We consider the following generalized version of the impulsive control problem presented in §1, extended for $\Omega = D \times \mathbb{R}^n$, where the evolution of the system takes the form:

$$(d,y)(s) = \left(d_\nu, y(\theta_\nu^+) + \int_{\theta_\nu}^{s} g(y(t), d_\nu) dt \right)$$

valid for $\theta_\nu \le s < \theta_{\nu+1}$. The jumps of the generalized state at instant θ_ν concern only the first component of the state, i.e. $(\tilde{d}, y(\theta_\nu)) = (d, y(\theta_\nu)) + (\tilde{d} - d, 0)$
It is evident that optimal switching problems can be considered as particular cases of impulsive control problems, where the dynamics associated to the impulsive part is not expansive. So we can announce the following rate of convergence of the discrete solution:

$$|u^d(x) - u_k^{d,h}(x)| \le C(h + \frac{k}{\sqrt{h}})^\gamma$$

where $\gamma = 1$ if $\lambda > L_g$, $\gamma \in (0,1)$ if $\lambda = L_g$, $\gamma = \lambda / L_g$ if $\lambda < L_g$. Note that this last inequality give a smaller estimate than (15) because $L = L_g$.

7 Conclusions

We have studied the fully discrete solution of an impulsive control problem. We have obtained the rate of convergence of the discrete solutions to the real solution, with a delay scheme and with a scheme that considers instantaneous impulsions. We have obtained an estimate of type $(h + \frac{k}{\sqrt{h}})^\gamma$. When h is of order k and $\gamma = 1$, that is when u is Lipschitz continuous, we have obtained an estimate of type $k^{\frac{1}{2}}$, which improves the estimate of type $k^{\frac{1}{2}}|ln(k)|$ obtained in [7] and [9].

After performing the time optimization with respect to h, k being fixed, we have obtained a bound that depends only on the parameter k, which is of order $k^{\frac{2\gamma}{3}}$.

We have also proved that the optimal switching problem is a special case of an impulsive control problem.

We have developed a simple numerical example, where the exact solution is known, in order to show the error between the real and the approximate solution for different relations between the time and the space discretization.

References

1. Bardi, M., Falcone, M., Soravia, P.: Fully discrete schemes for the value function of pursuit-evasion games. Advances in dynamic games and applications **1** (1994) 89–105
2. Bensoussan, A., Lions, J. L.: Contrôle impulsionnel et inéquations quasi-variationnelles. Dunod, Paris (1982)
3. Camilli, F., Falcone, M.: An approximation scheme for the optimal control of diffusion processes. Report Università degli Studi di Roma La Sapienza (1992)
4. Capuzzo Dolcetta, I., Evans, L. C.: Optimal switching for ordinary differential equations. SIAM J. of Control and Optimization **22** (1984) 143–161
5. Capuzzo Dolcetta, I., Ishii, H.: Approximation solutions of the Bellman equation of deterministic control theory. Appl. Math. Optim. **11** (1984) 161–181
6. Crandall, M.G., Lions, P. L.: Viscosity solutions of Hamilton-Jacobi equations. Trans. AMS **277** (1983) 1–42
7. Cortey-Dumont, P.: Approximation numérique d'une inéquation quasi variationnelle liée à des problèmes de gestion de stock. R.A.I.R.O. Analyse numérique **14** (1973) 17–31
8. González, R.: Sur la résolution de l'équation de Hamilton-Jacobi du contrôle déterministe. Cahiers de Mathématiques de la Décision, **8029** Ceremade Université de Paris-Dauphine (1980)
9. González, R., Rofman, E.: On deterministic control problem - An approximation procedure for the optimal cost. SIAM J. on Control and Optimization **23** (1985) 242–285
10. González, R., Tidball, M.: Sur l'ordre de convergence des solutions discrétisées en temps et en espace de l'équation de Hamilton-Jacobi. Note CRAS Paris **314** (1992) 479–482
11. Hanouset, B., Joly, J. L.: Convergence uniforme des iterés definisant la solution d'une inéquation quasi variationnelle abstraite. Note CRAS Paris **286** (1978) 735–738
12. Strang, G., Fix, G.: An Analysis of the finite elements method. Prentice-Hall, Englewood Cliffs, NJ (1973)
13. Tidball, M., Gonzâlez, R.L.V.: Zero sum differential games with stopping times. Some results about its numerical resolution. Advances in dynamic games and applications, **1** (1993).

Coarse-Grain Parallelisation of Multi-Implicit Runge-Kutta Methods

Roman Trobec[1], Bojan Orel[2], Boštjan Slivnik[1]

[1] Jožef Stefan Institute,* Slovenia
[2] University of Ljubljana, Slovenia
E-mail: roman.trobec@ijs.si

Abstract. A parallel implementation for a multi-implicit Runge-Kutta method (MIRK) with real eigenvalues is decribed. The parallel method is analysed and the algorithm is devised. For the problem with d domains, the amount of work within the s-stage MIRK method, associated with the solution of system, is proportional to $(sd)^3$, in contrast to the simple implicit finite difference method (IFD) where the amount of work is proportional to d^3. However, it is shown that s-stage MIRK admits much greater time steps for the same order of error. Additionally, the proposed parallelisation transforms the system of the dimension sd to s independent sub-systems of dimension d. The amount of work for the sequential solution of such systems is proportional to sd^3. The described parallel algorithm enables the solving of each of the s subsystems on a separate processor; finally, the amount of work is again d^3, but the profit of a larger time step still remains. To test the theory, a comparative example of the 3-D heat transfer in a human heart with 64^3 domains is shown and numerically calculated by 3-stage MIRK.

1 Introduction

With an increase in computer power, particularly with parallel computers, many numerical methods gained in importance. Let us mention the heat transfer simulation as a research area where numerical methods are an important solution tool. The basic physical equation which describes the heat transfer is the *diffusion equation*:

$$\frac{\partial T}{\partial t} = D\nabla^2 T.$$

D stands for the diffusion coefficient and ∇^2 is Laplacian operator. Heat transfer is an example of a *propagation problem* i.e., an *initial-value problem* in an *open domain* (open with respect to one of the independent variables). In this case, time t is taken as an independent variable with an open domain, because the initial heat distribution at time t_0 develops forward in time. Thus, we are looking for the time evolution of a heat distribution. However, the domains of the independent spatial variables x, y, z are closed, and the *boundary conditions* must, therefore,

* This work was supported in part by The Ministry of Science and Technology of The Republic of Slovenia, grant J2-5092-106-95.

be provided. For example, if the values of x, lie on the interval (x_{lo}, x_{hi}), values of $T_{x_{lo},y,z}(t)$ and $T_{x_{hi},y,z}(t)$, for any $t > t_0$, $y \in (y_{lo}, y_{hi})$ and $z \in (z_{lo}, z_{hi})$, must be given.

The exact solution of the diffusion equation can be computed only for simple cases, i.e.: for geometrical bodies consisting of one material. However, in practice, we are faced with more difficult problems: with bodies of irregular shapes which can hardly be modelled using a geometrical approach, with non-uniform initial heat distribution for which no simple mathematical function can be found, and with a number of different diffusion coefficients in the model [1]. In these cases, the diffusion equation should be rewritten as

$$\frac{\partial T}{\partial t} = D(x,y,z)\left[\frac{\partial^2 T}{\partial x^2} + \frac{\partial^2 T}{\partial y^2} + \frac{\partial^2 T}{\partial z^2}\right], \tag{1}$$

where $D(x,y,z)$ is a function of the coordinates x, y and z. Equation (1) can be semi-discretised for spatial variables. For example, the interval (x_{lo}, x_{hi}) can be divided into $\sqrt[3]{d}+1$ subintervals of length $l = (x_{hi} - x_{lo})/(\sqrt[3]{d}+1)$ with points $x_i = x_{lo} + il$. The term $\frac{\partial^2 T}{\partial x^2}$ can be replaced by

$$\frac{\partial^2 T(x,y,z,t)}{\partial x^2} \approx \frac{T_{i-1,j,k}(t) - 2T_{i,j,k}(t) + T_{i+1,j,k}(t)}{h^2},$$

where $T_{i,j,k}$ stands for $T(x_i, y_j, z_k)$ for all $i,j,k = 1,2,\ldots,\sqrt[3]{d}$. If similar procedure is applied for y and z coordinates, (1) is transformed to an initial-value problem (IVP)

$$\frac{dv(t)}{dt} = Mv(t) + v_b(t);\ v(0) = v_0, \tag{2}$$

where $v(t)$ denotes $[\ldots, v_m(t), \ldots]^T$, for $m = 1,2,\ldots,d$, with $v_m(t) = T_{i,j,k}$ and $m = i + (j-1) * \sqrt[3]{d} + (k-1) * \sqrt[3]{d^2}$, v_0 represents the initial temperature distribution, and $v_b(t)$ stands for a vector of the boundary conditions. Matrix M represents the system, incorporating the diffusion coefficients

$$M = \frac{D}{l^2}\begin{bmatrix} A & I & 0 & \ldots & 0 & 0 \\ I & A & I & \ldots & 0 & 0 \\ \vdots & & \ddots & & & \vdots \\ 0 & 0 & 0 & \ldots & A & I \\ 0 & 0 & 0 & \ldots & I & -A \end{bmatrix},$$

where

$$A = \begin{bmatrix} B & I & 0 & \ldots & 0 & 0 \\ I & B & I & \ldots & 0 & 0 \\ \vdots & & \ddots & & & \vdots \\ 0 & 0 & 0 & \ldots & B & I \\ 0 & 0 & 0 & \ldots & I & B \end{bmatrix} \text{ and } B = \begin{bmatrix} -2 & 1 & 0 & \ldots & 0 & 0 \\ 1 & -2 & 1 & \ldots & 0 & 0 \\ \vdots & & \ddots & & & \vdots \\ 0 & 0 & 0 & \ldots & -2 & 1 \\ 0 & 0 & 0 & \ldots & 1 & -2 \end{bmatrix}$$

are $\sqrt[3]{d} \times \sqrt[3]{d}$ and $\sqrt[3]{d^2} \times \sqrt[3]{d^2}$ square matrices, respectivelly, and I stands for the identity matrix of the appropriate dimension.

The IVP (2) can be efficiently solved by means of an implicit finite difference method (IFD); the time step is bounded by stability. Alternativelly, a more complicated method — MIRK [3] can be used; simple parallelisation and a larger timestep may give an important benefit in some particular applications. In the next section, the parallel definition of the MIRK is devised. In Sect. 3, the parallel algorithm is shown and analysed according to the amount of work required. Then, comparative example of the 3-D heat transfer on the human heart with 64^3 domains for 3-stage MIRK is given. In the conclusion, the results are discussed and evaluated.

2 Parallelisation of Runge-Kutta Method

2.1 Sequential Definition

The integration of IVP (2) may be solved numerically with the s-stage implicit Runge-Kutta method [2] expressed as

$$Y_i = y_{n-1} + h \sum_{j=1}^{s} a_{ij}(MY_j + y_b); \quad i = 1, 2, \ldots, s, \tag{3}$$

$$y_n = y_{n-1} + h \sum_{i=1}^{s} b_i(MY_i + y_b). \tag{4}$$

where h is a time step-size and y_n denotes the numerical approximation to $v(nh)$ and y_b is the discretization of boundary-conditions $v_b(t)$.

The method is usually represented by the vectors $b = (b_i), i = 1, \ldots, s$ and $c = (c_i), i = 1, \ldots, s$ and the matrix $A = (a_{ij}, i, j = 1, \ldots, s)$, which can be arranged into the Butcher tableau [2]

$$\begin{array}{c|cccc} c_1 & a_{1,1} & a_{1,2} & \cdots & a_{1,s} \\ c_2 & a_{2,1} & a_{2,2} & \cdots & a_{2,s} \\ \vdots & \vdots & \vdots & & \vdots \\ c_s & a_{s,1} & a_{s,2} & \cdots & a_{s,s} \\ \hline & b_1 & b_2 & \cdots & b_s, \end{array} \tag{5}$$

where the c_i are defined as $c_i = \sum_{j=1}^{s} a_{ij}$ for $i = 1, \ldots, s$.

System (3) implicitly defines the internal approximations $Y_i = [Y_i^1, \ldots, Y_i^d]$ that can be interpreted as an approximation of the exact solution $y(t)$ at the intermediate point $t = t_n + c_i h$. Consequently, a system of ds equations with ds unknowns must be solved at each step to compute quantities Y_i, which are combined into a solution value y_n at the next time step according to (4). In Fig. 1. the internal approximations Y_1, Y_2, Y_3 are given for the solution y_n, and the discretization in the (x, t) plane are shown.

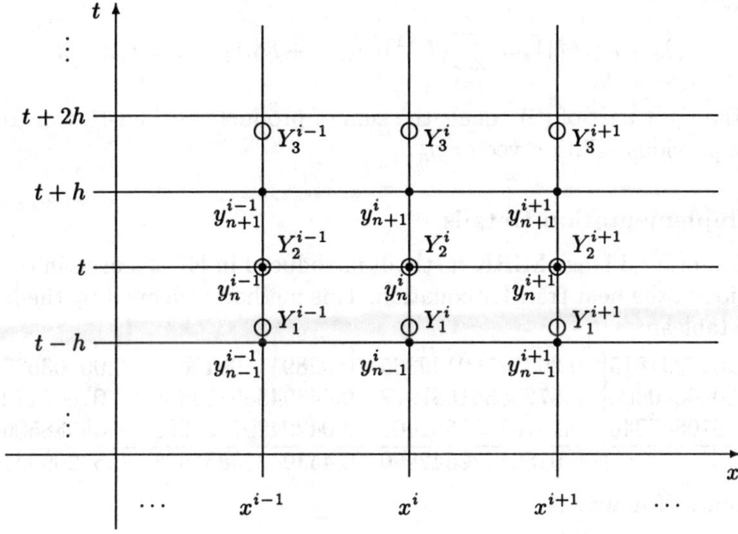

Fig. 1. Graphical presentation of the internal approximations for the solution y_n, and the 3-stage MIRK discretization.

2.2 Parallel Definition

System (3) can be written in tensor notation as

$$Y = e \otimes y_{n-1} + h(A \otimes M)Y + hA \otimes y_b, \tag{6}$$

where $Y = [Y_1^T, Y_2^T, ...Y_s^T]^T$, the s-dimensional vector e has unit entries and $\otimes$ is the tensor product (in general $A \otimes B = [a_{ij}B]$). If the eigenvalues γ_i of A are all real and distinct and T is the matrix of the eigenvectors of A, then the

$$D = T^{-1}AT$$

is a diagonal matrix with the eigenvalues γ_i on the diagonal. By introducing transforming vectors

$$\tilde{Y} = (T^{-1} \otimes I_d)Y, \text{ and } \tilde{y}_b = (T^{-1} \otimes I_d)y_b \tag{7}$$

where I_d is the unit matrix, (6) is transformed to

$$\tilde{Y} = T^{-1}e \otimes y_{n-1} + h(D \otimes M)\tilde{Y} + hD \otimes \tilde{y}_b,$$

which is a set of s subsystems with d equations and d unknowns in each subsystem. These subsystems are of the form

$$\tilde{Y}_i = \sum_{j=1}^{s}(T^{-1})_{ij}y_{n-1} + h\gamma_i M\tilde{Y}_i + h\gamma_i\tilde{y}_b; \quad i = 1, 2, ..., s \tag{8}$$

or

$$(I_d - h\gamma_i M)\tilde{Y}_i = \sum_{j=1}^{s}(T^{-1})_{ij}y_{n-1} + h\gamma_i\tilde{y}_b; \quad i = 1, 2, ...s. \tag{9}$$

where the right part of (9) equals the sum of products of the i-th row from T^{-1} and the previous solution vector y_{n-1}.

2.3 Implementation Details

The third-order 3-stage MIRK method, introduced in [4] was used in our implementation of the heat transfer equation. This method is defined by the following Bucher tableau:

0.17731047291815	0.20318569149365	−0.02891149910657	0.00303628053107
1.00000000000000	0.57206641687972	0.44394830854688	−0.01601472542660
2.83374670845340	0.10181357682466	2.04204807121208	0.68988506041666
	−0.01601472542660	0.44394830854688	0.57206641687972

Eigenvalues of matrix A are:

$$[\gamma_1, \gamma_2, \gamma_3] = [0.535870000,\ 0.46527906045718,\ 0.335870000].$$

The corresponding matrix of eigenvector of A is:

$$T = \begin{bmatrix} -0.99698590138860 & -0.99359718232641 & 0.98300104580292 \\ 0.07597739963378 & 0.11046776579844 & -0.17342135765532 \\ -0.01570182082594 & -0.02369624427039 & 0.06028247389871 \end{bmatrix}$$

and it's inverse is given by:

$$T^{-1} = \begin{bmatrix} -3.347643243690 & -48.055650853970 & -83.658467850145 \\ 2.438127219799 & 58.641197577395 & 128.942194784559 \\ 0.086432480687 & 10.533989095837 & 45.483459690559 \end{bmatrix}$$

3 Parallel Algorithm

The parallel 3-stage MIRK algorithm can be now represented as:

<u>*set-up*</u>
    $y_{old} = y_0$;   *set d initial values*
<u>*loop*</u> *for each time step*
<u>*loop*</u> *for* $i = 1, 2, 3$ *MIRK stages*
    $M_i = I_d - (h\gamma_i M)$;   *left part of (9)*
    $Right = \sum_{j=1}^{3}(T^{-1})_{i,j}y_{old} + h\gamma_i\tilde{y}_b$;   *right part of (9)*
    $\tilde{Y}_i = M_i^{-1}\ Right$;   *the solution of (9)*
<u>*end*</u>
    $Y = T\tilde{Y}$;   *from (7)*
    $Difference = h\sum_{i=1}^{3} b_i(MY_i + y_b)$;   *difference from previous time step*
    $y_{new} = y_{old} + Difference$;   *new value in time from (4)*
    $y_{old} = y_{new}$;   *save the calculated value*
<u>*end*</u>

It is obvious that the calculation of $M_i = I_d - (h\gamma_i M)$, and the multiplication $b_i M$ can be performed outside the loops because all the values are known; therefore, it can be put into the *set-up* segment.

If the amount of work associated with the solution of (6) for $s = 3$ is proportional to $(3d)^3$, the amount of work for the solution of the above algorithm is proportional to $3d^3$. Each subsystem can be solved on a separate processor, and the complexity of the calculation can be finally reduced to d^3 per processor, which is the original value for IFD.

4 Comparative Example

To obtain some experimental data on the behaviour of the presented method, a computation of a heat transfer during an open heart surgery was considered. A discretized model of 64^3 domains was used representing a cube of 128^3 mm^3. It contains a model of the human heart surrounded by a cooling liquid and a part of an open chest. The model is nonhomogenous with diffusion coefficents and initial temperatures lying in the intervals $[1.0596 \cdot 10^{-7}, 2.1287 \cdot 10^{-5}]$m/s^2 and $[0, 30]^o$C, respectivelly. A simplified simulation of the heat transfer using the diffusion equation instead of diffusion–convection equation was computed for 60 minutes, which is approximately twice as long as the surgery can last.

Because the diffusion equation cannot be solved exactly for the given model, a solution obtained by a method of explicit finite differences (EFD) using the largest time step permited by EFD method (0.03s) was used as a reference. The MIRK method for $s = 3$, described in Sect. 2, was tested. The solutions of the three systems of linear equations in step 1 of each RK iteration were computed by an iterative solver, with a modest convergence criterion $|(\Delta(\tilde{Y}_j)_i)_{max}| < 0.1$, for $j = 1, 2, 3$.

The absolute error of the MIRK method is shown in Fig. 2 for various time steps. In the particular medical application the model of the heart surgery is taken from, the absolute error which starts below 0.1^oC and later falls below 0.01^oC, is less than what can accurately be measured during the surgery. The precision of this method is satisfactory for the further medical modelling even with 10000 times longer time step as it is permited by the EFD method.

5 Conclusion

The above principle offers a straightforward possibility for the parallelisation of s-stage Runge-Kutta methods. We can simply take s processors and solve s systems in parallel. But, there are two problems: first, the dimension of the original system d may be too complex to be performed on a single processor; second, the number of stages cannot be arbitrarily increased. It follows that the above parallelisation offers a good choice for smaller problem domains on parallel processors with a relative low number of processors. The parallel algorithm needs almost no communication; therefore, message passing systems implemented with PVM can

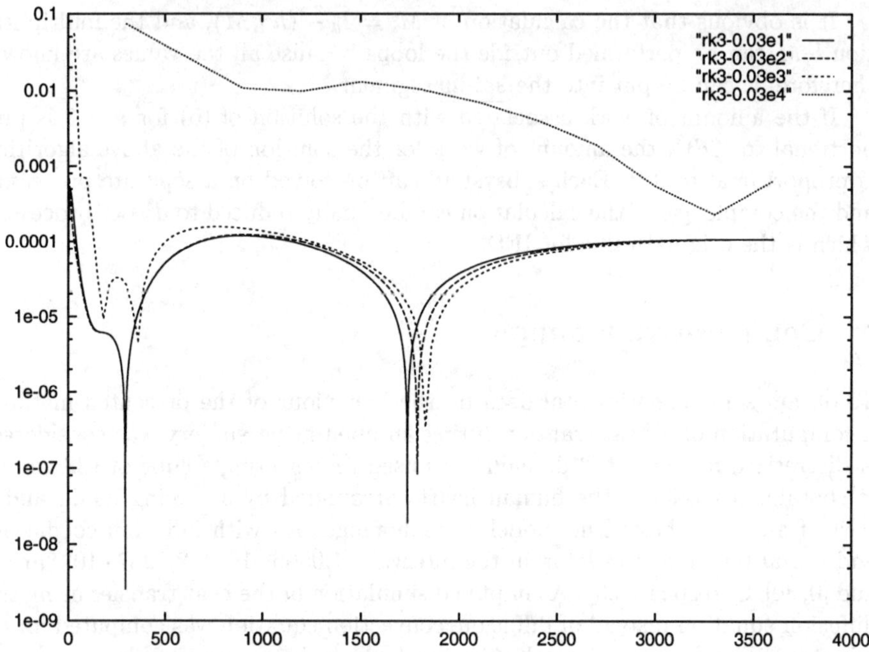

Fig. 2. The absoulte error of the MIRK method ($s = 3$) for various time steps ($h = 0.03 \cdot 10^1, 0.03 \cdot 10^2, 0.03 \cdot 10^3, 0.03 \cdot 10^4$s). The solution obtained by the EFD method with the time step $h = 0.03$s was taken as a reference.

also be used. In this way, the described parallel s-stage Runge-Kutta method becomes competitive with other finite-difference based numerical methods.

References

1. Gersak, B., Trobec, R., Gabrijelcic, T., Slivnik, B.: The Model of Topical Heat Cooling During Induced Hypothermic Cardiac Arrest in Open Heart Surgery. Proc. of the Int. Conf. Computer in Cardiology, Wien, Austria, (1995) 597-600.
2. Butcher, J. C.: On the implementation of implicit Runge-Kutta methods. BIT **16**(1976) 237–240.
3. Burrage, K.: A special family of Runge-Kutta methods for solving stiff differential equations. BIT **18**(1978) 22–41.
4. Orel, B.: Parallel Runge-Kutta methods with real eigenvalues. Appl. Num. Math. **11**(1993) 241–250.

Mosaic Ranks and Skeletons

Eugene Tyrtyshnikov

Institute of Numerical Mathematics,
Russian Academy of Sciences,
Leninski Prospect 32a, Moscow 117334, Russia
E-mail: tee@inm.ras.ru

Abstract. The fact that nonsingular coefficient matrices can be covered by blocks close to low-rank matrices is well known probably for years. It was used in some cost-effective matrix-vector multiplication algorithms. However, it has been never paid a proper attention in the matrix theory. To fill in this gap we propose a notion of mosaic ranks of a matrix which reduces a description of block matrices with low-rank blocks to a single number. A general algebraic framework is presented that allows one to obtain some theoretical estimates on the mosaic ranks. An algorithm for computing upper estimates of the mosaic ranks is given with some illustrations of its efficiency on model problems.

1 Introduction

For a dense $n \times n$ matrix, the matrix-vector multiplication requires $2n^2$ arithmetic operations. Some particular cases are yet known (Fourier matrices, Toeplitz matrices, and so on) for which special algorithms were invented to cut back the multiplication costs. It seems hard to believe that there would be a sufficiently general class of dense unstructured matrices allowing for a fast multiplication by a vector. In fact, such a class exists.

In most applications, the matrix entries are computed somehow, and usually all the entries can be partitioned into large enough groups within which they are produced by a common formula. Even in the case when all the entries are given by a single common formula, we may look for some partitioning in the hopes to have simpler approximate formulas for each group. If we add to this some natural and rather general assumptions about those formulas, we may get a firm ground for the design of advanced algorithms. For example, using piecewise smooth approximations to the kernel producing the entries we can partition the matrix into blocks, most of them having a small ε-rank. This observation (though in a somewhat weaker form) was made by V. V. Voevodin [10] as early as in 1979. Later the same was used (rather implicitly) in the matrix-vector multiplication algorithms proposed by Rokhlin [5], Hachbusch and Nowak [8], Myagchilov and Tyrtyshnikov [7].

In this paper, we consider the block skeleton approximations lying behind the above-mentioned approaches (cf. [9]). We begin with the notion of a skeleton (in the simplest case, this is a 1-rank matrix). Then we introduce the notion of the mosaic rank (and ε-rank) related to a given block partitioning of a matrix.

We present some algebraic tools that allow one to estimate the mosaic ranks both theoretically and numerically.

The algorithms computing the block skeleton approximations are usually referred to as compression algorithms. In contrast to [5, 7, 8] we claim that compression algorithms should not use anything but the entries of a matrix. Explicit form of the operator and the digitization technique comes in only when a rigorous proof of efficiency is derived.

2 Compression and skeletons

A matrix of the form uv^T, where u and v are column vectors, is called a *skeleton*.

If A is a sum of r linearly independent skeletons then the (classical) rank of A is equal to r. Using the skeleton expansion

$$A = \sum_{i=1}^{r} u_i v_i^T$$

we can calculate $y = Ax$ very efficiently in the two stages:

1) compute the scalars $\alpha_i = v_i^T x$;

2) compute the vector $y = \sum_{i=1}^{r} \alpha_i u_i$.

If A is $n \times n$ then we need only $2rn$ coupled operations instead of n^2 and we need only $2rn$ memory locations instead of n^2.

In the general case, when A is $m \times n$, the compressed memory is defined by the formula

$$\text{MEMORY} = \text{RANK} \cdot (m + n),$$

and we thus have

$$\text{THE COMPRESSION FACTOR} = \text{RANK} \cdot \frac{m+n}{mn}.$$

3 Compression and mosaic ranks

Unfortunately, we customarily deal with matrices which do not have a small rank. For such cases we want to introduce the mosaic rank (expecting it to be small for many nonsingular matrices) so that the formula

$$\text{MEMORY} = \text{MOSAIC RANK} \cdot (m + n)$$

be still valid.

3.1 Definitions

If B is a submatrix for a $m \times n$ matrix A then $\Gamma(B)$ denotes the $m \times n$ matrix with the same block B and zeroes elsewhere. A system of blocks A_i is called a *covering* of A if

$$A = \sum_i \Gamma(A_i),$$

and a *mosaic partitioning* of A if the blocks have no common elements.

For any given covering, the mosaic rank of A is defined as

$$\mathrm{mr}\, A = \sum_i \mathrm{mem}\, A_i \ / \ (m+n), \tag{1}$$

where

$$\mathrm{mem}\, A_i = \min\{m_i n_i\ ,\ \mathrm{rank}\, A_i(m+n)\}. \tag{2}$$

Similarly to the classical notion of ε-rank we introduce the notion of the mosaic ε-rank which is defined as the minimal mosaic rank over all ε-perturbations of A (usually with respect to the 2-norm).

The minimal mosaic rank over all mosaic partitionings will be called the optimal mosaic rank. Usually we need not know its exact value, and settle for sufficiently neat upper estimates.

3.2 Example

Let A_n be a $n \times n$ matrix with units on and below the main diagonal and zeroes in the upper triangular part. Then

$$\mathrm{rank} A_n = n \quad \text{whereas} \quad \mathrm{mr} A_n \le \log_2 2n.$$

The proof is easy by induction. For simplicity, assume that n is even and consider the partitioning

$$A = \begin{bmatrix} A_{n/2} & 0 \\ 1 & A_{n/2} \end{bmatrix}.$$

Using the inductive assumption, we have

$$\frac{2(\frac{n}{2}+\frac{n}{2})\log_2(2\frac{n}{2}) + (\frac{n}{2}+\frac{n}{2})}{n+n} \le \log_2 n + \frac{1}{2} \le \log_2 2n.$$

3.3 Mosaic versus standard partitionings

The logarithmic estimate for the above-considered example can not be achieved when using only standard partitionings of the form

$$A_n = \begin{bmatrix} A_{11} \dots A_{1k} \\ \dots \dots \dots \\ A_{k1} \dots A_{kk} \end{bmatrix}.$$

If each block is $\frac{n}{k} \times \frac{n}{k}$ then

$$\operatorname{mr} A_n = \frac{2\frac{n}{k}\frac{k^2-k}{2} + (\frac{n}{k})^2 k}{2n} = \frac{k-1}{2} + \frac{n}{2k} \sim \frac{1}{2}\left(k + \frac{n}{k}\right),$$

and for a proper k we have at best

$$\operatorname{mr} A_n = O(\sqrt{n}).$$

Therefore, standard partitionings can provide some compression (see the results reported in [3]), but they can be not as good as mosaic partitionings.

4 Some estimates

It is not casual that matrices in application problems have low mosaic ranks. We present here some theoretical results of the two types: algebraic and functional.

4.1 Algebraic preliminaries

Consider a matrix $A \in \mathbb{C}^{m \times n}$ written in the block form

$$A = \begin{bmatrix} A_{11} \; A_{12} \\ A_{21} \; A_{22} \end{bmatrix}, \quad A_{11} \in \mathbb{C}^{m_1 \times n_1}, \; A_{22} \in \mathbb{C}^{m_2 \times n_2}. \tag{3}$$

Lemma 1. *Given a matrix A of the form (3), assume that*

$$|(m_1 + n_1) - (m_2 + n_2)| \le q(m+n), \quad q < 1, \tag{4}$$

and there exist mosaic partitionings such that

$$\operatorname{mr} A_{ii} \le c \log_2(m_i + n_i), \; i = 1, 2; \quad \operatorname{mr} A_{ij} \le r, \; i \ne j; \tag{5}$$

and, moreover,

$$c \ge \frac{r}{\log_2 \frac{2}{q^2+1}}.$$

Then for A, there exists a mosaic partitioning such that

$$\operatorname{mr} A \le c \log_2(m+n). \tag{6}$$

Proof. Let

$$\alpha = m_1 + n_1,\ \beta = m_2 + n_2,\ \gamma = m_1 + n_2,\ \delta = m_2 + n_1.$$

Since $\alpha + \beta = \gamma + \delta$ and due to the concavity of the logarithm we have

$$\begin{aligned} \operatorname{mem} A &\le c\alpha \log_2 \alpha + c\beta \log_2 \beta + r(\alpha + \beta) \\ &\le c(\alpha + \beta) \log_2 \left(\frac{\alpha^2 + \beta^2}{\alpha + \beta}\, 2^{r/c} \right). \end{aligned}$$

From the identity

$$\frac{\alpha^2 + \beta^2}{\alpha + \beta} = \frac{\alpha + \beta}{2} + \frac{(\alpha - \beta)^2}{2(\alpha + \beta)}$$

it follows that the expression under the logarithm sign is less than or equal to $\alpha + \beta$ if and only if

$$\frac{(\alpha - \beta)^2}{(\alpha + \beta)^2} + 1 \le \frac{2}{2^{r/c}}.$$

The latter holds true whenever

$$q^2 + 1 \ \le\ \frac{2}{2^{r/c}} \quad \Longleftrightarrow \quad \frac{r}{\log_2 \frac{2}{q^2+1}} \ \le\ c,$$

and this completes the proof. □

Lemma 2. *Assume that all the hypotheses of Lemma 1 are fulfilled but instead of (5) for some $k \ge 0$ we have*

$$\operatorname{mr} A_{ii} \le c \log_2^{k+1}(m_i + n_i),\ i = 1, 2; \quad \operatorname{mr} A_{ij} \le r \log_2^{k}(m_i + n_j),\ i \ne j. \tag{7}$$

Then for A, there exists a mosaic partitioning such that

$$\operatorname{mr} A \le c \log_2^{k+1}(m + n). \tag{8}$$

Proof. Using the monotonicity property, we have

$$\begin{aligned} \operatorname{mem} A &\le c\alpha \log_2^{k+1} \alpha + c\beta \log_2^{k+1} \beta + r\gamma \log_2^{k} \gamma + r\delta \log_2^{k} \delta \\ &\le c\alpha \log_2^{k+1} \alpha + c\beta \log_2^{k+1} \beta + r(\gamma + \delta) \log_2^{k}(\gamma + \delta) \\ &\le (c\alpha \log_2 \alpha + c\beta \log_2 \beta + r(\alpha + \beta)) \log_2^{k}(\gamma + \delta), \end{aligned}$$

and it remains to have recourse to Lemma 1. □

Lemma 3. *Assume that*

$$A = \begin{bmatrix} A_{11} & \dots & A_{1k} \\ \dots & \dots & \dots \\ A_{k1} & \dots & A_{kk} \end{bmatrix}, \quad A_{ij} \in \mathbb{C}^{m_i \times n_j}, \tag{9}$$

where $\operatorname{mr} A_{ij} \le r$ for $i, j = 1, \dots, k$. Then

$$\operatorname{mr} A \ \le\ rk.$$

The proof is trivial.

4.2 Piecewise constant matrices

We call a pair $(i,j),(k,l)$ the (positively) ordered pair if one of the following three possibilities is valid:

$$(1)\ k=i,\ l=j+1; \quad (2)\ k=i+1,\ l=j; \quad (3)\ k=i+1,\ l=j+1.$$

Then we call a finite sequence $(i_1,j_1),\ \ldots,\ (i_k,j_k)$ the shearing line if each pair of neighbors is the ordered pair. In a matrix, any shearing line subdivides the entries into two "connected components", say, G_1 and G_2. We now consider a matrix $A=[a_{ij}]$ such that $a_{ij}=c_1$ for $(i,j)\in G_1$ and $a_{ij}=c_2$ for $(i,j)\in G_2$. The elements along the shearing line may take arbitrary values. We call A and A^T the piecewise constant matrices.

Theorem 1. *If $A\in\mathbb{C}^{m\times n}$ is a piecewise constant matrix then there is a mosaic partitioning such that*

$$\operatorname{mr} A \le c\log_2(m+n).$$

The previous Lemma 1 allows for an easy proof by induction.

Theorem 2. *If A is of the form (9), where each block is a piecewise constant matrix, then*

$$\operatorname{mr} A \le ck\log_2(m+n).$$

The proof follows from Theorem 1 and Lemma 3.

4.3 Smoothness and skeletons

Consider a sequence of matrices A_n generated by a function $f(x,y)$, where

$$x\in X\subset \mathbb{R}^m,\ y\in Y\subset\mathbb{R}^m,$$

in the following sense:

$$[A_n]_{ij}=f(x_i^{(n)},y_j^{(n)}),$$

where

$$x_1^{(n)},\ldots,x_n^{(n)}\in X,\quad y_1^{(n)},\ldots,y_n^{(n)}\in Y.$$

Under rather general assumptions such matrices appear to be close to matrices of low mosaic rank.

Let y have components $y_1,\ldots,y_m$ and ∂_i denote the partial differentiation with respect to y_i. If f is sufficiently smooth we can approximate it by a truncated Taylor expansion at some point $z\in Y$:

$$f(x,y)\approx f_N(x,y)=\sum_{k=0}^{N}\frac{(\delta_1\partial_1+\ldots+\delta_m\partial_m)^k}{k!}f(x,z),$$

where

$$\delta_1=y_1-z_1,\ldots,\delta_m=y_m-z_m.$$

Then

$$A_n \approx \tilde{A}_n \equiv [f_N(x_i^n, y_j^n)].$$

To be more precise, assume that

$$|(\delta_1 \partial_1 + \ldots + \delta_m \partial_k)^k f| \leq M|\delta|^k \frac{k!}{|x-y|^k}, \tag{10}$$

where

$$|\delta| = \sqrt{|\delta_1|^2 + \ldots + |\delta_m|^2}.$$

Consequently, if d is the diameter of Y and ρ is the distance between X and Y, then

$$\frac{|y-z|}{|x-z'|} \leq q \equiv \frac{d}{\rho} \quad \text{for any} \quad z' \in Y.$$

If $q < 1$ then

$$|f(x_i^{(n)}, y_j^{(n)}) - f_N(x_i^{(n)}, y_j^{(n)})| = O(q^{N+1}).$$

For usual digitization techniques we have

$$\varepsilon \sim q^N \sim 1/n^\gamma$$

for some $\gamma > 0$; hence, $N \sim \log n$. Taking into account that $f_N(x,y)$ is in fact a sum of $O(N^m)$ two-term products where one function depends only on x and the other one depends only of y we conclude that

$$\operatorname{rank} \tilde{A}_n = O(\log^m n).$$

This means that A_n is approximated by a sum of $\sim \log^m n$ skeletons.

In more general cases the logarithmic growth can be proven only for mosaic ranks.

Assume that $f(x,y)$ is *asymptotically smooth* (we adopt somewhat the definition used by Brandt [2]):

$$|\partial^p f(x,y)| \ \leq \ c_p |x-y|^{g-p}, \tag{11}$$

where ∂^p is any p-order derivative with respect to y or x. To be safe from infinite values, we set $f(x,y) = 0$ when $x = y$.

Note that c_p may grow with p, but we will assume that

$$c_p \ \leq \ c\, p^\alpha\, p! \qquad \text{for some} \qquad \alpha > 0,\ c > 0. \tag{12}$$

For example, if

$$f(x,y) = \frac{1}{|x-y|^\gamma}, \qquad \gamma > 0,$$

then

$$g = -\gamma, \qquad \alpha = \gamma - 1.$$

Consider now a model (but very instructive) case:

$$x_i^{(n)} = y_i^{(n)} = \frac{a}{n} i, \quad i = 1, \ldots, n. \tag{13}$$

Theorem 3. *Assume that matrices A_n are generated on the mesh points (13) by a sufficiently smooth function $f(x,y)$ satisfying (11) and (12). Then for any $\varepsilon > 0$ there exists a sequence of matrices $\tilde{A}_n$ such that*

$$\operatorname{mr} \tilde{A}_n = O(\log n \log \varepsilon^{-1}) \tag{14}$$

and

$$||A_n - \tilde{A}_n||_F = O(n\varepsilon). \tag{15}$$

Proof. Denote by $|\mathcal{M}|$ the greatest lower bound of the distance between the points in $\mathcal{M}$ and the origin. Suppose that X and Y are disjoint: $|X - Y| \neq 0$. Let $x \in X$, $y, z \in Y$, and $p = N + 1$. Then

$$|f(x,y) - f_N(x,y)| \leq cp^{\alpha} \frac{|y - z|^p}{|x - Y|^{p-g}} \equiv R_p.$$

We subdivide the interval $\Omega = (0, a]$ into four subintervals:

$$\Omega = \bigcup_{k=1}^{4} \Omega_k, \qquad \Omega_k = (\frac{k-1}{4} a, \ \frac{k}{4} a].$$

and denote by n_k the number of the mesh points (13) on Ω_k. It is easy to see that

$$n_i \leq \frac{n}{4} + 1, \quad n_1 + n_2 + n_3 + n_4 = n.$$

Let

$$A_n = [A_{n;\,kl}]_{kl=1}^{2},$$

where $A_{n;\,11}$ is $(n_1 + n_2) \times (n_1 + n_2)$ and $A_{n;\,22}$ is $(n_3 + n_4) \times (n_3 + n_4)$.

Now, build up approximations to the corner blocks and estimate their mosaic ranks.

Consider first the block $A_{n;\,12}$. Divide it into 2×2 blocks so that the partitioning squares with the above subintervals. Consider the following three cases:

(1) $X = \Omega_1$, $Y = \Omega_3$;
(2) $X = \Omega_1$, $Y = \Omega_4$;
(3) $X = \Omega_2$, $Y = \Omega_4$.

Let z be the center of Y in each case. Then we have

$$R_p \ \leq \ cp^{\alpha} \frac{(a/8)^p}{a^{p-g}} \ \leq \ c_0 q^p$$

for some c_0 and q such that

$$c_0 \ > \ ca^g, \qquad \frac{1}{8} \ < \ q \ < \ 1.$$

Therefore,

$$R_p \leq \varepsilon$$

whenever

$$p \geq \log_{q^{-1}} c_0 + \log_{q^{-1}} \varepsilon^{-1}.$$

Thus, all the blocks in $A_{n;\,12}$ save for the left-bottom one are approximated by the blocks with the ranks of the order $\log \varepsilon^{-1}$.

The left-bottom block is still to be partitioned in the same manner using finer subintervals of Ω_k. We can proceed like this recursively. It seems quite clear that in the end we come to some $\tilde{A}_{n;\,12}$ for which

$$\mathrm{mr}\, \tilde{A}_{n;\,12} = O(\log \varepsilon^{-1})$$

and

$$||\tilde{A}_{n;\,12} - A_{n;\,12}||_F = O(n\varepsilon).$$

The same can be done for the block $A_{n;\,21}$.

We are now in a position to apply Lemma 3, which obviously completes the proof. □

Similar results can be obtained for other meshes. Moreover, much the same is valid for the matrices arising in the method of moments (see Mikhailovski [6]).

The results can be carried over to the multidimensional case. Using the Taylor series as above, usually we can prove that the mosaic ε-rank of A_n increases as $\log n \log^m \varepsilon^{-1}$. If $\varepsilon \sim 1/n^\gamma$ then it behaves as $O(\log^{m+1} n)$.

However, in the multidimensional case it seems expedient to work with spherical harmonics rather than with the Taylor series. Eventually, we can use less skeletons and have smaller mosaic ranks.

5 Mosaic rank estimator

We present an algorithm that builds up a proper mosaic partitioning and thus estimates the optimal mosaic ε-rank for a given matrix. The idea of the algorithm is close to the ideas behind the adaptive quadrature procedures (see [1]).

The algorithm will generate only *regular mosaic partitionings*, where each block belongs to a standard partitioning for some other block.

5.1 Estimation of elementary ranks

Given a mosaic partitioning, it seems quite natural to proceed with evaluating the mosaic ranks for each block. However, it could be an expensive recursive procedure, and we propose to confine ourselves to estimating some 'elementary' ranks for each block. For a $m_i \times n_i$-block A_i the elementary rank is calculated as $\mathrm{mem}\, A_i \;/\; (m_i + n_i)$ where $\mathrm{mem}\, A_i$ is given by (2). Everywhere the ranks are understood as ε-ranks, where ε is an input parameter.

For small-size blocks the ε-rank is easily computed from the singular value decomposition (SVD) of the block. For larger blocks we fall back on the Lanczos bidiagonalization algorithm with a natural stopping criterion (cf [3]). The number of Lanczos iterations is never permitted to be large, because, roughly speaking, it should be equal to the wanted ε-rank. Consequently, the estimation procedure works with reasonable efficiency even for large blocks.

In the present algorithm, to find an approximate skeleton decomposition for a block we need to have all its entries. However, it is shown in [2] that this can be done with using only a small part of the entries.

5.2 The algorithm

1. Form the list of blocks of an initial mosaic partitioning.

2. Estimate the elementary ranks for each block.

3. Search for the 'worst' block in the list of blocks (presently the one of maximal elementary rank).

4. Cut off the 'worst' block into smaller blocks (presently with 2×2 standard partitioning) and temporarily update the list of blocks.

5. Estimate the elementary ranks for the new blocks and then estimate the mosaic rank for the new mosaic partitioning.

6. If the mosaic rank increases, then return to the previous mosaic partitioning and do not allow further to pick up that 'worst' block.

7. Check the stopping criteria (we set a limit on the total number of blocks and/or a threshold for the mosaic rank).

8. Go to Action 3 to proceed (if necessary).

We consider this algorithm as a useful research tool to acquire more knowledge about optimal mosaic ranks of A_n, the rationale for this being that some modest values of n may give insight into what kind of mosaic partitionings should be advocated for larger n.

5.3 Accuracy estimates

Note that the algorithm quite naturally produces a reliable *a posteriori* estimate on $||A_n - \tilde{A}_n||_2$, where $\tilde{A}_n$ is the matrix which possesses the computed mosaic rank.

If A is covered by blocks A_i then, obviously,

$$||A||_2 \leq \sum_i ||A_i||_2.$$

However, for regular mosaic partitionings there is a better estimate:

$$||A||_2 \leq \sqrt{\sum_i ||A_i||_2^2}. \qquad (*)$$

The inequality $(*)$ holds true for any regular mosaic partitioning of A. To prove this, it is sufficient to consider the following two cases:

$$(1)\ A = \begin{bmatrix} A_1 \\ A_2 \end{bmatrix}; \quad (2)\ A = [A_1\ \ A_2]\,.$$

The first case is trivial. The second case can be reduced to the first one by the transition to the transposed matrix A^T.

5.4 Numerical examples

We consider the following "piecewise constant" test matrices:

$$(1):\ a_{ij} = \begin{cases} 1\,,\ i \geq j, \\ 2\,,\ i < j; \end{cases}$$

$$(2):\ a_{ij} = \begin{cases} 1\,,\ i+j < n+1,\ i > j, \\ 2\,,\ i+j < n+1,\ i \leq j, \\ 3\,,\ i+j \geq n+1,\ i \leq j, \\ 4\,,\ i+j \geq n+1,\ i > j; \end{cases}$$

$$(3):\ a_{ij} = \begin{cases} 1\,,\ i > j,\quad i-j > n/2, \\ 2\,,\ i > j,\quad i-j \leq n/2, \\ 3\,,\ i \leq j,\ i-j > -n/2, \\ 4\,,\ i \leq j,\ i-j \leq -n/2, \end{cases}$$

and the following "smooth" test matrices with a_{ij} equal to

$$(4):\ \frac{1}{i-j+\frac{1}{2}};\quad (5):\ \frac{-1}{(i-j)^2-\frac{1}{4}};\quad (6):\ \frac{1}{i+j-1};\quad (7):\ \frac{1}{|i-j|+\frac{1}{10}}.$$

We tried the above-described mosaic rank estimator on these $n \times n$-matrices for $n = 512$ and obtained the following mosaic ε-ranks and compression factors:

	(1)	(2)	(3)	(4)	(5)	(6)	(7)
Mosaic rank	10.98 4%	19.48 8%	19.23 8%	41.25 16%	32.00 13%	8.88 3%	35.13 14%
Relative accuracy	10^{-6}	10^{-6}	10^{-6}	10^{-5}	10^{-5}	$6 \cdot 10^{-6}$	10^{-5}

6 Acknowledgments

This work was partially supported by the Russian Foundation for Basic Research under Grant 94-01-00989 and also by the Volkswagen-Stiftung.

The author is grateful to S. Goreinov, N. Mikhailovski, and N. Zamarashkin for fruitful discussions of the concept of mosaic ranks.

References

1. Bakhvalov, N. S.: Numerical Methods. Nauka, Moscow (1973)
2. Brandt, A.: Multilevel computations of integral transforms and particle interactions with oscillatory kernels. Computer Physics Communications **65** (1991) 24–38
3. Goreinov, S. A., Tyrtyshnikov, E. E., and Zamarashkin, N. L.: Pseudo-skeleton approximations of matrices. Dokladi Rossiiskoi Akademii Nauk **343** **(2)** (1995) 151–152
4. Goreinov, S. A., Tyrtyshnikov, E. E., and Yeremin, A. Yu.: Matrix-free iteration solution strategies for large dense linear systems. EM-RR 11/93 (1993)
5. Hackbusch, W., and Nowak, Z. P.: On the fast matrix multiplication in the boundary elements method by panel clustering. Numer. Math. **54** **(4)** (1989) 463–491
6. Mikhailovski, N.: Mosaic approximations of discrete analogues of the Calderon-Zigmund operators. Manuscript, INM RAS (1996)
7. Myagchilov, M. V., and Tyrtyshnikov, E. E.: A fast matrix-vector multiplier in discrete vortex method. Russian Journal of Numerical Analysis and Mathematical Modelling **7** **(4)** (1992) 325–342
8. Rokhlin, V.: Rapid solution of integral equations of classical potential theory. J. Comput. Physics **60** (1985) 187–207
9. Tyrtyshnikov, E. E.: Matrix approximations and cost-effective matrix-vector multiplication. Manuscript, INM RAS (1993)
10. Voevodin, V. V.: On a method of reducing the matrix order while solving integral equations. Numerical Analysis on FORTRAN. Moscow University Press (1979) 21–26

Numerical Techniques for Solving Convection-Diffusion Problems

Petr N.Vabishchevich

Institute for Mathematical Modeling, 4 Miusskaya Square, Moscow 125047, Russia

Abstract. Many applied problems of continuum mechanics are connected with the necessity to solve problems of mass transfer induced not only by diffusion but convection too. Mathematical models of such processes are based on elliptic and parabolic equations with terms containing first-order derivatives. The corresponding problems with non-selfadjoint elliptic operators are very difficult for numerical study. In the present work there is presented a review of the-state-of-the-art in numerical solving convection-diffusion problems.

1 Introduction

In problems of continuum mechanics the convection-diffusion equation involving diffusion of a substance as well as its transport by means of a moving medium can be considered as the basic equation [1, 2]. Peculiarities of such a mathematical model are connected primarily with the non-selfadjoint property of the corresponding elliptic operators, i.e. with existence of first-order spatial derivatives (convection, drift).

The theory of difference schemes [3] as well as projection-difference schemes (finite element methods) has been developed only for selfadjoint problems. The results obtained include issues of stability (well-posedness) of difference schemes [4] and problems of iterative solution of the corresponding grid equations [5]. As for problems with non-selfadjoit operators (such as the convection-diffusion equation) which are of great practical importance, unfortunately, today there is not any essential results for them. That is why different aspects of numerical solving convection-diffusion problems are so popular in the current scientific publications.

In the present report on the basis of fundamental results of the general theory of finite difference schemes we try to provide an analysis of principal issues resulted from the non-selfadjoint property of the problem (i.e. from the convective transport). First of all, it is necessary to recognize various types of convection-diffusion problems, where convective terms have divergent, nondivergent and skew-symmetric (the half-sum of two previous) forms. In constructing discrete analogs it is reasonable to be oriented to fulfillment of the corresponding properties of the differential problem — *a priori* estimates in the corresponding spaces, subordination of the convective transport operator to the diffusion one, the maximum principle *etc.* Here there are considered 2D model steady-state as well as unsteady problems of convection-diffusion on rectangular grids based on central as well as and on upwind differences for convective terms.

The material presentation is very close to the book [6]. Using the difference schemes developed here for convection-diffusion problems it is easy to construct numerical algorithms for solving problems of convective heat and mass transfer considered both in the primitive variables (pressure-velocity) and in the transformed formulation (stream function-vorticity variables) (see, e.g. refs [7, 8]).

2 Model problems of convection-diffusion

It is convenient to consider issues of construction, study and solution of convection-diffusion problems on the simplest problems which do have the main peculiarity of general multidimensional problems of continuum mechanics.

In rectangular

$$\Omega = \{x \mid x = (x_1, x_2), \quad 0 < x_\alpha < l_\alpha, \quad \alpha = 1, 2\}$$

we search the solution of elliptic equation

$$\sum_{\alpha=1}^{2} v_\alpha(x) \frac{\partial u}{\partial x_\alpha} - k \sum_{\alpha=1}^{2} \frac{\partial^2 u}{\partial x_\alpha^2} = f(x), \quad x \in \Omega. \tag{1}$$

Here $k = \text{const}$ — diffusivity, $v_\alpha(x), \alpha = 1, 2$ — velocity components, and convective terms are written in the nondivergent form. Let us supplement equation 1) with the simplest homogeneous boundary conditions of the Dirichlet type

$$u(x) = 0, \quad x \in \partial\Omega. \tag{2}$$

Equation (1) with nondivergent convective terms is the basic one in consideration of boundary value problems for elliptic equations of second-order. Concerning to applied mathematical modeling, more attention should be given to problems with convective terms in the divergent form:

$$\sum_{\alpha=1}^{2} \frac{\partial (v_\alpha(x) u)}{\partial x_\alpha} - k \sum_{\alpha=1}^{2} \frac{\partial^2 u}{\partial x_\alpha^2} = f(x), \quad x \in \Omega. \tag{3}$$

In the divergent (conservative) form (3) the equation of convection-diffusion is nothing but the corresponding conservation law. In problems of hydrodynamics it is often possible to employ the approximation of an incompressible medium. Using the incompressibility constraint

$$\text{div} \mathbf{v} \equiv \sum_{\alpha=1}^{2} \frac{\partial v_\alpha}{\partial x_\alpha} = 0, \quad x \in \Omega, \tag{4}$$

it is easy to rewrite equation (1) in equivalent form (3) and vice versa. In these problems special consideration should be given to the skew-symmetric form where the convection-diffusion equation is like this

$$\frac{1}{2}\sum_{\alpha=1}^{2}\left(v_\alpha(x)\frac{\partial u}{\partial x_\alpha}+\frac{\partial(v_\alpha(x)u)}{\partial x_\alpha}\right)-k\sum_{\alpha=1}^{2}\frac{\partial^2 u}{\partial x_\alpha^2}=f(x), \quad x\in\Omega. \tag{5}$$

In this case a combination of divergent and nondivergent forms is used for convective terms.

The solution of difference problems should inherit the primary properties of the original differential problem (1),(2) (problems (2),(3), (2),(5)), and difference operators must have just the same basic properties as the differential ones. First of all it means stability in the corresponding norms. Therefore we start our consideration from deriving the simplest *a priori* estimates for the differential problem.

Let $\mathcal{H}=L_2(\Omega)$ be a Hilbert space with the dot product

$$(v,w)=\int_\Omega v(x)w(x)dx$$

for functions $v(x)$ and $w(x)$ equal to zero on $\partial\Omega$, and the norm $\|v\|^2=(v,v)$. Let us rewrite convection-diffusion problem (1),(2) as a problem for the operator equation of the first kind

$$\mathcal{C}u+\mathcal{D}u=f. \tag{6}$$

The diffusive transport operator

$$\mathcal{D}u=-k\sum_{\alpha=1}^{2}\frac{\partial^2 u}{\partial x_\alpha^2} \tag{7}$$

is selfadjoint and positive definite in $\mathcal{H}$:

$$\mathcal{D}=\mathcal{D}^*\geq k\lambda_0 E, \tag{8}$$

where E — the identity operator, $\lambda_0>0$ — the minimal eigenvalue of the Laplace operator.

For the convective transport operator we have

$$\mathcal{C}u=\mathcal{C}_1 u=\sum_{\alpha=1}^{2}v_\alpha(x)\frac{\partial u}{\partial x_\alpha}. \tag{9}$$

Usage of the convection-diffusion equation in form (3) leads to

$$\mathcal{C}u=\mathcal{C}_2 u=\sum_{\alpha=1}^{2}\frac{\partial(v_\alpha(x)u)}{\partial x_\alpha}. \tag{10}$$

For the convective transport operator in the skew-symmetric form (equation (5)) we have

$$\mathcal{C} = \mathcal{C}_0 = \frac{1}{2}(\mathcal{C}_1 + \mathcal{C}_2). \tag{11}$$

Taking into account homogeneous boundary conditions (2), it is easy to see that the divergent and nondivergent forms of the convective transport operator are adjoint one to another and that the skew-symmetric form does provide the skew-symmetric property, i.e.

$$\mathcal{C}_1^* = \mathcal{C}_2, \quad \mathcal{C}_0 = -\mathcal{C}_0^*. \tag{12}$$

Moreover, at the fulfilment of the incompressibility constraint (4) the skew-symmetric property holds both for nondivergent (9) and for divergent (10) forms. The principle issue in constructing discrete approximations for the convective transport operator is the fact that the skew-symmetric property of operator $\mathcal{C}_0$ takes place at any $v_\alpha(x), \alpha = 1,2$, including those for which the incompressibility constraint (4) is not satisfied.

Thus, for the operators of convective transport introduced in correspondence with (9)–(11) we have

$$|(\mathcal{C}u, u)| \le \mathcal{M}_1 \|u\|^2, \tag{13}$$

where constant $\mathcal{M}_1$ depends only on div**v**. Let us present the estimate of subordination of the convective transport operator with respect to the operator of diffusion

$$\|(\mathcal{C}u\|^2 \le \mathcal{M}_2(\mathcal{D}u, u), \tag{14}$$

with constant $\mathcal{M}_2$ which depends on the velocity. The above estimates (13),(14) serve us as the check points in investigation of difference analogs for the convective transport operator.

To derive *a priori* estimate, let us perform the scalar product of equation (7) in *calH* by u taking into account the above estimates for the convective transport operator. For the skew-symmetric form of the convective transport operator it immediately follows that in $W_2^1(\Omega)$ and in $L_2(\Omega)$ we have:

$$\|u\|_{\mathcal{D}} \le \|f\|_{\mathcal{D}^{-1}}, \tag{15}$$

$$\|u\| \le \frac{1}{k\lambda_0}\|f\|. \tag{16}$$

For the divergent and nondivergent forms of convective terms it is necessary, in general, to use some additional restrictions on div**v** in order to obtain *a priori* estimates.

3 Difference schemes for steady-state problems of convection-diffusion

In rectangular Ω let us introduce a uniform (for simplicity) in both directions difference grid with spacing $h_\alpha, \alpha = 1, 2$. Let ω be the set of internal grid points:

$$\omega = \{x \mid x = (x_1, x_2), x_\alpha = i_\alpha h_\alpha, \quad i_\alpha = 1, 2, ..., N_\alpha - 1,$$

$$N_\alpha h_\alpha = l_\alpha, \quad \alpha = 1, 2\},$$

and $\partial\omega$ — the set of boundary points. The difference solution of the convection-diffusion problem at the time level t will be referred to as $y(x, t), x \in \omega \cup \partial\omega, t > 0$.

Using the standard index-free notations of the theory of difference schemes [3, 6] we have for the forward and backward derivatives

$$w_x = \frac{w(x+h) - w(x)}{h}, \quad w_{\bar{x}} = \frac{w(x) - w(x-h)}{h},$$

whereas the central difference derivative is defined via the following relation

$$w_{\overset{\circ}{x}} = \frac{1}{2}(w_x + w_{\bar{x}}) = \frac{w(x+h) - w(x-h)}{2h}.$$

For grid functions which are equal to zero at $\partial\omega$ let us define a Hilbert space H where the scalar product and the norm are introduced as follows

$$(y, w) = \sum_{x \in \omega} y w h_1 h_2, \quad \|y\| = \sqrt{(y, y)}.$$

For $R = R^* > 0$ via H_R we define space H with the dot product $(y, w)_R = (Ry, w)$ and the norm $\|y\| = \sqrt{(Ry, y)}$.

The difference analog for the diffusive transport at the set of functions $y \in H$ we define by means of the expression

$$Dy = -k \sum_{\alpha=1}^{2} y_{\bar{x}_\alpha x_\alpha}. \tag{17}$$

Operator D is selfadjoint in H and the estimate of type (8) takes place:

$$D = D^* \geq k\delta E. \tag{18}$$

It seems to be useful [3, 5] to mention the upper estimate for the difference operator D:

$$D \leq 4k(h_1^2 + h_2^2)E. \tag{19}$$

Let us approximate with the second order the convective terms in the convection-diffusion equation on the basis of the central differences. In accordance with (9) let us define

$$Cy = C_1 y = \sum_{\alpha=1}^{2} b_\alpha y_{\overset{\circ}{x}_\alpha}. \tag{20}$$

In the simplest case of smooth enough velocity components and the solution of the differential equation let us suppose, for instance, $b_\alpha(x) = v_\alpha(x), \alpha = 1, 2$. For the convection-diffusion equation with convective terms in the divergent form the following difference operator (see (10)) is employed

$$Cy = C_2 y = \sum_{\alpha=1}^{2} (b_\alpha y)_{\overset{\circ}{x}_\alpha}. \tag{21}$$

Similar to (11) we have

$$C = C_0 = \frac{1}{2}(C_1 + C_2),$$

i.e.

$$C_0 y = \frac{1}{2} \sum_{\alpha=1}^{2} (b_\alpha y_{\overset{\circ}{x}_\alpha} + (b_\alpha y)_{\overset{\circ}{x}_\alpha}). \tag{22}$$

Let us highlight the basic properties of the introduced difference operators for convective transport in H. Just as in the continuous case (see (12)) we have

$$C_1^* = -C_2, \quad C_0 = -C_0^*. \tag{23}$$

It is easy to check that in the difference case there is no full analog for the upper estimate for difference operators of convective transport in the divergent and nondivergent forms. In this case constant M_1 in the inequality

$$|(Cy, y)| \leq M_1 \|y\|^2 \tag{24}$$

depends on the grid spacing.

For difference operators of convective transport the following estimates of subordination take place

$$\|Cy\|^2 \leq M_2 (Dy, y) \tag{25}$$

with constants M_2 which are fully consistent with the continuous case under the condition that central differences are used to approximate the operator of divergence.

For equation (6) let us put into the correspondence the following operator-difference equation

$$Cy + Dy = \varphi, \quad x \in \omega, \tag{26}$$

where, for instance, $\varphi(x) = f(x), x \in \omega$. For equation (26) there do exist *a priori* estimates consistent with the estimates for the differential problem. For example, choosing $C = C_0$ we obtain (see (15),(16))

$$\|y\|_D \leq \|\varphi\|_{D^{-1}}, \tag{27}$$

$$\|y\| \leq \frac{1}{k\delta} \|\varphi\|. \tag{28}$$

Considering the problem for the error $z(x) = y(x) - u(x), x \in \omega$, from estimates (27),(28) it is easy to derive [9] convergence of difference schemes for the convection-diffusion problem in grid spaces $W_2^1(\omega)$ and $L_2(\omega)$. For smooth enough equation coefficients and the solution itself the difference scheme (26) at $C = C_0$ converges (and converges unconditionally) with the second order.

4 Monotone schemes

It is well-known that for convection-diffusion problems with convective terms in nondivergent form (1) the maximum principle holds. The similar fact can also be obtained for equations (3) and (5). It seems to be natural to require fulfilment of this property for the solution of the discrete convection-diffusion problem too. In this case we shall say about monotonicity of a difference scheme.

At first let us consider monotonicity of difference schemes for the convection-diffusion equation with convective terms in nondivergent form. For boundary value problem (1),(2) let us put into the correspondence the following difference problem

$$\sum_{\alpha=1}^{2} b_\alpha(x) y_{\overset{\circ}{x}_\alpha} + Dy = \varphi(x), \quad x \in \omega, \tag{29}$$

$$y(x) = 0, \quad x \in \partial\omega. \tag{30}$$

Monotonicity of difference scheme (ref4.7),(30) will be satisfied at

$$\frac{1}{k} h_\alpha \max_{x \in \omega} |b_\alpha(x)| < 2, \quad \alpha = 1, 2. \tag{31}$$

Values $\theta_\alpha(x) = 0.5k^{-1} h_\alpha b_\alpha(x), \alpha = 1, 2$ characterize the ratio between convective and diffusive transports on the grid employed.

Among unconditionally monotone difference schemes for convection-diffusion problems it is necessary to highlight the simplest schemes with directed differences. Let us define grid functions $b_\alpha^\pm(x), \alpha = 1, 2$ in the following way

$$b_\alpha(x) = b_\alpha^+(x) + b_\alpha^-(x), \quad b_\alpha^+(x) = \frac{1}{2}(b_\alpha(x) + |b_\alpha(x)|) \geq 0,$$

$$b_\alpha^-(x) = \frac{1}{2}(b_\alpha(x) - |b_\alpha(x)|) \leq 0.$$

Difference scheme

$$\sum_{\alpha=1}^{2} (b_\alpha^+(x) y_{\bar{x}_\alpha} + b_\alpha^-(x) y_{x_\alpha} + Dy = \varphi(x), \quad x \in \omega \tag{32}$$

is unconditionally monotone. It should be mentioned that this scheme in contrast to scheme (29),(30) has only the first order of approximation.

Construction of monotone schemes [10] can be performed on the basis of the regularization principle for difference schemes [3]. Let us start from some initial difference scheme which does not satisfy the maximum principle unconditionally. Concerning to the above problem (1),(2) it is natural to use as the initial scheme the scheme with central differences (29),(30). It is monotone with some restrictions (see restriction on grid spacing (31)).

A regularized scheme has the following form

$$\sum_{\alpha=1}^{2} b_\alpha(x) y_{\overset{\circ}{x}_\alpha} - k \sum_{\alpha=1}^{2} (1 + R_\alpha(x)) y_{\bar{x}_\alpha x_\alpha} = \varphi(x), \quad x \in \omega. \tag{33}$$

Scheme (30),(33) corresponds to regularization by means of increasing of local diffusion. An alternative approach is to perturb the velocity and to utilize instead of (33) the following difference equation

$$\sum_{\alpha=1}^{2} \frac{b_\alpha(x)}{1+R_\alpha(x)} y_{\overset{\circ}{x}_\alpha} + Dy = \varphi(x), \quad x \in \omega.$$

For scheme (30),(33) the monotonicity conditions lead to restrictions

$$1 + R_\alpha(x) > |\theta_\alpha(x)|, \quad \alpha = 1,2, \quad x \in \omega. \tag{34}$$

Due to the fact that $\theta_\alpha(x) = O(h_\alpha)$, it is sufficient to put $R_\alpha(x) = O(h_\alpha^2), \alpha = 1,2$ in order to preserve the second order.

Another class of regularized schemes is defined by the following selection of regularizers

$$R_\alpha(x) = \eta\theta_\alpha^2(x), \quad \alpha = 1,2. \tag{35}$$

The sufficient condition of monotonicity at this selection will be satisfied for (34) if in (35) we have $\eta > 0.25$. The schemes developed in [11, 12] also belong to the class of regularized unconditionally monotone difference schemes (30),(33),(35).

It is clear that instead of (35) we can employ and more complicated regularizers. Such an example is the exponential scheme [13, 14] where regularizers are as follows

$$R_\alpha(x) = \theta_\alpha(x)\mathrm{cth}\theta_\alpha(x) - 1, \quad \alpha = 1,2. \tag{36}$$

The primary shortcomings of exponential scheme (30),(33),(36) are connected with the fact that evaluation of coefficients for this scheme includes multiple calculation of exponents. Thus, it is natural to simplify coefficients of this scheme but retain its quality (see [2, 15]).

Hybrid monotone schemes are derived on the basis of discontinuous regularizers $R_\alpha(x), \alpha = 1,2$. For instance, well-known scheme [16] corresponds to the choice of the simplest discontinuous regularizers

$$R_\alpha(x) = \begin{cases} 0, & |\theta_\alpha(x)| \le 1, \\ |\theta_\alpha(x)|, & |\theta_\alpha(x)| > 1, \end{cases} \quad \alpha = 1,2,$$

It is possible to construct simple regularized schemes do not using the scheme with central differences as the initial one. An example is the following

$$1 + R_\alpha(x) = |\theta_\alpha(x)| + \frac{1}{1+|\theta_\alpha(x)|}, \quad \alpha = 1,2. \tag{37}$$

Scheme (30),(33),(37) is unconditionally monotone [3, 18].

For convection-diffusion problem with convective terms in divergent form (equation (3)) regularized unconditionally monotone schemes are constructed in a similar way. In this case it is reasonable to employ new sufficient conditions for fulfilment of the maximum principle for difference schemes [6, 10, 17].

For equation (3) difference equation with central differences has the following form

$$\sum_{\alpha=1}^{2}(b_\alpha(x)y)_{\overset{\circ}{x}_\alpha} + Dy = \varphi(x), \quad x \in \omega. \tag{38}$$

A regularized scheme can be derived as follows

$$\sum_{\alpha=1}^{2}(b_\alpha(x)y)_{\overset{\circ}{x}_\alpha} - k\sum_{\alpha=1}^{2}((1+R_\alpha(x))y)_{\bar{x}_\alpha x_\alpha} = \varphi(x), \quad x \in \omega. \tag{39}$$

For regularized scheme (30),(39) the monotonicity condition has the form (34). Further consideration [10] is conducted similar to problems with nondivergent convective terms.

5 Difference schemes for transient problems

For transient problems of convection-diffusion

$$\frac{du}{dt} + \mathcal{C}u + \mathcal{D}u = f, \tag{40}$$

$$u(x,0) = u_0(x), \quad x \in \Omega \tag{41}$$

the emphasis is on constructing difference schemes unconditionally stable in the corresponding norms. For problem (40),(41) let us put into the correspondence the following differential-difference problem

$$\frac{dy}{dt} + Cy + Dy = \varphi, \tag{42}$$

$$y(x,0) = u_0(x), \quad x \in \omega \tag{43}$$

with $C = -C^*, \quad D = D^* > 0$.

Let $\mathbf{y}^n$ be the difference solution at the time moment $t^n = n\tau$, where $\tau > 0$ is a time step. For problem (42),(43) let us consider the class of two-level weighted schemes

$$\frac{y_{n+1} - y_n}{\tau} + C\left(\sigma_1 y_{n+1} + (1-\sigma_1)y_n\right) + D\left(\sigma_2 y_{n+1} + (1-\sigma_2)y_n\right) = \varphi_n. \tag{44}$$

Stability of scheme (44) with equal weights

$$\sigma_1 = \sigma_2 = \sigma \tag{45}$$

is proved on the basis of the general theory of stability for difference schemes [3, 4]. In the canonical form

$$B\frac{y_{n+1} - y_n}{\tau} + Ay_n = \varphi_n, \quad n = 0, 1, \ldots \tag{46}$$

for scheme (44),(45) we have

$$B = E + \sigma\tau A, \quad A = C + D. \tag{47}$$

In $L_2(\omega)$ the stability restriction for scheme (46),(47) has the following form

$$A + \tau(\sigma - \frac{1}{2})A^* A \geq 0.$$

In particular, schemes with $\sigma \geq 0.5$ are unconditionally stable.

Particular attention should be paid to the explicit-implicit scheme

$$\frac{y_{n+1} - y_n}{\tau} + D y_{n+1} + C y_n = \varphi_n,$$

$$B = B^* = E + \tau D, \quad A = C + D.$$

This scheme also belongs to the class of unconditionally stable schemes [19]. The estimates which guarantees ρ-stability with respect to the initial data is like this

$$\|y_{n+1}\|_D \leq (1 + \tau \frac{M_2}{4}) \|y_n\|_D,$$

where M_2 — the constant from the inequality of subordination of the convection operator with respect to the operator of diffusion.

In using the corresponding approximations with the directed differences, unconditionally stable are fully implicit schemes ($\sigma = 1$ in (44),(45)). For the nondivergent form of convective transport

$$Cy = C_1 y = \sum_{\alpha=1}^{2} (b_\alpha^+(x) y_{\bar{x}_\alpha} + b_\alpha^-(x) y_{x_\alpha})$$

the estimate of stability is as follows

$$\|y_{n+1}\|_\infty \leq \|y_n\|_\infty + \tau \|\varphi_n\|_\infty,$$

where

$$\|y\|_\infty = \max_{x \in \omega} |y(x)|.$$

For divergent operator of convection

$$Cy = C_2 y = \sum_{\alpha=1}^{2} ((b_\alpha^+(x) y)_{\bar{x}_\alpha} + (b_\alpha^-(x) y)_{x_\alpha})$$

we have

$$\|y_{n+1}\|_1 \leq \|y_n\|_1 + \tau \|\varphi_n\|_1,$$

where now

$$\|y\|_1 = \sum_{x \in \omega} |y(x)| h_1 h_2.$$

These stability estimates can be obtained in the most simple way using the concept of the logarithmic norm of an operator [20].

References

1. Roache P.J.: Computational Fluid Dynamics. Hermosa, Albuquerque, N.M., 1982.
2. Patankar S.V.: Numerical Heat Transfer and Fluid Flow. Hemisphere, Washington, DC, 1980.
3. Samarskii A.A.: Theory of Difference Schemes. Nauka, Moscow, 3rd edn, 1989, in Russian.
4. Samarskii A.A., Gulin A.V.: Stability of Finite Difference Schemes. Nauka, Moscow, 1973, in Russian.
5. Samarskii A.A., Nikolaev E.S.: Numerical Methods for Grid Equations. Birkhauser Verlag, Basel, 1989.
6. Samarskii A.A., Vabishchevich P.N.: Computational Heat Transfer. Wiley, Chichester, 1995.
7. Churbanov A.G., Pavlov A.N., Vabishchevich P.N.: Operator-splitting methods for the incompressible Navier-Stokes equations on non-staggered grids. Part 1: First-order schemes. Int. J. Numer. Methods Fluids **21** (1995) 617–640.
8. Chudanov V.V., Popkov A.G., Churbanov A.G., Vabishchevich P.N., Makarov M.M.: Operator-splitting schemes for the stream function-vorticity formulation. Computers & Fluids **24** (1995) 771–786.
9. Vabishchevich P.N.: Difference schemes with central differences for problems of convection-diffusion. Institute for Mathematical Modeling RAS, Moscow **17** (1993), in Russian.
10. Vabishchevich P.N.: Monotone difference schemes for problems of convection-diffusion. Differentsial'nye Uravnenija **30** (1994) 503–513, in Russian.
11. Briggs D.G.: A finite difference scheme for the incompressible advection-diffusion equation. Comput. Meths. Appl. Mech. Engrg. **6** (1975) 233–241.
12. Joseph M.: Finite difference representations of vorticity transport. Comput. Meths. Appl. Mech. Engrg. **39** (1983) 107–116.
13. Allen D.N.De G., Southwell R.V.: Relaxation methods applied to determine the motion in two dimensions of a viscous fluid past a fixed cylinder. Quart. J. Mech. Appl. Math. **8** (1955) 129–145.
14. Doolan E.P., Miller J.J.H., Schilders W.H.A.: Uniform Numerical Methods For Problems with Initial and Boundary Layers. Dublin, Boole Press, 1980.
15. Chien J.C.: A general finite-difference formulation with application to Navier-Stokes equations. Comput. & Fluids **5** (1977) 15–31.
16. Spalding D.B.: A novel finite difference formulation for differential equations involving both first and second derivatives. Int. J. Numer. Methods Engrg. **4** (1972) 551–559.
17. Karetkina N.V.: Unconditionally stable difference scheme for parabolic equations involving the first derivatives. Zhurnal Vychislitel'noy Matematiki i Matematicheskoy Fiziki **20** (1980) 236–240, in Russian.
18. Samarskii A.A.: On monotone difference schemes for elliptic and parabolic equations in case of non-selfadjoint elliptic operator. Zhurnal Vychislitel'noy Matematiki i Matematicheskoy Fiziki **5** (1965) 548–551, in Russian.
19. Vabishchevich P.N.: Difference schemes for transient problems of convection-diffusion. Institute for Mathematical Modeling RAS, Moscow **3** (1994), in Russian.
20. Hairer E., Norsett S.P., Wanner G.: Solving Ordinary Differential Equations. Nonstiff Problems. Springer-Verlag, Berlin, 1987.

Explicit Fifth Order Runge-Kutta Methods with Five Stages for Quadratic ODEs

M. Van Daele, G. Vanden Berghe and H. De Meyer*

Vakgroep Toegepaste Wiskunde en Informatica
Universiteit–Gent
Krijgslaan 281 - S9, B9000 Gent, Belgium

Abstract. It is well-known that there exist no general purpose explicit fifth order five stage Runge-Kutta methods. One may however wonder whether there exist explicit five stage methods which are of fifth order for special kinds of problems, for instance problems $y' = f(x, y)$ where f is an m dimensional vector of polynomials of degree at most d in each of the arguments. It is shown that there exist explicit five stage Runge-Kutta methods for quadratic ODEs ($d = 2$). Solutions are given for the non-confluent case and a numerical example is incorporated.

1 Introduction

In recent years, there has been a great interest in Runge-Kutta methods solving Hamiltonian problems. In many cases, the methods that were studied preserved the symplecticness of the problems considered. These methods are presented in detail in Ref. [1].
Sometimes however, also the special form of the problem itself was taken into account. One such effort for quadratic ODEs was proposed by G. Ramaswami [2]. Indeed, frequently the Hamiltonian can be written as a polynomial that is at most quadratic in its arguments. For Runge-Kutta methods that are developed for this purpose there is a serious drawback : since all derivatives of order at least three of the r.h.s. identically vanish, some order conditions do not need to be fulfilled.

On the other hand, it is a well-known result that explicit fifth order Runge-Kutta methods for solving $y' = f(x, y)$ require a least six stages. Indeed, the 17 order conditions can not simultaneously be fulfilled by the 15 coefficients of a consistent explicit five stage method.

These two observations have led us to the following question : is it possible to obtain explicit five stage methods of order five that solve quadratic ODEs, i.e. problems for which each component ${}^kf$ of f (in a homogeneous form) is given by

$$ {}^kf(y) = \sum_{i=1}^{m}\sum_{j=1}^{i} {}^ku_{ij}\, {}^iy\, {}^jy + \sum_{i=1}^{m} {}^kv_i\, {}^iy + {}^kw, \qquad k = 1, 2, \ldots, m, \qquad (1) $$

* Research Director at the National Fund for Scientific Research (N.F.W.O. Belgium)

where ${}^k u_{ij}$, ${}^k v_i$ and ${}^k w$ are constants and m is the dimension of the problem? Of course, we do not look for symplectic methods since it is known that no such explicit methods exist [1].

Our investigations have lead to several results. First of all, it turns out that all solutions must have $c_5 = 1$ and that they must obey the so-called column simplifying assumption $D(1)$ (following the standard notation used in Ref. [3, 4]). Secondly, it turns out that some of the order conditions that need not be considered for solving problems of the form (1) are satisfied anyhow. Thirdly, we consider the non-confluent case and show that there exists a two parameter family solution in case $b_3 \neq 0$ and a one parameter solution in case $b_3 = 0$. Finally, we give an example method to sustain the theory.

2 Methods for Quadratic ODEs

Suppose an explicit five stage Runge-Kutta method with Butcher-representation

$$\begin{array}{c|c} c & A \\ \hline & b^T \end{array}, \qquad c_i = \sum_{j=1}^{i-1} a_{ij}\,, \qquad i = 1, 2, \ldots, 5\,,$$

is applied to a problem of the form (1), the order conditions $\Phi(t) = \dfrac{1}{\gamma(t)}$ have to be fulfilled. Here we will always use the standard notation of Butcher [3] : t is any rooted tree and Φ and γ are the elementary weight and the density function. Further, we will use τ to denote the rooted tree with only one vertex. Further, $\circ$ will be used to denote the Butcher product of rooted trees :

$$\tau \circ u = [u] \qquad \text{and} \qquad [t_1\, t_2 \ldots t_k] \circ u = [t_1\, t_2 \ldots t_k\, u]\,.$$

Written down explicitly, these order conditions are

$$\begin{array}{ll}
\text{order 1 } \sum_i b_i = 1, & \text{order 5 } \sum_{i,j} b_i\, c_i\, a_{ij}\, c_j^2 = \frac{1}{15}, \\
\text{order 2 } \sum_i b_i\, c_i = \frac{1}{2}, & \sum_i b_i \left(\sum_j a_{ij}\, c_j\right)^2 = \frac{1}{20}, \\
\text{order 3 } \sum_i b_i\, c_i^2 = \frac{1}{3}, & \sum_{i,j,k} b_i\, c_i\, a_{ij}\, a_{jk}\, c_k = \frac{1}{30}, \\
\sum_{i,j} b_i\, a_{ij}\, c_j = \frac{1}{6}, & \sum_{i,j,k} b_i\, a_{ij}\, c_j\, a_{jk}\, c_k = \frac{1}{40}, \\
\text{order 4 } \sum_{i,j} b_i\, a_{ij}\, c_j^2 = \frac{1}{12}, & \sum_{i,j,k} b_i\, a_{ij}\, a_{jk}\, c_k^2 = \frac{1}{60}, \\
\sum_{i,j} b_i\, c_i\, a_{ij}\, c_j = \frac{1}{8}, & \sum_{i,j,k,l} b_i\, a_{ij}\, a_{jk}\, a_{kl}\, c_l = \frac{1}{120}. \\
\sum_{i,j,k} b_i\, a_{ij}\, a_{jk}\, c_k = \frac{1}{24}, &
\end{array} \tag{2}$$

We will use the symbol S to denote the set of equations (2).

From the traditional set of 17 equations, the following equations no longer have to be fulfilled :

$$\sum_i b_i\, c_i^3 = \frac{1}{4}, \tag{3}$$

$$\sum_i b_i\, c_i^4 = \frac{1}{5}, \tag{4}$$

$$\sum_{i,j} b_i\, a_{ij}\, c_j^3 = \frac{1}{20}, \tag{5}$$

$$\sum_{i,j} b_i\, c_i^2\, a_{ij}\, c_j = \frac{1}{10}. \tag{6}$$

Lemma 1. *Each five stage explicit Runge-Kutta method for which the coefficients satisfy set S has $c_5 = 1$.*

Proof. It S is fullfilled, then it is clear that

$$(5\,c_2 - 2) \sum_{i,j,k,l} b_i\, a_{ij}\, a_{jk}\, a_{kl}\, c_l - c_2 \sum_{i,j,k} b_i\, a_{ij}\, a_{jk}\, c_k + \sum_{i,j,k} b_i\, a_{ij}\, a_{jk}\, c_k^2 \tag{7}$$

$$= \frac{5\,c_2 - 2}{120} - \frac{c_2}{24} + \frac{1}{60} = 0\,.$$

Writing down explicitly (7), one thus finds that

$$b_5\, a_{54}\, a_{43}\, [a_{32}\, c_2\, (5\,c_2 - 2) - c_3\, (c_2 - c_3)] = 0\,.$$

Now, $b_5\, a_{54}\, a_{43} \neq 0$ since

$$\sum_{i,j,k,l} b_i\, a_{ij}\, a_{jk}\, a_{kl}\, c_l = b_5\, a_{54}\, a_{43}\, a_{32}\, c_2 \neq 0\,, \tag{8}$$

and one thus obtains

$$a_{32}\, c_2\, (5\,c_2 - 2) - c_3\, (c_2 - c_3) = 0\,. \tag{9}$$

On the other hand

$$\sum_{i,j} b_i\, (c_5 - c_i)\, a_{ij}\, c_j\, (c_2 - c_j) - (2 - 5\,c_2) \sum_{i,j,k} b_i\, (c_i - c_5)\, a_{ij}\, a_{jk}\, c_k \tag{10}$$

$$= c_2\, c_5 \sum_{i,j} b_i\, a_{ij}\, c_j - c_2 \sum_{i,j} b_i\, c_i\, a_{ij}\, c_j - c_5 \sum_{i,j} b_i\, a_{ij}\, c_j^2 + \sum_{i,j} b_i\, c_i\, a_{ij}\, c_j^2$$

$$-(2 - 5\,c_2) \left(\sum_{i,j,k} b_i\, c_i\, a_{ij}\, a_{jk}\, c_k - c_5 \sum_{i,j,k} b_i\, a_{ij}\, a_{jk}\, c_k \right)$$

$$= \frac{c_2\, c_5}{6} - \frac{c_2}{8} - \frac{c_5}{12} + \frac{1}{15} - (2 - 5\,c_2) \left(\frac{1}{30} - \frac{c_5}{24} \right)$$

$$= \frac{c_2\, (1 - c_5)}{24}, \tag{11}$$

while the explicit form of expression (10) is given by

$$b_4\,(c_5 - c_4)\,a_{43}\,\left[c_3\,(c_2 - c_3) - a_{32}\,c_2\,(5\,c_2 - 2)\right]$$

which is equal to zero due to (9). From (11) and (8) it thus follows that $c_5 = 1$.

Lemma 2. *Each consistent five stage explicit Runge-Kutta method, for which the coefficients satisfy set S, satisfies the column simplifying condition*

$$\sum_{i=1}^{5} b_i\,a_{ij} = b_j\,(1 - c_j)\,, \qquad j = 1, 2, \ldots, 5\,.$$

Proof. For $j = 5$, the proof follows from Lemma 1.
For $j = 2$, 3 and 4, we rely on the relation

$$\Phi(t) - \Phi(t \circ \tau) - \Phi(\tau \circ t) = \sum_{i=1}^{s} \Phi_i(t)\,\left[b_i\,(1 - c_i) - \sum_{j=1}^{s} b_j\,a_{ji}\right]$$

where t is replaced by τ, $[\tau]$ and $[_2\tau]_2$. If S is fulfilled, then the l.h.s. identically vanishes. Since $\Phi_1(t) \equiv 0$ and $b_5\,(1 - c_5) = \sum_{i=1}^{5} b_i\,a_{i5}$, one obtains a homogeneous linear system of 3 equations in $b_j\,(1 - c_j) = \sum_{i=1}^{5} b_i\,a_{ij}$, $j = 2, 3, 4$, for which the determinant is given by $\Phi_2(\tau)\,\Phi_3([\tau])\,\Phi_4([_2\tau]_2) = a_{43}\,a_{32}^2\,c_2^3 \neq 0$ due to (8).

Finally, the result for $j = 1$ follows from the consistency relation $c_i = \sum_{j=1}^{i-1} a_{ij}$:

$$\begin{aligned}
\sum_{i=1}^{5} b_i\,a_{i1} &= \sum_{i=1}^{5} b_i\,(c_i - \sum_{j=2}^{5} a_{ij}) \\
&= \sum_{i} b_i\,c_i - \sum_{j=2}^{5} b_j\,(1 - c_j) \\
&= \frac{1}{2} - (1 - b_1) + \frac{1}{2} = b_1 = b_1\,(1 - c_1)\,.
\end{aligned}$$

With the help of these two lemmas we can now state the following theorem :

Theorem 3. *For each five stage explicit Runge-Kutta method for which the coefficients satisfy set S the coefficients also satisfy (3) and (6) while neither (4) nor (5) hold.*

Proof. If $D(1)$ holds, then it is known [3] that the order condion for $\tau \circ t$ is a direct consequence of the order conditions for t and $\tau \circ t$ since

$$\Phi(t) - \Phi(t \circ \tau) - \Phi(\tau \circ t) = \frac{1}{\gamma(t)} - \frac{1}{\gamma(t \circ \tau)} - \frac{1}{\gamma(\tau \circ t)} = 0\,.$$

Now apply this identity to $t = [_2\tau]_2$ and to $t = [\tau[\tau]_2$ to prove the first part. To prove the second part, we use $t = [_3\tau]_3$: the identity can no longer hold since otherwise we would obtain an explicit 5 stage method of order 5.

3 The Non-confluent Case

We consider the case, i.e. $c_i \neq c_j$ if $i \neq j$. The following set of equations remains to be solved :

$$\sum_{j=1}^{i-1} a_{ij} = c_i, \qquad i = 2, 3, 4, 5, \tag{12}$$

$$\sum_{j=i+1}^{5} b_j\, a_{ji} = b_i\,(1 - c_i), \qquad i = 2, 3, 4, \tag{13}$$

$$\sum_{i=1}^{5} b_i\, c_i^{k-1} = \frac{1}{k}, \qquad k = 1, 2, 3, 4, \tag{14}$$

$$\sum_{i=3}^{5} b_i\, c_i^{k-1} \sum_{j=2}^{i-1} a_{ij}\, c_j^{l-1} = \frac{1}{l\,(k+l)}, \qquad 1 < k \leq k + l - 2 < 4, \tag{15}$$

$$b_5\, a_{54}\, a_{43}\, a_{32}\, c_2 = \frac{1}{120}, \tag{16}$$

$$\sum_{i=3}^{5} b_i \left(\sum_{j=2}^{i-1} a_{ij}\, c_j \right)^2 = \frac{1}{20}. \tag{17}$$

Further, it is known that $c_1 = 0$ and $c_5 = 1$.

Eq. (12) can be used to determine the coefficients a_{i1} once all other coefficients have been determined.

The set (14) allows one to express b_1, b_2, b_3 and b_4 as a function of c_2, c_3 and c_4 and b_5. One thus finds

$$b_3 = \frac{-3 + 4\,(c_2 + c_4) - 6\,c_2\,c_4 + 12\,(c_2 - 1)\,(c_4 - 1)\,b_5}{12\,c_3\,(c_3 - c_2)\,(c_4 - c_3)}. \tag{18}$$

– In case $b_3 \neq 0$, the system (13), (15) determines a_{32}, a_{42}, a_{43}, a_{52}, a_{53} and a_{54} uniquely in terms of c_2, c_3, c_4, b_3, b_4 and b_5, i.e. due to the first step in terms of c_2, c_3, c_4 and b_5. Substituting these expressions in (16), b_5 can be expressed in terms of c_2, c_3 and c_4, such that every a and b-coefficient can be determined uniquely in terms of the three remaining c-values.
One thus finds :

$$b_2 = \frac{6 - 5\,c_2 - 10\,c_3 - 10\,c_4 + 5\,c_2\,c_4 + 20\,c_3\,c_4}{120\,(c_2 - 1)\,c_2\,(c_3 - c_2)\,(c_2 - c_4)},$$

$$b_3 = \frac{(5\,c_2 - 2)\,(3 - 5\,c_4)}{120\,(c_2 - c_3)\,(c_3 - 1)\,c_3\,(c_4 - c_3)},$$

$$b_4 = \frac{6 - 15\,c_2 - 10\,c_3 + 20\,c_2\,c_3 + 5\,c_2\,c_4}{120\,(c_2 - c_4)\,(c_3 - c_4)\,(1 - c_4)\,c_4},$$

$$b_5 = \frac{1}{120\,(1-c_4)\,\gamma},$$
$$a_{32} = \frac{(c_2-c_3)\,c_3}{c_2\,(5\,c_2-2)},$$
$$a_{42} = \frac{c_4\,(c_4-c_2)\,\left(-3\,c_2+5\,c_3-5\,{c_3}^2-2\,c_4+5\,c_2\,c_4\right)}{c_2\,(c_3-c_2)\,(6-15\,c_2-10\,c_3+20\,c_2\,c_3+5\,c_2\,c_4)},$$
$$a_{43} = \frac{(5\,c_2-2)\,(c_2-c_4)\,(c_3-c_4)\,c_4}{(c_2-c_3)\,c_3\,(6-15\,c_2-10\,c_3+20\,c_2\,c_3+5\,c_2\,c_4)},$$
$$a_{52} = \frac{1}{c_2\,(1-c_3)\,(c_3-c_2)\,(c_4-c_2)\,(c_4-c_3)}\times$$
$$\left((1-c_2)^2\,(5\,c_4-3)+(1-c_2)\,((1-c_4)(13-15\,c_3-10\,c_4)+5\,c_3\,(c_3-c_4))\right.$$
$$\left.+(1-c_3)\,(1-c_4)\,\left(20\,c_3\,c_4-5\,c_3-4\right)\right)\,\gamma,$$
$$a_{53} = \frac{(2-5\,c_2)\,\left(-3+c_3+7\,c_4-5\,{c_4}^2\right)}{c_3\,(c_3-c_2)\,(c_4-c_3)}\,\gamma,$$
$$a_{54} = \frac{(1-c_4)\,(6-15\,c_2-10\,c_3+20\,c_2\,c_3+5\,c_2\,c_4)}{c_4\,(c_4-c_2)\,(c_4-c_3)}\,\gamma,$$

where

$$\gamma = \frac{(1-c_2)\,(c_3-1)}{25\,c_2+30\,(c_3+c_4)-40\,c_3\,(c_2+c_4)-35\,c_2\,c_4+60\,c_2\,c_3\,c_4-24}.$$

The coefficients b_1 and a_{i1} (i = 2, 3, 4, 5) follow from (14) with $k = 1$ and (12).

From (17), a final relation between c_2, c_3 and c_4 is found :

$$\begin{aligned}0 &= 72-6\,(39\,c_2+35\,c_3+27\,c_4)\\ &+15\,(9\,c_2^2+31\,c_2\,c_3+31\,c_2\,c_4+10\,c_3^2+32\,c_3\,c_4+6\,c_4^2)\\ &-5\,(30\,c_2\,c_3^2-9\,c_2^2\,c_3+30c_2^2\,c_4+180\,c_2\,c_3\,c_4+70\,c_3^2\,c_4+51c_2\,c_4^2+54\,c_3\,c_4^2)\\ &-5\,(60\,c_2^2\,c_3^2+91\,c_2^2\,c_3\,c_4-60\,c_2\,c_3^2\,c_4-15\,c_2^2\,c_4^2-111\,c_2\,c_3\,c_4^2-40\,c_3^2\,c_4^2)\\ &-100\,c_2\,c_3\,c_4\,(c_2\,c_4-7\,c_2\,c_3+3\,c_3\,c_4)\,. \end{aligned}\tag{19}$$

This relation, that contains terms that are at most quadratic in each of the parameters c_2, c_3 and c_4 can be looked at as a quadratic equation in c_4. As is shown in Fig. 1, it turns out that most of the times, there exist real solutions for c_4 when c_2 and c_3 are given as input values. We can thus state that we have found a two-parameter solution of the set S.

– If $b_3 = 0$, if follows from (18) that

$$b_5 = \frac{3-4\,(c_2+c_4)+6\,c_2\,c_4}{12\,(c_2-1)\,(c_4-1)}\,. \tag{20}$$

This equation, together with (14) determines b_1, b_2, b_4 and b_5 in terms of c_2 and c_4. Now, (13), (15) form a system of six linear equations in five unknowns a_{42}, a_{43}, a_{52}, a_{53} and a_{54}. A solution can only be found for $c_4 = 3/5$. These results can now be substituted in (16) to find a solution for a_{32}. Finally, (17) requires $c_2 = \frac{1}{4}$, such that the following one parameter family of solutions of S is found :

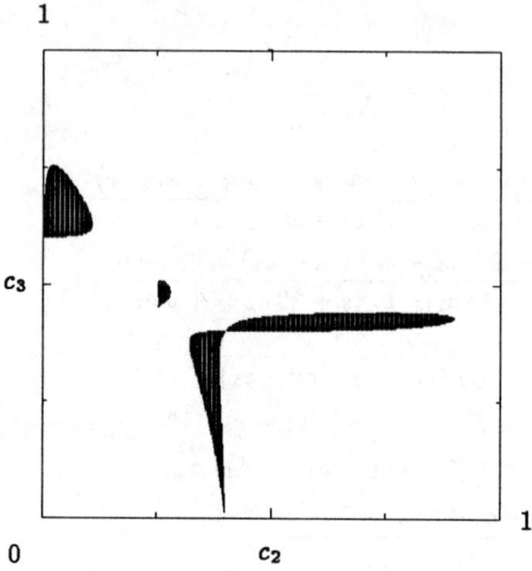

Fig. 1. The shaded area identifies couples (c_2, c_3) for which (3.8) has no real roots c_4.

$$
\begin{array}{c|ccccc}
0 & & & & & \\
\frac{1}{4} & \frac{1}{4} & & & & \\
c_3 & \frac{c_3\,(7-16\,c_3)}{3} & \frac{4\,c_3\,(4\,c_3-1)}{3} & & & \\
\frac{3}{5} & \frac{3\,(21-40\,c_3)}{500\,c_3} & \frac{84\,(5\,c_3-2)}{125\,(4\,c_3-1)} & \frac{63}{500\,c_3\,(4\,c_3-1)} & & \\
1 & \frac{3\,(8\,c_3-3)}{20\,c_3} & \frac{12\,(10-19\,c_3)}{35\,(4\,c_3-1)} & \frac{9}{20\,c_3\,(1-4\,c_3)} & \frac{10}{7} & \\
\hline
 & \frac{1}{9} & \frac{16}{63} & 0 & \frac{125}{252} & \frac{5}{36}
\end{array}
$$

4 An Example

As an example, we consider the member of the one parameter family with $c_3 = 1/2$. We apply this method to two test problems which are integrated to the point x_{end} :

(i) $y' = y^2$ with $y(0) = 1$ and $x_{\text{end}} = 0.5$. The solution is $y(x) = (1-x)^{-1}$.
(ii) $y' = -x^2\,y$ with $y(0) = 1$ and $x_{\text{end}} = 1$. The solution is $y(x) = \exp(-x^3/3)$.

According to the theory, the method used will produce a numerical solution in x_{end} with a global error that is proportional to h^5 for problem (i), and to h^4 for problem (ii). Figure 2 sustains these theoretical findings. This figure shows

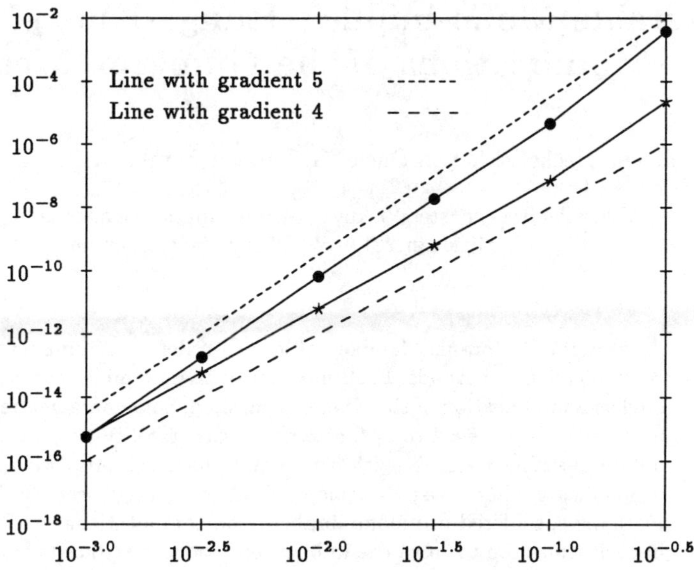

Fig. 2. A double logarithmic plot of stepsize .vs. global error in x_{end} for problem (i) (filled circles) and problem (ii) (stars).

a double logarithmic plot of the stepsize versus the absolute value of the global error in x_{end}. These errors were computed for constant step sizes $h = 10^{-q/2}$ for $q = 1, 2, \ldots, 6$. For each problem, the points thus found are connected. For problem (i) a straight line with gradient approximately 5 is obtained. The line corresponding to problem (ii) on the other hand is parallel to the dotted line with gradient 4.

References

1. Sans-Serna, J.M., Calvo, M.P., Applied Mathematics and Mathematical Computation 7 (ed. R.J. Knops & K.W. Morton), Numerical Hamiltonian Problems, (Chapman & Hall, London, 1994)
2. Ramaswami, G., Iserles, A., Sofroniou, M., Numerical methods for quadratic ODEs (Private communication and Abstracts of the 16th Biennal Conference on Numerical Analysis, 27 June – 30 June, 1995)
3. Butcher, J.C., The Numerical Analysis of Ordinary Differential Equations: Runge-Kutta and General Linear Methods , (J. Wiley, Chichester), (1987)
4. Hairer, E., Nørsett, S.P., Wanner, G., Solving Ordinary Differential Equations I : Nonstiff Problems, (Springer-Verlag, Berlin, 1987).

P-stable Mono-implicit Runge-Kutta-Nyström Modifications of the Numerov Method

T. Van Hecke*, M. Van Daele, G. Vanden Berghe and H. De Meyer**

Vakgroep Toegepaste Wiskunde en Informatica, Universiteit–Gent,
Krijgslaan 281 – S9, B9000 Gent, Belgium

Abstract. We present families of fourth-order mono-implicit Runge-Kutta-Nyström methods. Each member of these families can be considered as a modification of the Numerov method. Some parameters of these new methods are used to optimize the linear stability properties, i.e. to obtain P-stable methods with a minimal phase-lag. Also we show that in some cases there exist P-stable methods with stage-order 3. Since the methods considered are mono-implicit, the computational work needed in each time-step to solve the implicit equations is reduced seriously.

1 Introduction

One of the most popular codes for solving problems of the form $y'' = f(x, y)$, $y(x_0) = y_0$, $y'(x_0) = y'_0$ is given by the Numerov method

$$y_{k+2} - 2\,y_{k+1} + y_k = \frac{h^2}{12}\,(f_{k+2} + 10\,f_{k+1} + f_k)\ , \tag{1}$$

whereby $f_i = f(x_i, y_i)$. This method is of fourth order, its phase-lag is of fourth order and the interval of periodicity is $(0, H_p^2) = (0, 6)$. In the last decades, several authors (see $[1-4, 7, 9-13, 15-17]$ for example) considered modifications of the Numerov method (making it an explicit method, a hybrid method or using exponential fitting for instance) to raise the order, to extend the interval of periodicity or to make the phase-lag of the newly constructed method smaller. In this paper, we will propose new P-stable methods that satisfy the following requirements : (i) they can be regarded as a modification of Numerov's method, (ii) their order is at least four, (iii) they are mono-implicit RKN (MIRKN) methods, i.e. the methods are implicit in only one variable, and (iv) the methods only involve function evaluations at points $x = x_0 + n\,h$, $n = 0, 1, \ldots$.

The last two requirements can be motivated a s follows. In [14], it was shown that there exist L-stable MIRK methods with four stages of order four for solving first-order initial value problems where the c-vector of the RK method is given by $c = (0, 1, 2, 3)^T$. In [17], the stability properties were studied of MIRKN methods that are included within the class of methods currently studied. Within the families considered, P-stable methods were found.

---

* Research assistant of the University of Gent

** Research Director at the National Fund for Scientific Research (N.F.W.O. Belgium)

2 Mono-implicit Runge-Kutta-Nystrom Methods

The general form of an s-stage RKN method is

$$y_{k+1} = y_k + h\, y'_k + h^2 \sum_{i=1}^{s} \bar{b}_i\, f(x_k + c_i\, h, Y_i) \ ,$$

$$y'_{k+1} = y'_k + h \sum_{i=1}^{s} b_i\, f(x_k + c_i\, h, Y_i) \ ,$$

whereby

$$Y_i = y_k + c_i\, h\, y'_k + h^2 \sum_{j=1}^{s} a_{ij}\, f(x_k + c_i\, h, Y_j) \ , \qquad i = 1, 2 \ldots, s \ . \qquad (2)$$

Clearly, a RKN method is a one-step method that provides approximate values for the unknown solution y and its first derivative y'. As is the case for RK methods, these methods can be presented in a more compact way as

$$\begin{array}{c|c} c & A \\ \hline & \bar{b}^T \\ & b^T \end{array}$$

In general, a RKN method involves the solution of a system of non-linear equations. The dimension of this system is s times the dimension of the problem to be solved.

A RKN method can however also be rewritten in a parametrized form. Indeed, we rewrite (2) as

$$Y_k = (1 - v_i)\, y_k + v_i\, y_{k+1} + (c_i - v_i - w_i)\, h\, y'_k + w_i\, h\, y'_{k+1} + h^2 \sum_{j=1}^{s} x_{ij}\, f(x_k + c_i\, h, Y_j)$$

and this leads to the tableau

$$\begin{array}{c|c|c|c} c & v & w & X \\ \hline & & & \bar{b}^T \\ & & & b^T \end{array}$$

where v en w are vectors with components v_i en w_i and where X is a matrix with elements x_{ij}. The relation between the two forms is $A = X + v.\bar{b}^T + w.b^T$

If X is strictly lower triangular then one obtains a system of equations that is implicit in the variables y_{n+1} and y'_{n+1}. If in addition also $v \equiv 0$ or $w \equiv 0$, then the resultant methods are called mono-implicit RKN (MIRKN) methods. In each time-step, the dimension of the system of non-linear equations to be solved is then reduced to the dimension of the problem to be solved, as is the case for the Numerov method itself.

3 Mono-implicit Extentions of the Numerov Method

Starting from the LMMs

$$y_{k+1} = y_k + h\,y'_k + h^2\left(\frac{7}{24}\,f_k + \frac{1}{4}\,f_{k+1} - \frac{1}{24}\,f_{k+2}\right)\,, \tag{3}$$

$$y'_{k+1} = y'_k + h\left(\frac{3}{8}\,f_k + \frac{19}{24}\,f_{k+1} - \frac{5}{24}\,f_{k+2} + \frac{1}{24}\,f_{k+3}\right)\,, \tag{4}$$

one obtains, after eliminating the y', the Numerov method (1), so (3) and (4) are suited to build modified Numerov schemes upon. Our idea is to use these equations as parts of a RKN method, i.e. we propose RKN methods M_{23} en M_{32} which have the following structure :

$$M_{23} \qquad\qquad\qquad\qquad M_{32}$$

$$\begin{array}{c|c|c|cccc}
0 & 0 & 0 & 0 & 0 & 0 & 0\\
1 & 1 & 0 & 0 & 0 & 0 & 0\\
2 & v_3 & w_3 & x_{31} & x_{32} & 0 & 0\\
3 & v_4 & w_4 & x_{41} & x_{42} & x_{43} & 0\\
\hline
 & & & \frac{7}{24} & \frac{1}{4} & -\frac{1}{24} & 0\\
 & & & \frac{3}{8} & \frac{19}{24} & -\frac{5}{24} & \frac{1}{24}
\end{array}
\qquad \text{or} \qquad
\begin{array}{c|c|c|cccc}
0 & 0 & 0 & 0 & 0 & 0 & 0\\
1 & 1 & 0 & 0 & 0 & 0 & 0\\
3 & v_3 & w_3 & x_{31} & x_{32} & 0 & 0\\
2 & v_4 & w_4 & x_{41} & x_{42} & x_{43} & 0\\
\hline
 & & & \frac{7}{24} & \frac{1}{4} & 0 & -\frac{1}{24}\\
 & & & \frac{3}{8} & \frac{19}{24} & \frac{1}{24} & -\frac{5}{24}
\end{array}$$

Since the method is also supposed to be mono-implicit, we also have to require that $w_3 = 0$ and $w_4 = 0$. In both cases, the remaining parameters are chosen in such a way that the resulting RKN method is of order four, as is the case for the Numerov method. Solving the order conditions [5, 6, 8] 4 degrees of freedom remain : v_3, v_4, x_{42} and x_{43}.

	M_{23}	M_{32}
x_{31}	$\dfrac{47 - 10\,v_3 - v_4 - 6\,x_{42} - 12\,x_{43}}{30}$	$\dfrac{40 - 2\,v_3 - 5\,v_4 - 30\,x_{42} - 90\,x_{43}}{6}$
x_{32}	$\dfrac{13 - 5\,v_3 + v_4 + 6\,x_{42} + 12\,x_{43}}{30}$	$\dfrac{-13 - v_3 + 5\,v_4 + 30\,x_{42} + 90\,x_{43}}{6}$
x_{41}	$\dfrac{9 - v_4 - 2\,x_{42} - 2\,x_{43}}{2}$	$\dfrac{4 - v_4 - 2\,x_{42} - 2\,x_{43}}{2}$

Since the order cannot be raised from 4 to 5, we will use these parameters to optimize the stability properties of the methods considered.

4 Linear Stability Analysis

To investigate the stability properties, the scalar test equation $y'' + \lambda^2\,y = 0$ is introduced [10, 12, 16]. The application of a RKN method to this equation leads to

$$\begin{pmatrix} y_{k+1} \\ y'_{k+1} \end{pmatrix} = M \begin{pmatrix} y_k \\ y'_k \end{pmatrix} , \tag{5}$$

where

$$M = \begin{pmatrix} 1 - H^2\,\bar{b}^T.(I + H^2\,A)^{-1}.e & h\,[1 - H^2\,\bar{b}^T.(I + H^2\,A)^{-1}.c] \\ -\lambda H\,b^T.(I + H^2\,A)^{-1}.e & 1 - H^2\,b^T.(I + H^2\,A)^{-1}.c \end{pmatrix} ,$$

with $H = \lambda\,h$. Eliminating y'_k and y'_{k+1}, one obtains the following relation

$$y_{k+2} - \mathrm{tr}(M)\,y_{k+1} + \det(M)\,y_k = 0 . \tag{6}$$

Let r_1 and r_2 be the roots of its characteristic equation.

Definition 1. A numerical method has an interval of periodicity $(0,\,H_p^2)$ if, for all $H^2 \in (0,\,H_p^2)$, r_1 and r_2 satisfy $r_1 = e^{i\,\theta(H)}$ and $r_2 = e^{-i\,\theta(H)}$, where $\theta(H)$ is a real function of H. If the interval of periodicity is $(0,\,\infty)$, the method is said to be P-stable.

Suppose the method has a non-vanishing interval of periodicity. In that case, for a range of values of H, the characteristic equation associated with (6) becomes

$$r^2 - \mathrm{tr}(M)\,r + 1 = 0 . \tag{7}$$

Following [16], we give the next definition :

Definition 2. For any method corresponding to the characteristic equation (7), the phase-lag is defined as the leading term in the expansion of $\phi(H) = H - \theta(H)$. The order of the phase-lag is said to be q if $\phi(H) = \mathcal{O}(H^{q+1})$ as $H \to 0$.

In order to have complex conjugate roots r_1 and r_2 with modulus 1, the necessary and sufficient conditions are given by

$$\det(M) = 1 \qquad \text{and} \qquad D(H^2) = \mathrm{tr}(M)^2 - 4 < 0 .$$

In both cases, there are two possibilities to satisfy $\det(M) = 1$.

M_{23}

Case A : $x_{42} = 3\,(1 - 2\,x_{43})$ and $v_4 = 3\,(1 + 8\,x_{43})$.

Case B : $21 - v_4 - 6\,x_{42} - 12\,x_{43} \neq 0$ and v_3 given by

$$\frac{129 + 20\,v_4 - {v_4}^2 - 42\,x_{42} - 6\,v_4\,x_{42} - 732\,x_{43} + 12\,v_4\,x_{43} + 144\,x_{42}\,x_{43} + 288\,{x_{43}}^2}{5\;(21 - v_4 - 6\,x_{42} - 12\,x_{43})}$$

M_{32}

Case A : $v_3 = 3 + 360\,x_{43}$ and $v_4 = 2 + 192\,x_{43}$.

Case B : $7 + v_3 - 5\,v_4 + 600\,x_{43} \neq 0$ and x_{42} given by

$$\frac{44 + 8\,v_3 - 41\,v_4 - v_3\,v_4 + 5\,{v_4}^2 + 3522\,x_{43} - 18\,v_3\,x_{43} - 510\,v_4\,x_{43} - 10800\,{x_{43}}^2}{6\,(7 + v_3 - 5\,v_4 + 600\,x_{43})} \tag{8}$$

The remaining parameters can be used to determine a region of P-stable methods in the plane or the three-dimensional space. This requires the examination of the roots of $D(H^2)$. For $H^2 > 0$, $D(H^2)$ can each time be written als the product of three functions in H^2 : a nonnegative function, a quadratic polynomial D_2 and a cubic polynomial D_3, whereby $D_2(0)\,D_3(0) < 0$.

M_{23}

Case A : Since the highest degree coefficient of $D_2(H^2)\,D_3(H^2)$ is $15\,(v_3 - 2)^2$, we can only consider methods for which $v_3 = 2$. We are thus left with a one-parameter family of methods, which all turn out to be P-stable, and for which

$$\phi(H) = \frac{1}{90}\,H^5 - \frac{143}{120960}\,H^7 + \mathcal{O}(H^9)\ .$$

Case B : Since the highest degree coefficient of $D_2(H^2)\,D_3(H^2)$ is $(-12 + v_4 + 6\,x_{42} + 12\,x_{43})^2\,(3 - v_4 + 24\,x_{43})^2$, the only methods to be considered are those for which this term identically vanishes. If $-12 + v_4 + 6\,x_{42} + 12\,x_{43} = 0$, then it turns out that $D(H^2)$ contains a factor $H^2 - 12$, i.e. for these methods $H_p^2 \leq 12$. If on the other hand $x_{43} = (v_4 - 3)/24$, then one finds P-stable methods for (see Fig. 1)

$$v_4 > \frac{59 - 24\,x_{42}}{6}\ . \tag{9}$$

The order of the phase-lag for each member of this two-parameter family is four since

$$\phi(H) = \frac{-13 + 3\,v_4 + 12\,x_{42}}{2880}\,H^5 + \frac{29 - 21\,v_4 - 84\,x_{42}}{241920}\,H^7 + \mathcal{O}(H^9)\ .$$

Based on these observations, the best stability properties are obtained near $v_4 = (59 - 24\,x_{42})/6$. For this one-parameter family, one finds

$$\phi(H) = \frac{11}{1920}\,H^5 - \frac{71}{96768}\,H^7 + \mathcal{O}(H^9)\ .$$

To conclude our investigation of M_{23}, we want to remark that within the family of P-stable methods, there are also methods for which the order of the internal stages is raised to 3. The stage-order of a RKN method is defined to be $r + 2$ if $C(r)$ is fulfilled with

$$C(r)\ :\qquad A.c^q = \frac{c^{q+2}}{(q+2)\,(q+1)}\ ,\qquad q = 0,\,1,\,2,\ldots,\,r\ . \tag{10}$$

In that case $Y_i = y(x_k + c_i\,h) + \mathcal{O}(h^{r+3})$, $i = 1, 2, \ldots, s$. Indeed, for M_{23} condition $C(0)$ is automatically fulfilled, while $C(1)$ implies $v_4 = 3\,(9 - 2\,x_{42} -$

$4\,x_{43}$). Since also $x_{43} = (v_4 - 3)/24$ is required for P-stability, this condition becomes $v_4 = 19 - 4\,x_{42}$. As can be seen in Fig. 1, this one-parameter family lies in the P-stable region.

M_{32}

Case A : Expressing $D(H^2)$ in terms of x_{42} and x_{43}, it is easy to verify that there are P-stable methods for (see Fig. 2)

$$\begin{cases} x_{42} > \dfrac{1}{24}(13 - 912\,x_{43} + 1296\,x_{43}^2) \ , & -\dfrac{1}{36} < x_{43} \leq 0 \ , \\ x_{42} \geq \dfrac{1}{2}(1 - 82\,x_{43}) \ , & x_{43} \leq -\dfrac{1}{36} . \end{cases} \tag{11}$$

Again, it is impossible to raise the order of the phase-lag of the P-stable methods, since

$$\phi(H) = \frac{-4 + 15\,x_{42} + 375\,x_{43}}{720}\,H^5 + \frac{5 - 42\,x_{42} - 966\,x_{43} + 1512\,x_{42}\,x_{43} + 27720\,{x_{43}}^2}{24192}\,H^7 + \mathcal{O}(H^9) \ .$$

At the border of the region of P-stable methods, the phase-lag is minimal for $x_{43} = 0$ and $x_{42} = 13/24$ (which is not a P-stable method, in fact $H_p^2 = 24$) :

$$\phi(H) = \frac{11}{1920}\,H^5 - \frac{71}{96768}\,H^7 + \mathcal{O}(H^9) \ .$$

Finally, we look again at the methods for which the stage-order is 3. Condition $C(0)$ is again fulfilled identically, while $C(1)$ is given by

$$x_{42} = \frac{1}{6}\,(8 - v_4 - 18\,x_{43}) = 1 - 35\,x_{43} \ .$$

From (11), this means that for $-1/12 \leq x_{43} \leq 0$ there exist P-stable methods with stage-order 3.

Case B : We are now confronted with the examination of a quadratic and a cubic polynomial in H^2, for which the coefficients depend on three parameters. Since this is a (too) complex problem to handle, we simplify things by looking only at methods for which the stage order is 3, i.e. for which

$$x_{42} = \frac{8 - v_4 - 18\,x_{43}}{6} \ .$$

With (8), this means that $x_{42} = \dfrac{262 - 35\,v_4}{192}$ and $x_{43} = \dfrac{v_4 - 2}{192}$.

We are now left with a two-parameter family of methods of stage order 3 and a detailed analysis shows that these methods are P-stable whenever

$$\begin{cases} \dfrac{72\,(v_4 - 1)}{26 - v_4} \leq v_3 \leq \dfrac{6\,(2 + v_4)}{10 - v_4} \ , & v_4 \leq 2 \ , \\ \dfrac{72\,(v_4 - 1)}{26 - v_4} \leq v_3 \ , & 10 < v_4 < 26 \ . \end{cases} \tag{12}$$

The corresponding area is shown in Fig. 3. The phase-lag is given by

$$\phi(H) = \frac{834 + 40\, v_3 - 125\, v_4}{46080}\, H^5 - \frac{3992 + 140\, v_3 - 1162\, v_4 - 70\, v_3\, v_4 + 175\, v_4^2}{1548288}\, H^7 + \mathcal{O}(H^9) \ .$$

Phase-lage order 6 is again impossible since the line $40\, v_3 - 125\, v_4 = -834$ has no intersection with the P-stable region. The phase-lag is minimal if $v_3 = 3$ and $v_4 = 2$. In that case one obtains

$$\phi(H) = \frac{11}{720}\, H^5 - \frac{37}{24192}\, H^7 + \mathcal{O}(H^9) \ .$$

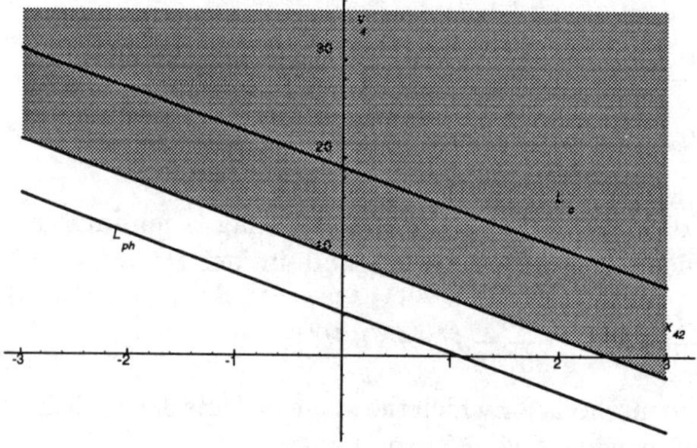

Fig. 1. M_{23}, Case B : The shaded area corresponds to P-stable methods. L_c is the line corresponding to methods for which $C(1)$ holds, L_{ph} corresponds to methods for which the phase-lag is of order 6.

5 Overview

An overview of the P-stable methods considered is presented in Table 1.

6 Illustration

We will try to solve the IVP [4]

$$y" = \begin{pmatrix} \mu - 2 & 2\,\mu - 2 \\ 1 - \mu & 1 - 2\,\mu \end{pmatrix} y, \qquad x \in [0, 10], \tag{13}$$

$$y(0) = \begin{pmatrix} 2 \\ -1 \end{pmatrix}, \qquad y'(0) = \begin{pmatrix} 0 \\ 0 \end{pmatrix}.$$

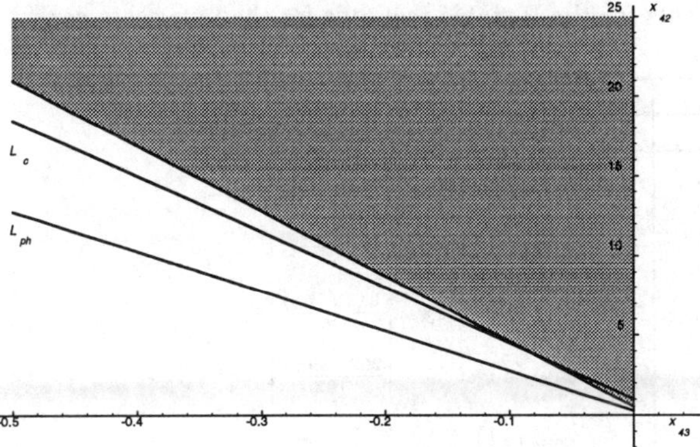

Fig. 2. M_{32}, Case A : The shaded area corresponds to P-stable methods. L_c is the line corresponding to methods for which $C(1)$ holds, L_{ph} corresponds to methods for which the phase-lag is of order 6.

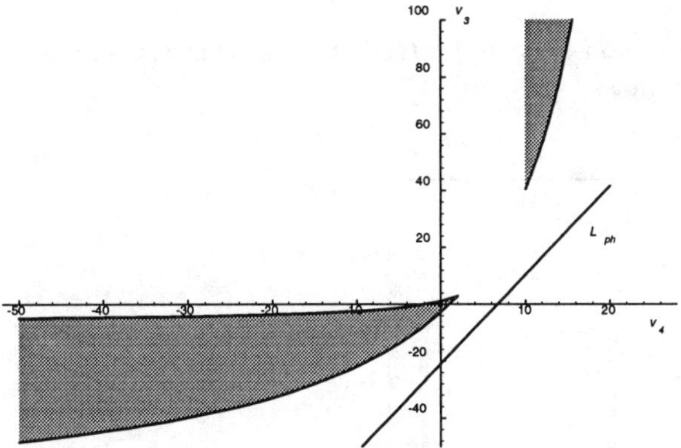

Fig. 3. M_{32}, Case B : The shaded area corresponds to P-stable methods for which $C(1)$ holds. L_{ph} corresponds to methods for which the phase-lag is of order 6.

which has the exact solution $y(x) = (2\cos x, -\cos x)^T$ for all real $\mu > 0$. The eigenvalues of $\partial f/\partial y$ are -1 and $-\mu$, so for large values of μ, methods with a large interval of periodicity are required.

We consider the following methods :

1. the Numerov method, for which y_1 instead of y'_0 is given as input.
2. M_{23}, Case A with $x_{43} = 0$.
3. M_{32}, Case A with $x_{42} = 1$ and $x_{43} = 0$.

In Fig. 4, which presents the maximum norm errors in logarithmic scale for the three methods considered, we use the value $\mu = 2500$ and fixed stepsize $h = \pi/60$. The Numerov method, for which $H_p^2 = 6$, fails to converge. On the other hand, the two other methods, that are both P-stable, behave quite

Table 1. Values of the coefficients for the P-stable methods for the four cases considered.

	M_{23}		M_{32}	
	Case A	Case B	Case A	Case B
x_{31}	$\frac{1}{5}$	$\frac{19-v_4-4\,x_{42}}{20}$	$4-5\,x_{42}-295x_{43}$	$-\frac{v_3}{3}$
x_{32}	$\frac{4}{5}$	$\frac{1+v_4+4\,x_{42}}{20}$	$-1+5\,x_{42}+115\,x_{43}$	$\frac{27-v_3}{6}$
x_{41}	$-7\,x_{43}$	$\frac{37}{8}-\frac{13\,v_4}{24}-x_{42}$	$1-x_{42}-97\,x_{43}$	$\frac{31\,(2-v_4)}{96}$
x_{42}	$3\,(1-2\,x_{43})$	see (9)	see (11)	$\frac{262-35\,v_4}{192}$
x_{43}		$\frac{-3+v_4}{24}$	see (11)	$\frac{-2+v_4}{192}$
v_3	2	2	$3\,(1+120\,x_{43})$	see (12)
v_4	$3\,(1+8\,x_{43})$	see (9)	$2\,(1+96\,x_{43})$	see (12)

well. Initially, the Numerov method behaves better than the other two methods considered due to its smaller phase-lag.

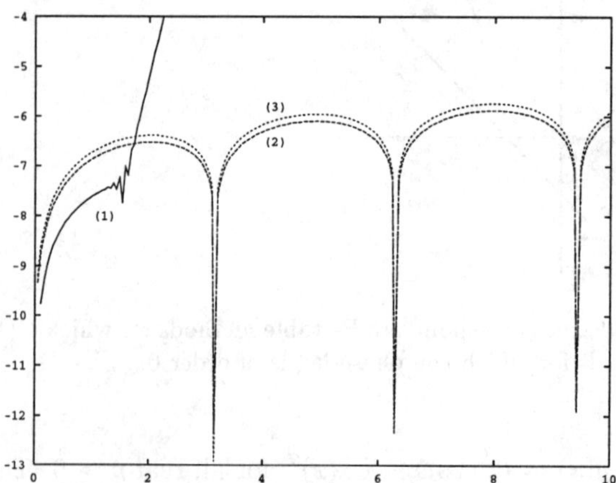

Fig. 4. The maximum norm errors in logarithmic scale for the three methods considered.

References

1. Cash, J.R., High order P-stable formulae for the numerical integration of periodic initial value problems, Num. Math. **37** (1981) 355–370
2. Chawla, M.M., P.S. Rao, P.S., A Numerov-type method with minimal phase-lag for the integration of second order periodic initial-value problems, J. Comp. Appl. Math. **11** (1984) 277–281
3. Coleman, J.P., Numerical methods for $y'' = f(x, y)$ via rational approximations for the cosine, IMA J. Numer. Anal. **9** (1989) 145–165
4. Franco, J.M., An explicit hybrid method of Numerov type for second-order periodic initial-value problems, J. Comp. Appl. Math. **59** (1995) 79–90
5. Hairer, E., Wanner, G., A theory for Nyström methods, Num. Math. **25** (1976) 383–400
6. Hairer, E., Méthodes de Nyström pour l'équation différentielle $y' = f(x, y)$, Num. Math. **27** (1977) 283–300
7. Hairer, E., Unconditionally stable methods for second order differential equations, Num. Math. **32** (1979) 373–379
8. Hairer, E., Nørsett, S.P., Wanner, G., Solving Ordinary Differential Equations I, Nonstiff Problems, (Springer-Verlag, Heidelberg, 1987)
9. Ixaru, L.Gr., Rizea, M., A Numerov-like scheme for the numerical solution of the Schrödinger equation in the deep continuum spectrum of energies, Comput. Phys. Commun. **19** (1980) 23–27
10. Lambert, J.D., Watson, I.A., Symmetric multistep methods for periodic initial value problems, J. Inst. Math. Applics. **18** (1976) 189–202
11. Simos, T.E., Raptis, A.D., Numerov-type methods with minimal phase-lag for the numerical integration of the one-dimensional Schrödinger equation, Computing **45** (1990) 175–181
12. Thomas, R.M., Phase properties of high order almost P-stable formulae, BIT **24** (1984) 225-238
13. Van Daele, M., De Meyer, H., Vanden Berghe, G., A modified Numerov integration method for general second order initial value problems, Intern. J. Computer Math. **40** (1991) 117-127
14. Van Daele, M., Vanden Berghe, G., De Meyer, H., A general theory of stabilized extended one-step methods for ODEs, Intern. J. Computer Math. **60** (1996) 253-263
15. Vanden Berghe, G., De Meyer H., Vanthournout, J., A modified Numerov integration method for second order periodic initial-value problems, J. Computer. Math. **32** (1990) 233-242
16. van der Houwen, P.J., Sommeijer, B.P., Explicit Runge-Kutta(-Nyström) methods with reduced plase errors for computing oscillating solutions, SIAM J. Numer. Anal. **24** (1987) 595–617
17. Van Hecke, T., Van Daele, M., Vanden Berghe, G., De Meyer, H., A mono-implicit Runge-Kutta-Nyström modification, of the Numerov method (to appear)

The V-Ray Technology of Optimizing Programs to Parallel Computers

Valentin V.Voevodin[1], Vladimir V.Voevodin[2]

[1] Institute of Numerical Mathematics, RAS, Russia
[2] Moscow State University, Russia

Abstract. This paper gives a brief overview of the V-Ray technology, based on the rigorous mathematical theory of analysis and transformation of programs, and intended for optimization of programs to parallel computers. This technology provides a basis for resolving the whole scope of problems related to mapping of applications onto parallel computers, starting from the commonly adopted control-flow and data-flow analysis up to the optimization of data distribution and data locality. High efficiency of the V-Ray technology is illustrated by successful optimization of the TRFD Perfect Benchmark to vector/parallel CRAY Y–MP M90/C90 as well as to massively parallel CRAY T3D supercomputers.

1 Introduction

The V-Ray technology is a set of software tools and mathematical methods designed for exploration of parallel properties of algorithms and programs. The technology was developed on the basis of the strict theory and it is mostly oriented to creating standalone software systems to match structure of programs to requirements of a parallel computer architecture. An urgent need for such tools is evident and it has been caused by serious difficulties of the matching process as well as low quality of the current generation of compilers in optimizing programs to parallel machines. On the principles of the V-Ray technology we have implemented the first version of the system that we call 'V-Ray system'.

Many recent results in the theory of informational structure of programs have been greatly improved before to be included into V-Ray. Depending on goals, this technology makes it possible either to detect and describe parallel properties of programs or to transform source codes in accordance with peculiarities of a particular parallel computer architecture. It should be noted that the V-Ray technology provides a mathematical guarantee that all suggested transformations are strictly equivalent (even taking into account round-off errors).

Having serious mathematical background the technology provides the real basis for solving the problem of code portability. It means that programs written in accordance with V-Ray's requirements can be relatively easy ported to a variety of different parallel computer platforms while keeping reasonable performance. In many cases the process of porting can be done almost automatically, since to use the V-Ray technology one need not at all any substantial information about a program under analysis except its source code on a high level language, for example, a Fortran code.

Many key features of the technology are described in [1]. In the paper we give a brief overview of the V-Ray technology (unfortunately without comparison with existing approaches [7]-[10] due to the limited space of the paper) and consider its application for optimizing TRFD Perfect Benchmark to CRAY Y–MP M90/C90 and CRAY T3D supercomputers.

The Perfect Club Benchmark suite has been used for many years for evaluating of real performance of parallel computers when solving industrial applications. By now a lot of papers have been published [11],[12] which describe its general organization, structure of each code, key features of some important subroutines, performances for a wide range of computers and etc. However, just that high popularity of the suite seemed attractive to us: it is much more interesting to produce a nontrivial result in the field where many researchers have worked very hard and, moreover, this is the only way to show to industrial users potentials of the V–Ray technology as compared with existing approaches to analysis and restructuring of codes.

To this end we have chosen the TRFD Perfect Benchmark. The original version of the code delivers the near 56 Mflop/s (baseline) performance on the CRAY Y–MP M90 computer independently of the actual number of processors available (we remind that the peak performance of the CRAY Y–MP M90 computer equals 333 Mflop/s per processor). The best hand optimization of the code enables one to speedup execution up to 82 Mflop/s for one Y–MP processor [3]. It is clear that the original TRFD code is not only poorly vectorized, but even hand transformations do not lead to any considerable improvement of the performance. Obviously two questions arise immediately:

- What is the reason of such poor performance?
- Is it possible at all to improve performance of this benchmark using only equivalent transformations?

Running a little ahead we give the answer right now: the V–Ray technology enables us, first, to improve performance of this code for one processor of CRAY Y–MP computer up to 247 Mflop/s, second, to keep significant growth of performance for multiprocessor Y–MP configurations, and at last to find a way to a really scalable program with performance 1.7 Gflop/s on 256 processors of massively parallel CRAY T3D computer.

2 Three stages in optimization of programs

The central part of the V-Ray technology is comprehensive investigation of parallel structure of "linear" programs. Class of these programs is organized quite simple and at the same time it is representative. The most essential restrictions of this class are the next requirements: all index expressions of array references, loop bounds and conditions of branch statements must be linear functions of

[3] "baseline" as well as "hand" optimization performance data are provided by Jeff Brooks from CRAY Research, Inc., Eagan MN, USA.

surrounded loop parameters and external integer parameters (variables) of a program; loops of the program must be rigorously nested (without partial overlapping) and have unit stride; linear programs must contain neither CALLs to functions nor subroutines (it should be noted that if an program under analysis does not satisfy a particular requirement of this class than it could be analyzed after some special transformations).

Application of the V-Ray technology to investigation and transformation of a particular program consists of three stages.

The main purpose of the first stage — to reduce investigation of parallel properties of an original program to investigation of parallel properties of a set of linear programs. In the simplest case when the original program is linear itself this set contains this program only. As practice has shown the most important fragments of codes are often linear programs.

Linear programs might be modifications of parts of the original code or even dummy programs built upon selected fragments. Modified and dummy linear programs help to detect parallel properties of parts of codes but they are not usually involved in practical parallel computations. One of the reasons for modifications is, for example, the need for explicit representation of computable index expressions. Dummy programs might appear in more complicated cases, for example while analyzing parallel structure of code with CALL statements (hereinafter we assume that analyzed programs are written on Fortran language). For such codes each CALL statement is substituted for a simple linear program that uses and produces data identical to input and output data of a routine invoked by this CALL.

The second stage is comprehensive investigation of parallel properties within each of the linear programs directed towards detecting the total resource of potential parallelism and key features of data locality. At the end of this stage linear programs are converted by equivalent transformations to a new form where massive parallelism is explicitly expressed in terms of ParDo statements. To optimize data management on distributed memory parallel system new representation of original programs contains information about possible distributions of arrays among memory modules and tasks among processors.

The third stage is fine tuning of the parallel program to match better to peculiarities of a target computer architecture and a particular compiler. These transformations exploit extensively information gathered on the two previous stages to optimize, for example, vector processing and using of cache, to balance use of multiple functional units and improve many other items.

The advantage of the V-Ray technology is an attempt to formalize all these stages to minimize (or at least decrease) human's efforts during parallel optimization process. By now formalization of the second stage is almost completed. Other stages are formalized to a less extend because of absence of suitable formal descriptions of parallel computer architectures as well as weak formalization of general structure and style of codes.

3 Comprehensive investigation of parallel properties of programs

The second stage is significantly more difficult than other stages since it assumes resolving a whole set of very difficult mathematical problems. It consists of five basic steps described briefly in this section.

Step 2.1. First of all we need to detect all types of dependences for an analyzed linear program. Many existing approaches use dependences between points of an iteration space to analyze and describe structure of programs. Usually a subset of points are associated with each point of iteration space in accordance with exactness of methods in use for discovering dependences. However we need to know more exact information about dependences to determine the real structure of the program. In particular, for each point A of an iteration space we should find the only point B generating dependence. This point is either lexicographical maximum or minimum of just mentioned subset depending on the type of analyzed dependence. Hence we have to deal with oriented graphs where nodes are points of the iteration space and arcs connect two lexicographically nearest points referenced the same data. Graphs are similar to each other but distinguished by definition of starting and end points of arcs and by memory access mode (read/write). As a rule arcs of these graphs represent lexicographical ascending or descending within iteration space.

Data-flow dependence graph has a particular meaning for exploring parallel structure so we use similar object and call it an algorithm graph. Its algebraic as well as minimax properties have been investigated in detailes [2] and the book [3] shows clearly such an important role this notion plays for resolving many difficult problems, for example, for designing systolic arrays. For the first time an algorithm for building of this graph is appeared in [4] and later in [1]. The papers [5],[6] contains many useful aspects of algorithm graph use. In this paper we would like to concentrate on parallel optimization of programs based on algorithm graph analysis.

From mathematical point of view all types of dependence graph are identical therefore we apply one and the same algorithm for their construction. The final formal description of these graph may be treated as a finite set of triples (F_i, Δ_i, N). N is a vector of unknown input integer variables of a program, the last entry of the vector is 1. Δ_i is a linear polyhedron within iteration space whose size and location depend on the vector N. The polyhedron is defined by a system of linear inequalities. The function F_i is linear and has the form $F_i = J_i x + \Phi_i N$. The matrices J_i, Φ_i are constant. The domain of the function F_i is the polyhedron Δ_i. The polyhedrons Δ_i may be intersected and may not cover the entire iteration space. The values of the functions F_i are points of iteration space. The points x and $J_i x + \Phi_i N$ are the start and end points of an arc respectively.

We have developed an efficient and fast algorithm to construct the complete set of triples by source code, i.e. to compute coefficients of linear inequalities describing the polyhedrons Δ_i and coefficients of the matrices J_i, Φ_i. Unlike many other algorithms it does not apply time-consuming integer programming technique. The key feature is using an important characteristic property of lexicographical order that we call L-property [4].

Dependence graphs of some linear programs might be more complicated. In the general case domain of the functions F_i is an intersection of polyhedrons Δ_i with some regular lattices. Parameters of the lattices can be determined also by source code in accordance with general theory (the actual version of the V-Ray system does not support lattice analysis because in practice we have met them only twice and found a way to process them by much simpler methods). It should be noted that if due to some reasons a dependence graph can not be expressed as a set of triples, than we use triples for its minimal extension keeping the further analysis without changes.

The result of the first step is explicit symbolic representation of all dependences. Exactly this form of representation enables us to solve many problems necessary for efficient exploration of parallel structure of algorithms and programs.

Step 2.2. The next step is detection of ParDo loops. Taking into account explicit formulas for data dependences this problem can be solved easily. To examine ParDo property one may use a few criteria (sufficient as well as sufficient and necessary) testing some joint properties of matrices J_i and Φ_i, for instance, for a loop to be the ParDo loop it is sufficient that the first entry of the vector function F_i for all triples of an algorithm graph for this loop to be the form $F_{i1} = x_1$. The actual number of loops and nesting structure do not affect seriously complexity of the criteria.

One simple example. Innermost ParDo loops are better suited for vector processing while outermost ones for parallel execution on multiprocessor computers. Hence sometimes for optimal mapping of programs onto a particular computer we would like to use loop interchange. To test whether it possible to apply this technique or not we make use a statement requiring quite trivial analysis of triples.

The second step mostly deals with general properties of programs which on the one hand do not depend on computer architecture (the output of this stage could be marked original code), but on the other hand can be easily detected after simple analysis of explicit formulas representing data dependences. An interesting observation: for the large part of realworld applications this information suffices for efficient target optimization. Notice, that advanced analysis of dependences should be continued on the third stage if we want to adjust well program structure with peculiarities of a particular computer and compiler.

Step 2.3. ParDo loops describe coordinate parallelism within iteration space. However resource of coordinate parallelism could be not sufficient therefore the main purpose of this step is discovering "skew" parallelism based on the notion of schedule.

Let an oriented graph be given in iteration space. A strict (generalized) schedule of the graph is a real function defined on iteration space and strictly increasing (nondecreasing) along arcs of the graph. Schedules have numerous very important properties. Suppose, for example, the graph be an algorithm graph. Let us take any schedule and examine its level surfaces. These surfaces split iteration space into nonintersecting sets, where each set corresponds to one level. The schedule gives equal value for all points of one set. Since the schedule does not decrease along arcs of the algorithm graph one may execute operations set-by-set according to growth of schedule values. If the schedule is strict then points of any set can not be linked by arcs and hence these operations can be run simultaneously. Moreover, it means that search for computational parallelism in a program we reduce to investigation of schedules for its algorithm graph!

The third step is completely intended to determining and analysis of schedules represented similar to graph description as a finite sets of triples (f_i, ∇_i, N). N is a vector of input integer variables of a program, ∇_i is a linear polyhedron within iteration space whose size and location depend on the vector N. The linear functional f_i sets a schedule and has the form $f_i = < j_i, x > + < \varphi_i, N >$, where $< *, * >$ means a dot product and the vectors j_i, φ_i are constant. The domain of the functional f_i is polyhedron ∇_i. Unlike triples of dependence graphs polyhedrons ∇_i are nonintersecting and cover the entire iteration space.

There is certain freedom in determining of polyhedrons ∇_i. In particular, we may consider them identical to domains of loop parameters or take into account that arcs of dependence are actually piecewise and so on. Assume that we fix polyhedrons ∇_i somehow. A condition of nondecreasing of schedules along arcs is equivalent to inequality $At \leq 0$, where t is a vector of direct sum of all vectors j_i and φ_i. The matrix A is computed unambiguously by triples of the graph, polyhedrons ∇_i and domains of loop parameters. Hence the schedule search problem we reduce to searching solutions of the system of linear inequalities $At \leq 0$. We call this system as a system of informational closure.

At the end of the third step we should find quite representative set of solutions of the system $At \leq 0$ in view of one more property of schedules. If do not go into details it can be stated as: if we have found S independent schedules than by means of these schedules we can transform program to a new form where each nest of loops with depth no less than s will have at least $s-1$ ParDo loops. On this step we detect possibility of this transformation but it will be done later.

Step 2.4. Proper data distribution is a crucial point for efficient use of detected parallelism on distributed memory parallel computers. A criterion of the optimal data distribution is minimum of switchings of data transfer channels between PEs and memory modules. The total number of switchings depends on distribution of tasks between processors so we have to find both proper data distribution and schedule for processors simultaneously.

Let us pick out some subset of occurrences of array references in a code. Assume that iteration space as well as arrays are splitted into nonintersecting sets so that each selected occurrence addresses data from the only 'data' set across all points of one 'space' set. Let us refer to both partitions as eigensets for the given subset of occurrences.

Suppose that different memory modules keep different eigensets of data and a particular processor executes one by one operations from a eigenset in iteration space. Then by definition of eigensets this processor will not require additional data from other modules up to the end of processing. The more eigensets of data, the more memory modules can be efficiently used inside parallel computer. The more eigensets within iteration space, the more considerable benefit can be gained from absence of switchings.

If initial subset contains all occurrences of a realworld program it might happen that there is only one eigenset of data coinciding with the entire data of the program. To find suitable data distributions and schedules for processors we have to make trade-off decision and try to determine eigensets for a part of occurrences. On the one hand the subset should cover as much memory references as possible, but on the other hand the total number of eigensets should be sufficient for efficient implementation of a program.

From mathematical point of view the problem of searching for eigensets is quite similar to spectral problem: given matrices B_k^{ij}, it is necessary to find nonzero vectors $\gamma^i = (\gamma_1^i, \ldots, \gamma_{n_i}^i)$ and $l^j = (l_1^j, \ldots, l_{m_j}^j)$ so that satisfy the equations $(\gamma_1^i B_1^{ij} + \ldots + \gamma_{n_i}^i B_{n_i}^{ij}) l^j = 0$. In these formulas n_i is a dimension of the i-th array, m_j is a dimension of the j-th nest of loops, the vector γ^i is a directing vector for eigensets in i-th array, the vector l^j is a directing vector for eigensets in j-th nest, the matrices B_k^{ij} are constructed from coefficients of index expressions of selected occurrences.

Since switching overheads depend on peculiarities computer architecture then requirements to solutions of this problem should be different. Let, for example, we use a computer with programmable structure of data transfer channels. For this type of computer switching time does not depend on the actual size of transferring data. Therefore it might be sufficient to find any nonzero solution for the given subset of occurrences.

Now assume that each memory module is attached to its own processor so switchings are equivalent to message passing. This is the typical case for the major part of massively parallel computers. Let us assume that a processor is going from one eigenset within iteration space to another. Then just in this moment at least one occurrence is going from one eigenset of data to another and hence we have to send (receive) some data from one processor to (from) another. Rate of data transfer is usually much less than rate of data processing so for interprocessor data transfers do not affect asymptotically the principal part of execution time, for example on massively parallel computer, it is necessary to ensure absence of switchings for all occurrences constituting the principal part of the total amount of memory references. If we intend to realize parallelism of outermost ParDo loops then the last condition is equivalent to equalities $\gamma_1^i \tilde{B}_1^{ij} + \ldots + \gamma_{n_i}^i \tilde{B}_{n_i}^{ij} = 0$. The matrices $\tilde{B}_k^{ij}$ are derived from the matrices B_k^{ij} by deleting first columns corresponding to the outermost ParDo loops. We see that requirements to solutions of stated above "spectral" problem are very rigid.

After the fourth step we would like to obtain quite representative set of solutions of the system $(\gamma_1^i B_1^{ij} + \ldots + \gamma_{n_i}^i B_{n_j}^{ij}) l^j = 0$.

Step 2.5. The meaning of this step is transformation of an original program to present results of previous analysis in explicit and easy-of-use form. For each type of computers, for instance vector, parallel, massively parallel, with shared or distributed memory and so on, transformations are different. All transformations make use solutions of problems from previous steps 2.2 — 2.4. All detected coordinate as well as skew parallelism is converted into ParDo loops. Indicated possible data distributions and schedules for processors are adjusted with computational parallelism.

4 Analysis and optimization of the TRFD code

Let us examine first the original structure of code. It is not difficult to determine that the major part of arithmetic costs fall on the subroutine OLDA. The first brief analysis reveals three interesting facts:

- independence of iterations inside all innermost loops of OLDA is easily detectable by very simple conventional methods;
- parallel structure of the outermost loop of the original routine can not be easily detected by conventional formal methods (that was the reason why the baseline performance does not improved with the number of available processors) but become evident when looking at the code and thus it is possible to improve performance with multiple CPUs using the manual optimization;
- combination of the complicated loop structure, the large depth of loop nesting and the compact one dimensional form of the main multidimensional array give rise to principal difficulties when performing any nontrivial equivalent transformation of the code.

With no doubt the third fact spoils a general impression about the code. Let us first analyze the performance of the straightforward implementation of TRFD on vector/parallel computer CRAY Y–MP which takes into account structure of the innermost and outermost loops only. On the one hand, the source code has the commonly adopted "optimal" form of a code with vectorizable innermost loops and parallelizable outermost ones. In other words, almost all operations of the code are executed in the vector mode and almost full time all processors are busy. But from the another hand, even in the case of manual optimization the performance does not exceed 25% of its peak value! It is clear that we need to change something, however with no deeper analysis and understanding of the code structure any substantial transformation may become very difficult or even useless. Indeed, we have only detected structure of the outermost and innermost loops of the original code. Relying on such a poor information we can not resolve any nontrivial problem:

- it is impossible to estimate the potential performance of the code;
- many of the code properties are hidden, therefore it is not clear what kind of transformations should be performed and what bottlenecks they should resolve;
- the current form of the code puts very severe constraints on the scope of possible equivalent transformations;

Fig .1. The potential coordinate parallelism of the OLDA subroutine

– it is possible that we shall not be able to exceed 82 Mflop/s but before giving up we must provide evidence that the performance can not be improved in principle.

All above problems were successfully resolved using the V–Ray mathematical technology. The first stage is reduction of the code to an analyzable form. We mentioned already that the serious difficulty when analyzing the TRFD code is related to use of the compact one dimensional form of the main multidimensional array. To overcome this difficulty we used the technique for reconstructing multidimensional arrays according to their compact one dimensional form. Although sometimes this problem may possess an arbitrary number of solutions we can take any of them which converts index expressions to a linear form. The new form of the arrays may not coincide with their original form before the memory optimization and this is not necessary since any acceptable form can clarify structure of the code. After array reconstruction we get modified version of the OLDA subroutine from linear class.

After determining of necessary dependence graphs (step 2.1) and detection of ParDo loops we realize the total resource of coordinate parallelism of OLDA. Figure 1 shows this resource in the form of loop profile, where each horizontal bracket represents one loop of OLDA and ParDo loops are additionally marked by a dot. The situation is principally changed because now we possess almost complete information about parallel structure of the code. It is clear what we can rely on and how large our reserves are. In particular, the ParDo loops of the second level can be used to increase length of vector operations and at the same time sequential loops might be useful for improving data locality.

Step 2.3 reveals additional reserve of skew parallelism however further transformations will not use it. Shared memory computers like CRAY Y–MP usually do not required step 2.4 so we proceed immediately to a target transformation (step 2.5). Taking into account that all loops of OLDA are quite short and have no more than 40 iterations, one should move two outermost loops inside, collapse and use for vectorization. In the same way the original innermost loop should be moved outside and used for parallel processing. To understand whether it possible to perform these transformations one need to test the triples of dependence graphs. The third stage is fine tuning of the code, where, for example, one can improve data locality by means of loop unrolling to use better vector registers.

The final results of optimization for CRAY Y–MP M90/C90 computers are shown in table 1 which adopts the following notation:

– 'Baseline perf.': corresponds to the original (nontransformed) code;
– 'Manual opt.': manual optimization was performed;

– 'V–Ray opt.': performance after optimization of the TRFD code using the V–Ray technology.

Table 1. The final results for vector/parallel computers CRAY Y–MP M90 and C90

Y–MP M90 CPUs	Baseline Mflop/s	Manual opt. Mflop/s	V-Ray opt. Mflop/s
1	56.19	81.72	247
4	54.86	261.64	822
8	54.34	481.03	954
Y–MP C90 CPUs	**Baseline Mflop/s**	**Manual opt. Mflop/s**	**V-Ray opt. Mflop/s**
1	89.5	139.71	579.7
8	89.6	962.68	2440.5

We would like to especially note that all TRFD optimizations were performed using equivalent Fortran transformations with no CRAY Assemble Language (CAL) fragments or straightforward calls of CAL coded library routines.

For distributed memory massively parallel CRAY T3D computer step 2.4 is of great importance. After analysis of the set of possible data distributions among processors by means of the V-Ray technology we could prove that:

1. Within each individual nest of loops, a data distribution strategy exists that supports two-dimensional parallelism. Two-dimensional parallelism implies independence of all iterations of two loops inside a nest. By using a consistent data distribution strategy within the loop nest, we can avoid the need for data transfer between iterations.
2. A data distribution strategy that could be used across all the nests that would also be consistent with the use of two-dimensional parallelism within each nest does NOT exist.

Having this information we had to use different data distribution strategies within each of the nests and perform the redistribution of data between them.

Fine tuning (the third stage) consists of improving of data and computational locality to use more efficient cache memory. The final results of optimization for CRAY T3D are shown in table 2.

The parallel program for CRAY T3D uses PVM version 3.2.6 and a few low-level communication routines. As a comment, we would like to note that original program includes operation similar to matrix transposition (the second nest of OLDA) widely recognized as matched badly to distributed memory massively parallel computers. Nevertheless selected final form of a parallel code has been found very efficient and presence of transpositions do not affect dramatically the performance.

Table 2. The final results for massively parallel computer CRAY T3D

CRAY T3D PEs	V-Ray opt. seconds	V-Ray opt. Mflop/s
4	6.013	71.66
8	3.215	134.01
16	1.702	253.43
32	0.896	480.85
64	0.494	871.44
128	0.301	1435.96
256	0.253	1700.93

References

1. Voevodin, V.V.: Mathematical foundations of parallel computing. World Sci. Publ. Co. Computer Science Series **33** (1992)
2. Voevodin, V.V.: Parallel Structures of Algorithms and Programs. Moscow DNM RAS (1987) (in Russian)
3. Voevodin, V.V.: Mathematical Models and Methods for Parallel Processes. Moscow Nauka (1986) (in Russian)
4. Voevodin, V.V., Pakulev, V.V.: Determination of agorithm graph arcs. Moscow DNM RAS Tech. Report **228** (1989) (in Russian)
5. Voevodin, Vl.V.: Theory and Practice of Parallelism Detection in Sequential Programs. Programming and computer software **18** (1992) n.**3**
6. Voevodin, V.V., Voevodin, Vl.V.: Why Do We Use an Algorithm Graph for Analysis of Program Structure? Research Report EM-RR 5/1992 Elegant Mathematics, Inc. (USA) (1992)
7. Feautrier, P.: Dataflow analysis of array and scalar references. Int. J. of Parallel Programming **20** (1991)
8. Feautrier, P.: Some efficient solutions to the affine scheduling problem, part I/II. Tech. Rep. 92.28/78 IBP/MASI (1992)
9. Kelly, W., Pugh, W.: A framework for unifying reordering transformations. Tech. Rep. CS-TR-2995.1 Dept. of Comp.Science Univ. of Maryland College Park (1993)
10. Maydan, D., Amarasinghe, S., Lam, M.: Array data-flow analysis and its use in array privatization. In ACM'93 Conf. on Principles of Programming Languages (1993)
11. Cybenko, G., Kipp, L., Pointer, L., Kuck, D.: Supercomputer performance evaluation and the Perfect Benchmarks. Tech. Rep. 965 CSRD Univ. of Illinois at Urbana Technical Report (1990)
12. Grassl, C.M.: Parallel Performance of Applications on Supercomputers. Parallel Computing **17** (1991) 1257-1273

Applications of Steklov-Type Eigenvalue Problems to Convergence of Difference Schemes for Parabolic and Hyperbolic Equations with Dynamical Boundary Conditions

Lubin G. Vulkov

Center of Applied Mathematics and Informatics
University of Rousse, 7017 Rousse, Bulgaria.
e-mail: vulkov@ami.ru.acad.bg

Abstract. Parabolic and hyperbolic equations with dynamical boundary conditions, i.e which involve first and second order time derivatives respectively, are considered. Convergence and stability of weighted difference schemes for such problems are discussed. Norms arising from Steklov-type eigenvalues problems are used, while in previously investigations, norms corresponding to Neumann's or Robin's boundary conditions are used. More exact stability conditions are obtained for the difference schemes parameters.

1 Introduction

We consider two model problems:

$$\frac{\partial u}{\partial t} = \frac{\partial^2 u}{\partial x^2}, \quad 0 < x < l, \quad 0 < t < T, \tag{1}$$

$$u(0,t) = u_1(t), \quad 0 < t < T, \tag{2}$$

$$K\frac{\partial u\,(l,t)}{\partial t} + \frac{\partial u\,(l,t)}{\partial x} = du\,(l,t) + u_2(t), \quad 0 < t < T, \tag{3}$$

$$u(x,0) = \varphi(x), \quad 0 < x < l, \tag{4}$$

and

$$\frac{\partial^2 u}{\partial t^2} = \frac{\partial^2 u}{\partial x^2}, \quad 0 < x < l, \quad 0 < t < T, \tag{5}$$

$$M\frac{\partial^2 u\,(l,t)}{\partial t^2} + \frac{\partial u\,(l,t)}{\partial x} = 0, \quad 0 < t < T, \tag{6}$$

$$u\,(x,0) = \varphi\,(x)\,, \quad \frac{\partial u\,(x,0)}{\partial t} = \psi\,(x)\,, \quad 0 < x < l, \tag{7}$$

and (2). Further, we shall reffer these problems as (P) and (H), respectively. Here, K and M are positive constants. (P) and (H) are formulated in [1]; the first one describes heat conduction with concentrated capacity K at the boundary

$x = l$, and the second - vibration of a string with concentrated mass M at the end $x = l$.

Let in problem (P) $u_1 \equiv 0$, $u_2 \equiv 0$. For obtaining an energy estimate for the solution of problem (P), we multiply (1) by $u(x,t)$ and integrate the result in bounds $(0,l)$. Taking into account the boundary and initial conditions, we obtain:

$$\frac{dE}{dt} = -\int_0^l \left(\frac{\partial u}{\partial x}\right)^2 dx \left(1 - d\frac{u^2(l,t)}{\int_0^l \left(\frac{\partial u}{\partial x}\right)^2 dx}\right), \tag{8}$$

$$E(t) = \frac{1}{2}\int_0^l u^2 dx + \frac{K}{2}u^2(l,t).$$

Let define μ and λ by

$$\frac{1}{\mu} = \max_{v \in H} \frac{v^2(l,t)}{\int_0^l (v')^2 dx}, \; ' \equiv \frac{d}{dx}, \tag{9}$$

$$\frac{1}{\lambda} = \max_{v \in H} \frac{\int_0^l v^2 dx + Kv^2(l,t)}{\int_0^l (v')^2 dx}, \tag{10}$$

where H is the set of functions:

$$H = \left\{v | v \in C^2(0,l) : v(0) = 0\right\}. \tag{11}$$

The solution of the varational problem $(9), (11)$ is the solution $(v(x), \mu)$ of the following spectral problem:

$$v'' = 0, \quad v(0) = 0, \quad v'(l) - \mu v(l) = 0, \tag{12}$$

i.e

$$v(x) = Cx, \quad C = const, \quad \mu = \frac{1}{l}.$$

This is a typical Steklov's eigenvalue problem - the eigenvalue μ is only at the boundary.

The solution of the variational problem (10), (11) is the pair $(v_1(x), \lambda_1)$, where λ_1 is the smallest eigenvalue and $v_1(x)$ - the corresponding eigenfunction of the spectral problem.

$$v'' + \lambda v = 0, \quad v(0) = 0, \quad v'(l) = \lambda K v(l). \tag{13}$$

Therefore, for each function $v \in H$, the following Steklov-Poincare-Friedrichs-type inequallity holds:

$$\|v\|_K = \int_0^l v^2 dx + Kv^2(l) \le \lambda_1 \int_0^l (v')^2 dx, \quad \lambda_1 = \beta_1^2, \tag{14}$$

where β_1 is the smallest positive root of the trancedental equation $ctg\beta l = K\beta$.

Now, from (8), the estimate follows for the energy $E(t)$:

$$E(t) \le e^{-kt} E(0), \quad k = 2\lambda_1 (1 - ld) \tag{15}$$

$$E(0) = \frac{1}{2} \int_0^l \varphi^2 dx + \frac{K}{2} \varphi^2(l).$$

Let $0 \le d < \frac{1}{l}$. Since $\lambda_1 > 0$, then $E(t) \to 0$ as $t \to \infty$, which means that the zero solution of (1)– (4) is asymptotically stable.

As usual, the differential problems cannot be solved analytically. When one applies numerical methods, the basic question is the convergence of the approximate solution, which for linear problems is equivalent to the stability. Here we study for stability of weighted difference schemes for problems (P) and (H) using a norm which is discrete analog of $\|.\|_K$. In ([1], p.396) and ([2], p.177) norms, corresponding to Neumann's or Robin's boundary conditions are used. Our choise of norm enable us to reduce stabilty requirments for the weight.

2 One-dimensional Steklov-type eigenvalue problems

When studying parabolic equations with variables coefficients one comes to more general spectral problems. A Sturm-Liouiville equation is a second-order homogeneous linear differential equation of the form:

$$(p(x)v')' - q(x)v + \lambda rv = 0, \quad 0 < x < l. \tag{16}$$

Here λ is a parameter, while p, q and r are real-valued functions of x; the functions p and r are positive. The equation (16) and two separated endpoint conditions

$$\alpha_0 v'(0) + \beta_0 v(0) = 0, \quad \alpha_0 + \beta_0 > 0,\ \alpha_0,\ \beta_0 \ge 0, \tag{17}$$

$$\alpha_1 v'(l) + \beta_1 v(l) = 0, \quad \alpha_1 + \beta_1 > 0, \alpha_1,\ \beta_1 \ge 0, \tag{18}$$

forms a Sturm-Liouville system. The following boundary conditions are nonstandard:

$$\alpha_0 v'(0) + \beta_0 v(0) = \lambda\gamma_0 v(0)\,, \; \alpha_0 + \beta_0 > 0, \tag{19}$$

$$\alpha_1 v'(l) - \beta_1 v_1(l) = \lambda\gamma_1 v(l)\,, \; \alpha_1 + \beta_1 > 0, \tag{20}$$

$$\alpha_i, \beta_i \geq 0, \quad \gamma_i \geq 0, \quad i = 0, 1, \quad \gamma_0 + \gamma_1 > 0.$$

We concern with Sturm-Lioville systems of the following type: (16), (17), (20) or (16), (18), (19) at $\gamma_0 \neq 0$ and (16), (19), (20) at $\gamma_1 \neq 0$. It was found that the spectral properties of the last problems are analogeous of those of the classical Sturm-Liouiville problem (16), (17), (18). Namely, each of the spectral problems (16), (17), (20) or (16), (18), (19) and (16), (19), (20) has the following properties:

1) It has an infinite sequence of simple real eigenvalues $0 < \lambda_1 < \lambda_2 < ...$ to which correspond the eigenfunctions $v_1(x), v_2(x), ...$;

2) The eigenfunctions $\{v_n(x)\}$ form an orthogonal system with respect to a convenient norm (see (14) about the norm).

3) There exist constants $c_1 > 0, c_2$ which depend only on the functions p, q, r and parameters $\alpha_i, \beta_i, \gamma_i, \quad i = 0, 1$ such that, $c_1 n^2 < \lambda_n < c_2 n^2$ when $n \to \infty$.

4) The eigenfunctions and their derivatives satisfy the inequalities:

$$|v_n(x)| \leq c_3, \quad |v_n'(x)| \leq c_4\sqrt{\lambda_n} \leq c_5 n, \quad c_4, c_5 - const$$

The oscillatory properties of all regular Sturm-Liouiville system are the same: the trigonometric equation $v'' + \lambda v = 0$ is typical. For problem (13) we have:

a) $\lambda_k = \beta_k^2,\ ctg\beta_k l = K\beta_k, \quad k = 1, 2, ...$;

$v_k(x) = \sqrt{2}\frac{\sqrt{1+K^2\beta_k^2}}{\sqrt{K+l(1+K^2\beta_k^2)}} \sin\beta_k x, \quad k = 1, 2, ...$

b) The eigenfunctions $\{v_k\}$ forms orthogonal basis:

$(v_k, v_m)_K \equiv \int\limits_0^l v_k(x)\, v_m(x)\, dx + K v_k(l)\, v_m(l) = \delta_{km} = \begin{cases} 0, & k \neq m, \\ 1, & k = m. \end{cases}$

If a function $f(x), \quad 0 \leq x \leq l$, satisfies $f(0) = 0,\ \|f\|_K^2 =< \infty$, then $f(x) = \sum\limits_{k=1}^{\infty} f_k v_k(x)$, where $f_k = (f, v_k)_K\,, \quad k = 1, 2, ...$

c) The derivatives of the eigenfunctions $v_k(x)$ satisfies the equalities

$v_k'(x) = \beta_k \overline{v}_k(x) = \sqrt{2}\beta_k \frac{\sqrt{1+K^2\beta_k^2}}{\sqrt{K+l(1+K^2\beta_k^2)}} \cos\beta_k x.$

It follows from here that the system $\{\overline{v}_k(x)\}$ is also orthogonal.

The following discrete spectral problem approximates (13) with $O(h^2)$ local truncation error:

$$y_{\overline{x}x,i} = (y_{\overline{x}})_{x,i} \equiv \frac{y_{i-1} - 2y_i + y_{i+1}}{h^2} = -\lambda y_i, \quad i = 1, ..., N-1 \tag{21}$$

$$y_0 = 0, \quad y_{\overline{x},N} = \frac{y_N - y_{N-1}}{h} = (K + 0.5h)\,\lambda y_N, \tag{22}$$

where $Nh = l$. As usual, the norm is used:

$$\|y\|_0^2 = \sum_{i=1}^{N-1} hy_i^2 + 0.5hy_N^2.$$

Here we introduce the mesh norm:

$$\|y\|_K^2 \equiv \sum_{i=1}^{N-1} hy_i^2 + (K + 0.5h)\, y_N^2 = \|y\|_0^2 + Ky_N^2.$$

Then the problem (21), (22), has the following properties.
(i) The eigenvalues and eigenfunctions of (21), (22) are:

$$0 < \lambda_1 = \frac{4}{h^2}\sin^2\frac{\alpha_1 h}{2} < \lambda_2 < ... < \lambda_N = \frac{4}{h^2}\sin^2\frac{\alpha_N h}{2},$$
$$y_k(x) = \sqrt{2}\frac{\sqrt{1+K^2\alpha_k^2}}{\sqrt{K + l(1+K^2\alpha_k^2)}}\sin\alpha_k x, \quad x = kh, \quad k = 1, ..., N,$$

where $\alpha_i,\ i = 1, \ldots, N$ are the first N positive roots of the equation

$$ctg\alpha l = K\frac{2}{h}ctg\frac{\alpha h}{2}.$$

(ii) The eigenfunctions $\{y_k(x)\}$ form orthogonal basis in the l^2-space with norm $\|.\|_K$ .If a mesh function satisfies: $f_0 = 0$ then

$$f(x_i) = \sum_{k=1}^{N} f_k v_k(x), \quad f_k = (f, v_k)_K, \quad k = 1, ..., N.$$

A direct consequence of (i), (ii) is the following lemma.

Lemma 1. *For an arbitrary mesh function y, such that $y(0) = 0$, the following inequalities hold*

$$\lambda_1\|y\|_K^2 \le \|y_{\overline{x}}\|^2 \equiv \sum_{i=1}^{N} y_{\overline{x},i}^2 \le \lambda_N\|y\|_K^2$$

3 Convergence and stability of difference schemes for problems (P), (H)

First we approximate the problem (H) by the scheme:

$$\begin{aligned}
&y_{\overline{t}t} = \sigma\widehat{y}_{\overline{x}x} = (1-2\sigma)\, y_{\overline{x}x} + \sigma\, \overset{\vee}{y}_{\overline{x}x}, \quad \overset{\vee}{y} = y^{j-1}, \ 0 < \sigma < 1, \\
&y_0 = y(0) = 0, \quad (M + 0.5h)\, y_{\overline{t}t} + \sigma\widehat{y}_{\overline{x}} + (1-2\sigma)\, y_{\overline{x}} + \sigma\overset{\vee}{y}_{\overline{x}}, \ x = l, \\
&y(x, 0) = \varphi(x), \quad y_t(x, 0) = \psi(x)
\end{aligned} \tag{23}$$

Theorem 2. *Let the following condition is fulfilled*

$$\sigma \geq \frac{1}{4} - \frac{1}{\lambda_N \tau^2}. \tag{24}$$

Then the scheme (23) *is stable with respect to initial data.*

Proof. By a technique similar to those in ([1], p. 316) we obtain the following energy equality

$$\mathrm{E}^{j+1} = \mathrm{E}^j = \left|\left|y_{\bar{t}}^j\right]\right|_M + \left(\sigma - \frac{1}{4}\right) \left\|y_{\bar{t}\bar{x}}^j\right\|^2 + \frac{1}{4}\left\|y_{\bar{x}} + y_{\bar{x}}^{j-1}\right\|. \tag{25}$$

The identity (25) is a discrete analog of the energy conservation for the differential problem (H) :

$$E(t) = \int_0^l \left(\frac{\partial u}{\partial t}\right)^2 dx + M\left(\frac{\partial u(l,t)}{\partial t}\right)^2 + \int_0^l \left(\frac{\partial u}{\partial x}\right)^2 dx = E(0),$$

$$E(0) = \int_0^l \psi^2 dx + M\psi^2(0) + \int_0^l (\varphi')^2 dx.$$

We seek the value of σ for which $\mathrm{E}^j \geq 0$ at each y^j and y^{j-1}. The well known inequality

$$\sum_{i=1}^{N} v_{\bar{x},i} h \leq \frac{4}{h^2}\left(\sum_{i=1}^{N-1} v_i^2 h + 0.5 h v_N^2\right) = \frac{4}{h^2}\|v\|_0^2, \quad v_0 = 0$$

is not effective here. We apply the Lemma 1 to (25) and obtain

$$\left|\left|y_{\bar{t}}^j\right]\right|_M^2 + \left(\sigma - \frac{1}{4}\right)\left\|y_{\bar{t}\bar{x}}^j\right\|^2 \geq \left(\frac{1}{\lambda_N} + \left(\sigma - \frac{1}{4}\right)\tau^2\right)\|y_{t\bar{x}}\|^2$$

It follows from here the inequality (24). The equality (25) means stability of (23) with respect to initial data in the seminorm $\left(\mathrm{E}^j\right)^{1/2}$. □

Now we approximate the problem (P) at $d = 0$ by the scheme:

$$\begin{aligned}
&y_t = \sigma \widehat{y}_{\bar{x},x} + (1-\sigma)\, y_{\bar{x},x}, \\
&y = y_{\bar{t}}^j = y\left(x_i^j\right), \quad \widehat{y} = y_i^{j+1}, \quad y_t = \frac{\widehat{y} - y}{\tau} \\
&y(x,0) = \varphi(x), \quad y(0,t) = \psi_0(t), \\
&(K + 0.5h)\, y_t + \sigma \widehat{y}_{\bar{x}} + (1-\sigma)\, y_{\bar{x}} = \psi_1(t) \ \ at\ x = l.
\end{aligned} \tag{26}$$

Just this problem is treated in ([2], p.177).

Theorem 3. *Let*

$$\sigma = \sigma_0 \geq \frac{1}{2} - \frac{1}{\tau\lambda_N} \tag{27}$$

Then (26) is stable with respect to initial data. Let

$$\sigma \geq \frac{1}{2} - \frac{1-\varepsilon}{\tau\lambda_N}, \ \varepsilon > 0. \tag{28}$$

Then the scheme is stable with respect to right side of the boundary conditions and uniformly convergent with rate of convergence $\mathcal{O}\left(\tau^{m_\sigma} + h^2\right)$.

Proof. Let $\psi_0 = \psi_1 \equiv 0$. Now we obtain the equality

$$2\tau\left(\|y_t\|^2 + (K+0.5h)\, y_{t,N}^2 + (\sigma - 0.5)\tau\, \|y_{t\bar{x}}\,]|^2\right) + \|\widehat{y}_{\bar{x}}\,]|^2 = \|y_{\bar{x}}\,]|^2 .$$

We conclude after applying the Lemma that when $\sigma \geq \sigma_0$, $\left\|y_{\bar{x}}^{j+1}\,]\right| \leq ... \leq \|y_{\bar{x}}^{\cdot}\,]|$, i.e the scheme (26) is stable with respect to initial data in the norm $\|y_{\bar{x}}\,]|$.

We turn to investigate (26) for stability with respect to right side of the boundary conditions. For the error $z = y - u$ we obtain the problem

$$\begin{aligned}
&z_t = \sigma \widehat{z}_{\bar{x}x} + (1-\sigma)\, \widehat{z}_{\bar{x}x} + \psi, \\
&z(x,0) = 0, \quad z(0,t) = 0, \\
&(K+0.5h)\, z_t + \sigma \widehat{z}_{\bar{x}} + (1-\sigma)\, z_{\bar{x}} = v,
\end{aligned} \tag{29}$$

where $\psi = O\left(\tau^{m_\sigma} + h^2\right)$, $\quad v = O\left(\tau^{m_\sigma} + h^2\right)$, $\quad \tau^{m_\sigma} = 1$ at $\sigma \neq 0.5$, $m_\sigma = 2$ at $\sigma = 0.5$ are the approximation errors. This problem can be written in operator form as follows:

$$Bz_t + Az = \Psi,$$

where

$$\begin{aligned}
&B = E + \tau\sigma A, \\
&(Av)_i = -v_{\bar{x}x,i},\ i = 1, \ldots, N-1 \\
&(Av)_N = \frac{1}{K+0.5h} v_{\bar{x},N}, \\
&\Psi_i = \psi_i, \quad 1 \leq i \leq N-1, \quad \Psi_N = \frac{v}{K+0.5h}.
\end{aligned}$$

The operator A is selfadjoint and positive in the mesh space with norm$\|.]|_K$. Indeed, using the discrete Green formula, we have

$$\begin{aligned}
(Ay, v]_K &= (Ay, v) + (K+0.5h)(Ay)_N v_N = \\
&= (v_{\bar{x}}, y_{\bar{x}}] = \sum_{i=1}^{N} v_{\bar{x}} y_{\bar{x}} = (Av, y]_K .
\end{aligned}$$

Applying Lemma 1 we conclude that A is positive operator. Since

$$(By, y]_K = \|y]|_K^2 + \tau\sigma(Ay, y]_K,$$

the operator B is also selfadjoint and positive.

Now we apply Theorem 5 , p.172 in [2], and conclude that the solution of problem(29) satisfies the inequality

$$\|z(t_{n+1})\|_A = \left(\sum_{i=1}^{N} h z_{\bar{x},i}(t_{n+1})\right)^{\frac{1}{2}} \leq \frac{1}{\sqrt{2\varepsilon}} \left(\sum_{i=0}^{n} \tau \|\Psi(t_i)]|_K^2\right)^{\frac{1}{2}}, \tag{30}$$

if $B \geq \varepsilon E + 0.5\tau A$, $\varepsilon > 0$. In our case this operator inequality takes the form

$$\|y]|_K^2 + \sigma\tau(Ay, y]_K \geq \varepsilon(y, y]_K + \frac{\tau}{2}(Ay, y]_K. \tag{31}$$

But, the Lemma 1 gives

$$\|y]|_K^2 \geq \frac{1}{\lambda_N}(Ay, y]_K.$$

Inserting the last inequality in(31), we obtain the condition (28).

Putting in (30) $\|\Psi]|_K^2 = (\psi, \psi) + (K + 0.5h)v^2$ and taking into account that

$$\|z\|_C = \max_{1 \leq i \leq N} |z_i| \leq \left(\sum_{i=1}^{N} z_{\bar{x},i}^2 h\right)^{\frac{1}{2}},$$

we establish uniform convergence of the sheme(26) with rate of convergence $\mathcal{O}\left(\tau^{m_\sigma} + h^2\right)$ at condition (28). □

The achieved results may be generalised to parabolic and hyperbolic equations with variable discontinuous coefficients. Also, difference schemes with variable weights [3], can be studied in norms similar to these introduced in the present paper.

Acknowledgment

This work is supported by Ministry of Science and Education of Bulgaria under Grant # MM–524/95.

References

1. Samarskii, A. A.: Theory of difference schemes. Nauka, Moscow, 1977 (in Russian)
2. Samarskii, A. A., Goolin A. V.: Stability of difference schemes. Nauka, Moscow, 1973 (in Russian)
3. Goolin, A. V.: On the stability of symmetrizable difference schemes. Mathematical Modeling **6** (1994) 9–13

Two-Step P-stable Methods with Phase-Lag of Order Infinity for the Numerical Solution of Special Second Order Initial Value Problems

P.S. Williams[1] **and T.E. Simos**[2]

[1] Department CISM, London Guildhall University, London EC3N 1JY, U.K.
[2] Laboratory of Applied Mathematics and Computers, Technical University of Crete, Kounoupidiana, 73100 Hania, Greece.

Abstract. Two 2-step P-stable methods for the numerical solution of special second order initial value problems are developed in this paper. One is of the Numerov type and of algebraic order 4 and the other is of the Runge-Kutta type and of algebraic order 6. Each of these methods has free parameters which may be chosen so that they are P-stable and have phase-lag of order infinity. The methods are used on problems with oscillatory solutions. The results indicate that these techniques are more efficient than other well known methods.

1 Introduction

In the last ten years there has been much activity concerned with the numerical solution of second order initial value problems of the type

$$y'' = f(x, y),\ y(x_0) = y_0,\ y'(x_0) = y'_0 \tag{1}$$

involving second order ordinary differential equations in which the first derivative does not appear explicitly. The equations that are investigated possess periodic solutions and model phenomena that occur in celestial mechanics, quantum mechanical scattering theory, theoretical physics and electronics (see [1], [2]).

There are two main groups of methods for the numerical solution of problems of the form (1) with periodic solutions. Category I consists of methods in which the coefficients that determine the numerical scheme are dependent on the particular problem to be solved. The application of these methods is possible when the frequency of the solution of the problem is known a priori. Category II consists of methods with constant coefficients, i.e. with coefficients independent of the problem. These latter methods can be applied to any problem with a periodic solution, even if the frequency of the problem is unknown initially.

Several methods have been developed for the solution of (1) belonging to Category I. We mention the works of Raptis and Allison [3], Cash, Raptis and Simos [4]. These works belong to the special subcategory of exponentially fitted methods. A more general subcategory of methods are the techniques with phase-lag of order infinity (called phase-fitted methods). In this area we mention the works of Simos [5], [6]. A weakness of these methods is that in the investigation of their stability it is assumed that the fitted frequency is the same as the test frequency. The main areas of application of these methods are problems for which the above requirement is possible, such as the Schrödinger equation.

Numerous category II numerical techniques have been obtained for the solution of (1). Methods of this category must be P-stable, and this applies especially in the case of problems with highly oscillatory solutions. The P-stability property was first introduced by Lambert and Watson [7]. We refer to the works of Cash [8] and Chawla and Rao [9]. In these works sixth order P-stable methods have been obtained. An important contribution for the P-stable methods is the work of Hairer [10] in which lower order P-stable methods have been developed.

The purpose of this paper is to develop two 2-step P-stable methods of algebraic order four and six with phase-lag of order infinity. We will assume here that the fitted frequency is different from the test frequency. The numerical tests show that these new methods are more efficient than the other well known P-stable methods. In section 2 we will describe the basic theory of the stability and phase-lag of symmetric multistep methods. In section 3 we will develop the P-stable methods of order four and six. Finally, in section 5 an application of the new method to a well known problem is presented.

2 Basic Theory

To investigate the stability properties of methods for solving the initial-value problem (1) Lambert and Watson [7] introduce the scalar test equation

$$y'' = -u^2 y \tag{2}$$

and the **interval of periodicity**.

When we apply a symmetric two-step method to the scalar test equation (2) we obtain a difference equation of the form

$$y_{n+1} - 2C_{H,V}(P)y_n + y_{n-1} = 0, \tag{3}$$

where $P = uh, h$ is the step length, $C_{H,V}(P) = \frac{B_{H,V}(P)}{A_{H,V}(P)}$ where $A_{H,V}(P)$ and $B_{H,V}(P)$ are polynomials in P and y_n is the computed approximation to $y(nh)$, $n = 0, 1, 2, \ldots$.

The characteristic equation associated with (3) is

$$s^2 - 2C_{H,V}(P)s + 1 = 0. \tag{4}$$

Bruca and Nigro [12] introduced the frequency distortion as an important property of methods for solving special second order initial-value problems. For frequency distortion other authors [5], [6], [11] use the terms of *phase-lag, phase error or dispersion*. From now on we use the term **phase-lag**.

Following Coleman [11] when we apply a symmetric two-step method to the scalar test equation $y'' = -u^2 y$ in which case we have the difference equation (3). The characteristic equation associated with (3) is given by (4). The roots of the characteristic equation (4) are denoted as s_1 and s_2.

The following definitions are pertinent to the basic theory.

Definition 1 *(see [13] and [14]) The method (3) is unconditionally stable if* $|s_1| \leq 1$ *and* $|s_2| \leq 1$ *for all values of* P.

Definition 2 *Following Lambert and Watson [7] we say that the numerical method (3) has an interval of periodicity* $(0, P_0^2)$ *if, for all* $P^2 \in (0, P_0^2)$, s_1 *and* s_2 *satisfy:*

$$s_1 = e^{i\theta(P)} \text{ and } s_2 = e^{-i\theta(P)}, \tag{5}$$

where $\theta(P)$ *is a real function of* P. *For any method corresponding to the characteristic equation (4) the phase-lag is defined (see [11]) as the leading term in the expansion of*

$$t = P - \theta(P) = P - \cos^{-1}[C_{H,V}(P)]. \tag{6}$$

If the quantity $t = O(P^{q+1})$ *as* $P \to 0$, *the order of phase-lag is* q.

Definition 3 *([7]) The method (3) is* **P -stable** *if its* **interval of periodicity** *is* $(0, \infty)$.

Theorem 1 *(for the proof see [14]) A method which has characteristic equation (4) has an interval of periodicity* $(0, P_0^2)$, *if for all* $P^2 \in (0, P_0^2)$ $|C_{H,V}(P)| < 1$.

Theorem 2 *(for the proof see [14]) For a method which has an interval of periodicity* $(0, P_0^2)$ *we can write:*

$$\cos[\theta(P)] = C_{H,V}(P), \text{ where } P^2 \in (0, P_0^2). \tag{7}$$

Using these concepts Coleman [11] makes the following remark.

Remark 1 *If the phase-lag order is* $q = 2r$ *then :*

$$\begin{aligned} t = cP^{2r+1} + O(P^{2r+3}) &\Rightarrow \cos(P) - C_{H,V}(P) = \cos(P) - \cos(P - t) \\ &= cP^{2r+2} + O(P^{2r+4}). \end{aligned} \tag{8}$$

Definition 4 *A symmetric two-step method phase-fitted if has phase-lag of order infinity, i.e. if* $\cos(P) = C_{H,V}(P)$.

3 Construction of P-stable phase-fitted schemes

3.1 A fourth order scheme

Consider the two parameter family of two-step methods

$$\begin{aligned} \overline{y}_n &= y_n - ah^2(f_{n+1} - 2f_n + f_{n-1}) \\ \overline{\overline{y}}_n &= y_n - bh^2(f_{n+1} - 2\overline{f}_n + f_{n-1}) \\ y_{n+1} - 2y_n + y_{n-1} &= \frac{h^2}{12}\left(f_{n+1} + 10\overline{\overline{f}}_n + f_{n-1}\right) \end{aligned} \tag{9}$$

where $f_q = f(x_q, y_q)$, $q = n-1(1)n+1$, $\overline{f}_n = f(x_n, \overline{y}_n)$, $\overline{\overline{f}}_n = f(x_n, \overline{\overline{y}}_n)$, a and b are free parameters and

$$L.T.E. = -\frac{h^6}{240}\left[y_n^{(6)} + (200b-1)y_n^{(4)}F_n\right], \tag{10}$$

where $F_n = \left(\frac{\partial f}{\partial y}\right)_n$.

Applying the above method to the scalar test equation

$$y'' = -w^2 y, \tag{11}$$

we get (3), with $P = H = wh, h$ is the step length, $C_{H,V}(H) = \frac{B_{H,V}(H)}{A_{H,V}(H)}$, where

$$A_{H,V}(H) = 1 + \frac{H^2}{12} + \frac{5bH^4}{6} - \frac{5abH^6}{3},$$
$$B_{H,V}(H) = 1 - \frac{5H^2}{12} + \frac{5bH^4}{6} - \frac{5abH^6}{3}. \tag{12}$$

From the definition 4 we see that the above method (9) is phase-fitted if $cos(H) = C_{H,V}(H) = \frac{B_{H,V}(H)}{A_{H,V}(H)}$ i.e.

$$b = \frac{(H^2+12)cos(H) + 5H^2 - 12}{10H^4(2aH^2-1)(cos(H)-1)}. \tag{13}$$

To find the interval of periodicity we apply the method (9) to the test equation

$$y'' = -s^2 y, \tag{14}$$

where $s \neq w$. Again we have the following difference equation

$$y_{n+1} - 2C_{H,V}(V)y_n + y_{n-1} = 0, \tag{15}$$

where $V = sh, h$ is the step length, $C_{H,V}(V) = \frac{B_{H,V}(V)}{A_{H,V}(V)}$, where

$$A_{H,V}(V) = 1 + \frac{V^2}{12} + \frac{5bV^4}{6} - \frac{5abV^6}{3},$$
$$B_{H,V}(V) = 1 - \frac{5V^2}{12} + \frac{5bV^4}{6} - \frac{5abV^6}{3}. \tag{16}$$

The characteristic equation associated with (15) is

$$s^2 - 2C_{H,V}(V)s + 1 = 0. \tag{17}$$

We make the following definition

Definition 5 *(see [15]) A family of phase fitted methods with the stability function $C_{H,V}(H,V)$, where $H = wh$ and $V = sh$, is* **P-stable** *if, for each value of w, the quantity $\mid C_{H,V}(V) \mid < 1$ is satisfied for all values of s and for all h, except possibly a set of exceptional values of h determined by the chosen value of w.*

Using the above definition we have, assuming that $A_{H,V}(V) > 0$,

$$| C_{H,V}(V) |< 1 \Rightarrow A_{H,V}(V) - B_{H,V}(V) > 0 \textit{ and } A_{H,V}(V) + B_{H,V}(V) > 0. \tag{18}$$

With the help of (16) we obtain

$$A_{H,V}(V) - B_{H,V}(V) = \frac{V^2}{2} > 0 \textit{ for all } h,\ w,\ s,$$
$$A_{H,V}(V) + B_{H,V}(V) = 2 - \frac{V^2}{3} + \frac{5bV^4}{3} - \frac{10abV^6}{3}. \tag{19}$$

The inequality $A_{H,V}(V)+B_{H,V}(V) > 0$ is satisfied if $-\frac{V^2}{3}+\frac{5bV^4}{3}-\frac{10abV^6}{3} > 0$. One way to obtain the previous condition is to assume that

$$-\frac{V^2}{3} + \frac{5bV^4}{3} - \frac{10abV^6}{3} = c(V)V^2, \tag{20}$$

where $c(V)$ is a non-negative function of V (i.e. for example $c(V) = 1+V^2+V^4$).
Using (20) and (13) we have

$$a = \frac{[2H^4(1+3c) + V^2S_1]cos(H) - 2H^4(1+3c) + V^2S_2}{2\{[2H^6(1+3c) + V^4S_1]cos(H) - 2H^6(1+3c) + V^4S_2\}},$$
$$b = \frac{[2H^6(1+3c) + V^4S_1]cos(H) - 2H^6(1+3c) + V^4S_2}{10V^2H^4(H^2 - V^2)(cos(H) - 1)}, \tag{21}$$

where $c = c(V)$, $S_1 = H^2 + 12$ and $S_2 = 5H^2 - 12$. It is obvious now that for a and b given by (21), $A_{H,V}(V) = 1 + \frac{V^2(2c+1)}{4} > 0$ for all $V \in (0, \infty)$.

3.2 A sixth order scheme

Consider the two parameter family of two-step methods

$$\overline{y}_n = y_n - ah^2(f_{n+1} - 2f_n + f_{n-1})$$
$$\overline{y}_{n\pm\frac{1}{2}} = \frac{1}{2}(y_n + y_{n\pm1}) - h^2\left[b\overline{f}_n + \left(\frac{1}{8} - b\right) f_{n\pm1}\right]$$
$$\overline{\overline{y}}_{n\pm\frac{1}{2}} = \frac{1}{2}(y_n + y_{n\pm1}) - \frac{h^2}{96}\left(f_{n\pm1} + 10\overline{f}_{n\pm\frac{1}{2}} + f_n\right) \tag{22}$$
$$y_{n+1} - 2y_n + y_{n-1} = \frac{h^2}{60}\left[f_{n+1} + 26f_n + f_{n-1} + 16\left(\overline{\overline{f}}_{n+\frac{1}{2}} + \overline{\overline{f}}_{n-\frac{1}{2}}\right)\right]$$

where $f_q = f(x_q, y_q)$, $q = n - 1(1)n + 1$, $\overline{f}_s = f(x_s, \overline{y}_s)$, $s = n,\ n \pm \frac{1}{2}$, $\overline{\overline{f}}_{n\pm\frac{1}{2}} = f(x_{n\pm\frac{1}{2}}, \overline{\overline{y}}_{n\pm\frac{1}{2}})$, a and b are free parameters and

$$L.T.E. = \frac{h^8}{1209600}\left(-10y_n^{(8)} - 21y_n^{(6)}F_n - 2975y_n^{(4)}F_n'F_n + 33600by_n^{(4)}F_n'F_n\right), \tag{23}$$

where $F_n = \left(\frac{\partial f}{\partial y}\right)_n$, $F_n' = \left(\frac{dF}{dx}\right)_n$.

Applying the above method to the scalar test equation (11) we get (3), with $P = H = wh, h$ is the step length, $C_{H,V}(H) = \frac{B_{H,V}(H))}{A_{H,V}(H)}$, where

$$A_{H,V}(H) = 1 + \frac{3H^2}{20} + \frac{H^4}{60} + \frac{H^6(1-8b)}{288} + \frac{abH^8}{18},$$
$$B_{H,V}(H) = 1 - \frac{7H^2}{20} - \frac{H^4}{60} - \frac{bH^6}{36} + \frac{abH^8}{18}. \quad (24)$$

Based on the definition 4 we have that the above method is phase-fitted if $cos(H) = C_{H,V}(H) = \frac{B_{H,V}(H)}{A_{H,V}(H)}$ i.e. for

$$b = \frac{(5H^6 + 24H^4 + 216H^2 + 1440)cos(H) + 24(H^4 + 21H^2 - 60)}{40H^6(1 - 2aH^2)(cos(H) - 1)}. \quad (25)$$

To find the interval of periodicity we apply the method (22) to the test equation (14). So, we have the difference equation (15) with $V = sh, h$ is the step length, $C_{H,V}(V) = \frac{B_{H,V}(V)}{A_{H,V}(V)}$, where

$$A_{H,V}(V) = 1 + \frac{3V^2}{20} + \frac{V^4}{60} + \frac{V^6(1-8b)}{288} + \frac{abV^8}{18},$$
$$B_{H,V}(V) = 1 - \frac{7V^2}{20} - \frac{V^4}{60} - \frac{bV^6}{36} + \frac{abV^8}{18}. \quad (26)$$

The characteristic equation associated with (15) is the equation (17)
Based on the definition 5 we have (18). With the help of (26) we obtain

$$A_{H,V}(V) - B_{H,V}(V) = \frac{V^2}{2} + \frac{V^4}{30} + \frac{V^6}{288} > 0 \ for\ all\ h,\ w,\ s,$$
$$A_{H,V}(V) + B_{H,V}(V) = 2 - \frac{V^2}{5} + \frac{V^6(1-16b)}{288} + \frac{abV^8}{9}. \quad (27)$$

The inequality $A_{H,V}(V) + B_{H,V}(V) > 0$ is satisfied if $-\frac{V^2}{5} + \frac{V^6(1-16b)}{288} + \frac{abV^8}{9} > 0$. One way to obtain the previous condition is to assume that

$$-\frac{V^2}{5} + \frac{V^6(1-16b)}{288} + \frac{abV^8}{9} = c(V)V^2, \quad (28)$$

where $c(V)$ is a non-negative function of V.
Based on (28) and (25) we have

$$a = \frac{[288H^6(1+5c) + V^4S_1]cos(H) + V^4S_2}{2[S_6cos(H) - 288H^8(1+5c) + S_7]},$$
$$b = \frac{S_8cos(H) - H^8S_5 + 48V^6(H^4 + 21H^2 - 60)}{80V^4H^6(V^2 - H^2)(cos(H) - 1)}, \quad (29)$$

where $c = c(V)$, $S_1 = 5H^6 + 48H^4 + 432H^2 + 2880$, $S_2 = 5H^6 + 24H^4 + 216H^2 + 1440$, $S_3 = 5H^6 + 48H^4 + 1008H^2 - 2880$, $S_4 = H^4 + 21H^2 - 60$, $S_5 = 1440c - 5V^4 + 288$, $S_6 = 288H^8(1+5c) + 2V^6S_3 - 5V^4H^8$, $S_7 = 48V^6S_4 + 5V^4H^8$, $S_8 = H^8S_5 + 10H^6V^6 + 48H^4V^6 + 432H^2V^6 + 2880V^6$, . It is obvious now that for a and b given by (21), $A_{H,V}(V) = 1 + \frac{V^2(2c+1)}{4} + \frac{V^4}{60} + \frac{V^6}{576} > 0$ for all $V \in (0, \infty)$.

4 Numerical Illustration

We have applied the new methods to the problem studied by Kramarz [16]

$$y_1'' = 2498y_1 + 4998y_2, \; y_1(0) = 2, y_1' = 0$$
$$y_2'' = -2499y_1 - 4999y_2, \; y_2(0) = -1, y_2' = 0 \tag{30}$$

with the analytical solution given by

$$y_1(x) = 2\cos x, \; y_2(x) = -\cos x \tag{31}$$

The eigenvalues of the matrix of the coefficients of equations (30) are equal to -1 and -2500. Hence it is possible to transform the above system of equations to the form of the scalar equation (14) with $s = 1$ and $s = 50$. Thus the new methods are applicable because we have two frequencies defined. Consequently, we set $w = 1$ and $s = 50$. In the exponentially fitted methods used for comparison purposes we set $w = 1$. We have also investigated the case $w = 50$ and the results are similar. For comparison purposes some well known P-stable methods are used.

We have solved the above problem in the interval $(0, 10\pi)$ using several step-sizes and the following methods

Method 1: The exponentially fitted methods of Cash et. al. [4].
Method 2: The sixth order phase fitted method of Simos [6].
Method 3: The sixth-order P-stable method of Cash [8].
Method 4: The sixth-order P-stable method of Chawla and Rao [9].
Method 5: The P-stable method (Case $m = 4$) of Simos [17].
Method 6: The fourth order P-stable phase-fitted method developed above.
Method 7: The sixth order P-stable phase-fitted method developed above.

Method	$h=\frac{\pi}{2}$	$h=\frac{\pi}{5}$	$h=\frac{\pi}{10}$	$h=\frac{\pi}{20}$
1		8.2e-3	7.3e-4	5.5e-5
2		7.2e-3	1.2e-4	3.4e-5
3	1.1e-2	4.7e-5	7.8e-7	1.2e-8
4	2.4e-1	3.4e-4	2.7e-6	5.4e-7
5	4.5e-4	5.8e-7	6.3e-9	9.3e-10
6	4.7e-12	3.4e-12	2.2e-12	1.3e-12
7	8.6e-13	6.5e-13	3.2e-13	1.1e-13

Table 1. Maximum absolute error in the interval $(0, 10\pi)$ using the methods 1-7.

In Table 1 we present the maximum absolute error in the interval $(0, 10\pi)$ using the methods 1-7. The blank sections indicate that the error is greater than 1.

From the results presented in Table 1 it is evident that the new methods 6 and 7 are much more efficient for the problems with stiff oscillatory solutions. We should note however, that Methods 1 and 2 were not developed for problems such as (30) but for problems such as the Schrödinger type, for which they give very accurate results. Note also, that the methods 3-5 are constant coefficient techniques.

References

1. Landau, L.D., Lifshitz, F.M.: Quantum Mechanics. Pergamon, New York, 1965
2. Liboff, R.L.: Introductory quantum mechanics. Addison-Wesley Publishing Company, 1980
3. Raptis, A.D., Allison, A.C.: Exponential-fitting methods for the numerical solution of the Schrödinger equation. Comput. Phys. Commun. **14** (1978) 1–5
4. Cash, J.R., Raptis, A.D., Simos, T.E.: A sixth-order exponentially fitted method for the numerical solution of the radial Schrödinger equation. J. Comput. Phys. **91** (1990) 413–423
5. Simos, T.E.: A two-step method with phase-lag of order infinity for the numerical integration of second order periodic initial value problem. Inter. J. Comput. Math. **39** (1991) 135–140
6. Simos, T.E.: Two-step almost P-stable complete in phase methods for the numerical integration of second order periodic initial value problems. Inter. J. Comput. Math. **46** (1992) 77–85
7. Lambert, J.D., Watson, I.A.: Symmetric multistep methods for periodic initial value problems. J. Inst. Math. Applic. **18** (1976) 189–202
8. Cash, J.R.: High order P-stable formulae for the numerical integration of periodic initial value problems. Numer. Math. **37** (1981) 355–370
9. Chawla, M.M., Rao, P.S.: High accuracy methods for $y'' = f(x, y)$. IMA J. Numer. Anal. **5** (1985) 215–220
10. Hairer, E: Uncoditionally stable methods for second order differential equations. Numer. Math. **32** (1979) 373–379
11. Coleman, J.P.: Numerical methods for $y'' = f(x, y)$ via rational approximation for the cosine. IMA J. Numer. Anal. **9** (1989) 145–165
12. Brusa, L., Nigro, L.: A one-step method for direct integration of structural dynamical equations. Int. J. Numer. Methods Engrg. **15** (1980) 685–699
13. Thomas, R.M.: Phase properties of high almost P-stable formulae. BIT **24** (1984) 225–238
14. Simos, T.E., Mousadis, G.: A two-step method for the numerical solution of the radial Schrödinger equation. Comput. Math. Appl. **29** (1995) 31–37
15. Coleman, J.P., Ixaru, L.Gr.: P-stability and exponential-fitting methods for $y'' = f(x, y)$. University of Durham, Numerical Analysis Report NA-94/05, Durham 1994
16. Kramarz, L.: Stability of colocation methods for the numerical solution of $y'' = f(x, y)$. BIT **20** (1980) 215–222
17. Simos, T.E.: New variable-step procedure for the numerical integration of the one-dimensional Schrödinger equation. J. Comput. Phys. **108** (1993) 175–179

Notes on the Classification of Numerical Algorithms with Respect to Their Stability to Roundoff Errors*

Plamen Y. Yalamov[1]

Center of Applied Mathematics and Informatics
University of Rousse, 7017 Rousse, Bulgaria

Abstract. First a review of previous results on dependence graphs and roundoff error analyses is presented. Then it is shown that allowing perturbations of outputs does not change the bounds on the forward error essentially in two important cases. A new condition number of the algorithm is introduced. Finally, a classification of all numerical algorithms is given.

1 Introduction

In this work we are concerned with the stability of direct (not iterative) algorithms with respect to roundoff errors. We assume the standard model of computation with a guard digit, i.e.,

$$fl(x * y) = (x * y)(1 + \sigma_*), \quad |\sigma_*| \leq \rho_0, \quad * \in \{+, -, \times, /\},$$

where ρ_0 is the machine roundoff unit.

There are two main approaches to assess the roundoff error propagation. These are forward [6] and backward [8] roundoff error analyses. The forward analysis bounds the errors in the outputs of the algorithm, while the backward analysis represents the errors from all the operations in a given algorithm as perturbations in inputs, and finds bounds on the perturbations. Sometimes, it is useful also to obtain the so called mixed analysis, where for some operations the errors are propagated backwards, and for the rest the errors are propagated forwards.

In this paper we first review in brief a new approach to roundoff error analysis, which is a generalization of all existing analyses. This approach was developed for the case of absolute errors and absolute perturbations by Voevodin and Yalamov in [7], and then extended to the case of relative errors and relative perturbations by Yalamov in [10]. Based on the results of [10] here we propose a classification of all numerical algorithms. The proofs are ommited, and a detailed version of this paper [12] can be obtained from the author.

* This work was supported by Grant MM-434/94 from the Bulgarian Ministry of Education, Science and Technology

2 Dependence graphs and roundoff error analysis

Let us suppose that the algorithm consists of arithmetic operations only, and that it is represented by its dependence graph. This means that 1) a vertex corresponds to each operation, and 2) if two vertices are connected by an arc the result of the first vertex is an argument for the second one. Denote the dependence graph by $G = (V, E)$, where V is the set of vertices, and E is the set of arcs.

Let us suppose that all the data (input, intermediate, and output) is kept in a vector A, and that $f_k,\ k = 1, ..., l$, are all the operations of the algorithm. Then the algorithm can be given as

$$A_k = f_k\left(A_{k_1}, A_{k_2}\right), \quad k = 1, ..., l,$$

where A_{k_1} and A_{k_2} are the inputs for the k-th operation. In computation with roundoff errors we have

$$\widetilde{A}_k = \widetilde{f}_k\left(\widetilde{A}_{k_1}, \widetilde{A}_{k_2}\right) = f_k\left(\widetilde{A}_{k_1}, \widetilde{A}_{k_2}\right)(1+\sigma_k), \quad k = 1, ..., l, \tag{1}$$

and $|\sigma_k| \leq \rho_0$. In [10] the computation is represented as follows (if possible):

$$\widetilde{A}_k(1+\rho_k) = f_k\left(\widetilde{A}_{k_1}(1+\rho_{k_1}), \widetilde{A}_{k_2}(1+\rho_{k2})\right), \quad k = 1, ..., l. \tag{2}$$

The quantities ρ_k are called relative equivalent perturbations, because if the algorithm is executed in exact arithmetic but with perturbed inputs $A_j(1+\rho_j)$ then we get the equations (2). The adjective "equivalent" comes from the fact that if the output ρ_k are zeros, for example, then the outputs from (1) and (2) are equal, and the introduction of perturbations is equivalent in some sense to the computation with roundoff errors. Now if we have bounds on the input ρ_k we can get bounds on the error in the solution by using perturbation analysis. This is the purpose also of the classical Wilkinson's backward analysis [8], i.e. to obtain bounds on the input equivalent perturbations. The difference is that in [7] and [10] we introduce equivalent perturbations not only of outputs, but also of intermediate and output results. This fact allows to propose a construction for calculating the bounds on input equivalent perturbations. In fact, the equivalent perturbations of the intermediate results exist implicitly but they were not distinguished in the existing literature. An example given in [12] show this fact.

To define all the equivalent perturbations we eliminate the quantities $\widetilde{A}$ from (1) and (2) to obtain

$$f_k\left(\widetilde{A}_{k_1}(1+\rho_{k_1}), \widetilde{A}_{k_2}(1+\rho_{k_2})\right) - f_k\left(\widetilde{A}_{k_1}, \widetilde{A}_{k_2}\right)(1+\rho_k)(1+\sigma_k) = 0, \tag{3}$$

$$k = 1, ..., l.$$

The simultaneous consideration of all these equations leads to a system with respect to the unknowns ρ_k. If we can find a solution to system (3), then we can get bounds on the equivalent perturbations.

If f_k are arithmetic operations then (3) is one of the following type (see [10]):

$$\tilde{x}\rho_x \pm \tilde{y}\rho_x - (\tilde{x} \pm \tilde{y})(1+\sigma_\pm)\rho_z = (\tilde{x} \pm \tilde{y})\sigma_\pm, \tag{4}$$

$$\rho_x + \rho_y + \rho_x\rho_y - (1+\sigma_\times) f_z = \sigma_\times, \tag{5}$$

$$\rho_x - (1+\sigma_/)\rho_y - (1+\sigma_/)\rho_y\rho_z - (1+\sigma_/)\rho_z = \sigma_/, \tag{6}$$

where the first equation corresponds to an addition (subtraction), the second to a multiplication, and the third to a division.

It is clear that the coefficients in (4) may depend on the growth of intermediate results in contrast to (5) and (6). Also it is well-known that there are stable algorithms (e.g. back substitution for triangular systems), for which the bounds of the equivalent perturbations do not depend on the growth of intermediate results. A natural question is, what are the conditions under which we have backward stability. The answer to this question is given below. The proof can be found in [10].

Definition 1. Let us consider the dependence graph of a given algorithm. Each vertex of the graph represents some local operation. All of the arcs carry results but some of these may be replicas of the others. In this case we assume that the arc (or equivalently the result) is *replicated*, or we say that there is *replication* in the vertex. This means that some result is used as an argument in more than one operation.

Let us denote by V_m, the subset of the set of vertices V, consisting of *vertices in which the outputs are obtained and of vertices the results of which are replicated.*

Theorem 2. *Let the following conditions be fulfilled:*

1. *the operands in all the additions and subtractions are not replicated;*
2. *at least one operand in every multiplication or division is not replicated;*
3. *the algorithm consists of arithmetic operations only;*

Suppose that all the data is non zero, and the equivalent perturbations of outputs are equal to zero. Then there exists an equivalent relative perturbation of every input a, the bound of which does not depend on the data, and can be given explicitly as

$$\rho_a = \sum_{j=1}^{r(a)} \sigma_j, \tag{7}$$

by neglecting terms of higher order in ρ_0, where $|\sigma_j| \leq \rho_0$, and $r(a)$ is the length of the minimal path leading from the vertex, where a is stored, to the subset V^*.

The algorithms satisfying Theorem 2 are very stable. Their bounds do not depend on the growth of intermediate results. Theorem 2 gives only sufficient conditions which are easy to check. But these are not the only algorithms which have such a nice property. Now we shall sketch the definition of a wider class of algorithms.

Let us suppose that

$$\rho_x = \rho_y \tag{8}$$

in equation (4). Then from (4) we obtain

$$\rho_x - (1 + \sigma_\pm)\, \rho_z = \sigma_\pm, \tag{9}$$

and (9) has the nice property that its coefficients do not depend on the growth of the intermediate results. After applying the basic step (8) to all additions (subtructions) we get the system of equations

$$B\rho + Q(\rho) = \sigma, \tag{10}$$

where each scalar equation is one of the type (5), (6), (9), ρ is the vector of relative equivalent perturbations, and σ is the vector of all local relative roundoff errors ($|\sigma_j| \le \rho_0$). $B\rho$ is the linear part, and $Q(\rho)$ is clearly quadratic with respect to the entries of ρ. In [10] it is shown that the linearization

$$B_1\rho = \sigma \tag{11}$$

gives quite a good approximation to ρ where B_1 is obtained from the matrix B by dropping the terms $\sigma_\pm, \sigma_\times, \sigma_/$ in all its coefficients.

In the usual backward analysis we set the equivalent perturbations of outputs equal to zero. This was done in [10] when deriving the wider class of stable algorithms. In this paper we shall use directly system (11) to define this class. This gives more degrees of freedom when solving system (11). In this way we get a mixed analysis. We show in the next section that this does not lead to big differences with the well known bounds derived from backward analysis in the two of the most spread cases - solution of linear systems and computation of eigenvalues.

From (11) we have the following

Theorem 3. *Suppose that the algorithm consists of arithmetic operations only, and matrix B_1 defined above has full row rank. Then there exist first order approximations of the equivalent relative perturbations which bounds do not depend on data, and the algorithm is backward stable.*

3 Bounds on the forward error

In this section we show that the introduction of equivalent perturbations to all results leads to similar bounds on the forward error as those coming from the usual backward analysis. We do this for the two of the most spread cases - solution of linear systems and simple eigenvalue computation.

First, we consider solution of linear systems with square matrices. It is clear that our analysis, which exhibits the perturbations in all the data, leads to a system

$$(A+\varepsilon_A)(\widetilde{x}+\varepsilon_x)=b+\varepsilon_b, \tag{12}$$

where $\varepsilon_A, \varepsilon_x$, and ε_b are the absolute equivalent perturbations of A, x, and b, respectively. We shall use the Skeel condition numbers [5]

$$\operatorname{cond}(A,x)=\frac{\left\| |A^{-1}|\,|A|\,|x| \right\|_\infty}{\|x\|_\infty}, \quad \operatorname{cond}(A)=\operatorname{cond}(A,e).$$

Theorem 4. *Let us suppose that in* (12) $|\varepsilon_A|\le\delta\,|A|, |\varepsilon_b|\le\delta\,|b|, |\varepsilon_x|\le\delta\,|\widetilde{x}|$, *i.e. the relative equivalent perturbation in each entry is bounded by* δ. *If* $cond(A)\delta<1$ *then*

$$\frac{\|\widetilde{x}-x\|_\infty}{\|x\|_\infty}\le\left(\frac{2\,cond\,(A,x)}{1-cond\,(A)\,\delta}+1\right)\frac{\delta}{1-\delta}.$$

In the case of a simple eigenvalue computation our analysis would give

$$(A+\varepsilon_A)\left(\widetilde{\lambda}+\varepsilon_\lambda\right)=\left(\widetilde{\lambda}+\varepsilon_\lambda\right)(\widetilde{x}+\varepsilon_x), \tag{13}$$

where $\varepsilon_A, \varepsilon_x$, and ε_λ are the absolute equivalent perturbations of A, x, and λ respectively. Here we shall use the componentwise condition number proposed by Geurts [1]:

$$\kappa(\lambda,E)=\frac{|y^T|\,E\,|x|}{|y^T x|\,|\lambda|},$$

and the absolute condition number

$$K(\lambda)=|y^T|\,|x|/|y^T x|,$$

where x is the right eigenvector corresponding to λ, and E is a matrix of the same size as A with nonnegative entries.

Theorem 5. *Let us suppose that in* (13) $|\varepsilon_A|\le\delta\,|A|, |\varepsilon_x|\le\delta\,|\widetilde{x}|, |\varepsilon_\lambda|\le\delta|\widetilde{\lambda}|$, *i. e. the relative equivalent perturbation in each entry is bounded by* δ. *If* $K(\lambda)\,\delta<1$, *then*

$$\frac{\left|\widetilde{\lambda}-\lambda\right|}{|\lambda|}\le\left[\frac{\frac{|y^T||A||\widetilde{x}|}{|y^T\widetilde{x}||\lambda|}}{1-\frac{|y^T||\widetilde{x}|}{|y^T\widetilde{x}|}\delta}(1+\delta)+1\right]\frac{\delta}{1-\delta}=$$

$$(\kappa(\lambda,|A|)+1)\,\delta+O\left(\delta^2\right).$$

The second bound is true in case δ *is small enough.*

These theorems show that allowing perturbations of all the data does not change essentially the perturbation analysis. But this gives more degrees of freedom when doing roundoff analysis. The proofs of Theorems 4 and 5 are given in [12].

4 Conditioning of the algorithm

In this section we introduce a new condition number which characterizes the stability of the algorithm.

It is clear that the stability of the algorithm depends on the solutions of system (3). In practice, system (11) has a lot of solutions. For simplicity we consider the linearized system (3) which is written as follows:

$$B_0\rho = \sigma, \tag{14}$$

i.e. matrix B in (10) is obtained from B_0 after applying the basic step (8). We are looking for solutions of (14) which are small with respect to some norm. For example, we can consider the following solution:

$$\min \|\rho\|_2\,, \text{ subject to } B_0\rho = \sigma, \tag{15}$$

or,

$$\min \|\rho\|_\infty\,, \text{ subject to } B_0\rho = \sigma. \tag{16}$$

Let us suppose the $\rho^{(1)}$ is a solution to (15), and $\rho^{(2)}$ is a solution to (16). Then we can write

$$\rho_1^{(1)} = B_0^+\sigma, \quad \rho^{(2)} = B_0^{++}\sigma,$$

where B_0^+ and B_0^{++} are some pseudoinverses of B.

Definition 6. We call the norm $\|B_0^+\|$ (or $\|B_0^{++}\|$, respectively) a condition number of the algorithm, where $\|.\|$ is some subordinate norm.

It is clear that

$$\left\|\rho^{(1)}\right\|_\infty \le \|B_0^+\|_\infty \rho_0, \quad \left\|\rho^{(2)}\right\|_\infty \le \|B_0^{++}\|_\infty \rho_0,$$

for example. So, the possible instabilities of the algorithm would be displayed by large equivalent perturbations, and large norms of B_0^+ (or B_0^{++}), respectively.

Let us note that $\|B_0^+\|\,\rho_0$ (or $\|B_0^{++}\|\,\rho_0$) is a bound for the perturbation δ in the theorems from the previous section.

It is not so easy to compute $\|B_0^+\|$ (or $\|B_0^{++}\|$) but in some cases we can get bounds on this condition number.

5 Classification of algorithms

Here we give a possible classification on numerical algorithms with respect to their stability to roundoff errors.

5.1 Unconditionally stable algorithms

Definition 7. We say that a given algorithm is unconditionally stable, if it satisfies the conditions of Theorem 2 or Theorem 3.

In other words, for unconditionally stable algorithms the bounds of the equivalent perturbations do not depend on the growth of intermediate results, and are independent of the input data.

Theorem 8. *For any algorithm satisfying the conditions of Theorem 2 we have*

$$\|B_0^{++}\|_\infty \leq r,$$

where r is the length of the maximal path in the dependence graph of the algorithm.

Theorem 9. *For any algorithm satisfying the conditions of Theorem 3 we have*

$$\|B_0^{++}\| \leq c_n,$$

where c_n is a constant depending on the problem size n, but not on the growth of intermediate results.

The proofs of these theorems are evident from Theorem 2, Theorem 3 and the previous section. It is clear that the condition numbers of the unconditionally stable algorithms are relatively small.

5.2 Conditionally stable algorithms

Definition 10. We say that a given algorithm is conditionally stable if it does not satisfy the conditions of Theorem 3.

From this definition it follows that matrix B_1 for a conditionally stable algorithm is not of full row rank. So, it is not possible to apply the basic step (8) to system (14). Then clearly the condition number of the algorithm depends on the intermediate results. Depending on the inputs, for some inputs we would have small $\|B_0^+\|$ (or $\|B_0^{++}\|$), and large $\|B_0^+\|$ (or $\|B_0^{++}\|$) for other inputs. For this reason we call such algorithms conditionally stable.

An example of this type of algorithms is the Gaussian eliminations without pivoting. For example, it is well-known that this algorithm is stable for diagonally dominant matrices, and can be unstable for arbitrary matrices (see [3]). Another example is the cyclic reduction for tridiagonal systems. In [11] it is shown that the equivalent perturbations are bounded by small constants when the tridiagonal matrix is diagonally dominant, s.p.d., totally nonnegative, or M-matrix. This algorithm can be unstable for arbitrary matrices. In the notations of this paper we have

$$\|B_0^{++}\|_\infty \leq O(n^\gamma), \quad \gamma > 0,$$

where $\gamma = \log_2 1.5$ for diagonally dominant matrices, and $n^\gamma = \log_2^2 n$ for symmetric positive definite, nonnegative, and M-matrices.

The possible instabilities can be analyzed from the entries of matrix B_0. Then one can take certain measures to stabilize the algorithm. Up to now there is no general way to overcome all possible instabilities. But there are cases in which we can propose stabilization of certain algorithms. One such case is the division by small numbers (or zeros) in some algorithms. These algorithms can be stabilized by adding small perturbations to the small numbers. This can be done under certain conditions. This approach was successfully applied to a fast Toeplitz solver in [2], then to the tridiagonal cyclic reduction in [11], and to a partitioning algorithm for tridiagonal systems in [4]. We expect that it could be applied to other (especially parallel) algorithms as well.

References

1. Chatin-Chatelin, F., Frayssé, V.: Lectures in finite precision computations. SIAM, 1996
2. Hansen, P. C., Yalamov, P. Y.: Stabilization by perturbation of a $4n^2$ Toeplitz solver. Preprint N25, Technical University of Russe, January 1995 (submitted to SIAM J. Matrix Anal. Appl.)
3. Higham, N. J.: Accuracy and stability of numerical algorithms. SIAM, 1996
4. Pavlov, V. T., Todorova, D.: Stabilization and experience with a partitining method for tridiagonal systems. Preprint N30, University of Rousse, 1996 (in this volume)
5. Skeel, R. D.: Scaling for numerical stability in Gaussian elimination. J. Assoc. Comput. Mach. **26** (1979) 494–526
6. Stummel, F.: Perturbation theory for evaluation algorithms of arithmetic expressions. Math. Comp. **37** (1981) 435–473
7. Voevodin, V. V., Yalamov, P. Y.: A new method of roundoff error estimation. Parallel and Distributed Processing (K. Boyanov, ed.), Elsevier, Amsterdam,1990, 315-333
8. Wilkinson, J. H.: The algebraic eigenvalue problem. Clarendon Press, Oxford, 1965
9. Yalamov, P. Y.: On the stability of the cyclic reduction without back substitution for tridiagonal systems. BIT **35** (1994) 428–447
10. Yalamov, P. Y.: On some classes of backward stable algorithms. Preprint N27, Center of Applied Mathematics and Informatics, University of Rousse, 1995
11. Yalamov, P. Y., Pavlov, V. T.: Stabilization by perturbation of ill-conditioned cyclic reduction. Preprint N28, Center of Applied Mathematics and Informatics, University of Rousse, 1996
12. Yalamov, P. Y.: Notes on the classification of numerical algorithms with respect to their stability to roundoff errors. Preprint N31, Center of Applied Mathematics and Informatics, University of Rousse, 1996

New IMGS-based Preconditioners for Least Squares Problems

Tianruo Yang
Department of Computer Science
Linköping University
581 83 Linköping
SWEDEN

Abstract. Convergence acceleration by preconditioning is usually essential when solving the standard least squares problems by an iterative method. IMGS, is an incomplete modified version of Gram-Schmidt orthogonalization to obtain an incomplete orthogonal factorization preconditioner $M = \bar{R}$, where $A = \bar{Q}\bar{R} + E$ is an approximation of a QR factorization, $\bar{Q}$ is an orthogonal matrix and $\bar{R}$ is upper triangular matrix respectively. Based on the IMGS orthogonalization, a relaxed Incomplete Modified Gram-Schmidt preconditioning and a new recursive selecting strategy of incomplete orthogonal preconditioning which updates the drop tolerance step by step and decides the corresponding value according to the recursive relation are proposed. The numerical experiments show clearly the robustness of this recursive selecting strategy. For the relaxed IMGS preconditioning approach, a suitable relaxation parameter influences the performance and quality of the preconditioner.

1 Introduction

Convergence acceleration by preconditioning the matrix A is usually essential when solving the standard least squares problems by an iterative method. Many preconditioners that have been proposed and analyzed include diagonal and SSOR preconditioners [3], incomplete Cholesky factorization [6], and incomplete orthogonal factorization [5, 8, 9, 11].

Our study will focus on the class of IMGS orthogonal factorization, an incomplete modified version of Gram-Schmidt orthogonalization to obtain an incomplete orthogonal factorization preconditioner $M = \bar{R}$, where $A = \bar{Q}\bar{R} + E$ is an approximation of a QR factorization, $\bar{Q}$ is an orthogonal matrix and $\bar{R}$ is upper triangular matrix respectively. In [9], the complete analysis has been given to the following members of the family: Incomplete Classical Gram-Schmidt(ICGS), Incomplete Modified Gram-Schmidt(IMGS) and Compressed Incomplete Modified Gram-Schmidt(CIMGS) factorizations. The numerical properties of each of these methods have been considered as well as the relationships between the methods concerning the preservation of sparsity, computational efficiency and the quality of the preconditioner. From his numerical experiments, the IMGS-based members were found to be more robust and to produce higher quality preconditioners than variants based on ICGS.

This paper is organized as follows. In section 2, a relaxed IMGS preconditioning technique is proposed. A new recursive selecting strategy of incomplete orthogonal preconditioning is presented in section 3. Finally the numerical experiments show clearly the robustness of this recursive selecting strategy and relaxed IMGS preconditioning technique.

2 Relaxed IMGS preconditioner

CIMGS also available in [10] is an alternative more compact way of computing the IMGS preconditioner. The basic idea of Compressed Incomplete Gram-Schmidt is to compress the information in the column vectors of A into a inner product form without losing the information needed for the computation of the factor. It can be shown that in exact arithmetic this produces the same incomplete factor R as IMGS, and therefore inherits the robustness of IMGS.

The approach of relaxed IMGS is based on CIMGS extended by a relaxation parameter $\omega \in [0,1]$. The idea is similar with RILU, Relaxed Incomplete LU factorization proposed by Axelsson and Lindskog in [1, 2]. For a fixed sparse pattern, the parameter value $\omega = 0$ in relaxed IMGS reproduces the corresponding CIMGS factorization.

Algorithmically, let $B = A^T A$. When A is a real matrix with full rank, B is symmetric positive definite. Given a drop set $P \subseteq P_n$, Relaxed IMGS generates the upper triangular matrix $R \in \Re^{n \times n}$ as follows:

ALGORITHM Relaxed IMGS Factorization

for $k = 1, 2, \ldots, n$ do

if $b_{kk} \neq 0$ then

$b_{kk} = \sqrt{b_{kk}}, \quad r_{kk} = b_{kk};$

for $j = k+1, k+2, \ldots, n$ do

$b_{kj} = b_{kj} / \sqrt{b_{kk}};$

$$r_{kj} = \begin{cases} 0 & (k,j) \in P \\ b_{kj} & \text{otherwise} \end{cases}$$

endfor

for $j = k+1, k+2, \ldots, n$ do

for $i = k+1, k+2, \ldots, n$ do

if $j = i$ then

$$b_{ij} = b_{ij} - b_{ki} b_{kj} + \omega \sum_{p=k+1}^{n} (b_{ip} - b_{ki} b_{kp});$$

else

$$b_{ij} = \begin{cases} 0 & (i,j) \notin P \\ b_{ij} - b_{ki}b_{kj} & \text{otherwise} \end{cases}$$

endif; endfor;

endfor; endif;

endfor

The performance of relaxed IMGS in exact arithmetic applied to $A^T A$ compared with the triangular factor as IMGS applied to A and the rounding error incurred by relaxed IMGS are being investigated right now. The main problem for relaxed IMGS is how to choose an optimal parameter ω.

3 Recursive IMGS preconditioner

The effects of varying the dropping strategy are discussed in detail in [9]. Not surprisingly, the choice of P can be crucial not only to the quality of the preconditioner but also to the combined efficiency of the two phases of the algorithm: preconditioner computation and iterative method. Wang [9] compared three different selection strategies, namely dynamic pattern selection, static pattern selection and semi-dynamic pattern selection.

Static pattern selection determines the positions in the drop set P before the factorization starts. This specifies exactly which elements of R must be evaluated in each step of IMGS and makes the dynamic nature of the data structure holding the evolving Q much simpler. This strategy allows us to arrange the computations in a favorable way. The disadvantage of this static strategy is that with the lack of knowledge of elements of R we may not be able to produce a good preconditioner.

By dynamic we mean that the set P is unknown until the factorization is complete. Certain criteria are imposed on the value of a matrix element and at some point during the factorization a decision is made to keep or drop it based on the criteria. Such a strategy has the advantage that it can adapt to numerical values encountered during the factorization and not just precondition based on its sparsity pattern. The main drawback with the dynamic approach is that there is no clear way of selecting the drop tolerance τ.

Saad [8], proposed a dropping strategy in ILQ factorization which allows static allocation of the data structures but does not determine in advance which elements are kept. We refer to this method of pattern selection as the semi-dynamic strategy. There are advantages to this method in addition to static space. This can also improve the quality of the preconditioner. On the other hand, it is difficult to anticipate the number of elements that should be kept.

From the numerical tests presented in [9], the dynamic pattern selection strategy tends to produce the best preconditioner for IMGS given that a good value of τ is known. How to find a good value of τ? Here we propose a recursive selection strategy which avoids requiring a good τ value and updates the drop tolerance step by step and decides the corresponding value according to the recursive relation where we assume $r_{ij} = \tau * \|a_i\|_2$ where $\|a_i\|_2$ is the i-th column of

A and apply this into the MGS factorization procedure to get the corresponding updated expressions of τ. The algorithm can be described as follows:

ALGORITHM Recursive IMGS Factorization

$$
\begin{array}{l}
\text{Set} \quad \tau = 0.02 \\
\text{for} \quad i = 1, 2, \ldots, n \quad \text{do} \\
\quad d_i = a_i, \qquad r_{ii} = \|a_i\|_2, \qquad q_i = a_i / r_{ii}; \\
\quad \text{for} \quad j = i+1, \ldots, n \quad \text{do} \\
\qquad r_{ij} = q_i^T a_j, \quad aold_j = a_j; \\
\qquad \text{if} \quad r_{ij} < \tau * \|a_i\|_2 \quad \text{then} \\
\qquad\quad r_{ij} = 0; \\
\qquad \text{else} \\
\qquad\quad a_j = a_j - r_{ij} q_i; \\
\qquad\quad \tau = \dfrac{a_i^T (aold_j - \tau * d_i)}{\|a_i\|_2^2}; \\
\qquad \text{endif} \\
\quad \text{endfor} \\
\text{endfor}
\end{array}
$$

The numerical study presented later of this recursive IMGS preconditioner not only shows that it is robust and efficient, but also promises potential for future research. Basic properties of this method will be explored including existence and numerical error properties in my future study.

4 Numerical tests

The following tests were performed in MATLAB on a SUN workstation using double precision. The first set of test problems denoted $P(m, n, d, p)$ was generated, as described in Paige and Saunders [7], as

$$A = Y \begin{pmatrix} D \\ 0 \end{pmatrix} Z^T \in \mathbf{R}^{m \times n}, \qquad Y = I - 2yy^T, \quad Z = I - 2zz^T.$$

Here y and z are Householder vectors of appropriate dimension, generated by

$$y_i = \sin(4\pi i/m), \qquad z_i = \cos(4\pi i/n),$$

followed by normalization so that $\|y\| = \|z\| = 1$. Taking $n = q$ the singular values are taken to be

$$D = q^{-p} \operatorname{diag}(q^p, \ldots, 3^p, 2^p, 1),$$

i.e., p is a power factor, and $n = qd$ leads to d copies of each singular value.

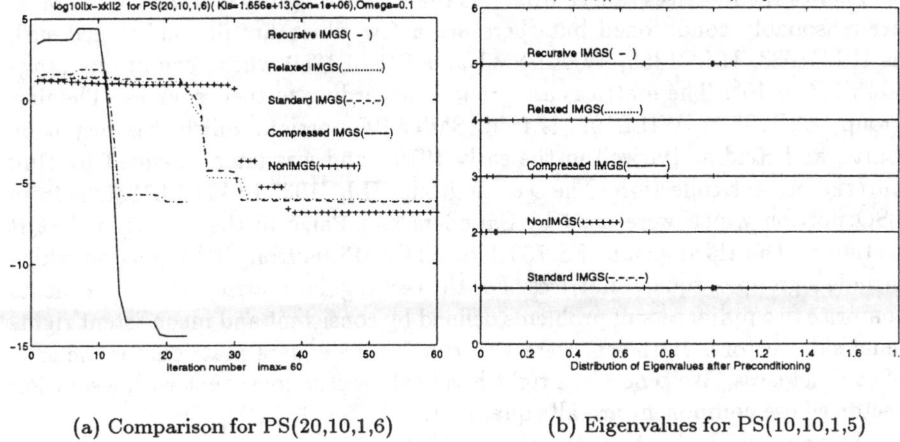

(a) Comparison for PS(20,10,1,6) (b) Eigenvalues for PS(10,10,1,5)

Fig. 1. The performance and eigenvalue distribution of IMGS-based preconditioners

The solution is taken to be $x = (n-1, \ldots, 2, 1, 0)^T$, and right hand side is

$$b = Ax + \rho r, \qquad r = Y \begin{pmatrix} 0 \\ c \end{pmatrix},$$

where

$$c = \frac{1}{m}(1, -2, 3, \ldots, \pm(m-n))^T, \qquad \|r\| = \rho\|c\|.$$

Thus taking $m > n$, and $\rho > 0$ gives an incompatible system. For reference, this gives $\|x\| \simeq n(n/3)^{1/2}$ and cond(A)=cond(D) $\simeq (\sigma_n/\sigma_1)^p = q^p$. The corresponding stopping rules are described fully in [7].

We use the set of problem to compare the different IMGS-based approaches as the preconditioner and CGLS as the basic iterative tool. The relaxed parameter for relaxed IMGS $\omega = 0.1$. From the numerical tests in Fig.1, the relaxed IMGS is not so good as expected because it is not so easy to find an optimal parameter. It seems that the performance of the recursive IMGS preconditioner is best. It reduces the iteration number sharply. Also we show the comparison of the spectrum of A and $M^{-1}A$ obtained for the first set of problems among IMGS-based approaches in Fig.1.

The second set of test problems are from Harwell-Boeing collection [4]. Characteristics of the matrices, including the number of row(m), the number of columns(n), the number of nonzeros(nnz(A)) and the density of the matrix are given in [4]. The density of a matrix $A \in \Re^{m\times n}$ is denoted dense(A) and expresses, as a percentage, the ratio of actual non-zero elements to the maximum possible, i.e.,100(nnz(A)/mn). The collection includes 15 matrices, of which 10 are square. The size of the matrices vary considerably both in term of dimensions $54 \leq m, n \leq 3564$ and number of non-zero elements $291 \leq nnz(A) \leq 22316$. The sparsity of the matrices ranges from less than 1% to slightly less than 10%. Most are reasonably conditioned but there are a few which are ill conditioned such as ILLC1033, ILLC1850, WELL1033 and WELL1850 whose conditions range from 10^2 to 10^4. The matrices are grouped according to their sources. The first group, ASH292 to WILL199, is from SMTAPE portion, which was begun by Curtis and Reid at Harwell in the early 1970's and was later extended by Duff into the present collection. The second group, ILLC1033 to WELL1850, is from LSQ portion which were used by Saunders and Paige in the testing of LSQR methods. The third group, FS_760_1 to STEAM3, is from RUA portion which include a group of square matrices. For the rectangular matrices the experiments comprise two major sets of problems defined by consistent and inconsistent right-hand side vectors. Both sets were also run for the square matrices for the sake of completeness. We generate a right-hand side vector consistent with a solution vector whose component are all equal to 1.

As we mentioned before, the choice of relaxed parameter ω influences significantly the quality of Relaxed IMGS preconditioner. From the limited experiments in Table 1 whose identifiers range from ILLC1033 to WELL1850, it seems that $\omega = 0.1$ is a better choice and we will focus on the theoretical studies of selecting an optimal relaxed parameter for Relaxed IMGS preconditioners.

Table 1. Influence of ω in Relaxed CIMGS

Identifier	$\omega = 1$	$\omega = 0.1$	$\omega = 0.01$	$\omega = 0$
ILLC1033	88	78	82	82
ILLC1850	92	82	84	87
WELL1033	38	30	36	38
WELL1850	40	34	37	37

The performance comparison of IMGS-based preconditioners can be seen in Table 2. Here we compare non-IMGS, IMGS with dynamic pattern of $\tau = 0.02$, Compressed IMGS with dynamic pattern of $\tau = 0.02$, Relaxed IMGS with dynamic pattern of $\omega = 0.1$ and $\tau = 0.02$ and Recursive IMGS preconditioners while CGLS acts as the basic iterative tool. If a problem does not converge within the given iteration limit l, then the iteration count is recorded as $> l$ and the corresponding measures of the time count reflects the activity of l iterations. In this table, we use ITER as the number of iterations, PT as preconditioning time in seconds and NN(R) as the number of non-zero elements of R. From the experiments, firstly we can see clearly the effect of the preconditioner. Secondly CIMGS can perform efficiently compared with IMGS with the smaller preconditioning time and almost same number of non-zero elements of R. Relaxed IMGS needs more preconditioning time than CIMGS, as we expected, but it works well for FS_760_3 while CIMGS and IMGS fail for this case. From the results on ILLC1033, WELL1850 and SAYLOR4, Relaxed IMGS performs extremely well with comparable number of non-zero elements of R. So, Relaxed IMGS is still

Table 2. Comparison of IMGS-based preconditioners

Identifier	NonIMGS	IMGS			CIMGS			RelIMGS			RecIMGS		
	ITER	ITER	PT	NN(R)	ITER	PT	NN(R)	ITER	PT	NN(R)	ITER	PT	NN(R)
ASH292	32	17	1.62	126	17	0.16	126	17	0.28	126	10	1.88	124
CURTIS54	18	4	0.54	22	4	0.11	22	4	0.23	22	4	0.72	22
GENT54	32	13	0.38	45	12	0.09	46	13	0.20	46	8	0.52	44
LUNDA	28	8	2.62	129	8	0.28	129	9	0.58	129	8	2.77	127
LUNDB	32	8	2.64	128	8	0.26	128	9	0.43	127	8	2.92	128
WILL199	28	10	2.44	49	10	0.24	49	10	0.50	49	10	2.51	50
ILLC1033	142	82	6.21	2630	82	0.18	2631	78	0.48	2601	44	8.25	2645
ILLC1850	160	87	51.32	6654	87	1.30	6655	82	2.55	6645	41	60.44	6672
WELL1033	88	30	7.60	2983	38	0.26	2984	30	0.74	2988	19	9.70	2984
WELL1850	101	37	65.20	6971	37	1.43	6972	34	3.04	6968	21	66.42	6988
FS_760_1	38	7	3.34	852	7	0.94	853	7	2.04	856	3	4.64	867
FS_760_3	> 1520	> 1520	71.30	8036	>1520	3.55	8037	1032	8.02	8006	1328	88.24	8022
SAYLR4	2820	2279	1371.88	12204	2277	21.15	12205	1760	40.26	12086	2034	1420.50	12098
STREAM1	30	10	2.12	378	10	0.72	379	10	1.88	376	4	2.28	374
STREAM2	68	28	2.54	1177	28	0.63	1178	26	1.53	1154	15	3.12	1126

a nice preconditioner. How to get the optimal relaxed parameter will be a very interesting topic. From the experiments, the numerical performance of Recursive IMGS is very promising. It reduces the number of iterations significantly with comparable number of non-zero elements of R. The only disadvantage we can see now is its relatively high preconditioning time because of its recursive relations.

References

1. Axelsson, O.: Incomplete block matrix factorization preconditioning methods. Journal of Compututational and Applied Mathematics. **12/13** (1985) 3–18
2. Axelsson, O., Lindskog, G.: On the eigenvalue distribution of a class of preconditioning methods. Numerische Mathematik. **48** (1986) 479–498
3. Björck, Å., Elfving, T.: Accelerated projection methods for computing pseudoinverse solutions of systems of linear equations. BIT. **19** (1979) 145–163
4. Duff, I., Grimes, R., Lewis, J.: Users' guide for Harwell-Boeing sparse matrix collection. CERFACS, Toulouse Cedex, France, October (1992)
5. Jennings, A., A. Ajiz, M.: Incomplete methods for solving $A^T Ax = b$. SIAM Journal on Scientific and Statistical Computing. **5** (1984) 978–987
6. A. Meijerink, J., and A. van der Vorst, H.: An iterative solution method for linear system of which the coefficient matrix is a symmetric M-matrix. Mathematics of Computation. **31** (1977) 148–162
7. C. C. Paige and M. A. Saunders.: LSQR: An algorithm for sparse linear equations and sparse least squares. ACM Transactions on Mathematical Software. **8** (1982) 43–71
8. Saad, Y.: Preconditioning techniques for nonsymmetric and indefinite linear systems. Journal on Computational and Applied Mathematics. **24** (1988) 89–105
9. Wang, X.: Incomplete factorization preconditioning for linear least squares problems. PhD thesis, Department of Computer Science, University of Illinois at Urbana-Champaign. (1993)
10. Wang, X., A. Gallivan, K., Bramley, R.: CIMGS: An incomplete orthogonal factorization preconditioner. Technical Report 393, Computer Science, Indiana University, Bloomington, Indiana, December 1994.
11. Zlatev, Z., Nielsen, H.: Solving large and sparse linear least-squares problems by conjugate gradient algorithms. Computer and Mathematics with Applications. **15** (1988) 185–202

Numerical Algorithm for Studying Heat Transfer in a Glass Melting Furnace

I. Zheleva, V. Kambourova, P. Dimitrov and R. Rusev

Institute of Chemical Technology and Biotechnology, 7200 Razgrad, POB 110, Bulgaria

Abstract. A numerical method for integration of the equation that describes heat-transfer processes in a glass melting furnace is proposed for different boundary conditions in a non-rectangular geometrical area. Irregular mesh is used which allows to obtain more precise results in the regions of large temperature gradients. The proposed algorithm is tested for different boundary conditions and for different embedded meshes. The results show that the algorithm is a reliable and efficient tool for investigating heat-transfer processes in a glass melting furnace.

1 Introduction

Processes taking place in a glass melting furnace producing flat glass are very complicated. In fact there are five relatively separated physico-chemical processes—silication, refining fusion, degassing, homogenization and cooling, which are closely interconnected at very high temperature and practically occur simultaneously.

Nowadays it seems that mathematical modeling is the most suitable and not so expensive investigation of these complicated processes. As a first stage of an adequate study of the above mentioned processes in the glass melting furnace only the heat transfer process is to be modeled and examined. In this paper we shall develop a numerical algorithm for solving Laplace equation, describing the stationary heat transfer for continuum media with isomorphic properties which do not depend of spatial coordinates. The geometric area corresponds to the real flat glass furnace, working in Diamond Ltd in Razgrad (Fig. 1).

2 Posing of the Mathematical Problem

We have to find a solution of the Laplace equation for the temperature T of the continuum medium, written in a Cartesian coordinate system (Fig. 1):

$$\Delta T = \frac{\partial^2 T}{\partial x^2} + \frac{\partial^2 T}{\partial y^2} = 0\,. \tag{1}$$

This solution has to satisfy the following boundary conditions:

— on the top surface (GH in Fig. 1):

$$K_{\text{eff}}\,\frac{\partial T}{\partial y} = q_T\,, \tag{2}$$

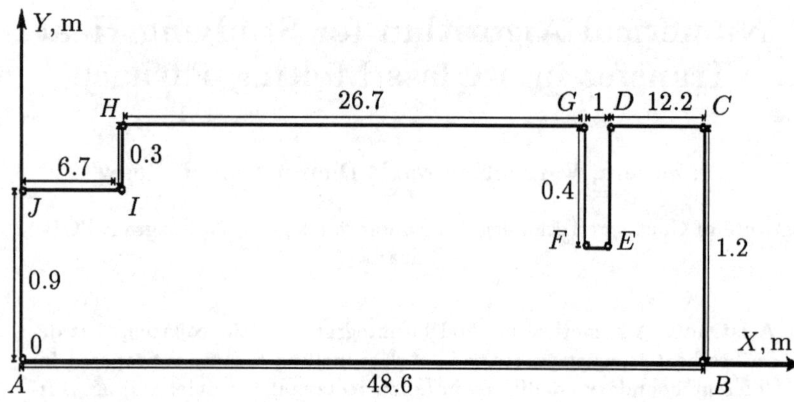

Fig. 1. Scheme of the glass melting furnace and the coordinate system

where K_{eff} is the effective thermal conductivity and q_T is the heat flux entering the glass melt surface.

— on the top surface in the cooling zone and for the shield assembly (CD, DE, EF, GH):

$$\frac{\partial T}{\partial N} = 0\,, \tag{3}$$

i.e. there is not heat exchange between glass melt and gas space.

— on the solid surface (the walls and the bottom JA, AB and BC) the heat transfer is expressed by means of the equation:

$$K_{\text{eff}}\,\frac{\partial T}{\partial N} = U_i\,(T - T_a)\,, \tag{4}$$

where T_a is the ambient temperature, N—the inward normal (i.e. the normal directed from the refractory surface into glass melt), U_i the heat transfer coefficient of solid surfaces which, depending on the surface has different values, namely, U_{W1} for the front wall, U_{W2} for the back wall and U_b for the bottom. Note that the heat transfer coefficients U_i may also be different in different sections of the delivery function.

It is assumed that the area JIH is occupied by the batch and the conditions at the interface between the batch and the glass melt are:

$$\frac{\partial T}{\partial N} = 0\,. \tag{5}$$

3 Numerical Algorithm

To solve the problem (1)–(5) a finite differences method is used. To this end it is necessary that a non-uniform grid, shown in Fig. 2, be introduced.

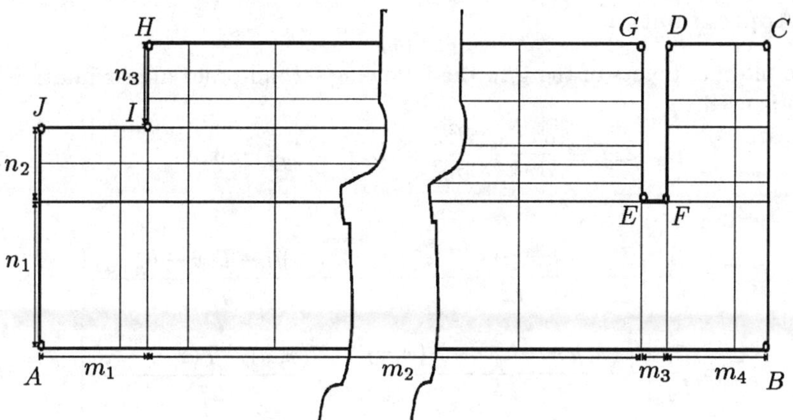

Fig. 2. Construction of the grid

3.1 Construction of the Grid

The points A, B, C, D, E, F, G, H, I and J of the boundary of the area, given in Fig. 2, have to be nodes of the grid. For describing the grid, the area is divided into four parts in the x-direction (m_1, m_2, m_3 and m_4) by the points I, F and E, and into three parts in the y-direction (n_1, n_2 and n_3) by the points F and I. Each of these parts is divided into k_i ($k_i > 0$, k_i integer) equal subintervals. If the values of k_i are well chosen, the step in each part will equal the step in other parts ($k_1/m_1 = k_2/m_2 = \cdots$). An uniform grid in the x-direction, in y-direction or both and in each of the mentioned parts can be developed in this way. The grid can be modified as the number of points is increased in the critical areas (near the walls, the bottom and the top surface), as shown in Fig. 3.

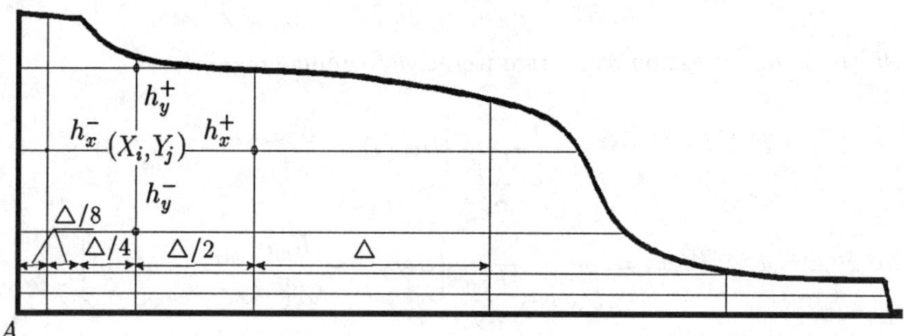

Fig. 3. The non-uniform grid used in the numerical solution

3.2 Approximation

For the internal points of the grid the following second-order approximation for Eq. (1) is used:

$$\Delta T \equiv \frac{\partial^2 T}{\partial x^2} + \frac{\partial^2 T}{\partial y^2} \approx L_x T + L_y T = 0\,. \tag{6}$$

Here

$$L_x T = \frac{1}{\hbar_x}\left[\frac{T(x-h_x^+, y) - T(x,y)}{h_x^+} - \frac{T(x,y) - T(x-h_x^-, y)}{h_x^-}\right],$$

$$L_y T = \frac{1}{\hbar_y}\left[\frac{T(x, y+-h_y^+) - T(x,y)}{h_y^+} - \frac{T(x,y) - T(x, y-h_x^-)}{h_y^-}\right], \tag{7}$$

$$\hbar_x = \frac{1}{2}(h_x^+ + h_x^-)\,, \quad \hbar_y = \frac{1}{2}(h_y^+ + h_y^-)\,,$$

see also Fig. 3. In turn, a first-order approximation is used for the boundary conditions.

3.3 Iterative Method

The employed technique is of Seidel type with acceleration procedure SOR ("successive over relaxation"). If $T^0(x,y)$ equals the initial temperature for all points of the grid, the successive values for each point of the grid is calculated by the following iterative formulae:

$$T^{n+1}(x,y) = \frac{M}{\hbar_x\, h_x^-}\, T^n(x-h_x^-, y) + \frac{M}{\hbar_x\, h_x^+}\, T^n(x-h_x^+, y-h_y^-)$$

$$+\frac{M}{\hbar_y\, h_y^-}\, T^n(x, y-h_y^-, y) + \frac{M}{\hbar_y\, h_y^+}\, T^n(x, y+h_y^+)\,, \tag{8}$$

$$M = \left(\frac{1}{\hbar_x\, h_x^+} + \frac{1}{\hbar_x\, h_x^-} + \frac{1}{\hbar_y\, h_y^+} + \frac{1}{\hbar_y\, h_y^-}\right).$$

On the boundaries the respective iterative formulae read:

On AJ: $$T^{n+1}(x,y) = \frac{1}{U_{W1} - \dfrac{K_{\text{eff}}}{h_x^+}}\left[U_{W1} T_0 - \frac{K_{\text{eff}}}{h_x^+}\, T^n(x-h_x^-, y)\right];$$

On BC: $$T^{n+1}(x,y) = \frac{1}{U_{W2} - \dfrac{K_{\text{eff}}}{h_x^-}}\left[U_{W2} T_0 - \frac{K_{\text{eff}}}{h_x^-}\, T^n(x-h_x^-, y)\right];$$

On AB: $$T^{n+1}(x,y) = \frac{1}{U_b - \dfrac{K_{\text{eff}}}{h_y +}}\left[U_b T_0 - \frac{K_{\text{eff}}}{h_y^+}\, T^n(x, y+h_y^+)\right];$$

On HG: $T^{n+1}(x,y) = q_T h_y^- T^n(x, y - h_y^-)$;

On DC: $T^{n+1}(x,y) = T^n(x, y - h_y^-)$;

On FG: $T^{n+1}(x,y) = T^n(x - h_x^-, y)$;

On DE: $T^{n+1}(x,y) = T^n(x + h_x^+, y)$;

On HI: $T^{n+1}(x,y) = T^n(x - h_x^-, y)$;

On EF: $T^{n+1}(x,y) = T^n(x, y - h_y^-)$;

On IJ: $T^{n+1}(x,y) = T^n(x, y - h_y^-)$.

This iteration procedure continues until the the relative error becomes small enough, more precisely, until the following condition is satisfied

$$\varepsilon = \max_{i,j} \left| \frac{T_{ij}^{n+1} - T_{ij}^n}{T_{ij}^n} \right| \leq 4 \cdot 10^{-4},$$

which needs, as a rule about 400 iterations. Here T_{ij}^{n+1} is the value of the temperature calculated from $(n+1)th$ iteration and T_{ij}^n is that of nth iteration.

The over relaxation procedure (SOR) [1] is used for the acceleration of the iterative procedure. The $(n+1)\ th$ value is calculated by the formula:

$$\overline{T}^{n+1}(x,y) = (1-w)\overline{T}^n(x,y) + wT^n(x,y),$$

where w is the SOR parameter, appropriately chosen, see Table 1.

4 Numerical Results

4.1 Test Examples

Several tests are first provided to verify the numerical algorithm. Let us denote by Ω and $\partial\Omega$ the geometrical area of the furnace and its boundary, respectively, and by Ω_1 its top surface $HGDC$, see Fig. 1. We study first several problems for the Laplace equation for which the exact analytical solution is known. The results of these test examples are summarized in Table 1.

The results of Table 1 clearly indicate that the developed algorithm works well for the test examples.

Table 1. Input data and results from tests

Boundary conditions	Analytical solution	Accuracy of the num. solution	Number iterations	SOR parameter
$T\big\|_{\partial\Omega} = 1$	$T = 1$	$4 \cdot 10^{-4}$	47	1.8
$T\big\|_{\partial\Omega\setminus\Omega_1} = x + y$	$T = x + y$	$4 \cdot 10^{-4}$	35	1.7
$T\big\|_{\partial\Omega} = x + y$	$T = x + y$	$4 \cdot 10^{-4}$	35	1.7
$\frac{\partial T}{\partial y}\big\|_{\Omega_1} = 1$		$4 \cdot 10^{-6}$	63	1.7
$T\big\|_{\partial\Omega} = xy$ $\frac{\partial T}{\partial y}\big\|_{\Omega_1} = x$	$T = xy$	$4 \cdot 10^{-6}$	69	1.7

4.2 Results of the Calculations for the Real Problem

The real problem is solved for the values of the parameters, given in Table 2 [2].

Table 2. Physical parameters used in calculation

Parameters	Values
Effective conductivity K_{eff}	$5.386 W/(mK)$
Melting temperature T_m	1100 K
Ambient temperature T_a	350 K

For this calculation non-uniform grid is used in the areas near the walls and the bottom. The reason is that the temperature gradients are considerable there. The grid is modified as the number of points increases in these areas as shown in Fig. 3.

The results for the temperature distribution are shown in Fig. 4.

5 Concluding Remarks

A reliable numerical algorithm for investigating complicated physico-chemical processes taking place in glass melting furnace is developed in the study. The algorithm is tested and the results from provided tests are acceptable and promising for this stage of the investigation.

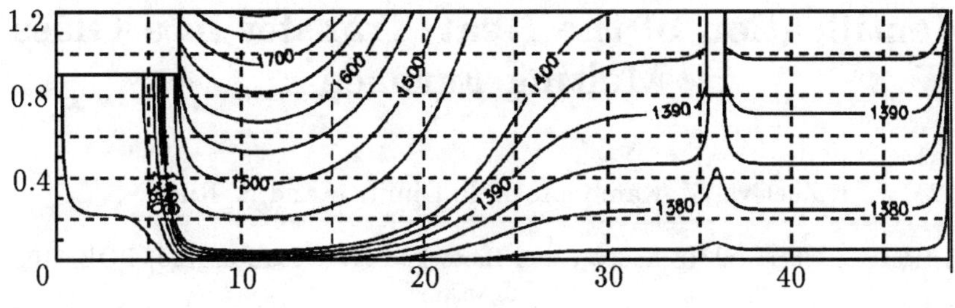

Fig. 4.

Acknowledgements. The support of the Bulgarian National Fund of Scientific Research under Grant No MM510/1995 is gratefully acknowledged. We thank Prof. M. Kaschiev for stimulating and very helpful discussions.

References

1. Marchuk, G. I.: Methods of numerical mathematics. Moscow, "Nauka" 1980, 535. (in Russian.)
2. Ungan, A., Viskanta, R.: Three dimensional numerical modeling of circulation and heat transfer in a glass melting tank. Glastech. Ber. **60**(3) (1987) 71–78; **60**(4) (1987) 115–124

Identification of the Heat–Transfer in a Glass Melting Furnace

I. Zheleva, V. Kambourova, P. Dimitrov and R. Rusev

Institute of Chemical Technology and Biotechnology, 7200 Razgrad, POB 110, Bulgaria

Abstract. A simplified mathematical model of heat-transfer processes in a glass melting furnace is studied. The model is built up on the basis of the laws of mass conservation, the energy and the moment of a continuous 2-D medium. It is assumed that the liquid glass melt is a Newtonian fluid. The boundary conditions describe the heat exchange on the front and back walls and the bottom of the furnace as well as the heat flow on the glass melting surface. The obtained system of partial differential equations is very complicated and nonlinear and no analytic solutions are known. For simplification very slow processes are only considered assuming also that the thermal problem is stationary. For solving the simplified system a numerical method is proposed. The latter allows to investigate numerically the influence of the heat flow and the insulation on the temperature distribution within the furnace. The obtained results can be used for minimizing the heat consumption of the furnace.

1 Introduction

The melting of glass batch and the production of homogeneous glass melt is a complicated physico-chemical process, taking place in a glass furnace at high temperature. This process usually incorporates the following relatively separated five parts: silication, refining fusion, degassing, homogenization and cooling. All these individual phases are closely interconnected and practically they occur almost simultaneously.

The kinetics of chemical transformation of batch in glass melt includes the following phases—refining fusion, degassing and homogenization and defines the temperature regime of the glass melt, going for drawing. The temperature regime is an important prerequisite for quality of drawing glass. Recall that the quality of glass is characterized by a constant thickness and smooth flat surface (without any waveness).

The outlet product of the furnace is a glass melt suitable for drawing. It goes to drawing machines in excess and depends on the equipment of the furnace. The basic characteristic of a glass melt is its quality which is defined by thermal and chemical homogeneity in the drawing volume and at the time. Therefore it is especially important for quality of the flat glass that the temperature regime in the furnace be within given limits. Hence the automatic control of glass' quality is directly connected with the control of temperature distribution in the furnace.

It is well known that information for heat and mass transfer processes, taking place in the furnace, can be obtained in three different ways:

— experimental measurements of temperature and velocity of the glass melt in a working furnace;

— physical study of reduced low temperature models;

— mathematical modeling and simulation.

Experimental investigations in a working furnace are very difficult and expensive due to the large volume of the latter, great inertia of the processes and high temperature. Especially difficult is the measurement of the temperature in a glass melting furnace. The temperature near the walls and the bottom of the furnace is usually measured using special thermocouples. The glass surface temperature is measured using pyrometers, which lead to considerable errors. As a matter of fact, the temperature within the glass melt cannot be measured and it is practically impossible to have reliable information about it.

Usage of mathematical models gives a possibility to decrease expenses for investigation and to simulate furnace work in different conditions. The mathematical modeling of the processes includes development of mathematical model of the phenomena, choice of proper numerical methods, creation of reliable software and numerical experiments. When complicated technological processes such as production of flat glass is under study, a hierarchy of models is used as a rule, which allows to elucidate the role of each factor. In this paper, as a first stage of the investigation, the diffusion heat transfer and its influence on the temperature distribution is only examined, on the example of the real flat glass furnace, working in Diamond Ltd in Razgrad.

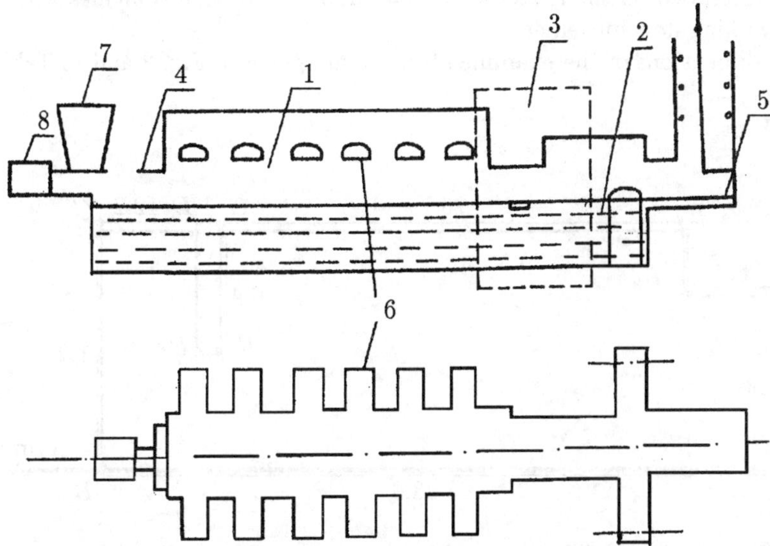

Fig. 1. Scheme of the flat glass melting furnace

2 Scheme of the Model Furnace

In this paper a glass melting continuous furnace for producing of flat glass is considered. The scheme of the furnace is given in Fig. 1. It consists of tank and combustion chambers. The tank is divided into two zones from the shield assembly (3), namely, the melting (1) and cooling (2) zones.

The following processes occur in the melting zone—batch routing, silication, refining fusion, degassing and homogenization, and in thc cooling zonc cooling, thermal homogenization and glass drawing. The batch is routed to the furnace from a doghouse (4) and the glass melt is drown from four drawing chambers (5). The parting of the tank from the shield assembly does not permit passing of the upset glass melt layer to the cooling zone. In this way the penetration of sandy and not seedy-free glass melt for drawing is prevented. Thus the glass melt of poor quality is moved off to the surface because its specific weight is less than the specific weight of the quality glass melt.

The melting zone of the tank is rectangular and the ratio between its length and width is 4:1 for the examined furnace. The insulation of the walls and the bottom is made by shamot bricks.

The combustion chamber is formed from walls, crown and steel constructions. The walls and the crown are built from dinas refractory. There are disposed burners (6) in the combustion chamber. The direction of the burner fire is changed periodically in the opposite direction. That is why the burners are disposed in couples and while fuel and air through the first burner are passing for combustion, the combustion gases are lead off the furnace from the other. The flame is always unscrewed on the glass melt surface and its direction coincides with that of the working flow movement.

The dimensions of the examined furnace are given in Fig. 2 and in Table 1.

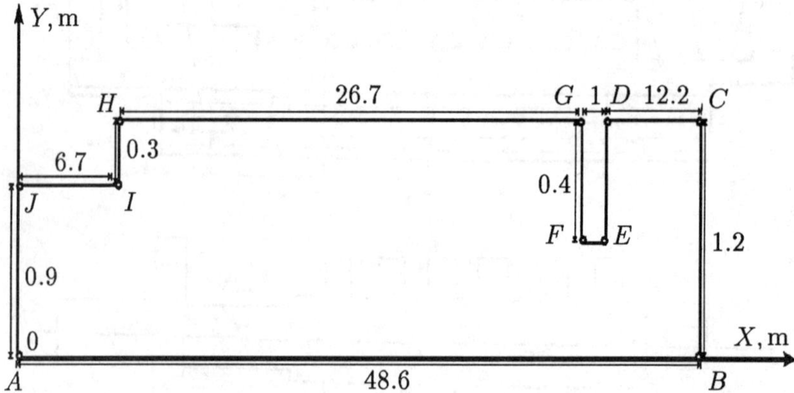

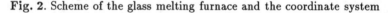

Fig. 2. Scheme of the glass melting furnace and the coordinate system

3 Mathematical Formulation of the Problem

3.1 Basic Equations

The glass melt is considered as a homogeneous Newtonian fluid, which interacts with the batch, the flame space, the walls and the bottom of the furnace. Heat and mass exchange processes in the furnace are described by the equations of conservation of the mass, the energy and the momentum. Two and three-dimensional models of these processes are examined in the literature [1, 2, 4] and their results show that there are no pronounced differences in the temperature and velocity fields in 2-D and 3-D cases. That is why in this paper only a two-dimensional model is studied.

Fluid flow and heat transfer in the system can be described by the equation of continuity, the Navier - Stokes equation and the differential energy balance:

$$\begin{gathered} \nabla \cdot \mathbf{V} = 0 \,, \\ \rho(\mathbf{V} \cdot \nabla)\mathbf{V} = \nabla P - \nabla \cdot \boldsymbol{\tau} + \rho\beta(T - T_0)g \,, \\ \rho C_p \mathbf{V} \cdot \nabla T = \nabla \cdot (K_{\text{eff}} \nabla T) \,, \end{gathered} \tag{1}$$

where ρ is the mass density, $\mathbf{V}$—the velocity vector, P—the pressure (which includes the effect of gravitational head), $\boldsymbol{\tau}$—the viscous stress tensor, β—the volumetric coefficient of expansion, T and T_0—the glass temperature and its reference value, g—the gravity acceleration, C_p—the specific heat and K_{eff}—the effective thermal conductivity.

The first stage of our investigation is to examine only heat transfer in the above mentioned furnace which corresponds to the conditions that glass melt is not moving, the coefficient K_{eff} is a constant and the heat transfer is stationary. On this case the system (1) reduces to:

$$\nabla \cdot \nabla T = \Delta T = \frac{\partial^2 T}{\partial x^2} + \frac{\partial^2 T}{\partial y^2} = 0 \tag{2}$$

in the Cartesian coordinate system Oxy, shown in Fig. 2.

3.2 Boundary Conditions

The solution of the Laplace equation (2) requires specification of the boundary conditions along the boundaries of the glass melt.

The top surface is partially exposed to combustion gases and the rest is covered by the batch blanket. The heat flux to the free glass melt surface in the melting zone (GH in Fig. 2) is:

$$K_{\text{eff}} \frac{\partial T}{\partial y} = q_T \,, \tag{3}$$

where q_T is the heat flux entering the glass melt surface. The model for calculating q_T is discussed in the next section.

The heat transfer in the cooling zone and on the boundaries of thc shield assembly (CD, DE, EF, GF) is:

$$\frac{\partial T}{\partial N} = 0\,, \tag{4}$$

i.e. there is no heat exchange between glass melt and gas space.

The glass melt from the walls and from the bottom is surrounded by the refractories. At the solid surface (the walls and the bottom JA, AB and BC) the heat transfer is expressed by means of the equation:

$$K_{\text{eff}}\,\frac{\partial T}{\partial N} = U_i\,(T - T_a)\,, \tag{5}$$

where T_a is the ambient temperature, N—the inward normal (i.e. the normal directed from the refractory surface into the glass), U_i—the heat transfer coefficient of solid surfaces which, depending on the surface has different values, namely, U_{W1} for the front wall, U_{W2} for the back wall and U_b for the bottom. Note that the heat transfer coefficients U_i may also be different in different sections of the delivery function. The values for U_i, used for investigation are given in Table 2.

It is assumed finally that the area JIH is occupied by the batch and the conditions at the interface between the batch and the glass melt are:

$$\frac{\partial T}{\partial N} = 0\,. \tag{6}$$

3.3 Heat Flow at the Glass Surface

For the present purposes we will formulate a model for heat flow on the basis of experimental data. For calculating the heat flow the distribution of the fuel flux from the burners is used. The heat losses with outlet combustion gases are taken into account. The distribution of the heat flow along the x-direction is given in Fig. 3. It is approximated by the function:

$$\text{Log}\, q_T = -7.76\cdot 10^{-5}\,X^2 + 1.83\cdot 10^{-2}\,X + 0.94 \tag{7}$$

and it is plotted in Fig. 3 as well. This model is chosen from 12 different functions. The criterion for the choose is maximal correlation coefficient and minimal mean-square error. The correlation coefficient of this approximation is 0.97.

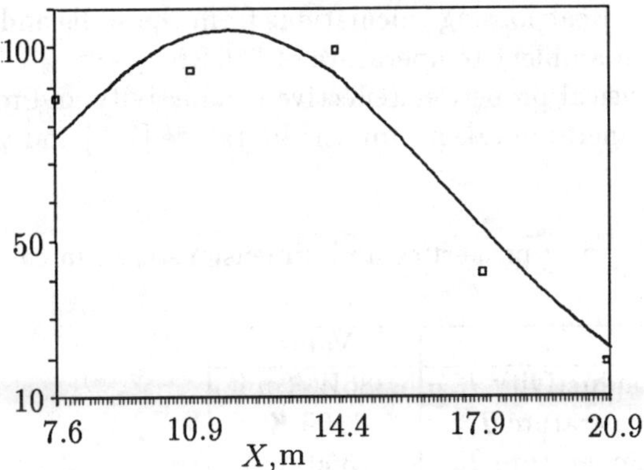

Fig. 3. Distribution of the heat flow in the x-direction and its approximation

4 Numerical Procedure

A finite differences method of Seidel type with acceleration procedure SOR ("successive over relaxation") is used for numerical solving of the Laplace equation with given boundary conditions [3]. A 5-point approximation is used for solution. The number of points used in the numerical solution is 501 in the x-direction and 27 in y-direction. The grid is non-uniform and it is concentrated in critical areas (near the walls, the bottom and the top surface). The solution procedure is iterative, and it converges to an acceptably accurate results if a proper grid and parameters are specified. In the calculations, the solution is considered to be converging if

$$\varepsilon = \max_{i,j} \left| \frac{T_{ij}^{n+1} - T_{ij}^{n}}{T_{ij}^{n}} \right| \leq 4 \cdot 10^{-4} \,. \tag{8}$$

Here $T_{ij}^{n+1}$is the result of the $(n+1)th$ iteration and is T_{ij}^{n} that of the nth iteration.

5 Results and Discussion

5.1 Model Parameters

The tank size considered (Fig. 2) is 48.6 m long and 1.2 m depth. The geometrical dimensions of the melting and cooling zones are given in Table 1. The front and back walls refractory construction are considered to be the same, and a value of 4.0 W/(m^2 K) is assigned to their heat transfer coefficients for the basic simulation. The heat transfer coefficient at the bottom is 4.0 W/(m^2 K) for the basic solution, too. Heat loosing calculations from the walls and the bottom are performed using an ambient temperature of 350 K.

The thermophysical properties (effective conductivity and melting temperature) for the glass melt are taken from the literature [1, 2] and they are given in Table 1.

Table 1. Physical properties and dimensions used in calculations

Parameters	Value
Effective conductivity K_{eff}	$5.386 W/(m\,K)$
Melting temperature T_m	1100 K
Ambient temperature T_a	350 K
Melting zone: length	35.4 m
depth	1.2 m
Length of the batch	0.3 m
Depth of the batch	6.7 m
Cooling zone: length	12,2 m
depth	1.2 m
Shield assembly: length	1.0 m
depth	0.35 m

5.2 Results of Basic Simulation

The basic simulation calculated for parameters given in Table 2 (simulation 1) is shown in Fig. 4. The maximal temperature is at the top surface in the melting zone for $x = 11\,\text{m}$ and it equals 1809 K. The maximal temperature gradient is in the melting zone under the maximal heat flow. The temperature at the top surface in the cooling zone is 1398 K and at the bottom is 1375 K. The minimal temperature in the tank is 1375 K and it is near the front wall and the bottom. Neglecting the convectional heat exchange leads to lower temperature in the area under the batch. The results from the energy balance model could be compared with measurements in the working furnace which is described in this study. The results for the top surface in the cooling and the melting zones agree very well with the measurements in the working furnace for the same fuel flux.

5.3 Influence of the Heat Transfer Coefficient of the Walls and the Bottom

Increasing the heat transfer coefficient (simulation 2 and 3 in Table 2) for the walls leads to decreasing the temperature near the front and the back walls and the bottom. The minimal temperature for simulation 2 in Table 2 (Fig. 6) is 1353 K near the point A (Fig. 2). The temperature in the drawing zone is 1395 K. In simulation 3 (Fig. 6) this temperature decreases to 1395 K. The minimal temperature in this case attains the value of 1320 K near the bottom and the front wall. The change of the heat transfer coefficient of the refractories does not lead to a change in the temperature field near the top surface in the melting zone.

Table 2. Summary of the parameters used in the simulations

Simulation	U_{W1} $W/(\mathrm{m}^2\,\mathrm{K})$	U_{W2} $W/(\mathrm{m}^2\,\mathrm{K})$	U_b $W/(\mathrm{m}^2\,\mathrm{K})$ W/(m2K)	Q% I – III couples burners	Q% IV – VI couples burners
1 (Fig. 4)	4	4	4	100	100
2 (Fig. 5)	8	8	4	100	100
3 (Fig. 6)	12	12	12	100	100
4 (Fig. 7)	4	4	4	100	80
5 (Fig. 8)	4	4	4	100	120

5.4 Influence of the Heat Flow

For the automatic control of the furnace the change of the fuel consumption from the last three couples burners is used. The decrease of the heat flow from the last three couples burners with 20% is investigated (simulation 4 in Table 2, corresponding to Fig. 7). The maximal temperature decreases by 25 K to 1778 K for $x = 11\,\mathrm{m}$. The temperature near the front wall and the bottom is changes slightly. The temperature in the drawing zone does not change.

The increase of the heat flow of the last couples of burners with 20 % (simulation 5 in Table 2, corresponding to Fig. 8) leads to a change of the temperature field only near the top surface. In this case the maximal temperature is increased with 40 K to 1843 K under the burner with maximal fuel flow. The influence of the change of the heat flow of three last couples burners is slight in the cooling zone and in the end of the melting zone because the movement of the glass melt is neglected.

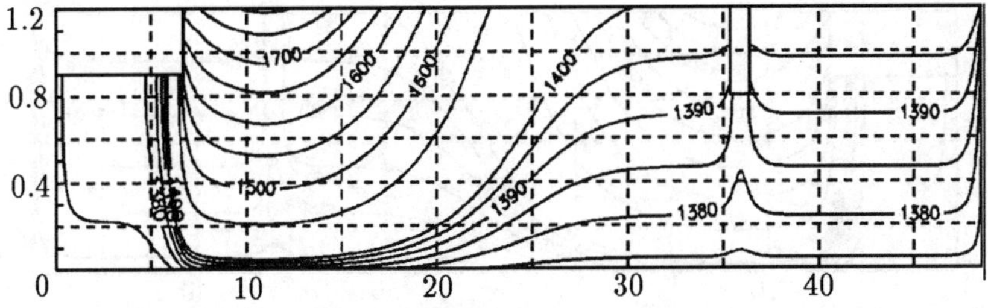

Fig. 4. Temperature distribution for the simulation 1 in Table 2

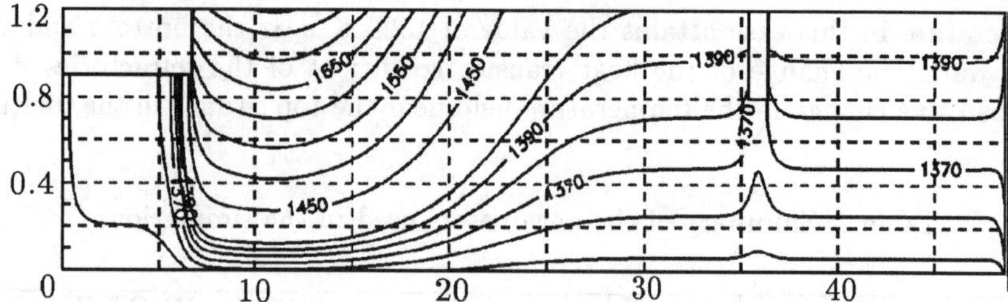

Fig. 5. Temperature distribution for the simulation 2 in Table 2

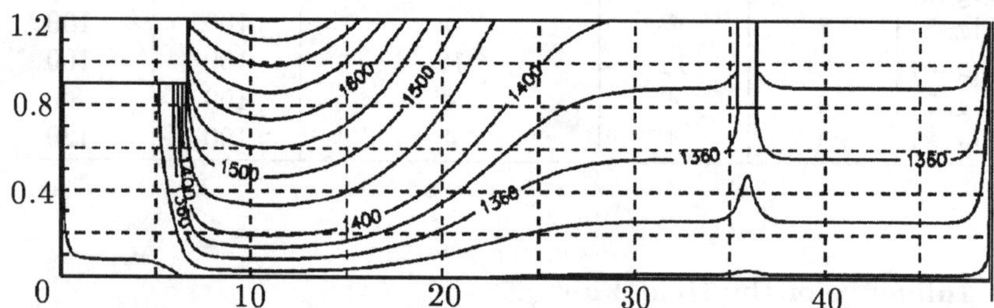

Fig. 6. Temperature distribution for the simulation 3 in Table 2

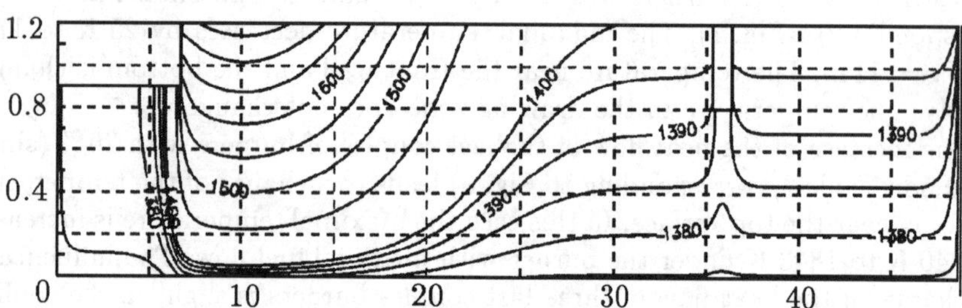

Fig. 7. Temperature distribution for the simulation 4 in Table 2

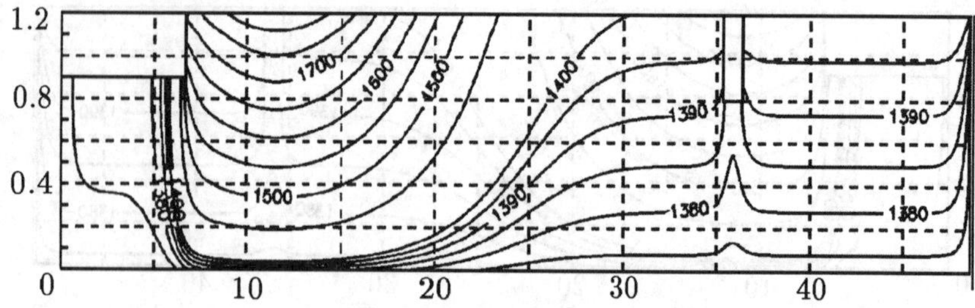

Fig. 8. Temperature distribution for the simulation 5 in Table 2

6 Concluding Remarks

A simplified mathematical model for flat glass furnace is presented as a first step of the investigation of the temperature and velocity fields. It is possible to study the energetical behaviour of the furnace and the influence of changes of technological parameters on the temperature field. The numerical solution uses the finite difference method with "over relaxation" (SOR).

The investigation has to be continued with a development of a numerical procedure for solving Navier-Stokes equations. This will give a possibility for simulating the convectional heat transfer and velocity fields.

The approach presented here can be also modified in order to include a model of the heat flow and heat transfer in the combustion chamber for calculating more precisely the boundary condition and the temperature at the glass melt surface.

Acknowledgements. The support of the Bulgarian National Fund of Scientific Research under Grant No MM510/1995 is gratefully acknowledged.

References

1. Ungan, A., Viskanta, R.: Three dimensional numerical modeling of circulation and heat transfer in a glass melting tank. Glastech. Ber. **60**(3) (1987) 71–78; **60**(4) (1987) 115–124
2. Horvath, Z., Hilbig G.: Mathematical model for fuel-heated glass melting tanks. Glastech. Ber. **61**(10) (1988) 277–282
3. Marchuk, G. I.: Methods of numerical mathematics. Moscow, "Nauka" 1980, 535. (in Russian.)
4. Zheleva, I., Kambourova, V.: Investigation on the influence of the heat flow upon the temperature field in a glass melting furnace. In Proc. II International Ceramic Congress, Vol. 1, pp. 359–365, Istanbul, 1994

Author Index

Lecture Notes in Computer Science

For information about Vols. 1–1122

please contact your bookseller or Springer-Verlag

Vol. 1123: L. Bougé, P. Fraigniaud, A. Mignotte, Y. Robert (Eds.), Euro-Par'96. Parallel Processing. Proceedings, 1996, Vol. I. XXXIII, 842 pages. 1996.

Vol. 1124: L. Bougé, P. Fraigniaud, A. Mignotte, Y. Robert (Eds.), Euro-Par'96. Parallel Processing. Proceedings, 1996, Vol. II. XXXIII, 926 pages. 1996.

Vol. 1125: J. von Wright, J. Grundy, J. Harrison (Eds.), Theorem Proving in Higher Order Logics. Proceedings, 1996. VIII, 447 pages. 1996.

Vol. 1126: J.J. Alferes, L. Moniz Pereira, E. Orlowska (Eds.), Logics in Artificial Intelligence. Proceedings, 1996. IX, 417 pages. 1996. (Subseries LNAI).

Vol. 1127: L. Böszörményi (Ed.), Parallel Computation. Proceedings, 1996. XI, 235 pages. 1996.

Vol. 1128: J. Calmet, C. Limongelli (Eds.), Design and Implementation of Symbolic Computation Systems. Proceedings, 1996. IX, 356 pages. 1996.

Vol. 1129: J. Launchbury, E. Meijer, T. Sheard (Eds.), Advanced Functional Programming. Proceedings, 1996. VII, 238 pages. 1996.

Vol. 1130: M. Haveraaen, O. Owe, O.-J. Dahl (Eds.), Recent Trends in Data Type Specification. Proceedings, 1995. VIII, 551 pages. 1996.

Vol. 1131: K.H. Höhne, R. Kikinis (Eds.), Visualization in Biomedical Computing. Proceedings, 1996. XII, 610 pages. 1996.

Vol. 1132: G.-R. Perrin, A. Darte (Eds.), The Data Parallel Programming Model. XV, 284 pages. 1996.

Vol. 1133: J.-Y. Chouinard, P. Fortier, T.A. Gulliver (Eds.), Information Theory and Applications II. Proceedings, 1995. XII, 309 pages. 1996.

Vol. 1134: R. Wagner, H. Thoma (Eds.), Database and Expert Systems Applications. Proceedings, 1996. XV, 921 pages. 1996.

Vol. 1135: B. Jonsson, J. Parrow (Eds.), Formal Techniques in Real-Time and Fault-Tolerant Systems. Proceedings, 1996. X, 479 pages. 1996.

Vol. 1136: J. Diaz, M. Serna (Eds.), Algorithms – ESA '96. Proceedings, 1996. XII, 566 pages. 1996.

Vol. 1137: G. Görz, S. Hölldobler (Eds.), KI-96: Advances in Artificial Intelligence. Proceedings, 1996. XI, 387 pages. 1996. (Subseries LNAI).

Vol. 1138: J. Calmet, J.A. Campbell, J. Pfalzgraf (Eds.), Artificial Intelligence and Symbolic Mathematical Computation. Proceedings, 1996. VIII, 381 pages. 1996.

Vol. 1139: M. Hanus, M. Rogriguez-Artalejo (Eds.), Algebraic and Logic Programming. Proceedings, 1996. VIII, 345 pages. 1996.

Vol. 1140: H. Kuchen, S. Doaitse Swierstra (Eds.), Programming Languages: Implementations, Logics, and Programs. Proceedings, 1996. XI, 479 pages. 1996.

Vol. 1141: H.-M. Voigt, W. Ebeling, I. Rechenberg, H.-P. Schwefel (Eds.), Parallel Problem Solving from Nature – PPSN IV. Proceedings, 1996. XVII, 1.050 pages. 1996.

Vol. 1142: R.W. Hartenstein, M. Glesner (Eds.), Field-Programmable Logic. Proceedings, 1996. X, 432 pages. 1996.

Vol. 1143: T.C. Fogarty (Ed.), Evolutionary Computing. Proceedings, 1996. VIII, 305 pages. 1996.

Vol. 1144: J. Ponce, A. Zisserman, M. Hebert (Eds.), Object Representation in Computer Vision. Proceedings, 1996. VIII, 403 pages. 1996.

Vol. 1145: R. Cousot, D.A. Schmidt (Eds.), Static Analysis. Proceedings, 1996. IX, 389 pages. 1996.

Vol. 1146: E. Bertino, H. Kurth, G. Martella, E. Montolivo (Eds.), Computer Security – ESORICS 96. Proceedings, 1996. X, 365 pages. 1996.

Vol. 1147: L. Miclet, C. de la Higuera (Eds.), Grammatical Inference: Learning Syntax from Sentences. Proceedings, 1996. VIII, 327 pages. 1996. (Subseries LNAI).

Vol. 1148: M.C. Lin, D. Manocha (Eds.), Applied Computational Geometry. Proceedings, 1996. VIII, 223 pages. 1996.

Vol. 1149: C. Montangero (Ed.), Software Process Technology. Proceedings, 1996. IX, 291 pages. 1996.

Vol. 1150: A. Hlawiczka, J.G. Silva, L. Simoncini (Eds.), Dependable Computing – EDCC-2. Proceedings, 1996. XVI, 440 pages. 1996.

Vol. 1151: Ö. Babaoğlu, K. Marzullo (Eds.), Distributed Algorithms. Proceedings, 1996. VIII, 381 pages. 1996.

Vol. 1152: T. Furuhashi, Y. Uchikawa (Eds.), Fuzzy Logic, Neural Networks, and Evolutionary Computation. Proceedings, 1995. VIII, 243 pages. 1996. (Subseries LNAI).

Vol. 1153: E. Burke, P. Ross (Eds.), Practice and Theory of Automated Timetabling. Proceedings, 1995. XIII, 381 pages. 1996.

Vol. 1154: D. Pedreschi, C. Zaniolo (Eds.), Logic in Databases. Proceedings, 1996. X, 497 pages. 1996.

Vol. 1155: J. Roberts, U. Mocci, J. Virtamo (Eds.), Broadbank Network Teletraffic. XXII, 584 pages. 1996.

Vol. 1156: A. Bode, J. Dongarra, T. Ludwig, V. Sunderam (Eds.), Parallel Virtual Machine – EuroPVM '96. Proceedings, 1996. XIV, 362 pages. 1996.

Vol. 1157: B. Thalheim (Ed.), Conceptual Modeling – ER '96. Proceedings, 1996. XII, 489 pages. 1996.

Vol. 1158: S. Berardi, M. Coppo (Eds.), Types for Proofs and Programs. Proceedings, 1995. X, 296 pages. 1996.

Vol. 1159: D.L. Borges, C.A.A. Kaestner (Eds.), Advances in Artificial Intelligence. Proceedings, 1996. XI, 243 pages. (Subseries LNAI).

Vol. 1160: S. Arikawa, A.K. Sharma (Eds.), Algorithmic Learning Theory. Proceedings, 1996. XVII, 337 pages. 1996. (Subseries LNAI).

Vol. 1161: O. Spaniol, C. Linnhoff-Popien, B. Meyer (Eds.), Trends in Distributed Systems. Proceedings, 1996. VIII, 289 pages. 1996.

Vol. 1162: D.G. Feitelson, L. Rudolph (Eds.), Job Scheduling Strategies for Parallel Processing. Proceedings, 1996. VIII, 291 pages. 1996.

Vol. 1163: K. Kim, T. Matsumoto (Eds.), Advances in Cryptology – ASIACRYPT '96. Proceedings, 1996. XII, 395 pages. 1996.

Vol. 1164: K. Berquist, A. Berquist (Eds.), Managing Information Highways. XIV, 417 pages. 1996.

Vol. 1165: J.-R. Abrial, E. Börger, H. Langmaack (Eds.), Formal Methods for Industrial Applications. VIII, 511 pages. 1996.

Vol. 1166: M. Srivas, A. Camilleri (Eds.), Formal Methods in Computer-Aided Design. Proceedings, 1996. IX, 470 pages. 1996.

Vol. 1167: I. Sommerville (Ed.), Software Configuration Management. VII, 291 pages. 1996.

Vol. 1168: I. Smith, B. Faltings (Eds.), Advances in Case-Based Reasoning. Proceedings, 1996. IX, 531 pages. 1996. (Subseries LNAI).

Vol. 1169: M. Broy, S. Merz, K. Spies (Eds.), Formal Systems Specification. XXIII, 541 pages. 1996.

Vol. 1170: M. Nagl (Ed.), Building Tightly Integrated Software Development Environments: The IPSEN Approach. IX, 709 pages. 1996.

Vol. 1171: A. Franz, Automatic Ambiguity Resolution in Natural Language Processing. XIX, 155 pages. 1996. (Subseries LNAI).

Vol. 1172: J. Pieprzyk, J. Seberry (Eds.), Information Security and Privacy. Proceedings, 1996. IX, 333 pages. 1996.

Vol. 1173: W. Rucklidge, Efficient Visual Recognition Using the Hausdorff Distance. XIII, 178 pages. 1996.

Vol. 1174: R. Anderson (Ed.), Information Hiding. Proceedings, 1996. VIII, 351 pages. 1996.

Vol. 1175: K.G. Jeffery, J. Král, M. Bartošek (Eds.), SOFSEM'96: Theory and Practice of Informatics. Proceedings, 1996. XII, 491 pages. 1996.

Vol. 1176: S. Miguet, A. Montanvert, S. Ubéda (Eds.), Discrete Geometry for Computer Imagery. Proceedings, 1996. XI, 349 pages. 1996.

Vol. 1177: J.P. Müller, The Design of Intelligent Agents. XV, 227 pages. 1996. (Subseries LNAI).

Vol. 1178: T. Asano, Y. Igarashi, H. Nagamochi, S. Miyano, S. Suri (Eds.), Algorithms and Computation. Proceedings, 1996. X, 448 pages. 1996.

Vol. 1179: J. Jaffar, R.H.C. Yap (Eds.), Concurrency and Parallelism, Programming, Networking, and Security. Proceedings, 1996. XIII, 394 pages. 1996.

Vol. 1180: V. Chandru, V. Vinay (Eds.), Foundations of Software Technology and Theoretical Computer Science. Proceedings, 1996. XI, 387 pages. 1996.

Vol. 1181: D. Bjørner, M. Broy, I.V. Pottosin (Eds.), Perspectives of System Informatics. Proceedings, 1996. XVII, 447 pages. 1996.

Vol. 1182: W. Hasan, Optimization of SQL Queries for Parallel Machines. XVIII, 133 pages. 1996.

Vol. 1183: A. Wierse, G.G. Grinstein, U. Lang (Eds.), Database Issues for Data Visualization. Proceedings, 1995. XIV, 219 pages. 1996.

Vol. 1184: J. Waśniewski, J. Dongarra, K. Madsen, D. Olesen (Eds.), Applied Parallel Computing. Proceedings, 1996. XIII, 722 pages. 1996.

Vol. 1185: G. Ventre, J. Domingo-Pascual, A. Danthine (Eds.), Multimedia Telecommunications and Applications. Proceedings, 1996. XII, 267 pages. 1996.

Vol. 1186: F. Afrati, P. Kolaitis (Eds.), Database Theory - ICDT'97. Proceedings, 1997. XIII, 477 pages. 1997.

Vol. 1187: K. Schlechta, Nonmonotonic Logics. IX, 243 pages. 1997. (Subseries LNAI).

Vol. 1188: T. Martin, A.L. Ralescu (Eds.), Fuzzy Logic in Artificial Intelligence. Proceedings, 1995. VIII, 272 pages. 1997. (Subseries LNAI).

Vol. 1189: M. Lomas (Ed.), Security Protocols. Proceedings, 1996. VIII, 203 pages. 1997.

Vol. 1190: S. North (Ed.), Graph Drawing. Proceedings, 1996. XI, 409 pages. 1997.

Vol. 1191: V. Gaede, A. Brodsky, O. Günther, D. Srivastava, V. Vianu, M. Wallace (Eds.), Constraint Databases and Applications. Proceedings, 1996. X, 345 pages. 1996.

Vol. 1192: M. Dam (Ed.), Analysis and Verification of Multiple-Agent Languages. Proceedings, 1996. VIII, 435 pages. 1997.

Vol. 1193: J.P. Müller, M.J. Wooldridge, N.R. Jennings (Eds.), Intelligent Agents III. XV, 401 pages. 1997. (Subseries LNAI).

Vol. 1196: L. Vulkov, J. Waśniewski, P. Yalamov (Eds.), Numerical Analysis and Its Applications. Proceedings, 1996. XIII, 608 pages. 1997.

Vol. 1197: F. d'Amore, P.G. Franciosa, A. Marchetti-Spaccamela (Eds.), Graph-Theoretic Concepts in Computer Science. Proceedings, 1996. XI, 410 pages. 1997.

Vol. 1198: H.S. Nwana, N. Azarmi (Eds.), Software Agents and Soft Computing: Towards Enhancing Machine Intelligence. XIV, 298 pages. 1997. (Subseries LNAI).

Vol. 1199: D.K. Panda, C.B. Stunkel (Eds.), Communication and Architectural Support for Network-Based Parallel Computing. Proceedings, 1997. X, 269 pages. 1997.

Vol. 1200: R. Reischuk, M. Morvan (Eds.), STACS 97. Proceedings, 1997. XIII, 614 pages. 1997.

Vol. 1201: O. Maler (Ed.), Hybrid and Real-Time Systems. Proceedings, 1997. IX, 417 pages. 1997.

Vol. 1202: P. Kandzia, M. Klusch (Eds.), Cooperative Information Agents. Proceedings, 1997. IX, 287 pages. 1997. (Subseries LNAI).

Vol. 1203: G. Bongiovanni, D.P. Bovet, G. Di Battista (Eds.), Algorithms and Complexity. Proceedings, 1997. VIII, 311 pages. 1997.

Vol. 1204: H. Mössenböck (Ed.), Modular Programming Languages. Proceedings, 1997. X, 379 pages. 1997.

Vol. 1206: J. Bigün, G. Chollet, G. Borgefors (Eds.), Audio- and Video-based Biometric Person Authentication. Proceedings, 1997. XII, 450 pages. 1997.